Biocatalysis

A. S. Bommarius, B. R. Riebel

Related Titles

S. Brakmann, K. Johnsson

Directed Molecular Evolution of Proteins

or How to Improve Enzymes for Biocatalysis

February 2002

ISBN 3-527-30423-1

M. C. Flickinger, S. W. Drew

Encyclopedia of Bioprocess Technology

Fermentation, Biocatalysis, and Bioseparation

April 1999

ISBN 0-471-13822-3

I. T. Horvath

Encyclopedia of Catalysis

6 Volume Set

February 2003

ISBN 0-471-24183-0

S. M. Roberts, G. Casy, M.-B. Nielsen, S. Phythian, C. Todd, U. Wiggins

Biocatalysts for Fine Chemicals Synthesis

October 1999

ISBN 0-471-97901-5

K. Drauz, H. Waldmann

Enzyme Catalysis in Organic Synthesis

A Comprehensive Handbook

April 2002

ISBN 3-527-29949-1

U. T. Bornscheuer, R. J. Kazlauskas

Hydrolases in Organic Synthesis

Regio- and Stereoselective Biotransformations

October 1999

ISBN 3-527-30104-6

B. Cornils, W. A. Herrmann, R. Schlögl, C.-H. Wong

Catalysis from A to Z

A Concise Encyclopedia

March 2000

ISBN 3-527-29855-X

Biocatalysis

A. S. Bommarius, B. R. Riebel

WILEY-VCH Verlag GmbH & Co. KGaA

Prof. Dr. Andreas Sebastian Bommarius
School of Chemical and Biomolecular Engineering
Parker H. Petit Biotechnology Institute
Georgia Institute of Technology
315 Ferst Drive, N. W.
Atlanta, GA 30332-0363
USA

Dr. Bettina Riebel
Emory University School of Medicine
Whitehead Research Building
615 Michael Street
Atlanta, GA 30322
USA

Library of Congress Card No.: applied for
A catalogue record for this book is available from the British Library.

Bibliographic information published by Die Deutsche Bibliothek
Die Deutsche Bibliothek lists this publication in the Deutsche Nationalbibliografie; detailed bibliographic data is available in the internet at http://dnb.ddb.de.

Printed in the Federal Republic of Germany.
Printed on acid-free paper.

Composition Manuela Treindl, Laaber
Printing betz-druck GmbH, Darmstadt
Bookbinding Großbuchbinderei J. Schäffer GmbH & Co. KG, Grünstadt

ISBN 3-527-30344-8

Preface

The field of biocatalysis is at a crossroads. On one hand, the frontier of research races ahead, propelled by advances in the database-supported analysis of sequences and structures as well as the designability of genes and proteins. Moreover, the "design rules" for biocatalysts have emerged from vague images on the horizon, to come into much clearer view. On the other hand, experienced practitioners from other areas as well as more and more students entering this field search for ways to obtain the level of knowledge in biocatalysis that advances their own agenda. However, both groups find a rapidly growing field with too little *guidance* towards the research front and too little *structure* in its guiding principles. In this situation, this book seeks to fill the gap between the research front and the area beyond basic courses in biochemistry, organic synthesis, molecular biology, kinetics, and reaction engineering. Students and practitioners alike are often left alone to bridge the gulf between basic textbooks and original research articles; this book seeks to cover this intermediate area.

Another challenge this book strives to address results from the interdisciplinary nature of the field of biocatalysis. Biocatalysis is a synthesis of chemistry, biology, chemical engineering, and bioengineering, but most students and practitioners enter this field with preparation essentially limited to one of the major contributing areas, or at best two. The essence of biocatalysis, as well as most of its current research, however, is captured in the interdisciplinary overlap between individual areas. Therefore, this work seeks to help readers to combine their prior knowledge with the contents and the methods in this book to make an integrated whole.

The book is divided into three parts:

- Chapters 1 through 7 cover *basic tools*. Many readers have probably encountered the contents of some chapters before; nevertheless, we hope to offer an update and a fresh view.
- Chapters 8 through 14 expand on *advanced tools*. While command of such advanced concepts is indispensable in order to follow, much less to lead, today's developments in biocatalysis, the mastering of such concepts and tools cannot necessarily be expected of all practitioners in the field, especially if their major course of study often did not even touch on such topics.
- Chapters 15 through 20 treat *applications* of all the tools covered in previous chapters. "Applications" here encompass not just industrial-scale realization of bio-

catalysis but also new intellectual frontiers in biological catalysis that are possible with today's technologies, such as rapidly expanding DNA databases or comprehensive coverage of three-dimensional structure analysis for many enzymes.

In the early part of the book, several chapters have a fairly clear emphasis on chemistry, biology, or chemical engineering. Chapters on the isolation of microorganisms (Chapter 3), molecular biology tools (Chapter 4), protein engineering (Chapter 10), or directed evolution (Chapter 11) have a distinct biological flavor. Chemistry is the main topic in the chapters on applications of enzymes as products (Chapter 6), in bulk and fine chemicals (Chapter 7), and in pharmaceuticals (Chapter 13). Chemical engineering concepts predominate in the chapters on biocatalytic reaction engineering (Chapter 5) or on processing steps for enzyme manufacture (Chapter 8). Other chapters contribute a perspective from biochemistry/enzymology, such as characterization of biocatalysts (Chapter 2) and methods of studying proteins (Chapter 9), or from informatics, most notably bioinformatics (Chapter 14).

Finally, a word on the history of this book: the idea for the present work originated during a lectureship of one of us (A.S.B.) as an adjunct faculty member at the Rheinisch-Westfälische Technische Hochschule (RWTH) Aachen in Aachen, Germany, for nine years while he was working at Degussa in Wolfgang, Germany. Time and time again, students enjoyed the interdisciplinary nature and coverage of biocatalysis but lacked adequate preparation in those basic tools that were not provided during their courses for their respective major, be it chemistry, biology, or chemical engineering. Similar observations were made when teaching biocatalysis or related subjects at the Georgia Institute of Technology in Atlanta/GA, USA. One of the aims of this book is to take readers back to scientific fundamentals often long forgotten, to let them to participate in the joy of discovery and understanding stemming from a multi-faceted picture of nature. While scientific fundamentals are a source of immense satisfaction, applications with an impact in the day-to-day world are just as important. Two of the biggest challenges facing mankind today (and not exclusively the industrial societies) are maintenance and improvement of *human health*, and maintenance and improvement of *the environment*. Biocatalysis aids the first of these goals through its selectivity in generating ever more complex pharmaceutically active molecules, and the second goal by opening new routes to both basic and performance chemicals with the aim of achieving sustainable development.

We hope that you enjoy reading this book. We encourage you to contact us to voice your opinion, gripe, laud, discuss aspects of the book, point out errors or ambiguities, make suggestions for improvements, or just to let us know what you think. The easiest way to do this is via email at bommariu@bellsouth.net or andreas.bommarius@alum.mit.edu.

We wish you pleasant reading.

Andreas S. Bommarius and Bettina R. Riebel
Atlanta/GA, USA
December 2003

Acknowledgments

For more than a decade, one of us (A.S.B.) had the good fortune to be associated with Degussa, one of the early players, and currently still strong, in the area of biocatalysis, in its R&D center in Wolfgang, Germany. While several factors were responsible for Degussa's venture into biocatalysis, certainly the most influential was the steadfast support of biocatalysis by Degussa's former board member and Head of Research, Professor Heribert Offermanns. His unconventional and far-sighted way of thinking remains an example and A.S.B. thanks him warmly for his attitude and encouragement. A.S.B. is also grateful to Professor Karlheinz Drauz, himself an accomplished author with Wiley-VCH, for sustained support and also for supporting biocatalysis at Degussa during difficult times. A.S.B. also fondly remembers co-workers at Degussa and its many subsidiaries. He thanks Wolfgang Leuchtenberger, his predecessor and representing a group too numerous to acknowledge individually, and encourages Harald Gröger, his successor.

The origin of this book stems from a biweekly lectureship that A.S.B. held at the RWTH Aachen (in Aachen, Germany) from 1991 to 2000, first at the Institute of Biotechnology under the late Harald Voss, then in the Institute of Technical Chemistry and Petroleum Chemistry under Wilhelm Keim. A.S.B. expressly thanks Wilhelm Keim for his continued support and advice, not just with the lectureship but also during his habilitation.

Both of us have several reasons to thank Professors Maria-Regina Kula at the University of Düsseldorf, Germany, and Christian Wandrey at the Research Center Jülich, Germany. While both of them have left a huge impact on the field of biocatalysis in general (acknowledged, among other honors, by the German Technology Transfer Prize in 1983 and the Enzyme Engineering Award in 1995 to both of them), they influenced each of us markedly. One of us (B.R.R.) thanks her advisor Maria-Regina Kula and, specifically, her direct mentor, Werner Hummel, for sustained support and interest during her formative thesis years and beyond. A.S.B. gladly acknowledges both of them and Christian Wandrey for many years of fruitful collaboration. The impact of their views on both of us is evident in many parts of this book.

One of us (A.S.B.) gratefully acknowledges the support from Georgia Tech, from the higher administration to the laboratory group, for getting his own research group started. As representatives for a much more numerous group, A.S.B. thanks Dr. Ronald Rousseau, his School Chair, himself an author of one of the most influential textbooks on chemical engineering, for his trust and his support of the area of biocatalysis in chemical engineering, as well as Dr. Phillip Gibbs, his first postdoctoral associate, for countless discussions on the research front in the field.

We thank our publisher, Wiley-VCH, in Weinheim, Germany, for their continual support and enthusiasm. The publishing team, including Karin Dembowsky, Andrea Pillmann, Eva Wille, Karin Proff, and Hans-Jochen Schmitt, had to put up with quite a scheduling challenge, not to mention the pain resulting from the need for both authors to relocate to Atlanta/GA, USA, and establish their careers there. Both of us thank the publishers for exemplary support and the high quality of workmanship reflected in the layout of this book.

Last but not least, we could write this book because we enjoyed countless interactions with other scientists and engineers who shaped our view of the field of biocatalysis. A representative, but certainly not exhaustive, list of these individuals, besides those already mentioned above, includes Frances Arnold, Uwe Bornscheuer, Stefan Buchholz, Mark Burk, Robert DiCosimo, David Dodds, Franz Effenberger, Uwe Eichhorn, Wolfgang Estler, Andreas Fischer, Tomas Hudlicky, Hans-Dieter Jakubke, Andreas Karau, Alexander Klibanov, Andreas Liese, Oliver May, Jeffrey Moore, Rainer Müller, Mark Nelson, David Rozzell, Roger Sheldon, Christoph Syldatk, Stefan Verseck, and George Whitesides. We thank all of them for their contribution to our view of the field.

Andreas S. Bommarius and Bettina R. Riebel
Atlanta/GA, USA
December 2003

Contents

1
Introduction to Biocatalysis

Summary

Over the last 20 years, many reservations with respect to biocatalysis have been voiced, contending that: (i) enzymes only feature limited substrate specificity; (ii) there is only limited availability of enzymes; (iii) only a limited number of enzymes exist; (iv) protein catalyst stability is limited; (v) enzyme reactions are saddled with limited space–time yield; and (vi) enzymes require complicated cosubstrates such as cofactors.

Driven by the discovery of many novel enzymes, by recombinant DNA technology which allows both more efficient production and targeted or combinatorial alterations of individual enzymes, and by process development towards higher stability and volumetric productivity, synthesis routes in which one or all of the steps are biocatalytic have advanced dramatically in recent years. Design rules for improved biocatalysts are increasingly precise and easy to use.

Biocatalysts do not operate by different scientific principles from organic catalysts. The existence of a multitude of enzyme models including oligopeptidic or polypeptidic catalysts proves that all enzyme action can be explained by rational chemical and physical principles. However, enzymes can create unusual and superior reaction conditions such as extremely low pK_a values or a high positive potential for a redox metal ion. Enzymes increasingly have been found to catalyze almost any reaction of organic chemistry.

Biotechnology and biocatalysis differ from conventional processes not only by featuring a different type of catalyst; they also constitute a new technology base. The *raw materials base* of a biologically-based process is built on sugar, lignin, or animal or plant wastes; in biotechnology, unit operations such as membrane processes, chromatography, or biocatalysis are prevalent, and the product range of biotechnological processes often encompasses chiral molecules or biopolymers such as proteins, nucleic acids or carbohydrates.

Cost and margin pressure from less expensive competitors and operation with emphasis on a clean (or less polluted) environment are two major developments. Fewer processing steps, with higher yields at each step, lower material and energy costs, and less waste are the goals. Biotechnology and biocatalysis often offer unique technology options and solutions, they act as *enabling technologies*; in other cases, biocatalysis has to outperform competing technologies to gain access. In the pharmaceutical industry, the reason for the drive for enantiomeric purity is that the vast majority of novel drugs are chiral targets, favoring biocatalysis as the technology with the best selectivity performance.

Biocatalytic processes increasingly penetrate the chemical industry. In a recent study, 134 industrial-scale biotransformations, on a scale of > 100 kg with whole cells or enzymes starting from a precursor other than a C-source, were analyzed. Hydrolases (44%), followed by oxido-reductases (30%), dominate industrial biocatalytic applications. Average performance data for fine chemicals (not pharmaceuticals) applications are 78% yield, a final product concentration of 108 g L^{-1}, and a volumetric productivity of 372 g $(L \cdot d)^{-1}$.

1.1
Overview: The Status of Biocatalysis at the Turn of the 21st Century

1.1.1
State of Acceptance of Biocatalysis

Over the last 20 years, many reservations with respect to biocatalysis have been voiced. The critics, often focusing on the drawbacks, have contended that

- enzymes only feature limited substrate specificity,
- there is only limited availability of enzymes,
- only a limited number of enzymes exist,
- protein catalyst stability is limited,
- enzyme reactions are saddled with limited space–time yield, and
- enzymes require complicated co-substrates such as cofactors.

The renaissance of biocatalysis, supported by the advent of recombinant DNA, is only about 20 years old. Recently, several publications have appeared which deal specifically with the attitudes listed above (Rozzell, 1999; Bommarius, 2001; Rasor, 2001). Most of the points above can either be refuted or they can be levied against any novel catalytic technology; the situation with some competing technologies such as chiral homogeneous catalysts is similar to that with enzymes (Chapters 18 and 20).

- *Enzymes only feature limited substrate specificity.* Often, enzymes designed to convert small molecules such as hydrogen peroxide, urea, fumaric acid, or L-aspartic acid feature extremely narrow substrate specificity; the corresponding enzymes catalase, urease, fumarase or aspartase, and L-aspartate decarboxylase take either few other substrates, such as alkyl peroxides in the case of catalase, or no other substrate, such as urease, which only converts urea. On the other hand, very large enzymes acting as multi-enzyme complexes such as nonribosomal peptide synthetases (NRPSs) (Kleinkauf, 1996) are often highly specific. Ordinary-sized enzymes working on medium-sized substrates, however, in most cases feature broad substrate specificity, a fact already noted by Rasor and Voss (Rasor, 2001).

- *There is only limited availability of enzymes.* Until very recently, limited availability of enzymes was indeed a major problem. About ten years ago, with 3196 different enzymes already listed in Enzyme Nomenclature (Moss, 1992), only about 50 enzymes were fully characterized and only about a dozen enzymes available commercially on a regular basis. However, recombinant DNA technology, discovered in 1978 by Cohen and Boyer (Stanford University, Palo Alto, CA, USA), over the next 20 years allowed enzymes to be produced much more efficiently, in higher purity, and more inexpensively (Baneyx, 1999), so that today a multitude of enzymes are available not only from established suppliers such as Sigma–Aldrich–Fluka (Milwaukee, WI, USA), E. Merck (Darmstadt, Germany), Mercian (Tokyo, Japan), or Roche Diagnostics (Mannheim, Germany) but increasingly also from smaller, more focused suppliers such as Biocatalytics (Pasadena, CA, USA) or Jülich Fine Chemicals (Jülich, Germany). The argument of unavailability or scarcity will be less and less justified in the future.

- *Only a limited number of enzymes exist.* This criticism, while depending on the observer's position, is indeed a drawback at the moment. Although enzymes have been found for every conceivable organic chemical reaction except the hetero-Cope rearrangement (Table 1.4, below), there are enzymes sought for many more reactions than there are enzymes available. If enzymes were inferior catalysts this situation would not arise, of course. In fact, enzymes are often superior catalysts (see the next section), so superior is fact, that the community seeks plenty more of them. Chapter 3 treats the discovery of novel enzymes, whereas Chapters 10 and 11 cover improvement of existing enzymes through rational (protein engineering) and combinatorial random mutagenesis (directed evolution).

- *Protein catalyst stability is limited.* This is one of major drawbacks of enzymes. They commonly require temperatures around ambient to perform (15–50°C), pH values around neutral (pH 5–9), and aqueous media. In addition, any number of system components or features such as salts, inhibitors, liquid–gas or liquid–solid interfaces, or mechanical stress can slow down or deactivate enzymes. Under almost any condition, native proteins, with their Gibbs free enthalpy of stability of just a few kilojoules per mole, are never far away from instability. In this book, we cover inhibitors (Chapter 5, Section 5.3) or impeding system parameters (Chapter 17) and successful attempts at broadening the choice of solvents (Chapter 12).

- *Enzyme reactions are saddled with limited space–time yield.* The notion that biocatalysts are slow catalysts is false. Slow catalysts, applied at low concentrations, certainly lead to low space–time yields. However, optimized syntheses not only produce very good selectivities or total turnover numbers but also satisfactory to excellent space–time yields. Examples with such good s.t.y. values are
  - in commodity biochemicals, the synthesis of L-aspartate from fumaric acid and ammonia with space–time-yields of up to 60 kg $(L \cdot d)^{-1}$ (Rozzell, 1999), and

– in advanced pharmaceutical intermediates, kinetically-controlled peptide synthesis to kyotorphin (Tyr-Arg) catalyzed by α-chymotrypsin from maleyl-L-Tyr-OEt and Arg-OEt, employing a highly soluble protecting group at the electrophile (Fischer, 1994). Space–time-yields of 1.34 kg $(L \cdot d)^{-1}$ have been achieved.

The question of high volumetric productivity is coupled to the solubility of substrates. High space–time-yields have been demonstrated to be correlated with high solubilities of substrates (Bommarius, 2001).

- *Enzymes require complicated co-substrates such as cofactors.* Much has been made of the requirement of some enzymes for cofactors, such as nicotinamide-containing compounds, NAD(P)(H), for dehydrogenases; flavin compounds, FMN or FAD, for oxidases; pyridoxylphosphate, PLP, for transaminases and decarboxylases; thiamine pyrophosphate, TPP, for carboligases, and vitamin B12 for glycerate dehydratase, among others. The scale-up of L-aspartate decarboxylation to L-alanine with the help of PLP-requiring L-aspartate decarboxylase, or of reductive amination of trimethylpyruvate to L-tert-leucine with the help of NADH-requiring leucine dehydrogenase demonstrates the feasibility of industrial processing with cofactor-requiring enzymes. The implementation also gives credence to the suggestion that cofactors are no longer the dominating cost component, as was believed until recently. Requirements for cofactors constitute a technological challenge but one that has been met successfully and so should not be regarded as impeding the use of biocatalysts in processing.

1.1.2 Current Advantages and Drawbacks of Biocatalysis

1.1.2.1 Advantages of Biocatalysts

The biggest advantage of enzymes is their often unsurpassed selectivity. While enzymes are used beneficially to increase chemical selectivity or regioselectivity of a reaction, their biggest advantage lies in the differentiation between enantiomeric substrates, a pair of substrates with Gibbs free enthalpy differences between the *R*- and the *S*-enantiomer ΔG_{RS} of around 1–3 kJ mol^{-1}. With enzymes, enantioselectivities of > 99% e.e. can be achieved routinely, although by no means in every case. This fact becomes increasingly important for using biocatalysts in the synthesis of advanced pharmaceutical intermediates, as regulatory agencies require separate toxicological studies for every impurity comprising above 1% of the content (Chapter 13, Section 13.1.4) (Crossley, 1995).

The fact that enzymes are active mostly at mild, near-ambient conditions of temperature and pH and preferentially in aqueous media is often regarded as an advantage rather than a drawback nowadays. Goals for industrial processing such as "sustainable development", "green chemistry", or "environmentally benign manufacturing", an increasingly important boundary condition for industrial activity in a large part of the world, would be much harder to attain without the availability of biocatalysts which tolerate and require such conditions.

Biocatalysts are able to catalyze an increasing breadth of reactions (see Table 1.4). This breadth translates into an increasing number of applications of biocatalysts on [and] industrial scale (Liese, 1999, 2000; Zaks, 2001; Straathof, 2002). Increasingly, biocatalysts are combined with chemical catalysts (Chapter 18) or utilized in a network of reactions in the cell ("metabolic engineering", Chapters 15 and 20).

1.1.2.2 Drawbacks of Current Biocatalysts

There are three essential drawbacks of today's biocatalysts:

1. biocatalysts are often not sufficiently stable in the desired media,
2. too few biocatalysts exist for the desired reactions from available substrates to targeted products, and
3. development cycles are too long for new and improved biocatalysts.

ad 1) *Biocatalysts are often not sufficiently stable in the desired media.* As mentioned above, this still is an essential drawback of biocatalysts. As even conformational changes of less than an Ångstrom can cause a precipitous decline in activity (Carter, 1988), retention of activity is a stringent criterion for the integrity of a protein molecule. Enzymes deactivate under a range of conditions such as extremes of temperature or pH value, physical forces such as cavitation by pumps and aqueous–organic or gas–liquid interfaces (Caussette, 1998), or specific covalent interactions (Quax, 1991; Kelly, 1994; Slusarczyk, 2000).

ad 2) *Too few biocatalysts exist for the desired reactions from available substrates to targeted products.* This argument can be approached from two vantage points. (We do not count the one that a higher demand than supply for biocatalysts speaks for itself.) As Table 1.4 (see below) reveals, there are biocatalysts for almost any reaction. However, most biocatalysts (there are now more than 4000 known) are either not well characterized, or proprietary, or at least not commercially available. The situation, though, improves steadily: just ten years ago, only about a dozen enzymes were available commercially, whereas nowadays the number has increased more than tenfold. Further rapid progress is to be expected.

ad 3) *Development cycles are too long for new and improved biocatalysts.* The development of some flagship biocatalytic processes of today took between 10 and 20 years: development of the acrylamide process took 20 years, of the Lonza process to L-carnithine 15 years. Compared to patent lifetimes around the same length, such durations can easily be deemed far too long for many applications. One reason for such timelines is the as-yet incomplete knowledge base of biotechnology and biocatalysis. With an improved knowledge base stemming from intensive research efforts, development times will certainly be decreased. Shortening the development cycle time for biocatalyts should therefore be a topic of active research. It is interesting, however, to pose the question whether long development times are particular for biocatalysis. Even a cursory inspection of the situation of homogeneous chiral chemical catalysis reveals that the situation is no different than for biocatalysis. While the work of some outstanding researchers such as Ryori Noyori (Nagoya University, Nagoya, Japan), K. Barry Sharpless (Scripps Institute, La Jolla, CA, USA), who both shared the Nobel Prize for 2001, or Eric Jacobsen (Harvard University, Cambridge, MA, USA) has been scaled up and used in industrial processes, widely applicable design rules cannot be laid down yet for homogeneous chiral chemical catalysts.

1.2 Characteristics of Biocatalysis as a Technology

1.2.1 Contributing Disciplines and Areas of Application

From different disciplines, biotechnology and biocatalysis are seen from very different angles and perspectives (Figure 1.1). Chemistry and chemists emphasize a *molecularly-oriented* perspective dominated by compounds and transformations, whereas chemical engineering and thus chemical engineers favor a *process-oriented* perspective of reactions and processes; lastly, biology and its practitioners contribute a *systems-oriented* perspective of description at the organism level as well as in their view of evolution.

Different parts of each of the three disciplines are needed for the successful practice of biocatalysis: biochemistry and organic chemistry from chemistry; molecular biology, enzymology, and protein (bio)chemistry from biology; and catalysis, transport phenomena, and reaction engineering from chemical engineering are indispensable. Both biotechnology and biocatalysis are interdisciplinary areas; as most practitioners tend to hail from one of the three major contributing disciplines, hardly anybody has an equally strong command of all the sub-disciplines of biocatalysis.

There are not only many contributing disciplines for biotechnology and biocatalysis, but also many *areas of application:*

- production and transformation of compounds, mainly in the chemical and pharmaceutical industry,
- analytics and diagnostics, mainly in medicine, and
- environmental protection and bioremediation (reconstruction of the environment).

The areas of application differ from the *industries* which apply them; the most important ones are the pharmaceuticals, food, fine chemicals, basic chemicals, pulp and paper, agriculture, medicine, energy production, and mining industries (Figure 1.1).

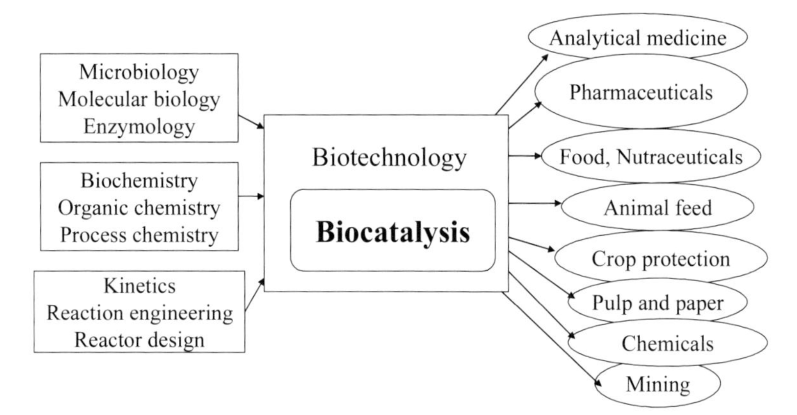

Figure 1.1 Central role of biocatalysis and biotechnology between interdisciplinary feeder sciences (biology, chemistry, chemical engineering science) and multiple user industries.

1.2.2
Characteristics of Biocatalytic Transformations

Biotechnological transformations include a broad range of processes, ordered according to the number of biologically performed process steps and the complexity of the substrates (Figure 1.2):

- *Fermentations* transform raw materials such as sugar, starch or methanol, often in industrial mixtures such as molasses or corn sleep liquor as carbon sources with living cells to more complex target products.
- *Precursor fermentations* start with defined educts and transform these, again with living cells, to the desired target products.
- *Biotransformations* transform defined precursors in a series of defined (not always known) steps with enzymes or resting cells to a desired target product.
- In *enzyme catalysis,* frequently *crude extracts* or *partially purified enzymes,* which only have to be free of side activities, are utilized for the transformation from defined substrate to target product.
- Purified enzymes are rarely used for the production of chemicals, possibly only for the production of highly priced fine chemicals such as pharmaceutical actives.

The limits between the areas are blurred: biotransformations and enzyme catalyses with crude extracts or pure enzymes are often summarized under the term "biocatalysis". "Biocatalytic processes" are taken to mean transformations of a defined substrate to a defined target product with one or several enzyme-catalyzed steps.

Biotechnology and biocatalysis not only differ from conventional processes by featuring a different type of catalyst but they constitute a new technology base:

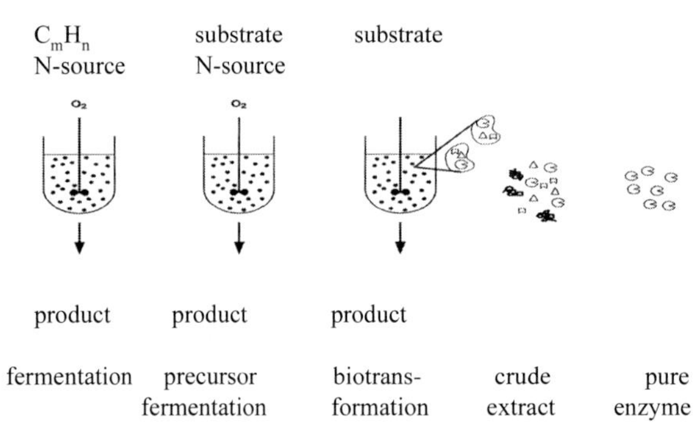

Figure 1.2 Biocatalysis as a continuum between fermentation and transformations with pure enzymes.

- Whereas the *raw materials base* of a conventional process is built on coal or oil, the base for biotechnology is sugar, lignin, or animal or plant wastes.
- Whereas conventional processes are dominated by unit operations such as distillation, often at high pressure or temperature, in biotechnology unit operations, such as membrane processes, chromatography or biocatalysis are prevalent.
- Whereas conventional *products* are often achiral organic molecules or polymers, the product range of biotechnological processes often encompasses chiral molecules or biopolymers such as proteins, nucleic acids, or carbohydrates.

1.2.2.1 Comparison of Biocatalysis with other Kinds of Catalysis

Compared with other kinds of catalysts, for example *homogeneous catalysts,* in which ligands are responsible for specificity, and *heterogeneous catalysts,* in which catalytically active centers are attached to solid carriers such as zeolites or metal oxides, enzymes feature the advantages and disadvantages listed in Table 1.1.

Whereas enzymes often feature great advantages in terms of selectivity, their stability is often insufficient. Additionally, long development times of new biocatalysts owing to an insufficient knowledge base of biocatalysis and biotechnology remain a problem and a challenge.

Table 1.1 Advantages and disadvantages of biocatalysts and enzymes.

Advantages	*Disadvantages*
Very high enantioselectivity	often low specific activity
Very high regioselectivity	instability at extremes of T and pH
Transformation under mild conditions	availability for selected reactions only
Solvent often water	long development times for new enzymes

1.2.3
Applications of Biocatalysis in Industry

1.2.3.1 Chemical Industry of the Future: Environmentally Benign Manufacturing, Green Chemistry, Sustainable Development in the Future

Owing to two very strong and important driving forces the chemical industry of the future will look considerably different from today's version:

- cost and margin pressure resulting from competition in an increasingly open market-oriented economy, and
- operation of the industry in a societal framework which puts emphasis on a clean (or at least less polluted) environment.

Processing with a view towards this new set of conditions focuses on the development of production routes with fewer processing steps, with higher yields on each step, to save material and energy costs. Less waste is generated, and treatment and disposal costs go down. Both pressures come together in the cases of environmental compliance costs.

In many cases, such as high-fructose corn syrup, or biotechnology and biocatalysis offer technology options and solutions that are not available through any other technology; in such situations such as acrylamide, nicotinamide or intermediates for antibiotics, biotechnology and biocatalysis act as "enabling technologies". In the remaining situations, biotechnology and biocatalysis offer one solution among several others, which all have to be evaluated according to criteria developed in Chapter 2: yield to product, selectivity, productivity, (bio)catalyst stability, and space–time-yield.

In this context, the three terms in the title are to a good extent synonymous; nevertheless, they have been developed in a slightly different context:

- *environmentally benign manufacturing* is a movement towards manufacturing systems that are both economically and environmentally sound;
- *sustainable development* is a worldwide Chemical Industry movement and represents a set of guidelines on how to manage resources such that non-renewables are minimized as much as possible;
- *green chemistry* is the design of chemical products and processes that reduce or eliminate the use and generation of hazardous substances.

"Green chemistry is an overarching approach that is applicable to all aspects of chemistry" (Anastas, 2002). Green chemistry methodologies can be viewed through the framework of the "Twelve Principles of Green Chemistry" (Anastas, 1998):

1. It is better to prevent waste than to treat or clean up waste after it is formed.
2. Synthetic methods should be designed to maximize the incorporation of all materials used in the process into the final product.
3. Wherever practicable, synthetic methodologies should be designed to use and generate substances that possess little or no toxicity towards human health and the environment.

4. Chemical products should be designed to preserve efficacy of function while reducing toxicity.
5. The use of auxiliary substances (e.g., solvents, separation agents, etc.) should be made unnecessary wherever possible, and should be innocuous when used.
6. Energy requirements should be recognized for their environmental and economic impacts and should be minimized. Synthetic methods should be conducted at ambient temperatures and pressures.
7. A raw material or feedstock should be renewable rather than depleting wherever technically and economically practicable.
8. Unnecessary derivatization (blocking group, protection/deprotection, temporary modification of physical/chemical processes) should be avoided wherever possible.
9. Catalytic reagents (as selective as possible) are superior to stoichiometric reagents.
10. Chemical products should be designed so that at the end of their function they do not persist in the environment and they do break down into innocuous degradation products.
11. Analytical methodologies need to be further developed to allow for real-time, in-process monitoring and control prior to the formation of hazardous substances.
12. Substances and the form of a substance used in a chemical process should be selected so as to minimize the potential for chemical accidents, including releases, explosions, and fires.

Catalysis offers numerous advantages for achieving green chemistry goals: novel, high-yield, shorter process routes; increased selectivity; and lower temperatures and pressures. Biocatalysis combines the goals of all three topics above. Biocatalysts, as well as many of the raw materials, especially those for fermentations, are themselves completely renewable and for the most part do not pose any harm to humans or animals. Through the avoidance of high temperatures and pressures and of large consumptions of metals and organic solvents, the generation of waste per unit of product is drastically reduced.

1.2.3.2 **Enantiomerically Pure Drugs or Advanced Pharmaceutical Intermediates (APIs)**

The most important property of a catalyst for application in a process is not its activity but rather its selectivity, followed by its stability, which is just activity integrated over time. In comparison with other catalysts, enzymes often feature superior selectivity, especially regio- and enantioselectivity. Enzymes are destined for selective synthesis of molecules with several similar functional groups or chiral centers. A growing emphasis is laid on the synthesis of enantiomerically pure compounds (EPCs). The interest in EPCs shown by all areas of the life science industries (e.g., pharmaceuticals, food and agriculture), stems from the challenge to develop structurally optimized inhibitors, almost always containing chiral centers.

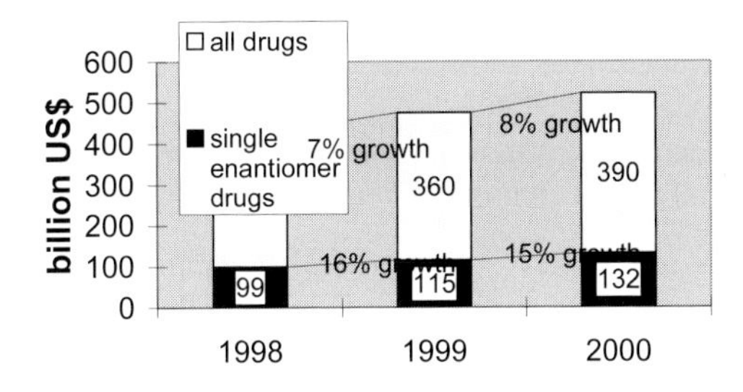

Figure 1.3 Relevance of chiral drugs (May, 2002).

The importance of the drive for enantiomeric purity for biocatalysis can be discerned from the fact that fully 88% of all biocatalytic processes above a scale of 100 kg involve chiral targets (Straathof, 2002). Of the new drugs introduced in 2000, 76% were single enantiomers compared with 21% in 1991 (Agranat, 2002). As Figure 1.3 reveals, the revenues from single-enantiomer drugs have exceeded $ (or €) 100 billion in 2000, compared to $ (€) 500 billion total, and feature a higher growth rate than non-single-enantiomer drugs (155 vs. 8% annual growth) (May, 2002).

The whole of Chapter 13 is devoted to the application of biocatalysis in pharmaceutical syntheses.

1.3 Current Penetration of Biocatalysis

1.3.1 The Past: Historical Digest of Enzyme Catalysis

Table 1.2 provides an overview of historic events in biocatalysis and biotechnology. It demonstrates that biotechnology is an old science, or even an old art. The big ideas and driving forces for biocatalysis in the 20th century were twofold: first, the idea of catalysis as transition-state complementarity in 1944 and second, the development of molecular biology after 1978 to allow the design of enzymes and their production vehicles.

1.3.2 The Present: Status of Biocatalytic Processes

Even a few years ago, biocatalytic processes on an industrial scale were few. As with so many novel technologies, however, the time lag between research, development, and large-scale application just had to pass before we could witness a range of such processes in industrial practice today. In a review summarizing the status of biocatalytic processing in industry, Straathof (2002) records that the number of industrial-scale biocatalytic processes has more than doubled over the last 10 years.

Straathof included 134 industrial-scale biotransformations, counting processes on a scale of > 100 kg with whole cells or enzymes starting from a precursor other than a C-source. Among the highlights of this review were the facts that:

Table 1.2 Historical digest of enzyme catalysis and biotechnology.

Year(s)	*Who?*	*Where?*	*What?*
BC	Unknown	Old World	Chymosin from calf and sheep stomach utilized for production of cheese
1783–1836	Spallanzani		verifies in vitro "digestion" of meat in stomach juice: factor called "pepsin"
1876	Kühne		term "enzyme" for catalysts not bound to living cells ("unorganized ferments")
1877	Eduard Buchner (Nobel prize 1907)	Berlin Agricultural College, Germany	1st alcoholic respiration with cell-free extract: vital force, *vis vitalis*, does not exist
1893	Wilhelm Ostwald	Leipzig Univ., Germany	definition of term "catalyst" (Nobel prize 1909)
1894	Emil Fischer	Berlin Univ., Germany	"lock-and-key" concept (Nobel prize 1902)
1903	Henry D. Dakin	London, UK	1st enantioselective synthesis, with oxynitrilase
1908	Otto Röhm	Darmstadt, Germany	Patent for enzymatic treatment of leather (with trypsin)
1913–1915	Röhm Company	Darmstadt, Germany	1st laundry detergent with enzyme (pancreatin): "Burnus"
1926	James B. Sumner	Cornell Univ., Ithaca, NY, USA	1st enzyme crystallized: urease from jack beans (Nobel prize 1946)
1936	Ernst Sym		lipase reaction in organic solvent
1944	Linus Pauling	Caltech, Pasadena, CA, USA	1st attempt to explain catalysis as transition-state complementarity
1950	Pehr Edman	Univ. of Lund, Sweden	protein degradation developed
1951	Frederick Sanger and Hans Tuppy	Univ. of Cambridge, UK	sequence determination of insulin β-chain: each protein is characterized by a sequence (Nobel prize 1978)
1960		Novo (Bagsvaerd, Denmark)	large-scale protease production from *Bacillus licheniformis* in submerged culture
1963	Stanford Moore and William Stein	Rockefeller Univ., New York, USA	amino acid sequence of lysozyme and ribonuclease elucidated (Nobel prize 1972)
1978	Stanley Cohen and Herbert Boyer	Stanford, CA, USA	method of recombination of DNA developed
1985	Michael Smith	Univ. of British Columbia, Canada	site-directed gene mutagenesis to change enzyme sequence (Nobel prize 1993)
1988	Kary B. Mullis	Cetus Corp., CA, USA	invention of PCR (Nobel prize 1993)
2000	Celera Genomics	Gaithersburg, MD, USA	sequencing of human genome announced (3 billion basepairs)

Table 1.3 Biotransformations on an industrial scale.

Production scale [tpy]	*Product*	*Enzyme*	*Reactor*	*Company*
> 1 000 000	high-fructose corn syrup (HFCS)	glucose isomerase	fixed-bed, IME	various
> 100 000	lactose-free milk	lactase	fixed-bed, IME	various
> 10 000	acrylamide	nitrilase	batch reactor	Nitto Co.
	cocoa butter*	lipase (CRL)	fixed-bed, IME	Fuji Oil
≥ 1,000				
	nicotinamide	nitrilase	3-stage batch	Lonza Guangzhou
	D-pantothenic acid	aldonolactonase		Fuji Pharmaceuticals
	(S)-chloropropionic acid	lipase		Dow Chemical
	6-aminopenillanic acid	penicillin amidase	fixed-bed, IME	various
	7-aminocephalosporanic acid	glutaryl amidase	Kundl/Hoechst	
	aspartame®	thermolysin	soluble enzyme	Tosoh/DSM
	L-aspartate	aspartase	fixed-bed, IME	various
	D-phenylglycine	hydantoinase/ (carbamoylase)	resting cells	Kanegafuchi
	D-*p*-OH-phenylglycine	hydantoinase/ carbamoylase	resting cells	Recordati
> 100	ampicillin	penicillin amidase	stirred IME	DSM–Gist Brocades
	L-methionine, L-valine	aminoacylase	EMR	Degussa (Rexim)
	L-carnitine	dehydrase/ hydroxylase	whole cells	Lonza
	L-dopa	β-tyrosinase		Ajinomoto
	L-malic acid	fumarase	fixed-bed, IME	Tanabe
	(*S*)-methoxyisopropylamine	lipase	repeated batch	BASF
	(*R*)-HPOPS	hydroxylase	batch reactor	BASF
	(*R*)-mandelic acid	nitrilase	batch reactor	BASF
	L-alanine	L-aspartate-β-decarboxylase	IME	various

tpy: (metric) tons per year; IME: immobilized enzyme reactor; EMR: enzyme membrane reactor; L-dopa: 3,4-dihydroxyphenylalanine; (*R*)-HPOPS: (*R*)-2-(4-hydroxyphenoxy)propionic acid.

* Operation depends on the price of substrate palm oil vis-à-vis competing sources.

- hydrolases (44%), followed by oxidoreductases (30%), dominate industrial biocatalytic applications, and
- average performance data for fine chemicals (not pharmaceuticals) applications are 78% yield, a final product concentration of 108 g L^{-1}, and a volumetric productivity of 372 g $(L \cdot d)^{-1}$.

Performance data for biocatalytic reactions and processes will be discussed throughout this book, especially in Chapters 2, 7, 18, and 19. Table 1.3 lists some of the largest-scale biocatalytic processes, categorized by size stated as of tons per year (tpy).

1.4 The Breadth of Biocatalysis

1.4.1 Nomenclature of Enzymes

According to the first report of the Enzyme Commission from 1961, all enzymes are classified in six enzyme classes, depending on the reaction being catalyzed. Within the scheme of identification , each enzyme has a number ("E.C." stands for "Enzyme Commission"):

E.C. a.b.c.d

a: enzyme class (of six)	**b**: enzyme-subclass
c: enzyme sub-subclass	**d**: enzyme sub-sub-subclass (running number)

The number of known enzymes has risen significantly, from 712 in the first edition of *Enzyme Nomenclature* of 1961 through 2477 in 1984 to 3196 in 1992, the year of the third edition. It is important to note that this classification scheme does not organize enzymes according to amino acid sequence or type of three-dimensional structure, and in principle not even according to chemical mechanism.

1.4.2 Biocatalysis and Organic Chemistry, or "Do we Need to Forget our Organic Chemistry?"

The question is often posed of whether biocatalysts operate by different scientific principles from organic catalysts. Careful analysis reveals that they do not (Knowles, 1991; Menger, 1993). Enzymes are "not different, just better" (Knowles, 1991). The multitude of enzyme models including oligopeptidic or polypeptidic catalysts (Chapter 18) proves that all enzyme action can be explained by rational chemical and physical principles. However, enzymes can create unusual and superior reaction conditions, such as extremely low pK_a values for a lysine residue (Westheimer, 1995) or a high positive potential for a redox metal ion (Wittung-Stafshede, 1998).

Enzymes increasingly have been found to catalyze almost any reaction of organic chemistry. Table 1.4 provides examples for a series of reactions.

The principles of how and why enzymes work are discussed in the next chapter.

Table 1.4 Chemical reactions and their chemical and enzymatic manifestations.

Reaction	*E.C. number*	*Enzyme*	*Reference*
Meerwein–Ponndorff–Verley reduction	1.1.1.1	alcohol dehydrogenase	
Oppenauer oxidation	1.1.1.1	alcohol dehydrogenase	
Baeyer–Villiger oxidation	1.14.13.22	BV monooxidase (CHMO)	Taschner, 1988
Ether cleavage	1.14.16.5	glyceryl etherase	Kaufmann, 1997
Disproportionation	1.15.1.1	superoxide dismutase	McCord, 1969
Etherification	2.1.1.6	COMT	Mannisto, 1998
Transamination	2.6.1.x	aminotransaminase	Stirling, 1992
Hydrolysis Phosphate hydrolysis	3.1.–.– 3.1.3.2/1	lipase, esterase phosphatase, acid/alkali	
Esterification	3.4.21.14	subtilisin	
Transesterification	3.4.21.14	subtilisin	
Oximolysis	3.1.1.3	lipase	Gotor, 1990
Electrophilic addition	4.2.1.2	fumarase	
Aldol reaction	4.1.2.x	aldolase	Wymer, 2001
Decarboxylation (β-elimination)	4.1.1.12	L-asp. decarboxylase	Chibata, 1984
Mannich reaction	4.1.99.1 4.1.99.2	tryptophanase β-tyrosinase	
Diels–Alder reaction		Diels–Alderase	Oikawa, 1995
Amadori rearrangement	4.1.3.27	PRAI	Jürgens, 2000
Claisen rearrangement	5.4.99.5	chorismate mutase	Pawlak, 1989
Racemization	5.1.2.2	mandelate racemase	Kenyon, 1995
Isomerization	5.3.1.5	glucose isomerase	
Kolbe–Schmitt reaction	6.4.–.–	phenol carboxylase	Aresta, 1998
Ligation	6.5.1.1	DNA ligase	

BV monooxygenase: Baeyer–Villiger monooxygenase; CHMO: cyclohexanone monooxygenase; COMT: catechol *O*-methyltransferase; PDC: pyruvate decarboxylase; PRAI: phosphoribosyl anthranilate isomerase.

Suggested Further Reading

Reports

The Application of Biotechnology in Industrial Sustainability, OECD, **2001**.

Books

K. Drauz and H. Waldmann (eds.), *Enzyme Catalysis in Organic Synthesis*, Wiley-VCH, Weinheim, 2nd edition, **2002**.

K. Faber, *Biotransformations in Organic Chemistry*, Springer-Verlag, Berlin, 4th edition, **2000**.

A. Liese, K. Seelbach, and C. Wandrey, *Industrial Biotransformations: A Collection of Processes*, VCH, Weinheim, **2000**.

R. N. Patel (ed.), *Stereoselective Biocatalysis*, Marcel Dekker, Amsterdam, **1999**.

A. J. J. Straathof and P. Adlercreutz (eds.), *Applied Biocatalysis*, Taylor & Francis, 2nd edition, **2000**.

Review Articles

B. C. Buckland, D. K. Robinson, and M. Chartrain, Biocatalyis for pharmaceuticals – status and prospects for a key technology, *Metabolic Eng.* **2000**, *2*, 42–48.

A. Liese and M. V. Filho, Production of fine chemicals using biocatalysis, *Curr. Opin. Biotechnol.* **1999**, *10*, 595–603.

R. N. Patel, Biocatalytic synthesis of intermediates for the synthesis of chiral drug substances, *Curr. Opin. Biotechnol.* **2001**, *12*, 584–604.

J. P. Rasor and E. Voss, Enzyme-catalyzed processes in pharmaceutical industry, *Appl. Catal.: General* **2001**, *221*, 145–158.

A. Schmid, J. S. Dordick, B. Hauer, A. Kiener, M. Wubbolts, and B. Witholt, Industrial biocatalysis today and tomorrow, *Nature (London)*, **2001**, *409*, 258–268.

A. Zaks, Industrial biocatalysis, *Curr. Opin. Chem. Biol.* **2001**, *5*, 130–136.

References

I. Agranat, H. Caner, and J. Caldwell, Putting chirality to work: the strategy of chiral switches, *Nat. Rev. Drug Discov.* **2002**, *1*, 753–768.

P. T. Anastas and T. C. Williamson, Frontiers in green chemistry, *Green Chem.* **1998**, 1–26.

P. T. Anastas and M. M. Kirchhoff, Origins, current status, and future challenges of green chemistry, *Acc. Chem. Res.* **2002**, *35*, 686–694.

M. Aresta, E. Quaranta, R. Liberio, C. Dileo, and I. Tommasi, Enzymic synthesis of 4-hydroxybenzoic acid from phenol and carbon dioxide: the first example of a biotechnological application of a carboxylase enzyme, *Tetrahedron* **1998**, *54*, 8841–8846.

F. Baneyx, Recombinant protein expression in *Escherichia coli*, *Curr. Opin. Biotechnol.* **1999**, *10*, 411–421.

A. S. Bommarius, M. Schwarm, and K. Drauz, Factors for process design towards enantiomerically pure L- and D-amino acids, *Chimia* **2001**, *55*, 50–59.

P. Carter and J. A. Wells, Dissecting the catalytic triad of a serine protease, *Nature* **1988**, *332*, 564–568.

M. Caussette, A. Gaunand, H. Planche, P. Monsan, and B. Lindet, Inactivation of enzymes by inert gas bubbling: a kinetic study, *Ann. N.Y. Acad. Sci.* **1998**, *864*, 228–233.

I. Chibata, T. Tosa, and S. Takamatsu, Industrial production of L-alanine using immobilized *Escherichia coli* and *Pseudomonas dacunhae*, *Microbiol. Sci.* **1984**, *1*, 58–62.

R. Crossley, *Chirality and the Biological Activity of Drugs*, CRC Press, Boca Raton, **1995**.

U. Eichhorn, A. S. Bommarius, K. Drauz, and H.-D. Jakubke, Synthesis of dipeptides by suspension-to-suspension conversion via thermolysin catalysis – from analytical to preparative scale, *J. Peptide Sci.* **1997**, *3*, 245–251.

A. Fischer, A. S. Bommarius, K. Drauz, and C. Wandrey, A novel approach to enzymatic peptide synthesis using highly solubilizing N^{α}-protecting groups of amino acids, *Biocatalysis* **1994**, *8*, 289–307.

V. Gotor and E. Menendez, Synthesis of oxime esters through an enzymatic oximolysis reaction, *Synlett* **1990**, *(11)*, 699–700.

C. Jürgens, A. Strom, D. Wegener, S. Hettwer, M. Wilmanns, and R. Sterner, Directed evolution of a (beta alpha)(8)-barrel enzyme to catalyze related reactions in two different metabolic pathways, *Proc. Natl. Acad. Sci.* **2000**, *97(18)*, 9925–9930.

F. Kaufmann, G. Wohlfarth, and G. Diekert, Isolation of *O*-demethylase, an ether-cleaving enzyme system of the homoacetogenic strain MC, *Arch. Microbiol.* **1997**, *168*, 136–142.

S. T. Kelly and A. L. Zydney, Effects of intermolecular thiol–disulfide interchange reactions on BSA fouling during microfiltration, *Biotech. Bioeng.* **1994**, *44(8)*, 972–982.

G. L. Kenyon, J. A. Gerlt, G. A. Petsko, and J. W. Kozarich, Mandelate racemase: structure–function studies of a pseudosymmetric enzyme, *Acc. Chem. Res.* **1995**, *28*, 178–186.

H. Kleinkauf and H. von Doehren, A nonribosomal system of peptide biosynthesis, *Eur. J. Biochem.* **1996**, *236*, 335–351.

J. R. Knowles, Enzyme catalysis: not different, just better, *Nature* **1991**, *350*, 121–124.

A. Liese and M. V. Filho, Production of fine chemicals using biocatalysis, *Curr. Opin. Biotechnol.* **1999**, 10, 595–603.

A. Liese, K. Seelbach, and C. Wandrey, *Industrial Biotransformations: A Collection of Processes*, VCH, Weinheim, **2000**.

P. T. Mannisto, Catechol *O*-methyltransferase: characterization of the protein, its gene, and the preclinical pharmacology of COMT inhibitors, *Adv. Pharmacol.* **1998**, *42*, 324–328.

J. M. McCord and I. Fridovich, Superoxide dismutase. Enzymic function for erythrocuprein (hemocuprein), *J. Biol. Chem.* **1969**, *244*, 6049–6055.

F. M. Menger, Enzyme reactivity from an organic perspective, *Acc. Chem. Res.* **1993**, *26*, 206–212.

G. P. Moss, *Enzyme Nomenclature*, Academic Press, San Diego, CA, **1992**, and later Supplements.

H. Oikawa, K. Katayama, Y. Suzuki, and A. Ichihara, Enzymic activity catalyzing exo-selective Diels–Alder reaction in solanapyrone biosynthesis, *J. Chem. Soc., Chem. Commun.* **1995**, 1321–1322.

J. L. Pawlak, R. E. Padykula, J. D. Kronis, R. A. Aleksejczyk, and G. A. Berchtold, Structural requirements for catalysis by chorismate mutase, *J. Am. Chem. Soc.* **1989**, *111*, 3374–3381.

W. J. Quax, N. T. Mrabet, R. G. M. Luiten, P. W. Schuurhuizen, P. Stanssens, and I. Lasters, Enhancing the thermostability of glucose isomerase by protein engineering, *Bio/Technology* **1991**, *9*, 738–742.

J. P. Rasor and E. Voss, Enzyme-catalyzed processes in pharmaceutical industry, *Appl. Catal. A: General* **2001**, *221*, 145–158.

J. D. Rozzell, Biocatalysis at commercial scale: myths and realities, *Chimica Oggi* **1999**, *(6/7)*, 42–47.

H. Slusarczyk, S. Felber, M. R. Kula and M. Pohl, Stabilization of NAD-dependent formate dehydrogenase from *Candida boidinii* by site-directed mutagenesis of cysteine residues, *Eur. J. Biochem.* **2000**, *267*, 1280–89.

D. I. Stirling, The use of aminotransferases for the production of chiral amino acids and amines, in *Chirality in Industry*, A. N. Collins, G. N. Sheldrake, and J. Crosby (eds.), Wiley, New York, **1992**, pp. 209–222.

A. J. J. Straathof, S. Panke, and A. Schmid, The production of fine chemicals by biotransformations, *Curr. Opin. Biotechnol.* **2002**, *13*, 548–556.

M. J. Taschner and D. J. Black, The enzymatic Baeyer–Villiger oxidation: enantioselective synthesis of lactones from mesomeric cyclohexanones, *J. Am. Chem. Soc.* **1988**, *110*, 6892–6893.

F. H. Westheimer, Coincidences, decarboxylation, and electrostatic effects, *Tetrahedron* **1995**, *51*, 3–20.

P. Wittung-Stafshede, E. Gomez, A. Ohman, R. Aasa, R. M. Villahermosa, J. Leckner, B. G. Karlsson, D. Sanders, J. A. Fee, J. R. Winkler, B. G. Malmstrom, H. B. Gray, and M. G. Hill, High-potential states of blue and purple copper proteins, *Biochim. Biophys. Acta* **1998**, *1388*, 437–443.

N. J. Wymer, L. V. Buchanan, D. Henderson, N. Mehta, C. H. Botting, L. Pocivavsek, C. A. Fierke, E. J. Toone, and J. H. Naismith, Directed evolution of a new catalytic site in 2-keto-3-deoxy-6-phosphogluconate aldolase from *Escherichia coli*, *Structure* **2001**, 9, 1–10.

A. Zaks, Industrial biocatalysis, *Curr. Opin. Chem. Biol.* **2001**, 5, 130–136.

2
Characterization of a (Bio-)catalyst

Summary

Enzymes are a class of macromolecules with the ability both to bind small molecules and to effect reaction. Stabilizing forces such as hydrophobic effects only slightly dominate destabilizing forces such as Coulombic forces of equal polarity; thus the Gibbs free enthalpy of formation of proteins, $\Delta G_{formation}$, is only weakly negative.

In an enzyme reaction, initially free enzyme E and free substrate S in their respective ground states initially combine reversibly to an enzyme–substrate (ES) complex. The ES complex passes through a *transition state,* $\Delta G_{tr,cat}$, on its way to the enzyme–product (EP) complex and then on to the ground state of free enzyme E and free product P. From the formulation of the reaction sequence, a *rate law,* properly containing only observables in terms of concentrations, can be derived. In enzyme catalysis, the first rate law was written in 1913 by Michaelis and Menten; therefore, the corresponding kinetics is named the Michaelis–Menten mechanism. The rate law according to Michaelis–Menten features saturation kinetics with respect to substrate (zero order at high, first order at low substrate concentration) and is first order with respect to enzyme.

Important milestones in the rationalization of enzyme catalysis were the lock-and-key concept (Fischer, 1894), Pauling's postulate (1944) and induced fit (Koshland, 1958). Pauling's postulate claims that enzymes derive their catalytic power from *transition-state stabilization;* the postulate can be derived from transition state theory and the idea of a thermodynamic cycle. The Kurz equation, $k_{cat}/k_{uncat} \approx K_S/K_T$, is regarded as the mathematical form of Pauling's postulate and states that transition states in the case of successful catalysis must bind much more tightly to the enzyme than ground states. Consequences of the Kurz equation include the concepts of effective concentration for intramolecular reactions, cooperativity of numerous interactions between enzyme side chains and substrate molecules, and diffusional control as the upper bound for an enzymatic rate.

Enzymes demonstrate *rate accelerations* (k_{cat}/k_{uncat} ratios) of $\leq 1.4 \cdot 10^{17}$ and *proficiencies* [$k_{cat}/(k_{uncat}\ K_M)$ ($= 1/K_T$)], of $\leq 10^{23}$ M^{-1}. Surprisingly, for a range of enzymatic reactions k_{cat} is within two orders of magnitude whereas k_{uncat} varies by more than six orders of magnitude; the highest rate accelerations are observed for reactions with low k_{uncat}.

Every (bio)catalyst can be characterized by the three basic dimensions of merit – *activity*, *selectivity* and *stability* – as characterized by turnover frequency (tof) (= $1/k_{cat}$), enantiomeric ratio (E value) or purity (e.e.), and melting point (T_m) or deactivation rate constant (k_d). The dimensions of merit important for determining, evaluating, or optimizing a *process* are (i) product yield, (ii) (bio)catalyst productivity, (iii) (bio)catalyst stability, and (iv) reactor productivity. The pertinent quantities are turnover number (TON) (= [S]/[E]) for (ii), total turnover number (TTN) (= mole product/mole catalyst) for (iii) and space–time yield [kg $(L \cdot d)^{-1}$] for iv). Threshold values for good biocatalyst performance are $k_{cat} > 1\ s^{-1}$, $E > 100$ or e.e. > 99%, TTN > 10^4–10^5, and s.t.y. > 0.1 kg $(L \cdot d)^{-1}$.

2.1 Characterization of Enzyme Catalysis

2.1.1 Basis of the Activity of Enzymes: What is Enzyme Catalysis?

Enzymes are a class of multifunctional, multivalent macromolecules with the ability to bind small molecules and, much more importantly, subsequently to effect reaction. Stabilizing forces such as hydrophobic effects (Tanford, 1961) only slightly dominate destabilizing forces such as Coulombic forces of equal polarity; thus the Gibbs free enthalpy of formation, $\Delta G_{formation}$, is weakly negative. Without exception, enzymes are proteins and consist of one or more linear chains of, with rare exceptions, the common 20 proteinogenic amino acids. The fairly frequent disulfide bonds between cysteines are formed through crosslinking of side-chain interactions but do not constitute branching of the backbone chain. Extremely rarely, an additional genetically coded 21st amino acid is found, such as selenocysteine in formate dehydrogenase or glutathione peroxidase (Chambers, 1986; Zinoni, 1986), or pyrrolysine, the "22nd" amino acid, in methanogens (belonging to the domain of archaea (Chapter 3; Hao, 2002; Srinivasan, 2002), or *p*-fluorophenylalanine in a redesigned *E. coli* cell (Mehl, 2003). All three examples used a stop codon to code for the unusual amino acid.

Non-proteinogenic amino acids are found much more commonly in non-enzyme proteins. One prominent example is collagen, from gelatin, which mainly consists of triplets of hydroxyproline-glycine-proline, in which the hydroxyproline is generated via posttranslational hydroxylation. Other proteins apart from enzymes exist that can bind substrates and catalyze reactions, such as catalytic antibodies (Chapter 18, Section 18.1; Pollack, 1986; Tramontano, 1986) or ribozymes (Chapter 18, Section 18.2; Cech, 1986, 1993). On the other hand, there are many proteins that do not catalyze any reaction, and thus are not enzymes, but function as oxygen storage and shuttle units (hemoglobin and myoglobin), receptors (signal transduction proteins such as the EGF-receptor) or as structural proteins (myelin).

2.1.1.1 Enzyme Reaction in a Reaction Coordinate Diagram

The idea of an enzyme reaction consisting of both a binding step and a subsequent reaction step is embodied in a free enthalpy-reaction coordinate (ΔG–ξ) diagram depicting the change in Gibbs free enthalpy ΔG over the extent of reaction ξ (zeta) (Figure 2.1). In an uncatalyzed reaction, one or more substrates or reactants initially are in their respective *ground states;* during the reaction, the reactants pass the point of maximum free enthalpy $\Delta G_{tr,uncat}$, termed the *"transition state"*, and continue to the ground state of the product(s).

In an enzyme reaction, initially free enzyme E and free substrate S likewise are in their respective ground states. From the ground state, enzyme and substrate combine reversibly to an enzyme–substrate (ES) complex. If [S] > K_M, the ES complex forms an *"intermediate"*, i.e., a local minimum of free enthalpy. (If [S] < K_M, the free enthalpy of the ES complex is higher than the enzyme's ground state; at [S] = K_M, both free enthapy levels are the same.) The ES complex passes through another transition state, $\Delta G_{tr,cat}$, on its way to the enzyme–product (EP) complex and then on to the ground state of free enzyme E and free product P.

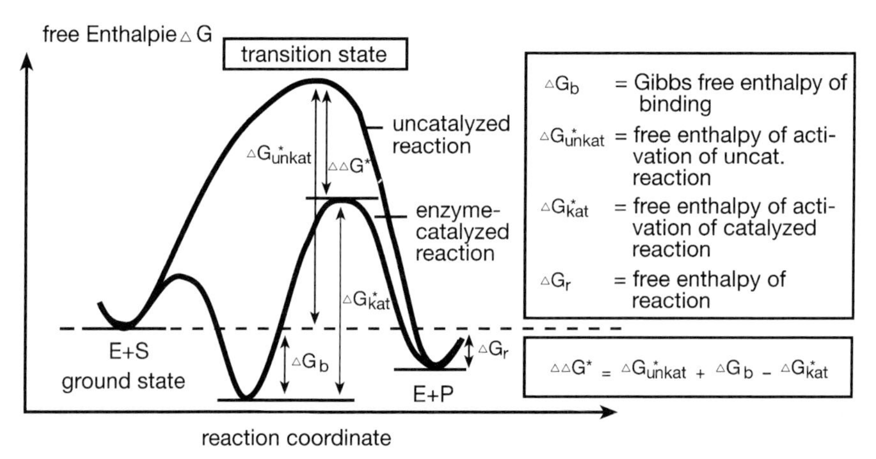

Figure 2.1 Free enthalpy reaction coordinate diagram for an enzyme reaction.

The maximum of the Gibbs free enthalpy between the ground states of substrate and product forms the Gibbs free enthalpy of activation with the energy difference $\Delta G^{\neq}$, which determines the rate constant of the reaction. Like every catalyst, an enzyme decreases the value of $\Delta G^{\neq}$ and thus accelerates the reaction. (An agent increasing the value of $\Delta G^{\neq}$ is termed an "anti-catalyst".)

2.1.2 Development of Enzyme Kinetics from Binding and Catalysis

From the idea of enzyme kinetics as a binding and a reaction step with the corresponding course of the energy curve in the Gibbs free enthalpy–reaction coordinate ($\Delta G - \xi$) diagram, the reaction scheme represented by Eq. (2.1) can be drawn.

$$E + S \Leftrightarrow ES \rightarrow EP \rightarrow (\Leftrightarrow) E + P \quad (2.1)$$

Among the several ways of verifying or disproving such a reaction scheme (Chapter 9, Section 9.2), the derivation of a *rate law* linking a product formation rate or substrate consumption rate with pertinent concentrations of reactants, products, and auxiliary agents such as catalysts probably has the greatest utility, as conversion to product can be predicted. A proper rate law contains only *observables,* and no intermediates or other unobservable parameters. In enzyme catalysis, the first rate law was written in 1913 by Michaelis and Menten (the corresponding kinetics is therefore aptly named the Michaelis–Menten (MM) mechanism).

The kinetic schemes for the description of enzyme reactions are similar to those for homogeneous and heterogeneous catalysis and were developed roughly contemporaneously. In heterogeneous catalysis, a mechanism postulating a reaction of a mobile substrate molecule on a solid surface is modeled by a rate law named after Langmuir and Hinshelwood, and later also after Hougen and Watson (Hougen, 1947) (LHHW). LHHW and MM both assume a limiting number of active sites which can be saturated with reacting molecules from a continuous phase.

The kinetic scheme according to Michaelis–Menten for a one-substrate reaction (Michaelis, 1913) assumes three possible elementary reaction steps: (i) formation of an enzyme–substrate complex (ES complex), (ii) dissociation of the ES complex into E and S, and (iii) irreversible reaction to product P. In this scheme, the product formation step from ES to E + P is assumed to be rate-limiting, so the ES complex is modeled to react directly to the free enzyme and the product molecule, which is assumed to dissociate from the enzyme without the formation of an enzyme–product (EP) complex [Eq. (2.2)].

$$E + S \Leftrightarrow ES \rightarrow E + P \quad (2.2)$$

In the derivation according to Michaelis and Menten, association and dissociation between free enzyme E, free substrate S, and the enzyme–substrate complex ES are assumed to be at equilibrium, $K_S = [ES]/([E] \cdot [S])$. [The Briggs–Haldane derivation (1925), based on the assumption of a steady state, is more general; see Chapter 5, Section 5.2.1.] With this assumption and a mass balance over all enzyme components ($[E]_{total} = [E]_{free} + [ES]$), the rate law in Eq. (2.3) can be derived.

$$v = v_{max}[S]/(K_M + [S]) = k_{cat}[E]_{total}\,[S]/(K_M + [S]) \quad (2.3)$$

In contrast to $[E]_{free}$, $[E]_{total}$ is observable. Eq. (2.3) is written with K_M, the Michaelis constant, instead of the equilibrium binding constant K_S; unless the enzyme reaction is very fast (Section 2.3.3.); i.e., in almost all cases, $K_M \approx K_S$. In Eq. (2.3), the reaction rate is traditionally denoted by v [concentration/time] and k_{cat} is the reaction rate constant [time^{-1}]. The equation describes a two-parameter kinetics, with a monotonically rising reaction rate with respect to substrate concentration and saturation at high substrate concentration. The maximum reaction rate at saturation is denoted by v_{max}, with $v_{max} = k_{cat}[E]$. The K_M value corresponds to the substrate concentration at half saturation ($v_{max}/2$) and is a measure of binding affinity of the substrate to the enzyme: a high K_M value corresponds to loose binding, and a low value to tight binding between enzyme and substrate. At *low* substrate concentration, more precisely if $[S] \ll K_M$, Eq. (2.3) simplifies to $v = v_{max}[S]/K_M$ and consequently is *first* order with respect to the substrate concentration [S]. In contrast, at *high* substrate concentration, more precisely if $[S] \gg K_M$, Eq. (2.3) simplifies to $v = v_{max}$ and consequently is *zeroth* order with respect to [S]. In all situations, v is proportional (first-order with respect) to [E].

2.2 Sources and Reasons for the Activity of Enzymes as Catalysts

2.2.1 Chronology of the Most Important Theories of Enzyme Activity

Just one year after Ostwald's hypothesis about the existence of catalysts in 1893, when nobody yet had a clear idea of the structure and composition of enzymes, Emil Fischer voiced the idea for the first time that a substrate molecule fits into the pocket of an enzyme, the "lock-and-key hypothesis" (Fischer, 1894). Both the lock (enzyme) as well as the key (substrate) were regarded as rigid.

This hypothesis was modified later in many ways. According to Haldane (1965), catalysis of a reaction occurs only if a catalyst in the active center is complementary to the *transition state* of the substrate during the reaction. Therefore, the transition state between substrate and products fits best into a pocket close to the enzyme. The substrate molecule is subject to *strain* upon binding to the active center and changes its conformation to fit into the active center; the key (the substrate) does not fit completely into the lock but is strained and bent.

In a further modification of this concept, a reaction is accelerated if a catalyst stabilizes the transition state; in contrast, stabilization of the ground state leads to a deceleration of the reaction. This concept of *transition-state stabilization*, formulated first by Haldane, was expanded later by Linus Pauling (Pauling 1946, 1948). The notion of lowering of $\Delta G^{\neq}$ attributed to the stabilization of the transition state by the catalyst or to the destabilization of the ground state in comparison to the transition state is generally accepted nowadays.

Instead of assuming a "solid" enzyme, in which active center the substrate molecule is "bent" (the concept of *substrate strain*), the idea was developed (Koshland 1958, 1966) that enzymes can embrace the substrate molecule flexibly in the active center and effect reaction by the formation of specific interactions, the so-called *"induced fit"*. This picture is especially appropriate with allosterically activated enzymes or in situations in which part of the enzyme molecule has to turn or move over longer distances to effect catalysis *(hinge movement)*, as for instance with most NAD(P)(H)-dependent enzymes (Stillman, 1999).

An equation for the acceleration of the reaction rate by a deliberate catalyst and the relative strength of the binding to the transition state and ground state was formulated in 1963 by Kurz, who combined the ideas of the thermodynamic cycle and transition-state theory and whose result, unrecognized by Kurz, can be regarded as a quantification of the Pauling postulate (Kurz, 1963). Just a few years later Jencks originated the idea of transition-state analogs as inhibitors and cited references in the literature (Jencks, 1969). Finally, it was observed (Kraut, 1977) that the best inhibitors for serine proteases such as subtilisin, chymotrypsin, trypsin or elastase all led to geometrically similar transition states for the hydrolysis reaction and all bound in the complementary oxyanion hole. All four proteases belonged to the small number of enzymes whose three-dimensional structure was known by then.

2.2.2 Origin of Enzymatic Activity: Derivation of the Kurz Equation

The key equation resulting from the application of two known theories, the Born–Haber cycle and the transition-state theory, was formulated for any catalyst and without reference to enzymes (Kurz, 1963); a good derivation can be found in the article by Kraut (1988).

Transition-state theory is based on two assumptions, the existence of both a dynamic bottleneck and a preceding equilibrium between a transition-state complex and reactants. Eq. (2.4) results, where k denotes the observed reaction rate constant, κ the transmission coefficient, and ν the mean frequency of crossing the barrier.

$$k = \kappa \cdot \nu \cdot K^{\neq} \qquad (2.4)$$

All correction factors such as tunneling, back-crossing of the barrier, and solvent frictional effects are captured by κ, which is between 0.1 and 1 for the reactions in solution in question here. The equilibrium constant $K^{\neq}$ is expressed by partition functions, where the mode normal to the reaction coordinate ν is approximated by the term $k_B T/h\nu$. In the resulting Eq. (2.5), ν cancels out.

$$k = \kappa \cdot (k_B T/h) \cdot K^{\neq\prime} \qquad (2.5)$$

Note that $K^{\neq\prime}$ is not the thermodynamic equilibrium constant and $k_B T/h$ is not a universal frequency for the decomposition of the transition-state complexes into products.

The thermodynamic cycle relevant for further discussion is shown in Figure 2.2.

If one compares the first-order rate constant for the enzyme-catalyzed single-substrate elementary reaction k_e with the one for the uncatalyzed reaction k_u, Eq. (2.6) is obtained with Eq. (2.4).

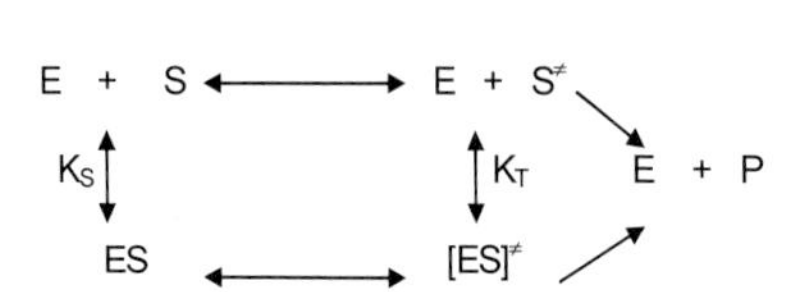

Figure 2.2 Thermodynamic (Born–Haber)-cycle.

$$k_e/k_u = (\kappa_e \cdot \nu_e \cdot K_e^{\neq})/(\kappa_u \cdot \nu_u \cdot K_u^{\neq}) \tag{2.6}$$

In the case of a simple elementary enzyme reaction, k_e is identical to k_{cat}. By utilizing the thermodynamic cycle the quotient of the constants of formation of the transition state $K_e^{\neq}/K_u^{\neq}$ can be equated with the quotient of the dissociation constants for substrate K_S and transition state K_T [Eq. (2.7)].

$$k_{cat}/k_u = (\kappa_e \cdot \nu_e \cdot K_S)/(\kappa_u \cdot \nu_u \cdot K_T) \tag{2.7}$$

Although only a few data are available for the comparison of $\kappa_e \cdot \nu_e$, and $\kappa_u \cdot \nu_u$, it is assumed that the two terms do not differ much in magnitude, so that Eq. (2.8) holds approximately.

$$k_{cat}/k_u \approx K_S/K_T \tag{2.8}$$

This is the equation derived by Kurz, expressed here for the case of an enzyme reaction. The ratio of enzyme-catalyzed to uncatalyzed reaction k_{cat}/k_u or k_{cat}/k_{uncat}, i.e., the *rate acceleration,* often reaches orders of magnitude of 10^{10} to 10^{12} (Radzicka, 1995), and in the best case so far $1.4 \cdot 10^{17}$ (orotidine monophosphate decarboxylase, OMPD), which translates into a proficiency $k_{cat}/(k_{uncat} \cdot K_M)$ of 10^{23} M^{-1} (Lee, 1997). This underscores the quality of enzymes as catalysts. In comparison, the best value for the rate acceleration of catalytic antibodies is $2.3 \cdot 10^8$ (Zhong, 1999), and typical values are 10^3 to 10^5 (Hasserodt, 1999). Radzicka (1995) found that k_{cat} for the reactions in question is often within one or two orders of magnitude; interestingly, the highest efficiencies are observed with reactions of very low k_{uncat}. Thus, Eq. (2.8) means that transition states in the case of successful catalysis must bind much more tightly to the enzyme than ground states; in this way, Eq. (2.8) is the mathematical form of the Pauling postulate (Pauling 1944, 1948). It is emphasized again that the action of enzymes through the stabilization of the transition state is not an independent concept but is derived from transition-state theory and the idea of a thermodynamic cycle.

2.2.3
Consequences of the Kurz Equation

The Kurz equation [Eq. (2.8)] has simplified and channeled many discussions in enzymology. Many other explanations of enzyme action and many phenomena observed with enzyme reactions can be reduced to the validity of Eq. (2.8), as shown in the following paragraphs.

- The principle of transition state stabilization can explain why many enzymes show higher activity against larger, sterically more demanding substrate molecules than against smaller ones. Paradoxically, the cause of the enhanced activity was an increased k_{cat}, not a decreased K_M value, i.e., an enhanced maximum rate and not a tighter substrate binding. This phenomenon led to the postulation of "induced fit". But if the enzyme template binds the transition state more tightly than the ground state, it can be expected that peripheral parts of the substrate molecule have an important role for the binding in the transition state but not necessarily for the binding in the ground state. Conformational changes of the enzyme template do not need to be postulated. Frequently, however, conformational changes during binding and catalysis are observed which are then also called "induced fit" by several authors.

- The exact loci of binding and catalysis cannot be distinguished exactly. In subtilisin, the amino acids forming the catalytic triad, Ser221–His64– were replaced in all combinations by Ala. Even the triple mutant without any amino acid from the original catalytic center displayed a 1000-fold higher reaction rate than the uncatalyzed reaction, and the remainder of the enzyme molecule bound the substrate better in the transition state than in the ground state (Carter, 1988).

- The numerous interactions between enzyme side chains and substrate molecules act synergistically, so the fit is quite exact. This cooperativity enhances substrate specificity, for a small change of substrate can effect a large change in the binding of the transition state. Triose phosphate isomerase is only 1/1000 as effective if the residue Glu165 is moved away from the substrate by even 1 Å, as was demonstrated with the exchange of Glu165 for Asp165 (Joseph-McCarthy, 1994). The change of only one amino acid of the triad in subtilisin causes an activity decrease of 10^{-6}, whereas the exchange of all three causes a decrease of only 7×10^{-7} and not of 10^{-18} (Carter, 1988). The results seem to be explained by an all-or-nothing situation rather than additivity.

- Intramolecular reactions proceed much more rapidly than intermolecular ones owing to entropic stabilization. Page and Jencks have calculated the entropies of translation, rotation, and vibration (Page, 1971). The entropy loss of the reactants already realized in the enzyme–substrate (ES) complex allows for a much faster reaction than in the situation without complexation, binding, or geometric orientation. Introduction of an *effective concentration* $[S]_{eff}$ alleviates the trouble of comparing results of *inter*molecular catalysis with rate constants of dimension $[\text{concentration} \cdot \text{time}]^{-1}$ (second-order reaction) with those of *intra*molecular catalysis with constants of dimension $[\text{time}]^{-1}$. The high values often achieved for $[S]_{eff}$ (often > 10 M) reflect the efficiency of intramolecular catalysis. In this context, enzymes are often referred to as "entropy traps", because many contributions to entropy are "frozen in" after binding of the substrate to the enzyme molecule.

- Many enzymatic reaction mechanisms are accelerated or even only made possible by the absence of water or other solvents. The enzyme supports desolvation by redirecting part of the possible binding energy liberated from the binding between enzyme and substrate.

- Upper boundary of rate constant of an enzyme reaction: diffusion control

The upper boundary of the reaction rate is reached when every collision between substrate and enzyme molecules leads to reaction and thus to product. In this case, the Boltzmann factor, $\exp(-E_a/RT)$, is equal to 1in the transition-state theory equations and the reaction is *diffusion-limited* or *diffusion-controlled* (owing to the difference in mass, the reaction is controlled only by the rate of diffusion of the substrate molecule). The reaction rate under diffusion control is limited by the number of collisions, the frequency Z of which can be calculated according to the Smoluchowski equation [Smoluchowski, 1915; Eq. (2.9)].

$$-dc/dt = -8\pi RD \cdot c^2 \tag{2.9}$$

R is the normalized radius of both participating particles $(1/r_1 + 1/r_2)$, D the normalized diffusion coefficient $(1/D_1 + 1/D_2)$. The collision frequency Z is then calculated from Eq. (2.10).

$$Z = 2RT/(3000\eta \cdot [(r_1 + r_2)^2/r_1r_2]) \tag{2.10}$$

As the enzyme molecule is much larger than the substrate or product molecule ($r_E \gg r_S$) and thus diffuses much more slowly ($D_E \ll D_S$), Eqs. (2.9) and (2.10) can be simplified to Eqs. (2.11) and (2.12).

$$-dc/dt = -8\pi RD_E \cdot c^2 \tag{2.11}$$

$$Z = 2RT/(3000\eta \cdot r_E) \tag{2.12}$$

In the case of a highly viscous solution the influence of viscosity η can dominate (Eichhorn, 1997). For simple, uncharged particles in water at 25 °C the second-order rate constant is $3.2 \times 10^9\ (\mathrm{M} \cdot \mathrm{s})^{-1}$ (Gerischer, 1969; Creutz, 1977). Some cases of wholly or partially diffusion-controlled enzyme reactions are listed in Table 2.1. Rearrangement of Eq. 2.7 results in Eq. 2.13.

$$k_{cat}/K_S = (k_u \cdot \kappa_e \cdot \nu_e)/(\kappa_u \cdot \nu_u \cdot K_T) \tag{2.13}$$

Eq. (2.13) is the manifestation of the second-order rate constant for a reaction between free enzyme and free substrate to give free enzyme and free product; thus, the rate constant cannot exceed the maximum value for k_{diff}, i.e., $3.2 \times 10^9\ \mathrm{M}^{-1} \cdot \mathrm{s}^{-1}$.

Table 2.1 Wholly or partially diffusion-controlled enzyme reactions (adapted from Fersht, 1985).

Substrate	***Enzyme***	k_{cat} ***[s^{-1}]***	K_M ***[M]***	k_{cat}/K_M ***[s^{-1} · M^{-1}]***
$CO_2 \Leftrightarrow HCO_3^-$	Carbonic anhydrase	1×10^6	0.012	8.3×10^7
H_2O_2	Catalase	4×10^7	1.1	4×10^7
fumarate	Fumarase	800	5×10^{-6}	1.6×10^8
benzylpenicillin	*β*-Lactamase	2×10^3	2×10^{-5}	1×10^8

A perfectly evolved enzyme therefore features a decreased binding constant K_s, which means a stronger binding in the transition state, possibly up to the thermodynamic limit.

Limitation of a reaction by translational diffusion in solution is a rather rare case. Much more frequently the limitation of the observed overall reaction rate is by *external mass transfer* (through a laminar film around a solid macroscopic carrier) (Chapter 5, Section 5.5.1) or *internal mass transfer* (diffusion of substrate or product through the pores of a solid carrier or a gel network to an enzyme molecule in the interior of the carrier) (Chapter 5, Section 5.5.2).

2.2.4
Efficiency of Enzyme Catalysis: Beyond Pauling's Postulate

Let us return one more time to the Kurz equation [Eq. (2.8)], which is regarded as the quantification of Pauling's postulate that transition-state stabilization, i.e., the tighter binding between enzyme and transition state, expressed by K_T, as compared to binding between enzyme and ground state, expressed by K_S, is the source of catalytic rate enhancement.

$$k_{cat}/k_{uncat} \approx K_S/K_T \tag{2.8}$$

Identifying K_S with K_M, K_T approximately equals the ratio $(k_{uncat} \cdot K_M)/k_{cat}$, which is the inverse of termed the *"proficiency"*, which has been determined experimentally for a number of systems (see Section 2.3.4 below). If the Kurz equation captured the whole essence of enzyme catalysis, K_T should be proportional to $(k_{uncat} \cdot K_M)/k_{cat}$. However, Figure 2.3 reveals that the correlation of K_T (≡ "K_{TS}") for some enzyme reactions is much better with k_{uncat} (≡ "k_{non}") than with k_{cat} (Bruice, 2000).

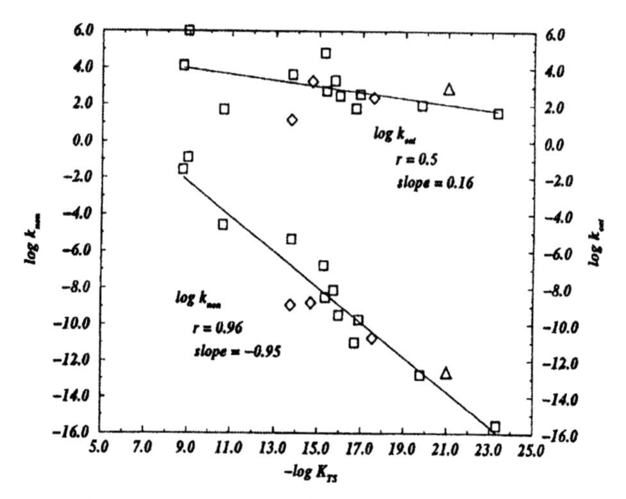

Figure 2.3 Comparison of the relationship of K_{TS} with the enzymatic reaction rate constant, k_{cat}, and with the uncatalyzed solution or reference reaction rate constant k_{non} (Bruice, 2000).

This seeming contradiction to Pauling's postulate and the notion of the transition-state energy and conformation as the sole source of enzymatic catalytic power cast the geometry of the substrate ready to interact with the enzyme, i.e., substrate *preorganization*, into a new and more important role. While the discussion on this subject is far from settled, a novel concept based on molecular mechanics calculations of the stability of different substrate conformations, termed substrate *conformers*, has emerged which emphasized substrate preorganization. The substrate conformers which allow reactions to occur most easily clearly resemble the transition state and are called *"near-attack conformers"* (NACs) (Bruice, 2000, 2002). The relative rate constants of anhydride formation k_{rel} from mono-*p*-bromophenyl esters were found to correlate with the mole fractions (P) of ground-state conformers that could be classified as NACs according to Eq. (2.14) (Bruice, 2002).

$$\log k_{rel} = 0.94 \log P + 7.48 \quad (r^2 = 0.915) \tag{2.14}$$

The results are depicted in Figure 2.4.

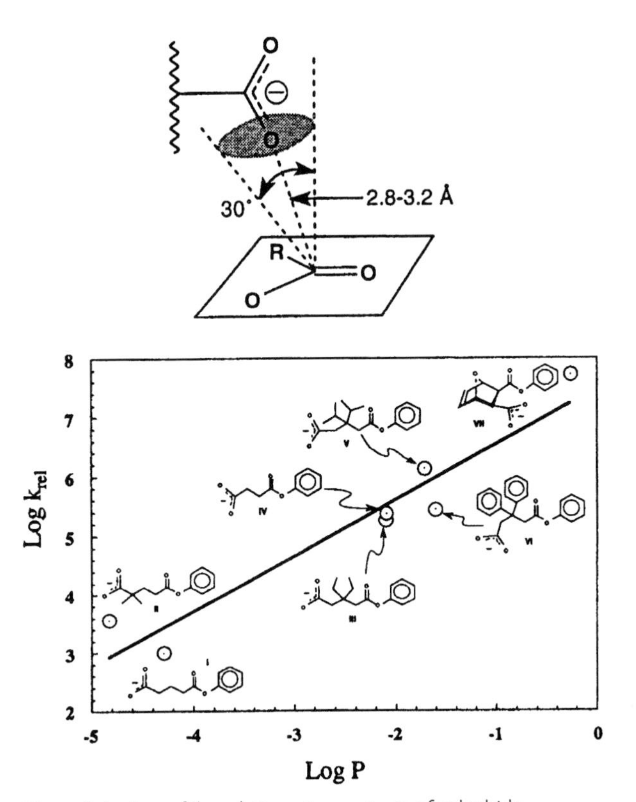

Figure 2.4 Log of the relative rate constants of anhydride formation k_{rel} from mono-*p*-bromophenyl esters vs. the log of the probability (P) of NAC formation of each monophenyl ester of the various dicarboxylic acids (Bruice, 2000).

2.3
Performance Criteria for Catalysts, Processes, and Process Routes

2.3.1
Basic Performance Criteria for a Catalyst: Activity, Selectivity and Stability of Enzymes

Every catalyst, and thus also every biocatalyst, can be characterized by the three basic dimensions of merit, namely *activity*, *selectivity*, and *stability*. Additional, but not frequently employed, performance criteria beyond the basic dimensions are discussed in Section 2.3.3.

1. Biocatalysts often feature much better *selectivity* than non-biological catalysts, so they are developed for use because of their selectivity, be it enantioselectivity, chemo- or regioselectivity.

2. As *activity* is straightforward to measure and necessary to know for even the basic experimental protocol, enzyme activity is often well studied. In contrast to other catalysts, most enzymes are only active and stable in a very limited temperature and pH range (mostly between 15 and 50 °C or pH 5 and 10).

3. Unlike activity, *stability* of enzymes is often interpreted simplistically as *thermal* stability, i.e., a temperature beyond which the enzyme loses stability. Although this quantity is important, first every statement of stability at a certain temperature depends on exposure time and thus is often ambiguous; and second, for biocatalytic process applications, a more important quantity is the *process* or *operational stability*, which is the long-term stability under specified conditions.

2.3.1.1 Activity

The value of the (overall) enzyme activity is usually provided in "International Units" or "Units":

$$1 \text{ Unit} \equiv 1 \text{ IU} \equiv 1 \text{ U} \equiv 1\ \mu\text{mol min}^{-1}$$

In most cases a value for the overall activity of enzyme is not very interesting. Much more important are both the *specific activity*, scaled to the mass of catalyst, and the *volumetric activity*, based on the activity per unit volume:

$$1 \text{ U (mg protein)}^{-1} = 1\ \mu\text{mol (min mg protein)}^{-1} \equiv \text{specific activity}$$

and

$$1 \text{ U mL}^{-1} = 1\ \mu\text{mol (min mL)}^{-1} \equiv \text{volumetric activity}$$

All activity data are meaningless without a specification of the conditions of measurement. Connonen, the activity of enzymes is provided at nearly physiological conditions at 30 °C and pH 7.5, unless specified otherwise. Any enzyme sold commercially comes with information about the detailed assay conditions used to arrive at the specific (solid preparation) or volumetric (liquid preparation) activity. To judge the quality of a biocatalyst, its specific activity can serve as a guideline: the threshold for a useful biocatalyst is 1 U (mg pure protein)$^{-1}$; the threshold for a biocatalyst developed for plant-scale level, however, lies much higher, rather closer to 100 U (mg pure protein)$^{-1}$.

The volumetric activity of a (bio)catalyst can be enhanced by simple addition of more catalyst to a system. Specific activity, however, must be improved through optimization of the reaction conditions, or through variation of the structure of the carrier (Chapter 5) or even of the enzyme (Chapters 10 and 11).

A mass-independent quantity of activity, superior to specific activity, is the turnover frequency (tof), defined as:

$$\text{turnover frequency (tof)} = \frac{\text{number of catalytic events}}{\text{time} \times \text{number of active sites}}$$

The turnover frequency allows performance comparison between different catalyst systems, biological and/or non-biological. Its threshold is at 1 event per second per active site. According to the definition, a turnover frequency can be determined only if the number of active sites is known (Chapter 9, Section 9.2.3). For an enzyme reaction obeying Michaelis–Menten kinetics, Eq. (2.15) holds.

$$\text{tof} = 1/k_{cat} \tag{2.15}$$

2.3.1.2 **Selectivity**

The notion of selectivity needs to be specified further: *point selectivity* is the incremental selectivity, usually towards product, at a specific degree of conversion, whereas *integral selectivity* is the overall average selectivity at the same specific degree of conversion. Owing to the importance of enantiomeric purity of target product molecules in life science applications and the pre-eminent position biocatalysts enjoy with respect to the achievement of that goal, enantioselectivity is the most important kind of selectivity in the context of biocatalysis.

The *enantiomeric ratio* or E value (Chen, 1982) serves as a measure of enantioselectivity at a certain degree of conversion (for the derivation of E, see Chapter 5, Section 5.7) [Eq. (2.16)].

$$E = (k_{cat}/K_M)_A/(k_{cat}/K_M)_B = \ln([A]/[A]_0)/([B]/[B]_0) \tag{2.16}$$

Although the E value is measured at a specific degree of conversion, nevertheless it is an integral selectivity measure. This is even more obvious in the case of the other measure commonly employed, the enantiomeric excess, e.e. [Eq. (2.17)], which reflects overall selectivity up to the point of isolation of the product.

$$\text{e.e.} = ([A] - [B])/([A] + [B]) \times 100\% \tag{2.17}$$

The *threshold* for an enantioselective (bio)catalyst is either at $E = 100$ or at 99% e.e., depending on whether one is evaluating either the reaction itself or a reaction product. [These values are linked to the requirement by the Food and Drug Administration of the United States (FDA), the most important body in the world for setting guidelines for both procedures for novel drug approval and drug performance evaluation, that a separate toxicological study be conducted for any impurity, including an enantiomer of the active drug, exceeding a concentration level of 0.5% (Chapter 13).]

Besides enantioselectivity, regioselectivity (Chapter 12, Section 12.6) and to a lesser extent, chemoselectivity (Chapter 7, Section 7.3.2) are also important issues of enzymatic reaction selectivity.

2.3.1.3 Stability

Stability of an enzyme is usually understood to mean temperature stability, although inhibitors, oxygen, an unsuitable pH value, or other factors such as mechanical stress or shear can decisively influence stability (Chapter 17). The *thermal stability* of a protein, often employed in protein biochemistry, is characterized by the melting temperature T_m, the temperature at which a protein in equilibrium between native (N) and unfolded (U) species, N ⇔ U, is half unfolded (Chapter 17, Section 17.2). The melting temperature of a protein is influenced on one hand by its amino acid sequence and the number of disulfide bridges and salt pairs, and on the other hand by solvent, added salt type, and added salt concentration. Protein structural stability was found to correlate also with the Hofmeister series (Chapter 3, Section 3.4; Hofmeister, 1888; von Hippel, 1964; Kaushik, 1999) [Eq. (2.18)].

$$T_m = T_0 + K[S] \tag{2.18}$$

In Eq. (2.18), T_0 is the melting temperature in the absence of salt, [S] the salt concentration, and K the corresponding coefficient.

Storage stability over time under fixed conditions of temperature, pH value and concentration of additives often can be expressed by a first-order decay law (analogous to radioactive decay) [Eq. (2.19)].

$$[E]_t = [E]_0 \exp(-k_d \cdot t) \tag{2.19}$$

The validity of a first-order decay law over time for the activity of enzymes according to Eq. (2.19), with $[E]_t$ and $[E]_0$ as the active enzyme concentration at time t or 0, respectively, and k_d as the deactivation rate constant, is based on the suitability of thinking of the deactivation process of enzymes in terms of Boltzmann statistics. These statistics cause a certain number of active protein molecules to deactivate momentarily with a rate constant proportional to the amount of active protein [for evidence for such a “catastrophic “decomposition, see Craig (1996)].

The relevant stability criteria for process applications (the *process stability* or *operating stability*) is discussed in the next section.

2.3.2
Performance Criteria for the Process

In evaluating a biocatalyst for a given processing task, there are performance criteria to be met not only for the biocatalyst but also for the *process*. The dimensions of merit important when determining, evaluating, or optimizing a route for a process are (i) *product yield*, (ii) *(bio)catalyst productivity*, (iii) *(bio)catalyst stability*, and (iv) *reactor productivity*.

2.3.2.1 Product Yield

Chemical yield to product is most important for the economics of the process. Product yield is inversely proportional to the amount of reactants required per unit of product output. As the key substrate and other raw materials in most mature processes make up more than 50% of the variable cost, a high product yield is indispensable for an economic process. The yield of product y is linked to selectivity σ and the degree of conversion x by Eq. (2.20).

$$y = x \cdot \sigma \tag{2.20}$$

so that the product concentration [P] can be calculated from Eq. (2.21).

$$[\mathrm{P}] = [\mathrm{S}]_0 \cdot y = [\mathrm{S}]_0 \cdot x \cdot \sigma \tag{2.21}$$

Yields much less than 100% are no longer acceptable either economically or ecologically. In the separation of a racemate, where the yield per run is limited to 50% per pass, this means that internal or external racemization is necessary unless either the substrate is inexpensive enough to lose up to 50% or the co-product can be marketed in similar amounts. Both of the latter situations are highly unlikely.

A threshold for a sufficiently high yield is hard to define, as the threshold value seems to correlate inversely with the unit value of the product. While for basic, large-volume chemicals yields of 98 or 99% are absolutely essential, the situation in fine chemicals calls for 90–95% yield, and in the initial stage of production of extreme performance chemicals, such as pharmaceuticals, yields of > 80% are very acceptable; sometimes values down to 50% have to be encountered. Acceptable yields depend on the number of process steps, including product isolation. If all the steps are assumed to fetch 90% yield, the overall yield depends on the number of steps n as in Eq. (2.22).

$$y_{\mathrm{overall}} = (y_{\mathrm{step}})^n = (0.9)^n \tag{2.22}$$

so that for a one-step process $y_{\mathrm{overall},1} = 90\%$, for three-step process $y_{\mathrm{overall},3} = 73\%$, and for a ten-step process $y_{\mathrm{overall},10} = 35\%$!

2.3.2.2 (Bio)catalyst Productivity

If a large amount of (bio)catalyst is added to a substrate solution better results are achieved than if just a small amount of a highly productive catalyst is used. To even call an agent a catalyst, one condition is that the agent be added, as a minimum, in substoichiometric amounts. The smaller the amount of catalyst that has to be added for the same result, the better its performance. The relevant criterion is the dimensionless turnover number, TON [Eq. (2.23)].

$$\mathrm{TON} = \frac{\text{amount of product}}{\text{amount of catalyst}} = \frac{\text{substrate}}{\text{catalyst ratio}} = [\mathrm{S}]/[\mathrm{C}] \tag{2.23}$$

The turnover number is not used frequently in biocatalysis, possibly as the molar mass of the biocatalyst has to be known and taken into account to obtain a dimensionless number, but it is the decisive criterion, besides turnover frequency and selectivity, for evaluation of a catalyst in homogeneous (chemical) catalysis and is thus quoted in almost every pertinent article. Another reason for the low popularity of the turnover number in biocatalysis, apart from the challenge of dimensionality, is the focus on reusability of biocatalysts and the corresponding greater emphasis on performance over the catalyst lifetime instead of in one batch reaction as is common in homogeneous catalysis (Blaser, 2001). For biocatalyst lifetime evaluation, see Section 2.3.2.3.

The turnover number of a catalyst does not refer to the timescale of activity of that catalyst in a process. If the timescale of activity is taken into consideration, the *productivity* of the catalyst is recovered, expressed as a *productivity number*, PN, defined in Eq. (2.24).

$$\mathrm{PN} = \frac{\text{mass of product}}{\text{unit catalyst} \times \text{time}} \tag{2.24}$$

2.3.2.3 (Bio)catalyst Stability

As in all catalytic processes, catalyst stability is a key process criterion. In contrast to mere temperature or storage stability, which refer to the catalyst independently of a process, the *operating stability* or *process stability* is the relevant and decisive dimension of merit. It is determined by comparing the amount of product generated with the amount of catalyst spent. The relevant quantity, also sometimes found in homogeneous catalysis, is the total turnover number (TTN) [Eq. (2.25)].

$$\mathrm{TTN} = \frac{\text{moles of product produced}}{\text{mole of catalyst spent}} \tag{2.25}$$

Quite common in applied biocatalysis, where the purity of biocatalyst often is not known, is the expression of biocatalyst stability as an the enzyme consumption number (e.c.n.) [Eq. (2.26)].

$$\text{e.c.n.} = \frac{\text{amount of enzyme preparation spent}}{\text{amount of product generated}} = \frac{\text{g of enzyme}}{\text{kg (or lb) of product}} \tag{2.26}$$

The e.c.n. value depends on process parameters such as temperature, pH value, and concentrations of substrate(s) and product(s). With the molar masses of enzyme and product, e.c.n. and TTN can be interconverted [Eq. (2.27)].

$$\mathrm{TTN} = (1000/\mathrm{e.c.n.}) \times (\mathrm{MW}_{\mathrm{enzyme}}/\mathrm{MW}_{\mathrm{product}}) \qquad (2.27)$$

It should be emphasized that the TTN is not a completely suitable quantity for the evaluation of operating stability because the number of moles of biocatalyst is not a suitable reference for the complexity and cost of its manufacture. However, the values for both numerator and denominator in Eq. (2.27) are usually known and can be expressed in monetary terms. For checking the application of such a biocatalyst, the contribution of the biocatalyst to the overall cost can be assessed readily.

The relevant parameter for studies of operating stability of enzymes is the product of active enzyme concentration $[E]_{active}$ and residence time τ, $[E]_{active} \cdot \tau$. In a continuously stirred tank reactor (CSTR) the quantities $[E]_{active}$ and τ are linked by Eq. (2.28), where $[S_0]$ denotes the initial substrate concentration, x the degree of conversion and $r(x)$ the conversion-dependent reaction rate (Wandrey, 1977; Bommarius, 1992).

$$([E]_{active} \cdot \tau)/[S_o] = x/r(x) \qquad (2.28)$$

For a comparison and discussion of the concepts of stability at rest [Eq. (2.19)] and operating stability [Eq. (2.28)], see Chapter 5, Sections 5.6.1–5.6.3.

The threshold value for sufficient biocatalyst stability depends on the application, as does the value for product yield. For any application in synthesis the TTN should exceed 10 000, and for large-scale processing a value of > 1 000 000 is preferred.

2.3.2.4 **Reactor Productivity**

For the assessment of reactor productivity that is independent of catalyst, the space–time yield (s.t.y.) [Eq. (2.29)]. is considered, as in any chemical process.

$$\text{s.t.y.} = \text{mass of product generated}/(\text{reactor volume} \times \text{time})\ [\text{kg}\ (\text{L} \cdot \text{d})^{-1}] \qquad (2.29)$$

An *increase in s.t.y. in the same reactor is equivalent to an increase in reaction rate.* When considering the Michaelis–Menten equation [Eq. (2.3): $v = v_{max}[S]/(K_M + [S]) = k_{cat}[E][S]/(K_M + [S])$], there are three possible ways to achieve a maximum reaction rate:

1. High substrate concentration (increase of space–yield). Enzyme reactions can be accelerated by increasing the substrate concentration up to the limit of saturation ($\approx 10K_M$). If $[S] \gg K_M$, the enzyme is saturated and Eq. (2.3) is reduced to Eq. (2.30).

$$v = v_{max} = k_{cat} \cdot [E] \qquad (2.30)$$

At substrate saturation, the reaction is zeroth-order with respect to substrate. With a loosely binding substrate, i.e., a high K_M value, the enzyme is often saturated only above the solubility limit of the substrate. To improve the reaction rate or the s.t.y. further, according to Eq. (2.29) the yield per unit time must be increased, which means enhancing either the biocatalyst concentration and/or the time constant k_{cat}.

2. High enzyme concentration. The reaction rate and s.t.y. can be enhanced by increasing the catalyst concentration [E]; in practice, however, in contrast to the formalism of Eq. (2.3), owing to an either excessive viscosity increase or excess of deactivated protein in the reactor, a maximum limit of enzyme concentration is reached.

3. Increase in the time constant. According to Eq. (2.3), k_{cat} signifies the time constant of the enzyme reaction [time^{-1}], and the corresponding reaction is performed over a time scale $1/k_{cat}$ [time]. An increase can be effected through the temperature (Arrhenius behavior) but also for instance through changing a protecting group in peptide synthesis (Fischer, 1994). An increase in k_{cat} also increases the acceleration ratio k_{cat}/k_{uncat} (Radzicka, 1995) of the enzyme catalyst, in contrast to case 2) (see Section 2.3.3).

A minimum threshold value for reactor productivity can be set at a space–time-yield of about 100 g $(L \cdot d)^{-1}$, a value which tends to be compromised more by lack of substrate solubility than by biocatalyst reactivity. Well-developed biocatalytic process often feature space–time yields of > 500 g $(L \cdot d)^{-1}$ or even > 1 kg $(L \cdot d)^{-1}$ (Fischer, 1994; Rozzell, 1999; Bommarius, 2001).

2.3.3 Links between Enzyme Reaction Performance Parameters

2.3.3.1 Rate Acceleration

While users of biocatalysts are often concerned first and foremost about a high maximum rate v_{max}, a high k_{cat} value, and possibly also a low K_M value, a good (bio)catalyst is one that enhances the chemical background rate by as high a factor as possible, at least 10^5, possibly 10^{10} or even more. The dimensionless ratio of the enzyme catalytic rate constant over the uncatalyzed (i.e., the chemical background) rate constant, k_{cat}/k_{uncat}, is called the *rate acceleration*. An alternative nomenclature used in the literature, the *efficiency* or *proficiency*, should be reserved for the term $k_{cat}/(k_{uncat} \cdot K_M)$ quantifying the total energetic cost of enzymatic catalysis (Taylor, 2001). Surprisingly, when analyzing the source of good rate accelerations, i.e., high k_{cat}/k_{uncat} values, or the source of high proficiencies, i.e., high $k_{cat}/(k_{uncat} \cdot K_M)$ values, k_{cat} was found to be within two orders of magnitude for a variety of enzymes whereas k_{uncat} varied by several orders of magnitude, irrespective of the value of k_{cat} (Figures 2.5 and 2.6; Radzicka, 1995; Wolfenden, 2001).

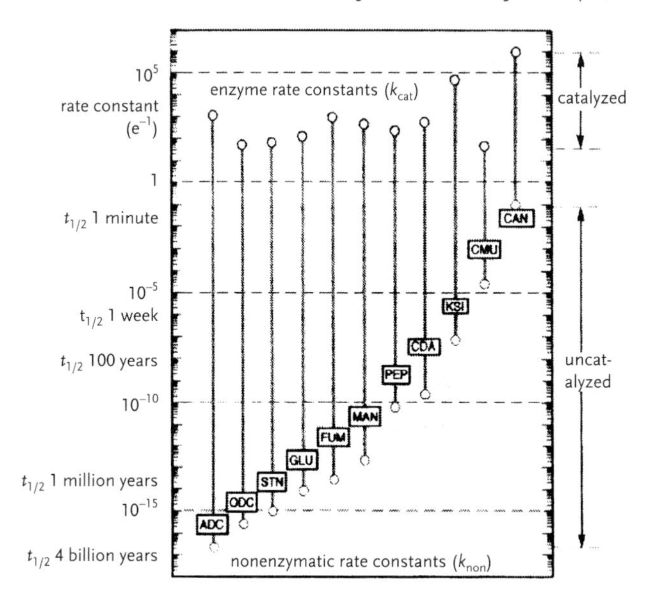

Figure 2.5 Logarithmic scale comparison of k_{cat} and k_{uncat} ($\equiv k_{non}$) for some representative reactions at 25 °C. The length of each vertical bar represents the rate enhancement. (Wolfenden, 2001). ADC: arginine decarboxylase; ODC: orotidine 5′-phosphate decarboxylase; STN: staphylococcal nuclease; GLU: sweet potato β-amylase; FUM: fumarase; MAN: mandelate racemase; PEP: carboxypeptodase B; CDA: *E. coli* cytidine deaminase; KSI: ketosteroid isomerase; CMU: chorismate mutase; CAN: carbonic anhydrase.

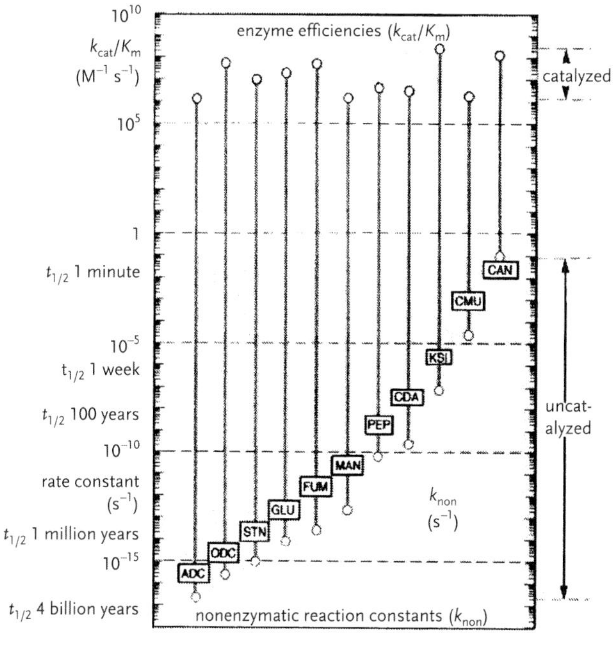

Figure 2.6 Logarithmic scale comparison of k_{cat}/K_M and k_{uncat} ($\equiv k_{non}$) for some representative reactions at 25°C. The length of each vertical bar represents the transition-state affinity or catalytic proficiency (its reciprocal). For abbreviations, see Figure 2.5 (Wolfenden, 2001).

The data provides clear guidance that the biggest improvement in enzyme catalysts can be achieved for reactions with very low chemical background rate constants and not by optimizing rate constants or specificities which are already fairly high.

2.3.3.2 Ratio between Catalytic Constant k_{cat} and Deactivation Rate Constant k_d

In Chapter 17, Section 17.4, we will encounter the rate equation for a deactivating enzyme in a batch reactor [Eq. (2.31)].

$$x \cdot [S]_0 - K_M \ln(1-x) = [E]_0 \cdot (k_{cat}/k_d) \cdot \{1 - \exp(-k_d \cdot t)\} \tag{2.31}$$

The influence of deactivation depends linearly on the dimensionless ratio k_{cat}/k_d, which might serve as a ratio to assess quickly the potential of a deactivating enzyme for synthesis.

2.3.3.3 Relationship between Deactivation Rate Constant k_d and Total Turnover Number TTN

Specific enzyme consumption [U per kg product], a hands-on parameter, can be obtained from Eq. (2.32) (Kragl, 1996), where $a_{vol,0}$ [U L^{-1}]is the initial volumetric activity).

$$\text{specific enzyme consumption} = (a_{vol,0} \cdot k_d)/(\text{s.t.y.})\ [\text{U (kg product)}^{-1}] \tag{2.32}$$

A high value for s.t.y. and low values for [E]′ (a modified enzyme concentration [g enzyme L^{-1}]), $a_{vol,0}$, or k_d all lead to a favorable, i.e., low, specific enzyme consumption.

To convert the specific enzyme consumption into the total turnover number TTN, the specific enzyme consumption, sp.e.c. [U (kg product)$^{-1}$] [Eq. (2.32)] first has to be converted into the enzyme consumption number e.c.n. [g enzyme (kg product)$^{-1}$] [Eq. (2.26)].

$$\text{e.c.n.} = (\text{sp.e.c.} \cdot [E]')/a_{vol,0}\ \ [\text{g enzyme (kg product)}^{-1}] \tag{2.33}$$

If Eq. (32) is inserted into Eq. (2.33) and the resulting equation into Eq. (2.27), the result is Eq. (2.34), which with $[E]'/MW_{enzyme} = [E]_0$ [mol L^{-1}] becomes Eq. (2.35).

$$\text{TTN [mol product (mol catalyst)}^{-1}] = (MW_{enzyme}/MW_{product})\ (1000 \times \text{s.t.y.})/([E]' \cdot k_d) \tag{2.34}$$

$$\text{TTN} = (1000 \times \text{s.t.y.})/(MW_{product} \cdot [E]_0 \cdot k_d) \tag{2.35}$$

A favorable, i.e., high, TTN is achieved by driving up the s.t.y. of a pertinent experiment while keeping both $[E]_0$ and k_d low. (The conversion factor of 1000 does not appear if s.t.y. is expressed in [g (L · d)$^{-1}$] and not [kg (L · d)$^{-1}$].)

2.3.4
Performance Criteria for Process Schemes, Atom Economy, and Environmental Quotient

Roger Sheldon, of TU Delft (Delft, The Netherlands), has compiled a list of trends for chemical and biochemical process technology (Sheldon, 1994); they are towards

- catalytic reactions instead of stoichiometric ones,
- reactions with 100% selectivity at 100% conversion,
- high substrate concentrations,
- no detrimental solvents, and no change of solvent during the process,
- enhanced use of solid or volatile acids and bases as well as pH-stat techniques.

This set of trends is easier to quantify than the "Twelve Principles of Green Chemistry" (Anastas, 1998) discussed in Chapter 1. We recognize the importance of high conversion, high selectivity, and thus high product yields; and of high substrate concentration, often leading to high s.t.y. and TTN. However, there are several additional guidelines which help to optimize a complete process.

Some of those criteria have been discussed above in the context of the catalyst or the process, such as enantioselectivity or diastereoselectivity instead of simple selectivity to product, and catalyst performance data such as turnover frequency (tof), turnover number (TON), total turnover number (TTN), and space–time-yield (s.t.y.).

Others have to deal with inputs and outputs of the process apart from reactants, catalysts, and products, such as:

1. the *atom economy*, i.e., the fraction of carbon (or other elements) of the substrate that is utilized or recovered in the product
2. the *specific consumption* (consumption per kg product) *of solvents, salts, and other auxiliaries* such as work-up materials (active carbon, filter aids), and
3. the degree of *recovery of all materials*, ranging from main process components to solvents, salts, and traces as well as contents of trace streams, often termed ancillary losses.

ad 1) *Atom economy*. The degree of utilization of inputs in the final product is termed the atom economy (Trost, 2002). An atom-economic process uses as many atoms as possible in the reaction stoichiometry. Examples of simple, but atom-economic, chemical processes are the manufacture of maleic anhydride by air-oxidation of butane/butene instead of benzene, or of propylene oxide by air-oxidation of propylene (propene) instead of by ring-closing substitution of epichlorohydrin (with loss of HCl). An example of a process with low atom economy is any synthesis with protection/deprotection steps, such as those that occur during peptide synthesis (while possibly still featuring very high product yields).

ad 2) *Specific consumption of solvents, salts, and auxiliaries*. This can be captured by evaluating the environmental quotient, EQ (Sheldon, 1994), consisting of a relative mass of by-product versus target product [kg kg^{-1}] (or for our American readers, lb lb^{-1}) multiplied by a measure of the environmental unfriendliness. Water or even NaCl might carry a factor of 1 or close to 1, whereas mercury, halogenated solvents, or other toxic by-products might carry a factor of 10 or even 100.

ad 3) *Recovery of all materials.* Eco-balances have become increasingly popular with regulators to gauge the impact of discharges. Increasingly, such instruments replace simple levying of costs on discharges or setting inflexible upper limits for concentrations of a range of ecologically unfriendly agents.

When aiming for scale-up or when actually already running a process on a pilot or a full industrial scale instead of trying to gain knowledge about a problem under controlled laboratory conditions, one should ensure that the different goals are reflected in different measures taken to obtain data or to develop a catalyst or process for an ultimate large scale. Table 2.2 contrasts typical laboratory practices with conditions prevailing on production scale. It is prudent, however, for processes to be practiced on a large scale to be developed accordingly, even during the test phase on lab scale.

Table 2.2 Comparison of conditions for laboratory- and industrial-scale processes.

	Laboratory-scale processes	*Industrial-scale processes*
Medium	dilute solutions	(highly) concentrated media
Substrate	often natural	often unnatural
pH control	with buffer	titration with acid/base; solid acids/bases
Time	sufficient for reaction	fast reactions desired
Temperature, pH gradients	none	heat and mass transfer influence reaction

Suggested Further Reading

ALAN FERSHT, *Structure and Mechanism in Protein Science*, Freeman, New York, **1999**.

WILLIAM P. JENCKS, *Mechanisms in Chemistry and Enzymology*, Dover, New York, **1975**.

References

P. T. ANASTAS and T. C. WILLIAMSON, Frontiers in green chemistry, *Green Chemistry* **1998**, 1–26.

H. U. BLASER, F. SPINDLER, and M. STUDER, Enantioselective catalysis in fine chemicals production, *Appl. Catal. A: General* **2001**, *221*, 119–143.

A. S. BOMMARIUS, K. DRAUZ, U. GROEGER, and C. WANDREY, **1992**, Membrane bio-reactors for the production of enantiomer ically pure α-Amino acids, in: *Chirality in Industry*, A. N. COLLINS, G. N. SHELDRAKE, AND J. CROSBY (eds.), Wiley & Sons Ltd., London, Chapter 20, pp. 371–397.

A. S. Bommarius, M. Schwarm, and K. Drauz, Comparison of different chemo-enzymatic process routes to enantiomerically pure amino acids, *Chimia* **2001**, *55*, 50–59.

T. C. Bruice and S. J. Benkovic, Chemical basis for enzyme catalysis, *Biochemistry* **2000**, *39*, 6267–6274.

T. C. Bruice, A View at the Millenium: the efficiency of enzymatic catalysis, *Acc. Chem. Res.* **2002**, *35*, 139–148.

P. Carter and J. A. Wells, Dissecting the catalytic triad of a serine protease, *Nature*, **1988**, *332*, 564–568.

T. R. Cech, RNA as an enzyme, *Sci. Am.* **1986**, *255(5)*, 64–75.

T. R. Cech, RNA. Fishing for fresh catalysts, *Nature*, **1993**,*365*, 204–205.

I. Chambers, J. Frampton, P. Goldfarb, N. Affara, W. McBain, and P. R. Harrison, The structure of the mouse glutathione peroxidase gene: the selenocysteine in the active site is encoded by the 'termination' codon, TGA, *EMBO J.* **1986**, *5*, 1221–1227.

C. S. Chen, Y. Fujimoto, G. Girdaukas, and C. Sih, Quantitative analyses of biochemical kinetic resolutions of enantiomers, *J. Am. Chem. Soc.* **1982**, *104*, 7294–7299.

D. B. Craig, E. A. Arriaga, J. C. Y. Wong, H. Lu, and N. J. Dovichi, Studies on single alkaline phosphatase molecules: reaction rate and activation energy of a reaction catalyzed by a single molecule and the effect of thermal denaturation – the death of an enzyme, *J. Am.Chem. Soc.* **1996**, *118*, 5245–5253.

C. Creutz and N. Sutin, Vestiges of the inverted region for highly exergonic electron-transfer reactions, *J. Am.Chem. Soc.* **1977**, *99*, 241–243.

U. Eichhorn, A. S. Bommarius, K. Drauz, and H.-D. Jakubke, Synthesis of dipeptides by suspension-to-suspension conversion via thermolysin catalysis – from analytical to preparative scale, *J. Peptide Sci.* **1997**, *3*, 245–251.

A. Fersht, *Enzyme Structure and Mechanism*, 2nd edition, Freeman, New York, **1985**.

A. Fischer, A. S. Bommarius, K. Drauz, and C. Wandrey, A novel approach to enzymatic peptide synthesis using highly solubilizing N^{α}-protecting groups of amino acids, *Biocatalysis* **1994**, *8*, 289–307.

E. Fischer, Einfluss der Configuration auf die Wirkung der Enzyme, *Ber. dtsch. chem. Ges.* **1894**, *27*, 2985–2993.

H. Gerischer, J. F. Holzwarth, D. Seifert, and L. Strohmaier, Flow method with integrating observation for very fast irreversible reactions, *Ber. Bunsenges. Physik. Chem.* **1969**, *73*, 952–955.

J. B. S. Haldane, *Enzymes*, M.I.T. Press, Cambridge/MA, USA, **1965**.

B. Hao, W. Gong, T. K. Ferguson, C. M. James, J. A. Krzycki, and M. K. Chan, A new UAG-encoded residue in the structure of a methanogen methyltransferase, *Science* **2002**, *296*, 1462–1466.

J. Hasserodt, Organic synthesis supported by antibody catalysis, *Synlett* **1999**, *12*, 2007–2022.

F. Hofmeister, Zur Lehre von der Wirkung der Salze. Zweite Mitteilung, *Arch. Exp. Pathol. Pharmakol.* **1888**, *24*, 247–260.

O. A. Hougen and K. M. Watson, *Chemical Process Principles*, Part III, Wiley, New York. **1947**.

W. P. Jencks, in: *Catalysis in Chemistry and Enzymology*, McGraw-Hill, New York, **1969**, pp. 288–291.

D. Joseph-McCarthy, L. E. Rost, E. A. Komives, and G. A. Petsko, Crystal structure of the mutant yeast triosephosphate isomerase in which the catalytic base glutamic acid 165 is changed to aspartic acid, *Biochemistry* **1994**,*33*, 2824–2829.

J. K. Kaushik and R. Bhat, A mechanistic analysis of the increase in the thermal stability of proteins in aqueous carboxylic acid salt solutions, *Protein Sci.* **1999**, *8*, 222–233.

D. E. Koshland Jr., Application of a theory of enzyme specificity to protein synthesis, *Proc. Natl. Acad. Sci. U.S.A.*, **1958**, *44*, 98–104.

D. E. Koshland Jr., G. Neméthy, and D. Filmer, Comparison of experimental binding data and theoretical models in proteins containing subunits, *Biochemistry* **1966**, *5*, 365–385.

U. Kragl, D. Vasic-Racki, and C. Wandrey, Continuous production of L-*tert*-leucine in series of two enzyme membrane reactors, *Bioproc. Eng.* **1996**, *14*, 291–297.

J. Kraut, Serine proteases: structure and mechanism of catalysis, *Ann. Rev. Biochem.* **1977**, *46*, 331–358.

J. Kraut, How do enzymes work?, *Science* **1988**, *242*, 533–540.

J. L. Kurz, Transition state characterization for catalyzed reactions, *J. Am. Chem. Soc.* **1963**, *85*, 987–991.

J. K. Lee and K. N. Houk, A proficient enzyme revisited: the predicted mechanism for orotidine monophosphate decarboxylase, *Science* **1997**, *276*, 942–945.

R. A. Mehl, J.C . Anderson, S. W. Santoro, L. Wang, A. B. Martin, D. S. King, D. M. Horn, and P. G. Schultz, Generation of a bacterium with a 21 amino acid genetic code, *J. Am. Chem. Soc.* **2003**, *125*, 935–939.

L. Michaelis and M. L. Menten, Die Kinetik der Invertinwirkung, *Biochem. Z. 49*, **1913**, 333–369.

M. I. Page and W. P. Jencks, Entropic contributions to rate accelerations in enzymic and intramolecular reactions and the chelate effect, *Proc. Natl. Acad. Sci. USA* **1971**, *68*, 1678–1683.

L. Pauling, Molecular architecture and biological reactions, *Chem. Eng. News* **1946**, *24*, 1375–1377.

L. Pauling, Chemical achievement and hope for the future, *Am. Sci.* **1948**, *36*, 51–58.

S. J. Pollack, J. W. Jacobs, and P. G. Schultz, Selective chemical catalysis by an antibody, *Science* **1986**, *234*, 1570–1573.

A. Radzicka and R. Wolfenden, A proficient enzyme, *Science* **1995**, *267*, 90–93.

J. D. Rozzell, Biocatalysis at commercial scale: myths and realities, *Chimica Oggi* **1999**, *(6/7)*, 42–47.

R. A. Sheldon, Consider the environmental quotient, *CHEMTECH* **1994**, *3*, 38–47.

M. von Smoluchowski, Versuch einer mathematischen Theorie der Koagulationskinetik kolloidaler Lösungen, *Z. physik. Chem.* **1917**, *92*, 129–168.

G. Srinivasan, C. M. James, and J. A. Krzycki, Pyrrolysine encoded by UAG in *Archaea*: charging of a UAG-decoding specialized tRNA, *Science* **2002**, *296*, 1459–1462.

T. J. Stillman, A. M. B. Migueis, X.-G. Wang, P. J. Baker, L. Britton, P. C. Engel, and D. W. Rice, Insights into the mechanism of domain closure and substrate sof glutamate dehydrogenase from *Clostridium symbosium, J. Mol. Biol.* **1999**, *285*, 875–885.

C. Tanford, *Physical Chemistry of Macromolecules*, Wiley, New York, **1961**.

E. A. Taylor, D. R. J. Palmer, and J. A. Gerlt, The lesser 'burden borne' by o-succinylbenzoate synthase: an 'easy' reaction involving a carboxylate carbon acid, *J. Am. Chem. Soc.* **2001**, *123*, 5824–5825.

A. Tramontano, K. D. Janda, and R. A. Lerner, Chemical reactivity at an antibody binding site elicited by mechanistic design of a synthetic antigen, *Proc. Natl. Acad. Sci. USA* **1986**, *83*, 6736–6740.

B. M. Trost, On inventing reactions for atom economy, *Acc. Chem. Res.* **2002**, *35*, 695–705.

P. H. von Hippel and K.-Y. Wong, Neutral salts: the generality of their effects on the stability of macromolecular conformation, *Science* **1964**, *145*, 577–580.

C. Wandrey, Reaction engineering studies on enzyme catalysts for the development of continuous processes, Habilitation Thesis, TH Hannover, Germany, **1977**.

R. Wolfenden and M. J. Snyder, The depth of chemical time and the power of enzymes as catalysts, *Acc. Chem. Res.* **2001**, *34*, 938–945.

G. Zhong, R. A. Lerner, and C. F. Barbas III, Enhancement of the repertoire of catalytic antibodies with aldolase activity by combination of reactive immunization and transition state theory, *Angew. Chem. Int. Ed.* **1999**, *111*, 3957–3960.

F. Zinoni, A. Birkmann, T. C. Stadtman, and A. Boeck, Nucleotide sequence and expression of the selenocysteine-containing polypeptide of formate dehydrogenase (formate-hydrogen-lyase-linked) from *Escherichia coli, Proc. Natl. Acad. Sci. USA* **1986**, *83*, 4650–4654.

3
Isolation and Preparation of Microorganisms

Summary

Microorganisms are predominant as a source for novel enzymes, whereas animal organs and plant materials contribute much less than 10% each to the total amount of enzymes processed.

Microorganisms can be obtained either through screening or through strain collections. While genetic engineering methods now occupy a firm place for improving existing enzyme activity or stability, screening of cultures of organisms still remains the principal way of finding novel activities.

In the still-valid taxonomic model, microorganisms can be classified into the three domains of eubacteria, archaea, and eukaryotes. The three domains branched off very early in evolution, united by a common ancestor; the corresponding picture of the relationships between the domains and kingdoms is termed the "universal phylogenetic tree".

To gain industrially useful properties, the original isolate of a microorganism is greatly modified in the laboratory; among the techniques used are both non-genetic methods, such as mutation with unspecific agents or recombination during propagation, and genetic engineering with in-vitro mutations and recombinations, in-vivo selection pressure, and repression or elimination of minor metabolic pathways with resulting metabolic imbalance.

As a result of this modification, progressive improvements in the yields of the desired products have been achieved, such as the increase in penicillin production from 1–10 $\mu g\ mL^{-1}$ in the 1940s to 85 $mg\ mL^{-1}$ 50 years later, all of this without genetic engineering. One of the best examples of metabolic engineering in amino acid production, the formation of L-lysine from glucose, includes by-passing feedback regulation through auxotrophic mutants or screening for toxin-resistant mutants. The wild type of *Brevibacterium lactofermentum* does not produce any lysine, but incorporation of both these measures increased the yield to 50 $g\ L^{-1}$:

Owing to the increased ability to handle large amounts of samples by *high throughput screening*, HTS, conventional screening methods such as isolation of microorganisms in the neighborhood of specific habitats, selection of strains that are taxonomically close to prior successes, and enrichment cultures are increasingly replaced or supplemented by PCR and rRNA methods, by cloning for high expression with highly transformable hosts, or by using related primer pairs in PCR based on known enzyme functions aided by increased genome sequencing for similar enzymes in different organisms.

3.1 Introduction

The most prominent source of enzymes is microorganisms. Even back in the 1980s, animal organs and plant materials contributed only 8% and 4%, respectively, to the total amount of enzymes processed. Since the advent of recombinant DNA technology and improved requirements for uniform quality, microorganisms have gained even more ground as a source of enzymes.

Microorganisms can be classified into three domains: eubacteria, archaea, and eukaryotes (fungi). The domain accounting for the most numerous representatives is the eubacteria. Figure 3.1. elucidates the connection between the three domains, which branched off very early in evolution, in all likelihood more than a billion years ago.

Molecular structures and sequences generally reveal more of evolutionary relationships than do classical phenotypes, particularly so among microorganisms. Consequently, the basis for the definition of taxa has shifted progressively from the organismal level in Darwin's time to the cellular and then to the molecular level. Molecular comparisons show that life on this planet divides into three primary groupings, commonly known as the eubacteria, the archaebacteria, and the eukaryotes, each containing two or more kingdoms (Woese, 1990). In this model (see, however, Chapter 16, for very new views on this topic), all domains are united at the base and thus all derive from a common ancestor. Relationships between the domains or the kingdoms can be pictured as a tree, termed the "universal phylogenetic tree". The universal phylogenetic tree not only spans all extant life, but its root and earliest branchings represent stages in the evolutionary process before modern cell types had come into being (Woese, 2000). The evolution of the cell is an interplay between vertically derived and horizontally acquired variation. Primitive cellular entities were necessarily simpler and more modular in design

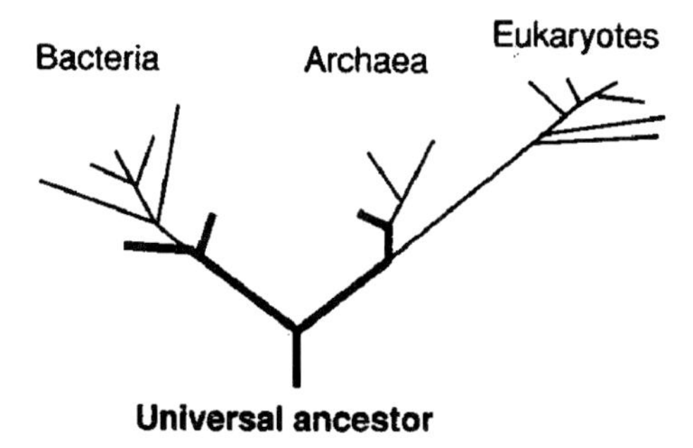

Figure 3.1 Relationship between the three domains (Woese, 1990).

than are modern cells. Consequently, horizontal gene transfer early on was pervasive, dominating the evolutionary dynamic. The root of the universal phylogenetic tree represents the first stage in cellular evolution when the evolving cell became sufficiently integrated and stable to the erosive effects of horizontal gene transfer for true organismal lineages to exist.

Screening or Culture Collection?

Microorganisms can be obtained either through *screening* or through a *strain collection*. Strain collections have been gathered and prepared in many laboratories dealing with enzyme technology. In addition, many countries have built up national collections administered by government agencies (Table 3.1); these collections originally grew out of the need for deposition of strains claimed in patent applications. In the United States, the ATCC (American Type Culture Collection) in Rockville/MD (American Type Culture Collection, P.O. Box 1549, Manassas, VA 20108, 703-365-2700) acts as the national strain collection; the corresponding agency in Germany, for instance, is the DSMZ (Deutsche Sammlung von Mikroorganismen und Zellkulturen e.V., Mascheroder Weg 1b, D-38124 Braunschweig/Germany). Each newly deposited strain is numbered and a (usually freeze-dried) sample can be ordered for a small payment, today often via the Internet (in the case of non-pathogenic organisms and if ordered from an institutional address; regulations have tightened owing to security concerns in regard to potential bioterrorism). (Alternatively, in the case of more common organisms, cDNA or genomic DNA can be ordered instead of the whole organism.)

Table 3.1 A selection of national culture collection agencies.

Abbreviation	*Name*	*Location*
ATCC	American Type Culture Collection	Rockville, Maryland/USA
CBS	Centraalbureau voor Schimmelcultur	Baarn/The Netherlands
CCM	Czechoslovak Collection of Microorganisms	J. E. Purkyne University, Brno/Czech Republic
CDDA	Canadian Department of Agriculture	Ottawa/Canada
CIP	Collection of the Institut Pasteur	Paris/France
DSMZ	Deutsche Sammlung von Microorganismen und Zellkulturen	Braunschweig/Germany
JCRB	Japanese Collection of Research Biosources	Tokyo/Japan
NCTC	National Collection of Type Cultures	London/UK

3.2
Screening of New Enzyme Activities

There is a range of techniques for obtaining new enzyme activities:

- screening for new activities in different environments (soil, polluted areas, deep vents) (this chapter),
- discovery of novel unnatural activities of existing enzymes
- utilization of novel reaction conditions, of altered reaction media (Chapter. 12), or of novel effectors such as metal ions,
- applications of genetic engineering techniques such as protein engineering (Chapter 10) or directed evolution (Chapter 11), and
- combination of chemical and enzymatic catalysis to obtain new catalysts (Chapter 18)

In the age of recombinant DNA technology and genetic engineering-based methodologies, why is the search for new microorganisms or new enzyme activities still so important?

From the economic point of view the pharmaceutical industry is probably the most important industry using microorganisms. Virtually all of the medically important antibiotics are produced by microbes as well as many steroid hormones, such as genetically engineered bacteria for the production of human insulin and human growth hormone, antiviral and antitumor agents such as interferons, blood products such as blood-clotting factors and erythropoietin, and a variety of vaccines and monoclonal antibodies for diagnostics.

In general, microorganisms offer a practical solution to the challenge of procuring and developing enzyme activities: they are easy to maintain, feature rapid growth, and can be coaxed to focus on the production of just one major desired compound. They are extremely versatile and nowadays it is believed that for every target product a microorganism or enzymatic catalyst derived from a microorganism can be found. A good example of the direction of a microorganism to the production of one major product is *Corynebacterium glutamicum*. This bacterium was engineered in the 1950s to produce massive amounts of L-glutamic acid, used especially in East Asia as a taste enhancer. Interestingly, the same organism a generation later has been employed in the 1980s and 1990s for metabolic engineering towards the large-scale production of another amino acid, L-lysine, used in animal feed to supplement diets rich in corn-based feed which is lysine-poor.

Genetic engineering methods, on the other hand, despite all progress, feature a range of challenges, such as (i) incorrect folding which has to be circumvented by urea denaturation with subsequent renaturation; (ii) missing or incorrect patterns of glycosylation of proteins from eukaryotic organisms in bacteria such as *E. coli*; and (iii) missing or incomplete cleavage of the N-terminal methionine residue by prokaryotes. Most importantly, such techniques now occupy a firm place for improving existing enzyme activity or stability. However, screening of cultures of organisms still remains the principal way of finding novel activities. Organisms can be obtained through culture collections as described above or, at least at an early stage, from samples in soil, from aqueous habitats, or from animal or plant material. Obtaining microbes from their natural habitats belongs to the field of microbial ecology (see below).

The criteria for success of a screening process with the goal of are:
1. The organism must deliver the desired enzyme in good yield and within a reasonably short time frame if possible, in submerged culture.
2. The organism must produce the enzyme with inexpensive and accessible nutrients.
3. The organism should be separable from the fermentation medium, and the enzyme should be excreted extracellularly, and easily separable from the fermentation broth.
4. The organism should be non-pathogenic, it should be genetically stable, and it should not produce undesired or even toxic compounds.

3.2.1 Growth Rates in Nature

The natural habitats of microorganisms are exceedingly diverse. Any habitat suitable for the growth of higher organisms will also permit microbial growth, but there are also many habitats, such as hot or sulfurous springs, that are unfavorable to other organisms but where microorganisms (archaea) can exist and flourish.

Despite adsorption to any tiny microenvironment or surface, nutrient concentrations are usually much lower (frequently 100 to 1000 times) in nature than they are in the usual laboratory culture conditions. Thus microorganisms in natural habitats are often subjected to nutrient-poor conditions, resulting in much lower counts regarding an individual species. Thus organisms in nature must be able to adapt quickly to a rapid change in nutrient availability. This requirement is probably one of the main reasons for the existence of highly sophisticated regulatory mechanisms and also provides the diversity of diets upon which microorganisms can feed.

Extended periods of exponential growth (see Chapter 8, Section 8.2) in nature are rare; growth more often occurs in spurts, when substrate becomes available, followed by an extended stationary phase, after the substrate is used up. For example, the generation time of *Escherichia coli* in the intestinal tract is about 12 h, whereas in culture it grows much faster, with a division rate approximately every 20–30 min.

3.2.2 Methods in Microbial Ecology

Rarely does a natural environment contain only a single type of microorganism. In most cases, a variety of organisms are present and it is particularly challenging for the microbiologist to devise methods and procedures to permit isolation and culture of organisms of interest. The most common technique is *enrichment culture.* In this method, a medium and a set of incubation conditions are used that are selective for the desired organism. Enrichment cultures are frequently established by placing the inoculum directly into a highly selective medium; many common prokaryotes can be isolated in this way. Such an enrichment culture selects either for the highest growth rate (see Chapter 8, Section 8.2) or for organisms that can utilize the carbon (C) or nitrogen (N) source offered in the medium. By application of an appropriate *selection pressure* (Chapter 11, Section 11.2), the desired activity can be filtered from a series of organisms, for instance by providing a sole N-source in the medium which is transformed by the growing organism by breaking the same bond as desired later in the biotransformation. A good example is screening for hydantoinases by supplying hydantoins to the media (Chapter 7, Section 7.2.2.3).

Frequently the objective of an enrichment culture study is to obtain a pure culture. In this context, "pure" means free from foreign living elements, and hence containing only one kind of organism. In examining a culture microscopically, it is almost impossible to detect a stray contaminant because the sensitivity of the light microscope is low. With the average 100× lens, the field size is such that if the bacterial count is 10^6 cells mL^{-1}, on average there will be only one cell/field. If a contaminant numbered 10^4 cells mL^{-1}, one would have to examine 100 fields to detect a single contaminant. Pure cultures can be obtained in many ways; the most frequently used are the streak plate, the agar shake, and liquid dilution methods. By repeated picking and restreaking of a well-isolated colony, one usually obtains a pure culture.

3.3 Strain Development

Industrially useful microbes are a unique subset of all the microorganisms that are available: whereas microbes isolated from nature exhibit cell growth as their main physiological property, industrial microorganisms are most often organisms which have been selected carefully for specific and optimum product formation (Chapter 8, Section 8.2.). Even if the industrial microbe is one which has been isolated by traditional techniques, it becomes a highly modified organism before it enters large-scale production.

3.3.1 Range of Industrial Products from Microorganisms

Microbial products of industrial interest can be of several major types. The cells themselves can be the source of interest, as in yeast for baking, or mushrooms cultivated for food. Another source are bacteria used for inoculation in food products, such as lactobacilli in dairy and sausage products. As stated above, in comparison to enzymes from animal or plant sources, enzymes produced by microorganisms are frequently the desired product and are becoming more and more important. A number of commercially important enzymes are produced in large-scale microbial processes, including starch-digesting enzymes (amylases), protein-digesting enzymes (proteases such as renin, subtilisin, thermolysin), and fat-digesting enzymes (lipases) (Table 3.2). An important industrial enzyme is glucose isomerase, used in large amounts for the production of fructose sweetener from glucose (Chapter 7, Section 7.3.1). Another important enzyme is penicillin acylase, used industrially in the manufacture of semi-synthetic penicillins (Chapter 7, Section 7.5.1). Industrial enzymes have now reached an annual market of US $ 1.6 billion (Demain, 2000).

Microbial metabolites contribute to the list of products as well, such as with fermentations to such major products as ethanol, acetic acid, *n*-butanol, or lactic acid; key growth factors such as amino acids or vitamins; or pharmacologically active compounds such as antibiotics, steroids, or alkaloids. Pharmacologically active agents are generally catagorized as *secondary metabolites,* which most often implies production in the stationary (non-growth-associated) phase of fermentation (Chap-

Table 3.2 Microbial enzymes and their application.

Enzyme	*Source*	*Application*	*Industry*
Amylase	fungi	bread	baking
	bacteria	starch coatings	paper
	fungi	syrup and glucose manufacture	food
	bacteria	cold-swelling laundry starch	starch
	fungi	digestive aid	pharmaceutical
	bacteria	removal of coatings (desizing)	textile
Protease	fungi	bread	baking
	bacteria	spot removal	dry-cleaning
	bacteria	meat tenderizing	meat
	bacteria	wound cleansing	medicine
	bacteria	desizing	textile
	bacteria	household detergent	laundry
Invertase	yeast	soft-center candies	food
Glucose oxidase	fungi	glucose removal, oxygen removal, test paper for diabetes	pharmaceutical
Pectinase	fungi	pressing clarification	wine, fruit juice
Rennin	fungi	coagulation of milk	cheese

Source: Brock, 1988.

ter 8, Section 8.2) and coincides with metabolic regulation which differs from that of primary metabolites. Secondary metabolites are not immediately necessary for bacterial growth and are not produced from the C- or N-sources in the medium but from metabolic intermediates themselves. Secondary metabolites can inhibit growth when they interfere with bacterial metabolism; fermentation processes have to be designed accordingly to circumvent such problems.

Through the years, as large-scale microbial processes have been perfected, a number of industrial strains have been deposited in culture collections. When a new industrial process is patented, the applicant for the patent is required to deposit a strain capable of carrying out the process in a recognized culture collection. Although these culture collections can serve as ready sources of cultures, it should be understood that most industrial companies will be reluctant to deposit their best cultures.

3.3.2 Strain Improvement

To obtain industrially useful properties, an original isolate of a microorganism is greatly modified in the laboratory. Most of the organisms are altered

- genetically by spontaneous mutation, mutation with unspecific agents such as chemicals (MNNG) or hard UV light, or recombination during propagation (this is not to be confused with genetically engineered organisms where mutations and recombinations are introduced in vitro), or
- by in-vivo selection pressure (see above), or
- by repression or elimination of minor metabolic pathways with resulting metabolic imbalance.

As a result of this modification, progressive improvement in the yield of the desired product should be achieved. One of the most dramatic early examples of such improvement is the production of penicillin, the antibiotic produced by the fungus *Penicillium chrysogenum*. When penicillin was first produced on a large scale at the end of the 1940s, yields of 1–10 $g \cdot L^{-1}$ were obtained. Over the years, as a result of strain improvement coupled with changes in media and growth conditions, the yield has been increased to about 85 $g \cdot L^{-1}$. Remarkably, this increase was obtained without genetic engineering, but just by selection for natural mutations and for mating.

If the *metabolic pathway* from a substrate such as glucose to a desired final product such as lysine is well understood, one can try to redirect a maximally high flux in the direction of the end product. The targeted change of flux rates in the metabolism for the purpose of optimization of product yield is called *metabolic engineering* (Chapters 15 and 20). Examples of metabolic engineering in amino acid production include bypassing feedback regulation by identifying auxotrophic mutants, and screening for mutants that are resistant to a toxic analog of the desired metabolite (Demain, 2000).

In the best-documented example of the formation of lysine, the product is formed from aspartate which reacts via aspartylphosphate and aspartate semialdehyde to lysine. The wild type of *Brevibacterium lactofermentum* does not produce any lysine. With the following steps the yield could be increased to 50 g L^{-1}:

1. *S*-(2-Aminoethyl)-L-cysteine (AEC), $H_2N\text{-}CH_2\text{-}CH_2\text{-}S\text{-}CH_2\text{-}CH(NH_2)\text{-}COOH$, a lysine analog, acts as a false feedback inhibitor on aspartokinase, which produces aspartylphosphate from aspartate. The inhibitor simulates, for aspartokinase, the absence of lysine and threonine , and as a consequence the AEC insensitive mutant is no longer inhibited by lysine and threonine. The result was a yield increase from 0 to 16 g L^{-1}.
2. Lysine is formed from aspartate and pyruvate; pyruvate, however, is also consumed for the synthesis of alanine. The discovery of an alanine auxotroph, a mutant which needs external alanine addition for growth because it cannot catalyze a precursor step to alanine, was responsible for a yield increase from 16 to 33 g L^{-1}.
3. α-Chlorocaprolactam and γ-methyl-L-lysine inhibit enzymes which are on the metabolic path to lysine. After the respective mutants were found, the yield could be increased to 43 g L^{-1}.
4. β-Fluoropyruvate inhibits an enzyme which is located on a bypass of lysine metabolism, with the result that more carbon flows towards lysine. The yield was increased to 50 g L^{-1}.

In the future, the engineering of pathways in strains will be even further facilitated by the availability of genome and gene expression data. Recently (April 2003), the sequencing of approximately 110 prokaryotic genomes has been completed; for the results, see www.ncbi.nlm.nih.gov/genomes. There are several driving forces for improvement in the field of metabolic engineering.

- the growing field of genomics (Chapter 15); genome sequencing, and the analysis of DNA microarrays for expression of certain genes will give insight in transcriptional controls of whole pathways, identifying genes responsible for high productivity as an example (Stafford, 2001).
- developments in applied molecular biology such as promoter optimization for specialized gene expression (Chapter 4), together with combinatorial methods such as random mutagenesis and gene recombination (Chapter 11) (Stafford, 2001); combining those methods can create artificial pathways with improved enzymes within the same target cell.
- The need for renewable resources for synthesis of chemicals with regard to environmental aspects (Chapter 20).

3.4 Extremophiles

A very interesting research area in biology and biotechnology is the topic of extremophiles. Life on Earth has adapted over the course of evolution, even to extreme habitats (although there is evidence that organisms had to adapt to ever cooler conditions on Earth over the course of early evolution). "Extreme" habitats encompass the range of conditions as described:

Temperature:	–5 to +110 °C
Hydrostatic pressure:	0.1 to 120 MPa
Water activity:	0 to 0.6 (corresponds to a salt concentration up to and beyond 6 M)
pH:	pH 1–12

The three strategies of nature for the adaptation to environmental stress factors are avoidance, tolerance, and adaptation. Many organisms are successful in the avoidance of stress under extreme conditions of pH and salt concentration by utilizing kosmotropes or compatible solutes to neutralize extreme pH values or to compensate high osmotic pressure between the inner core of cells and the outer membrane space. As kosmotropes (= "structure formers", the opposite is chaotropes = "structure breakers"). Molecules such as glycerol, betaine, proline, or *N*-acylamino acids can function. Ions can also act as kosmotropes (or chaotropes) whose potential can be read from their relative position in the Hofmeister series (Hofmeister, 1888):

← Increasing chaotropic effect

cations Al^{3+}, Ca^{2+}, Mg^{2+}, Li^{+}, Na^{+}, K^{+}, NH_4^{+}, $(CH_3)_4N^{+}$
anions SCN^{-}, I^{-}, ClO_4^{-}, Br^{-}, Cl^{-}, SO_4^{2-}, HPO_4^{2-}, citrate^{3-}

Increasing stabilization →

The isothermal and isobaric conditions in a habitat require adaptation as a strategy at unusual pressures and temperatures. Extremophilic microorganisms are adapted to survive in ecological niches, and they produce unique biocatalysts that function under extreme conditions comparable to those prevailing in various industrial processes.

Extremophiles can be divided into groups according to:

1. temperature tolerance,
2. salt concentration,
3. pH range,
4. or pressure conditions.

ad 1) *Temperature.* For psychrophilic microorganisms the temperature ranges between 0 °C < T_{opt} < 20 °C, for thermophilic 50 °C < T_{opt} < 80 °C, and for hyperthermophilic 80 °C < T_{opt} < 120 °C. Temperature creates challenges, low temperature causing formation of ice crystals within the cell and high temperature causing denaturation of biomolecules (Chapter 17) and increase in cell membrane fluidity to lethal levels. The most hyperthermophilic organisms can exist above 100 °C, with *Pyrolobus fumarii* (Crenarchaeota), a nitrate-reducing chemolithoautotroph capable of growing at temperatures up to 113 °C (Blochl, 1997).

ad 2) *Salt.* Halophilic microorganisms tolerate salt con=centrations of 0–6 M. Organisms live within a range of salinities, from water that is essentially distilled to saturated salt solutions. Halophily refers to the ionic requirements for life at high salt concentrations: a halophile must cope with osmotic stress. Some archaea can withstand long periods in saturated NaCl.

ad 3) *pH.* Acidophiles can survive at pH 0–5, alkaliphiles at pH 9–12. Biological processes tend to occur towards the middle range of the pH spectrum, and intracellular pH of the cytoplasm is usually between pH 5.0 and 7.5. Acidophilic archaea flourish under extreme acidity; for example, *Ferroplasma acidarmanus* has been described growing at pH 0 in acid mine drainage in Iron Mountain, CA, USA, thriving in a brew of sulfuric acid and high levels of copper, arsenic, cadmium, and zinc with only a cell membrane and no cell wall (Edwards, 2000).

ad 4) *Pressure.* Piezophilic or barophilic microorganisms tolerate pressures of 50–110 MPa. Pressure challenges life because it forces volume changes to occur. It compresses the packing of lipids, and membranes become less fluid. Growth occurs up to 70–80 MPa; piezophiles have been observed within the 50–80 MPa range. Piezophiles occur in the deep ocean, such as in the vicinity of deep sea vents. At the bottom of the Mariana Trench, the deepest known site in the ocean, the pressure measures 110 MPa.

Even extremophilic organisms and their proteins contain the same 20 amino acids with bonds similar to those in mesophiles. As the difference in free enthalpy between folded and unfolded states of globular proteins $\Delta G_{N \to G}$ is only about 45 ± 15 kJ mol^{-1} the sequence and structure of extremophilic proteins should differ from those of ordinary species. However, the main question, namely which properties cause the increase in denaturation temperature of thermostable proteins, is still debated (Rehaber, 1992). Theoretical and experimental analyses have shown that thermal stability is largely achieved by small but relevant changes at different locations in the structure involving electrostatic interactions and hydrophobic effects (Karshikoff, 2001). There is no evidence for a common determinant or for just one effect causing thermostability.

Recent structural comparisons between enzymes from mesophiles and thermophiles have validated numerous protein-stablizing effects, including hydrophobic interactions, packing efficiency, salt bridges, hydrogen bonds, loop stabilization

Table 3.3 Correlation of number of salt bridges and higher temperature resistance.

Protein	T_{opt}	N_s	N_r
Rubredoxin (*Pyroccocus furiosus*)	100	4	4.2
Glutamate dehydrogenase (*Pyrococcus furiosus*)	100	42	25
Glutamate dehydrogenase (*Thermatoga maritima*)	80	30	24
Ribosomal protein (*Thermus thermophilus*)	75	9	6.8
Superoxide dismutase (Mn, Fe) (*Thermus thermophilus*)	75	10	10.1
Malate dehydrogenase (*Thermus flavus*)	75	17	16.7
3-Isopropylmalate dehydrogenase (*Thermus aquaticus*)	75	29	19.5
Elongation factor Tu (*Thermus aquaticus*)	75	36	28.3
Seryl-tRNA synthetase (*Thermus thermophilus*)	75	36	28.6

N_S: number of salt bridges;
N_r: number of salt bridges statistically expected for that protein structure.
Source: Karshikoff (2001).

and resistance to covalent destruction, one cause maybe evolving from others (Zeikus, 1998).

With more and more crystal structures becoming available, opinions arise that electrostatic interactions are an important factor conferring thermostability on proteins. This opinion is supported by the increasing number of salt bridges found in many thermostable and hyperthermostable proteins (Table 3.3) (Karshikoff, 2001); comparison of the capsid surface residues in lumazine synthase from *Bacillus subtilis* (mesophilic) and *Aquifex aeolicus* (hyperthermophilic) show significantly more charged residues than polar residues in the hyperthermophilic protein.

3.4.1
Extremophiles in Industry

While extremophiles have provided data that are basic to biological science, including information on protein folding (see the previous section), extremophiles certainly have the potential to support multibillion dollar industries, including agriculture, chemical synthesis, laundry detergents and pharmaceuticals (Rothschild, 2001). Functional and stable enzymes are sought for economically preferable environments such as high temperatures, required for a high space–time-yield; or for a processable substrate solution or a chemically challenging environment, such as an alkaline laundry detergent solution.

Enzymes from extremophiles, such as thermozymes, have potential either as products themselves, or as catalysts, or they may be used as sources of ideas to modify mesophile-derived enzymes. Most of the thermozymes maintain their thermoresistant properties when expressed in a mesophilic organism such as

Escherichia coli (Niehaus, 1999). Whereas halophiles (Margesin, 2001) and piezophiles (Abe, 2001) have attracted attention, the main aspect of interest for biotechnological application remains thermal stability, focusing interest on thermophiles or even hyperthermophiles. Increase in temperature has a significant influence on the bioavailability and solubility of organic compounds (not on the solubility of O_2, however, which is nearly zero at extremophilic temperatures!). Current applications of extremozymes include:

- heat-stable amylases and glucoamylases from organisms such as *Pyrococcus woesei* or *Thermococcus profundus* for the degradation of starch. Thermostable enzymes can improve industrial starch bioconversion processes by reducing steps with temperature shifts and simplifying conditions through salt reduction and higher acidity (Niehaus, 1999).

- conversion of glucose to a 55% fructose–glucose mixture (high fructose corn syrup, HFCS) performed with glucose isomerase (GI) (see Chapter 7, Section 7.3.1, and Chapter 19, Section 19.1), a huge market. To achieve the desired 55% fructose concentration, equal in sweetness to the same weight of sucrose, without the currently required chromatographic concentration step, the reaction would have to be run at 110 °C. Despite development of an improved thermostable GI from mesophilic enzymes (Crabb, 1997), application of a thermophilic glucose isomerase such as GI from *Thermus maritima* (stable up to 100 °C with a half-life of 10 min at 115 °C) could move this vision closer to industrial application.

- protein degradation. Proteolytic enzymes account for the largest worldwide commercial enzyme application (Niehaus, 1999). Serine proteases are used in alkaline detergents for household laundry. A variety of heat-stable proteases, some with half-lives of 4 h at 95 °C, have been identified in hyperthermophilic archaea belonging to the genera *Desulphococcus, Sulfolobus, Staphylothermus, Thermococcus,* and *Pyrococcus* (Niehaus, 1999). Laundering at higher temperatures reduces the amount of detergent without diminishing laundering efficiency.

- solvent-resistant strains. Whereas organic solvents can be applied for biotransformations (Chapter 12) (Nicolova, 1993) and hydrophobic solvents especially are compatible with enzyme activity (Chapter 12, Section 12.1), hydrophobic organic solvents such as toluene or benzene are toxic for living organisms because they accumulate in and disrupt cell membranes. Interestingly, tolerance of organic solvents, which after all comprise another extreme environment, can largely be found in mesophiles. Most *Pseudomonas* strains tolerate toluene, heptanol, and dimethyl phthalate whereas only two deep-sea strains, a typical extremophilic source, show tolerance against benzene and toluene (Isken, 1998).

- last not least, DNA polymerases (E.C. 2.7.7.7). Instead of the heat-labile Klenow enzymes used in early versions of PCR (polymerase chain reaction, Chapter 4,

Section 4.3), which had to be replenished at every new cycle, the application of heat-stable polymerases from organisms such as *Thermus aquaticus* (Taq polymerase), *Pyrococcus woesii* (Pwo polymerase), *Pyrococcus furiosus* (Pfu polymerase), or *Thermococcus litoralis* (Vent polymerase) facilitated automation of the thermal cycling process.

3.5 Rapid Screening of Biocatalysts

Several recent publications have shown (Winson, 1997) that the method of screening of a large number of otherwise undetermined samples for certain targets with an assay for a specific activity or properties, the *random screening* approach, has grown by leaps and bounds owing to the increased ability to handle large amounts of samples by *high-throughput screening* (HTS).

- Instead of the conventional number of 96 wells on a plate, 384 well plates and even 1536 well plates are now in use.
- Robotic systems with abilities to localize single colonies on a plate, to grow colonies on a plate, and to inoculate in liquid culture allow screening of up to 1000 strains per week. Given the necessity to screen about 10–100 000 strains in order to find a useful biocatalyst, this means one could expect the process to take 10–100 weeks to a breakthrough.
- Pharmaceutical companies today can test up to 50 000 compounds per day for pharmacological activity (with up to 20 activity tests simultaneously), 30 times the capacity compared to even three years before.

The conventional screening methods, such as
- isolation of microorganisms in the neighborhood of habitats with enhanced concentration of the substrate such as alkanes in the vicinity of oil wells,
- selection of strains based on taxonomic closeness to prior successful strains, and
- enrichment cultures,

are increasingly being replaced or supplemented by novel methods.

Examples of such novel methods are:
- PCR and rRNA methods, based in part on published nucleotide sequences of proteins with a similar activity profile;
- cloning for high expression (Chapter 4) with highly transformable hosts such as *E. coli* or *Aspergillus oryzae*;
- single cell selection by identification of interesting cells, mostly on solid supports, with the desired activity, and picking with robotic arms for further cultivation; and
- together with increased genome sequencing, acceleration of the search for similar enzymes in different organisms by using related primer pairs in PCR.

The ideal assay system for HTS would include automated manipulation of single wells and each should be tested for a variety of activities simultaneously. One of the main challenges is to develop a simple assay sensitive enough to pick up enzyme activities that might be below their optimal level under the assay conditions. Nowadays novel colorimetric, luminescence, and fluorescence methods have been established in HTS with automated multiplex compound testing (typically 10–20 compounds per well) (Winson, 1997).

The same screening strategies can be applied to activity engineering using directed evolution (Chapter 11), as single cell selection is important for screening of mutants. Detecting genetic diversity through random mutagenesis can be achieved by using the Fluorescence Enzyme Bead Assay System (FEBAS), in which an evenly distributed substrate becomes a fluorescently localizable product after catalysis by enzymes attached to beads. With this method, volumes as small as 750 nL can be screened in drops of 0.5–3.5 nl, and cells presenting the product of interest can be identified by Confocal Nanospectroscopic Scanning (CNS) (Winson, 1997).

Another strategy exploits the nucleic acids from diverse organisms where it is not necessary to isolate the bacteria themselves. This approach has particular benefits because only a small fraction of organisms in a particular habitat is culturable (Table 3.4).

Uncomfortable culture conditions (as are often the case with extremophiles) or even the inability to culture at all can be avoided by cloning the nucleic acids from those organisms into *E. coli*-based plasmid and phage expression vectors, termed "genomic libraries" (Chapters 4 and 11). The Diversa Company (San Diego, CA, USA) specializes in improving the accessibility of enzymes of otherwise unculturable organisms. The procedure is strongly dependent on the source of the genomic DNA, as the DNA for constructing these libraries has to be totally pure. In special cases, an intermediate step of primary enrichment culture might be necessary, with subsequent purification of genomic DNA from such cultures. Screening these libraries consists of the following steps: (i) distribution of clones; (ii) screening for a desired activity; and (iii) purification of single clones through HTS. Figure 3.2 demonstrates the assay procedure.

Table 3.4 Fraction of culturable organisms (Amann, 1995).

Habitat	*Culturability [%]*
Sea water	0.001–0.1
Fresh water	0.25
Mesotrophic lake	0.1–1
Unpolluted estuarine waters	0.1–3
Activated sludge	1–15
Sediments	0.25
Soil	0.3

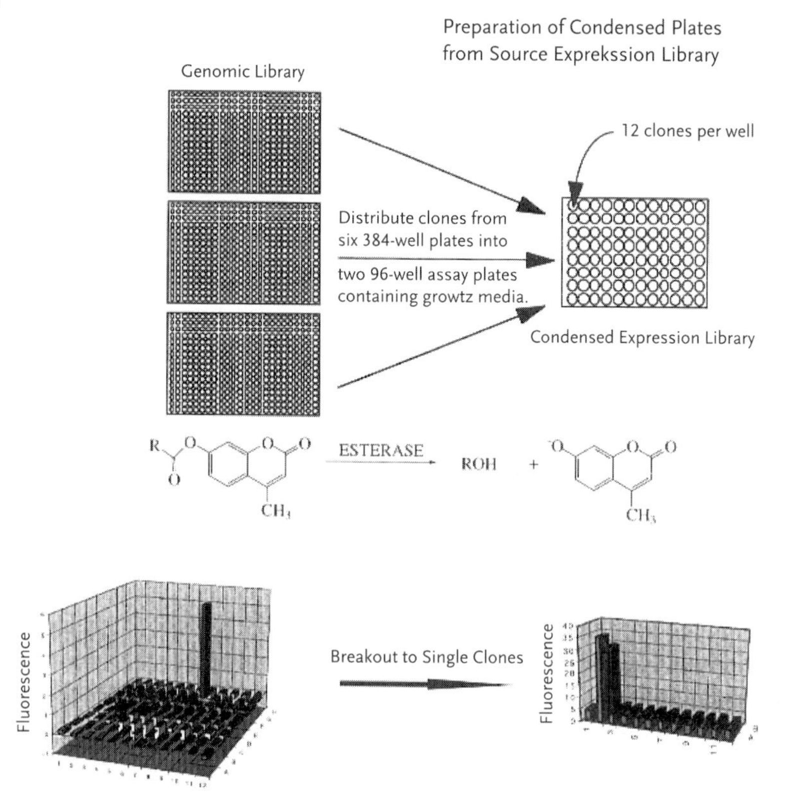

Figure 3.2 Condensation of genomic libraries and example assay of enzyme activity (Robertson, 1996).

Host cells are grown to achieve maximum expression of cloned inserts. A simple but highly sensitive fluorescence assay is employed, with a low-fluorescence substrate resulting in a highly fluorescent product. After suitable active clones have been obtained, the insert is sequenced, and the resulting open reading frame is identified. The target gene is subcloned (transferred to a more suitable vector, probably getting rid of non-coding regions in that genomic insert) for high-level expression. The recent application of molecular phylogeny (Chapter 14) to environmental samples has resulted in the discovery of an abundance of unique and previously unrecognized microorganisms. The vast majority of this microbial diversity has proved refractory to cultivation. Cells were encapsulated in gel micro droplets for massively parallel microbial cultivation under low nutrient flux conditions, followed by flow cytometry to detect micro droplets containing micro colonies (Zengler, 2002). The technique was applied to samples from both sea water and soil. The ability to grow and investigate previously uncultured organisms in pure culture serves two goals: (i) to provide a source of new microbial enzyme functions; and (ii) to enhance understanding of microbial physiology and metabolic adaptation.

Suggested Further Reading

J. Postgate, *The Outer Reaches of Life*, Cambridge University Press, Cambridge, UK, 1st edition, **1994**.

References

F. Abe and K. Horikoshi, The biotechnological potential of piezophiles, *Trends Biotechnol.* **2001**, *19*, 102–108.

R. I. Amann, W. Ludwig, and K.-H. Schleifer, Phylogenetic identification and in-situ detection of individual microbial cells without cultivation, *Microbiol. Rev.* **1995**, *59*, 143–169.

E. Blochl, R. Rachel, S. Burggraf, D. Hafenbradl, H. W. Jannasch, and K. O. Stetter, *Pyrolobus fumarii*, gen. and sp. nov., represents a novel group of archaea, extending the upper temperature limit for life to 113°C, *Extremophiles* **1997**, *1*, 14–21.

T. D. Brock and M. T. Madigan (eds.), *Biology of Microorganisms*, 5th edition, Prentice-Hall, New York, **1988**.

S. K. Chapman and G. A. Reid, Breaking the code, *Chemistry in Britain*, **1993**, *29*, 202–204.

P. S. J. Cheetham, Screening for novel biocatalysts, *Enzyme Microb. Technol.* **1987**, *9*, 194–213.

D. W. Crabb and C. Mitchinson, Enzymes involved in the processing of starch to sugars, *Trends Biotechnol.* **1997**, *15*, 349–352.

A. L. Demain, Microbial biotechnology, *Trends Biotechnol.* **2000**, *18*, 26–31.

K. J. Edwards, P. L. Bond, T. M. Gihring, and J. F. Banfield, An archaeal iron-oxidizing extreme acidophile important in acid mine drainage, *Science* **2000**, *287*, 1796–1799.

F. Hofmeister, Zur Lehre von der Wirkung der Salze. Zweite Mitteilung, *Arch. Exp. Pathol. Pharmakol.* **1888**, *24*, 247–260.

S. Isken and J. A. M. de Bont, Bacteria tolerant to organic solvents, *Extremophiles* **1998**, *2*, 229–238.

A. Karshikoff and R. Ladenstein, Ion pairs and the thermotolerance of proteins from hyperthermophiles: a 'traffic rule' for hot roads, *Trends Biochem. Sci.* **2001**, *26*, 550–556.

R. Margesin and F. Schinner, Potential of halotolerant and halophilic microorganisms for biotechnology, *Extremophiles* **2001**, *5*, 73–83.

F. Niehaus, C. Bertoldo, M. Kähler, and G. Anthranikian, Extremophiles as a source of novel enzymes for industrial application, *Appl. Microbiol. Biotechnol.* **1999**, *51*, 711–729.

P. Nikolova and O. P. Ward, Whole cell yeast biotransformations in two-phase systems: effect of solvent on product formation and cell structure, *J. Ind. Microbiol.* **1992**, *10*, 169–177.

P. Nikolova and O. P. Ward, Whole cell biocatalysis in nonconventional media, *J. Ind. Microbiol.* **1993**, *12*, 76–86.

V. Rehaver and R. Jaenicke, Stability and reconstitution of D-glyceraldehyde-3-phosphate dehydrogenase from the hyperthermophilic eubacterium *Thermotoga maritima*, *J. Biol. Chem.* **1992**, *267*, 10999–11006.

D. E. Robertson, E. J. Mathur, R. V. Swanson, B. L. Marrs, and J. M. Short, The discovery of new biocatalysts from microbial diversity, *SIM News* **1996**, *46*, 3–8.

L. J. Rothschild and R. L. Mancinelli, Life in extreme environments, *Nature* **2001**, *409*, 1092–1101.

D. E. Stafford and G. Stephanopoulus, Metabolic engineering as an integrating platform for strain development, *Curr. Opin. Microbiol.* **2001**, *4*, 336–340.

M. K. Winson and D. B. Bell, If you've got it, flaunt it – rapid screening for microbial biocatalysts, *TIBTECH* **1997**, *15*, 120–122.

C. R. Woese, O. Kandler, and M. L. Wheelis, Towards a natural system of organisms: proposal for the domains Archaea, Bacteria, and Eucarya, **1990**, *Proc. Natl. Acad. Sci. USA* **1990**, *87*, 4576–4579.

C. R. Woese, Interpreting the universal phylogenetic tree, *Proc. Natl. Acad. Sci. USA* **2000**, *97*, 8392–8396.

J. G. Zeikus, C. Vieille, and A. Savchenko, Thermozymes: biotechnology and structure–function relationships, *Extremophiles* **1998**, *2*, 179–183.

K. Zengler, G.Toledo, M. Rappe, J. Elkins, E. J. Mathur, J. M. Short, and M. Keller, Cultivating the uncultured, *Proc. Natl. Acad. Sci. USA* **2002**, *99*, 15681–15686.

4
Molecular Biology Tools for Biocatalysis

Summary

The basic dogma of biology states that information flows from genes, represented by DNA, via a messenger molecule, the mRNA, to proteins. For recombinant expression of proteins, bacterial hosts are often preferred: transcription (DNA to RNA) and translation (RNA to protein) are less complicated than in eukaryotes, and bacteria feature a limited number of expressed genes, short doubling times, no organelles for compartmentation, and quite a straightforward metabolism.

When molecular biology tools are being used there are six basic steps for obtaining recombinant proteins:

1. isolating the DNA of the organism of interest, and DNA purification;
2. finding the gene of interest within the isolated DNA;
3. defining primers and restriction sites for successful cloning;
4. cloning the gene into an appropriate vector and transforming a suitable heterologous host with it (such as an *E. coli* strain);
5. sequencing the clone to confirm the correct sequence of the inserted gene, and
6. expressing the protein of interest.

The sequence of a gene of interest is known either partially (most often the N-terminus from protein sequencing), or fully from databases. To obtain the gene from genomic DNA or cDNA, either the *polymerase chain reaction* (PCR) or, in the second case, a hybridization technique such as *Southern blot* can be applied. The success of the PCR depends on (i) the right choice of a primer set (with a length of 25–30 base pairs, the highest possible T_m, and a GC content not exceeding 40–50%), (ii) the correct Mg^{2+} concentration, and (iii) purity and the correct concentration of the DNA template. Southern blotting involves transfer of the target DNA from a gel-electrophoretically separated DNA template to a nitrocellulose or nylon membrane by means of capillary forces or electric current.

A *vector* can be described as a double-stranded *plasmid* (a circular DNA molecule propagated in a host) which includes the recombinant gene of interest, selection markers, an origin of replication, a promoter, and a multiple cloning site (MCS) to insert a gene. *Cloning* is the process of introducing new DNA into an organism and subsequently creating identical copies of the newly created organism. To incorporate fragments of foreign DNA into a plasmid vector, *restriction enzymes* for cutting and ligases for rejoining of double-stranded DNA are required. The action of restriction enzymes involves recognition of a specific DNA sequence, producing either *cohesive ends* with protruding overhangs on the 5′ or 3′ end, or *blunt ends* with no overhang.

Ligated plasmids should be checked for insertion of the desired fragment instead of just religation, most conveniently by *blue–white screening*. In vectors with the *lacZ′* gene sequence, insertion at any restriction site of the MCS results in disruption of that sequence so that no β-galactosidase is expressed and the synthetic X-gal substrate cannot form blue colonies.

To achieve the desired amount of recombinant protein product, *overexpression* is required. Success is still difficult to predict and depends on the expressed protein, host, vector, promoter, and culturing conditions. Since *constitutive expression* in high yields influences cell metabolism negatively, *inducible overexpression* with 10–40% yield of soluble target protein is the goal. Main obstacles include bad *codon usage*, the preference for specific triplets to code for an amino acid, and an insufficient number of repressors to retain tight control in the case of high-copy expression vectors. Alternatives to expression as soluble protein are (i) expression as inclusion bodies, (ii) expression as a fusion protein, or (iii) secretion into the medium.

4.1
Molecular Biology Basics: DNA versus Protein Level

When a researcher develops a biocatalyst, the first choice for a host is always a bacterial system. Bacteria represent the most versatile organisms as far as reproduction, growth rates, and handling efficiency are concerned. Since the 1990s, molecular biology has become a powerful tool for creating recombinant proteins, especially enzymes. *Escherichia coli* is the most commonly used host because it is such a well characterized system. Although *E. coli* cannot be used to produce all large, complex proteins, such as those with disulfide bonds or unpaired thiols, or proteins that require post-translational modification for activity, there are some examples of even higher eukaryotic proteins that have been produced successfully, such as interferons, insulin, and human serum albumin (Hodgson, 1993). This chapter provides an overview of today's possibilities when cloning and expression are required to obtain a biocatalyst.

Compared to higher organisms, bacteria seem to be rather simple when one considers the number of their genes (Table 4.1). Among unicellular organisms, *E. coli* has 4289 genes coding for proteins, *Agrobacterium tumefaciens* 1833, and *Bacillus subtilis* 4104; the archaeum *Aquifex aeolicus* has only 1529 genes encoded for functioning as a whole organism. Higher organisms feature more genes, such as the fruitfly *Drosophila* with 13 500 annotated genes, the earthworm *Caenorhabditis elegans* with 19 000 genes, or humans (*Homo sapiens*) with a predicted number of 35 000 genes coding for proteins (Celera databank). While the number of genes

Table 4.1 Total number of genes for different species with (nearly) completed genomes.

Species	***Gene number***
Aquifex aeolicus	1 529
Agrobacterium tumefaciens	1 833
Bacillus subtilis	4 104
Escherichia coli	4 289
Drosophila melanogaster	13 500
Homo sapiens	35 000–100 000

seems to correlate with organism complexity, one should keep in mind that a nematode such as *C. elegans* has almost half the genes of a human being.

Any living organism consists of a varying number of cells. Whereas bacteria are unicellular, humans as the most highly developed organism consist of several billions of cells, differentiated and organized in different tissues. All these cells have some common features:

- metabolism,
- procreation,
- movement, and
- information transfer from DNA into proteins functioning as enzymes or structural proteins.

Central dogma of biology

The basic dogma of biology states that information flows from genes, represented by DNA, via a messenger molecule, the mRNA, to proteins:

$$\text{DNA} \xrightarrow[\text{transcription}]{} \text{RNA} \xrightarrow[\text{translation}]{} \text{protein}$$

This dogma is valid for most living organisms; exceptions are found within viruses. The rule calls for transcription of DNA into RNA, which incorporates uracil (U) instead of thymine (T) as the fourth nucleic acid, and RNA is then translated into proteins using a translational machinery consisting of ribosomes. As bacterial transcription and translation are less complicated than the processes in eukaryotes, bacteria are commonly used as hosts for recombinant expression of proteins. For further study and details of transcription of DNA into RNA and translation of RNA into protein, biochemistry or molecular biology textbooks, such as listed under "Suggested Further Reading" are recommended.

1. *Transcription*. Transcription is the process of reading coding regions within the DNA molecule and transcribing them into an mRNA molecule, performed by

RNA polymerases, enzymes that are able to read the base pair information written in the DNA sequence. RNA polymerases use one strand of double-stranded DNA as a template to create the anti-sense strand represented by the developing RNA molecule.

2. *Translation.* The mRNA molecule, representing an ORF (open reading frame), a direct copy of the corresponding DNA coding region from which it originated, can be translated into an amino acid sequence with the genetic code. Each codon (triplet of base pairs) within a DNA sequence of an ORF represents one of the 20 proteinogenic amino acids or a "stop" event. As the code is degenerated, some amino acids are represented by more than one codon.

3. *Ribosomes.* The translation machinery, consisting of proteins and rRNA (ribosomal RNA) molecules, binds to the mRNA and translates their sequence into the amino acid sequence (the primary sequence of proteins) by using tRNAs (transfer RNAs), molecules which have again the codons corresponding to the amino acids they are carrying. tRNAs are transferred to the ribosomes, and the corresponding amino acid is incorporated into the growing amino acid chain at the exact location of the codon in the mRNA molecule.

These features differ in bacteria and eukaryotes and are illustrated in Figure 4.1.

In addition to the limited number of expressed genes, bacteria feature a highly reproductive cycle but no cell nucleus, no organelles for compartmentation, no cell-to-cell communication as in tissues, and quite a straightforward metabolism. Favored bacterial systems for production of recombinant proteins are various strains of *E. coli*, *Bacillus*, and *Pseudomonas*. Non-bacterial options such as yeast or mammalian cells are discussed briefly later in this chapter.

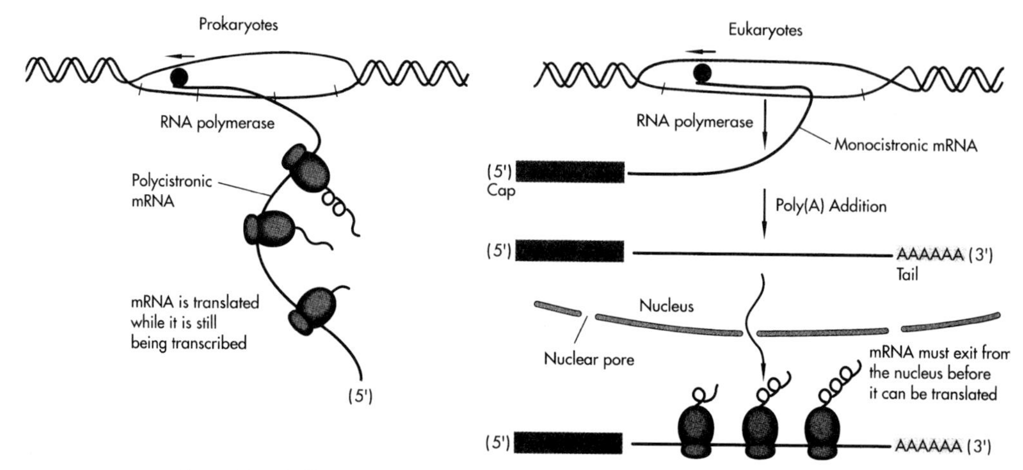

Figure 4.1 Differences in transcription and translation in bacteria and eukaryotes (Watson, 1992).

E. coli offers several strains, each with a slightly different genome (Brown, 1991). The most common are *E. coli* XL1Blue, JM105, JM109, and DH5α (all are derivatives of the strain K12), and BL21, HB101, which are also derivatives of the strain B. These strains can be combined with vectors in multiple ways. A vector can be described as a double-stranded plasmid which includes the recombinant gene of interest, selection markers, an origin of replication, a promoter, and a multiple cloning site where a gene can be inserted. One copy of a gene or genome codes for more than one corresponding protein molecule because (i) every DNA molecule can be transcribed into several identical mRNA copies, and (ii) every mRNA can be translated into several identical copies of protein. Thus, only a small amount of DNA, about one or two copies, is required for a reasonable amount of protein. The bacterial translation machinery is evolutionally designed for high efficiency to serve bacterial needs within a 20-min cell cycle. Higher organisms progressively develop increasing complexity, including cell differentiation, prolonged cell cycles, and regulation, which complicates optimization of single protein production.

When molecular biology tools are being used there are six basic steps for obtaining recombinant proteins:

1. isolating the DNA of the organism of interest, and DNA purification;
2. finding the gene of interest within the isolated DNA;
3. defining primers and restriction sites for successful cloning;
4. cloning the gene into an appropriate vector and transforming a suitable heterologous host with it (such as an *E. coli* strain);
5. sequencing the clone to confirm the correct sequence of the inserted gene; and
6. expressing the protein of interest.

4.2 DNA Isolation and Purification

Experimentally, nowadays, DNA isolation and purification can be regarded as straightforward, as different molecular biology kits, from a variety of vendors such as Qiagen, BioRAD, or Clontech, can be purchased for isolating DNA or RNA. With bacteria as the source of genomic DNA, one should distinguish between Gram-negative and Gram-positive bacteria, which feature different cell wall (murein sacculus) compositions which determine the way cells can be lysed and DNA extracted. Gram-negative bacteria (such as *E.coli*) are usually easier to handle owing to lower strength of the cell walls due to fewer peptide crosslinks within the peptidoglycan layer within the murein sacculus. Crucial steps in purification involve (i) cell lysis, (ii) degradation of cell proteins with Proteinase K, and (iii) subsequent phenol extractions for DNA purification.

There are different DNA purification protocols for different bacterial organisms, based on their different cell wall compositions. Purification of macromolecular DNA can be achieved through subsequent steps of phenol extraction (using Tris-saturated phenol to prevent the DNA from getting redissolved in the phenol). The

DNA is precipitated after phenol extractions using standard protocols with high salts and ethanol at –70 °C (Maniatis, 1989). Good results can be achieved using genomic DNA isolation kits, as of Molecular Research Center, Inc. (Cincinnati, OH, USA), resulting in a genomic DNA yield of up to 40 µg (50 mL bacterial culture)$^{-1}$. For isolating low molecular weight DNA, such as plasmids, purification using special kits is state-of-the-art. All kits follow a basic protocol of lysing the cells, denaturing debris and proteins, and binding the DNA located in the supernatant to a positively charged membrane. After washing steps, the DNA can be eluted from that membrane. One can choose purification either with membrane columns (more common nowadays) or with a powder to which DNA adheres during purification.

4.2.1 Quantification of DNA/RNA

DNA can be quantified either spectrophotometrically at 260 nm or through comparison of ethidium bromide- or SYBR orange-stained DNA samples to DNA ladders (markers of known molecular weights) of known quantity in an agarose gel. The lowest possible amount that can be visualized is about 20 ng (Lottspeich, 1998). Agarose polymer gels are commonly used to separate DNA samples of different sizes on the basis of electrophoretic mobility of the DNA molecule in the gel. As electrophoretic mobility depends on the size of the sample molecule (here, DNA) as well as the characteristic density (the persistence length ξ) of the gel (Chapter 9, Section 9.3.3), the lower the agarose concentration and the lower the molecular weight of DNA, the faster the molecule travels in the gel. Therefore, different gel concentrations are employed for different ranges of sample sizes.

Table 4.2 Spectrometrically determined DNA can be quantified with the estimated amounts of DNA shown.

	Equivalent to
$OD_{260\ nm} = 1$	50 µg mL^{-1} double-stranded DNA
	40 µg mL^{-1} single-stranded DNA
	30 µg mL^{-1} single-stranded RNA
Nucleotide	**Molar extinction coefficient ε [mM^{-1} cm^{-1}]**
dATP	15.4
dCTP	9.0
dGTP	13.7
dTTP	10.0
$\Sigma[\varepsilon(dNTP)_{oligonucleotide} = 1$ µmol mL^{-1}	

Source: Maniatis, 1989.

For spectrometry, relatively large amounts of DNA are required (> 0.25 μg mL^{-1}). The purity of nucleic acids can be determined by the ratio of extinction at 260 to that at 280 nm. Pure DNA has a ratio of 1.8, pure RNA one of 2.0. The DNA concentration can be determined using the approximate values in Table 4.2.

Libraries

A gene *library* is a collection of different DNA sequences or DNA fragments, each of which has been cloned into a vector for ease of purification, storage, and analysis. Depending on the source of DNA used, there are two types of gene libraries: (i) a *genomic* library, if the source of DNA is genomic; or (ii) a *cDNA* library, if the DNA is a copy of an mRNA population (see Section 4.3.3). A *representative* library contains all the original sequences. An *enriched cDNA* library lacks certain sequences but is enriched in others. Representation of the starting material is an important consideration when producing a library. One cause for a non-representative library might be the lack of appropriate restriction sites. Importantly in any case, however, the library must contain a *sufficient number of clones* to be representative.

4.3 Gene Isolation, Detection, and Verification

Considering a gene of interest, two possibilities can occur:

1. the gene sequence is fully known from databases (see Chapter 14 for how to get this information), or
2. only part of the sequence is known, such as the N-terminal sequence, if the full sequence of the protein or DNA has not been submitted yet.

For both scenarios the polymerase chain reaction (PCR) can be used. Alternatively, in the second case, a hybridization technique such as Southern blot (see later) can be applied.

4.3.1 Polymerase Chain Reaction

The polymerase chain reaction, PCR, the revolutionary technique for amplification of DNA first described by Mullis et al. (Mullis, 1986; Nobel Prize 1993), changed and is still changing the field of molecular biology rapidly. The principle relies on multiple repeating cycles using a thermostable polymerase and a set of gene-specific primers resulting in specific amplification of the DNA sequence that lies between the sense and anti-sense primer pair (Figure 4.2).

PCR utilizes the enzymatic properties of a DNA polymerase to duplicate DNA through extension of a short, existing, double-stranded piece of DNA, usually 15–30 bp. This short, double-stranded DNA piece can be mimicked through addition of short oligonucleotides (primer) to each of the DNA strands, the sense (5′–3′) and anti-sense (3′–5′) primer. First, in the denaturing step, the DNA has to be

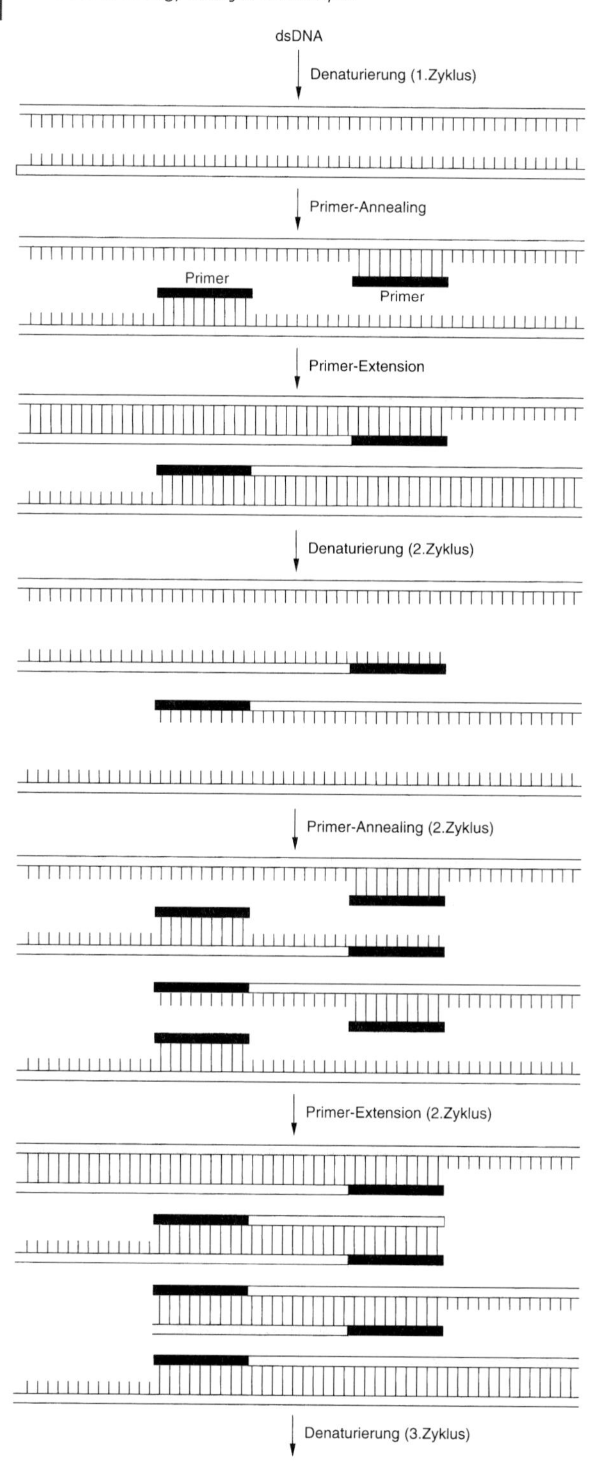

Figure 4.2 Polymerase chain reaction (PCR) (Lottspeich, 1998). The PCR cycles between a denaturing step to obtain single-stranded DNA, an annealing step for the primer attachment to the template DNA, and a polymerization step, in which the heat-stable polymerase elongates the corresponding strand using the primers as starting points.

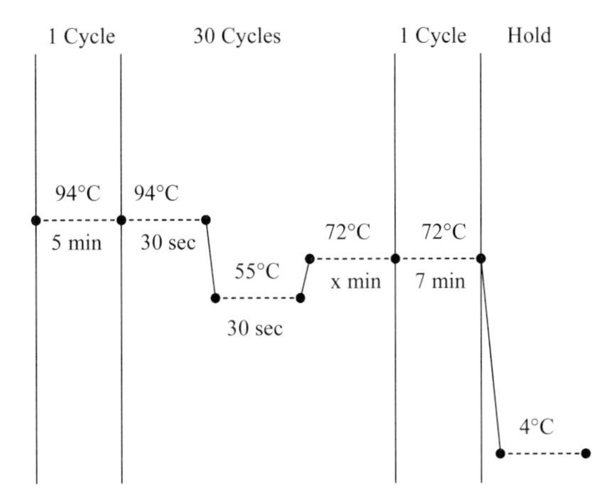

Figure 4.3 Typical profile of a PCR reaction. The *x* min in the elongation step is dependent on the length of the amplified DNA.

heated to 95 °C to obtain single strands. In the second step, the primers are able to bind to the corresponding part on the single-stranded DNA, a process termed "annealing". The third step involves an optimal polymerase temperature for primer extension (72 °C for the Taq polymerase from *Thermus aquaticus*). These three steps can be repeated in cycles because of the thermal stability of the polymerases which, after all, are obtained from extremophilic organisms (Chapter 3, Section 3.4). Each newly synthesized DNA strand and each one that already exists serves as a template for the next cycle, thereby creating an exponential number of DNA molecules. With each cycle, DNA amplification within the primers outruns the longer DNA present at the outset; in the end, practically all the amplified DNA lies within the boundaries of the chosen primers.

4.3.2 Optimization of a PCR Reaction

One of the crucial and time-consuming steps when establishing a PCR reaction of an unknown product is its optimization protocol. First, the *right choice of a primer set* should be considered a necessity. The primers should have an ideal length of 25–30 base pairs, a suitable T_m range for both of them (they should not differ in their T_m more than 2 °C) and the T_m should be as high as possible to achieve specificity in the PCR products. The primer GC content should not exceed 40–50% (if this is not possible, a different PCR kit for high GC content should be used), and the user should test the primers for primer-dimer production. A primer-dimer means (i) the annealing of one primer to different positions within the gene, causing short or long primer additions, or nonsense products that can be misleading; or (ii) the annealing of one primer to itself. There are several computer programs available for correct primer choice (such as Amplify for Macintosh, or on the Web, http://www.hgmp.mrc.ac.uk/GenomeWeb/nuc-primer.html).

Second, DNA polymerases need magnesium ions for activity; a *correct Mg^{2+} concentration* is crucial for successful amplification as well as correctness in sequence of the amplified product. This fact can be used as a tool for directed evolution to introduce sequence errors deliberately during amplification (error-prone PCR; see Chapter 11, Section 11.3) by adding geometrically similar Mn^{2+} to the PCR buffer already containing Mg^{2+}. In most cases, the concentration of the buffer provided in a commercial PCR kit is optimized for the corresponding polymerase, so adjustment of Mg^{2+} concentration should rarely have to be considered.

Third, the *purity and concentration of the DNA template* are very important for successful PCR. If possible, the DNA should be purified with high-quality kits (such as are available from Qiagen or Clontech), and no traces of inhibitory substances such as ethanol should be left after genomic DNA preparation. The amount of template DNA, based on the gene of interest and not on the whole genomic or plasmid DNA, ideally should lie within 10–100 pg and should be verified at the beginning of the optimization process. An excessive template DNA concentration can prevent amplification, whereas too little of the desired gene to be amplified results in excessive cycle time or poor resolution. At the outset of PCR optimization, different template concentrations covering orders of magnitudes should be tested to verify the optimal concentration range.

A short list of possible PCR kits and useful websites is given in Table 4.3.

For large PCR products or difficult templates (secondary structure or high GC content), high-fidelity (low error rate) polymerases such as the Advantage PCR (Clontech) or Failsafe PCR (Epicentre) are highly recommended.

If, even after consideration of all the PCR optimization techniques mentioned, still no PCR amplification is obtained, it is feasible to try a *touch-down PCR* with the same primers to establish the correct annealing temperature. A touch-down PCR implies a decrease of the annealing temperature in 0.1–0.5 °C increments in the first 10 cycles to determine the optimal annealing temperature. The starting temperature is the highest allowed for the primer pair. After about 10 cycles a "normal" PCR program commences, using the highest annealing temperature from the touch-down program for amplification. This procedure usually results in higher specificity of the PCR product. For challenging PCR reactions with difficult templates showing secondary structures or a high GC content, additives such as DMSO, formamide, Tween 20 or Triton-X-100 are sometimes needed to facilitate denaturing of the DNA template. Other compounds such as glycerol, BSA, or PEG are needed for stabilizing a fragile primer–DNA binding.

Table 4.3 Short list of polymerases for PCR.

PCR kit	*Supplier*	*Website*
taq polymerase	Qiagen	http://www.expasy.ch
Failsafe PCR	Epicentre	http://www.ncbi.nlm.nih.gov
Advantage PCR	Clontech	http://www.hgmp.mrc.ac.uk/GenomeWeb/nuc-primer.html
Pfu polymerase	Stratagene	http://srs.ebi.ac.uk/

4.3.3 Special PCR Techniques

4.3.3.1 Nested PCR

Nested PCR is used for eukaryotic gene amplification using cDNA (copy-DNA, being copied from mRNA through reverse transcription) libraries. The technique is useful for i) increasing specificity of amplification out of a template pool and ii) verifying the correct amplification product. A nested PCR uses two different sets of primers, an outer and inner pair. The outer primer pair can be complementary to the 5′ and 3′ regions of the mRNAs but normally anneals to 5′ and 3′ regions next to the gene of interest and should have a higher melting temperature (T_m) than the inner pair of primers, which can lie within the gene or mark the N- and C-terminal parts of the gene. A first PCR reaction uses the outer pair to generate a PCR product longer than the gene; the second PCR reaction is performed on the first amplification product using the second inner pair of primers. The two reactions can either be carried out in consecutive steps in two tubes or, if the melting temperatures of the two different primer pairs differ significantly from each other, the nested amplification can occur during the same reaction. In that case, the longer product with the higher Tm is amplified first, then, the annealing temperature is switched for the inner primer pair to amplify the shorter gene product.

4.3.3.2 Inverse PCR

Inverse PCR was originally developed as a method to rapidly amplify regions of unknown sequence flanking any characterized segment of the genome. The technique eliminates the need to construct and screen DNA libraries (see earlier in this chapter) to walk hundreds of base pairs into flanking regions and is useful mostly for large genomes (eukaryotic, starting from *C. elegans*) and screening of YACs (yeast artificial chromosomes). It can be adapted to generate linking fragments – pieces of DNA that identify adjacent restriction fragments (Collins, 1984) – for creating an ordered library and large-scale physical maps of complex genomes. In addition, Helmsley et al. (Helmsley, 1989) have devised a method for site-directed mutagenesis using inverse PCR. In this scheme, one of the oligonucleotide primers is synthesized with an alternative base reflecting the desired mutation.

4.3.3.3 RACE: Rapid Amplification of cDNA Ends

For some useful biocatalysts, bacteria may not be the only source, and retrieving the gene from eukaryotic organisms might be necessary.

To secure a gene from a eukaryotic genome, the most efficient method among several is to obtain cDNA (complementary DNA) from mRNA via reverse transcriptase. Use of cDNA avoids dealing with the problem of introns, where a complete gene can stretch over thousands of base pairs but is divided into several small exons, alternating with introns of sometimes remarkable size [several thousand base pairs (kB) for one intron]. As, however, only a few eukaryotic genes are fully sequenced and deposited in one of the databases, one has to rely on ESTs (expressed sequence tags) for the genome sequences available in the databases. These tags give limited information about the gene (mainly to identify and classify the type of enzyme or protein) but sufficient to perform a RACE, for which either self-synthesized cDNA (as described below) or commercially available cDNA libraries (common for tissues of higher organisms) can be used.

In essence, the RACE protocol generates amplified genes by using PCR to amplify copies of the region between a single point in the transcript (a primer generated against the known part of the sequence, such as the EST tag) and the 5′ or 3′ end (general primers, such as an adapter primer to the flanking UTR region of a gene). The self-generated primer provides specificity to the reaction (see the kits for further details). In this context, a follow-up amplification or verification of the correct DNA piece using a nested PCR (see above) can be applied. First, an appropriate cDNA library has to be chosen, depending on the tissue (in the case of mammalian genes), or total RNA isolation has to be performed (described below) to translate RNA into cDNA which serves as a template for the RACE.

Total RNA isolation

To obtain a template for the RACE, RNA isolation is necessary to rewrite it into cDNA. Precautions have to be taken, as RNA is a considerably more unstable molecule than DNA and, to add to the situation, its degrading enzymes are among the most stable known. Nowadays, most suppliers provide RNAse-free buffers and water, so DEPC (diethylpyrocarbonate) treatment to destroy RNAse activity is necessary only for home-made buffers. Kits for RNA preparation, either for total RNA or for mRNA isolation, do not require any DEPC treatment.

Regarding the isolation procedure itself, cells are lysed carefully using a special buffer commensurate with the RNA source and are then disrupted during centrifugation using a membrane column (Qiashredder from Qiagen). The lysate is mixed with an equal volume of 70% ethanol (RNAse-free) and then transferred to another column, in which the RNA binds to the column membrane. Subsequent washing steps purify the RNA, which is finally eluted with RNAse-free water (for details, consult http://www.qiagen.com/literature/BenchGuide/BG_RNA_lit.asp). At –80 °C, RNA is stable for a few months. Another method uses Trizol as a reagent with a protocol from Life Technologies (http://www.lifetechnologies.com/content/sfs/manuals/15596026.pdf).

Reverse transcriptase reaction

After obtaining pure total RNA from eukaryotic cells, the RNA needs to be rewritten into cDNA to serve as a template in PCR, as RNA cannot be amplified by PCR. The task of rewriting is accomplished with reverse transcriptase (RT), a viral enzyme used by retroviruses (whose name stems from harboring this enzyme and the ability to rewrite RNA into cDNA). The group of retroviruses has such members as the AIDS virus, Avian myeloblastosis virus, Murine leukemia virus (Frohmann, 1988; Kawasaki, 1988), and Adenovirus. Commonly employed reverse transcriptases stem from either the avian virus (AMV-RT) or the murine leukemia virus (MMLV, used in the Clontech RT-kit).

Today, the enzyme is from a recombinant source and forms part of RT kits (Clontech, Novagen, Qiagen). RT requires a short primer able to attach to mRNA and then rewrites the first strand of cDNA using dNTPs as well as mRNA as a template (Figure 4.4). During the second round, the single- stranded cDNA is complemented to a double-stranded DNA. The poly-A tail of eukaryotic mRNA can be exploited for synthesizing cDNA using an oligo-dT primer which attaches to the poly-A tail and serves as a starting point for the reverse transcriptase. For optimal results, the amount of total RNA used in the reaction should be within 1–2 μg and very pure, i.e., free from genomic DNA to avoid the genomic gene copy functioning as a template. RNA, together with the starting primer (oligo-dT), is first denatured at 70 °C to destroy secondary structure and then transferred to ice to allow the primer to attach to the single-stranded extended RNA. Optimum reaction conditions for RT are 42 °C and a reaction time usually of 1 h. The resulting cDNA can be used either for creating cDNA libraries, or for quantifying the amount of DNA expression of a target gene, or for determining unknown sequences on the 5′ or 3′ ends of the RNA (the RACE).

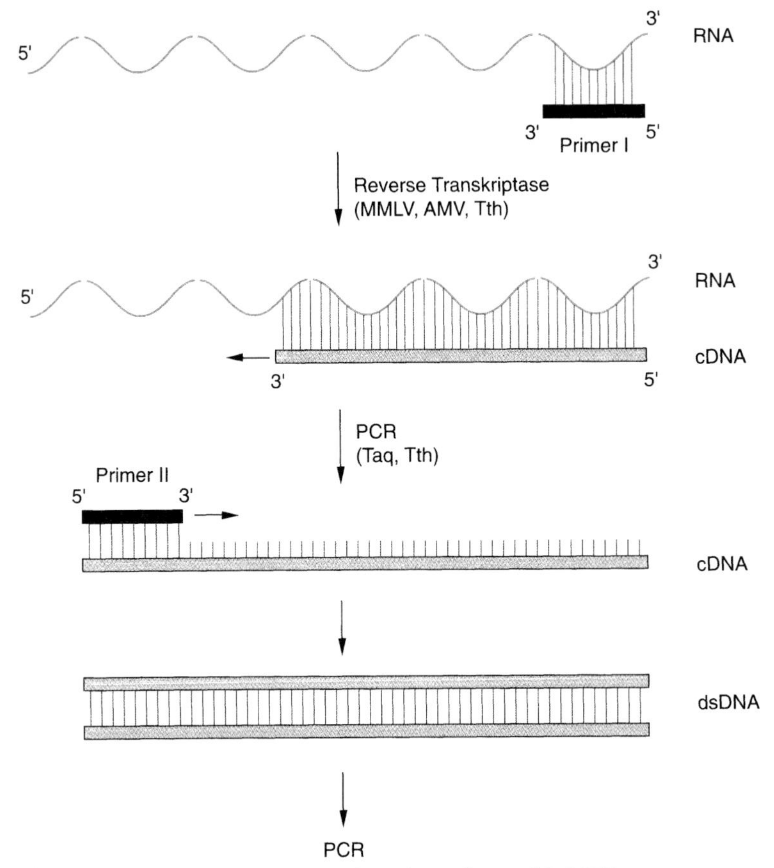

Figure 4.4 Reverse transcriptase reaction scheme (Lottspeich, 1998). Tth = *Thermus thermophilus*.

4.3.4
Southern Blotting

The Southern blot, developed by E. M. Southern (1974), is an alternative route to PCR to obtain a complete gene sequence when only parts of the gene sequence are known. It is an important tool for identifying genes within a pool of DNA templates to a known part of the target gene by the hybridization technique. Southern blotting involves transfer of the target DNA from a gel-electrophoretically separated DNA template to a nitrocellulose or nylon membrane (Table 4.4) by means of capillary forces or electric current. DNA needs to be denatured after electrophoresis so as to be single-stranded; the negatively charged single-stranded DNA is then transferred to a positively charged nylon membrane using high-salt buffer at pH 7.0. Various techniques for blot transfer exist, such as capillary force-driven transfer, dry blot or electroblotting. Capillary transfer, which is the most widely used and provides the highest sensitivity and reproducibility, describes passive transfer of the target in buffer through capillary forces, requiring depurination, denaturation, and neutralization of the DNA within the agarose gel. The dry blot works without transfer buffer and is useful for gel shift assays (an assay for analyzing DNA–protein associations). Electroblotting is only recommended for acrylamide gels, as are applied to small-size DNA down to oligonucleotides, and provides the most rapid transfer. The transfer efficiency depends on the length of transfer, buffer concentration and quality of the membrane used.

The transferred DNA is then crosslinked to the membrane using UV light or by heating for several hours at 80 °C. Crosslinking machines exploiting UV light (e.g., from Stratagene) optimize efficiency of yield and time. If the chosen membrane is a positively charged nylon membrane, crosslinking can even be neglected as DNA will stay bound to the membrane anyway.

Any single-stranded DNA on the membrane can then be hybridized to single-stranded probes which are identical to parts of the target gene.

Table 4.4 Overview of commercially available membranes for Southern blotting.

Membrane type	*Detection*	*Value*
Positively charged nylon	colorimetric or chemiluminescent	great variability in background (dependent on the vendor), higher target binding capacity, higher sensitivity, can work without crosslinking
Uncharged nylon	colorimetric or chemiluminescent	lower binding capacity, crosslinking necessary
Nitrocellulose	colorimetric	lower sensitivity, crosslinking not possible, less durable

Source: Lottspeich, 1998.

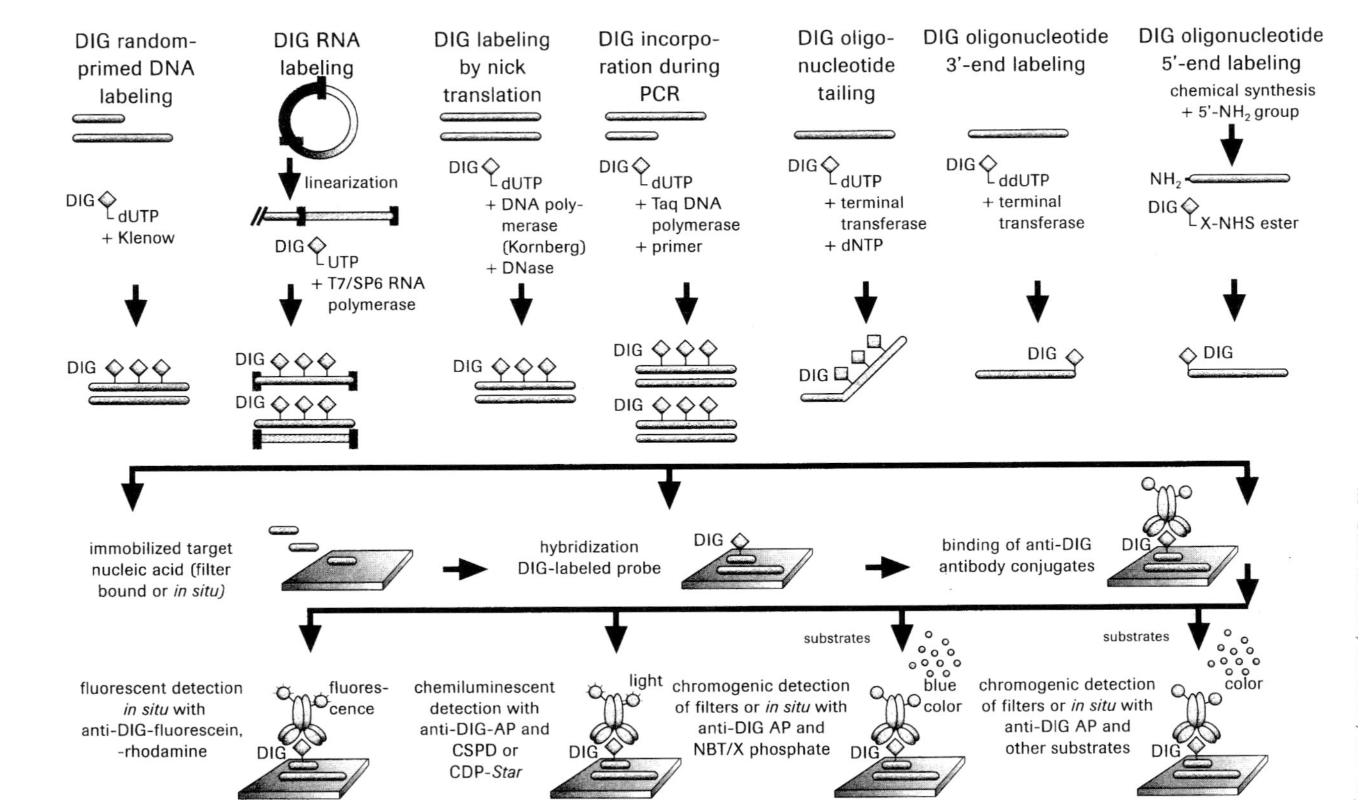

Figure 4.5 Digoxigenin (DIG) labeling of hybridization probes (from Roche hybridization manual). Of all the possibilities for labeling, the first, DIG random-primed DNA labeling, is usually preferred because of its high yield.

4.3.4.1 Probe Design and Labeling

A probe for Southern blotting represents part of the desired gen which is specifically labeled. Its length depends on prior knowledge and can vary from a few nucleotides (40–100 base pairs) to substantial parts (50–70%) of the gene. As a rule, a longer probe correlates with higher specificity and therefore a higher success rate in securing the correct gene. The probe can be a PCR product or an oligonucleotide synthesized against the N-terminal part of the gene, usually the minimum sequence information available from N-terminal sequencing. The probe can be labeled either radioactively (usually with ^{32}P) or non-radioactively, most commonly with digoxigenin (DIG; Roche kit). In the latter procedure, the PCR product or oligonucleotide, which is even available in labeled form from vendors, is incubated with desoxynucleotides, a part of which has been modified with digoxigenin. Using the Klenow enzyme (a polymerase without a 3′,5′ proofreading exonuclease activity) the DIG-modified nucleotide is randomly inserted into the template, while duplicating the second strand. The product is then purified and concentrated (for an overview from labeling to detection, see Figure 4.5).

4.3.4.2 Hybridization

When all the necessary preparations for a Southern blot have been completed, the genomic DNA can be hybridized against the labeled probe of the target sequence. The probe hybridizes to the corresponding gene within the single-stranded DNA during incubation in a hybridization oven at a temperature which is a function of the length and GC content of the probe. The less stringent the reaction, depending on the probe length and specificity, the longer the incubation times that are required. Following incubation, several washing steps remove unspecifically absorbed probe material.

The success of hybridization often depends on several factors that affect the rate of probe binding to the target DNA on the membrane. The most important ones include hybridization temperature, probe concentration, ionic strength, pH, and viscosity of hybridization solution (Wahl, 1987). As the hybridization temperature is one of the crucial parameters, its optimization has to be considered. Equation (4.1) is often used for estimating hybridization temperature, where n is the number of nucleotides in the probe and the Na^+ concentration of the hybridization buffer is 1 M or less.

$$T_m = 81.5 + 16.6\ (\log_{10}[Na^+]) + 0.41\ ([G + C]/100n\ x] - 600/n \tag{4.1}$$

Hybridization of nucleic acid probes to membrane-bound target sequences is a commonly used technique in gene structure and expression analysis, medical diagnostics, and gene targeting.

4.3.4.3 Detection

One of the most sensitive method for detecting probe–target hybrids involves an alkaline phosphatase-conjugated anti-DIG antibody and chemoluminescent alkaline phosphatase substrates (Roche, Molecular Diagnostics). Following hybridization, the membrane needs to be washed stringently before starting the detection procedure to eliminate unspecific probe binding to the membrane. Subsequently, the detection protocol uses steps similar to Western blots, as an antibody reaction is involved. First, the membrane needs to be blocked with either a blocking solution supplied with the kit or a PBS working solution with 0.1% Tween 20 and 5% milk powder for 1–3 h to minimize unspecific binding of the antibody. The next step involves anti-DIG-antibody incubation for 1 h at room temperature. After several washing steps with 0.1 M maleate buffer (+0.15 M NaCl, pH 7.5 with 0.3% Tween 20) the detection substrate can be applied to the membrane. An expected weak chemoluminescence reaction signal can be enhanced by incubation at 37 °C. The signal can be developed using X-ray films or an imaging system suitable for chemoluminescence signals (Alpha Innotech, UVP Imaging Systems, Stratagene).

4.3.5 DNA-Sequencing

Nowadays, DNA sequencing is usually performed by DNA service core facilities. In brief, the automated systems utilize the Sanger DNA sequencing method (Sanger, 1977). The method is based on the principle of randomly terminated DNA elongation through incorporation of didesoxynucleotides (chain termination). The DNA chain cannot be elongated at this nucleotide because the phosphate necessary for forming the phosphodiester bond of the DNA backbone on position C2 in the ribose ring is missing. By offering a mixture of the four nucleotides, each of which has a portion of the corresponding didesoxynucleotide, a mixture of randomly terminated DNA chains develops, statistically representing a stop at every nucleotide of the target sequence. The DNA molecules of different lengths are separated on a polyacrylamide gel and detected when passing a laser beam owing to fluorescence markers attached to the didesoxynucleotides.

4.4 Cloning Techniques

Cloning is the process of introducing new DNA into an organism and subsequently creating identical copies of the newly created organism. Nowadays, cloning techniques have been standardized to such a point that in most cases kits (from companies such as Stratagene, Clontech, Roche, Amersham–Pharmacia, or Qiagen) are available for this step, vastly improving the chances of success. Still, two crucial steps should be kept in mind when solving a cloning challenge: first, choosing the right combination of restriction enzymes including accurate design of primers for the PCR reaction; secondly, using an efficient ligation technique.

4.4.1 Restriction Mapping

To incorporate fragments of foreign DNA into a plasmid vector, methods for cutting and rejoining of double-stranded DNA are required. Identification and manipulation of restriction endonucleases in the 1960s constituted a key discovery for cloning. Today, commonly used restriction enzymes are produced recombinantly and distributed widely.

The action of restriction enzymes involves recognition of a specific DNA sequence, often a palindrome (an identical sequence when read backwards). They can produce cohesive ends, i.e., a protruding 5′ end with a phosphate group and a corresponding 3′ end with a free hydroxyl group. Alternatively, fewer enzymes generate blunt ends with no overhang on either the 5′ or the 3′ end. Cloning of a gene using blunt end-generating restriction enzymes is less likely to succeed because both orientations of the gene can occur. Provided the sequence of the gene of interest is known, a restriction map of the gene and the vector has to be generated. Vector maps are easily accessible, and are provided by the vendors over the Web. Restriction mapping helps to make choices for the flanking restriction sites at both sides of the gene for cloning, because the map will provide information about any enzyme that will or will not cut within the gene of interest.

If the sequence is not known, a vector system such as the TA cloning kit (Clontech) should be used. The vector provided has a protruding T-base and the resulting PCR product can be directly ligated into the vector, because Taq polymerases, generally used in PCR, generate a one-base A-overhang at the 3′ end on both strands. Restriction mapping can be performed either directly on the Web by visiting web pages such as www.ebi.srs.uk (the SRS retrieval system) or www.expasy.ch, or through programs such as "Lasergene" (DNAstar).

4.4.2 Vectors

Vectors are specially designed vehicles derived from naturally occurring bacterial plasmids. A plasmid is a circular DNA molecule that gets propagated in a host; a vector is a molecular biology tool that provides the necessary features for creating a recombinant plasmid. When ligating the gene of interest into a vector plasmid the most frequent unwanted product is the unchanged, re-ligated vector plasmid itself (see Section 4.4.4). Therefore, the ligated plasmid should be analyzed for insertion of the desired fragment to avoid just having produced re-ligated vector. One popular analysis method is blue–white selection, discussed below, followed by restriction analysis of positive clones within the blue–white selection.

One of the earliest vectors was named pBR322; it contains the gene to confer ampicillin resistance to the host. Derivative vectors, called the pUC series, also include a *multiple cloning site* (MCS) and contain an engineered version of the *lacZ′* gene which has multiple restriction enzyme recognition sites within the first part of the coding region of that gene. Insertion of a DNA fragment at any restriction site of this MCS results in disruption of the *lacZ'* gene sequence. This fact can be exploited for blue–white screening for the presence of a recombinant plasmid. Blue–white screening is based on the production of a blue compound catalyzed by β-galactosidase which is encoded by the *lacZ* gene and under the control of the *lacZ* promoter. The *E. coli* hosts employed only have an incomplete version of the *lacZ* gene, which must be completed by the *lacZ'* part from the vector to result in an active β-galactosidase, as well as a *lacZ* repressor. The *lacZ* repressor can be activated by induction with either lactose (the natural inductor) or its unnatural derivative IPTG (isopropyl-β-D-galactopyranoside). β-Galactosidase is able to cleave the synthetic substrate 5'-bromo-4'-chloro-3-indoyl-β-D-galactopyranoside (X-gal) to D-galactose and blue–greenish 5-bromo-4-chloroindole. The result is a blue colony if the *lacZ'* fragment on the vector is intact or a white colony if the *lacZ'* fragment has been disrupted by insertion of the desired gene. Therefore, white colonies on a transformation plate treated with X-gal and IPTG indicate success in inserting the gene of interest and should be selected for further experiments.

Vectors such as the pUC series (Molecular Diagnostics) or the Bluescript series (Stratagene) are ideal cloning vectors because of their relatively small size, good acceptance by the hosts, and weak promoters causing only little product to be expressed, which helps forming colonies. Figure 4.6 gives two examples of cloning vectors.

Cloning DNA fragments into bacteria can be also achieved through bacteriophage vectors, such as lambda. Non-essential portions of the linear 48.5 kB lambda genome may be replaced by up to 23 kB of foreign DNA; therefore, the vector is useful for cloning libraries. The phage particle injects its linear DNA into the host cell, where the *cos* sites at both ends of the linear phage particle allow ligation into a circular phage. The host cell may either replicate to form many phage particles, which are released from the cell by lysis and cell death (lytic phase), or the phage DNA may integrate into the host genome by site-specific recombination, where it

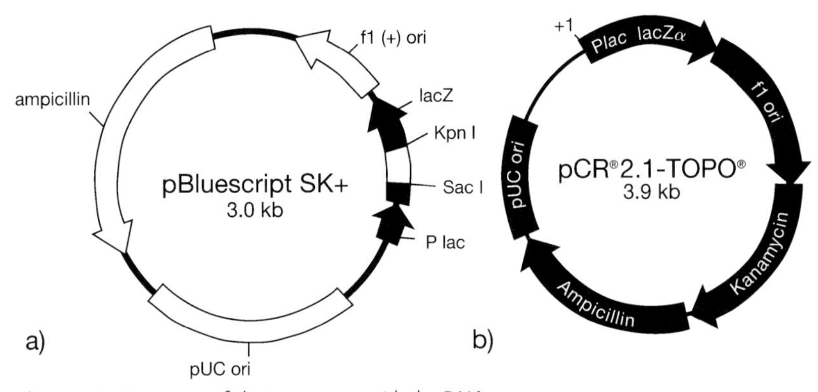

Figure 4.6 Two types of cloning vectors with the DNA parts necessary to function as a selectable, propagating tool for carrying foreign DNA:
(A) pBluescript vector from Stratagene;
(B) TA vector from Invitrogen.

remains for a long period (lysogenic phase). The target DNA fragments are ligated with the λ DNA ends, which provide the essential genes for infection of *E. coli*. An infected cell then produces recombinant phages, which can be screened for the desired DNA.

Eukaryotic vectors

There exist a variety of vectors for cloning into eukaryotic systems, ranging from yeast (*Saccharomyces* as well as *Pichia*) through insect cells (*Baculovirus*) and plants (Ti plasmid from *Agrobacterium tumefaciens*) to mammalian cells (transfected by viral or mammalian vectors). As expression in eukaryotic hosts is less efficient than bacterial expression in terms of yield and time and more complicated in terms of vector structure and culture conditions, such eukaryotic expression systems are only used for genes whose proteins require posttranslational modification which is not possible in bacteria. Yeast is the preferred option as a relatively easily culturable single-cell system but posttranslational modification capabilities is limited. The additional complexity can be circumvented in part by exploiting the ability of eukaryotic vectors to act as shuttle vectors, which can be shuttled between two evolutionarily different hosts. Thus, eukaryotic vectors can be replicated and analyzed in bacteria and transfected into eukaryotic cells for expression of the recombinant product.

Bacterial expression vectors

These vectors are dealt with in connection with Section 4.3.

4.4.3 Ligation

After restriction, the resulting cleaved phosphodiester bond has to be resealed to obtain circular DNA, with or without insertion of a target DNA fragment into a vector for the covalent linking of DNA molecules. Provided one of the double strands has a phosphate group at the 5′ end, DNA ligases can seal a cleaved phosphodiester bond in an annealed pair of cohesive ends of broken double-stranded DNA resulting from the action of restriction enzymes. Such a restricted vector can therefore be used to insert a DNA fragment, provided the fragment features the same cohesive ends as the now-linear vector. For efficient ligation there are several kits available from vendors such as Novagen or Roche. These "Rapid" ligation kits provide a genetically engineered T4 ligase which allows successful ligation at room temperature within 5 min, a considerable time saving as common protocols recommend 14 °C for 16 h. However, slower ligation kinetics at 14 °C often lead to more highly accurate profiles and should be preferred for certain substrates. Lack of ATP can cause failure of the ligation reaction, as the commonly used T4 ligases are ATP-dependent.

Before blunt-end ligation of an insert, the vector needs to be dephosphorylated because resealing of the two ends of the linear vector to form a circular plasmid representing an intramolecular reaction is entropically favored, and thus faster (Chapter 2, Section 2.2.3), than the intermolecular insertion of the blunt-end fragment into the vector. A dephosphorylated vector cannot be re-ligated into circular double-stranded DNA owing to the missing 5′-end phosphate groups crucial for the ligase reaction. Dephosphorylation can be performed using highly efficient shrimp alkaline phosphatase (Roche) in a one-pot reaction with the vector restriction reaction upon addition of appropriate buffer and enzyme; both processes are carried out at 37 °C.

4.4.3.1 Propagation of Plasmids and Transformation in Hosts

During transformation (bacterial uptake of foreign DNA) the components of the mixture of recombinant and native plasmid molecules formed by ligation are transferred to a host cell for replication (cloning). The most common hosts are strains of *E. coli* with specific genetic properties (see the *Molecular Biology Labfax* Series, e.g., Brown, 1991). Before transformation, *E. coli* cells have to be rendered *competent*, i.e., susceptible to DNA uptake. A simple method is pretreatment with Ca^{2+}, Rb^{+}, or Mn^{2+} solutions or a mixture of these ions. Another method renders cells electrocompetent and uses them in electroporation to transform foreign DNA. In general, competent cells can be either purchased or easily created (Hanahan, 1983).

DNA and competent cells are mixed and incubated on ice for 15–30 min to allow the DNA to be taken up. The mixture is then heat-shocked at 42 °C for 1–2 min, which induces DNA repair enzymes, allowing the cells to recover from the competent state and DNA uptake. Recovery is allowed to occur in growth medium at 37 °C before streaking on an agar plate with appropriate selection. The plates are incubated at 37 °C overnight; if successful, several colonies are formed.

Transfection, DNA uptake in eukaryotic systems, often is more problematic then bacterial transformation: the mode of DNA uptake is poorly understood and efficiency is much lower. In yeast, cell walls can be digested with degradative enzymes to yield fragile protoplasts, which are then able to take up DNA. Cell walls are resynthesized after removal of the degrading enzymes. Mammalian cells take up DNA after precipitation onto their surface with calcium phosphate [Fugene 6 (Roche); Lipofectin (Life Technologies);Effectene (Qiagen)]. Electroporation is often more efficient for transfection in eukaryotic cell systems, especially in yeasts.

4.5 (Over)expression of an Enzyme Function in a Host

4.5.1 Choice of an Expression System

Expression of genes at a regularly low level, resulting in a low level of product, occurs in every organism. For recombinant proteins, overexpression which leads to considerably enhanced amounts of the recombinant protein is desired. Success of an attempt at recombinant overexpression cannot be predicted as of yet and depends on many factors, such as the protein to be expressed, host, vector, promoter, culture conditions, and, last but not least, the experience of the investigator.

The two most important factors in the choice of an expression system should be (i) economy and efficiency, and (ii) the greatest possible closeness of the host to the DNA source if posttranslational modifications are required. However, in the case of eukaryotic genes, compromises are frequently necessary. The need for a close relationship often rules out overexpression, because often no efficient high-yield expression system exists among related eukaryotes; one of the best known systems employs CHO (Chinese hamster ovary) cells.

A few problems during overexpression, which have been dealt with, are mentioned here. Recombinant proteins can be toxic to the host; then overexpression results in low expression rates, lysis of the cells, or proteolysis of the recombinant protein. Options other than changing the host, which often is impossible, include regulation of culture conditions, change of inductors, or secretion of the recombinant protein. Proteolysis can be avoided by

- using protease-deficient (such as negative ompT, $DlgP^-$) *E. coli* strains (Murby, 1996),
- modifying the N-terminal amino acid to stabilize the protein (Met, Ser, Ala, Thr, Val, or Cys are stabilizing; Arg, Leu, Asp, Lys, or Phe are destabilizing) (Lottspeich, 1998), or by
- lower culture temperatures and short induction times.

Since *constitutive* expression in high yields negatively influences cell metabolism with consequences up to and including cell death, the commonly used expression systems are *inducible*. The goal of inducible overexpression is a yield of soluble target protein between 5 and 50%, typically about 20%, of total cell protein. If expression as soluble protein is deemed impossible, alternatives are (i) expression as inclusion bodies (insoluble protein fraction), (ii) expression as a fusion protein, or (iii) secretion into the medium. For a good ratio of soluble to insoluble recombinant protein (the ratio is hardly ever infinity), expression criteria have to be optimized.

4.5.2
Translation and Codon Usage in E. coli

E. coli requires a certain sequence for initiation of translation, the ribosome binding site (RBS or Shine–Dalgarno sequence) AGGA. This sequence involves binding of the mRNA to the 16S-rRNA of the bacterial ribosome (Figure 4.7). The gap between the RBS and the start codon ATG (Met) of the target protein should range from three to ten base pairs, or sometimes even 15 base pairs. Variation of this gap towards the same target protein can greatly influence overexpression. As an example, a strongly expressed protein can be shifted from insoluble to soluble expression by creating a weaker translation initiation signal by widening the gap between RBS and ATG. Provided the restriction sites within the multiple cloning site allow flexibility, the gap between the RBS and the start codon initially, when using a strong promoter, should be kept as wide as possible to increase soluble overexpression.

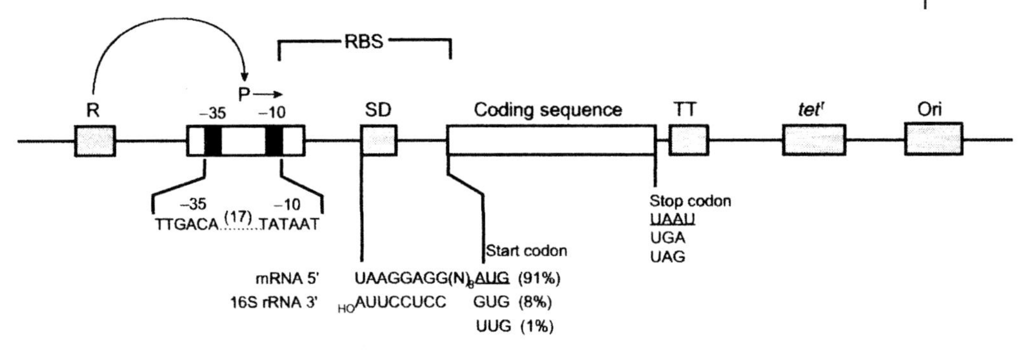

Figure 4.7 Schematic representation of the salient features of a prokaryotic expression vector (Hannig, 1998). The expression vector has an upstream promoter consisting of at least a –35 and –10 region; the space between them defines the weakness or strength of a promoter. Between the inserted target gene and promoter lies the ribosome binding site (RBS) with its Shine–Dalgarno sequence AGGA. The gene is usually inserted into an artificially constructed multiple cloning site. After the stop codon, sometimes to be added by the user (as is the start codon ATG), a termination signal follows to release the RNA polymerase from the DNA. Two other important features are the site of the origin of bacterial replication (ori), which insures that the plasmid is copied within each cell cycle, and a selection marker, in this case the *tet*^{r} gene.

Codon usage in different organisms, the preference of an organism for one or a few of several (in the case of degeneracy) triplet combinations to code for an amino acid, is one of the main obstacles to high-yield overexpression. The greater the relational distance between organisms, the greater the chance of problems with codon usage. In contrast to their use in eukaryotic systems, the corresponding tRNAs for the arginine codons AGG and AGA are very rare in *E. coli*; the same holds for leucine (CUA), isoleucine (AUA), and proline codons (CGG). This problem can be overcome by special strains (both Stratagene): BL21-CodonPlus-RIL cells contain extra *argU*, *ileY*, and *leuW* tRNA genes, BL21-CodonPlus-RP cells contain extra *argU* and *proL* tRNA genes. To overcome such problems, even drastic measures such as gene synthesis should be considered to adjust the codon usage.

When using high-copy expression vectors, which are able to generate up to 100 target protein molecules per cell, the number of copies of *lacI* repressor provided by the host cell, such as *E. coli*, is insufficient to insure tight control. Expression yields can be enhanced by supplying extra copies of repressor on a second plasmid.

High-copy plasmids might also express large amounts of products that are toxic to the host. Without selective pressure and if the product is not toxic, the losses are at low frequency (10^{-5}–10^{-6} per generation) but can increase tremendously when the expressed insert is toxic (Baneyx, 1999). This can even occur under selective pressure because the plasmid population contains a mixture of plasmids with and without insert. As cells carrying plasmids with inserts will not survive, such plasmids are often excised by the cell, a phenomenon termed "plasmid instability". To circumvent this problem, vectors that cause cell death upon loss of plasmid can be used, because an essential gene for the host metabolism gets lost together with the plasmid. This mechanism provides a better selection than antibiotic resistance, as the latter is found on vectors with and without target insert.

4.5.3 Choice of Vector

An overview of some of the most common vector systems is provided in Table 4.5. All the vectors chosen here are inducible expression systems. Induction can result from a change in temperature, from addition of metabolites, or from removal of a necessary carbon source.

One of the crucial parts of any vector system is the promoter, a piece of DNA sequence which controls the level of protein expression.

The hybrid tac promoter, a combination of lac and trp promoters, combines the high transcription efficiency of the trp promoter with the easy inducibility of the lac promoter. The lac promoter, like the other IPTG-inducible promoters, does not have tight repressor control, lacI^{q} *E. coli* strains with a large quantity of lacI repressor should always be used with this promoter; possibly extra copies of lacI should even be provided (such as rep4 plasmid from Qiagen).

Table 4.5 Common vector choices for *E. coli.*

Promoter	*Method of induction*	*Advantages*	*Disadvantages*
lac	IPTG	induction at low temperatures possible	low yields, no tight repression, repression through glucose
trp	scarcity of Trp	high yield, adjustable induction, many vectors available, low temperature possible	no tight repression
tac	IPTG	many vectors available, high yields, low temperature possible, adjustable induction	no tight repression
T7	IPTG	many vectors available, extremely high yields and low temperature possible	tight control combined with pLysS or pLysE plasmids only, useful only with BL21 (DE3) strains
araB	arabinose	adjustable induction, low temperature possible	only a few vectors available, catabolic repression through glucose
phoA	scarcity of phosphate	high yields, low temperature possible	induction not adjustable, only a few culture media, few vectors

Source: Weickert, 1996.

The highly transcription-efficient T7 promoter is recognized by the T7 RNA polymerase. This expression type originates from the bacteriophage T7, whose transcription apparatus successfully competes with the one from the bacterial host. The T7 expression system is perfectly embodied in the pET vectors (Novagen), of which a large variety is available, distinguishable by a number such as "pET30a". This system allows expression of the recombinant protein by itself, with tags (histidine), or as fusion proteins. Very good results can be achieved, with target protein up to 50% of total cell protein. Only *E. coli* BL21 (DE3) strains can be used with the T7 expression system because these strains have the necessary T7 RNA polymerase under the control of the lacUV5 promoter integrated into the genome. T7 polymerase production is induced by the lactose analog IPTG, which then identifies the minimal T7 promoter upstream of the target gene. The use of two promoters to express the target gene ensures tight control. However, as no expression system acts independently of the protein expressed, even in the T7 system promoter leakiness can occur and thus, in most cases, poor expression results.

Another tightly controlled promoter system (araB; Invitrogen) is based on arabinose. Induction of the promoter is totally repressed by sugars such as glucose or fructose, whereas transcription control can be leveled through control of the arabinose concentration.

4.5.3.1 **Generation of Inclusion Bodies**

High levels of expression of hydrophobic proteins can lead to insoluble aggregates, the so-called "inclusion bodies". They consist of mainly misfolded recombinant protein and are a telltale sign of overburden of the expression machinery unable to provide sufficient time or aids for protein folding (Chapter 17, Section 17.8). The generally undesirable advent of inclusion bodies can only be turned into an advantage if the protein can be easily resolubilized, as the purification process in this case is very simple and higher yields can result than for purification of soluble proteins. If the protein is not easily resolubilized such aggregates must be circumvented by any possible means.

Possible means include limiting the concentration of recombinant protein by cutting short induction by either lowering the temperature, choosing a weaker promoter, or adding a non-digestible sugar, such as desoxyglucose, as carbon source at the induction time point. Alternatively, utilization through coexpression of *E. coli* chaperones, such as GroEL or GroES (large and small subunits of a double-heptamer heatshock protein), can help correct folding of the target protein; however, success strongly depends on the target protein and cannot be predicted (Weickert, 1996; for a positive example, see Cole, 1996). Another option is expression of the recombinant protein as a fusion protein.

4.5.3.2 **Expression of Fusion Proteins**

As heterologous proteins can be easily degraded in or toxic to the host cell, fusing the target gene to the corresponding gene of a host-specific soluble protein can decelerate either detrimental process. A common challenge is expression of a hy-

Table 4.6 Overview of possible fusion proteins.

Fusion protein	*Origin*	*MW [kD]*	*Ligand for affinity purification*
β-Galactosidase	*E. coli*	116	APTG
Protein A	*Staphylococcus aureus*	31	IgG
Streptavidin	*Streptomyces avidinii*	13	biotin
Glutathione-S-transferase	*Schistosoma japonicum*	26	glutathione
Maltose-binding protein	*E. coli*	40	starch
Intein/chitin domain	*Saccharomyces cerevisiae/ Bacillus circulans*	55	chitin
Ubiquitin	*Saccharomyces cerevisiae*	8	–

APTG: *p*-aminophenyl-β-D-thiogalactoside; IgG: immunoglobulin G.
Source: Lottspeich, 1998.

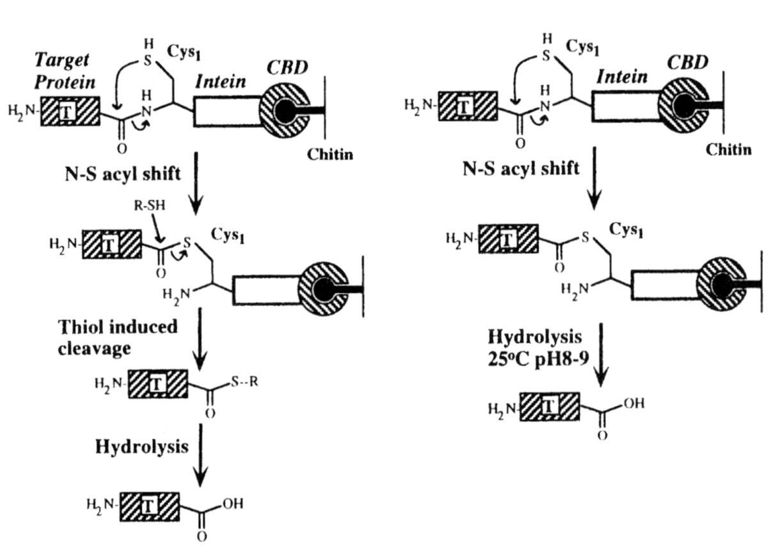

Figure 4.8 The intein system as used for fusion proteins (Xu, 2000). The fusion proteins undergo self-cleavage at the upstream splice junction. The amino acids that participate directly are shown, and the remainder of the intein and target protein are drawn as boxes. Usually, the intein is conjugated to a chitin-binding domain, so purification of the fusion protein can occur by using a chitin-conjugated column. Self-cleavage can occur overnight.

drophobic protein prone to aggregation; this can be met by fusing the gene of the hydrophobic protein to a hydrophilic domain to enhance solubility of the target protein. If fused to a domain enabling affinity purification, the fusion protein can be purified in one step. Examples of common fusion proteins are glutathione-S-transferase (GST) and the maltose-binding protein (MBP) (Table 4.6). A successful expression of a eukaryotic minimal receptor fused to bacterial GST protein (**G**lutathion-**S**-**T**ransferase) is described by Boesen (2000).

The separation of the target protein from the fusion protein can be performed chemically or enzymatically. The basis for chemical cleavage is acid or base stability of the target protein. Enzymatic separation by proteases is highly specific but its efficiency can be decreased by limited access to the part of the amino acid sequence required for proteolysis.

A special but elegant option as fusion protein is the intein domain, which exploits the self-splicing ability of this protein group (Figure 4.8). Self-cleavage is initiated by the N-terminal cysteine residue of the fused intein protein, which produces a thioester bond through nucleophilic substitution of the amide of the peptide bond with the C-terminal residue of the target protein. The thioester is hydrolyzed by the addition of high amounts of thiols, such as 25 mM DTT, thus achieving cleavage of the intein from the target protein. The whole process can be done on a single column by adding DTT overnight (for a good description of the procedure, see Xu, 2000).

4.5.3.3 **Surface Expression**

By the method of phage display, recombinant proteins can be expressed on the surface of bacteria, yeast, or bacteriophages. This procedure is useful mainly for extracting new protein or peptide variants with specialized functions. Phage display works through fusion of peptides or proteins with the capsule proteins of phages, which can be assembled in vivo. The selection process includes incubation of the phages with defined structures or substrates to achieve the desired new proteins (panning). Through panning, new enzyme functions can be found as part of in-vitro evolution.

4.5.4
Expression of Eukaryotic Genes in Yeasts

Expression of eukaryotic genes in yeast has two main advantages: (i) the yeast expression system contains many features of a eukaryotic expression system such as glycosylation or disulfide bond formation, and (ii) yeast is a very economical system. Yeasts are single cells, can be cultivated easily, feature fairly short doubling times, and require relatively inexpensive medium ingredients; in many aspects they resemble bacteria.

Saccharomyces cerevisiae

Just like *E. coli*, *S. cerevisiae* is a well-studied organism with a well-established genetic and molecular biology knowledge base. Commercial availability includes:

- several protease-deficient mutants, rendering the possibility of recombinant protein degradation small;
- several constitutively expressed or inducible promoters; examples of the former are constitutively expressed promoters of glycolytic enzymes such as phosphoglycerol kinase 1 or alcohol dehydrogenase 1, and the most common inducible promoter is GAL1 of the galactokinase gene.

An advantage of constitutive expression is a constant production of protein with high yield in the case of stable and soluble proteins. Induction of GAL1 occurs through addition of galactose to glucose-free cells. Although *S. cerevisiae* is capable of adding glycosylation for secreted proteins, excessive mannosylation common in this type of yeast can lead to adverse effects such as lower expression levels or misfolded proteins.

Methylotrophic yeasts

Yeasts such as *Pichia pastoris* (Invitrogen) or *Hansenula polymorpha* (Rhein–Braun Biotech), which were rare as expression systems even a few years ago, are becoming more and more established. *P. pastoris*, especially, is suitable for expression of glycosylated proteins owing to fewer incorporated mannose residues. The vector system of both yeasts features a strong, methanol-inducible promoter, derived from the alcohol oxidase gene (AOX1 and 2) for *P. pastoris* and the methanol oxidase gene for *H. polymorpha*. While both expression systems are still not completely understood, more and more publications describe successful expression with up to 30% overall expression yield. To avoid the need for optimizing separate bacterial and eukaryotic expression systems if the requirement or preference is not clear at the outset, a promising attempt to construct a dual expression vector for *Pichia pastoris* is given by Luecking (2000).

For eukaryotic genes, expression in insect cells (*Baculovirus* system) or mammalian cells (Chinese hamster ovary cells, cos7 monkey cells) is possible and is often more likely to lead to active protein than in bacteria or yeast owing to a more efficient posttranslational machinery. However, expression rates are very low and not feasible for production of biocatalysts.

Once expression has been achieved, a good screening system for positive clones should be available – in all likelihood a convenient and accurate activity test, if possible even suitable for high-throughput screening.

Suggested Further Reading

T. A. Brown (ed.), *Molecular Biology Labfax*, Academic Press, New York, **1991**.

D. M. Glover and B. D. Hames (eds.), *DNA Cloning I: Core Techniques. DNA Cloning II: Expression Systems* (Practical Approach Series, No. 149), IRL Press, 2nd edition, **1995**.

T. Maniatis, E. F. Fritsch, and J. Sambrook, *Molecular Cloning: A Laboratory Manual*, Cold Spring Harbor Laboratory, Cold Spring Harbour, 2nd edition, **1989**.

D. L. Nelson, Albert L. Lehninger, and M. M. Cox, *Principles of Biochemistry*, Worth Publishing; 3rd edition, **2000**.

L. Stryer, *Biochemistry*, W H Freeman, New York, 4th edition, **1995**.

J. D. Watson, M. Gilman, J. Witkowski, and M. Zoller, *Recombinant DNA*, Scientific American Books, W. H. Freeman, New York, 2nd edition, **1992**.

J. D. Watson and Joan Steitz, *Molecular Biology of the Gene*, Addison-Wesley, 4th edition, **2001**.

References

F. Baneyx, Recombinant protein expression in *Escherichia coli*, *Curr. Opin. Biotechnol.* **1999**, *10*, 411–421.

C. C. Boesen, S. A. Motyka, A. Patamawenu, and P. D. Sun, Development of a recombinant bacterial expression system for the active form of a human transforming growth factor β type II receptor ligand binding domain, *Prot. Expr. Purif.* **2000**, *20*, 98–104.

T. A. Brown (ed.), *Molecular Biology Labfax* (Labfax Series), Academic Press, New York, **1991**.

P. A. Cole, Chaperone-assisted protein expression, *Structure* **1996**, *4*, 239–242.

F. S. Collins and S. M. Weissman, Directional cloning of DNA fragments at a large distance from an initial probe: a circularization method, *Proc. Natl. Acad. Sci. USA* **1984**, *81(21)*, 6812–6.

M. A. Frohmann, M. K. Dush, and G. R. Martin, Rapid production of full-length cDNAs from rare transcripts: amplification using a single gene specific oligonucleotide primer, *Proc. Natl. Sci. USA* **1988**, *85*, 8998–9002.

D. Hanahan, Studies on transformation of *Escherichia coli* with plasmids, *J. Mol. Biol.* **1983**, *166*, 557–580.

G. Hannig and S. C. Makrides, Strategies for optimizing heterologous protein expression in *Escherichia coli*, *Tiptech* **1998**, *16*, 54–60.

A. Helmsley, N. Arnheim, M. D. Toney, G. Cortopassi, and D. J. Gales, A simple method for site-directed mutagenesis using PCR, *Nucleic Acid Res.* **1989**, *17*, 6545–6551.

J. Hodgson, Expression systems: a user's guide. Emphasis has shifted from the vector construct to the host organism, *Bio/Technology* **1993**, *11*, 887–893.

E. S. Kawasaki, S. S. Clark, M. Y. Coyne, S. D. Smith, R. Champlin, O. N. Witte, and F. P. McCormick, Diagnosis of chronic myeloid and acute lymphocytic leukemias by detection of leukemia-specific mRNA sequences amplified in vitro, *Proc. Natl. Sci. USA* **1988**, *85*, 5698–5702.

S. Y. Lee, High-cell density culture of *Escherichia coli*, *Trends Biotechnol.* **1996**, *14*, 98–105.

F. Lottspeich and H. Zorbas, *Bioanalytik, a Textbook*, Spektrum Akademischer Verlag, **1998**.

A. Lueking, C. Holz, C. Gotthold, H. Lehrach, and D. Cahill, A system for dual protein expression in *Pichia pastoris*, *Prot. Expr. Purif.* **2000**, *20*, 372–378.

T. Maniatis, E. F. Fritsch, and J. Sambrook, *Molecular Cloning: A Laboratory Manual*, Cold Spring Harbor Laboratory, Cold Spring Harbor, 2nd edition, **1989**.

K. Mullis, F. Faloona, S. Scharf, R. Saiki, G. Horn, and H. Ehrlich, Specific enzymatic amplification of DNA in vitro: the polymerase chain reaction, *Cold Spring Harbor Symp. Quant Biol.* **1986**, *51*, 263–273; republished in: *Biotechnology* **1992**, *24*, 17–27.

M. Murby, M. Uhlen, and S. Stahl, Upstream strategies to minimize proteolytic degradation upon recombinant production in *Escherichia coli, Prot. Expr. Purif.* **1996**, *7*, 129–136.

F. Sanger, G. M. Air, B. G. Barrel, N. L. Brown, A. R. Coulson, J. C. Fiddes, C. A. Hutchinson, P. M. Slocombe, and M. Smith, Nucleotide sequence of bacteriophage ϕX174, *Nature* **1977**, *265*, 678–695.

E. M. Southern, An improved method for transferring nucleotides from electrophoresis strips to thin layers of ion-exchange cellulose, *Anal Biochem.* **1974**, *62*, 317–318.

P. C. Turner, A. G. McLennan, A. D. Bates, and M. R. H. White, *Molecular Biology*, B. D. Hames (series ed.), Springer Verlag, 2nd editon.

G. M. Wahl, S. L. Berger, and A. R. Kimmel, Molecular hybridization of immobilized nucleic acids: theoretical concepts and practical considerations, *Methods Enzymol.* **1987**, *152*, 399–407.

J. D. Watson, M. Gilman, J. Witkowski, and M. Zoller, *Recombinant DNA*, Scientific American Books, W. H. Freeman, New York, 2nd edition, **1992**.

M. J. Weickert, D. H. Doherty, E. A. Best, and P. Q. Olins, Optimization of heterologous protein production in *Escherichia coli, Curr. Opin. Biotechnol.* **1996**, *7(5)*, 494–499.

M. Q. Xu, H. Paulus, and S. Chong, Fusions to self-splicing inteins for protein purification, *Methods Enzymol.* **2000**, *326*, 376–418.

5
Enzyme Reaction Engineering

Summary

The main goal of *enzyme reaction engineering* is a high space–time yield at a high degree of conversion of the substrate or substrates under the conditions of high chemical and enantiomeric selectivity. The goal of reactor design is the description of the behavior of both the pertinent degree of conversion x $(0 \le x \le 1)$ and of selectivity σ $(0 \le \sigma \le 1)$ over the residence time τ of the reactor.

To investigate the impact of kinetics on reactor design, we have to: (i) develop the pertinent kinetic rate equation; (ii) insert this rate law into the equation for the reactor we intend to operate; and (iii) integrate over all degrees of conversion.

Inhibition can become very detrimental if high degrees of conversions are sought, such as in any situation in synthesis or large-scale processing. The inhibition ratio K_M/K_I is particularly important as it is totally outside the control of the researcher or operator. An inhibition ratio $K_M/K_I < 1$ is not detrimental: the substrate binds more tightly than the inhibitor and the high degrees of conversion can still be achieved without much difficulty, maybe at longer reaction times. If $K_M/K_I > 1$, however, the reaction comes to a virtual stop at intermediate degrees of conversions and the reaction will not reach completion.

Biocatalysts are used advantageously in immobilized form if:
- the immobilized enzyme, e.g., glucose isomerase (GI) or penicillin G acylase (PGA), is more active and/or more stable than the soluble equivalent, or
- the packed-bed reactor configuration commonly employed with immobilized enzymes yields a higher degree of conversion or higher space–time yield than a CSTR; examples are the nitrile hydratase-catalyzed process to 5-cyanovaleramide or the decarboxylation of D,L-aspartate to D-aspartate, L-alanine, and CO_2.

Some rules for the choice of reactors are that:
- as deactivation with an activation energy $E_{a,deactivation}$ of 200–300 kJ mol^{-1} predominates over the enzyme reaction with $E_{a,reaction}$ around 100–120 kJ mol^{-1} at high temperature, an enzyme reaction should be run at *low temperature for minimization of deactivation;*
- if a *very high degree of conversion* ($x > 0.9$) or even complete conversion is the goal, a BR or PFR requires a much smaller reactor volume than a CSTR;

- at constant conditions of T and pH, an enzyme reaction should be run with the *smallest possible contribution by inhibition*. In the case of substrate inhibition, a CSTR minimizes substrate concentration at all points in the reactor and is therefore preferred. In the case of product inhibition, a BR or a PFR minimizes reactor volume significantly at high degrees of conversion.

5.1
Kinetic Modeling: Rationale and Purpose

The main goal of enzyme reaction engineering is a high space–time yield at a high degree of conversion of the substrate or substrates under the conditions of high chemical and enantiomeric selectivity. The approach towards that goal is illustrated in Figure 5.1.

At the start of optimization of the reaction system, suitable values for pH and temperature have to be chosen as a function of the properties of the reactants and enzymes. Fortunately, most enzyme reactions operate in a narrow band with respect to pH value (7–10) and temperature (30–50 °C). The initial substrate concentration and, in the case of two-substrate reactions, the stoichiometric ratio of the two reactants, have to be selected. The selected enzyme concentration influences both the achievable space–time-yield as well as the selectivity in the case of undesired parallel or consecutive side reactions. In the case of multi-enzyme systems, the optimal activity ratio has to be found. The activity and stability of all the enzymes involved have to be known as a function of the reaction conditions, before the kinetic measurements are made. Enzyme stability is an important aspect of biocatalytic processes and should be expressed preferably as an enzyme unit consumption number, with the dimension "unit of activity per mass of product" (such as mole, lb, or kg). In multi-enzyme systems the stability of all the enzymes has to be optimized so that an optimal reaction rate and space–time-yield result.

Reaction kinetics

Reaction kinetics have to be found under conditions relevant to the process, whereas most of the literature values refer to optimal values of pH and temperature for the enzyme in question, and mostly for very dilute solutions. To develop meaningful reaction kinetics, the influence of all the substrates, products, and inhibitors has to be known as a function of the degree of conversion; it does not suffice just to take initial rate kinetics. Also, the thermodynamics of an enzyme reaction has to be integrated into the kinetic model, as enzyme catalyze both, the forward and reverse reaction, so that both reactions have to find consideration in the same kinetic model. Advantageously, kinetic parameters such as v_{max}, k_{cat}, or K_M for kinetic models should be determined with non-linear regression models and not, as unfortunately is still relatively common, from linearized equations such as Lineweaver–Burk. Linearized equations obscure the measured data points through large errors at low and especially at high substrate concentrations. On the other hand, non-

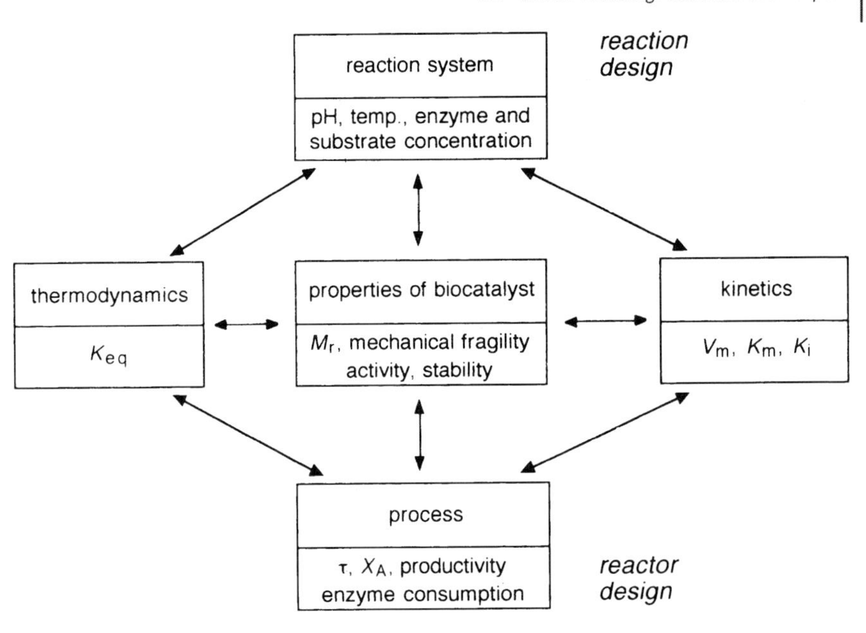

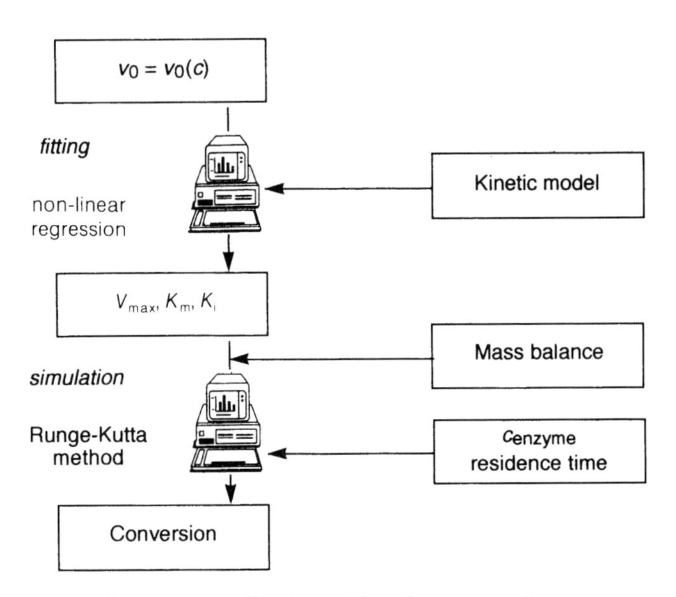

Figure 5.1 Approach to kinetic modeling of enzyme reactions: linkage of different elements of enzyme reaction engineering (top); parameter estimation and determination of operating points (bottom) (Bommarius, 1993).

linear regression models should not be used indiscriminately, just because otherwise ill-understood computer routines are available. The best evidence for a kinetic model is a concentration–time curve over the whole range of conversion in a batch reactor. If the resulting equations are difficult or impossible to integrate, numerical methods such as Runge–Kutta have to be utilized. Meanwhile such kinetic models have been developed even for multi-enzyme systems such as redox reactions with regeneration of cofactors.

Reactor kinetics

There are several phenomena which can interfere with the intrinsic kinetics and influence the overall reaction rate, degree of conversion, reactor productivity, or (bio)catalyst lifetime. Among those phenomena are:

- *partitioning* of substrate and products owing to the influence of charge or hydrophobicity in the vicinity of surfaces or networks, or in channels or pores of a solid carrier;
- *external diffusion* near a macroscopic surface owing to the formation of a boundary layer with only a diffusive flux instead of convective transport (Section 5.5.1);
- *internal diffusion* or *pore diffusion*, caused by hindered transport in narrow channels or pores of a carrier slowing down transport and/or even rendering parts of the carrier inaccessible to substrate (Section 5.5.2).

5.2 The Ideal World: Ideal Kinetics and Ideal Reactors

5.2.1 The Classic Case: Michaelis–Menten Equation

Even the simplest enzyme reaction consists of at least a binding and a catalytic step. The simplest kinetic representation resulting from a reversible binding step, followed by an irreversible catalytic step, is written in Eq. (5.1).

$$\mathrm{E} + \mathrm{S} \underset{k_{-1}}{\overset{k_1}{\Leftrightarrow}} \mathrm{ES} \overset{k_2}{\rightarrow} \mathrm{E} + \mathrm{P} \tag{5.1}$$

The rate law, containing just observables, following from the kinetics in Eq. (5.1) is obtained in the most general way in four steps:

1. identification of the product formation rate or substrate consumption rate [Eq. (5.2)]:

$$\mathrm{d[P]}/\mathrm{d}t = -\,\mathrm{d[S]}/\mathrm{d}t = k_2 \cdot [\mathrm{ES}] \tag{5.2}$$

2. writing steady-state balances for each species [Eqs. (5.3)]–(5.6)]:

For [E]: $$d[E]/dt = -k_1 \cdot [E][S] + k_{-1} \cdot [ES] + k_2 \cdot [ES] \quad (5.3)$$

For [S]: $$d[S]/dt = -k_1 \cdot [E][S] + k_{-1} \cdot [ES] \quad (5.4)$$

For [ES]: $$d[ES]/dt = k_1 \cdot [E][S] - k_{-1} \cdot [ES] - k_2 \cdot [ES] \quad (5.5)$$

For [P]: $$d[P]/dt = +k_2 \cdot [ES] \quad (5.6)$$

3. writing the mass balance of all catalytic species, assuming $[E]_{total}$ is observable, i.e., under the control of the operator [(Eq. (5.7)]:

$$[E]_{total} = [E] + [ES] \quad (5.7)$$

4. applying the Bodenstein approximation [Eq. (5.8)]. As [ES] is small, its derivative $d[ES]/dt \approx 0$:

$$d[ES]/dt \approx 0 = k_1 \cdot [E][S] - k_{-1} \cdot [ES] - k_2 \cdot [ES] \quad (5.8)$$

With Eq. (5.7) this gives:

$$0 = k_1 \cdot \{[E]_{total} - [ES]\}[S] - k_{-1} \cdot [ES] - k_2 \cdot [ES]$$

or:

$$[ES] = k_1 \cdot [E]_{total}[S]/(k_{-1} + k_2 + k_1[S])$$

or, by dividing by k_1, Eq. (5.9):

$$[ES] = [E]_{total}[S]/\{(k_{-1} + k_2)/k_1 + [S]) \quad (5.9)$$

and finally, with Eq. (5.2), Eq. (5.10):

$$d[P]/dt = k_2 \cdot [E]_{total}[S]/\{(k_{-1} + k_2)/k_1 + [S]) \quad (5.10)$$

which contains only observables, as required. If k_2 is identified with k_{cat} and the term $\{(k_{-1} + k_2)/k_1\}$ with the Michaelis constant K_M, the already familiar Michaelis–Menten equation is recovered [Eq. (5.11)]:

$$d[P]/dt = k_{cat} \cdot [E]_{total}[S]/\{K_M + [S]) = v_{max} \cdot [S]/\{K_M + [S]) \quad (5.11)$$

The Michaelis–Menten equation, especially if derived with the steady-state concept as above, is a rigorous rate law which not only fits almost all one-substrate enzyme kinetics, except in the case of inhibition (see Section 5.3), but also allows identification of the kinetic constants with the elementary steps in Eq. (5.1).

5.2.2
Design of Ideal Reactors

Ideal reactors behave as reactors for which very simplified assumptions are met. These assumptions concern the dimension(s) over which change occurs: these dimensions can either be time or concentration, or both. Three types of ideal reactors are known:

1. the batch reactor, BR,
2. the plug-flow reactor, PFR, and the
3. continuous stirred tank reactor, CSTR.

In addition, many mixed types, such as the cascade or the recycle reactor, are known which will not be discussed here but are covered in the "Suggested Further Reading". The goal of reactor design is the description of the behavior of both the pertinent degree of conversion x ($0 \leq x \leq 1$) and of selectivity σ ($0 \leq \sigma \leq 1$) over the relevant timescale, here the residence time τ of the reactor. The residence time τ is the mean amount of time a particle entering the reactor spends before exiting. The relationships of the three ideal reactors between residence time τ [time], degree of conversion x [–], reactor volume V [L], volumetric flux F_0 [volume · time^{-1}], initial concentration of the limiting substrate $[C]_0$ [mol·L^{-1}] and the reaction rate r [concentration time^{-1} = mol (L·s)$^{-1}$] are given in Eqs. (5.12)–(5.14):

Batch reactor (BR): $$V/F_0 = \tau = [C]_0 \int dx/r \qquad (5.12)$$

Plug-flow reactor (PFR): $$V/F_0 = \tau = [C]_0 \int dx/r \qquad (5.13)$$

Continuous stirred-tank reactor (CSTR): $$V/F_0 = \tau = (x_0 - x_e)/r \qquad (5.14)$$

It can be readily discerned that the reactor equation for the batch reactor (5.12) and the plug-flow reactor (5.13) are identical. In the former, the concentration changes with time, in the latter, with location. In contrast to the situation in the other two ideal reactors, the residence time τ in a CSTR is only an average, as every volume element has a different residence time throughout the reactor.

5.2.3
Integrated Michaelis–Menten Equation in Ideal Reactors

There are well-established methods for obtaining the type of inhibition and the value of the inhibition constants from initial-rate kinetics, often from linearized plots such as Lineweaver–Burk, Eadie–Hofstee, or Hanes. As these procedures are covered very well by a range of basic textbooks on biochemistry and kinetics (see the list of "Suggested Further Reading") we will not repeat these procedures here. Instead, we will discuss the situation in which an enzyme reaction is followed over more than just the initial range of conversion. Towards this end, the rate equation, in our discussion the Michaelis–Menten equation, must be inserted into the pertinent reactor equation, such as Eqs. (5.12)–(5.14) for the ideal reactors. As the last step, the reactor equation has to be integrated from time zero to time t and from the initial degree of conversion (usually zero) to the final degree of conversion (usually high, often 0.9, 0.99, or even 1.0). We will cover the cases of no inhibition and of substrate and product inhibition (Section 5.3.2).

5.2.3.1 Case 1: No Inhibition

The simplest case of a reactor is a cuvette, such as that in a photometer. From the Michaelis–Menten equation and the equation for the batch reactor [Eqs. (5.11) and (5.12)], respectively, as well as the definition for the degree of conversion x for the simple reaction $S \rightarrow P$, $x = 1 - [S]/[S_0] = [P]/[S_0]$, the integrated equation (5.15) for an enzyme reaction following a Michaelis–Menten law in a batch reactor is obtained.

$$r = k_{cat}[E][S]/(K_M + [S]) \quad (5.11)$$

$$\tau = [S]_0 \int (1/r)\, dx \quad (5.12)$$

$$\tau = [S]_0 \int (1/k_{cat}[E])\, (1 + K_M/[S])\, dx = \{1/k_{cat}[E]\} \int \{[S]_0 + K_M/(1 - x)\}\, dx \quad (5.13)$$

$$x[S]_0 + K_M \ln (1 - x) = k_{cat}[E]\tau \quad (5.14)$$

The same form of equation results from the equation for the PFR [Eq. (5.13)]. For a CSTR with the mass balance given by Eq. (5.14), the integrated rate equation (5.16) results from a procedure similar to that for the batch reactor.

$$\tau = x \cdot [S]_0/r \quad (5.15)$$

$$x[S]_0 + K_M \{x/(1 - x)\} = k_{cat}[E]\tau \quad (5.16)$$

Cases 2 and 3 will be discussed below in connection with inhibition.

5.3 Enzymes with Unfavorable Binding: Inhibition

5.3.1 Types of Inhibitors

Enzyme inhibitors can be classified into four categories:

1. classical inhibitors,
2. tight-binding inhibitors,
3. slowly binding inhibitors, and
4. slowly and tight-binding inhibitors.

Whereas the first two categories have been being investigated for a long time and corresponding solutions are a standard part of the literature (Morrison, 1982; Williams, 1979), solutions for the determination of the constants for the last category of slowly and tight-binding inhibitors have been developed just recently (Wang, 1993).

ad 1) *Classical inhibitors.* These act reversibly on their targets. The free inhibitor molecule, essentially at the total inhibitor concentration, can interact either with the free enzyme E (competitive inhibition), or with the enzyme–substrate complex ES (uncompetitive inhibition), or with both E and ES. While classical inhibitors inhibit reversibly, i.e., sometimes not very strongly, they do inhibit fast, i.e., on the timescale of or faster than catalytic turnover.

ad 2) *Tight-binding inhibitors.* When treating classical inhibitors the assumption was made that $[I] \gg [E]_t$, so that the free inhibitor concentration was approximately equal to the total added inhibitor concentration. However, many inhibitors are known to bind with such potency that the inhibitor concentration is significantly diminished. As a rule of thumb, $K_I \geq 1000[E]_t$ to justify the assumption that $[EI] \approx 0$ and that the steady-state approximation is valid; for *tight-binding inhibitors*, the assumption is usually not met.

ad 3 und 4) *Slowly binding inhibitors.* If inhibitors bind slowly to enzymes compared to the timescale of catalytic turnover, the initial velocity is going to change with time; such inhibitors are referred to as *slow-binding* or *time-dependent* inhibitors. Pertinent examples include *mechanism-based* or *suicide* inhibitors (Walsh, 1982). If neither the substrate is already near depletion (long timescale) nor the onset of inhibition is not yet discernible (short timescale), product formation can be divided into a linear part not influenced by inhibition and a curvilinear part reflecting slowdown by inhibition (Copeland, 1996) {Eq. (5.17), with v_i and v_s as the initial and final, i.e. steady-state, rate and k_{obs} as the rate constant converting the initial into the inhibited form].

$$[P] = v_s \cdot t + \{(v_i - v_s)/k_{obs}\}[1 - \exp(-k_{obs}\, t)] \tag{5.17}$$

Time-dependent inhibition leading to tight binding as compared to non-specific reversible binding made all the difference in the inhibition of cyclooxygenase-2 by aryl sulfonamides (Cox-2; E.C. 1.14.99.1) instead of Cox-1 (Copeland, 1995b). Only the Cox-2 isoform is thought to be the site of therapeutic action of non-steroidal anti-inflammatory drugs (NSAIDs), whereas stimulus of Cox-1 causes the well-known side effect of stomach problems.

5.3.2
Integrated Michaelis–Menten Equation for Substrate and Product Inhibition

In this section, we will discover that inhibition, which tends to be a nuisance at low conversion in initial-rate kinetics studies, can become very detrimental if high degrees of conversions are sought, such as in any situation in synthesis or large-scale processing.

To investigate the impact of inhibition on reactor design, the procedure again is as follows: (i) we have to develop the pertinent kinetic rate equation, (ii) insert this rate law into the equation for the reactor we intend to operate, and (iii) integrate over all applicable degrees of conversion. As the two most frequent cases of inhibition are the occurrence of substrate and (even more often) product inhibition, we will treat those two cases in the following section. We will just mention the equation for substrate inhibition, but we will develop step-by-step the equation for product inhibition.

5.3.2.1 Case 2: Integrated Michaelis–Menten Equation in the Presence of Substrate Inhibitor

The kinetic rate law for a substrate-inhibited enzyme is given by Eq. (5.18), which reveals that substrate inhibition is a special case of *uncompetitive* inhibition (Scheme 5.1).

$$r = v_{max}[S]/(K_M + [S] + [S]^2/K_S) \tag{5.18}$$

$$\begin{array}{ccccc} E & \Leftrightarrow & ES & \rightarrow & E + P \\ & I = S & \updownarrow K_I & & \\ & & ESI = ESS & & \end{array}$$

Scheme 5.1 Uncompetitive inhibition.

An additional substrate molecule interacts with the ES complex at high substrate concentration to form the unproductive complex ESI, or in this case ESS. With the respective mass balances for the ideal reactors the integral rate laws of Eqs. (5.19) and (5.20) are obtained.

$$\text{CSTR:} \quad x[S]_0 + K_M\, x/(1 - x) + ([S]_0^2/K_S)/(x - x^2) = k_{cat}[E]\tau \tag{5.19}$$

$$\text{BR and PFR:} \quad x[S]_0 + K_M \ln(1 - x) + ([S]_0^2/2K_S)(2x - x^2) = k_{cat}[E]\,\tau \tag{5.20}$$

To ameliorate the effects of substrate inhibition, running the reactor in semi-batch mode, i.e., with continuous feeding of substrate, is always an option.

5.3.2.2 Case 3: Integrated Michaelis–Menten Equation in the Presence of Inhibitor

Product inhibition of four different kinds can occur: (1) competitive, (2) uncompetitive, (3) non-competitive, and (4) mixed. The most frequent case, competitive inhibition, affects only the effective strength of binding of the substrate to the enzyme (Scheme 5.2).

E ⇔ ES → E + P
↕ P
EP

Scheme 5.2 Competitive binding.

Obtaining the integrated rate equation is still quite simple. The basic kinetic equation is Eq. (5.21), from which Eqs. (5.22) and (5.23) result.

$$v_{Pi} = v_{max}[S]/\{K_M\,(1 + [P]/K_{IP}) + [S]\} \quad (5.21)$$

1. BR and PFR: $(1 - K_M/K_{IP})\, x\, [S]_0 - (1 + [S]_0/K_{IP})K_M \ln(1 - x) = k_{cat}[E]\tau$ (5.22)

2. CSTR: $x[S]_0 + K_M \ln(1 - x) + (K_M[S]_0/K_{IP})(x^2/(1 - x)) = k_{cat}[E]\tau$ (5.23)

For the discussion of noncompetitive and mixed product inhibition, we turn to an actual case and its corresponding treatment in the literature (L. G. Lee, 1986). The oxidation of alcohols, especially diols, to ketones with dehydrogenases and concomitant reduction of NAD^+ to NADH is prone to product inhibition by the ketones. If the reaction

$$NAD^+ + RH\text{–}OH \rightarrow R{=}O + NADH + H^+$$

is represented by

$$A + B \rightarrow P + Q$$

the rate equation for the ordered bi–bi mechanism (A on first, then B, P off first, then Q) is Eq. (5.24).

$$v/v_{max} = [B]/\left\{K_{MB}\left(1 + \frac{K_{IA}}{[A]}\right)\left(1 + \frac{K_{MQ}[P]}{K_{IQ}K_{MP}}\right)\right\} + [B]\left\{1 + \frac{K_{MA}}{[A]} + \frac{[P]}{K_{IP}}\right\} \quad (5.24)$$

With the assumptions that [A] is saturating, $[A] \gg K_{MA}$, K_{IA}, and the definition of the ratio of the inhibition constants as R [Eq. (5.25)], Eq. (5.24) can be expressed as Eq. (5.26).

$$R = K_{IQ}K_{MP}/(K_{IP}K_{MQ}) \quad (5.25)$$

$$v/v_{max} = [B]/\left\{K_{MB}\left(1 + \frac{[P]}{R \cdot K_{IP}}\right) + [B]\left(1 + \frac{[P]}{K_{IP}}\right)\right\} \quad (5.26)$$

With the further assumption that substrate is converted only to product which simultaneously acts as (the only) inhibitor, $x = 1 - [S]/[S_0] = [P]/[S_0]$, and with $K_{MB} \equiv K_M$ and $K_{IP} \equiv K_I$, Eq. (5.27) is obtained from Eq. (5.26).

$$v/v_{max} = (1-x)/\left\{1 + \frac{K_M}{[S_0]} + x\left[\frac{[S_0]}{K_I}(1-x) + \frac{K_M}{R \cdot K_I} - 1\right]\right\} \quad (5.27)$$

In the (albeit not frequent) case of noncompetitive inhibition, $R = 1$; otherwise, inhibition is mixed and $R < 1$. If we now recall that $v = d[P]/dt = d(x[S_0])/dt$, Eq. (5.27) can be transformed into Eq. (5.28)

$$dt = \frac{[S_0]}{(1-x)\,v_{max}}\left\{\frac{-x^2 \cdot K_M \cdot [S]_0}{2 \cdot K_I \cdot K_M} + x\left(\frac{[S]_0}{K_M}\frac{K_M}{K_I} + \frac{K_M}{R \cdot K_I} - 1\right) + \left(\frac{K_M}{K_I} + \frac{K_M}{[S]_0}\frac{K_I}{K_M}\right) + 1\right\} dx \quad (5.28)$$

and integrated from $t = 0$ to t and $x = 0$ to x, to read:

$$t = \frac{[S_0]}{v_{max}}\left\{\frac{x^2 \cdot K_M \cdot [S_0]}{2\,K_I \cdot K_M} + \left(1 - \frac{K_M}{R \cdot K_I}\right)x - \left(\frac{K_M}{R \cdot K_I} + \frac{K_M}{[S_0]}\right)\ln(1-x)\right\} \quad (5.29)$$

Inspection of Eq. (5.29) reveals that there are two quantities besides R that control the degree of conversion x at a particular point in time, the *saturation ratio* $[S_0]/K_M$ and the *inhibition ratio* K_M/K_I. Figures 5.2–5.4 elucidate the importance of both ratios.

- *Saturation ratio.* In the case of competitive inhibition a high initial substrate concentration $[S_0]$ is always favorable, but in the cases of non-competitive or mixed inhibition too high a ratio is detrimental at high degrees of conversion: the reason is the inhibition by (plenty of) product at such degrees of conversion, and a ratio of the term $([S_0]/[K_M])\,(K_M/K_I)$ closer to unity seems optimal.

- *Inhibition ratio.* The inhibition ratio K_M/K_I is particularly important as it is totally outside the control of the researcher or operator. Figures 5.2–5.4 demonstrate that an inhibition ratio $K_M/K_I < 1$ is not detrimental: the substrate binds more tightly than the inhibitor and the high degrees of conversion can still be achieved with out much difficulty, maybe at longer reaction times. If $K_M/K_I > 1$, however, the reaction crawls to a virtual stop at intermediate degrees of conversion and the reaction will not reach completion.

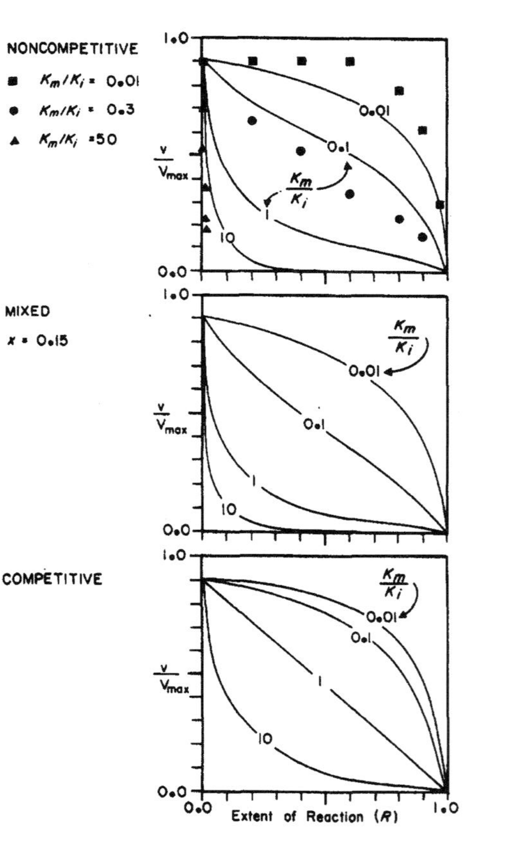

Figure 5.2 Relative rate of reaction v/v_{max} as a function of the extent of reaction (degree of conversion x) for competitive, noncompetitive ($R = 1$), and mixed ($R = 0.15$) product inhibition (L. G. Lee, 1986).

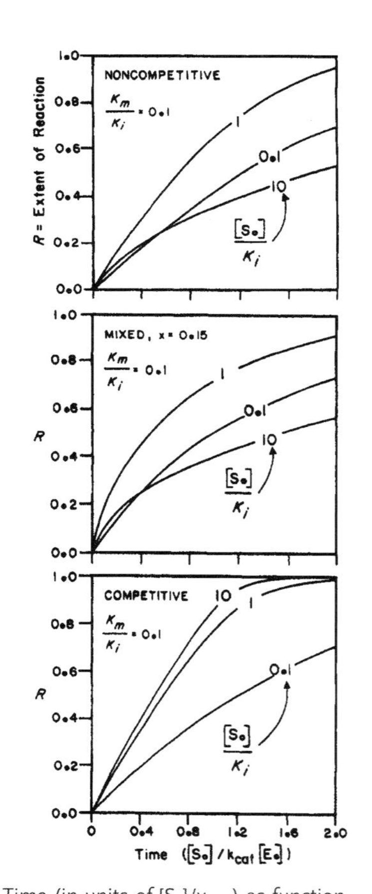

Figure 5.3 Time (in units of $[S_0]/v_{max}$) as function of the extent of reaction (degree of conversion x) for competitive, noncompetitive ($R = 1$), and mixed ($R = 0.15$) product inhibition; the value of the inhibition ratio K_M/K_I is 0.1 (L. G. Lee, 1986).

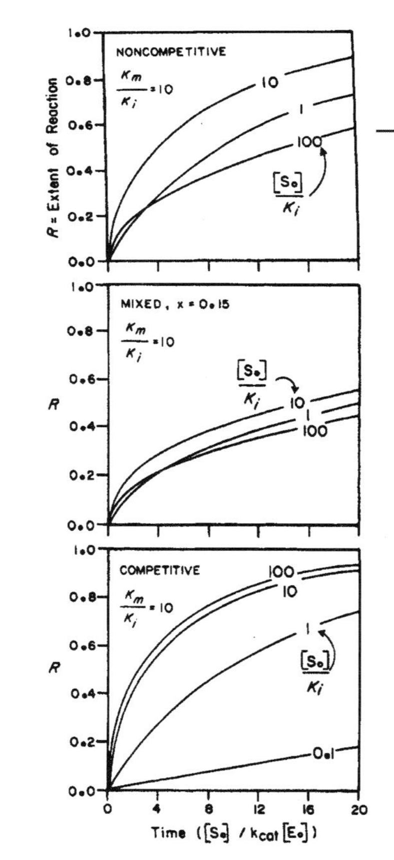

Figure 5.4 Time (in units of $[S_0]/v_{max}$) as a function of the extent of reaction (degree of conversion x) for competitive, noncompetitive ($R = 1$), and mixed ($R = 0.15$) product inhibition; the value of the inhibition ratio K_M/K_I is 10 (L. G. Lee, 1986).

5.3.3
The K_I –$[I]_{50}$ Relationship: Another Useful Application of Mechanism Elucidation

Determination of an inhibition constant K_I is often attempted by a short-cut procedure instead of the correct method, which entails the measurement of several v–[S] plots at different concentrations [I]. This short-cut procedure calls for measuring the reaction rate v in the absence of the inhibitor, subsequent addition of different concentrations [I] of inhibitor at constant substrate concentration and determination of IC_{50}, in other words, the inhibitor concentration at which the initial reaction rate is decreased to 50% of the rate at $[I] = 0$. Such a short-cut procedure can be misleading, however, as will be demonstrated in the following paragraphs.

It is assumed for the ensuing discussion that the Michaelis–Menten equation holds, that only one-substrate cases are treated, and that the inhibitor is a classical one (category 1 in the above list; Scheme 5.3).

Case 1: Competitve inhibition

$$\begin{array}{ccccc} E & \Leftrightarrow & ES & \rightarrow & E + P \\ \updownarrow I & & & & \\ EI & & & & \end{array}$$

Scheme 5.3 Case 1: competitive inhibition.

The basic kinetic equation is Eq. (5.30).

$$v_I = v_{max}[S]/\{K_M\ (1 + [I]/K_I) + [S]\} \tag{5.30}$$

If $[I] = I_{50}$ then $v_0 = 2 \cdot v_I$, leading to Eq. (5.31).

$$v_{max}[S]/(K_M + [S]) = 2\ v_{max}[S]/\{K_M\ (1 + I_{50}/K_I) + [S]\} \tag{5.31}$$

After rearranging, Eq. (5.32) is obtained.

$$[I]_{50} = K_I\ [1 + ([S]/K_M)] \tag{5.32}$$

Only if $[S] \gg K_M$, i.e., if the enzyme is completely saturated with substrate, does Eq. (5.33) hold.

$$[I]_{50} = K_I \tag{5.33}$$

Case 2: Non-competitive inhibition

$$\begin{array}{ccccc} E & \Leftrightarrow & ES & \rightarrow & E + P \\ I \updownarrow K_{IS} & & I \updownarrow K_{II} & & \\ EI & \Leftrightarrow & ESI & & \end{array}$$

Scheme 5.4 Case 2: non-competitive inhibition.

The basic kinetic equation for case 2 (Scheme 5.4) is Eq. (5.34).

$$v_I = v_{max}[S]/\{K_M (1 + [I]/K_{IS}) + [S] \cdot (1 + [I]/K_{II})\} \tag{5.34}$$

If $[I] = I_{50}$, then $v_0 = 2 \cdot v_I$ and Eq. (5.35) results.

$$v_{max}[S]/(K_M + [S]) = 2 \cdot v_{max}[S]/\{K_M (1 + [I]_{50}/K_{IS}) + [S] (1 + [I]_{50}/K_{II})\} \tag{5.35}$$

This becomes Eq. (2.36) after rearranging.

$$[I]_{50} = (K_M + [S])/\{(K_M + K_{IS}) + ([S]/K_{II})\} \tag{5.36}$$

If the affinity of the inhibitor for the enzyme (E) equals the affinity for the enzyme–substrate complex (ES) then $K_{II} = K_{IS} = K_I$, and Eq. (2.36) can be rearranged to yield Eq. (2.37).

$$I_{50} = K_I \tag{5.37}$$

If the enzyme is saturated with substrate ($[S] \gg K_M$), but $K_{IS} \neq K_{II}$, Eq. (5.36) can be rearranged to yield Eq. (5.38).

$$[I]_{50} = K_{IS}/\{(K_M/[S]) + (K_{IS}/K_{II})\} \tag{5.38}$$

If the substrate concentration is adjusted so that $K_M/[S] \ll K_{IS}/K_{II}$ holds, Eq. (5.38) can be transformed into Eq. (5.39).

$$[I]_{50} = K_{II} \tag{5.39}$$

Case 3: Uncompetitive inhibition

$$\begin{array}{ccccc} E & \Leftrightarrow & ES & \rightarrow & E + P \\ & & I \updownarrow K_I & & \\ & & ESI & & \end{array}$$

Scheme 5.5 Case 3: uncompetitive inhibition.

The basic kinetic equation for case 3 (Scheme 5.5) is Eq. (5.40).

$$v_I = v_{max}[S]/\{K_M + [S] \cdot (1 + [I]/K_I)\} \tag{5.40}$$

If $[I] = [I]_{50}$, then $v_0 = 2 \cdot v_I$ and Eq. (5.41) results, or after rearranging, Eq. (5.42).

$$v_{max}[S]/(K_M + [S]) = 2v_{max}[S]/\{K_M + [S] \cdot (1 + [I]_{50}/K_I)\} \tag{5.41}$$

$$[I]_{50} = (K_M + [S])/\{(K_M + K_{IS}) + ([S]/K_{II})\} \tag{5.42}$$

If the affinity of the inhibitor for the enzyme (E) equals the affinity for the enzyme–substrate complex (ES), then $K_{II} = K_{IS} = K_I$, and Eq. (5.42) can be transformed to yield Eq. (5.43).

$$[I]_{50} = K_I \, [1 + ([S]/K_M)] \tag{5.43}$$

If the enzyme is saturated with substrate ($[S] \gg K_M$) then Eq. (5.43) becomes Eq. (5.44).

$$I_{50} = K_I \tag{5.44}$$

So if $[S] \gg K_M$, then $[I]_{50}$ is independent of [S].

In summary, the quantity $[I]_{50}$ cannot simply be equated with the inhibition constant K_I. This can be done only if the following conditions are met: (i) existence of non-competitive or uncompetitive inhibition; *and* (ii) saturation of the enzyme by substrate.

5.4 Reactor Engineering

5.4.1 Configuration of Enzyme Reactors

As in the case of chemical processing, all enzyme reactors can be classified according to the degree of mixedness of the substrate and product solution; they belong to only a few basic types. The different reactor types used in biocatalysis are depicted in Figure 5.5.

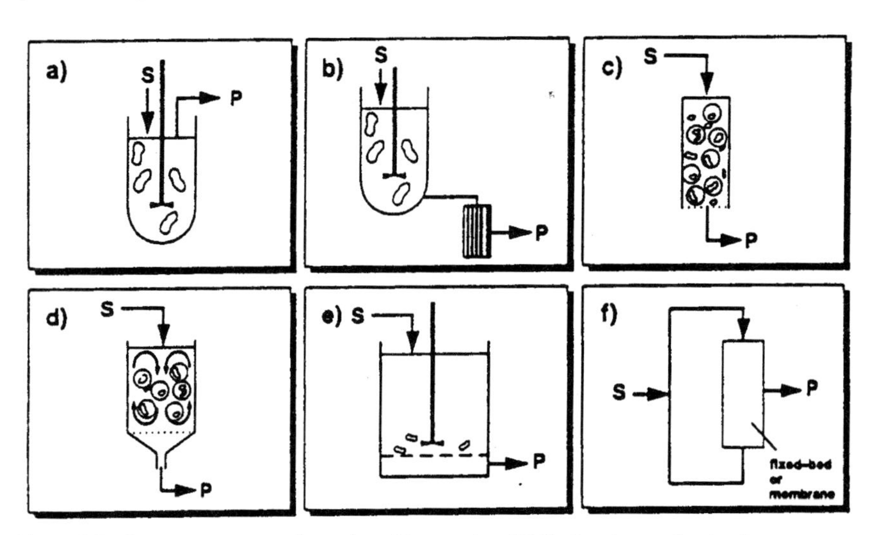

Figure 5.5 Enzyme reactor configurations (Bommarius, 1993). See the text for details.

Just like chemical processes, biocatalytic reactions are performed most simply in *batch reactors* (Figure 5.5a). On a lab scale and in the case of inexpensive or rapidly deactivating biocatalysts, this is the optimal solution. If the biocatalyst is to be recycled, but the mode of repeated batches is to be maintained, a batch reactor with subsequent ultrafiltration is recommended (*batch-UF reactor*; Figure 5.5b). The residence times of catalyst and reactants are identical in all batch reactor configurations.

With the increasing complexity of biocatalysis in organic synthesis and on a large scale, more methods were needed to retain the catalyst efficiently in the reaction system. Figure 5.6 provides an overview of the options. Binding of the enzyme to solid supports (immobilization) was the beginning of this development. Immobilized enzymes are bound mostly to spherical beads instead of to planar surfaces, so that they can be packed into a *fixed-bed reactor* operated as a PFR (Figure 5.5c). Undesired radial or axial pH or temperature gradients can be countered by an enhanced flow rate of the fluid phase in the column so that the reactor might then be operated as a *fluid-bed reactor* (Figure 5.5d).

Membrane reactors allow a different option for the separation of biocatalysts from substrates and products and for retention in the reactor. Size-specific pores allow the substrate and product molecules, but not the enzyme molecules, to pass the membrane. Membrane reactors can be operated as CSTRs with *dead-end filtration* (Figure 5.5e) or as *loop* or *recycle reactors* (Figure 5.5f) with *tangential* (crossflow) *filtration*.

Immobilized enzyme and enzyme membrane reactors each have advantages and disadvantages (Table 5.1).

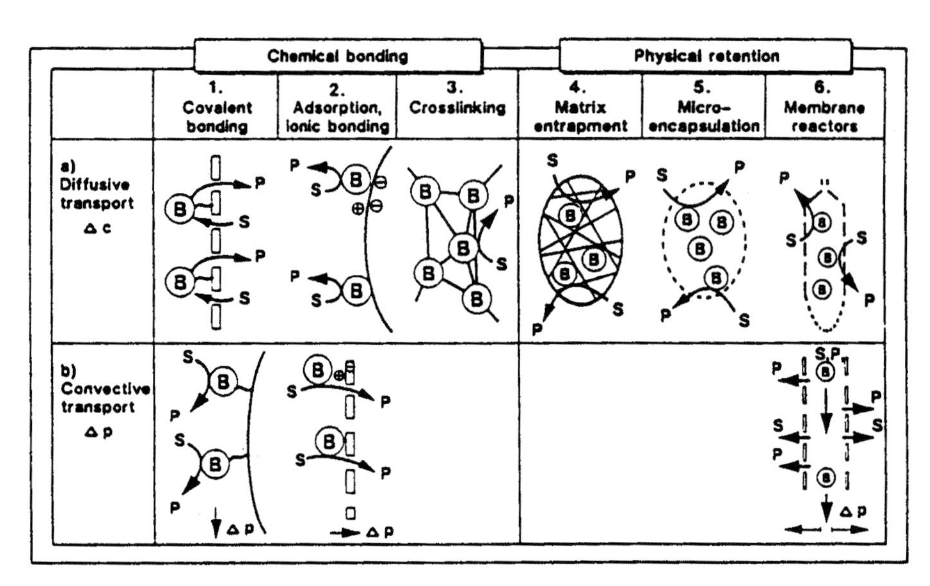

Figure 5.6 Configurations for the retention of biocatalysts (Bommarius, 1993).

Table 5.1 Advantages and disadvantages of membrane and immobilization reactors.

Reactor	*Advantages*	*Disadvantages*
Membrane reactors	– no mass transfer limitations – pyrogen-free products – scale-up simple	– pre-filtration required – no polymeric products possible – no product precipitation possible
Immobilized enzyme reactors	– potentially more stable enzymes – co-immobilization possible	– difficult to sterilize – mass transfer limitations – immobilization increases costs

Immobilization influences the activity and stability of biocatalysts much more than encapsulation by membranes; however, enhanced activity (rarely) or stability (often) can be an important reason to pick immobilization for the retention of enzymes.

Crucial advantages of membrane reactors are the absence of mass transfer limitations, even at higher catalyst concentrations (at least in purely aqueous solution), the convenient scale-up by adaptation of the available membrane surface area, and the simple replenishment of enzyme activity in the case of deactivation in a reactor run with a constant conversion policy. Additionally, the filtrates are usually free of microorganisms and pyrogens after passage through the membrane, which is increasingly important for the production of pharmaceuticals. Isolation and purification of products cannot occur through precipitation as this would block the membrane.

5.4.1.1 Characteristic Dimensionless Numbers for Reactor Design

To understand the relative importance of different rate processes such as reaction, mixing, diffusion, or dispersion, knowledge of the relevant timescales of these processes is invaluable. The ratios of such timescales form a set of dimensionless quantities. While we cannot possibly give an exhaustive account here (the reader is referred to reactor design texts), we mention the most important dimensionless quantities.

The characteristic timescales of the aforementioned rate processes are:

- the *residence time* τ for the reactor,
- the *inverse catalytic constant* $1/k_{cat}$ for the enzyme reaction (alternatively, the quantity $\theta = K_M/k_{cat}[E]$),
- the *inverse time constant* δ/k_f for external mass transfer (external diffusion),
- the *inverse time constant* L^2/D_{eff} for internal mass transfer (internal or pore diffusion),
- the *inverse time constant* $R^2/D_{ax,L}$ for axial dispersion in a fixed-bed or plug flow reactor.

(δ = boundary layer thickness, L = characteristic length of solid carrier, R = characteristic length of pipe)

The ratio of reactor and reaction timescale is the (dimensionless, of course) Damköhler number of the first kind: $Da_I = k \cdot \tau$ (for a first-order reaction). In a recycle reactor operated as a CSTR with defined residence time τ, the degree of conversion can be achieved by keeping Da_I constant [Eq. (5.45)].

$$Da_I = \frac{\text{time constant of reactor}}{\text{time constant of reaction}} = (\tau \cdot k\text{cat}[E])/K_M = \text{const} \times [E] \cdot \tau \quad (5.45)$$

Other factors limiting the overall rate can be external or internal mass transfer, or axial dispersion in a fixed-bed reactor. Pertinent dimensionless numbers are the Biot number Bi, the Damköhler number of the second kind Da_{II}, or the Bodenstein number Bo (Eqs. (5.46)–(5.48)].

$$Da_{II} = \text{timescale of pore diffusion/timescale of reaction} = (L^2 \times k_{cat})/D_{eff} \quad (5.46)$$

$$Bi = \text{external mass transfer/internal mass transfer} = (k_f \times R)/D_{eff} \quad (5.47)$$

$$Bo = \text{convective mass transfer/dispersive mass transfer} = (u \times R)/D_{ax,L} \quad (5.48)$$

As a general rule, the longest timescale always dominates the overall rate process.

5.4.2 Immobilized Enzyme Reactor (Fixed-Bed Reactor with Plug-Flow)

5.4.2.1 Reactor Design Equations

The reactor equation for a fixed-bed reactor with plug flow is Eq. (5.13).

$$\tau = [S]_0 \cdot \int dx/r(x) \quad (5.13)$$

The Reynolds number Re_p for the particles in the bed is calculated from the equivalent diameter of the particles D_p (for spheres, $D_p = d_p$), by Eq. (5.49).

$$Re_p = D_p v_{tube} \cdot \rho/(1 - \varepsilon_{total})\, \eta \quad (5.49)$$

With a given bed material of the column D_p and total porosity ε_{total}, and with a given substrate solution, the material properties ρ and η, can be regarded as constant so that the superficial velocity v_{tube} is particularly important for process design. This quantity is linked to the easily measurable flow rate Q and the area of the bed A by Eq. (5.50).

$$v_{tube} = Q/A = Q \times 4/(\pi \cdot d_{column}{}^2) \quad (5.50)$$

Equation (5.50) yields Eq. (5.51) for the residence time τ which has column height.

$$\tau = V/Q = \pi \cdot d_{column}{}^2 \cdot h\, \varepsilon_{total}/(4Q) = h\, \varepsilon_{total}/v_{tube} \quad (5.51)$$

Fixed-bed reactors on the lab scale have to be tested more severely than CSTRs because

- at $h/d_p < 100$ entrance effects occur but often in lab reactors $h/d_p \approx 25$;
- at $d_{column}/d_p < 10–20$ wall effects can be observed so with $d_p = 1 - 2$ mm, a lab reactor has to have a diameter of at least about 2–3 cm;
- axial dispersion can be relevant, especially if $h/d_p \approx 25$ and additionally Re $\ll 1$.

With immobilized enzymes in a fixed-bed reactor, a constant residence time τ in the fixed bed is the applicable design criterion for scale-up.

5.4.2.2 **Immobilization**

Without innovative process development, biocatalysis would have stopped at a few batch processes. Only through use of immobilization could the potential of biocatalysts be tapped for large-scale processing. Immobilization, of course, does not just find use for biologically-based synthesis but also for reactors in biomedicine (Langer, 1990) and for the coating of suitable medicinals in therapy (Chang, 1976). Besides achieving retention of (bio)catalysts, immobilization is often employed to achieve additional purposes such as (i) stabilization of enzymes, as translational motion as well as volume-enhancing unfolding of enzymes is restricted in immobilized mode; or (ii) a change in microenvironment with respect to pH or hydrophobicity.

Immobilized enzyme systems can be differentiated according to mode of immobilization, carrier properties, and rate-determining step.

- *Mode of immobilization.* Immobilization can be effected either chemically, by *covalent bonding* of the biocatalyst on a surface (Figure 5.6, option 1), by *adsorption*, or by ionic interactions between catalyst and surface (option 2), as well as by *cross-linking* of biocatalyst molecules for the purpose of enlargement (option 3), or physically by *encapsulation* in matrices or by embedding in a membrane (option 4).

- *Carrier properties.* Carriers can be shaped and configured as films, fibers, planar surfaces, or spheres. Surface morphology, i.e., surface texture and porosity, can exert a decisive influence as can carrier materials; the most important are inorganic materials such as ceramics or glass, synthetic polymers such as nylon or polystyrene, and polysaccharide materials such as cellulose, agarose, or dextran.

- *Rate-determining step.* In immobilized (bio)catalyst systems, especially in porous particles, the catalytic step is only one of several rate processes in sequence:
  1. film diffusion of the substrate molecule through the boundary layer close to the surface,
  2. pore diffusion from the surface to the active site,
  3. biocatalytic step(s),
  4. pore diffusion of the product molecule back to the surface, and
  5. film diffusion of the product molecule to the bulk.

Depending on the reaction conditions, each of the transport steps 1, 2, 4, or 5 can be rate-determining instead of the biocatalytic step 3. It is of the utmost importance to establish which step is rate-limiting under the prevailing conditions.

5.4.2.3 Optimal Conditions for an Immobilized Enzyme Reactor

For two reasons, biocatalysts are used advantageously in immobilized form:

1. The immobilized enzyme is more active and/or more stable than the soluble equivalent; the immobilization process retards unfolding. Examples are glucose isomerase (GI) (Chapter 7, Section 7.3.1.) or penicillin G acylase (PGA) (Section 7.5.1).

2. The packed-bed reactor configuration commonly employed with immobilized enzymes yields a higher degree of conversion or higher space–time yield than a CSTR, the typical configuration for soluble enzymes. Examples are the nitrile hydratase-catalyzed process to 5-cyanovaleramide (5-CVAM) (Chapter 7, Section 7.1.1.3) or the decarboxylation of D,L-aspartate to D-aspartate, L-alanine, and CO_2 (Section 7.2.2.5).

5.4.3
Enzyme Membrane Reactor (Continuous Stirred Tank Reactor, CSTR)

5.4.3.1 Design Equation: Reactor Equation and Retention

The enzyme membrane reactor (EMR), if the reactor contents are well-mixed so the reactor is operated as a CSTR as is commonly done, is characterized by a particularly simple reactor equation [(Eq. (5.52)].

$$\tau = (V/[S_0]) \cdot (x_e - x_{0i}) \quad (5.52)$$

The utilization of enzyme membrane reactors with soluble, homogeneous enzymes has been reviewed on several occasions (Bommarius, 1992; Kragl, 1992). The principal advantages and disadvantages of an enzyme–membrane reactor operated as a CSTR are listed in Table 5.2.

Complete retention of the enzyme in the reactor is essential for successful use of the EMR. In a CSTR Eq. (5.53) is valid as the wash-out equation (dilution equation), with R as degree of retention ($0 < R < 1$); this is completely analogous to the form of the deactivation equation $[E]_t = [E]_0 \exp(k_d\, t)$ [Eq. (2.19)].

$$[E]_t = [E]_0 \exp(R \cdot t) \quad (5.53)$$

Table 5.2 Advantages and disadvantages of enzyme membrane reactors.

Advantage	*Disadvantage*
No external mass transfer limitation	Immobilized enzymes often more stable
No activity loss upon immobilization	Bigger effort for enzyme preparation
Easy addition of fresh enzyme	Soluble substrates and products only
Constant space–time-yield possible	

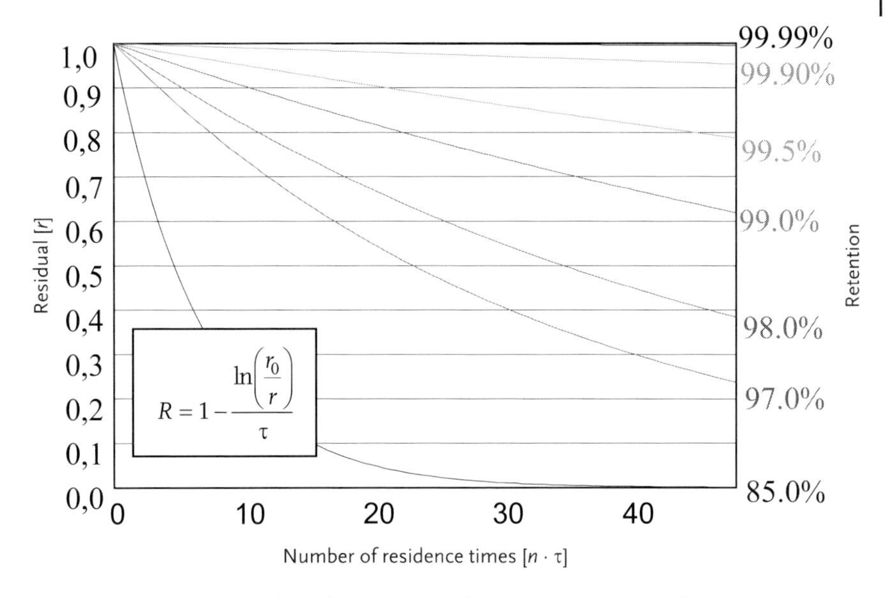

Figure 5.7 Dilution curves for different degrees of retention in a CSTR (Woltinger, 2001).

At long residence times, incomplete retention ($R < 0.99 - 0.999$) leads to large enzyme losses (Figure 5.7).

The UF membranes, which have been manufactured reproducibly since about 1980, with cut-offs between 5 to 50 kDa (typically 10 kDa), feature a basically perfect retention (> 99.99%) so that in most cases enzyme deactivation effects dominate over losses of enzyme through the UF membrane.

5.4.3.2 **Classification of Enzyme Membrane Reactors**

Membrane reactors became an option for the retention of biocatalysts when the processing of membrane materials had progressed sufficiently to control thickness and pore structure and to manufacture a membrane that was defect-free. Besides its function as a retainer the membrane also serves other functions such as (i) to stabilize the phase boundary in case of multi-phase reactions; (ii) as a consequence of (i), to transport dissolved O_2 preferentially over gaseous O_2; and (iii) to support purification and sterilization of air and other nutrients in fermentations.

A membrane can be generated by polymerization around a few biocatalyst molecules which surround a space of a few hundred micrometers (microencapsulation; Figure 5.6, option 5), or it can be of macroscopic dimensions (Figure 5.6, option 6). In the latter case, membrane reactors can be classified according to (i) driving force, (ii) pore structure and (iii) pore size.

Classification according to driving force

If separation is based on a concentration gradient Δc only, [part a) in Figure 5.6], the membrane acts as a barrier to *diffusive flux* only. (The process of diffusion against a water reservoir is termed *"dialysis"*, if the diffusion of charged species is aided by a charge gradient the process is termed *"electrodialysis"*.) In case of a pressure gradient ΔP across the membrane, separation relies on the effect of *convective* forces [part b) in Figure 5.6]. Configurations with membrane flux enforced by convection achieve much higher fluxes than those based exclusively on diffusive mass transfer.

Classification according to pore structure

Conventional filters, such as a coffee filter, termed *"depth filters"*, consist of a network of fibers and retain solute molecules through a stochastic adsorption mechanism. In contrast, most membranes for the retention of biocatalysts feature holes or pores with a comparatively narrow pore size distribution and separate exclusively on the basis of size or shape of the solute; such membranes are termed *"membrane filters"*. Only membrane filters are approved by the FDA for sterilization in connection with processes applied to pharmaceuticals. Table 5.3 lists advantages and disadvantages of depth and membrane filters.

Proteins can be viewed as hard spheres, and this property renders them easy to filter. Coiled polymers, on the other hand, can uncoil much more readily and thus pass through membrane pores with higher nominal cut-off.

Classification according to pore size

Membranes of interest in biocatalysis and biotechnology are divided into three categories:

1. *microfiltration* membranes (MF-membranes), with pore sizes in the range 0.1–10 μm (normally 0.1 or 0.22 μm), suitable for retention of cells ($d \approx 1$ μm) but not enzymes or salts,
2. *ultrafiltration* membranes (UF-membranes), with a molecular weight cut-off (MWCO) between 0.5 and 500 kDa (normally 10–100 kDa), suitable for retention of enzymes, but not salts and
3. *nanofiltration* or reverse osmosis membranes (NF- or RO-membranes), applicable for retention of low molecular weight molecules such as salts. RO membranes find wide use in desalting of seawater.

Table 5.3 Advantages and disadvantages of depth and membrane filters.

Filter type	*Advantages*	*Disadvantages*
Depth filter	high flux inexpensive apparatus	no absolute retention growth of microorganisms channeling observed leakage of filter material
Membrane filter	absolute retention pyrogen-free product stable phase boundary	easier blocking

Membrane materials often employed are hydrophobic polysulfone or hydrophilic regenerated cellulose or cellulose acetate; other materials are nylon, polytetrafluoroethylene (PTFE, Teflon), polyether ether ketone (PEEK) or poly(divinyl fluoride) (PDVF).

5.4.4 Rules for Choice of Reaction Parameters and Reactors

In this section only some rules of thumb are provided.

Choice of reaction temperature: activation versus deactivation

The question is often raised whether the reaction temperature should be lowered to decrease enzyme deactivation, even though enzyme activity suffers. If enzyme reaction and deactivation are treated as parallel reactions the answer can be obtained from classical chemical reaction engineering: whereas the activation energy of an enzymatic reaction $E_{a,reaction}$ is often around 100–120 kJ mol^{-1} the activation energy of the deactivation process $E_{a,deactivation}$ in most cases is considerably higher (200–300 kJ mol^{-1}). Since the reaction with the higher activation energy increasingly dominates with rising temperature, an enzyme reaction should be run at low temperature for minimization of deactivation.

Choice of reactor in the case of high degrees of conversion

As very high degrees of conversion are achieved in a PFR with much smaller reactor volumes than in a CSTR (Lilly, 1976; Vieth, 1976) a PFR with immobilized enzymes is most suitable if complete conversion has to be achieved, as in the case of isomerization of glucose or the decarboxylation of D,L-aspartate with L-aspartate-β-decarboxylase.

Choice of reactor in the case of inhibition

Independently of enzyme stability an enzyme reaction at constant temperature and pH should be run with the smallest possible contribution by inhibition. For this reason, a CSTR is most suitable in the case of *substrate* inhibition because the substrate concentration is evened out across the reactor volume and thus minimized. In the case of *product* inhibition, however, a batch reactor or a PFR is preferred, as the reactor volume required for complete or nearly complete conversion is much smaller than in the case of a CSTR.

5.5 Enzyme Reactions with Incomplete Mass Transfer: Influence of Immobilization

Immobilization can influence biocatalytic reactions in two different ways: (i) through a change in effective reaction parameters (pH value, charge density, and hydrophobicity); or (ii) through a change in the rate-limiting step away from the biocatalytic step towards either preceding or successive mass-transfer steps.

On ionic or inhomogeneous surfaces, the local charge density is altered in comparison with conditions in the bulk phase. In turn, the local charge density influences the concentration of H^+ and therefore the local pH value. Close to positively charged surfaces the local H^+ ion concentration is decreased; therefore, the apparent pH value is shifted into the alkaline region in comparison with the bulk phase. Correspondingly, negatively charged surfaces increase local concentrations of hydrogen ions so that an apparent pH shift near the surface towards more acidic values results. Similarly, hydrophobic surfaces lead to an apparent broadening of the pH scale.

At catalytically active centers in the center of carrier particles, *external mass transfer (film diffusion)* and/or *internal mass transfer (pore diffusion)* can alter or even dominate the observed reaction rate. External mass transfer limitations occur if the rate of diffusive transport of relevant solutes through the stagnating layer at a macroscopic surface becomes rate-limiting. Internal mass transfer limitations in porous carriers indicate that transport of solutes from the surface of the particle towards the active site in the interior is the slowest step.

5.5.1 External Diffusion (Film Diffusion)

If external diffusion dominates the overall rate, the process obviously reduces the observed enzyme activity. The flux N through the stagnating film at the surface can be expressed as in Eq. (5.54), where δ signifies the thickness of the stagnating layer and k_s is the *mass transfer coefficient* of the respective solute; k_s can be estimated by the simple relationship of Eq. (5.55).

$$N = (k_s/\delta) \cdot \Delta[S] = (k_s/\delta) \cdot ([S]_{bulk} - [S]_{surface}) \tag{5.54}$$

$$Sh = 2 + c \cdot Re^{1/2}\, Sc^{1/3} \text{ or } (k_s \cdot d_p/D) = 2 + c\,(d_p v/\nu)^{1/2}\,(\nu/D)^{1/3} \tag{5.55}$$

Sh, Re, and Sc are the dimensionless Sherwood, Reynolds, and Schmidt numbers, d_p is the relevant length scale, which here is the diameter of the particle; $c = 4.7$ for fixed beds and laminar flows. As usual, D is diffusion coefficient and v the dynamic viscosity (both [length · time^{-2}]).

5.5.2 Internal Diffusion (Pore Diffusion)

Internal diffusion in porous catalysts, if dominant, also reduces the observed activity of the biocatalyst. The decisive coefficient for mass transfer is the *effective diffusion coefficient* D_{eff}, which is defined in Eq. (5.56), where D_0 is the diffusion coefficient in solution, ε the porosity of the carrier, and τ the tortuosity factor.

$$D_{eff} = D_0 \cdot (\varepsilon/\tau)(K_{particle}/K_{pore}) \tag{5.56}$$

The ratio $K_{particle}/K_{pore}$ describes the partition coefficient of the diffusing particle between the bulk phase and pore. If electrostatic or hydrophobic interactions are unimportant and if only geometric restrictions in narrow pores are important, $K_{particle}/K_{pore}$ can be approximated by $1 - (r_{particle}/r_{pore})^4$. To examine whether internal diffusion is important in a given situation, the *effectiveness factor* η is determined; it is defined by Eq. (5.57).

$$\eta = \frac{\text{reaction rate under influence of internal diffusion effects}}{\text{intrinsic reaction rate without diffusional effects}} \tag{5.57}$$

If $\eta \sim 1$ the reaction is not, or not significantly, influenced by pore diffusion. If $\eta \ll 1$, pore diffusion is the sole dominating rate-limiting step. For the determination of η, the combined diffusion and reaction equation has to be solved. With a sequential model of the two rate phenomena, diffusion and reaction, and with the assumption of spherical geometry and validity of the Michaelis–Menten equation for the enzyme kinetics, $r = k_{cat}[E] \cdot [S]/(K_M + [S])$, Eq. (5.58) results.

$$d^2[S]/dr^2 + (2/r) \cdot d[S]/dr - k_{cat}[E][S]/D_{eff}(K_M + [S]) = 0 \tag{5.58}$$

Equation (5.59) gives the solution in the limiting case of a first-order reaction, where $\phi = L \cdot (k_{cat}[E]/(K_M D_{eff}))^{1/2}$ describes the dimensionless *Thiele modulus; L* is the characteristic length, which is half the width of a flat plate, $r/2$ of a long cyclindrical pellet, or $r/3$ of a sphere.

$$\eta = 3/(\phi \cdot \tan \phi) - 1/\phi \tag{5.59}$$

If $\phi < 0.5$, $\eta \sim 1$ results. If $\phi > 5$, then $\eta = 1/\phi$ and the substrate concentration decreases very rapidly towards zero in the pore; this case is termed *strong pore diffusion.*

In the case of other geometries, other solutions for the effectiveness factor η are obtained [Eqs. (5.60), (5.62)], and the respective solutions, in the limiting case of first-order reaction [Eqs. (5.61), (5.63)] read:

Plate:

$$d^2[S]/dr^2 + d[S]/dr - k_{cat}[E][S]/D_{eff}(K_M + [S]) = 0 \tag{5.60}$$

$$\eta = (\tanh \phi)/\phi \tag{5.61}$$

Cylinder:

$$d^2[S]/dr^2 + 1/r \cdot d[S]/dr - k_{cat}[E][S]/D_{eff}(K_M + [S]) = 0 \tag{5.62}$$

$$\eta = 2I_1(\phi)/(\phi \cdot I_0(\phi)) \tag{5.63}$$

(I_0 and I_1 are zeroth- and first-order Bessel functions.)

For different geometries and for kinetics of arbitrary order a general expression for the calculation of the Thiele modulus φ and the relationship between η and φ has been developed (Aris, 1957) [Eq. (5.64)].

$$\phi = V/A \cdot \{(n + \tfrac{1}{2})(k[C]^{n-1}/D_{eff}))^{1/2}\} \tag{5.64}$$

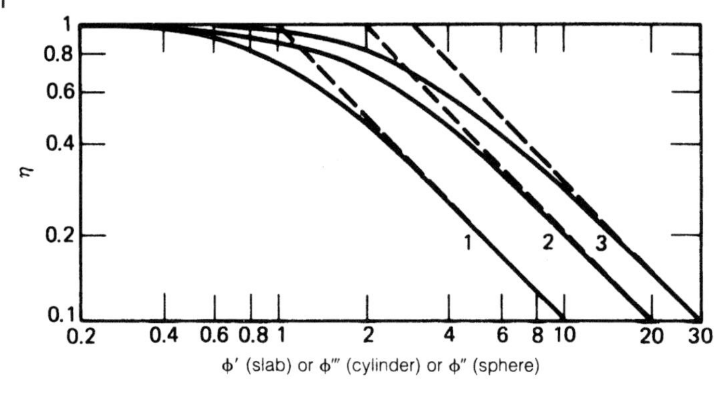

Figure 5.8 Influence of geometry on the link between the Thiele modulus ϕ and the effectiveness factor η (1: slab; 2: cylinder; 3: sphere) (Aris, 1969).

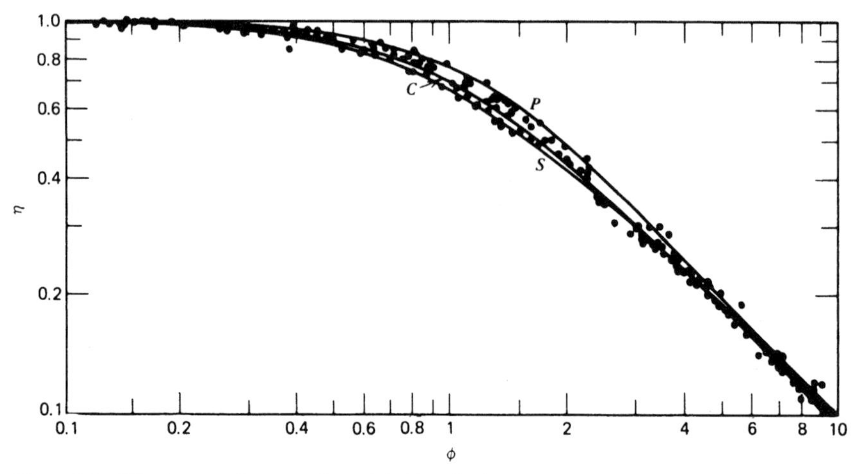

Figure 5.9 Effectiveness factors η for a slab (P), cylinder (C), and sphere (S) as functions of the Thiele modulus ϕ (Aris, 1965).

The solution to Eq. (5.64) for arbitrary geometry is Eq. (5.65).

$$\eta = 1/\phi \, \{[(3\phi) \cdot \coth(3\phi) - 1]/(3\phi)\} \tag{5.65}$$

Figures 5.8 and 5.9 illustrate the influence of geometry on the link between ϕ and η.

5.5.3
Methods of Testing for Mass Transfer Limitations

How is a carrier particle with biocatalytic activity tested for potential mass transfer limitations?

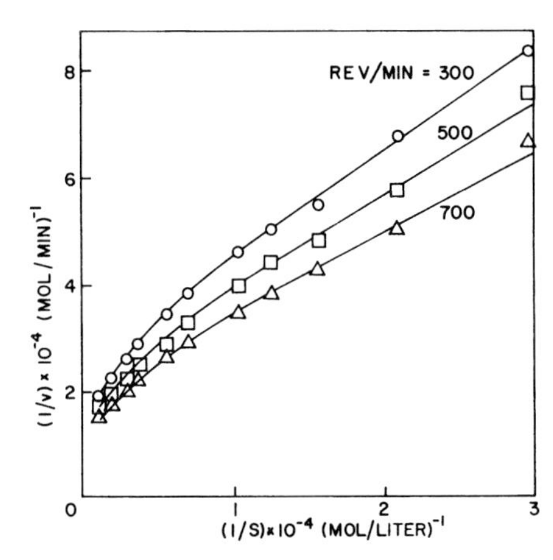

Figure 5.10 Influence of stirrer speed on film diffusion (Lilly, 1968).

External diffusion effects scale approximately with the inverse thickness of the stagnant film layer; they can be reduced by stronger stirring which leads to a compaction of the stagnant layer at the particle surface. If the observed reaction rate does not change upon enhanced stirring, external mass transfer is not rate-limiting. Figure 5.10 demonstrates a typical case in which the influence of film diffusion could be demonstrated by variation of stirring. Fixed-bed reactors cannot be controlled easily by stirring. They can be checked for diffusional limitations by measuring reaction rates at different bed heights and constant residence times or, even more simply, by establishing reaction rates at constant bed height and varying flow rates.

After the absence of film diffusion effects has been verified and if the reaction order n is known, the expression for the rate equation $r = \eta \cdot k_{cat}[E][S]_{bulk}/K_M$ (first-order reaction assumed) can be inserted into the definition for η and the unknown rate constant k can be eliminated (Weisz, 1954) [Eq. (5.66)].

$$-r \cdot L^2/(D_{eff}[S]_{bulk}) = \eta \cdot \phi^2 \tag{5.66}$$

If pore diffusion does not play an important role $\eta \sim 1$, and thus $\phi < 1$ and thence also Eq. (5.67) holds.

$$-r \cdot L^2/(D_{eff}[S]_{bulk}) < 1 \tag{5.67}$$

In the case of strong pore diffusion, $\eta = 1/\phi$ and $\phi > 1$ so that Eq. (5.68) results from Eq. (5.67).

$$-r \cdot L^2/(D_{eff}[S]_{bulk}) > 1 \tag{5.68}$$

Equations (5.67) and (5.68), termed the *Weisz-Prater modulus*, contain only observable quantities. A quick test for internal diffusion can be performed by varying L and observing the effect on r; if no effect is detected, pore diffusion cannot be important in this situation. Figure 5.11 elucidates the possible influence of pore diffusion on the observed reaction rate.

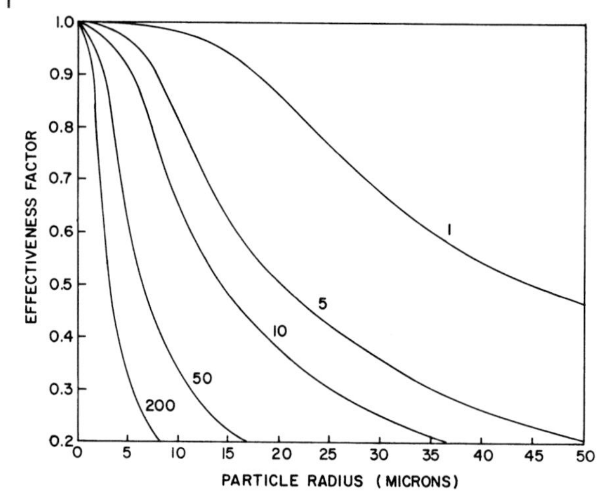

Figure 5.11 Influence of radius on the effectiveness factor of a spherical porous immobilized enzyme particle (enzyme contents [mg cm^{-3}] are listed besides curves) (specific activity: 100 IU, $D_{eff} = 4 \times 10^{-8}$ m^2 s^{-1}, $[S]/K_M = 10$) (Regan, 1974).

5.5.4
Influence of Mass Transfer on the Reaction Parameters

External diffusion

Mass transfer can alter the observed kinetic parameter of enzyme reactions. Hints of this are provided by non-linear Lineweaver–Burk plots (or other linearization methods), non-linear Arrhenius plots, or differing K_M values for native and immobilized enzymes. Different expressions have been developed for the description of apparent Michaelis constants under the influence of external mass transfer limitations by Hornby (1968) [Eq. (5.69)], Kobayashi (1971), [Eq. (5.70)], and Schuler (1972) [Eq. (5.71)].

$$K_M(\text{app}) = K_M + k_{cat}[\text{E}]/k_s \quad (5.69)$$

$$K_M(\text{app}) = K_M + \tfrac{3}{4}\, k_{cat}[\text{E}]/k_s \quad (5.70)$$

$$K_M(\text{app}) = K_M + k_{cat}[\text{E}] \cdot K_M/k_s(K_M + [\text{S}_0]) \quad (5.71)$$

Internal diffusion

Pore diffusion leads to a reduction by half of the apparent activation energy of enzyme reactions, as in the case of chemical reactions (Levenspiel, 1999). In the literature it has been reported that pore diffusion leads to enhanced apparent temperature and pH stability owing to broadening of the optimal activity regions as a function of temperature and pH activity (Karanth, 1978). Hindered pore diffusion of a substrate molecule in some cases also leads to an alleviation of the effects of substrate inhibition (Dickensheets, 1977; Tramper, 1978).

5.6 Enzymes with Incomplete Stability: Deactivation Kinetics

5.6.1 Resting Stability

If proteins are used as biocatalysts the reason for deactivation is often not quite as important as the rate of deactivation. The kinetics of deactivation over time under constant conditions of temperature, substrate, and product concentration can be determined analogously to other kinetics, most easily by following residual activity over time. In many cases, a simple negative exponential law [Eq. (5.72)] is found to describe deactivation.

$$-\,\mathrm{d}[\mathrm{E}]/\mathrm{d}t = k_1 \cdot [\mathrm{E}] \tag{5.72}$$

In this model, deactivation occurs regardless of the amount of catalysis performed until a certain time; as only time is the cause of deactivation, rather than stirring rate or catalysis, this type of deactivation is termed *"resting stability"*. This quantity is important for

- storage of the enzyme in solution,
- determination of the influence of a reaction on stability, and
- determination of rate constants of processes inherent to the protein, such as unfolding.

If the enzyme can be assumed to be transformed directly from the native (active) N to the deactivated state D, Eq. (5.73) holds.

$$\mathrm{N} \xrightarrow{k_1} \mathrm{D} \tag{5.73}$$

The solution of Eq. (5.73) is well known [Eq. (5.74)].

$$N_t = N_0 \cdot \exp\,(-k_1 \cdot t) \tag{5.74}$$

If the enzyme is deactivated in two sequential steps according to Eq. (5.75), the solution to Eq. (5.75) can be found by using Eq. (5.74) first to express N_1 as a function of N_0, i.e., $N_1 = N_0 \cdot \exp\,(-k_1 t)$; subsequently, this expression for N_1 is inserted into the rate law for N_2, giving Eq. (5.76).

$$\mathrm{N}_1 \xrightarrow{k_1} \mathrm{N}_2 \xrightarrow{k_2} \mathrm{D} \tag{5.75}$$

$$-\,\mathrm{d}N_2/\mathrm{d}t = k_2 \cdot N_2 - k_1 N_0 \cdot \exp\,(-k_1 t) \tag{5.76}$$

Equation (5.76) can be integrated with the help of the parameters $a_1 = N_2/N_1$ and $a_2 = D/N_1$. If N_t is a weighted average of the activities N_1 and N_2 with the definition $N_t = \{N_1 + aN_2\}/N_0$, where N_0 is the initial activity, then Eq. (5.77) gives the integrated rate law.

$$N_t/N_0 = 1/(k_2 - k_1) \cdot [k_2 \cdot \exp(-k_1 t) - k_1 \cdot \exp(-k_2 t)] \tag{5.77}$$

In a semi-logarithmic plot of log $\{N_t/N_0\}$ versus t, a single straight line is obtained in the case of the simple mechanism according to Eq. (5.73). In the case of sequential two-step deactivation, Eq. (5.75), two straight lines are obtained if the model is fitted well by the data. More complicated cases should not be treated by simple curve-fitting.

Even with precise measurements, models reflected in Eqs. (5.74) and (5.76) provide a reasonably exact description of the decrease in enzyme activity with time. Deactivation can be characterized in an especially simple manner by the half-life after which 50% of the initial activity remains [Eq. (5.77), with $t_{1/2}$ and k_1 in hours].

$$t_{1/2} = \ln 2/k_1 \tag{5.78}$$

It is important to understand that equations such as Eqs. (5.74) or (5.77) do not provide any mechanistic detail about the deactivation process.

The parameters to be held constant for measurements of resting temperatures are

- temperature: usually the parameter with the largest influence on stability;
- pH value: stability usually decreases dramatically with increasing distance from pH 6–8;
- concentration of substrates and products of the reaction to be catalyzed: usually, enzymes are stabilized by the presence of components of the reaction, especially if substrate or product inhibition occurs, in which case a reactant binds particularly strongly to the catalyst;
- ionic strength: a moderately high ionic strength often stabilizes an enzyme whereas high concentrations of denaturing agents such as urea or guanidinium hydrochloride cause unfolding of the enzyme with subsequent rapid deactivation.

5.6.2 Operational Stability

If the enzyme is utilized for catalysis of a desired reaction the *operational stability* is often a much more interesting parameter than the resting stability. Catalyzing a reaction can influence enzyme stability in both positive and negative fashions:

- Often in the presence of substrate and nearly always in the presence of inhibitor, a stabilization of the enzyme is observed as the enzyme molecule in the presence of a bound molecule commonly unfolds more slowly or at higher temperature.

- At high substrate concentration, chaotropic or kosmotropic effects can occur, depending on whether the substrate or added salts destroy or support water structures.

The topic of operating stability has not been treated adequately in enzyme reaction engineering. Even when investigating biochemical reaction systems, enzyme stability is often still expressed as a half-life $\tau_{1/2}$ after which 50% of initial activity remains. Such a half-life is determined by repeatedly measuring initial activity after incubation of the enzyme at a respective temperature (temperature stability) or by comparison of initial rates in a series of batch reactions with the permanent change in degree of conversion (repeated batch method). Both methods are inadequate for the characterization of enzyme stability under operating conditions. Temperature stability always refers to the stability of the protein molecule; as deactivation occurs even in the absence of a reaction; this property is not a measure of operating stability. Results of stability at varying degrees of conversion are not reproducible, unless the same conversion profile is repeated. Defined measurements of operating stability of enzymes have to be performed under a constant concentration of substrate, which is equivalent to a constant turnover number (reacted molecules per unit time and per active center) for a stable enzyme. A condition for the presence of a constant amount of substrate is the use of a continuous reactor. For immobilized enzymes, use of a fixed-bed plug flow reactor (PFR) is most suitable. For soluble enzymes, a continuous stirred tank reactor (CSTR) is more advantageous. Enzyme membrane reactors are thus extremely useful pieces of equipment for the determination of the operating stability of soluble biocatalysts.

The numerical value of the operational stability is expressed as the number of units of enzyme which have to be expended to generate a defined amount of product. The following conditions have to be specified: besides the obvious ones such as T, pH, and substrate concentration, these are the quantity $[E] \cdot \tau$ (see below) and the degree of conversion x.

The relevant parameter for the study of the operating stability of impure enzyme preparations is $[E]^* \cdot \tau$, the product of active enzyme concentration $[E]^* = m_E/V$ [g L^{-1}] and residence time τ. The following derivation illustrates the usefulness of this parameter.

The reaction rate is defined by Eq. (5.79), with the volumetric reaction rate r' [mol (L min^{-1})] and the specific (mass-based) reaction rate r [mol (g min)$^{-1}$].

$$-d[S]/dt = r' = r \cdot [E]^* \tag{5.79}$$

Equation (5.80) introduces the degree of conversion x.

$$[S_0] \cdot dx/dt = r \cdot [E]^* \tag{5.80}$$

Equation (5.81) results from rearrangement and integration over the degree of conversion from the initial value to residence time t.

$$\int_0^x (1/r)\,dx = \int_0^\tau [E]^*/[S_0]\,dt = ([E]^* \cdot \tau)/[S_0] \tag{5.81}$$

In a CSTR the integral $\int (1/r)\, dx$ is reduced to the algebraic equation [Eq. (5.82)] because the reaction rate r is uniform across the reactor.

$$([E]^* \cdot \tau)/[S_0] = x/r(x) \tag{5.82}$$

A very useful property of Eq. (5.82) is its independence from enzyme kinetics and enzyme mechanism although r still depends on x. With constant average residence time τ and initial substrate concentration $[S_0]$ at the entrance to the reactor, the degree of conversion x remains constant as long as the active enzyme concentration $[E]^*$ remains constant. If it is assumed that no enzyme is washed out of the reactor, $[E]^*$ can only decrease by deactivation. A lower value for $[E]^*$ through enzyme deactivation, however, results in a lower degree of conversion x. There are two strategies to revert back to the previous value for the parameter $[E]^* \cdot t$ and thus to the previously measured degree of conversion x:

- either the residence time τ is increased, commonly by decreasing the flow rate Q, to compensate for the decrease of $[E]^*$, or
- fresh enzyme is added into the reactor to raise the concentration of active enzyme $[E]^*$ back to its initial value.

In a PFR with immobilized enzymes, the increase in residence time t is achieved much more simply than the replenishment of spent enzyme. Especially in cases where a smooth enzyme concentration profile in both axial and radial dimensions of the PFR is aimed for, replacement of immobilized enzyme is difficult. Soluble enzymes, however, are best investigated in a CSTR according to the second method.

With both methods, the amount of enzyme added is compared to the amount of product produced ($[P] = x \cdot [S]$) to obtain a measure of the operating stability. At the beginning of a process design the prices for a unit of enzyme as well as for a mole (or pound or kilo or ton) of product are known, so an assessment of the influence of catalyst cost is possible.

5.6.3
Comparison of Resting and Operational Stability

Discussion of the concepts of resting and operational stability raises the question of their comparability and mutual convertibility. This subject has been discussed by Yamane et al. (Yamane, 1987). The validity of Eq. (5.74) for the description of enzyme deactivation is assumed.

For the discussion, *apparent half-lives* are defined: these are times after which either the degree of conversion or the flow rate is reduced to half its initial value. The half-lives obtained in this manner are then compared to the half-lives according to Eq. (5.74): $[E]_t = [E]_0 \cdot \exp(-k_d \cdot t)$. The half-life functions as a parameter for the characterization of stability. Two different operating policies for the determination of half-life in continuous reactors are investigated: (i) the *constant flow (Q) policy* with varying degree of conversion x, and (ii) the *constant conversion policy* at a constant degree of conversion x but a varying flow rate Q.

In summary, the results are quantified in Table 5.4 and are summarized as follows:

- Comparability depends on the policy of measuring half-lives: the mode of keeping a constant degree of conversion at variable flow rate or enzyme concentration (constant conversion policy) gives better results than the mode of constant flow rate accompanied by varying degree of conversion (constant flow rate policy).
- If pore diffusion is unimportant, i.e., if the effectiveness factor η is equal to 1, then with constant conversion policy both CSTR and PFR yield half-lives identical to Eq. (5.74) with arbitrary kinetics. With constant flow rate policy, the measured half-life is identical to that obtained through Eq. (5.74) only if the enzyme is saturated, i.e., $[S] \gg K_M$, and the reaction is zeroth order.

Under conditions of dominant pore diffusion ($\eta = 1/\varphi \ll 1$), the apparent (but not the actual!) half-life is always longer than that measured with native enzyme (where $\eta = 1$). The result depends on the reaction order, initial degree of conversion, and operational policy.

Figure 5.12 elucidates the contents of Table 5.4.

Table 5.4 Comparability of apparent and actual half-lives (Yamane, 1987).

Kinetics, Diffusion	*Continuous stirred tank reactor (CSTR)*		*Plug-flow reactor (PFR)*	
	Q = const.	x = const.	Q = const.	x = const.
Reaction order n arbitrary, $\eta = 1$	$X(t)/x(0) = E(t)/E(0)$ with $n = 0$; if $n = 1$: $x(t)(1-x(0))/[(1-x(t)) \cdot x(0)] = E(t)/E(0)$	$F(t)/F(0) = E(t)/E(0)$	$x(t)/x(0) = E(t)/E(0)$ with $n = 0$; if $n = 1$: $\ln(1-x(t))/\ln(1-x(0)) = E(t)/E(0)$	$F(t)/F(0) = E(t)/E(0)$
1st order, $\eta \ll 1$	$X(t)(1-x(0))/[(1-x(t)) \cdot x(0)] = [E(t)/E(0)]^{0.5}$, if $\phi \to \infty$; if $\phi \to 0$: $x(t)(1-x(0))/[(1-x(t)) \cdot x(0)] = E(t)/E(0)$	$F(t)/F(0) = [E(t)/E(0)]^{0.5}$, if $\phi \to \infty$; if $\phi \to 0$: $F(t)/F(0) = E(t)/E(0)$	$\ln(1-x(t))/\ln(1-x(0)) = E(t)/E(0)$ with $\phi \to 0$; $\ln(1-x(t))/\ln(1-x(0) = [E(t)/E(0)]^{0.5}$with $\phi \to \infty$	$F(t)/F(0) = [E(t)/E(0)]^{0.5}$, if $\phi \to \infty$; if $\phi \to 0$: $F(t)/F(0) = E(t)/E(0)$
Order $n \neq 1$, $\eta \ll 1$		With $n = 0$: if $\phi \leq 6^{0.5}$, $F(t)/F(0) \to E(t)/E(0)$; if $\phi \geq 6^{0.5}$, $F(t)/F(0) \to [E(t)/E(0)]^{0.5}$		With $n = 0$: if $\phi \leq 6^{0.5}$, $F(t)/F(0) \to E(t)/E(0)$; if $\phi \geq 6^{0.5}$, $F(t)/F(0) \to [E(t)/E(0)]^{0.5}$

Flow rate Q = const. ≡ constant flow rate policy;
degree of conversion x = const. ≡ constant conversion policy.

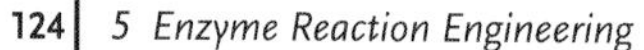

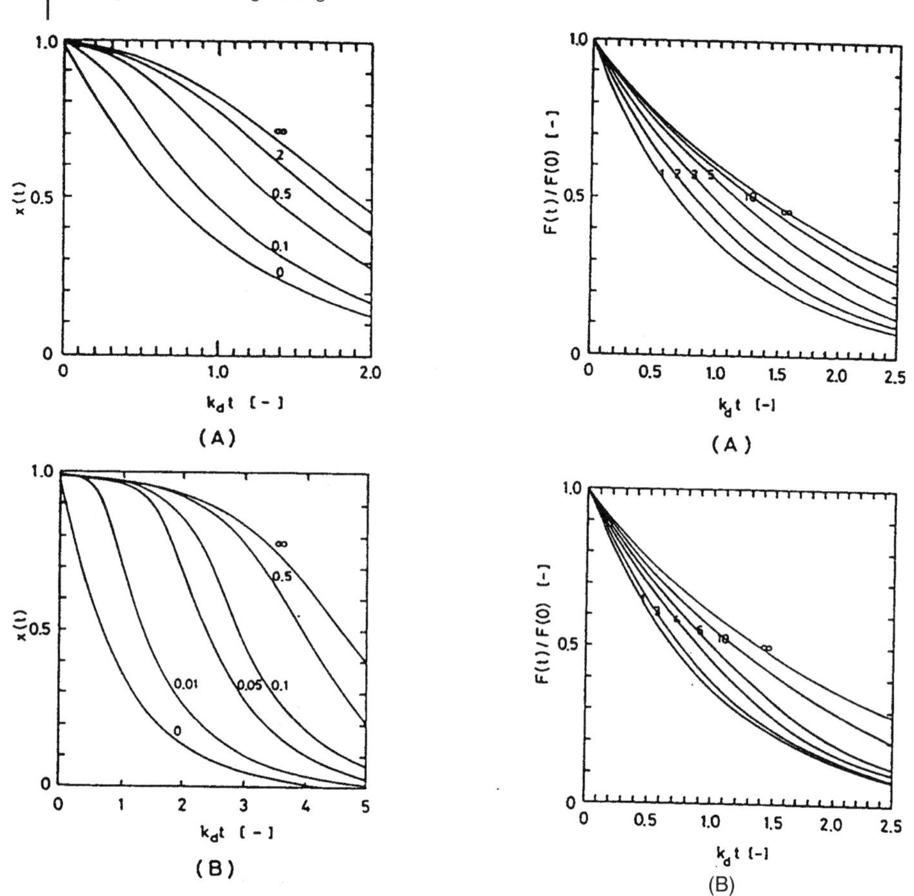

Figure 5.12 (A): PFR; (B): CSTR. Left: Profiles of decrease in $x(t)$ for Michaelis–Menten kinetics if flow rate F = const., $E(t)/E(0) = \exp(-k_d \cdot t)$, $\eta = 1$, and $x(0) = 0.99$; the parameter is K_M. Right: Profiles of decrease in $F(t)/F(0)$ for intraparticle diffusion-influenced zero-order reaction with spherical immobilized enzyme particles packed in the reactor operated under a constant conversion policy ($x = 0.99$). Enzyme activity decays as $E(t)/E(0) = \exp(-k_d \cdot t)$. Curves of $\varphi_0(0) = 1$ and $\varphi_0(0) = \infty$ correspond to $F(t)/F(0) = \exp(-k_d \cdot t)$ and $F(t)/F(0) = \exp(-k_d \cdot t/2)$, respectively.

5.6.4
Strategy for the Addition of Fresh Enzyme to Deactiving Enzyme in Continuous Reactors

In industrial reactors in most cases enzyme deactivation is important over the relevant time horizon of the reactor operating life, of the campaign, or the catalyst batch. Therefore, many attempts have been made to come to grips with the problem, especially in continuous reactors, or at least to alleviate its effects, by various addition strategies.

For the optimal strategy of maintaining operational stability, Lee et al. have calculated the optimal profile of addition of fresh, non-deactivated enzyme into a CSTR under different deactivation kinetics. If a CSTR is charged initially with an amount of enzyme of initial activity N_0, at time t under deactivation, the amount remaining is given by Eq. (5.83), where $k(t)$ denotes an arbitrary deactivation function (J. Y. Lee, 1990).

$$N(t) = N_0 \cdot k(t) \tag{5.83}$$

If enzyme is added continuously so that the initial activity N_0 is maintained, then enzyme molecules of different ages and with a specific age distribution are in the reactor. At time t' the added portion contributes the fraction of total activity $a_0 \cdot [\mathrm{E}](t') \cdot \mathrm{d}t'$, then at an arbitrary later point in time the total activity is $k(t - t') \cdot a_0 \cdot [\mathrm{E}](t') \cdot \mathrm{d}t'$. Then, $[\mathrm{E}](t)$ is given by Eq. (5.84).

$$N_0 \cdot k(t) + \int_0^t a_0[\mathrm{E}](t') \cdot k(t - t') \cdot \mathrm{d}t' = N_0 \tag{5.84}$$

With Eq. (5.83), Eq. (5.84) can then be rewritten as Eq. (5.85).

$$N_i(t) + (a_0 / N_0) \cdot \int_0^t [\mathrm{E}](t') \cdot N_i(t - t') \cdot \mathrm{d}t' = N_0 \tag{5.85}$$

The integral can be solved by Laplace transformation. With a given deactivation kinetics $N_i(t)$ is then fixed; at inversion of the solution of Eq. (5.85), the necessary addition profile is given by Eq. (5.86).

$$\hat{E}(s) = (N_0/a_0) \cdot (N_0 - s \cdot \grave{O}_i(s))/(s \cdot \grave{O}_i(s)) \tag{5.86}$$

In the case of first-order deactivation, $\mathrm{d}N/\mathrm{d}t = k_\mathrm{d} \cdot N$, one obtains Eq. (5.87) for $[\mathrm{E}](t)$.

$$[\mathrm{E}](t) = (N_0/a_0) \cdot k_\mathrm{d} \tag{5.87}$$

In the case of sequential deactivation, $\mathrm{N} \rightarrow \mathrm{N}^* \rightarrow \mathrm{D}$, with the deactivation kinetics $\mathrm{d}N/\mathrm{d}t = k_{\mathrm{d},1} \cdot N$ and $\mathrm{d}N^*/\mathrm{d}t = k_{\mathrm{d},1} \cdot N - k_{\mathrm{d},2} \cdot N^*$, one finally obtains Eq. (5.88) for $[\mathrm{E}](t)$.

$$[\mathrm{E}](t) = (N_0/a_0) \cdot [k_{\mathrm{d},1} \cdot \mathrm{k}_{\mathrm{d},2} \cdot /(k_{\mathrm{d},1} + k_{\mathrm{d},2})] \cdot \{1 - \exp - (k_{\mathrm{d},1} + k_{\mathrm{d},2}) \cdot t\} \tag{5.88}$$

For the optimum measurement strategy of operational stability, Lee et al. have calculated the time-dependent profile of addition of fresh, unactivated enzyme into a CSTR at different deactivation kinetics (Table 5.5).

Table 5.5 Time-dependent profile of addition of fresh, unactivated enzyme into a CSTR at different deactivation kinetics (J. Y. Lee, 1990).

Deactivation kinetics	*Time profile*
Single first-order [Eq. (5.52)]	$E(t) = (N_0/a_0)\ k_d$
Sequential [Eq. (5.88)]	$E(t) = (N_0/a_0)\ (k_1k_2/(k_1 + k_2))\ \{1 - \exp - [(k_1 + k_2)t]\}$

The results demonstrate that under conditions of deactivation according to the first-order law a constant amount of enzyme per unit time has to be added, whereas under conditions of sequential two-step first-order deactivation the amount of enzyme added has to be increased steadily.

5.7 Enzymes with Incomplete Selectivity: *E*-Value and its Optimization

5.7.1 Derivation of the E-Value

With the assumptions that A is the faster-reacting and B the slower-reacting enantiomer, that they both compete for the same active center of the enzyme, and that the reactions are irreversible and no product inhibition occurs, the ratio of the respective reaction rates v_A and v_B of the partial reactions (Scheme 5.6) can be written as in Eq. (5.89).

$$E + A \Leftrightarrow EA \rightarrow EP \rightarrow E + P$$

$$E + B \Leftrightarrow EB \rightarrow EQ \rightarrow E + Q$$

Scheme 5.6 Partial reactions of enantioners A (slow) and B (fast) with enzyme E: model for derivation of *E* value.

$$v_A/v_B = (v_{max,A}/K_{MA}) \cdot [A]/(v_{max,B}/K_{MB}) \cdot [B] \tag{5.89}$$

Integration of Eq. (5.89) yields Eq. (5.90).

$$\frac{\ln([A]/[A_0])}{\ln([B]/[B_0])} = \frac{v_{max,A}/K_{MA}}{v_{max,B}/K_{MB}} = E \tag{5.90}$$

The chiral discrimination between the enantiomers A and B is determined by the ratio of specificity constants $v_{max,i}/K_i$ or also $(k_{cat}/K_M)_i$, which is called the *enantiomeric ratio* or *E value*.

A different derivation starts with the total Gibbs free enthalpy of activation $\Delta G_T^{\ddagger}$ between substrate and enzyme, which consists of the energetically negative (i.e., advantageous) enthalpy of binding ΔG_S and the energetically positive (i.e., disadvantageous) activation barrier $\Delta G^{\ddagger}$ (Eq. (5.91).

$$\Delta G_T^{\ddagger\prime}\ \Delta G_S + \Delta G^{\ddagger} \tag{5.91}$$

According to the theory of transition states Eqs. (5.92) and hence (5.93) hold:

$$RT \cdot \ln k_{cat}/K_M = -\Delta G_T^{\ddagger} \tag{5.92}$$

$$RT \cdot \ln k_{cat}/K_M = RT \cdot \ln kT/h - \Delta G_S - \Delta G^{\ddagger} \tag{5.93}$$

and in the case of two different substrates A and B, Eq. (5.94) or (5.95):

$$RT \cdot (\ln k_{cat}/K_M)_A = RT \cdot (\ln k_{cat}/K_M)_B - (\Delta G_A^{\ddagger} - \Delta G_B^{\ddagger}) \tag{5.94}$$

or

$$RT \cdot \ln \{k_{cat}/K_M)_A\}/\{k_{cat}/K_M)_B\} = -\Delta\Delta G^{\ddagger} \tag{5.95}$$

and also, in analogy to Eq. (5.90), Eq. (5.96), from which Eq. (5.97) follows.

$$E = \{k_{cat}/K_M)_A\}/\{k_{cat}/K_M)_B\} = \exp(-\Delta\Delta G^{\ddagger}/RT) \tag{5.96}$$

$$-\Delta\Delta G^{\ddagger} = -RT \cdot \ln E \tag{5.97}$$

Additional important relationships are gained by defining the degree of conversion x and the enantiomeric excess of substrate and product, respectively, e.e.$_S$ and e.e.$_P$ [Eqs. (5.98)].

$$x = 1 - \frac{[A] + [B]}{[A]_0 + [B]_0} \tag{5.98a}$$

$$\text{e.e.}_S = \frac{[B] - [A]}{[A] + [B]} \tag{5.98b}$$

$$\text{e.e.}_P = \frac{[P] - [Q]}{[P] + [Q]} \tag{5.98c}$$

$$x = \text{e.e.}_S/(\text{e.e.}_S + \text{e.e.}_P) \tag{5.98d}$$

For a racemate e.e. = 0, whereas an enantiomerically pure component features an e.e. value of 1. If the separation of the racemate is achieved by reacting just one of the enantiomers while leaving the other enantiomer untouched, the relationships between the enantiomeric ratio E and the degree of conversion x and either e.e.$_S$ or e.e.$_P$ can be expressed by Eqs. (5.99) and (5.100).

$$\frac{\ln\{(1-x)(1-\text{e.e.}_S)\}}{\ln\{(1-x)(1+\text{e.e.}_S)\}} = E \tag{5.99}$$

$$\frac{\ln\{1-x(1+\text{e.e.}_P)\}}{\ln\{1-x(1-\text{e.e.}_P)\}} = E \tag{5.100}$$

Of course, one can also eliminate the degree of conversion x and express E in terms of e.e._S and e.e._P values:

$$\frac{\ln\{(1-\text{e.e.}_S)/(1+\text{e.e.}_S/\text{e.e.}_P)\}}{\ln\{(1+\text{e.e.}_S)/(1+\text{e.e.}_S/\text{e.e.}_P)\}} = E \tag{5.101}$$

It is important to repeat the assumptions for the derivation of Eqs. (5.99)–(5.101):
1. the separation of racemates is carried out in a homogeneous batch reactor;
2. the reaction can be described as a (pseudo) uni–uni-reaction;
3. the reaction must obey a (pseudo) Michaelis–Menten mechanism; and
4. the reaction is irreversible.

5.7.2 Optimization of Separation of Racemates by Choice of Degree of Conversion

5.7.2.1 Optimization of an Irreversible Reaction

The utility of Eqs. (5.99)–(5.101) becomes clear when either e.e._S or e.e._P is plotted against the degree of conversion x, with different E values as the parameter (Figure 5.13).

With increasing degree of conversion x the enantiomeric purity of the remaining substrates increases as well. Once the desired enantiomeric purity is reached

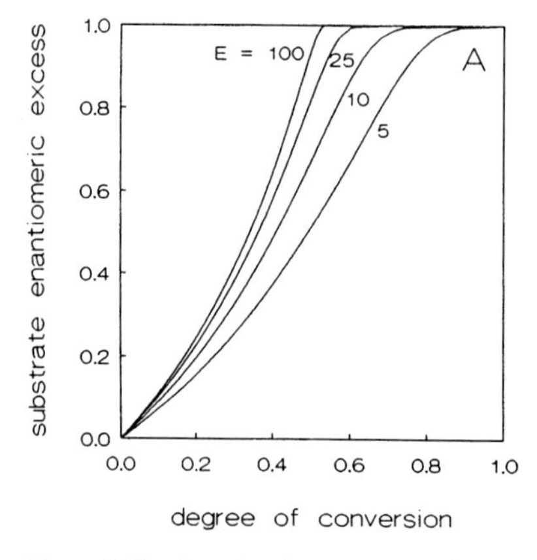

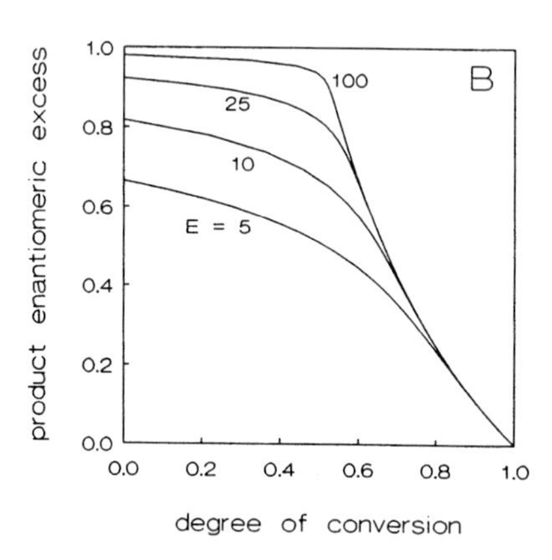

Figure 5.13 Enantioselectivity E as a function of degree of conversion x for substrate (left) and product (right) (Rakels, 1993).

(in most cases, at about 99% e.e.) the reaction can be stopped and the remaining substrate recovered. The higher the enantioselectivity of the reaction, the more substrate can be recovered, but this, in the best case, is only 50% of the total starting amount of substrate. For the product, however, the situation appears decidedly less promising: only at very high enantioselectivity of the reaction can a sufficiently great amount of product be recovered. At degrees of conversion above 40–50% the enantioselectivity is decreased drastically in any case.

5.7.2.2 Enantioselectivity of an Equilibrium Reaction

When considering a reversible reaction which runs to equilibrium with an equilibrium constant K (relaxing condition 4 at the end of Section 5.7.1), Eq. (5.102) can be written for the E value.

$$\frac{\ln\{1-(1+1/K)\,(x+\mathrm{e.e.}_S(1-x))\}}{\ln\{a-(1+1/K)\,(x-\mathrm{e.e.}_S(1-x))\}}=E \tag{5.102}$$

In the case of a reversible reaction, $K = \infty$ and Eq. (5.101) is recovered. Graphically the relationship between $\mathrm{e.e.}_S$ and $\mathrm{e.e.}_P$ can be depicted as in Figure 5.14.

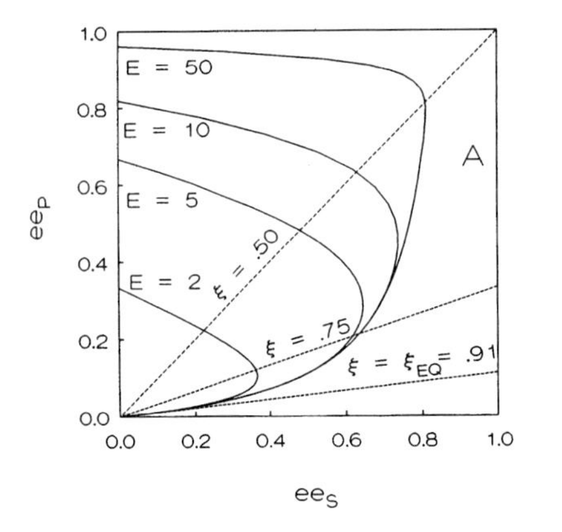

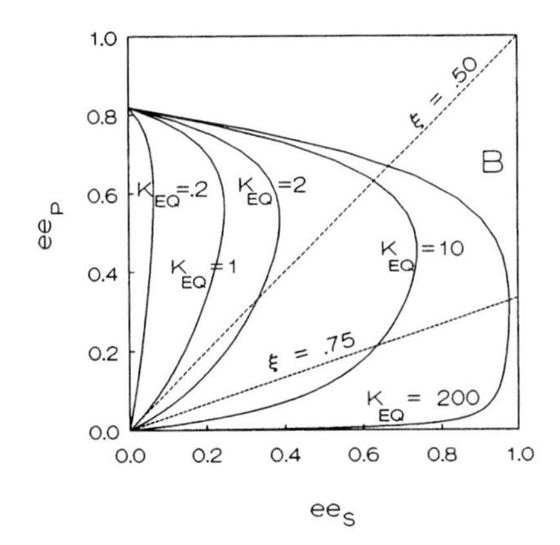

Figure 5.14 Enantioselectivity as a function of degree of conversion x and equilibrium constant (Rakels, 1993).
Left: parameter enantiomeric ratio E at different extents of reaction ξ; right: parameter equilibrium constant K at different extents of reaction ξ.

Whereas a high value for $\mathrm{e.e.}_S$ can always be achieved with irreversible reactions (maybe only in case of high degree of conversion), this is not possible in the case of reversible reactions. However, an optimal value for $\mathrm{e.e.}_S$ can be found which depends on E and K. As expected, the maximum $\mathrm{e.e.}_P$ value is reached at a zero degree of conversion and does not depend on K: at a degree of conversion of zero, the reaction is far from equilibrium.

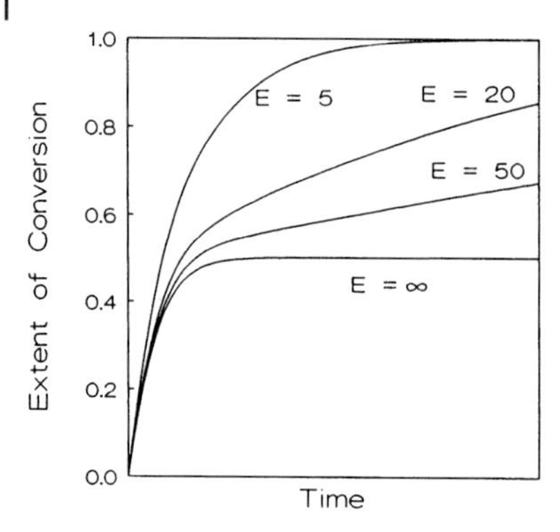

Figure 5.15 Conversion–time profiles at different enantioselectivities (Rakels, 1993).

5.7.2.3 **Determination of Enantiomeric Purity from a Conversion–Time Plot**

If the degree of conversion of a separation of racemate is followed over time, different conversion–time profiles, depending on enantioselectivity, can be measured as a function of time (Figure 5.15).

Figure 5.15 reveals that at $E = \infty$ the conversion–time profile bends sharply at a conversion of 50% and stays constant after that point. At lower E values, however, the discontinuity in the profile is not as discernible and the final degree of conversion is much higher than 50%. If only the racemate is available but none of the enantiomers, and thus e.e.$_S$ and/or e.e.$_P$ cannot be measured, the determination of the E value of the reaction can only be performed by numerical fitting of the conversion–time profile.

5.7.3 Optimization of Enantiomeric Ratio E by Choice of Temperature

5.7.3.1 **Derivation of the Isoinversion Temperature**

Many complex reactions consisting of several elementary steps feature a strongly temperature-sensitive overall selectivity as well as an inversion point with a maximum or minimum selectivity parameter. However, the empirical rule that stereoselective reactions should be performed at the lowest possible temperatures to achieve the highest selectivities is not always followed. Instead, the competition of enthalpy and entropy determines the overall selectivity, depending on the temperature range.

For separation of racemates, Eq. (5.103) holds, recounting the definition of the E value.

$$E = \{\ln([\mathrm{R}]/[\mathrm{R}_0])\}/\{\ln([\mathrm{S}]/[\mathrm{S}_0])\} = (k_{cat}/K_{\mathrm{M}})_{\mathrm{R}}/(k_{cat}/K_{\mathrm{M}})_{\mathrm{S}} \tag{5.103}$$

The difference of the Gibbs free enthalpy of activation for the enantiomers R and S, $\Delta\Delta G^{\ddagger} = \Delta G^{\ddagger}{}_R - \Delta G^{\ddagger}{}_S$, can thus be expressed according to Eq. (5.104).

$$\Delta\Delta G^{\ddagger} = \Delta G^{\ddagger}{}_R - \Delta G^{\ddagger}{}_S = -RT \cdot \ln E \tag{5.104}$$

$\Delta\Delta G^{\ddagger}$ is also defined by Eq. (5.105).

$$\Delta\Delta G^{\ddagger} = \Delta\Delta H^{\ddagger} - T \cdot \Delta\Delta S^{\ddagger} \tag{5.105}$$

If at a certain temperature no enantioselectivity can be observed, $E = 1$ and thus according to Eq. (5.104) $\Delta\Delta G^{\ddagger} = 0$. The inversion temperature T_i can then be calculated from Eq. (5.105) to read.

$$T_i = \Delta\Delta H^{\ddagger} / \Delta\Delta S^{\ddagger} \tag{5.106}$$

5.7.3.2 **Example of Optimization of Enantioselectivity by Choice of Temperature**

The results of the temperature dependence of the reaction rates of the enantiomers of secondary alcohols with a secondary alcohol dehydrogenase (SADH) from the thermophilic bacterium *Thermoanaerobacter ethanolicus* demonstrated a temperature-dependent reversal of stereospecificity (Pham, 1990) (Figure 5.16). At $T < 26$°C, (*S*)-2-butanol was a better substrate than (*R*)-2-butanol on the basis of k_{cat}/K_M values; however, at $T > 26$°C, (*R*)-2-butanol was a better substrate than (*S*)-2-butanol. (*S*)-2-Pentanol was the preferred substrate at $T < 60$°C; however, the data predict that (*R*)-2-pentanol would be preferred at $T > 70$°C. (*S*)-2-Hexanol was predicted to be the preferred enantiomer only at $T > 240$°C. Therefore, the concept of iso-inversion temperature is as valid for enzyme reactions as for others; only the range of catalytically accessible temperatures is smaller.

A mutant enzyme, C295A, shows no significant temperature dependence of stereospecificity, in contrast to the wild-type enzyme, and also demonstrates an increased ability to bind (*R*)-alcohols (Heiss, 2001). The data suggest that removal of

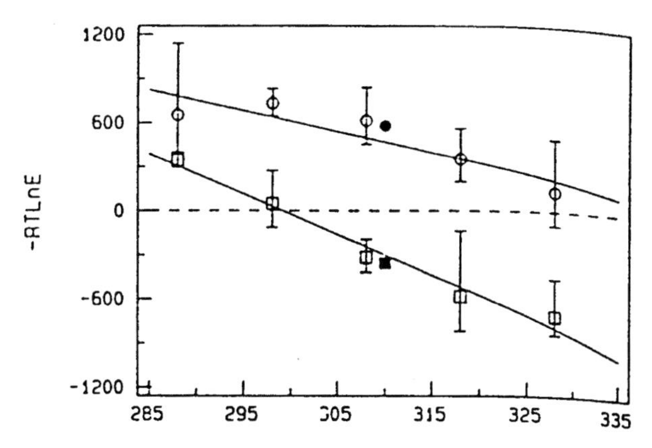

Figure 5.16 Enantioselectivity of the redox reaction of ADH from *T. ethanolicus* with different alcohols (Pham, 1990). Temperature dependence of free energy of activation differences for 2-butanol and 2-pentanol: open squares: 2-butanol; open circles: 2-pentanol; filled square: reduction of 2-butanone; filled circle: reduction of 2-pentanone.

the bulky Cys-295 sulfhydryl group from the small binding pocket in SADH increases the activation entropy for the (*R*)-enantiomers with large alkyl groups. The most likely explanation of this entropic behavior is a bound water molecule in the binding pocket of the wild type enzyme that must be displaced to bind (*R*)-alcohols, whereas (*R*)-alcohols bind more easily to the mutant enzyme, as the structured water is eliminated or disordered. Thus, the stereospecificity of SADH may be partly determined by the expulsion of bound water from the small alkyl binding pocket upon binding of (*R*)-alcohols.

Suggested Further Reading

H. W. Blanch and D. S. Clark, *Biochemical Engineering*, Marcel Dekker; New York, **1997**.

P. C. Engel (ed.), *Enzymology Labfax* (Labfax Series), Academic Press, **1996**.

H. S. Fogler, *Elements of Chemical Reaction Engineering*, Prentice Hall PTR, 3rd edition, **1998**.

G. G. Hammes, *Thermodynamics and Kinetics for the Biological Sciences*, John Wiley, **2000**.

O. Levenspiel, *Chemical Reaction Engineering*, John Wiley, ,3rd edition, **1999**.

D. L. Nelson, Albert L. Lehninger, and M. M. Cox, *Principles of Biochemistry*, Worth Publishing; 3rd edition, **2000**.

L. Stryer, *Biochemistry*, W H Freeman, 4th edition, **1995**.

References

R. Aris, Shape factors for irregular particles. I. The steady-state problem. Diffusion and reaction, *Chem. Eng. Sci.* **1957**, *6*, 262–268.

R. Aris, A normalization for the Thiele modulus, *Ind. Eng. Chem. Fundamentals* **1965**, *4*, 227–229.

R. Aris, *Elementary Chemical Reactor Analysis*, Prentice-Hall, Englewood Cliffs, NJ, **1969**.

A. S. Bommarius, K. Drauz, V. Grolger, and C. Wandrey, Membrane bioreactors for the production of enantiomerically pure α-amino acids, in: *Chirality in Industry*, eds.: A. N. Collins, G. N. Sheldrake, and J. Crosby, Wiley & Sons Ltd., London, New York, **1992**, Chapter 20, 371–397.

A. S. Bommarius, Biotransformations and enzyme reactors, in: *Bioprocessing*, G. N. Stephanopoulos (ed.), Vol. 3 in *Biotechnology* Series, H.-J. Rehm and G. Reed (series eds.), VCH, Weinheim, 2nd edition, **1993**, Chapter 17, pp. 427–466.

A. S. Bommarius, M. Schwarm and K. Drauz, The membrane reactor for the acylase process – from laboratory results to commercial scale, *Chim. Oggi* **1996**, *14(10)*, 61–64.

H. Buschmann, H.-D. Scharf, N. Hoffmann, and P. Esser, The isoinversion principle. A general selection model in chemistry, *Angew. Chem.* **1991**, *103*, 480–518; *Angew. Chem, Int. Ed. Engl.* **1991**, *30*, 477–515.

T. M. S. Chang, Methods for the therapeutic applications of immobilized enzymes, *Meth. Enzymol.* **1976**, *44*, 676–698.

C. S. Chen and C. J. Sih, Enantioselective biocatalysis in organic solvents. Lipase catalyzed reactions, *Angew. Chem.* **1989**, *101*, 711–724.

Y.-C. Cheng and W. H. Prusoff, Relationship between the inhibition constant (K_I) and the concentration of inhibitor which causes 50 per cent inhibition (I_{50}) of an Enzymatic reaction, *Biochem. Pharmacol* **1973**, *22*, 3099–3108.

R. A. Copeland, *Enzymes – A Practical Introduction to Structure, Mechanism, and Data Analysis*, Wiley-VCH, New York, **1996**, Chapter 9.1, p. 241.

R. A. Copeland, D. Lombardo, J. Giannaras, and C. P. Decicco, Estimating K_I values for tight binding inhibitors from dose–response plots, *Bioorg. Med. Chem.* **1995a**, 5, 1947–1952.

R. A. Copeland, J. M. Williams, N. L. Rider, D. E. Van Dyk, J. Giannaras, S. Nurnberg, M. Covington, D. Pinto, R. L. Magolda, and J. M. Trzaskos, Selective time dependent inhibition of cyclooxygenase-2, *Med. Chem. Res.* **1995b**, 5, 384–393.

P. A. Dickensheets, L. F. Chen, and G. T. Tsao, Characteristics of yeast invertase immobilized on porous cellulose beads, *Biotechnol. Bioeng.* **1977**, *19*, 365–375.

C. Heiss, M. Laivenieks, G. J. Zeikus, and R. S. Phillips, The stereospecificity of secondary alcohol dehydrogenase from *Thermoanaerobacter ethanolicus* is partially determined by active site water, *J. Am. Chem. Soc.* **2001**, *123*, 345–346.

W. E. Hornby, M. D. Lilly, and E. M. Crook, Some changes in the reactivity of enzymes resulting from their chemical attachment to water-insoluble derivatives of cellulose, *Biochem J.* **1968**, *107*, 669–674.

N. G. Karanth and J. E. Bailey, Diffusional influences on the parametric sensitivity of immobilized enzyme catalysts, *Biotechnol. Bioeng.* **1978**, *20*, 1817–1831.

T. Kobayashi and M. Moo-Young, Kinetic and mass transfer behavior of immobilized invertase on ion-exchange resin beads, *Biotechnol. Bioeng.* **1973**, *15*, 47–67.

Kragl, **1992**

R. Langer, H. Bernstein, L. Brown, and L. Cima, Medical reactors, *Chem. Eng. Sci.* **1990**, *45*, 1967–1978.

J. Y. Lee, A. Velayudhan, and M. R. Ladisch, Maintaining constant enzyme activity in a continuous flow reactor, *Chem. Eng. J.* **1990**, *45*, B1–B4.

L. G. Lee and G. M. Whitesides, Preparation of optically active 1,2-diols and α-hydroxy ketones using glycerol dehydrogenase as catalyst: limits to enzyme-catalyzed synthesis due to noncompetitive and mixed inhibition by product, *J. Org. Chem.* **1986**, *51*, 25–36.

O. Levenspiel, in: *Chemical Reaction Engineering*, John Wiley, New York, 3rd edition, **1999**, Chapter 18.

M. D. Lilly and A. K. Sharp, Kinetics of enzymes attached to water-insoluble polymers, *Chem. Engineer,* **1968**, *21k*, CE12–CE18.

M. D. Lilly and P. Dunnill, Immobilized-enzyme reactors, *Meth. Enzymol.* **1976**, *44*, 717–738.

J. F. Morrison, The slow-binding and slow, tight-binding inhibition of enzyme-catalyzed reactions, *Trends Biochem. Sci.* **1982**, *7*, 102–105.

V. T. Pham and R. S. Phillips, Effects of substrate structure and temperature on the stereospecificity of secondary alcohol dehydrogenase from *Thermoanaerobacter ethanolicus, J. Am. Chem. Soc.* **1990**, *112*, 3629–3632.

R. S. Phillips, Temperature effects on stereochemistry of enzymatic reactions, *Enzyme Microb. Technol.* **1992**, *14*, 417–419.

J. L. Rakels, A. J. Straathof, and J. J. Heijnen, A simple method to determine the enantiomeric ratio in enantioselective biocatalysis, *Enzyme Microb. Technol.* **1993**, 15, 1051–1056.

D. L. Regan, M. D. Lilly, and P. Dunnill, Influence of intraparticle diffusional limitation on the observed kinetics of immobilized enzymes and on catalyst design, *Biotechnol. Bioeng.* **1974**, *16*, 1081–1093.

M. L. Schuler, R. Aris, and H. M. Tsuchiya, Diffusive and electrostatic effects with insolubilized enzymes, *J. Theor. Biol.* **1972**, *35*, 67–76.

J. Tramper, F. Muller, and H. C. van der Plas, Immobilized xanthine oxidase: kinetics, (in)stability, and stabilization by coimmobilization with superoxide dismutase and catalase, *Biotechnol. Bioeng.* **1978**, *20*, 1507–1522.

W. R. Vieth, K. Venkatasubramanian, A. Constantinides and B. Davidson, Design and analysis of immobilized-enzyme flow reactors, *Appl. Biochem. Bioeng.* **1976**, *1*, 221–327.

C. Walsh, Suicide substrates: mechanism-based enzyme inactivators, *Tetrahedron* **1982**, *38*, 871–909.

Z.-X. Wang, A simple method for determining kinetic constants of slow, tight-binding inhibtion, *Anal. Biochem.* **1993**, *213*, 370–377.

S. R. Weijers and K. van't Riet, Enzyme stability in downstream processing Part 1: Enzyme inactivation, stability and stabilization, *Biotech. Adv.* **1992a**, *10*, 237–249; Part 2: Quantification of inactivation, 251–273.

S. R. Weijers and K. van't Riet, Enzyme stability in downstream processing Part 2: Quantification of inactivation, *Biotech. Adv.* **1992b**, *10*, 251–273.

P. B. Weisz and C. D. Prater, Interpretation of measurements in experimental catalysis, *Adv. Catal.* **1954**, *6*, 143–196.

J. W. Williams and J. F. Morrison, The kinetics of reversible tight-binding inhibition, *Methods Enzymol.* **1979**, *63*, 437–467.

J. Woltinger, K. Drauz, and A. S. Bommarius, The membrane reactor in the fine chemicals industry, *Appl. Catal. A: General* **2001**, *221*, 171–185.

T. Yamane, P. Siricote, and S. Shimizu, Evaluation of half-life of immobilized enzyme during continuous reaction in bioreactors: a theoretical study, *Biotech. Bioeng.* **1987**, *30*, 963–969.

6
Applications of Enzymes as Bulk Actives: Detergents, Textiles, Pulp and Paper, Animal Feed

Summary

Proteases against egg and blood stains as well as cellulases against grass, lipases against grease stains, and amylases against starchy residues are standard in most detergent formulations today. Laundry detergents offer several particularly hard challenges for an enzyme, such as elevated temperatures of 40–80 °C in alkaline pH values of 9–12, large amounts of mostly non-ionic surfactants, strongly alkaline soaps, and oxidizing compounds.

The application of enzymes in textile finishing becomes more and more important. Enzymes offer an ecologically benign alternative to common textile chemical processes, and totally new possibilities and chances to the textile finisher. Common applications of enzymes in the textile industry include improvement of softness, shine, and smoothness (biofinishing); creation of washout effects on blue denim through use of cellulases instead of pumice stones.

Driven by market and environmental demands for less chlorinated products and by-products, the pulp and paper industry is one of the fastest-growing markets for industrial enzymes. Today, the pulp and paper industry operates two pulping processes, the mechanical and the Kraft pulping process, and three bleaching processes: (i) the chlorine bleaching process; (ii) the elementary Cl_2-free (ECF) process; and (iii) the totally Cl_2-free (TCF) process. While Cl_2 bleaches most effectively and dissolves all lignin without attacking the cellulose, chlorinated lignin is very persistent in nature and thus is a major water and air pollutant. Using sodium chlorate and process modifications, the ECF process achieves the same quality of paper without the environmental drawbacks. The TCF process uses H_2O_2 but uses 10% more wood and generates shorter fibers.

Phytase offers significant promise as a means to reduce phosphorus levels in animal waste by 30–35%, while also reducing the cost of phosphorus supplementation. The enzyme hydrolyzes phytate (*myo*-inositol hexakisphosphate), the primary storage form of phosphorus in plant seeds and pollen, in several steps into inositol and inorganic phosphorus, which is readily bioavailable to the farm animals. Phytases can also have non-specific phosphorus monoester activity. Addition of phytases to farm animals' diets significantly enhances bioavailability of plant phosphorus for the animals while reducing phosphorus in the waste and simultaneously allowing a reduction of total phosphorus in the feed; 500 units of phytase per kg feed are equivalent to 0.76% crude protein. The nitrogen content of the feces is estimated to be reduced by 5% with this level of phytase added.

6.1 Application of Enzymes in Laundry Detergents

6.1.1 Overview

If viewed on the basis of tonnage, by far the main application of enzymes today is still in the laundry detergent sector, about 30–40% of the total. Around 30 years ago, proteases were first added to detergents to remove recalcitrant stains such as blood and egg protein from laundry. Not only is this application standard in most detergent formulations today but several other enzymes have been added over the years, such as cellulases in 1983 against grass, lipases in 1988 against grease stains, and amylases against starchy residues.

The application of enzymes, however, is not novel: already in 1913, Otto Röhm, founder of what would become the chemical giant Rohm and Haas, had added a proteolytic enzyme, mainly trypsin, from milled animal pancreas and dog waste to detergents. Marketed in Germany in 1914 as "Burnus", this innovation was way ahead of its time and ran into several problems: First, the very crude proteolytic enzyme contained many impurities which sometimes stained the very textile it was supposed to clean. Second, one tablet could be added to 10 liters of water to remove stains from clothes. Unfortunately, consumers were used to bulky washing powders which lathered a lot, and they did not believe that such a small tablet could work. Third, the process of enzyme extraction was not economical enough to include it in routine household detergents.

Laundry detergents offer several particularly hard challenges for an enzyme. Washing liquids often operate at elevated temperature of 40–80 °C (although the industry trend favors lower temperatures) in alkaline pH values of 9–12 in the presence of inhibitory or deactivating compounds. As may be discerned from Table 6.1, detergent preparations are routinely formulated with large amounts of mostly nonionic surfactants to allow water to contact hydrophobic dirt, strongly alkaline soaps to enhance washing effect, and oxidizing compounds such as hypochlorite or alternatively, in Europe, perborate or percarbonate, both slow H_2O_2 release agents, as well as builders (such as sodium tripolyphosphate) to counteract water hardness and prevent scaling. All of these compounds can easily deactivate enzymes. In addition, formulations containing bleach contain peroxides which work on enzymes even faster. The feat of stabilizing enzymes in such an environment is no small one.

Laundry detergents are available in granulated form for use in detergent powders, as liquid preparations in aqueous solution, and as liquid concentrates in "spotting" detergents, used for removing stubborn stains. Recent trends towards liquid detergents over granulated formulations raise the bar even further as enzymes in liquid environments tend to deactivate more rapidly than in dry ones.

Table 6.1 Composition of an enzyme detergent.

Constituent	*Composition [%]*
Sodium tripolyphosphate (water softener, loosens dirt)*	38.0
Sodium alkane sulfonate (surfactant)	25.0
Sodium perborate tetrahydrate (oxidizing agent)	25.0
Soap (sodium alkane carboxylates)	3.0
Sodium sulfate (filler, water softener)	2.5
Sodium carboxymethylcellulose (dirt-suspending agent)	1.6
Sodium metasilicate (binder, loosens dirt)	1.0
Bacillus protease (3% active)	0.8
Fluorescent brighteners	0.3
Foam-controlling agents	trace
Perfume	trace
Water	to 100%

* A recent trend is to reduce this phosphate content for environmental reasons. It may be replaced by sodium carbonate plus extra protease or by zeolites.

Table 6.2 Laundry detergent enzymes.

Commercial product	*Company*	*Enzyme*	*Organism*	*Function*
Alcalase	Novozymes	mainly subtilisin	*B. licheniformis*	10–65 °C and pH 7–10.5
Esperase	Novozymes		*B. licheniformis,* alkalophilic strain	pH < 12
Savinase	Novozymes		*B. amyloliquifaciens,* alkalophilic strain	pH < 11
Lipolase	Novozymes	lipase	*Humicola*, expr. in *Aspergillus oryzae*	
Maxatase	Gist Brocades	mainly subtilisin	*B. licheniformis*	10–65 °C and pH 7–10.5
Termamyl	Novozymes	α-amylase (same as for syrup)	*B. licheniformis*	Dishwashing, de-starching
Carezyme	Novozymes	cellulase		Removes fuzz on cotton clothes
Guardzyme	Novozymes	peroxidase		inhibits dye transfer
Purafect®, Purafect® L	Genencor	mainly subtilisin	*B. subtilis*	high-pH protease efficiency
Properase®	Genencor	high-pH protease		opt. for low-*T* wash conditions
Purafect®; Purafect® OxAm	Genencor	α-amylase	*B. licheniformis*	dishwashing, de-starching
Puradax®	Genencor	high-pH cellulase		removes fuzz on cotton clothes

Enzymes used in laundary detergents are almost all produced using *Bacillus* enzymes, most commonly *Bacillus subtilis* or *Bacillus licheniformis*, which all secrete the detergent enzymes into the fermentation broth. The market is highly concentrated; two producers cover more than 50% of the market share worldwide: Novozymes (Bagsvaerd, Denmark) and Genencor (Palto Alto/CA, USA). Table 6.2. lists several commercial products with their functions.

6.1.2
Proteases against Blood and Egg Stains

Protease is used in detergents to remove protein-based stains such as blood, mucus, feces, and various foods such as egg and gravy. These substances are almost insoluble and they tend to adhere to textiles and other surfaces. The protease hydrolysis products are peptides which are readily dissolved or dispersed in the washing liquor.

All proteolytic enzymes described are fairly non-specific serine endoproteases, cleaving peptide chains preferentially at the carboxyl side of hydrophobic amino acid residues. The enzymes convert their substrates into small, readily soluble fragments which can be removed easily from fabrics. Only serine protease can be used in detergent formulations, as thiol proteases such as papain would be oxidized by the bleaching agents, acidic proteases are not active at common laundry conditions, and metalloproteases such as thermolysin would lose their metal cofactors because of complexation with the water-softening agents or hydroxyl ions.

The most important enzyme in laundry detergents is subtilisin, which cleaves many peptide bonds, especially hydrophobic ones, and which is active in the desired alkaline pH range between 8 and 10.5, in which most washing cycles occur. The small, compact shape (MW 27.5 kDa; monomer) and the known 3D structure were additional reasons for developing this enzyme further. A protease such as Esperase has a pH activity optimum and also a high temperature optimum of 60–70 °C compatible with the high values of laundry cycle operation. This makes Esperase an effective means of improving the washing efficacy in industrial laundering systems with their relatively short washing time of 3–6 minutes. Industrial laundering systems often face special challenges such as cleaning of textiles heavily soiled with blood, e.g., from hospitals and slaughterhouses.

In the development process at Genencor (Palo Alto/CA, USA), subtilisin was found to be labile to oxidation: the cause was found to be a methionine side chain in position 222 (M222), directly adjacent to the catalytic serine (S221, part of the catalytic triad) (Oxender, 1987). Oxidation of the side-chain sulfur to the sulfoxide led to a 90% activity decrease. As is described in more detail in Chapter 10, Genencor solved the problem by site-directed mutagenesis in position 222; substitution of methionine by all other 19 proteinogenic amino acids yielded positive results in the cases of leucine, serine, and especially alanine, in that the mutated enzyme was stable for the required one hour in a 1 M H_2O_2 solution. This accomplishment stands as the first great success of protein engineering.

6.1.3
Lipases against Grease Stains

Lipase is used in detergent formulations to remove fat-containing stains such as those resulting from frying fats, salad oils, butter, fat-based sauces, soups, human sebum, or certain cosmetics. The enzyme hydrolyzes triglycerides into mono- and diglycerides, glycerol, and free fatty acids, all of which are more soluble than the original fats. In 1988, Novozyme launched Lipolase for the detergent industry – the first commercial enzyme developed by the application of genetic engineering and the first-ever detergent lipase. Lipolase has now been incorporated into a great number of major detergent brands around the world.

A protein-engineered variant (Lipolase Ultra) has since been developed with enhanced performance at low temperatures. At low temperatures, fatty stains become more troublesome to remove. Lipases therefore have an even more valuable role in cool washes. Although it is more expensive to produce, the manufacturer believes it to render more value, especially in low-temperature washes.

6.1.4
Amylases against Grass and Starch Dirt

Amylase facilitates the removal of starch-containing stains such as those from pasta, potato, gravy, chocolate, and baby food. Dried-up starch is difficult to remove from medium- to low temperatures. Amylase adheres to the surface of laundry, acting as an adhesive for other stain components. Starch acts as a kind of glue which binds particulate soil to the surface. Amylase hydrolyzes the starch into dextrins and oligosaccharides; the latter are readily dissolved in the washing liquor and thus successively diminish the stain. Likewise, dried-on food, in particular stains and films from starch-containing foods, may be difficult to remove in a dishwasher. Just like laundry detergents, modern automatic dishwashing detergents (ADDs) usually contain an amylase.

Starch also is very often present on laundry fabrics from the deliberate application of starch during fabric production. Complete removal of starch from cotton fibers is extremely difficult to achieve without detergent amylases.

6.1.5
Cellulases

As far back as 1983, alkali-stable fungal cellulase preparations were introduced into laundry detergent formulations for use in washing cotton fabrics. During use, small fibers are raised from the surface of cotton thread, resulting in a change in the "feel" of the fabric and, particularly, in the lowering of the brightness of colors. Cellulases also prevent pilling (the formation of small balls of fuzz on the fabric surface) (see also Section 6.2, below). Treatment with cellulase removes the small fibers without apparently damaging the major fibers, and restores the fabric to its "as new" condition. The cellulase also aids the cleaning process by decreasing the ability of cellulosic fibers to bind soil, and by the removal of soil particles from the wash by hydrolyzing associated cellulose fibers; this way, it acts as an anti-redeposition agent. Cellulases have been used for the same purpose in the textile industry for the treatment of new garments (see Section 6.2).

6.1.6 Bleach Enzymes

While enzymes can also be thought to support or replace the effect of oxidizing agents in laundry detergents, and development efforts have sought enzymes with this function, no such enzyme has been introduced into detergents yet. Opportunities could exist for enzymes such as glucose oxidase, lipoxygenase, and glycerol oxidase as means of generating hydrogen peroxide in situ. Additionally, added peroxidases may aid the bleaching efficacy of the peroxide in the wash liquor by supporting reaction between oxidizer and stain.

Researchers from Novozyme (Bagsvaerd, Denmark) recently described the directed evolution of a heme peroxidase (CiP), isolated from the inkcap mushroom *Coprinus cinereus*. CiP acts as a dye-transfer inhibitor in laundry detergent by decolorizing free dyes that have leached out of clothing, thereby preventing their reuptake by other garments (Cherry, 1999; Tobin, 1999). The combination of protein variations employed, made by both rational (see Chapter 10) and combinatorial (see Chapter 11) methods to increase the thermal and oxidative stability of CiP, has the potential to improve biocatalysts for a wide variety of applications. The wild-type peroxidase is rapidly inactivated under laundry conditions due to the high pH (10.5), high temperature (50 °C) and high peroxide concentration (5–10 mM). Peroxidase mutants created by site-directed mutagenesis based on structure–function considerations, error-prone PCR to create random mutations, and lastly, two rounds of in-vivo shuffling led to a mutant with 174 times the thermal stability and 100 times the oxidative stability of wild-type CiP.

6.2 Enzymes in the Textile Industry: Stone-washed Denims, Shiny Cotton Surfaces

6.2.1 Build-up and Mode of Action of Enzymes for the Textile Industry

The application of enzymes in textile finishing is becoming more and more important. Enzymes are not only an ecologically benign alternative to common textile chemical processes, they offer totally new possibilities and chances to the textile finisher. Enzymes are also not new in textile finishing. The possibility of using "amylases" for the desizing of starch, for example, has been known since the beginning of the 20th century. The first trials to treat wool with enzymes in order to improve the felting behavior were already started in 1910.

Table 6.3 Enzymes for textile finishing.

Enzyme	*Application*
Amylase	desizing of starch sizes (latest enzyme systems have wider pH and *T* profiles)
Cellulase	biofinishing, stonewashing of pigment and indigo-dyed fabrics, anti-pilling treatment
Peroxidase	antiperoxide (removal of peroxide residues after bleaching and before dyeing)
Proteases	modification of protein fibers (wool), improvement of felting properties, improvement of printability of wool, degumming of silk

In the last few years, further applications for enzymatic processes in textile finishing were developed and – very importantly – the necessary suitable enzymes were found. Large-scale production of such enzymes led to price reductions, which in turn led to more applications, as enzymes were found to be efficacious in the textile industry. Another major reason why enzymatic processes are becoming more popular is the increasing importance of environmentally friendly technologies due to customers' demands or restrictions by law.

Table 6.3 shows the types of enzymes which are already used in textile finishing or which could possibly be used in the near future.

Common applications of enzymes in the textile industry include the following:

- enzymatic singeing (removal of fibers protruding from yarns or fabrics)
- removal of hairiness prior printing to achieve very bright and sharp prints
- removal of pilling (twisting of fibers to form fuzzy ball-like structures)
- improvement of softness, shine, and smoothness (biofinishing)
- creation of washout effects on blue denim with its indigo surface dyed warp yarns
- use of cellulase enzymes for the partly or total exchange of pumice stones and therefore a more gentle treatment for textiles and machines.

6.2.2 Cellulases: the Shinier Look

In recent years, the use of cotton in garments has witnessed a revival. The higher comfort level of cotton fabric as compared to man-made fibers tilts the balance of customer favor towards cotton as the basic fiber in clothes. After many wash cycles, however, both the appearance and the soft feel of the cotton fabric are lost: the surface appears bland and the fabric touch has turned rough. This roughness stems at least in part from loose ends of cotton fibers which have separated from the bulk fabric and stick out on the surface. To combat this phenomenon termed pilling, cellulases have been developed to preserve both the appearance and feel of cotton fabric during washing.

Cellulases belong to the class of hydrolases and accelerate the degradation of cellulose by hydrolysis of the $\beta 1 \rightarrow 4$ glycosidic links. (Note that, in contrast, amylases hydrolyze the $\alpha 1 \rightarrow 4$ glycosidic links of starch.). Most of the cellulase enzymes nowadays used in practice are multi-enzymatic systems, consisting of endoglucanases, cellobiohydrolases, and β-glucosidases (Figure 6.1).

An important parameter influencing the mode of action of cellulases is the accessibility of the cellulose to the enzymes. The molecular weights of cellulases range between 30 and 80 kDa. A comparison of the size of cellulase (3–8 nm) and the pore size of cotton swollen in water (1–7 nm) shows very clearly that cellulases can penetrate the cellulose to a limited extent only. In addition, the enzyme reaction takes place preferentially on amorphous cellulose because the more compact, crystalline cellulose structures do not offer any space for such macromolecules. Thus – provided of enzyme and process parameters have been selected correctly – cellulases act mainly on the textile surface. In this way interesting effects on cellulosic fibers can be achieved.

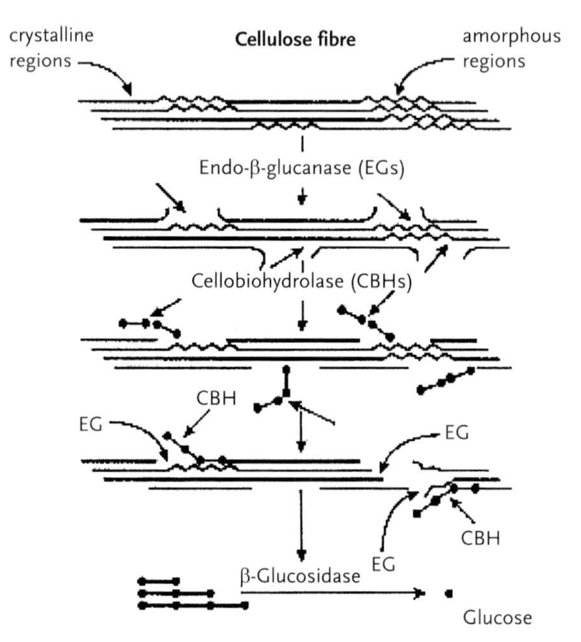

Figure 6.1 Mode of action of cellulase on textile fibers. To degrade long cellulose chains – such as cotton – down to the monomer glucose, many intermediate steps are necessary. Examinations have shown that endoglucanases, cellobiohydrolases and β-glucosidases have a synergistic effect on the degradation of cellulose, leading to both faster and more complete fiber disintegration. With textile applications, however, such synergy effects of multi-enzymatic systems are not always desirable since normally complete disintegration of the cellulosic fiber is not the aim.

An already common process is called "Biofinishing". This process leads not only to a smooth surface (as one can see in the Inset Figure at the end of this chapter), it also improves the quality of the treated fabric. "Sand wash", "silk look" or "peach skin touch" are well known terms for such finishing effects.

6.2.3
Stonewashing: Biostoning of Denim: the Worn Look

Not only have biocatalytic processes (see Chapter 20) been developed for the manufacture of indigo, the dye that gives denim its customary look, but also processes for the partial removal of indigo dye: many customers prefer denim clothing in a state which is attained usually only after several washing cycles, preferably as in the old days with brimstone ("stone-washed"). Instead of brimstone, cellulases are increasingly used which help to loosen the bond of the indigo molecule to the fiber (Figure 6.2); with the new process, clothes and washing machines suffer much less wear and tear and the dust load in laundries is reduced significantly.

In dyeing garments with indigo dye, the reduced indigo dye is applied to cotton yarn in a ball warp or warp sheet arrangement by dipping the clothes through boxes to layer the indigo dye on the outside of the yarn. The indigo dye is then air-oxidized back to its insoluble form. Because an efficient garment dyeing process had never been developed, it was hoped that creation of an improved process would give clothing manufacturers a valuable option in producing a variety of offerings, in various hues, that consumers would find appealing. Indigo dye, in its pigment form, will not dissolve in water, and has no natural affinity for cotton.

However, by reducing the dye and improving the dye-to-fiber affinity, a team at Cotton Incorporated (Cary/NC, USA) obtained positive results by dyeing and extracting cotton garments in an inert oxygen-free atmosphere, which prevents premature oxidation. Nitrogen was chosen to purge air from the dye machine, thereby providing an acceptable dyeing environment. After the dyeing cycle is complete, the nitrogen purge is again applied during the drain and extraction cycles where the excess dyebath is drained into a collection tank for re-use in other dyeing batches. The machine is then opened to the air and oxidation occurs, thus restoring the indigo color and promoting uniform dye application, improved dye yield, and colorfastness. Such a process seeks to allow manufacturers to re-use dyebaths instead

Cellulase enzymes break down the surface of the denim, thus releasing the indigo dye. The desired abrasion effect can be obtained totally without the use of purnice stones.

Figure 6.2 Effect of cellulase on indigo dye on denim cloth: stone-washing.

of releasing dyestuffs and accompanying chemicals into the sewage system, thereby simultaneously cutting process costs and achieving increased environmentally friendliness.

6.2.4 Peroxidases

Bleaching processes which in former times were done with chlorine are today in most cases replaced by a peroxide bleaching process (see Section 6.3). Hydrogen peroxide is used in more and more cases because of its good ecological properties. Especially when bleaching is followed by a dyeing process, one has to make sure that there is *no* residue of hydrogen peroxide on the fabric. Many dyestuffs are sensitive to oxidation and therefore a residue of hydrogen peroxide in the dye bath or on the fabric may cause problems with shade change. In the past there were two possible ways to solve this problem (Figure 6.3):

1. rinsing several times in order to wash off the remaining peroxide, or
2. adding an inorganic reducing agent in order to destroy the remaining peroxide.

The reaction of peroxide and inorganic salt is stoichiometric, so the dosage of reducing agent is not easy to adjust to the right value. If reducing salts remain in the dyebath this may cause problems with shade change. Therefore, after the addition of reducing inorganic salts more rinsing steps are necessary to remove the residues.

Traditional method

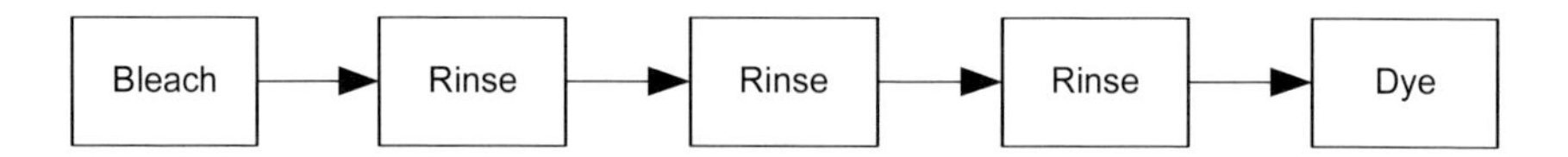

Reduction with inorganic salts

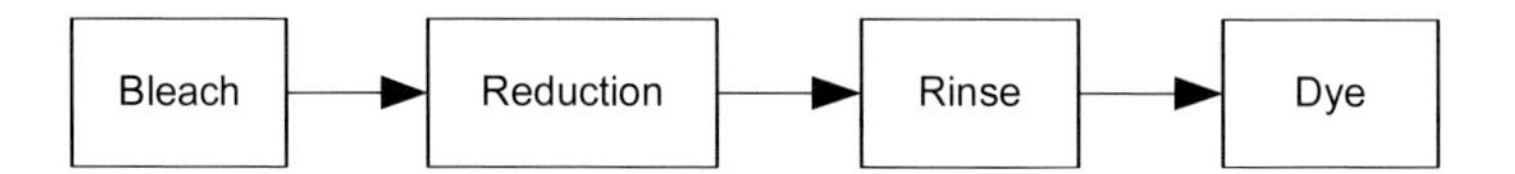

Process with PERIZYM RED (Peroxidase)

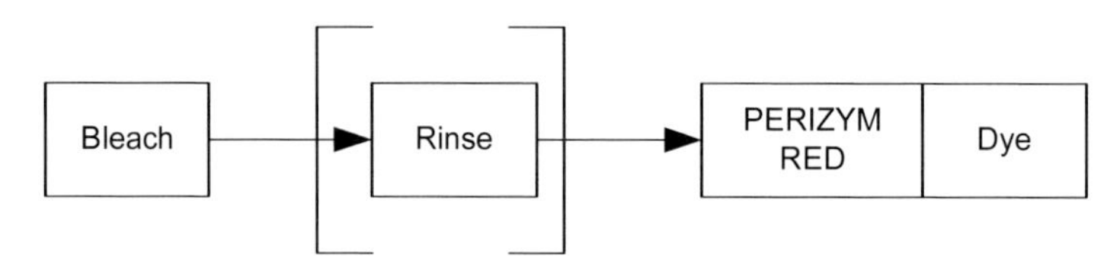

Figure 6.3 Comparison of the traditional and peroxidase routes to clean dyeing.

With peroxidases, a reduction process can be applied which shows the full benefits of an enzymatic reaction. The reaction products are ecologically innocuous – only oxygen and water are formed, and no salt. Peroxidase catalysts are in a position to destroy peroxide residues with much lower dosages than inorganic salts, and there is no reaction with dyes or with the textile itself at all. The peroxidase has been tested with great success in several textile finishing mills. The main benefits to the textile finisher consist in saving time and water and in improving the safety of the process.

6.3
Enzymes in the Pulp and Paper Industry: Bleaching of Pulp with Xylanases or Laccases

6.3.1
Introduction

The pulp and paper industry is very capital-intensive with small profit margins. A paper mill can easily cost more than $ or € 1 billion. It must meet increasing demands for pulp and paper and, at the same time, comply with increasingly stringent environmental regulations. Driven by market and environmental demands for fewer chlorinated products and by-products, the pulp and paper industry is one of the fastest-growing markets for industrial enzymes.

Today, the pulp and paper industry operates with two main pulping processes and three bleaching processes. The latter are the chlorine bleaching process, the elementary chlorine-free (ECF) process, and the totally chlorine-free (TCF) process; they are compared in Table 6.4.

Chlorine bleaching process. Elemental chlorine (Cl_2) is the most effective bleaching agent: all remaining lignin is dissolved, even without attacking the cellulose, which is bright white and remains white for many years. The drawback of this method is that dissolved chlorinated lignin cannot be degraded readily by bacteria and thus is very persistent in nature. The amount of dissolved chlorinated organics in the effluent can be measured as AOX (absorbable organic halogens), the amount of organochlorines which can be absorbed by active carbon. Elemental chlorine has also been targeted recently as a purported hazard to the atmosphere because it can be transformed into chlorohydrocarbons as well as aerosolic species such as chlorine radicals which can attack the ozone layer in the Earth's atmosphere.

Elementary chlorine-free (ECF) process. Using modifications to the standard process, such as alternatives to free chlorine in the form of chlorine dioxide (delivered by sodium chlorate), prolonged cooking times, and prebleaching with oxygen before the main bleaching with chlorine dioxide, about the same quality of paper was reached without the drawbacks of elementary chlorine (see Table 6.4).

Totally chlorine-free (TCF) process. The TCF process uses hydrogen peroxide in the bleaching process. As a drawback, about 10% more wood has to be used for the

Table 6.4 Comparison of bleaching processes.

Process	*Bleaching agent*	*AOX [g t^{-1}]*	*Organics in effluent [kg t^{-1}]*	*Dioxins [μg/ton pulp]*	*Dioxins in paper [rel. value]*	*Effluent toxic for fish?*
Cl_2	Cl_2	3–5000	50	1		yes
ECF	ClO_2	4–800	5	0	1	yes
TCF	H_2O_2	0		0	35	yes

same quality of paper, and the fibers remain shorter. Short fibers reduce recyclability as fibers that are too short have to be discarded and more new pulp added; each time paper or cardboard is recycled, the remaining fibers become shorter. The generally better results with respect to environmental compatibility were balanced by a very high dioxin level in fresh TCF paper.

6.3.2
Wood

The raw material for all types of paper manufacture is wood. Wood consists of about 40–55% of *cellulose*, i.e., β1→4 glycosidically bound glucose molecules (in contrast to the α1→4 glycosidically bound glucose molecules in starch) (Figure 6.4), of 25–35% of *hemicellulose* (short, branched polymers of pentoses with only a few hexoses), and finally of 15–30% of *lignin*, a polyphenolic network with more than ten different bonds (Figure 6.5), which renders wood hard ("concrete of nature") and which bonds can only be cleaved with difficulty.

6.3.2.1 Cellulose

Cellulose is a long-chain polymer with repeat units of D-glucose (Figure 6.4). The glucose units are in pyranose (six-membered ring) form, the oxygen link between units is between C-1 (acetal linkage) of one pyranose ring and the C-4 of the next ring. In cellulose, the C-1 oxygen is in the β-configuration (cellulose is poly[β-1,4-D-anhydroglucopyranose]), with all the functional groups in equatorial positions. The β-configuration causes the cellulose molecule to extend in a more-or-less straight line and renders cellulose a good fiber-forming polymer. (In contrast, amylose, a constituent of starch, is a related polymer of glucose, but with the C-1 oxygens in an α-configuration. This configuration forces the linkage to the next glucopyranose ring to assume an axial position, and the starch molecules tend to coil, rather than extend. Even though it often has long molecular chains, amylose is not a good fiber-former.)

The all-equatorial positions of the hydroxyls on the cellulose chain renders them available for hydrogen bonding, causing highly ordered (crystal-like) structures. The strong inter-chain hydrogen bonds in the crystalline regions provide the fibers with good strength, confer insolubility in most solvents, and also prevent cellulose from melting. As cellulose is very hygroscopic, it swells but does not dissolve in water.

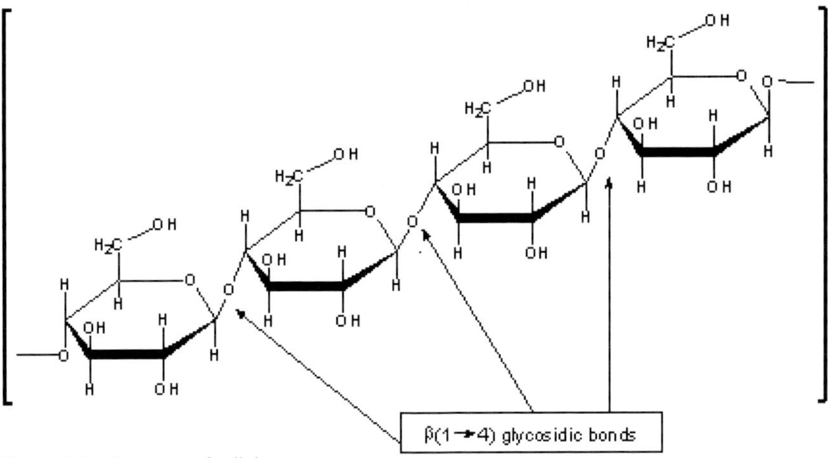

Figure 6.4 Structure of cellulose.

6.3.2.2 Hemicellulose

In contrast to cellulose, which is crystalline, strong, and resistant to hydrolysis, hemicellulose has a random, amorphous structure with little strength. It is easily hydrolyzed by dilute acid or base, but nature provides an arsenal of hemicellulase enzymes for its hydrolysis. Hemicellulases are commercially important because they open the structure of wood for easier bleaching and thus support the introduction of ECF or TCF methods. Many different pentoses are usually present in hemicellulose. Xylose, however, is always the predominating sugar. The pentoses are also present in rings (not shown) that can be five- or six-membered.

6.3.2.3 Lignin

Softwood lignin contents are on the order of 26–32% and hardwood lignin contents range from 20 to 28%. Lignin (Figure 6.5) is biosynthesized by a free-radical coupling reaction. Peroxidases and laccases initiate the process by homolytically breaking the covalent bond between the phenolic oxygen and its hydrogen. A series of resonance-stabilized free radicals in equilibrium with one another is formed which combine to create a new covalent bond and a dimer. The dimer still has a free phenolic hydroxyl, and an enzyme can remove the hydrogen and electron to make another free radical. The dimer reacts to a trimer or tetramer and this process continues until a polymeric structure results.

The free-radical polymerization process generates a series of interunit linkages. The predominant linkage is the so-called beta-O-4 linkage (Figure 6.6). The intermediate of this coupling is called a quinone methide, which usually rearranges (with the addition of water) to form the stable beta-O-4 ether bond. About 40–60% of all interunit linkages in lignin are formed as such ether bonds, and it is this bond that is usually broken during the delignification (pulping) processes.

How this process is initiated and controlled is poorly understood. Understanding the process of lignification is the first step required to manipulate it rationally to provide cell wall material more amenable to standard pulping processes. Some

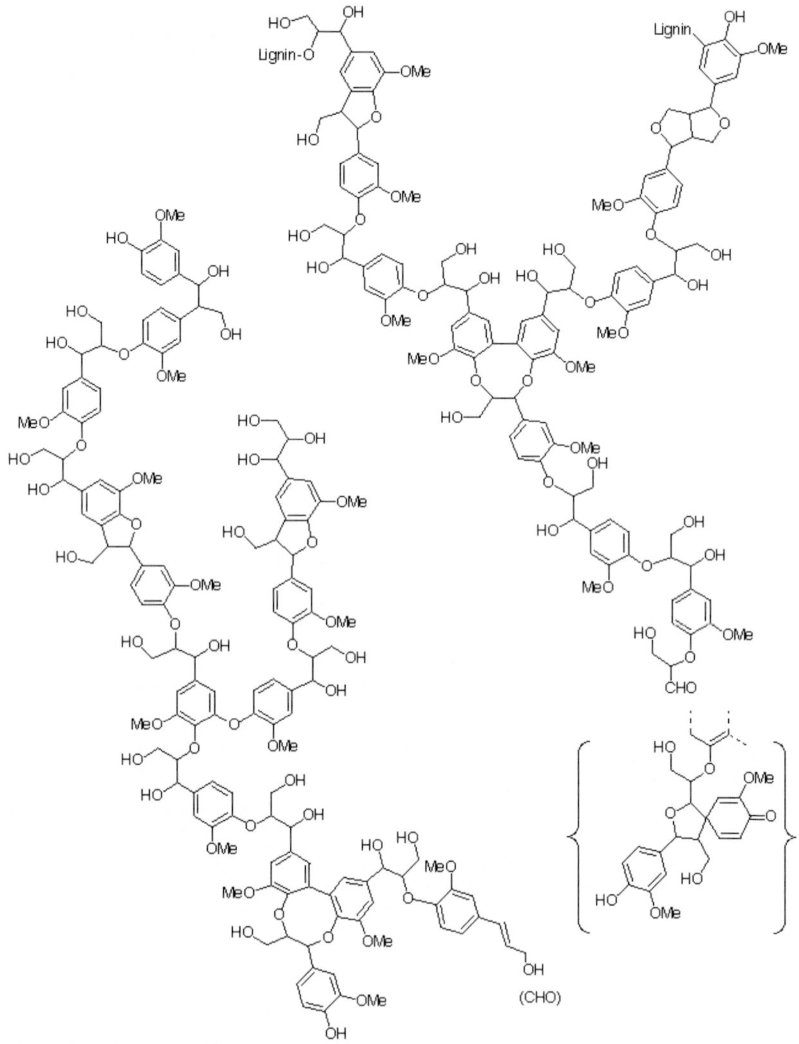

Figure 6.5 Structure of lignin.

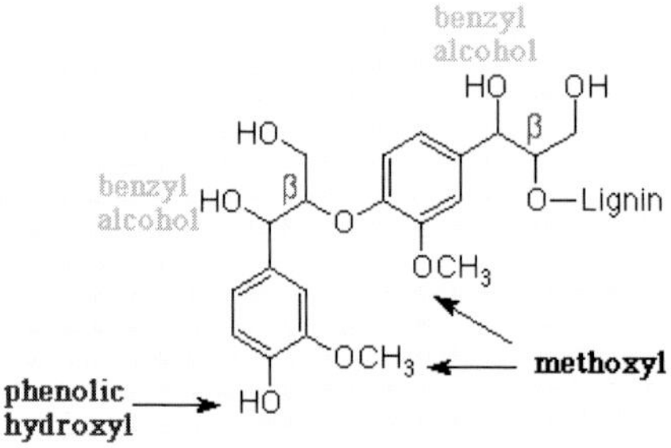

Figure 6.6 Key structural element of lignin.

scientists feel that there is no real control over the polymerization process (Ralph, 1997), whereas others feel that there is a high level of control (Davin, 1997). Owing to the small energetic differences in the different pathways and the considerable amount of radical species involved, there seems to be considerable "plasticity" in the polymerization process which amounts to a process which is hard to follow and control.

6.3.3
Papermaking: Kraft Pulping Process

The process of pulping involves the separation of useful cellulose fibers from lignin and other components. Pulping can be performed with mechanical or chemical processes. Mechanical pulping gives a high yield of pulp but it is of poor quality as the lignin components are not significantly solubilized. Such mechanical pulps are mainly used for the manufacture of newsprint.

The Kraft process combines good bleaching quality with an affordable price. In this process (Figure 6.7), wood is cut into chips which are cooked in caustic soda. This step dissolves and removes most of the lignin without excessively attacking the remaining cellulose. When the remaining pulp of the Kraft process is not further bleached, it is used as cardboard: the rather dark color is from the lignin remaining after the cooking process. The resulting darkly colored Kraft pulp must undergo substantial bleaching, i.e., unspecific oxidative cleavage, before being suitable for paper manufacture. The different methods of bleaching encompass chlorine, chlorine dioxide, oxygen, ozone, or hydrogen peroxide. Enzyme treatment of Kraft pulps has been shown to remove the hemicelluloses bound to the surface of fibers, which makes it easier to remove bound lignin components and thereby reduces the requirement for chlorine bleaching. This process has become known as "bleach boosting".

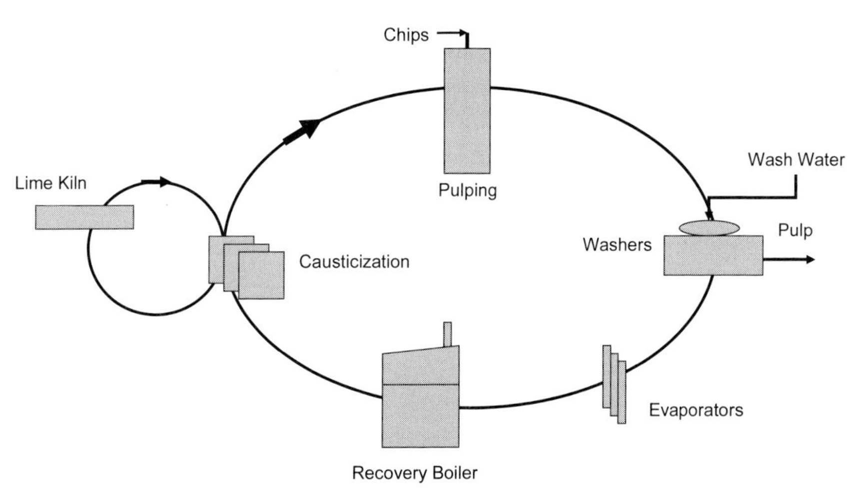

Figure 6.7 Simplified flow diagram of the Kraft pulping process.

Kraft pulping was originally based on the soda process, which necessitated the additional use of NaOH and, owing to its associated cost, could not compete with the sulfite process, which used Na_2SO_3. Around 1879, soda (Na_2CO_3) was replaced with the cheaper sodium sulfate (Na_2SO_4). Upon heating sodium sulfate, sodium sulfide (Na_2S) is formed [Eq. (6.1)] which, when mixed with water, provides a mixture of sodium hydroxide (NaOH) and sodium hydrosulfide (NaSH) [Eq. (6.2)]. With this innovation, the Kraft process was born.

$$Na_2SO_4 \rightarrow Na_2S \tag{6.1}$$

$$Na_2S + H_2O \leftrightarrow NaOH + NaSH \tag{6.2}$$

The heart of the Kraft pulping operation is actually chemical recovery, or liquor recycling. Sodium hydroxide and sodium hydrosulfide [see Eq. (6.2)] are the reagents that delignify. Typically, the pH of the solution is greater than 12 and cooking times range from 0.5 to 3 h at temperatures between 160 and 180 °C. There are a number of variables used to control the process, such as total alkali, active alkali, sulfidity, and the *H*-factor, which describes the rate of delignification. Pulp mills shoot for a certain *H*-factor, and this will provide a psulp of a specific lignin content. Lignin content is estimated by determination of the *"kappa number"*, which is a measure of the pulp's ability to reduce potassium permanganate ($KMnO_4$) to MnO_2. The remaining lignin is oxidized in the permanganate reaction: the higher the lignin content, the higher the kappa number. As a rough estimate, lignin content = 0.15 × kappa number.

While the Kraft process initially had two distinct disadvantages, it produced a dark pulp and required chemical recovery, the pulp is stronger and has better dimensional stability than in the old *soda* process. Chemical recovery systems rapidly became quite advanced and most advanced mills are now aiming for 98–99% chemical recovery.

6.3.4 Research on Enzymes in the Pulp and Paper Industry

6.3.4.1 Laccases

Research nowadays is focused on two types of enzymes, xylanases/cellulases and laccases, which are useful for fiber modification in both the pulp and paper and the textile industries. Attempts were made to circumvent the heterogeneity of lignin enzymatically by cleaving the hemicellulose chains adjacent to lignin with xylanases and thus extracting the lignin including the attached xylan residues from the wood structure. This approach has been taken by several companies but seems to have run into difficulty owing to the precipitation of extracted lignin residues onto the paper fiber as well to the separate heterogeneity of the hemicellulose structures.

For this reason, considerable interest has arisen in the pulp and paper industry concerning the potential replacement of traditional bleaching reagents with en-

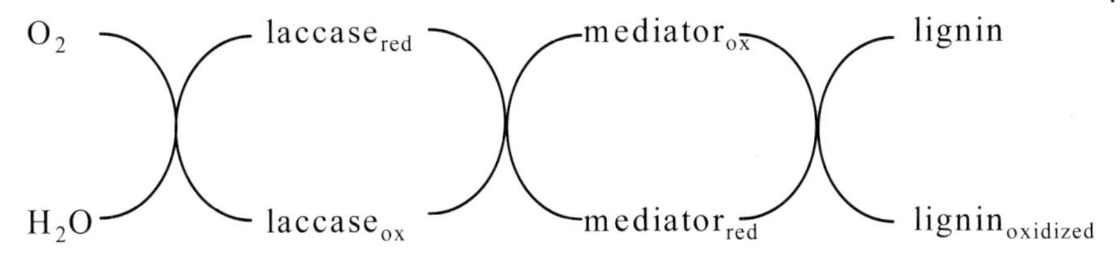

Figure 6.8 Laccase-mediator system.

zymes, in particular laccase, from white-rot fungi that are known to degrade lignin naturally, and are observed to be effective in bleaching kraft pulp through lignin degradation. Residual lignin in pulp causes discoloration and requires an aggressive bleaching process. Studies have shown that a laccase–mediator system (LMS) (Figure 6.8) can potentially remove 55% of the residual lignin from a Kraft pulp. LMS systems will have a lower impact on the environment than oxygen delignification, by eliminating the formation of chloride and absorbable, organically bound chlorides (AOX). They will also aid compliance with environmental regulations and a closed-loop water program. Their other expected benefits are a lower capital investment, reduced operating costs in mills, and a safe system for selectively removing lignin and improving pulp yields.

Goals in the development of a recombinant laccase are to obtain a laccase variant with increased reduction potential, thermal stability, and pH range and to identify new mediators and processes that improve the effectiveness of laccase in the degradation of lignin. Also, improved recombinant laccases and improved laccase mediators should result in higher strength fiber compared to conventional bleaching. Other potential applications of laccase include treatment of fabrics and catalysis of polymer synthesis.

Laccase is a blue copper enzyme (a polyphenol oxidase) that catalyzes the transfer of four electrons from various organic substrates to reduce dioxygen to water. Although laccase can be prepared from cultures of various white-rot fungi, no effective bacterial expression system for laccase has been reported. However, with the basidiomycete *Trametes versicolor*, a white-rot fungus, good production yields in submerged culture have been achieved by using homologous expression, where the laccase gene has been reintroduced into the former host strain at a higher copy number and optimized expression signals (Wich, *year unknown*). If such an expression system were successfully developed, then traditional techniques of protein engineering (see Chapters 10 and 11) could be employed to create laccase variants with increased thermal and chemical stability and even greater oxidizing capabilities than the existing wild-type enzyme. Modified laccases with these characteristics are expected to be more effective in a pulp mill setting than the wild-type enzyme.

6.3.4.2 **Xylanases**

Xylanase is an enzyme that selectively hydrolyzes xylan to yield xylose as a product. Xylanase does not remove lignin, but it allows it to be more easily removed in subsequent bleaching steps by other chemicals. Addition of xylanases therefore reduces the amount of bleaching chemicals such as chlorine and chlorine dioxide that are needed. This gives corresponding reductions in the amounts of harmful chlorinated organic compounds in the effluent.

6.3.4.3 Cellulases in the Papermaking Process

Enzymes also help to keep high-speed paper machines running. Cellulases and hemicellulases partially break down cellulose and can improve drainage of water from the pulp and thus the speed of paper machine operation, which depends in part on the drainage rate. As this rate tends to be lower for recycled fibers, the production rate declines as the recycled fiber content increases. These enzymes are now used commercially as drainage aids. The enzymes remove the finest fibers, allowing water to drain away faster, cutting the processing time and reducing the energy required to dry the paper.

6.4 Phytase for Animal Feed: Utilization of Phosphorus

6.4.1 The Farm Animal Business and the Environment

In the farm animal business, speed of growth is essential for rapid turnaround of delivery to market. An adequate diet containing all the required nutrients is essential not just for rapid growth but also for the prevention of disease. Phosphorus, essential for DNA synthesis and energy metabolism, but mainly also for proper skeletal growth, is supplied through plant components called phytates which are hexaphosphorylated inositols (or phosphorylated cyclic sugar alcohols); the acid is called phytic acid (Figure 6.9). As much as 80% of the phosphorus in cereal grains and oilseed meals is chemically bound in the form of phytate. This phosphorus has a very low bioavailability; pigs and chickens digest it poorly because they lack the enzyme phytase. All of this unavailable phosphorus is excreted in the feces. The US poultry industry, which contributes more than $ 12 billion annually to the economy, includes some 75 000 growers nationwide who produce more than 6.5 billion broilers, 285 million turkeys, and 20 million ducks. These operations generate 20 million tons of manure annually, which must be disposed of in an environmentally safe manner. Generally, the waste is applied to the land as a fertilizer. At the same time, however, producers must add inorganic phosphorus to the animals' rations to meet their needs – and part of this supplemental phosphorus is also excreted. This leads to a vast oversupply of the total amount of phosphorus in the animals' diet, causing a waste problem that places a strain on waters such as rivers and creeks in the vicinity of animal farms but ultimately also aquifers. Excess phosphorus in waterways can cause rapid algal growth and can affect the amount of oxygen in the water, leading to fish kills and other severe environmental, health, and economic problems.

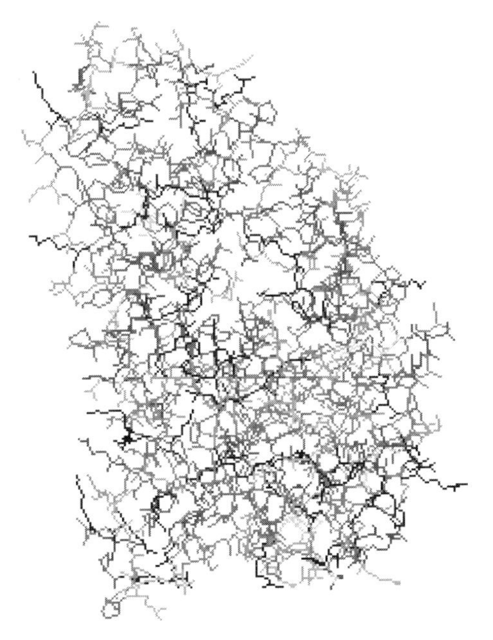

Figure 6.9 Structure of phytase from *Aspergillus ficum* (Kostrewa, 1997).

6.4.2
Phytase

Phytase offers significant promise as a means to reduce phosphorus levels in animal waste by 30–35%, while also reducing the cost of phosphorus supplementation (Vincent, 1992). In several steps, the enzyme hydrolyzes phytate (*myo*-inositol hexakisphosphate; Figure 6.10), the primary storage form of phosphorus in plant seeds and pollen, into inositol and inorganic phosphorus which is readily bioavailable to the farm animals. Phytases can also have non-specific phosphorus monoester activity. Addition of phytases to farm animals' diet significantly enhances bioavailability of plant phosphorus for the animals while reducing phosphorus in the waste and simultaneously allowing a reduction of total phosphorus in the feed.

From the crystal structure of an extracellular phytase enzyme from the fungus *Aspergillus ficuum*, we can see that phytase is a monomer with two domains, an alpha domain and an alpha/beta domain (Figure 6.9) (Kostrewa, 1997). The active site is in an indentation between these domains. The indentation is closed off at the back by an N-terminal lid. Basic amino acids at the active site help bind the negatively charged 3-phosphorus group on phytate. It is thought that a histidine (H59) makes the nucleophilic attack on the phosphorus group and an aspartate (D339) provides the proton for the leaving alcohol. This crystal structure was obtained through X-ray diffraction of the crystallized enzyme and has a resolution of 2.5 Å.

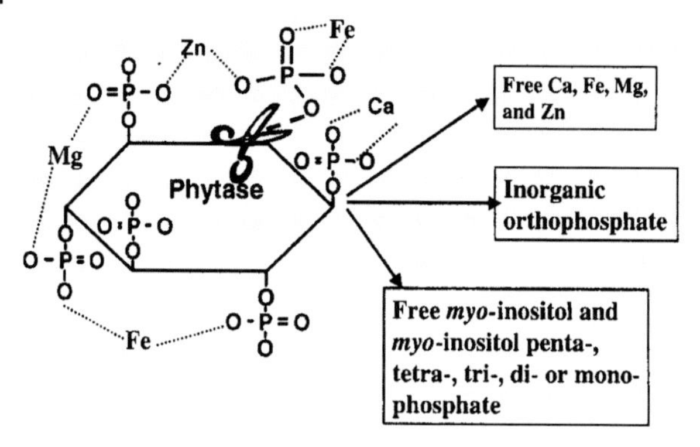

Figure 6.10 Structure of phytic acid (Lei, 2001).

6.4.3
Efficacy of Phytase: Reduction of Phosphorus

A trial was conducted with 2800 Ross–Hubbard male broiler chicks to check the efficacy of phytase addition to the feed (Harper, 1999). Diets were formulated to contain either the typical diet used in the industry with 0.425% non-phytate phosphorus (P) as control, or two reduced levels of non-phytate P (0.225% and 0.325%) and three levels of phytase (0, 300, and 600 phytase units per kilogram feed). Before treatments were assigned, all chicks until three weeks of age were fed a commercial starter mash adequate in all nutrients.

The results in Table 6.5 reveal that neither performance nor bone strength, important to help birds withstand stress during growth and processing, was significantly influenced by a reduction of non-phytate P to 0.325% as compared to the positive control of 0.425%. However, when non-phytate P was reduced to 0.225%, a significant negative impact on body weight, feed consumption, feed efficiency, and bone strength was observed.

Phytase significantly increased body weight at the lower non-phytate P level but not at the higher non-phytate P level. Phytase at 300 units per kilogram feed had a greater influence on bone mineral content, bone density, bone breaking strength, and livability in broilers fed 0.225% non-phytate P than in broilers fed 0.325% non-phytate P. This indicates that supplementing phytase in grower diets containing reduced levels of non-phytate P significantly improved performance and bone strength of broilers. The negative impact on bone strength associated with the deficient level of 0.225% non-phytate P was completely reversed by the inclusion of 300 phytase units (kg feed)$^{-1}$ in diets containing marginal to deficient levels of non-phytate P (Table 6.5). The addition of microbial phytase at 300 and 600 phytase units (kg feed)$^{-1}$ prevented P deficiency symptoms by allowing birds to utilize phytate P. However, increasing phytase levels from 300 to 600 phytase units (kg feed)$^{-1}$ provided no additional benefit.

These results demonstrate that addition of 300 units of phytase per kilogram diet allows producers to reduce dietary P by 0.1% in grower and finisher diets with

Table 6.5 Efficacy of added microbial phytase in broiler grower and finisher diets: control versus treatment comparisons at six weeks of age.

	Treatment diet				
Variable	1 (control)	2	3	4	5
Non-phytate P [%]	0.425	0.32 5	0.32 5	0.22 5	0.225
Phytase [units (kg diet)$^{-1}$]	0	0	300	0	300
Body weight at 6 wk [g]	2,430	2,46 2	2,43 1	2,38 4	2,392
Feed efficiency	2.11	2.07	2.10	2.12	2.11
Mortality [%]	0.46	0.46	0.21	0.79	0.33
Bone mineral content* [g cm^{-1}]	0.180	0.17 8	0.18 7	0.14 1**	0.181
Bone density* [g cm^{-2}]	0.195	0.188	0.200	0.160 **	0.202
Bone breaking strength* [kg]	39.37	37.65	39.88	31.05 **	42.22

Source: Harper, 1999.

* Bone criteria with higher values represent a stronger bone.

** Indicates significantly weaker bones compared to Diet 1.

no adverse effect on broiler performance. Use of phytase in poultry diets will significantly reduce environmental phosphorus pollution.

6.4.4 Efficacy of Phytase: Effect on Other Nutrients

In another study (Skaggs, 1999), animals showed linear improvements in phosphorus digestibility, average daily gain, feed intake, and feed efficiency as the level of phytase in the diet increased. At 750 or 1000 phytase units per kilogram of complete feed, the pigs on the low-phosphorus diets had performance and bone mineralization equal to or greater than the positive controls. Phosphorus concentrations in the feces decreased linearly as phytase levels increased. Based on both linear and nonlinear response equations, Skaggs (1999) reported that the average phosphorus equivalence of phytase was 500 units kg^{-1} for 1.06 g of phosphorus fed as mono- or dicalcium phosphate (Yi, 1996a).

The amount that fecal phosphorus levels decline with phytase supplementation will depend on the percentage of phosphorus that is available in the base ration (Yi, 1996b). With available phosphorus at 0.05% of the ration in this study, excretion of phosphorus was reduced by about 25% in pigs receiving phytase compared to the control animals. When the base diet in this study contained 0.16% available phosphorus, excretion was reduced by 50% in pigs receiving phytase but no inorganic phosphorus.

Phytase addition can have other effects in addition to helping to release phosphorus from phytate (Denbow, 1995; Yi, 1996d):

- *Calcium digestibility.* Phytate can bind other minerals in addition to acting as a mans for storage of phosphorus. Skaggs (1999) and Harper (1999) reported linear increases in calcium digestibility and decreased fecal concentrations of calcium with phytase supplementation. It has been suggested that dietary zinc fortification may also be decreased with phytase supplementation (Yi, 1996c).

- *Increased concentrations of hemoglobin.* Piglets also absorbed and used more iron for making hemoglobin. In experiments with 32 anemic piglets, it was found that animals fed dietary phytase had higher hemoglobin concentrations than those given feed without the enzyme (Harper, 1997). Also, hemoglobin concentrations in the phytase-supplemented pigs were similar to those in pigs fed supplemental iron. Iron, an essential component of hemoglobin, is derived from food and by recycling iron from old red blood cells. One cause of iron deficiency is poor absorption of iron by the body. Understanding the factors that improve and promote iron absorption should lead to better use of dietary iron.

- *Amino acid digestibility.* This may also be improved with phytase supplementation. In finishing pigs, the digestibility of all amino acids except proline and glycine increased linearly as phytase supplementation increased (Zhang, 1999). Based on those responses, the researchers calculated that 500 units (kg feed)$^{-1}$ of phytase was equivalent to 0.76% crude protein. They estimated that the nitrogen content of the feces was reduced by 5.1% with this level of phytase added.

Heat stability is a critical issue for commercial phytase products. Since phytase can be denatured and inactivated by the heat associated with feed processing, it is nec-

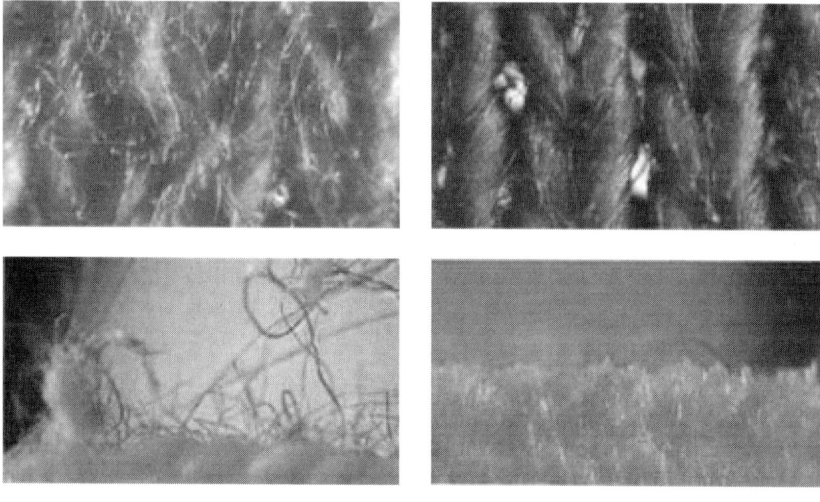

Dyed, untreated fabric | after biofinishing with cellulose

Inset Figure Biofinishing of cotton knitwear.

essary to use a thermostable product form or post-processing spraying of a liquid phytase product so the animals receive the required levels.

In summary, the savings from reduced phosphorus costs can virtually offset the cost of phytase itself, which would make the significant environmental benefits remarkably inexpensive to achieve.

Suggested Further Reading

Work on the application of phytase can be followed on the phytase.net site at: http://www.enzymes.co.uk/phytase/indexphytase.html

References

Anon., *Enzymes at Work*, Handbook of Novozyme A/S, Bagsvaerd, Denmark, **1997**.

J. R. Cherry, M. H. Lamsa, P. Schneider, J. Vind, A. Svendsen, A. Jones, and A. H. Pedersen, Directed evolution of a fungal peroxidase, *Nature Biotechnology* **1999**, *17*, 379–384.

L. B. Davin, H.-B. Wang, A. L. Crowell, D. L. Bedgar, D. M. Martin, S. Sarkanen, and N. G. Lewis, Stereoselective bimolecular phenoxy radical coupling by an auxiliary (dirigent) protein without an active center, *Science* **1997**, *275*, 362–366.

D. M. Denbow, V. Ravindran, E. T. Kornegay, Z. Yi, and R. M. Hulet, Improving phosphorus availability in soybean meal for broilers by supplemental phytase, *Poultry Sci.* **1995**, *74*, 1831–1842.

A. F. Harper, E. T. Kornegay, and T. C. Schell, Phytase supplementation of low-phosphorus growing–finishing pig diets improves performance, phosphorus digestibility, and bone mineralization and reduces phosphorus excretion, *J. Anim. Sci.* **1997**, *75*, 3174–3186.

A. F. Harper, J. H. Skaggs, H. P. Veit, and E. T. Kornegay, Efficacy and safety of Novo SP938 microbial phytase supplementation of a corn–soybean meal diet fed to growing pigs, *J. Anim. Sci.* **1999**, *77*, 174–175.

D. Kostrewa, F. Gruninger-Leitch, A. D'Arcy, C. Broger, D. Mitchell, and A. P. G. M. van Loon, Crystal structure of phytase from *Aspergillus ficuum* at 2.5 Å resolution, *Nature Struct. Biol.* **1997**, *4*, 185–190.

X. G. Lei and C. H. Stahl, Biotechnological development of effective phytases for mineral nutrition and environmental protection, *Appl. Microbiol. Biotechnol.* **2001**, *57*, 474–481.

D. L. Oxender and C. F. Fox (eds.), *Protein Engineering*, Alan R. Liss, New York, **1987**, Chapter 25, pp. 279–287.

J. Ralph, J. J. MacKay, R. D. Hatfield, D. M. O'Malley, R. W. Whetten, and R. R. Sederoff, Abnormal lignin in a Loblolly Pine mutant, *Science* **1997**, *277*, 235–239.

J. H. Skaggs and E. T. Kornegay, Dose response of Novo SP938 microbial phytase in weanling pigs fed a low-P corn–soybean diet, *J. Anim. Sci.* **1999**, *77*, 174.

M. Tobin, J. A. Affholter, W. P. C. Stemmer, and J. Minshull, Colorless green ideas, *Nature Biotechnology* **1999**, *17*, 333–334.

J. B. Vincent, M. W. Crowder, B. A. Averil, Hydrolysis of phosphate monoesters: a biological problem with multiple chemical solutions, *Trends in Biochemical Science* **1992**, *17*, 105–110.

G. Wich, Weissfäulepilze im Dienste der Biotechnologie: das Lignozym-Verfahren (in German), www.fh-weihenstephan.de/bt/berichte/wich.pdf

Z. Yi, E. T. Kornegay, V. Ravindran, M. D. Lindemann, and J. H. Wilson, Effectiveness of Natuphos phytase in improving the bioavailabilities of phosphorus and other nutrients in soybean metal-based semipurified diets for young pigs, *J. Anim. Sci.* **1996a**, *74*, 1601–1611.

Z. Yi, E. T. Kornegay, and D. M. Denbow, Supplemental microbial phytase improves zinc utilization in broilers, *Poultry Sci.* **1996b**, *75*, 540–546.

Z. Yi, E. T. Kornegay, V. Ravindran, and D. M. Denbow, Improving phytate phosphorus availability in corn and soybean meal for broilers using microbial phytase and calculation of phosphorus equivalency values for phytase, *Poultry Sci.* **1996c**, *75*, 240–249.

Z. Yi, E. T. Kornegay, and D. M. Denbow, Effect of microbial phytase on nitrogen and amino acid digestibility and nitrogen retention of turkey poults fed corn–soybean meal diets, *Poultry Sc.* **1996d**, *75*, 979–990.

Z. Zhang and E. T. Kornegay, Phytase effects on ileal amino acid digestibility and nitrogen balance in finishing pigs fed a low-protein plant-based diet, *J. Anim. Sci.* **1999**, *77*, 175.

7
Application of Enzymes as Catalysts: Basic Chemicals, Fine Chemicals, Food, Crop Protection, Bulk Pharmaceuticals

Summary

Applications of biocatalysis in large-scale processes in industry advance only slowly against established chemical processes, even with stoichiometry-based chemistry. Introduction of biocatalysis into existing processes often requires process modifications that are not economical in view of the short life span of the product and/or the low fixed costs of the existing process owing to written-off plants. It should be emphasized that the desire to reduce chemical wastes, imposed by either company policy or governmental measures, needs to be matched by favorable process economics. Therefore, the introduction of biocatalytic options at the very beginning of product and process development is of the utmost importance. Biocatalytic processes and technologies are spreading increasingly through all branches of the chemical process industries.

In basic chemicals, nitrile hydratase and nitrilases have been most successful. Acrylamide from acrylonitrile is now a 30 000 tpy process. In a product tree starting from the addition of HCN to butadiene, nicotinamide (from 3-cyanopyridine, for animal feed), 5-cyanovaleramide (from adiponitrile, for herbicide precursor), and 4-cyanopentanoic acid (from 2-methylglutaronitrile, for 1,5-dimethyl-2-piperidone solvent) have been developed. Both the enantioselective addition of HCN to aldehydes with oxynitrilase and the dihydroxylation of substituted benzenes with toluene (or naphthalene) dioxygenase, which are far superior to chemical routes, open up pathways to amino and hydroxy acids, amino alcohols, and diamines in the first case and alkaloids, prostaglandins, and carbohydrate derivatives in the second case.

In the fine chemicals industry, enantiomerically pure amino acids are mainly produced by the aminoacylase process, the amidase process, and the hydantoinase/carbamoylase process, all three of which are suitable for L- and D-amino acids. Dehydrogenases and transaminases are now becoming established for reduction processes.

The food industry is a fertile area for biocatalysis applications: high-fructose corn syrup (HFCS) from glucose with glucose isomerase, the thermolysin-catalyzed synthesis of the artificial sweetener Aspartame®, hydrolysis of lactose for lactose-intolerant consumers, and the synthesis of the nutraceutical L-carnitine in a two-enzyme system from γ-butyrobetaine all serve as examples.

In crop protection, the processes to herbicide intermediates (*R*)-HPOPS and (*S*)-MOIPA are described. A process catalyzed by an organometallic compound (Ir–ferrocene derivative) has turned out to be superior to the transaminase-catalyzed biocatalytic route to (*S*)-MOIPA whereas enzymatically-catalyzed aromatic hydroxylations with HPOPS as an example are the method of choice.

Enzymatic processes also advance in the area of large-scale pharma intermediates: β-lactam antibiotics can now be produced in a fully biotechnological process, including the semi-synthesis from the β-lactam core to the penicillin or cephalosporin. A precursor to ephedrine, long produced by a whole-cell process in yeast, can be obtained from benzaldehyde and acetaldehyde with the help of pyruvate dehydrogenase acting as a carboligase.

7.1
Enzymes as Catalysts in Processes towards Basic Chemicals

A productive role of biocatalysts in processes to basic chemicals is anything but obvious: for decades, it was believed that enzymes might be suitable catalysts but only for specialty applications. Such beliefs were guided by the assumption that enzymes in general are expensive and thus not suitable for processes to basic chemicals. Given sufficiently good stability and catalytic and volumetric productivity values (see Chapter 2), however, enzymes have proven to be the catalysts of choice for quite a few applications. New technologies such as protein engineering (Chapter 10), directed evolution (Chapter 11), and metabolic engineering (Chapters 15 and 20) are likely to open up further opportunities for biocatalytic routes, provided they can meet the stringent productivity and cost criteria set by processes to basic chemicals.

7.1.1
Nitrile Hydratase: Acrylamide from Acrylonitrile, Nicotinamide from 3-Cyanopyridine, and 5-Cyanovaleramide from Adiponitrile

7.1.1.1 Acrylamide from Acrylonitrile

In the polymerization of acrylamide, $H_2C{=}C{-}C(O)NH_2$, to polyacrylamide, very high degrees of polymerization can be achieved only if the monomer feed is very pure, especially with respect to metals and anions. The conventional chemical process to acrylamide from acrylonitrile, $H_2C{=}C{-}C{\equiv}N$, a solvent, is based on Cu catalysis. At Kyoto University (Kyoto, Japan), Hideaki Yamada and his group developed a whole-cell biocatalyst system, ultimately resting cells of *Rhodococcus rhodochrous* J1, in which a nitrile hydratase hydrolyzes acrylonitrile to acrylamide. Nitrile hydratases are Fe^{3+}- or Co^{3+}-containing hydrolases, leading to amides, in contrast to nitrilases which are adding two water molecules to the nitrile group to form the carboxylic acid (Figure 7.1). Nitrile hydratase was found not just to hydrolyze acrylonitrile but a range of aromatic nitriles to the corresponding amides (Table 7.1 and Section 7.1.2). The substrate concentrations that can be employed are so extremely high that, in case of solid nitriles and amides, precipitation of amide

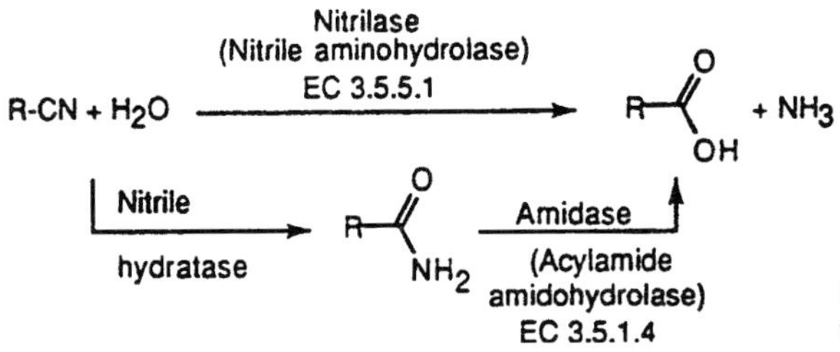

$$R\text{-}CN + H_2O \xrightarrow[\text{EC 3.5.5.1}]{\text{Nitrilase (Nitrile aminohydrolase)}} R\text{-}C(=O)OH + NH_3$$

$$R\text{-}CN + H_2O \xrightarrow{\text{Nitrile hydratase}} R\text{-}C(=O)NH_2 \xrightarrow[\text{EC 3.5.1.4}]{\text{Amidase (Acylamide amidohydrolase)}} R\text{-}C(=O)OH$$

Figure 7.1 Enzyme reactions with the cyano group.

Table 7.1 Aromatic amides and achievable concentrations with *Rhodococcus rhodochrous*.

Amide	*Concn. [g L^{-1}]*	*Amide*	*Concn. [g L^{-1}]*
nicotinamide	1465 g/L	0,0-difluorobenzoylamide	393 g/L
γ-picolylcarboxamide	1099 g/L	2-thienylcarboxamide	254 g/L
α-picolylcarboxamide	977 g/L	indole-3-carboxamide	697 g/L
pyrazine carboxamide	985 g/L	benzoylamide	848 g/L

Figure 7.2 Aggregation state of the reaction system after different reaction times [h] (Yamada, 1990).

product renders the system completely solid at intermediate degrees of conversion (Figure 7.2). In case of liquid nitriles, the reaction can be run in neat reactant (see 12.4.1). The extreme substrate concentrations possible with many nitriles (often more than 1000 g/L!) lead to very high space–time yields, a necessary criterion for low capital (i.e. fixed) cost for efficient processes in basic chemistry.

The biocatalytic acrylamide process is run by the Nitto Chemical Corp., now part of the Mitsubishi Rayon Corp., in Tokyo Bay on a scale of 30 000 tpy, in fed-batch mode up to 25–40% acrylamide at 0–10°C to complete conversion and with product yields > 99.9%, conditions under which a significant cost differential can be assumed with respect to the conventional chemical process.

7.1.1.2 Nicotinamide from 3-Cyanopyridine

In analogous fashion to hydrolysis of acrylonitrile by nitrile hydratase, 3-cyanopyridine (nicotine nitrile) can be hydrolyzed by the same catalyst to nicotinamide, which finds use as vitamin B_3 in animal feed mixtures and also as an agent against pellagra in humans. Lonza (Visp, Switzerland) has obtained a license from Kyoto University for use of nitrile hydratase from *Rhodococcus rhodochrous* J1 and in 2000 established this process on a large scale with a designed capacity of 3400 tpy (Thomas, 2002) at Lonza Guangzhou Fine Chemicals in Guangzhou, China (Kiener, 1995). The raw material 3-cyanopyridine is obtained from 2-methyl-1,5-diaminopentane (Dytek®), a by-product of the nylon-6,6 raw material adiponitrile (1,6-dicyanobutane, ADN) from DuPont (obtained upon addition of HCN at the 2-position of butadiene instead of the desired 1-position), which is first dehydrogenated to 3-methylpyridine and then ammoxidized to 3-cyanopyridine.

The process is configured as a series of three stirred-tank reactors with the substrate 3-cyanopyridine continuously fed at 10–20 wt.% concentration and the biocatalyst flowing countercurrently. Enzymatic hydrolysis yields the desired nicotinamide at > 99.3% selectivity, in contrast to the chemical alkaline hydrolysis process which results in about 3–5% nicotinic acid, an undesirable by-product because it causes diarrhea in farm animals (instead of supporting growth; for animal feed supplements, see Chapter 6, Section 6.4). Thus, the enzymatic process competes well with the chemical hydrolysis.

7.1.1.3 5-Cyanovaleramide from Adiponitrile

The same raw material as for nylon-6,6, adiponitrile (ADN), is the starting material for the nitrile hydratase-catalyzed process to 5-cyanovaleramide (5-CVAM), itself in turn a starting material for the synthesis of the novel herbicide azafenidin (Milestone®) from DuPont (Wilmington/DE, USA) (Hann, 1999). Screening of a variety of microorganisms for nitrile hydratase (E.C. 4.2.1.84) activity led to the selection of *Pseudomonas chlororaphis* B23, the same organism, incidentally, as is used in the second-generation Mitsubishi Rayon process to acrylamide described above.

The *Pseudomonas chlororaphis* B23 cells, immobilized in calcium alginate beads, catalyze hydration of adiponitrile to 5-cyanovaleramide with high regioselectivity (Figure 7.3). Fifty-eight consecutive batch reactions with biocatalyst recycle were run to convert a total of 12.7 metric tons of adiponitrile to 5-cyanovaleramide. At 97% adiponitrile conversion, the yield of 5-cyanovaleramide was 13.6 metric tons (93% yield, 96% selectivity), and the total turnover number of 5-cyanovaleramide produced per unit weight of catalyst was 3150 kg kg^{-1} (dry cell weight).

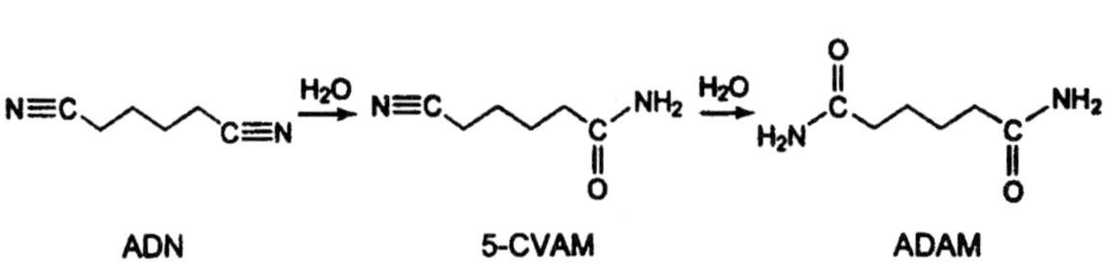

$$N{\equiv}C(CH_2)_4C{\equiv}N \xrightarrow{H_2O} N{\equiv}C(CH_2)_4C(=O)NH_2 \xrightarrow{H_2O} H_2NC(=O)(CH_2)_4C(=O)NH_2$$

ADN 5-CVAM ADAM

Figure 7.3 Process to 5-cyanovaleramide from adiponitrile.

Whereas chemical hydrolysis of adiponitrile with manganese dioxide resulted in significant by-products, 5% adipamide (ADAM) at 25% conversion of adiponitrile, and waste generation [1.25 kg waste (kg 5-CVAM)$^{-1}$], the enzymatic route generated about 5% adipamide upon full conversion of adiponitrile, and a fraction of the waste.

7.1.2
Nitrilase: 1,5-Dimethyl-2-piperidone from 2-Methylglutaronitrile

DuPont has developed another commercial process, based on catalysis by a nitrilase (E.C. 3.5.5.1), to the solvent 1,5-dimethyl-2-piperidone (1,5-DMPD) (Xolvone™) with applications in electronics and coatings (Thomas, 2002). The raw material is 2-methylglutaronitrile (MGN), a by-product during the manufacture of adipodinitrile (ADN) for nylon 6,6 discussed in the previous section. Such a raw material situation leads to coupling of nylon-6,6, 5-cyanovaleramide, and 1,5-dimethyl-2-piperidone production, a situation that most likely is specific to DuPont and thus not prone to much competition.

The route in point is a two-step chemoenzymatic process (Figure 7.4): first, 2-methylglutaronitrile is hydrolyzed to 4-cyanopentanoic acid ammonium salt with the help of immobilized whole-cell *Acidovorax facilis* 72W microbial catalyst at > 98% selectivity and 100% conversion; the dual-hydrolysis product 2-methylglutaric acid is the only by-product. Average productivity after 195 batches was 78 g (L h)$^{-1}$ (Hann, 2002; Chauhan, 2003). After removal of the immobilized cells for recycle, the residual aqueous phase containing the ammonium 4-cyanopentanoate is catalytically hydrogenated over Pd/C in the presence of methylamine. This straightforward process features a markedly improved yield of 1,5-DMPD over the 1,3-DMPD isomer and significantly less waste.

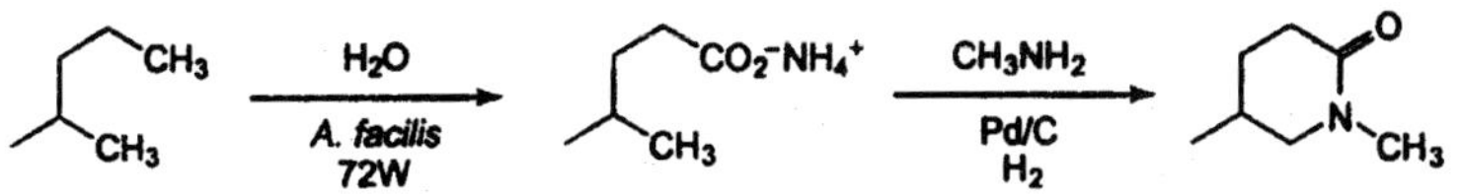

Figure 7.4 Chemoenzymatic process to 1,5-dimethyl-2-piperidone (1,5-DMPD) (Thomas, 2002).

7.1.3
Toluene Dioxygenase: Indigo or Prostaglandins from Substituted Benzenes via cis-Dihydrodiols

The discovery of an intermediate, benzene-*cis*-dihydrodiol, from the oxidation of benzene to CO_2 in a mutant of *Pseudomonas putida* (Gibson, 1968) led to a big effort to find corresponding new transformations and to utilize the diol products for other syntheses.

The reaction is catalyzed by toluene dioxygenase (E.C. 1.13.11) and has no simple equivalent in nature but opens a route to enantiomerically pure, stereochemically

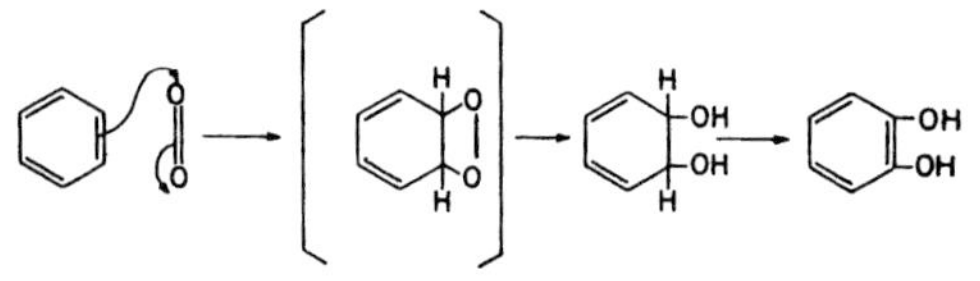

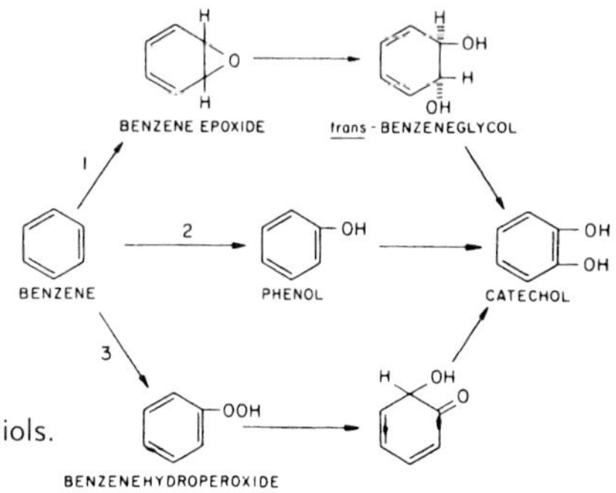

BENZENE EPOXIDE
trans-BENZENEGLYCOL
BENZENE
PHENOL
CATECHOL
BENZENEHYDROPEROXIDE

Figure 7.5 Pathway for oxidation of benzene to dihydrodiols.

predictable intermediates (Figure 7.6a) of a series of benzene-derivatives which can be further converted to carbohydrates, prostaglandins, alkaloids, and other products (Hudlicky, 1988; Carless, 1989; Hudlicky, 1989, 1990; Ley, 1990) (Figure 7.6b). Several reactor configurations have been proposed (Brazier, 1990; Woodley, 1991), among others a process which continuously extracts the diols (Van de Tweel, 1988). Typically, resting cells of *Pseudomonas* strains serve as carriers of enzyme activity in a batch process.

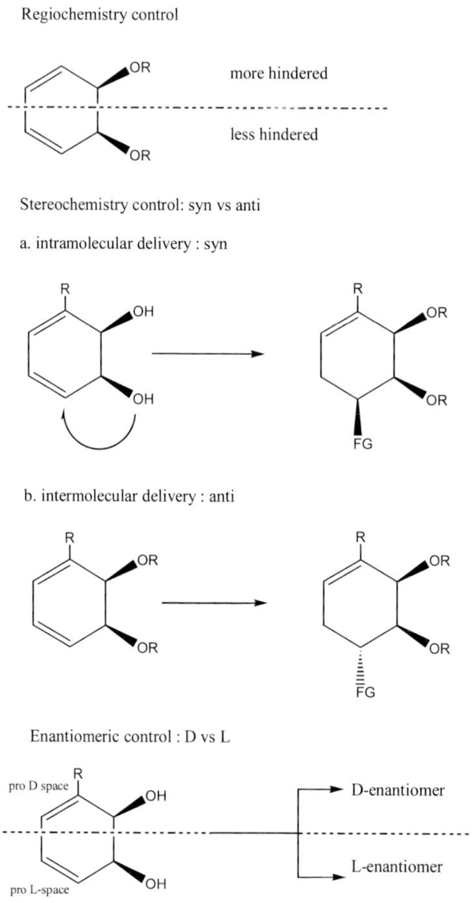

Regiochemistry control

more hindered

less hindered

Stereochemistry control: syn vs anti

a. intramolecular delivery : syn

b. intermolecular delivery : anti

Enantiomeric control : D vs L

pro D space

pro L-space

D-enantiomer

L-enantiomer

Figure 7.6a Stereochemical control with benzene *cis*-dihydrodiols.

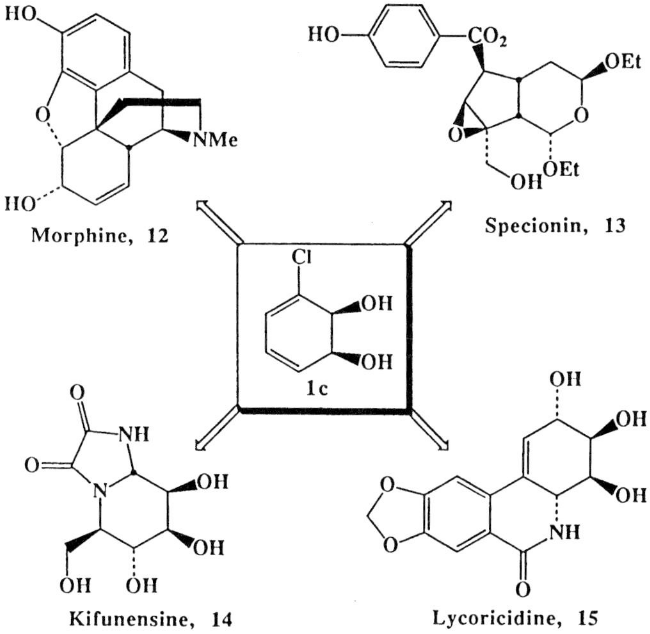

Figure 7.6b Conversions from chlorobenzene *cis*-dihydrodiol.

The blue dye indigo which provides the characteristic color for denims (or "jeans" in Europe) is produced up to this day according to the conventional chemical process that has been in use for almost 100 years. Owing to the vast amounts of waste produced, alternatives employing both chemical catalytic and biocatalytic routes are explored (Figure 7.7). A fermentation process to indigo via tryptophan starting from glucose containing a key dihydroxylation step of intermittently formed indole has been developed by Genencor (Palo Alto/CA, USA) (see Chapter 20). Several other companies, among them Chirotech (Cambridge, UK, now a subsidiary of Dow Chemical Company, previously Chiroscience, earlier Chiros, and even earlier Enzymatix), today offers several of the dihydrodiols in pilot quantities; larger quantities and increased use of these compounds can be expected.

Dioxygenases are important not only for the synthesis of indigo and other potentially large-scale processes but also for an impressive range of syntheses to such diverse products as carbohydrates, prostaglandins, and alkaloids. The status of the field has been reviewed by Hudlicky (1996).

Advantageous use of homochiral cyclohexadiene-*cis*-1,2-diol, available by means of biocatalytic oxidation of chlorobenzene with toluene dioxygenase, has enabled the synthesis of all four enantiomerically pure C_{18}-sphingosines (Nugent, 1998), which are known inhibitors of protein kinase C and important in cellular response mediation for tumor promoters and growth factors. The four requisite diastereomers of azido alcohol precursors were accessed by regioselective opening of epoxides with either azide or halide ions.

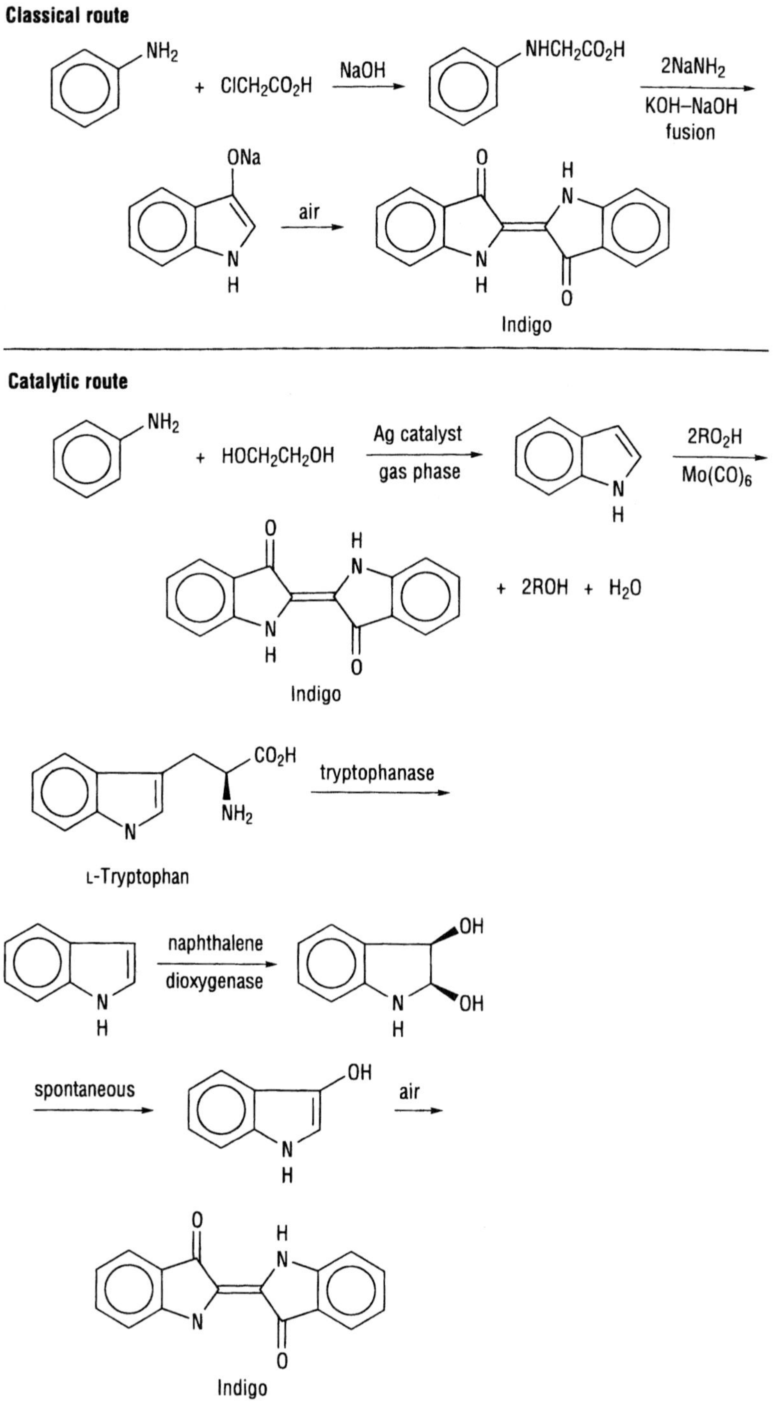

Classical route

NH_2 + $ClCH_2CO_2H$ $\xrightarrow{NaOH}$ $NHCH_2CO_2H$ $\xrightarrow[\text{KOH–NaOH fusion}]{2NaNH_2}$

ONa, N, H $\xrightarrow{air}$ O, H, N, N, H, O

Indigo

Catalytic route

NH_2 + $HOCH_2CH_2OH$ $\xrightarrow[\text{gas phase}]{\text{Ag catalyst}}$ N, H $\xrightarrow[Mo(CO)_6]{2RO_2H}$

O, H, N, N, H, O + 2ROH + H_2O

Indigo

CO_2H, NH_2, N $\xrightarrow{\text{tryptophanase}}$

L-Tryptophan

N, H $\xrightarrow[\text{dioxygenase}]{\text{naphthalene}}$ OH, N, H, OH

$\xrightarrow{\text{spontaneous}}$ OH, N, H $\xrightarrow{air}$

O, H, N, N, O

Indigo

Figure 7.7 Comparison of different processes to indigo (Sheldon, 1994).

The regioselective character of dioxygenase reactions was demonstrated by a whole repertoire of native and recombinant dioxygenases, including toluene dioxygenase, naphthalene dioxygenase, and biphenyl dioxygenase (Whited, 1994). Oxidation on 2-methoxynaphthalene as a test substrate in whole cells yielded (1*R*,2*S*)-dihydroxy-7-methoxy-1,2-dihydronaphthalene as a major product for naphthalene and biphenyl dioxygenase but (1*R*,2*S*)-dihydroxy-6-methoxy-1,2-dihydronaphthalene for toluene dioxygenase. The other compound in each case, and (1*S*,2*S*)-dihydroxy-3-methoxy-1,2-dihydronaphthalene, were detected as minor components. These findings indicate the very high degree of enantioselectivity of dioxygenases in general and very good but specific regioselectivity of each dioxygenase in particular.

7.1.4
Oxynitrilase (Hydroxy Nitrile Lyase, HNL): Cyanohydrins from Aldehydes

In the 1960s, the synthetic potential of (*R*)-oxynitrilase (hydroxyl nitrile lyase, HNL, E.C. 4.1.2.10), which can be isolated from bitter almonds (*Prunus amygdalus*), was recognized (Becker, 1965). The enzyme catalyzes the (*R*)-enantiospecific addition of HCN to a series of aldehydes, utilizing the reverse reaction of a protective mechanism of plants against predators (Table 7.2). In the late 1980s, it was demonstrated (Effenberger, 1987, 1990, 1991) that many aldehydes yield cyanohydrins with a high e.e. value. Enantiomerically pure cyanohydrins can be converted to α-amino and α-hydroxy acids, α-hydroxy aldehydes, β-amino alcohols and other compounds (Figure 7.8).

The achievement of high e.e. values during formation of the cyanohydrins is made difficult by the competitive chemical addition of HCN to the aldehydes as a side reaction, which renders racemic cyanohydrins. As the chemical addition is very fast above the p*K* value of 4.5 for cyanide, different reaction models have been proposed which are supposed to help suppress the unspecific reaction:

- reaction with immobilized enzyme in organic solvent (Brussee, 1990);
- use of acetone cyanohydrin as a donor of HCN (Ognyanov, 1991);
- batch reaction at low pH value in water (Niedermeyer, 1990);
- an enzyme membrane reaction with a high enzyme concentration and very short residence time at a pH value of 3–5 (Niedermeyer, 1989; Kula, 1990); and
- a two-phase system with a small amount of organic solvent (Solvay Duphar, 1991).

(*R*)- and (*S*)-cyanohydrins are both now accessible as a result of the availability, relatively high stability, and convenient handling of several hydroxynitrile lyases (HNLs) (Effenberger, 2000). The optimization of reaction conditions, employing protein engineering (site-directed mutagenesis) as well as reaction engineering (solvent, temperature), has enabled HNL-catalyzed preparations of optically active cyanohydrins on pilot scale. In comparison, the enantioselectivity of chiral metal-complex-catalyzed additions of trimethylsilyl cyanide to aldehydes has been improved, but is a long way from being competitive with the HNL-catalyzed reactions (North, 1993).

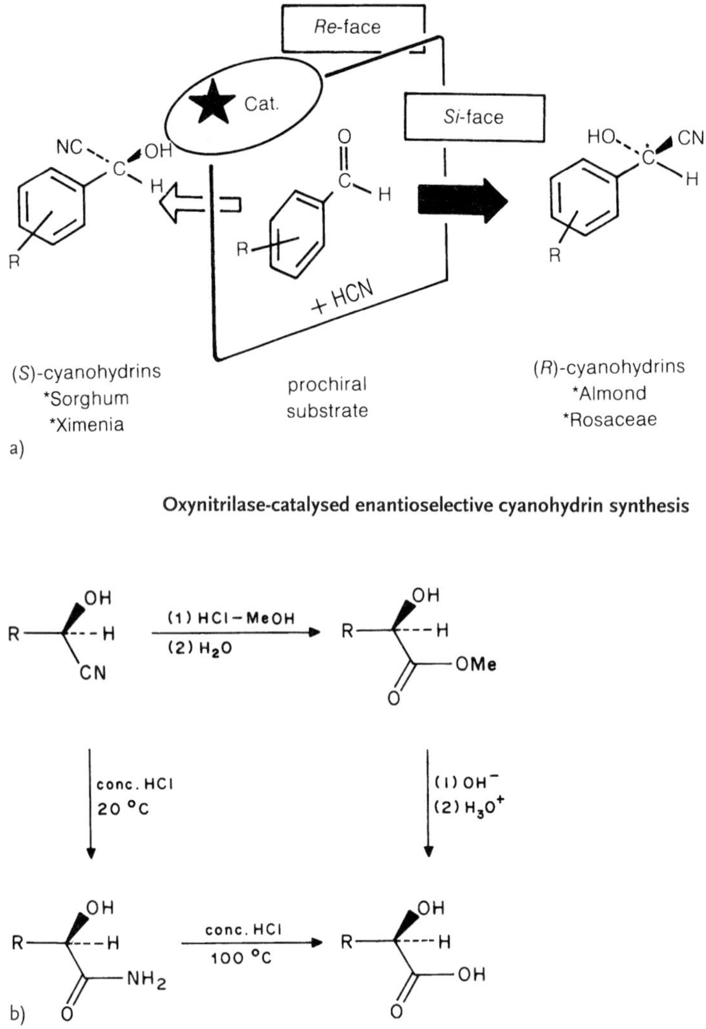

Oxynitrilase-catalysed enantioselective cyanohydrin synthesis

Figure 7.8 a) Enzymatic (oxynitrilase-catalyzed) formation of cyanohydrins from aldehydes. b) Chemical hydrolysis of cyanohydrins to hydroxy acids.

Table 7.2 Enzymatic formation of cyanohydrins from aldehydes (Griengl, 2000).

Structure	*Hnl*	*e.e. [%]*	*Yield [%]*	*Conditions*
(*R*)-Cyanohydrins	Pa	99	95	Ethanol/acetate buffer (ph 5.4), KCN/HOAc, 0 °C
	Pa	97	94	Pal-Inl adsorbed on Auricel, ethyl acetate/acetate buffer (pH 5.4), HCN, room temperature
	Pa	97	80	Pal-Inl adsorbed on Auricel, disopropylether/acetate buffer (pH 4.5), HCN, room temperature
	Pa	95	36	Diethyl ether/acetate buffer (pH 5.0), acetone cyanohydrin, room temperature
	Hb[a]	99	89	Disopropyl ether/citrate buffer (pH 4.0), HCN, 0 °C
(*S*)-Cyanohydrins	Pa[a]	99	71[c]	Pal-Inl adsorbed on Auricel, disopropyl ether/acetate buffer (pH 3.3), HCN, room temperature
	Hb	87	99	Methyl tert-butyl ether/citrate buffer (pH 5.5), HCN, room temperature
	Sb	87	99	Citrate buffer (pH 3.75), HCN, room temperature
	Me	98	92	MeHnl adsorbed on nitrocellulose, diisopropyl ether/citrate buffer (pH 4.3), HCN, room temperature
	Hb	99	94	Citrate buffer (pH 4.0), KCN, 0 °C
	Hb	96	99	Methyl tert-butyl ether/citrate buffer (pH 5.5), HCN, room temperature
	Hb	74	95	Citrate buffer (pH 4.0), 0 °C

Hnl: hydroxynitrile lyase; Pa: *Prunus amygdalus*; Hb: *Hevea brasiliensis*; Sb: *Sorghum bicolor*; Me: *Manihot esculenta*.

7.2
Enzymes as Catalysts in the Fine Chemicals Industry

Applications of biocatalysis in large-scale processes in industry advance only slowly against established chemical processes, even with stoichiometry-based chemistry. Introduction of biocatalysis into existing processes often requires process modifications that are not economical in view of the short life span of the product and/or the low fixed costs of the existing process owing to written-off plant. It should be emphasized that the desire to reduce chemical wastes, imposed by either company policy or governmental measures, needs to be matched by favorable process economics. Therefore, the introduction of biocatalytic options at the very beginning of product and process development is of the utmost importance.

7.2.1
Chirality, and the Cahn–Ingold–Prelog and Pfeiffer Rules

Chiralica are compounds which find application at least in part owing to their property of existence in enantiomerically pure form. As Figure 7.9 demonstrates, two enantiomers can feature completely different properties.

The vast majority of newly registered drugs contain one or several chiral centers (Stinson, 1994). As the wrong enantiomer can cause harmful side effects, very high enantiomeric purity of therapeutics is essential (Crossley, 1995). The observation that chirality can play a major role in the toxicity and specificity of therapeutic agents was first made by Carl Pfeiffer (Pfeiffer, 1956) and is commonly referred to as Pfeiffer's Rule, which states "...the greater the difference between the pharmacological activity of the D and L isomers the greater is the specificity of the active isomer for the response of the tissue under test." In retrospect, it is not surprising that, given the chiral nature of most biological components such as amino acids, sugars, or nucleotides, different enantiomers would produce different pharmacological effects.

Chirality is established by two configurations at a carbon atom (usually) which are not superimposable except by a plane of symmetry. Chirality is commonly caused by four different residues on a carbon atom, although there are many additional forms which cannot be discussed here. A chiral center has either (*R*)- or (*S*)-configuration, which can be determined by the Cahn–Ingold–Prelog rules: the residues are ordered according to priority; (*R*)-configuration is established if the residues of first to third priority yield a clockwise circle with the fourth-priority residue pointing perpendicularly away from the viewer, and correspondingly for the (*S*)-case. Priority is established by a bond from the chiral atom to a neighbor of higher molecular mass; if two have the same mass, the bond to the next-heaviest atom is considered. In contrast to the L/D system of nomenclature, the *R*/*S* system is always unambiguous.

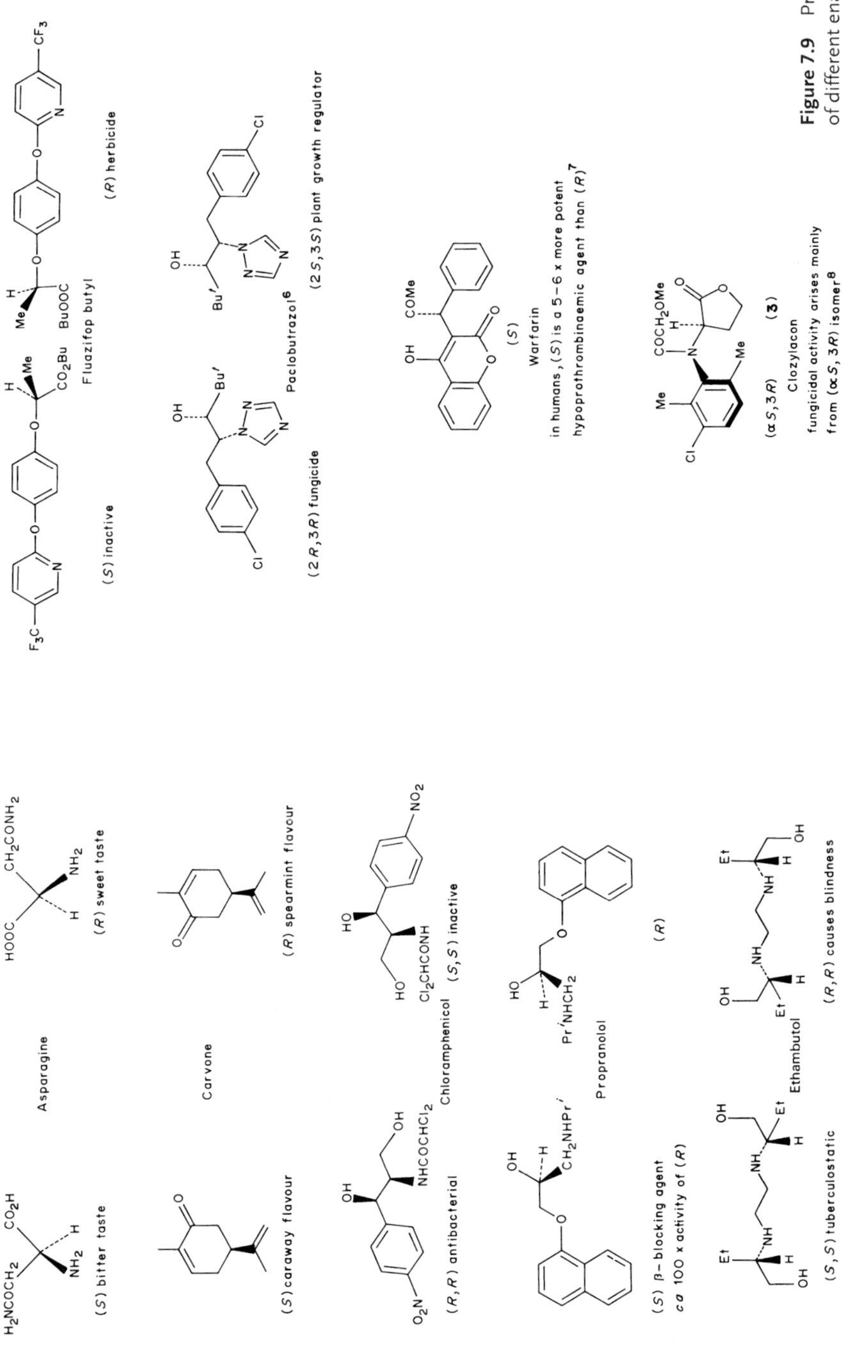

Asparagine

(*S*) bitter taste

(*R*) sweet taste

Carvone

(*S*) caraway flavour

(*R*) spearmint flavour

Chloramphenicol

(*R*,*R*) antibacterial

(*S*,*S*) inactive

Propranolol

(*S*) β-blocking agent
ca 100 x activity of (*R*)

(*R*)

Ethambutol

(*S*,*S*) tuberculostatic

(*R*,*R*) causes blindness

Fluazifop butyl

(*S*) inactive

(*R*) herbicide

Paclobutrazol[6]

(2*R*,3*R*) fungicide

(2*S*,3*S*) plant growth regulator

(*S*)
Warfarin
in humans, (*S*) is a 5–6 x more potent hypoprothrombinaemic agent than (*R*)[7]

(α*S*,3*R*) (**3**)
Clozylacon
fungicidal activity arises mainly from (α*S*, 3*R*) isomer[8]

Figure 7.9 Properties of different enantiomers.

7.2.2
Enantiomerically Pure Amino Acids

Enantiomerically pure amino acids owe their great importance among chiral compounds to the fact that not only are they among the most versatile building blocks with a rich and vast history of transformation to other products such as peptides, amino alcohols, amino aldehydes, and many others, but that most natural L-amino acids are important components of infusion solutions, health food, and animal feed preparations. For this reason, several processes exist on a large scale that are described in the following sections.

7.2.2.1 The Aminoacylase Process

The best-established method for the enzymatic production of L-amino acids is the separation of racemates of *N*-acetyl-DL-amino acids by acylase I (aminoacylase; E.C. 3.5.1.14). *N*-Acetyl-L-amino acid is cleaved and yields L-amino acids whereas *N*-acetyl-D-amino acid does not react (Figure 7.10).

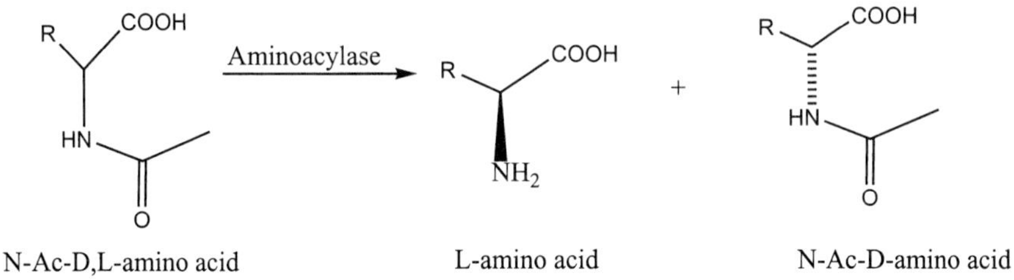

Figure 7.10 Acylase reaction.

After separation of the L-amino acid through ion exchange or by a crystallization step, the remaining *N*-acetyl-D-amino acid is recycled by thermal racemization under drastic conditions or by a racemase to achieve an overall yield of around 45% of L-amino acid (50% is the theoretical maximum) (Figure 7.11). D-Amino acids are also accessible by chemical hydrolysis of the *N*-acetyl-D-amino acid or directly by use of D-selective acylases.

With the aminoacylase process, Tanabe Seiyaku (Japan) in 1969 commercialized the first immobilized enzyme reactor system at all, after the process had been run in batch mode since 1954. Enzyme from *Aspergillus oryzae* fungus was immobilized through ionic bonding to DEAE-Sephadex. In a fixed-bed reactor at elevated temperature, L-methionine, L-valine and L-phenylalanine were produced. In 1982, Degussa (Frankfurt, Germany) introduced the enzyme membrane reactor, with which soluble enantioselective L-aminoacylase, also from *Aspergillus oryzae* fungus, catalyzes the reaction to the respective products at appreciable cost (Figure 7.12). Currently, several hundred tons per year of L-methionine are produced by this enzymatic conversion using an enzyme membrane reactor (Bommarius, 1992).

The starting material for the acylase process is a racemic mixture of *N*-acetyl amino acids which are chemically synthesized, most conveniently by acetylation of

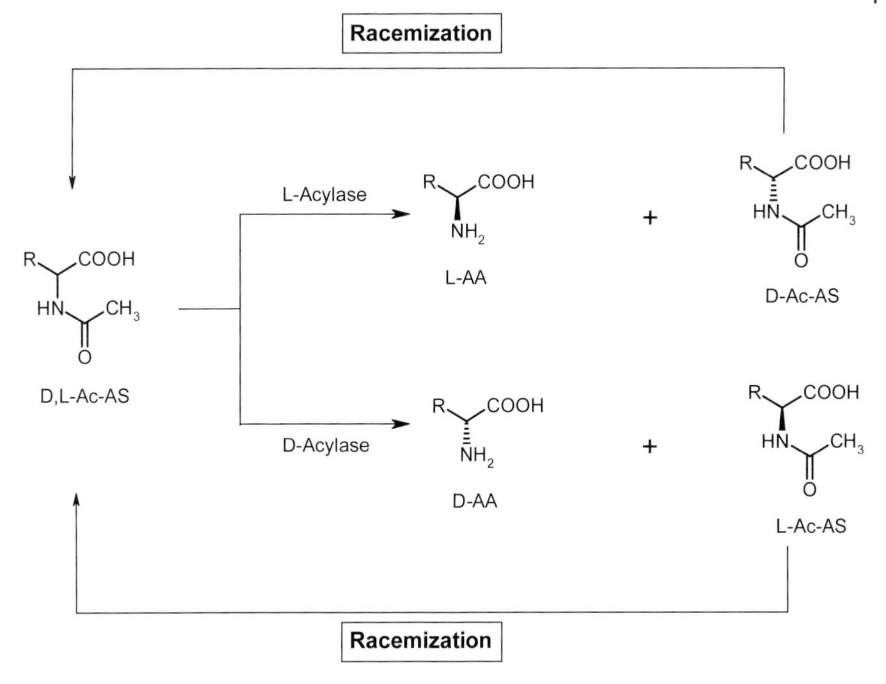

Figure 7.11 Acylase process with racemization (May, 2002). Conventional process: racemization at pH 3, 80–100 °C; novel process: racemization with enzyme (NAAR), pH 7–8, 25–40 °C.

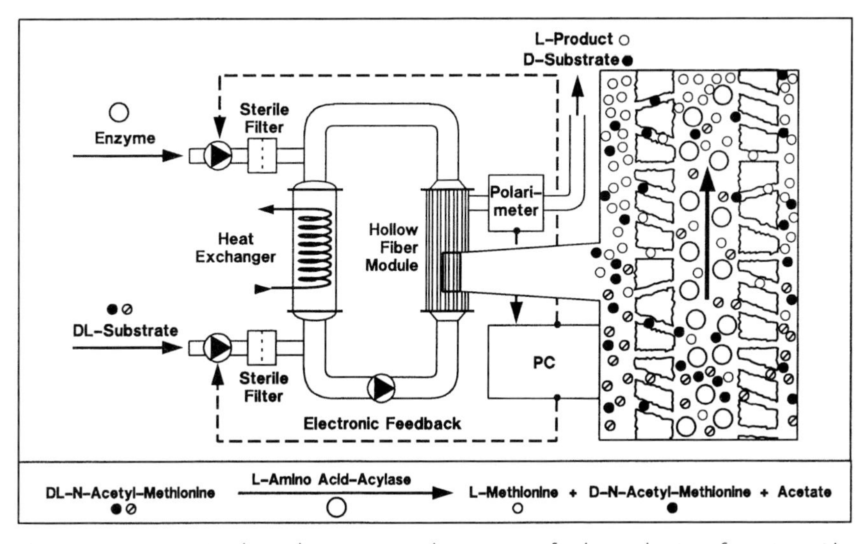

Figure 7.12 Degussa's pilot-scale enzyme membrane reactor for the production of L-amino acids.

DL-amino acids with acetyl chloride or acetic anhydride in alkali in a Schotten–Baumann reaction (Sonntag, 1954).

Selective in-situ racemization of *N*-acetyl amino acid could dramatically improve the acylase process by eliminating costly racemization and separation steps. Such an *N*-acylamino acid racemase (NAAR) activity was found by Takeda in the actinomycete *Amycolatopsis* sp. TS-1-60 (Tokuyama, 1994, 1995a,b) and by Degussa in *Amycolatopsis orientalis* subsp. *lurida* (Verseck, 2001). The latter enzyme was screened on the basis of the known consensus motif, KXK, in the active site (Babbitt, 1996) and the highly homologous N-terminus with other enolase superfamily members or the other known NAAR (Tokuyama, 1995a). Both NAAR genes have identities of 86% at the level of DNA, and 90% at the level of amino acids (Tokyama, 1995b; Verseck, 2001). Both enzymes require divalent metal ions for activation and feature side chain specificity similarly to L-aminoacylase. As a D-acylase has also been developed recently (Hibino, 1999; Isobe, 1999; Tokoyama, 1999), the process can now be driven to L- or D-amino acids, with in-situ racemization of the remaining *N*-acetyl amino acid enantiomer as an option.

7.2.2.2 **The Amidase Process**

Another process to L-amino acids starts from racemic α-H-α-amino acid amides, which can be obtained through alkaline hydrolysis of the cyano precursor, which in turn is available from Strecker synthesis of aldehyde, HCN, and ammonia (Figure 7.13). DSM (previously Dutch State Mines) in Geleen, The Netherlands, has optimized this process and operated it for years. Three enzymes, aminopeptidase from *Pseudomonas putida* (Hermes, 1993), amino acid amidase from *Mycobacterium neoaurum* (Schoemaker, 1992), and broad-range amidase from *Ochrobactrum anthropi* NCIMB 40321 (Van de Tweel, 1993) have been purified and characterized; the first enzyme is most widely employed, usually in whole-cell form. All enzymes yield L-amino acid and D-amides in enantiomeric purities e.e. of > 99% for a wide variety of amino acid amides. In contrast to the slightly thermodynamically limited acylase process, the amidase process reaches 50% conversion. Biocatalyst consumption was determined for the *Ps. putida* biocatalyst by automated sampling of conversion over time by a coupled reversed-phase HPLC analysis with fluorimetric detection (Duchateau, 1991).

7.2.2.3 **The Hydantoinase/Carbamoylase Process**

The hydantoinase process, consisting of a racemization reaction and hydrolyses of the hydantoin and the carbamoylic acid (Figure 7.14), has been enjoying much industrial success (> 1000 tpy) for almost 30 years in the production of D-amino acids such as D-phenylglycine and *p*-OH-phenylglycine which serve as side chains for β-lactam antibiotics ampicillin and amoxicilin (Cecere, 1976).

In contrast to acyl amino acids ($pK_a > 30$) or amides, most 5-monosubstituted hydantoins racemize comparatively easily; phenyl-substituted ones even racemize spontaneously at slightly alkaline conditions as their pK_a is around 8 (Kato, 1987). Under spontaneous or enzymatic racemization (Pietzsch, 1990), racemic hydantoins with the help of enantioselective D- or L-hydantoinases and the respective carb-

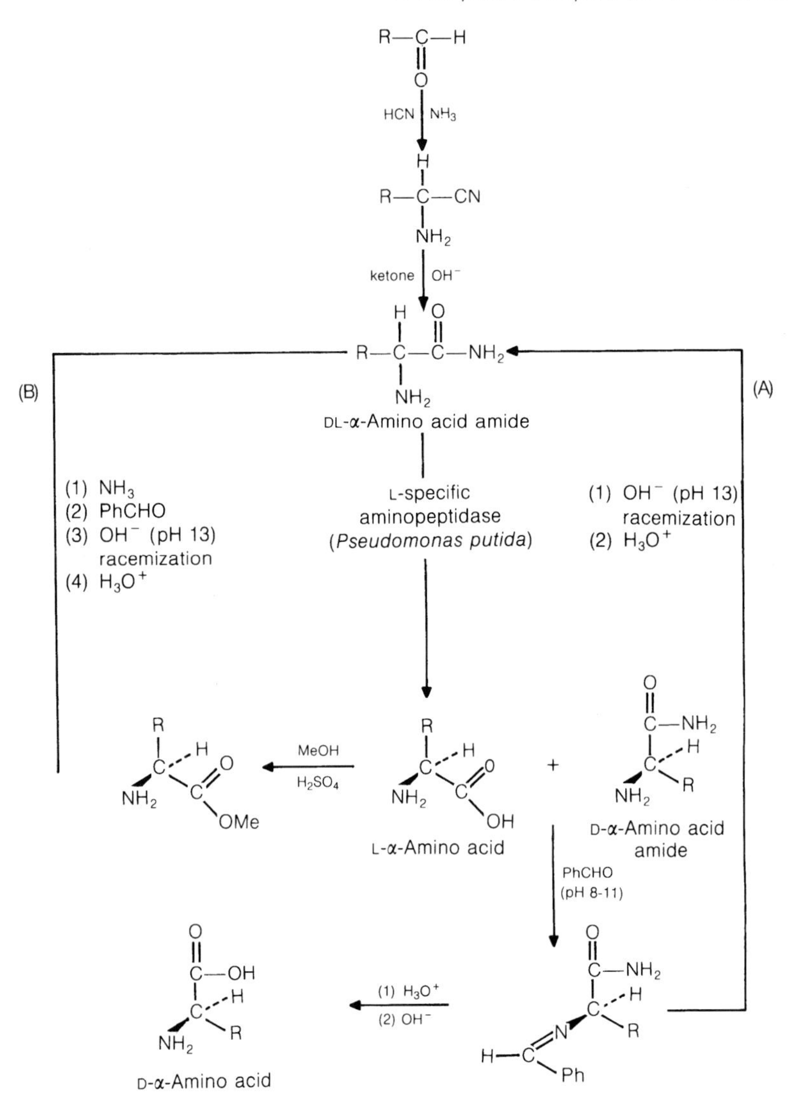

Figure 7.13 Amidase process from DSM to L-amino acids.

amoylases can be converted to enantioselective D- or L-amino acid products at close to 100% yield in mild conditions with only water and ammonia as by-products (Figure 7.14).

Thus, provided all necessary biocatalysts are or were available, either D- or L-amino acids can be synthesized. Straightforward synthesis of most racemic 5-monosubstituted hydantoins from inexpensive and often readily available starting materials through the following reactions provides another advantage for the process (Syldatk, 1999):

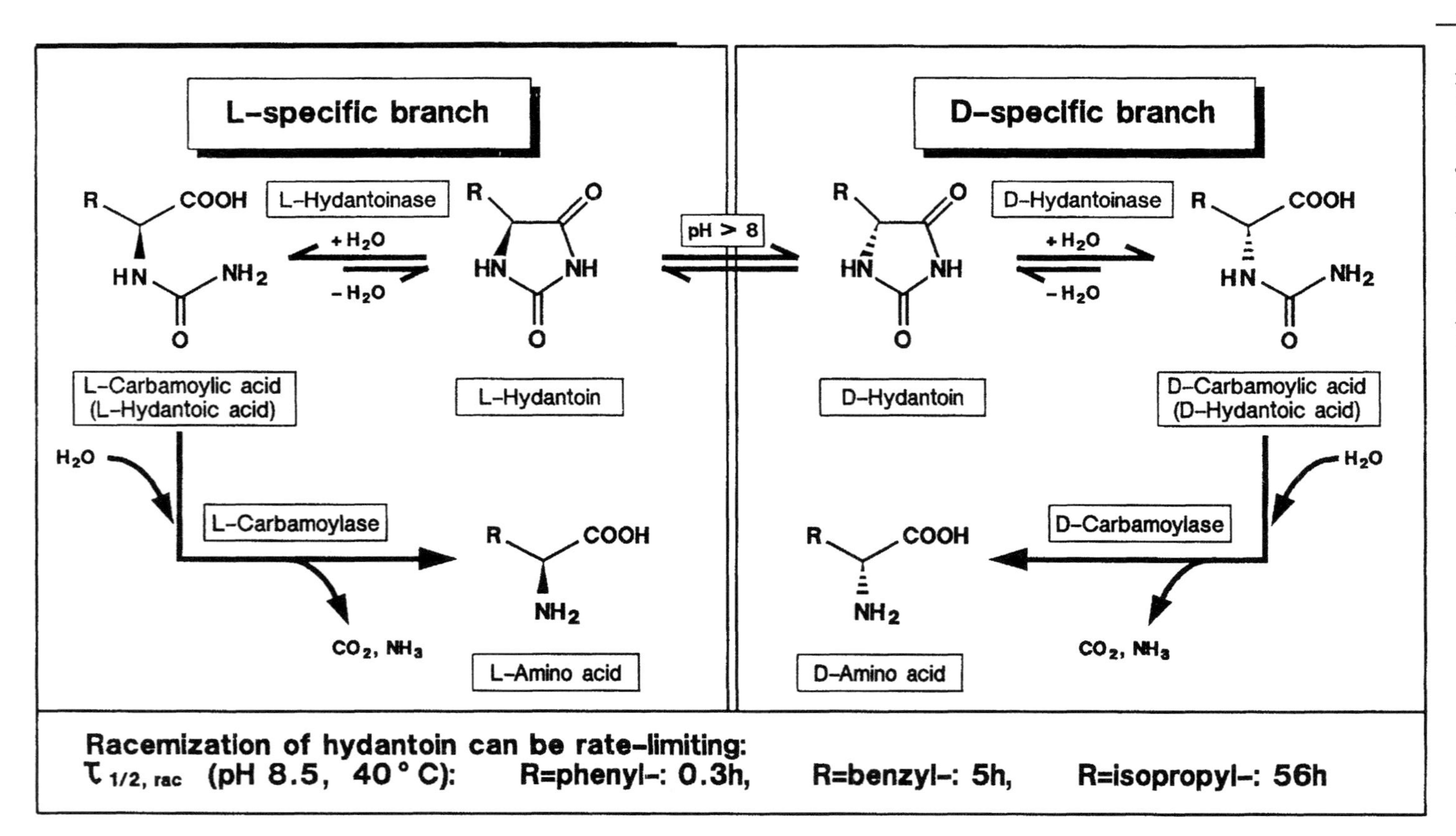

Figure 7.14 Hydantoinase/carbamolyase/(hydantoin racemase) process to D- or L-amino acids.

- aldol condensation between aldehydes and (unsubstituted) hydantoin with subsequent hydrogenation;
- Strecker reaction between an aldehyde, HCN, and ammonium carbonate; or
- carbamoylation of amino acids with cyanate (NaOCN or KOCN) and ring closure under acidic conditions.

As evidenced by the well-established industrial production of D-amino acids (mostly D-phenylglycine and *p*-OH-D-phenylglycine by the hydantoinase/carbamoylase route), both D-hydantoinases and D-carbamoylases are well developed. Enantioselectivity in most cases is not a problem (Gross, 1987).

In contrast to the D-branch, application of L-hydantoinases and L-carbamoylases has not gone beyond small pilot scale yet. The enantioselectivity of most L-hydantoinases varies depending on the substitution in the 5-position and can even cross over to the D-side (Nishida, 1987). In addition, L-carbamoylases are often very unstable (Cotoras, 1984). Over several years, the following steps were taken to improve the process to economically necessary levels (May, 2002):

- increasing the activity of the biocatalyst and improving the space–time yield (F. Wagner, 1996; T. Wagner, 1996);
- design of a whole-cell biocatalyst by overexpressing genes of hydantoinase, carbamoylase, and hydantoin racemase from *Arthrobacter* sp. DSM 9771 (F. Wagner, 1996; T. Wagner, 1996); and
- improving the enantioselectivity of the "L"-hydantoinase by directed evolution (May, 2000; see Chapter 11, Section 11.4.6).

Given the promising catalytic, economic, and ecological fundamentals of both the D- and the L-hydantoinase/carbamoylase technology, further technological advances are being actively sought. Thus, the perspective for this technology remains bright.

7.2.2.4 Reductive Amination of Keto Acids (L-tert-Leucine as Example)

Overview

L-*tert*-leucine is a non-proteinogenic amino acid with special properties owing to the close proximity of the sterically demanding tertiary butyl group to the amino and carboxyl moiety. As peptide bonds involving L-*tert*-leucine are only slowly hydrolyzed by proteases in the body, pharmaceuticals containing it or its derivatives are part of a variety of novel pharmaceuticals (Figure 7.15) (Bommarius, 1995; Whittaker, 1999). Probably owing to the structural singularity of L-*tert* leucine, all other synthesis strategies (such as the acylase, amidase, and hydantoinase/carbamoylase routes discussed above) failed, so reductive amination route of α-keto acids was developed to obtain the compound (Figure 7.16). Trimethylpyruvate, the substrate obtained through (i) oxidation of pinacolone, (ii) hydrolysis of pivaloyl cyanide (available from reaction of pivaloyl chloride with cyanide), or (iii) reaction of pivaloyl chloride with oxalic ester and hydrolysis, is reduced with incorporation of ammonia and catalysis by leucine dehydrogenase (LeuDH) to L-*tert*-leucine; the concomitant oxidation occurs with NADH to NAD^+. As the biological cofactor

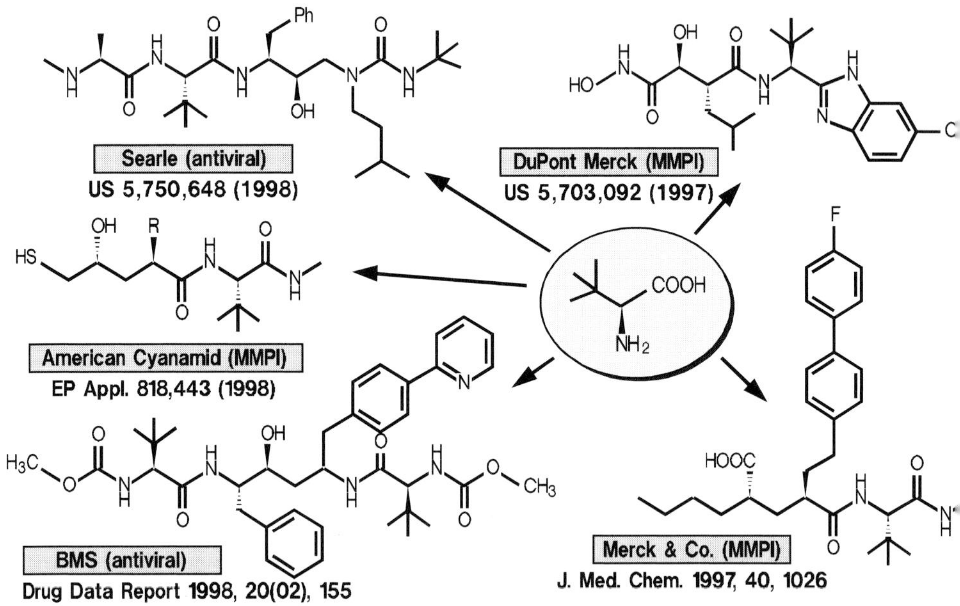

Searle (antiviral)
US 5,750,648 (1998)
DuPont Merck (MMPI)
US 5,703,092 (1997)
American Cyanamid (MMPI)
EP Appl. 818,443 (1998)
BMS (antiviral)
Drug Data Report 1998, 20(02), 155
Merck & Co. (MMPI)
J. Med. Chem. 1997, 40, 1026

Figure 7.15 L-*tert*-Leucine (L-Tle) as a building block of several novel pharmaceuticals.

NADH is much too expensive to be used in stoichiometric amounts, the reaction product NAD^+ has to be regenerated (i.e., reduced) to NADH by irreversible oxidation of formate to CO_2 catalyzed by formate dehydrogenase (FDH) (Schütte, 1976). Other, far less suitable, regeneration schemes for NADH from NAD^+ employ glucose, which is oxidized to gluconic acid with the help of glucose dehydrogenase (GluDH) (Hanson, 1999), or, for regeneration to NADPH from $NADP^+$, glucose-6-phosphate afforded by glucose-6-phosphate dehydrogenase (Wong, 1981; Hirschbein, 1982).

COOH
Trimethylpyruvate
LeuDH
COOH
NH2
L-Tle
NADH + H⊕
NAD⊕
CO2
HCOO⊖ NH4⊕
FDH

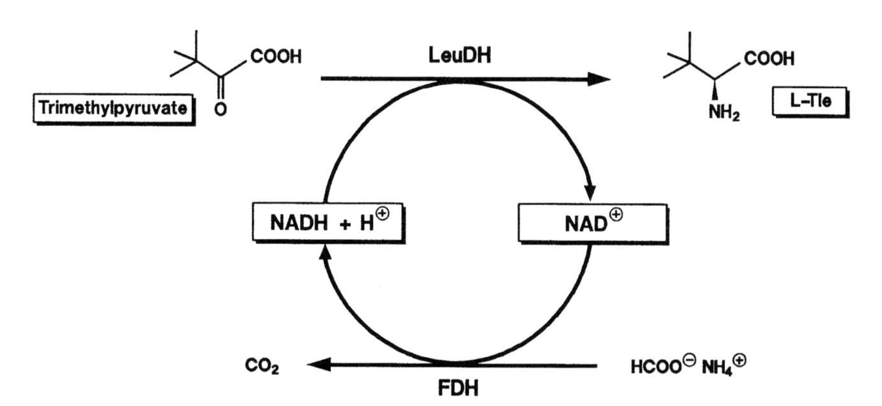

Figure 7.16 Reductive amination of trimethylpyruvate to L-*tert*-leucine (L-Tle).

Biocatalysts

Leucine dehydrogenase (LeuDH, E.C. 1.4.1.9) has been isolated and characterized with respect to substrate specificity from three bacilli, *B. sphaericus* (Hummel, 1981; Ohshima, 1989a), *B. cereus* (Schütte, 1985), and the moderately thermophilic *B. stearothermophilus* (Ohshima, 1985a,b; Nagata, 1988; Ohshima, 1989a,b; Bommarius, 1994; Krix, 1997). The sequential and structural relationship of the three LeuDHs is discussed in Chapter 14. LeuDH from *B. stearothermophilus* (Nagata, 1988; Oka, 1989) and from *B. cereus* (Ansorge, 2000a,b) have been cloned, overexpressed, and developed for production on a large scale. The substrate specificity of LeuDHs reveals a preference for aliphatic α-keto acids, both straight-chain and branched, and a similar pH profile. The kinetic parameters v_{max} and K_M show opposite behavior for good and bad substrates: the dimethyl-substituted substrates show K_M values above 10 mM (v_{max} values are between 0.2 and 30% of the reactivity of 2-oxo-4-methylpentanoic acid, the base case), whereas K_M values below 1 mM are typical for good substrates ($v_{max} \approx$ 100% compared to the base case, 2-oxo-4-methylpentanoic acid).

Formate dehydrogenase (FDH, E.C. 1.2.1.2) has been investigated most intensively on the enzyme from *Candida boidinii*, a homodimer (MW 40.6 kDa per subunit), a pI of 5.4, *T* and pH optima of 55 °C and 7.0, respectively, a K_M value of 5.0 mM towards formate and a specific activity of 6.5 U (mg protein)$^{-1}$ (Slusarczyk, 1997). The wild-type FDH is rather labile to oxidation in solution. Homology modeling of the *C. boidinii* FDH with that from *Pseudomonas putida* with a solved structure (Lamzin, 1992) and 40.5% amino acid sequence identity revealed a cysteine (Cys23) and a methionine (Met134) penetrating the surface as well as another cysteine (Cys262) in the interior. Thus, the *Candida* FDH was stabilized by the generation of partially or wholly cysteine-free mutants through site-directed mutagenesis (Chapter 10) (Slusarczyk, 2000). As the specific activity with 6.5 U mg^{-1} is very low, enhancement of specific activity has been conducted by directed evolution (see Chapter 11) (Slusarczyk, 2003).

Processing

While L-*tert*-leucine has been synthesized on a small scale (in this case, up to 20 kg) by a batch process (Bommarius, 1994), a continuous version was developed for a larger scale (Wichmann, 1981; Wandrey, 1986b). Instead of immobilization on beads, LeuDH and FDH are preferentially retained in an enzyme membrane reactor (EMR) as reactants and the products are highly soluble. To keep the cofactor from penetrating through the membrane, it can be enlarged optionally with polyethylene glycol (PEG) (Kula, 1988). The pros and cons of an EMR were discussed in Chapter 5, Section 5.3, and the enzyme membrane reactor design is covered in detail in Chapter 19, Section 19.2. Complementary α-keto acid substrate specificity between LeuDH, PheDH, and GluDH, high stability, and extremely high enantioselectivity in a coupled process with FDH demonstrate the possibilities of this process. Total turnover numbers (TTNs) of 600 000 and 125 000 have been achieved with phenylpyruvate/PheDH and trimethylpyruvate/LeuDH, respectively, as well as space–time yields of 638 g (L d)$^{-1}$ for trimethylpyruvate/LeuDH (Bommarius,

1992). While L-*tert* leucine has been the main target, reductive amination has been employed as a technique for a range of L-amino acids, such as L-leucine (Wandrey, 1986a,b; Kragl, 1992), L-phenylalanine (Hummel, 1987), L-neopentylglycine (Krix, 1997), L-β-hydroxyvaline (Hanson, 1990), or 6-hydroxy-L-norleucine (see Chapter 13, Section 13.3.4.2) (Hanson, 1999). The economics of the process is influenced decisively by the retention and regeneration of both production (amino acid DH) and regeneration enzyme (FDH).

7.2.2.5 **Aspartase**

Owing to the commercialization of Aspartame® the demand for L-aspartic acid increased steeply. L-Aspartate can be produced by enantioselective addition of ammonia to fumaric acid catalyzed by aspartase (E.C. 4.3.1.1) (Figure 7.17).

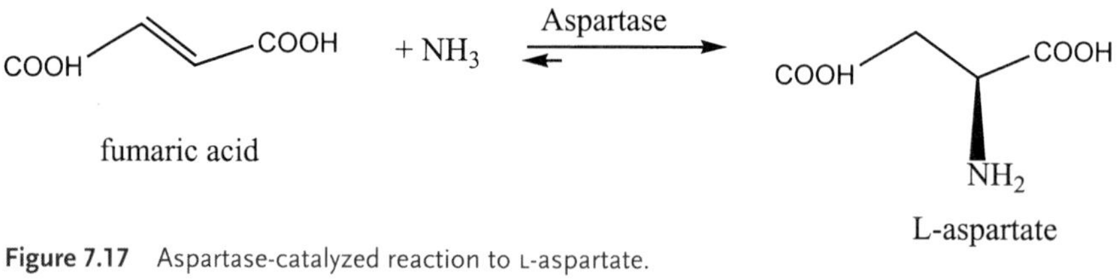

Figure 7.17 Aspartase-catalyzed reaction to L-aspartate.

From 1960, Tanabe was producing L-aspartate with living *E. coli* cells in a batch reactor until the company changed over to an immobilized enzyme system in 1973. Aspartase from *E. coli* cells was inactivated by a 1 M ammonium fumarate substrate solution (50% after 30 days at 30 °C); in contrast, the cells in polyacrylamide gels yielded a half-life of up to 120 days at 37 °C. By immobilization onto κ-carrageenan gel which had been hardened by treatment with glutaraldehyde and hexamethylenediamine, and by switching to a genetically modified *E. coli* strain, the stability was enhanced even further (in 1978, the half-life was two years). The process is run on a scale of 4000 tpy.

7.2.2.6 **L-Aspartate-β-decarboxylase**

Tanabe produces L-alanine from L-aspartate (see above) by decarboxylation with the help of L-aspartate-β-decarboxylase (E.C. 4.1.1.12) from *Pseudomonas dacunhae* (Figure 7.18).

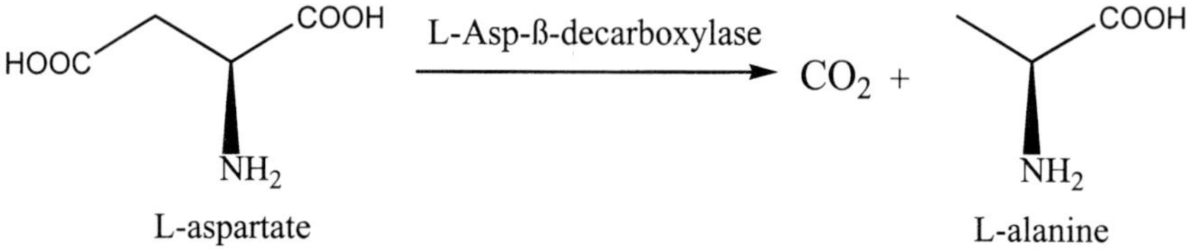

Figure 7.18 Decarboxylation of L-aspartate to L-alanine.

Because of the difficulties of stabilizing the pH at the pH optimum of 6.0 owing to liberation of CO_2, a loop reactor was developed which keeps the CO_2 dissolved at 10 bar and thus helps to stabilize the pH. Co-immobilization of *E. coli* and *Ps. dacunhae* cells for direct production of L-alanine from fumaric acid was not successful because *E. coli* cells work best at a pH of 8.5, in comparison with a pH of 6.0 for *Ps. dacunhae* cells and the decarboxylase. The sequential process has been run since 1982. The high enantioselectivity of L-aspartate-β-decarboxylase (ADC) led to a process, in 1989, in which inexpensive DL-aspartate was converted to L-alanine and D-aspartate [Eq. (7.1)]; the latter commands interest for synthetic penicillins:

$$\text{NH}_4\text{ fumarate} \xrightarrow{\text{aspartase}} \text{DL-aspartate} \xrightarrow{\text{L-Asp-}\beta\text{-decarboxylase}} \text{D-aspartate} + \text{L-alanine} + \text{CO}_2 \qquad (7.1)$$

7.2.2.7 L-2-Aminobutyric acid

In a clever utilization of unstable chemical intermediates of oxidized L-threonine, L-2-aminobutyric acid was synthesized in a transamination reaction from L-threonine and L-aspartic acid as substrates in a whole-cell biotransformation using recombinant *Escherichia coli* K12 (Fotheringham, 1999). The cells contained the cloned genes *tyrB*, *ilvA* and *alsS* which respectively encode tyrosine aminotransferase of *E. coli* (E.C. 2.6.1.5), threonine deaminase of *E. coli*, and α-acetolactate synthase of *B. subtilis* (E.C. 4.1.3.18). The 2-aminobutyric acid was produced by the action of the aminotransferase on 2-ketobutyrate and L-aspartate (Figure 7.19). The 2-ketobutyrate is generated in situ from L-threonine by the action of the deaminase, and the pyruvate by-product is eliminated by the acetolactate synthase. The concerted action of the three enzymes offers significant yield and purity advantages over the process using the transaminase alone, with an eight- to tenfold increase in the ratio of product to the major impurity.

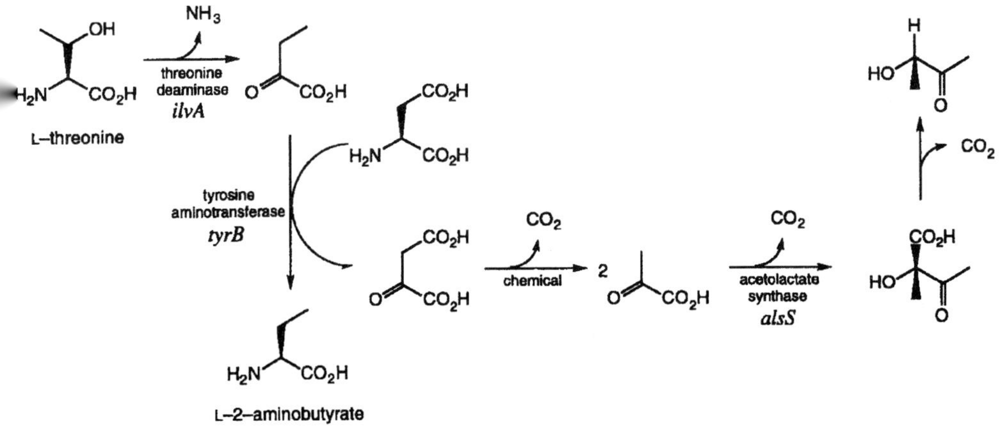

Figure 7.19 L-Aminobutyric acid from L-threonine.

7.2.3
Enantiomerically Pure Hydroxy Acids, Alcohols, and Amines

7.2.3.1 Fumarase

L-Malic acid is used in the form of a salt complementing basic amino acids and as an acidulant in the food industry, on a scale of 500 tpy. Analogously to L-aspartate, L-malic acid can be obtained from fumaric acid and water with the help of fumarase (E.C. 4.2.1.2) ($K = 4.4.$ at 25 °C) (Figure 7.20).

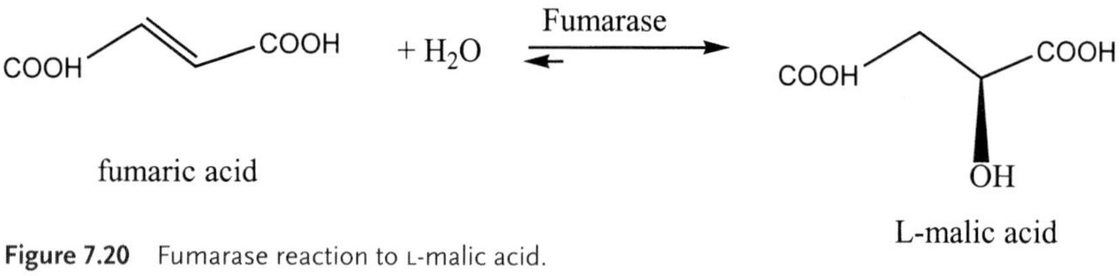

COOH–CH=CH–COOH (fumaric acid) + H_2O ⇌ (Fumarase) COOH–CH₂–CH(OH)–COOH (L-malic acid)

Figure 7.20 Fumarase reaction to L-malic acid.

Process development with *Brevibacterium ammoniagenes* (introduced in 1974), similarly to the development in the L-aspartate process led to 25-fold improvements, mainly through use of a *Brevibacterium flavum* strain immobilized on κ-carrageenan gel with polyethyleneimine (PEI) crosslinker, so that a half life of 310 days at 37°C has been achieved.

7.2.3.2 Enantiomerically Pure Amines with Lipase

Racemic amines can be resolved efficiently by using ethyl methoxyacetate, $MeOCH_2CO_2Et$, as acylating agent in a lipase-catalyzed reaction. As an example, treating racemic phenylethylamine, $PhCH(NH_2)CH_3$, with $MeOCH_2CO_2Et$ in the presence of a lipase from *Burkholderia plantarii* (E.C. 3.1.1.1) gave 41% (*R*)-PhCH $(NHC(O)CH_2OMe)CH_3$ and 45% (*S*)-phenylethylamine, (*S*)-$PhCH(NH_2)CH_3$, both with e.e. > 93% (Balkenhohl, 1997) (Figure 7.21). The methoxy group turned out

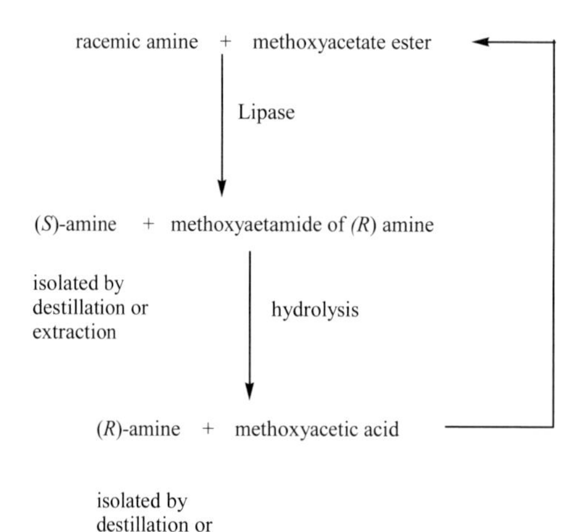

Figure 7.21 BASF's lipase process.

to be crucial for a sufficiently high enzymatic rate. The process can now be run at BASF (Ludwigshafen, Germany) on a scale of 1000 tpy, essentially using existing infrastructure as the reaction does not need much dedicated equipment and technology.

7.2.3.3 Synthesis of Enantiomerically Pure Amines through Transamination

While kinetic resolution with the help of lipases or esterases has seen the greatest success for the synthesis of enantiomerically pure amines, the same target can be reached by employing ω-transaminases (ω-TA) to reductively transaminate ketones to either (*S*)- or (*R*)-amines, depending on the transaminase. The reaction is shown in Figure 7.22 with acetophenone and (*S*)-transaminase as an example (Shin, 1998, 1999).

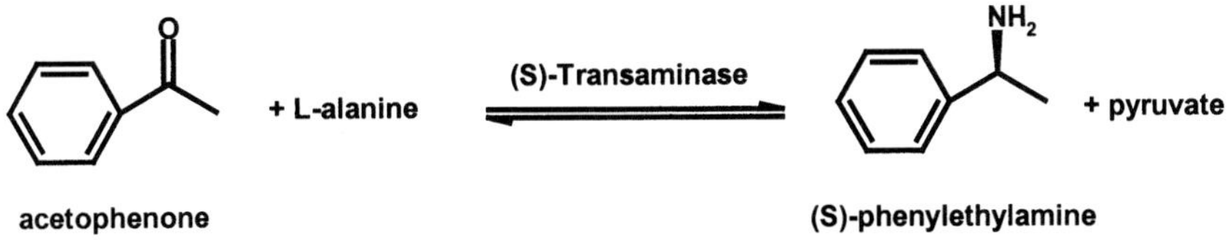

Figure 7.22 Enantiomerically pure (*S*)-amines via ω-transaminases (Drauz, 2002).

Each enantiospecific ω-transaminase can be applied for the synthesis of (*S*)- or (*R*)-amines (Stirling, 1992) by:

- reductive transamination of the ketone to the amine by a transaminase of the same specificity, or
- separation of the racemic mixture of the amine to obtain the remaining amine enantiomer of opposite enantiospecificity to the transaminase, in 50% yield maximally.

Both (*R*)- and (*S*)-aminotransferase are available for the synthesis of enantiomerically pure amines from racemic amines. Degrees of conversion were at or close to 50% for resolutions, and enantioselectivities customarily reached > 99% e.e. for the amine product from both resolutions or syntheses from ketones (Stirling, 1992; Matcham, 1996). The donor for resolutions of amine racemates was usually pyruvate whereas either isopropylamine or 2-aminobutane served as donors for reduction of ketones. The products range from L- and D-amino acids such as L-aminobutyric acid (see Section 7.2.2.6) and L-phosphinothricin (see Section 7.4.2) to amines such as (*S*)-MOIPA (see Section 7.4.2).

Challenge: driving the reaction to completion

The major disadvantage of the transamination technology is an equilibrium constant *K* often near unity. As $K \approx 1$ would limit the net conversion of substrates to around 50%, the key to efficient transamination technology lies in overcoming the problem of incomplete conversion of the 2-keto acid precursor to the desired amino acid product. The main approach to this problem is the coupling of the transamination reaction to a second reaction that consumes the by-product keto acid in an essentially irreversible step; this drives the transamination reaction to completion. When an aminotransferase that can utilize aspartic acid efficiently as the amino group donor is used, the corresponding by-product 2-keto acid is oxaloacetate, which, as a β-ketoacid, can be easily decarboxylated to pyruvate. This decarboxylation occurs spontaneously in aqueous solution (Figure 7.23), accelerated by base, various metal ions, and amines, or enzymatically using the enzyme oxaloacetate decarboxylase. The essentially irreversible decarboxylation of oxaloacetate to pyruvate drives the entire process to completion, allowing the transamination of 2-keto acids to amino acids in yields approaching 100% of theoretical (Rozzell, 1985, 1987; Crump, 1990, 1992). As even pyruvate often interferes with the production reaction (it is a keto acid like the substrate), pyruvate can be dimerized to acetolactate with the help of acetolactate synthase (ALS), which in turn spontaneously decarboxylates to acetoin. Importantly, all these methods of driving the reaction to completion may be used for the production of either D- or L-amino acids.

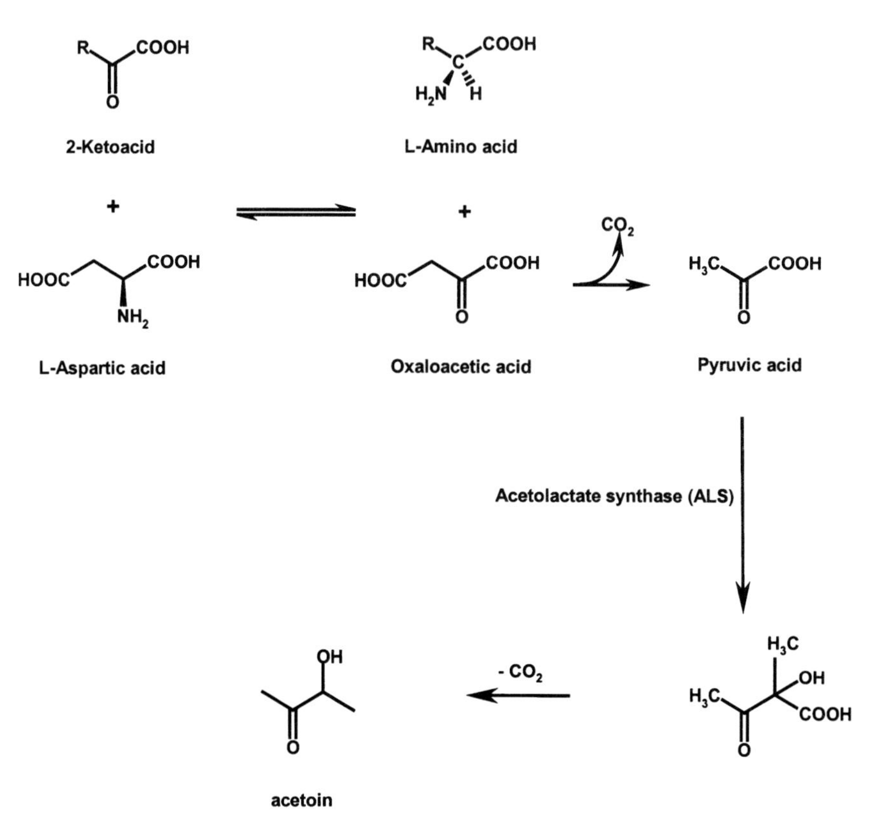

Figure 7.23 Driving the transamination reaction to completion (Drauz, 2002).

7.2.3.4 **Hydroxy esters with carbonyl reductases**

Enantiomerically pure alcohols are valuable building blocks in a range of pharmaceuticals; in particular, alkyl (*R*)-4-chloro-3-hydroxybutanoate is a convenient precursor of L-carnitine (see Section 7.3.4) whereas alkyl (*S*)-4-chloro-3-hydroxybutanoate and alkyl (*R*)-3-hydroxybutanoate are important intermediates in the synthesis of HMG-CoA reductase inhibitors (anticholestemics; Chapter 13) or angiotensin-converting enzyme (ACE) inhibitors. Carbonyl reductase from *Candida magnoliae* was found to convert the ethyl esters of 4-substituted 3-oxobutanoates efficiently to their (*S*)-3-hydroxy equivalents with very high enantioselectivity, > 99% (Yasohara, 2001). It should be noted that, owing to the changing priority of substituents according to the Cahn–Ingold–Prelog rules (see Section 7.2.1) the unsubstituted ethyl 3-oxobutanoate is reduced to the ethyl (*R*)-3-hydroxybutanoate, with > 99% e.e. as well! Carbonyl reductase from *Candida magnoliae* requires NADPH as cofactor; glucose dehydrogenase (GDH) was picked as regeneration enzyme, oxidizing glucose to gluconic acid. To circumvent the problem of weak cofactor regeneration in native whole-cell yeast systems, a transformant was constructed containing the genes of both carbonyl reductase and glucose dehydrogenase. As 4-substituted 3-oxobutanoates tend to be chemically unstable, a two-phase water/butyl acetate system was employed which did not compromise both enzymes' stabilities. Reaction conditions were 20 h at 30 °C in phosphate buffer at pH 6.5, 0.4 M, of substrate (3-oxoester) and glucose each, based on the aqueous volume, saturating concentrations of NADPH, and a water/butyl acetate phase ratio of of 1 : 1. The turnover number of $NADP^+$ was found to be 650. Ethyl (*S*)-4-bromo-3-hydroxybutanoate (0.38 M, $x = 0.95$) was recovered. In another configuration, a higher substrate concentration of 1.6 M yielded ethyl (*S*)-4-bromo-3-hydroxybutanoate in 94% chemical yield and 97% e.e. (Katoaka, 1999).

The relevance of dehydrogenases and transaminases has now been firmly established, but the search for suitable biocatalyst and reactor configuration continues. Asymmetric reductions from ketones to alcohols can be performed with growing or non-growing cells (maintenance only, $\mu = 0$) which provide both the reductase and the cofactor-recycling enzyme. The advantage is obvious: no separate cofactor-recycling system is required. On the other hand, the cofactor-recycling power is often not very strong, so that the whole system becomes limited by the recycling part. In addition, and often much more seriously, there are several reductases present in the yeast system which sometimes feature counteracting enantioselectivities. At least three enzymes with reducing activity towards β-keto esters have been isolated from baker's yeast, fatty acid synthase, aldo–keto reductase, and α-acetoxyketone reductase (Shieh, 1985, 1993; Sybesma, 1998; Stewart, 2001). There are essentially two methods to improve this situation: (1) partial (or whole) purification, and (2) genetic engineering of yeasts:

1. In *Geotrichum candidum*, both an (*R*)- and an (*S*)-dehydrogenase are present; this was found only when one of the two failed to utilize cyclopentanol to recycle NADP(H) (Matsuda, 2000). Preparation of an acetone powder has already improved enantioselectivity markedly.

2. In two successive generations of genetically engineered constructs, the genes of each of the three reductase-activity carrying enzymes were first knocked out or overexpressed. Subsequently, strains for (3*R*), (3*S*), and *syn*-(2*R*,3*S*) activity were prepared by tuning the activity level of the three genes mentioned (Rodriguez, 2000, 2001). Better selectivities, in accord with the design goal, were achieved.
   - Heterologous expression of genes for both the carbonyl reductase and glucose dehydrogenase in *E. coli* has been achieved as mentioned in the previous section (Kataoka, 1999; Kizaki, 2001) (see Section 19.3).

7.2.3.5 **Alcohols with ADH**

Among enantioselective organometallic catalysts (see Chapter 18), the Duphos series is one of the most successful. Precursors to Duphos, such as (*R,R*)- or (*S,S*)-2,5-hexanediol, are therefore much sought-after building blocks. One of the most promising routes to (2*R*,5*R*)-hexanediol is enzymatic reduction of 2,5-hexanedione with alcohol dehydrogenases. In a recent synthesis route (Haberland, 2002), the reduction of both oxo functions proceeded highly diastereoselectively, catalyzed by two different (*R*)-ADHs from *Lactobacillus kefir* (DSM 20587). (2*R*,5*R*)-Hexanediol was produced in quantitative yields starting from 2,5-hexanedione, with both enantiomeric and diastereomeric excesses > 99%. A maximum yield of (2*R*,5*R*)-hexanediol was achieved at pH 6, 30 °C, and with equimolar amounts of substrate and cosubstrate. The applicability in fed-batch mode, including the feed specific biomass concentration required to reach maximal yield and selectivity, was determined.

The ADH from *Lactobacillus brevis* (Riebel, 1997) has a broad substrate specificity and converts even bulky aromatic ketones with high activity (Hummel, 1999; Wolberg, 2001). In addition, the enzyme is the best characterized completely (*R*)-specific ADH. The enzyme belongs to the class of short-chain dehydrogenases and its 3D structure has recently been solved (Niefind, 2003). The recombinant form of *L. brevis* ADH in *E. coli* accepts a variety of β,δ-diketo esters as was determined in the synthesis of potential building blocks for HMG CoA reductase inhibitors (see also Chapter 13, Section 13.3.2) (Wolberg, 2001). *tert*-Butyl 3,5-dioxohexanoate and *tert*-butyl 3,5-dioxoheptanoate were reduced on a preparative scale to afford the corresponding (*R*)-δ-hydroxy-β-keto esters with 99.4% e.e. and 98.1% e.e., respectively.

A similar compound, *tert*-butyl 6-chloro-3,5-dihydroxyhexanoate, was synthesized via the key step of a highly regio- and enantioselective single-site reduction of *tert*-butyl 6-chloro-3,5-dioxohexanoate by two enantiocomplementary biocatalysts:

- ADH from *Lactobacillus brevis* afforded a 72% yield of enantiopure *tert*-butyl (*S*)-6-chloro-5-hydroxy-3-oxohexanoate;
- the enantiomer (*R*)-*tert*-butyl (*S*)-6-chloro-5-hydroxy-3-oxohexanoate was prepared with 90–94% e.e. by baker's yeast reduction in a biphasic system (50% yield).

Both biotransformations were performed on a gram scale. The β-keto group of the enantiomeric δ-hydroxy-β-keto esters thus obtained was reduced by syn- and anti-selective borohydride reductions (see Section 13.3.2).

7.3
Enzymes as Catalysts in the Food Industry

7.3.1
HFCS with Glucose Isomerase (GI)

The isomerization of glucose to fructose is a sought-after process in the food industry because the sweetening power of fructose is three times the power of sucrose for a smaller number of calories. Fructose is obtained most conveniently and least expensively by the isomerization of glucose. However, chemical isomerization with sulfuric acid leads to decomposition and brownish coloring of the glucose syrup. A much milder process was developed on the basis of glucose isomerase (GI, xylose isomerase, D-xylose ketoisomerase, E.C. 5.3.1.5), the natural function of which is the isomerization of xylose to xylulose. Whereas the process was already known and had been performed on a small scale for several decades, demand soared in the 1970s (1984: 5×10^6 tpy, 2000: 11×10^6 tpy worldwide) owing to rising sugar prices and a change in consumer preferences towards lower-calorie foods. Several process improvements led to decreased costs.

The GI process starts with 40–45% glucose syrup on a dry basis (≈ 95% glucose); this high concentration results in high viscosity and thus requires processing at high temperature. To prevent excessive loss of enzyme activity during isomerization, the solution is filtered, treated with active carbon, and ion-exchanged to remove inhibitory Ca^{2+}. Mg^{2+} is then added as an activator, and the reaction is run at pH 8.5 for 0.8–4 h. Using parallel trains of packed-bed reactors with differently deactivated enzyme batches in tube bundles, the age and temperature of each en-

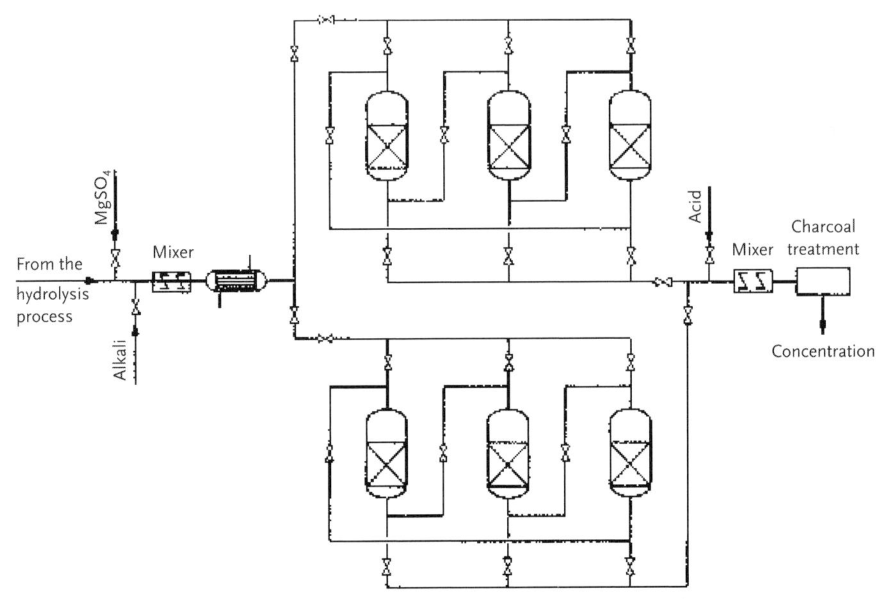

Figure 7.24 Process scheme for the glucose isomerase process (Crueger, 1989).

zyme batch are adapted to regulate and stabilize production capacity: a high capacity is achieved at 63–65°C with fresh enzyme whereas for a lower capacity 55–57°C and already partially deactivated enzyme suffice. The system shown in Figure 7.24 runs with two parallel rows, each of three 2.2 m^3 columns, and produces 100 t d^{-1} of fructose as a glucose/42% fructose mixture, termed High-Fructose Corn Syrup 42 (HFCS 42), which is close to the equilibrium limit. At the same degree of conversion and temperature, this continuous system produces a more consistent product, compared to the batch process, at fivefold improved space–time yield and only 4% of reactor volume. A more detailed packed-bed reactor design is described in Chapter 19, Section 19.1.

The 42% fructose solution has to be concentrated by chromatography to 55% fructose/45% glucose (HFCS 55), the same sweetness level as the equivalent weight of sucrose, to meet demand in the beverage industry, a major user. One of the open technological challenges of the HFCS process is the design of a sufficiently stable GI to run at temperatures where the equilibrium is more favorable to allow a degree to conversion to 55% fructose directly in the column.

7.3.2
Aspartame®, Artificial Sweetener through Enzymatic Peptide Synthesis

Aspartame is a dipeptide ester, α-L-aspartyl-L-phenylalanine-OMe, α-L-Asp-L-Phe-OMe, 200 times as sweet as sucrose, the sweetening power of which was discovered accidentally by researchers at G. D. Searle (Chicago/IL, USA) in 1965 during work on the intestinal hormone gastrin (Mazur, 1969). Aspartame® is utilized by now as a low-calorie sweetener in soft drinks, salad dressings, ready-made meals, table-top sweeteners, and pharmaceuticals and had reached a market volume of several thousand tons per year in 1992 and more than 12000 tpy in 1998. Since 1993 the patents on the application have expired in most countries. Among the several syntheses, one of the most interesting and successful is the Toya Soda enzymatic process which runs on and industrial scale in a joint venture with DSM (Dutch State Mines) at the Holland Sweetener Company in Geleen, The Netherlands (Oyama, 1984a,b).

Formation of the peptide bond is catalyzed by thermolysin, a neutral zinc protease (E.C. 3.4.24.4). The advantages of this process are:

- its complete enantiospecificity: depending on price, either L-phenylalanine-OMe or DL-phenylalanine-OMe can be used;
- its complete regiospecificity: there are no β-Asp bonds formed, in contrast to the situation with other processes;
- the thermal stability of the enzyme: its optimum lies at 50–60°C; and
- as a neutral protease, the absence of any esterase activity of thermolysin.

The kinetics of this equilibrium-controlled reaction have been worked out in detail (Oyama, 1981a,b). From all the possible ways to shift the equilibrium in favor of synthesis of dipeptide:

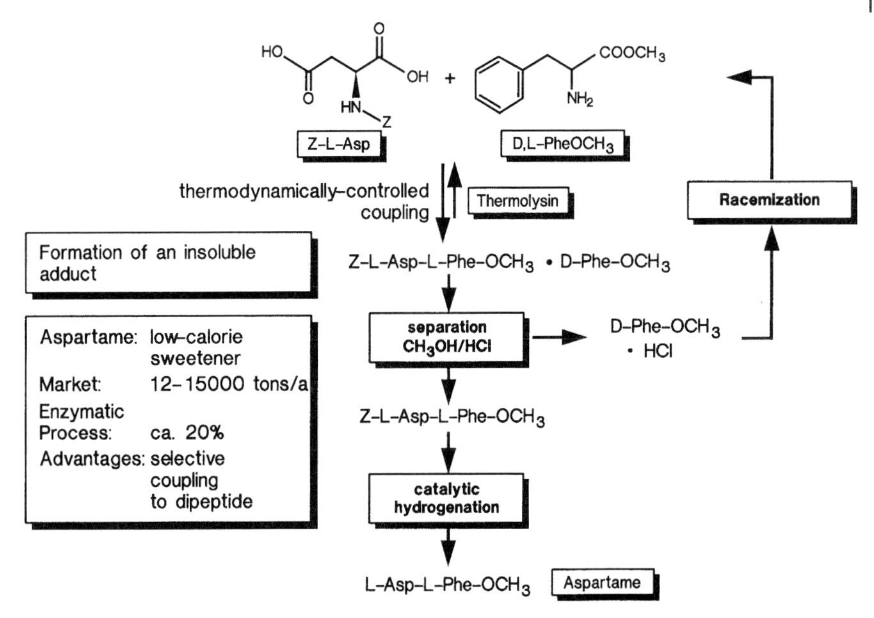

Figure 7.25 Toyo Soda/DSM process for the enzymatic production of aspartame.

1. addition of water-soluble organic solvents (Homandberg, 1978; Oyama, 1987),
2. addition of water-insoluble organic solvents (Oyama, 1981a,b),
3. precipitation of the product, or
4. complexation of product in an equilibrium-driven sequential reaction,

a combination of methods 3 and 4 was picked for process design (Oyama, 1984a,b). Given the solubility of the adduct of Z-L-Asp-L-Phe-OMe and D-Phe-OMe of 0.5 g L^{-1}, conversion can be shifted up to 96% dipeptide product (Figure 7.25).

Enzymatic peptide synthesis

In nature, the main purpose of proteases is to *cleave* peptide bonds. As the peptide and its amino acid components are in equilibrium, and as catalysts always catalyze both directions of a reversible reaction, the same reaction in principle can be used to *synthesize* peptide bonds.

Proteases can be classified according to the active center responsible for catalysis:

- *Serine proteases.* The β-hydroxy group of serine acts as a nucleophile under slightly alkaline conditions; examples are trypsin, chymotrypsin, and subtilisin.
- *Cysteine* (or thiol) *proteases.* The β-thiol group is an even better nucleophile in comparison with the serine side chain; thiol proteases operate under alkaline pH values as well; examples include papain and bromelain.
- *Metalloproteases.* These are also termed neutral proteases because the zinc center acting as a Lewis acid requires a neutral pH range; an example is thermolysin

- *Acid proteases.* Proteases featuring aspartate or glutamate in the active site need to operate under acidic conditions and are therefore called acid proteases; examples include gastrin and HIV protease.

Both alkaline proteases form an intermediate, the acyl–enzyme complex, on the reaction coordinate from the amino acid component to the dipeptide, which is formed by the triad Ser-(or Cys-)-His-Asp (or -Glu) (see Chapter 9, Section 9.5). The acyl–enzyme complex can be formed with the help of an activated amino acid component such as an amino acid ester. The complex can react either with water to the undesired hydrolysis product, the free amino acid, or with the amine of the nucleophile, such as an amino acid ester or amide, to the desired dipeptide. The particular advantage of enzyme-catalyzed peptide synthesis rests in the biocatalyst specificity with respect to particular amino acids in electrophile and nucleophile positions. Figure 7.26 illustrates the principle of kinetically and thermodynamically controlled peptide synthesis while Table 7.3 elucidates the specificity of some common proteases.

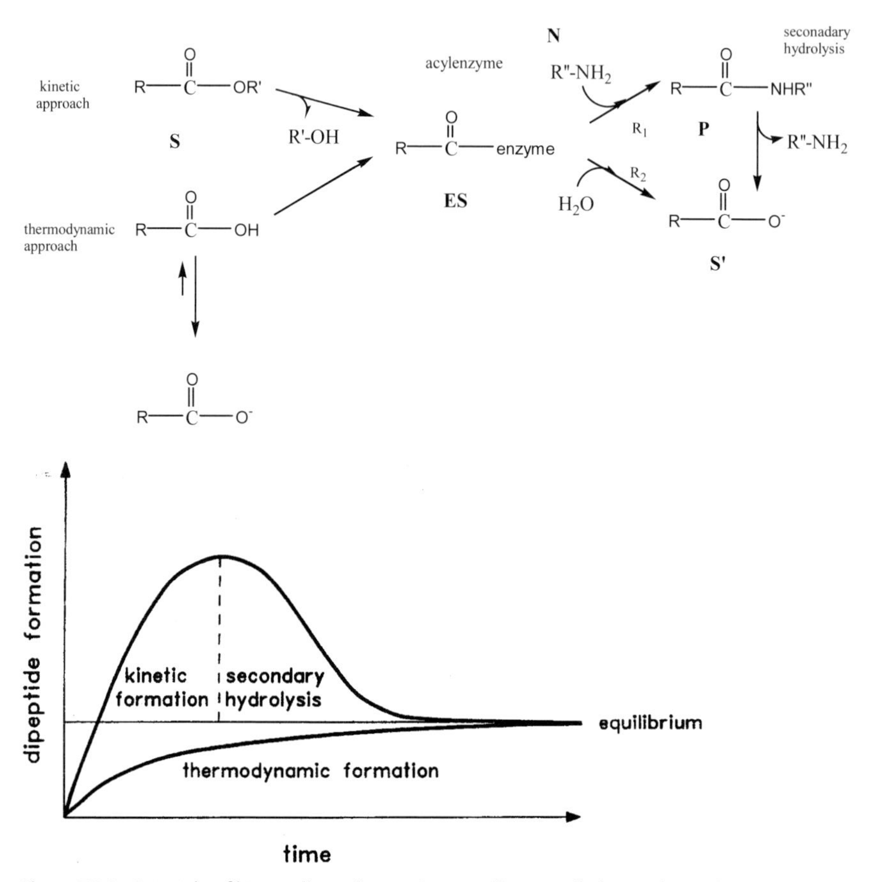

Figure 7.26 Principle of kinetically or thermodynamically controlled peptide synthesis.

Table 7.3 Specificity of proteases.

Enzyme	*Preferred cleavage sites (N-terminal → C-terminal)*
Bromelain	$-\text{Lys}\overset{\downarrow}{-}\text{Z};\ -\text{Arg}\overset{\downarrow}{-}\text{Z};\ -\text{Phe}\overset{\downarrow}{-}\text{Z};\ -\text{Tyr}\overset{\downarrow}{-}\text{Z}$
Chymotrypsin	$-\text{Trp}\overset{\downarrow}{-}\text{Z};\ -\text{Tyr}\overset{\downarrow}{-}\text{Z};\ -\text{Phe}\overset{\downarrow}{-}\text{Z};\ -\text{Leu}\overset{\downarrow}{-}\text{Z}$
Papain	$-\text{Phe}-\text{AA}\overset{\downarrow}{-}\text{Z};\ -\text{Val}-\text{AA}\overset{\downarrow}{-}\text{Z};\ -\text{Leu}-\text{AA}\overset{\downarrow}{-}\text{Z};\ -\text{Ile}-\text{AA}\overset{\downarrow}{-}\text{Z}$
Pepsin	$-\text{Phe (or Tyr,Leu)}\overset{\downarrow}{-}\text{Trp (or Phe,Tyr)}-$
Thermolysin	$-\text{AA}\overset{\downarrow}{-}\text{Leu}-;\ -\text{AA}\overset{\downarrow}{-}\text{Phe}-;\ -\text{AA}\overset{\downarrow}{-}\text{Ile}-;\ -\text{AA}\overset{\downarrow}{-}\text{Val}-$
Trypsin	$-\text{Arg}\overset{\downarrow}{-}\text{Z};\ -\text{Lys}\overset{\downarrow}{-}\text{Z}$

7.3.3 Lactose Hydrolysis

One of the largest-scale processes with enzymes is the hydrolysis of lactose to glucose and galactose (250 000 tpy) (Liese, 1999). As a significant fraction of consumers, especially in Asia, are lactose-intolerant, a condition which gives rise to rashes and stomach ailments, hydrolysis of lactose has considerable utility for the manufacture of lactose-free milk (Silk®).

7.3.4 "Nutraceuticals": L-Carnitine as a Nutrient for Athletes and Convalescents (Lonza)

L-Carnitine, (*S*)-3-hydroxy-4-trimethylaminobutyrate, belongs to the category of "shadow nutrients" or "nutraceuticals". Such compounds, while not being essential for a healthy grown-up person in the same way as vitamins, nevertheless aid metabolism and thus support a good state of health. For groups with special nutritional needs, including children, seniors, convalescents from illnesses, and active athletes, shadow nutrients are an important part of an optimum diet. An ever-growing part of the population in the developed countries (but still too small a fraction, as evidenced by the increasing problem of obesity in all developed countries) is very health-conscious and receptive to taking nutraceuticals as part of the diet. Nutraceuticals, together with growing trends towards self-medication and alternative medicines which probably often have similar effects to nutraceuticals, occupy an increasingly important position between the food and the pharmaceutical industries. Life-style drugs such as Viagra® and anti-obesity treatments (such as Xenical (Orlistat®), see Chapter 13, Section 13.3.5) belong to a similar category.

There are several routes to L-carnitine: (i) reduction of β-keto acids (see Section 7.2.2 and Chapter 19, Section 19.2); (ii) resolution of racemic carnitine by metabolism of D-carnitine; and (iii) two-step dehydrogenation/hydration formally resembling a hydroxylation from γ-butyrobetaine to L-carnitine. We discuss the last two in this section.

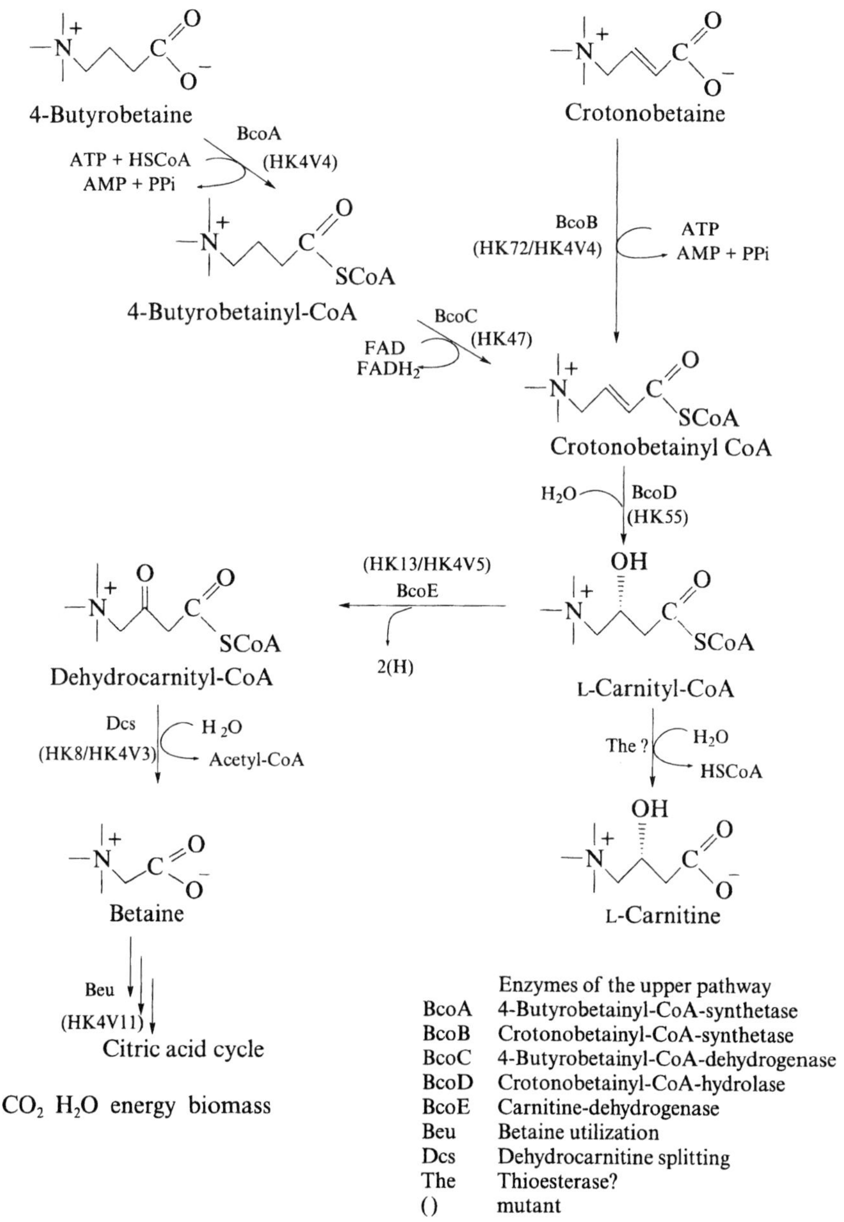

Figure 7.27 Reaction network to L-carnitine.

L-Carnitine of high optical purity was prepared via kinetic resolution using a mutant strain of *Acinetobacter calcoaceticus* ATCC 39647 (Figure 7.27) (Ditullio, 1994). This organism preferentially metabolized the D-enantiomer of the racemate to furnish L-carnitine. Recovery of L-carnitine was 93%, providing a total weight yield of 46.5% in 92% enantiomeric excess. The mode of degradation of carnitine was shown to proceed via a monooxygenase-catalyzed oxidative cleavage resulting in the formation of trimethylamine and malic acid. The data suggest that the stereoselective metabolism of DL-carnitine is probably the result of differential permeability of the cell membrane towards the optical antipodes.

Lonza (Visp, Switzerland) started strain development towards L-carnitine in 1983 with a small group of researchers; the product concentration was 0.1 g L^{-1} and the organism was *Agrobacterium/Rhizobium* HK4 from soil. After blocking L-carnitine degradation by the cell and after frame-shift mutagenesis, a 1–3 g L^{-1} yield was already achieved. Further selections of spontaneous mutants and a higher product tolerance led to 70–80 g L^{-1} of L-carnitine realized in the process. To reach 0.1–1 g L^{-1} Lonza took one year, to reach 1–10 g L^{-1} altogether 2.5 years, and to reach the range of 10–100 g L^{-1} a total of five years, and twice as long for implementation on a large scale. Meanwhile, the development of a biocatalyst on the basis of recombinant organisms has been started.

In the current process (shown schematically in Figure 7.28), the substrate concentration of γ-butyrobetaine is more than 80 g L^{-1}, probably between 80 and 100 g L^{-1}, and the productivity on the 15 m^3-scale is 5 g $(L\ h)^{-1}$; the yield and the product purity lie in the vicinity of > 99% and > 99% e.e., respectively. For desalination of the fermentation solution electrodialysis is utilized.

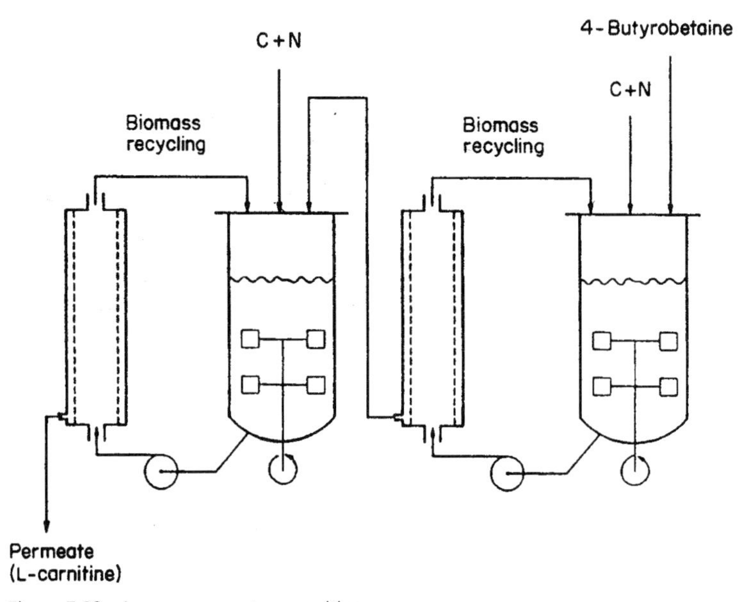

Figure 7.28 Lonza process to L-carnitine.

Particularly as a certain purity level is required for L-carnitine in its food and pharmaceutical applications, process optimization of biotransformation and downstream processing should be done in an integral way. The L-carnitine process offers an example where process integration considerations ultimately even drive the choice of C- and N-sources as raw materials for the biotransformation, and the choice of reactor configuration. The options for the biotransformation step were either a fed-batch or a cell recycle chemostat, the latter in either single or multistage mode. Blackman kinetics proved adequate to model product formation from γ-butyrobetaine to L-carnitine and the process options were investigated in a 450 L bioreactor for a run time of 700 h (Hoeks, 1992). With reactor productivity, simplicity of overall process scheme, product purity, and total cost as target criteria, the fed-batch biotransformation of γ-butyrobetaine to L-carnitine appeared to be superior to several modes of operation of the continuous process with cell recycling. In contrast, cell recycle demonstrated the highest volumetric productivity; but a lower number of downstream processing unit operations with resulting lower investment and operating costs for downstream processing tipped the balance in favor of the fed-batch process (Hoeks, 1996; Zimmermann, 1997).

7.3.5 Decarboxylases for Improving the Taste of Beer

As a side activity, many decarboxylases catalyze the formation of C–C bonds. In the reaction of two pyruvate molecules, catalyzed by pyruvate decarboxylase (PDC, E.C. 4.1.1.1), α-acetolactate is formed, an important intermediate of valine biosynthesis. In turn, α-acetolactate can be oxidatively decarboxylated by oxygen to diacetyl or enzymatically decarboxylated by acetolactate decarboxylase (ADC, E.C. 4.1.1.5) to (*R*)-acetoin (Figure 7.29).

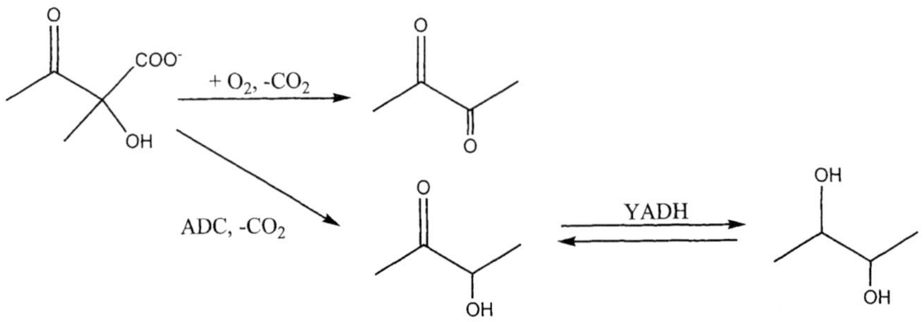

Figure 7.29 Reactions of α-acetolactate in beer.

In brewery processes, the bad taste of diacetyl spoils the beer. By addition of ADC (Novo, Bagsvaerd, Denmark) or by cloning and overexpression of ADC in brewery yeast (Kirin, Japan) acetolactate is first decarboxylated to acetoin and then reduced to innocuous 2,3-dihydroxybutane by yeast alcohol dehydrogenase (YADH).

7.4 Enzymes as Catalysts towards Crop Protection Chemicals

7.4.1 Intermediate for Herbicides: (*R*)-2-(4-Hydroxyphenoxypropionic acid (BASF, Germany)

BASF (Ludwigshafen, Germany) produces isomerically pure (*R*)-2-(4-hydroxyphenoxy)-propionic acid (HPOPS) from (*R*)-2-phenoxypropionic acid (POPS) on a 100 m^3 fermenter scale for use as a herbicide intermediate (Figure 7.30) (Cooper, 1990).

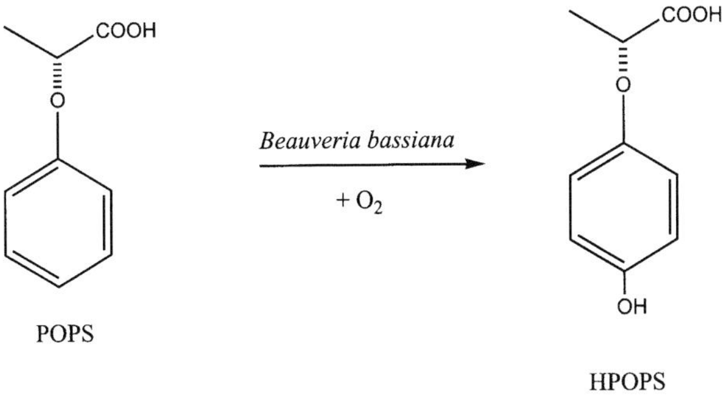

Figure 7.30 Hydroxylation of (*R*)-POPS to (*R*)-HPOPS.

Despite the good product (chemical composition > 99%, e.e. > 98%) the best strain from a collection of originally 7900 was too slow and also osmosensitive. Through classical mutation (UV light, MNNG (N-methyl-N′-nitro-N-nitroso guanidine)) the space–time yield could be improved from 0.5 g $(L\ d)^{-1}$ to 7.0 g $(L\ d)^{-1}$ and the substrate/product tolerance could be increased to 100 g L^{-1}. The isolation of the enzyme failed: even one re-use of the crude extract is not possible. Meanwhile two genes have been isolated (monooxygenase and reductase).

The process of developing biological routes covers three or four stages:

1. it starts with *biodiversity* (a source of organisms);
2. after subsequent enrichment culture one arrives at a *pure culture* (enrichment);
3. protein purification and strain improvement yield a *biocatalyst* (execution);
4. after immobilization and process optimization, *biotransformation* (optimization) is achieved.

For the illustration of a typical timescale of the development of a biotransformation at BASF, the hydroxylation of aromatics serves as an example: BASF took three years to increase the product titer from 1–5 g L^{-1} to 40 g L^{-1}, another four years after mutation of the parent strain to jump from 40 to 120 g L^{-1}. The hydroxylation of aromatics (POPS to HPOPS) today runs on a scale of > 100 tpy.

7.4.2
Applications of Transaminases towards Crop Protection Agents: L-Phosphinothricin and (S)-MOIPA

L-Phosphinothricin: herbicide building block

L-Phosphinothricin, the active ingredient of the broad-spectrum herbicide Basta (Hoechst, Frankfurt, Germany), can be obtained through enzymatic transamination of the corresponding oxo acid, 2-oxo-4-[(hydroxy)(methyl)phosphinoyl]butyric acid, in a coupled system (Figure 7.31) with aspartate aminotransferase (AAT, E.C. 2.6.1.1) and 4-aminobutyrate:2-ketoglutarate transaminase (E.C. 2.6.1.19) from *E. coli* (Figure 7.31) (Bartsch, 1996). In solutions containing 10% substrate, 85% conversion was reached with only < 3% of amino acid by-products. For this process, a new AAT from *B. stearothermophilus* had been screened and characterized (T_{opt} = 95°C, pH_{opt} = 8.0) before being cloned and overexpressed in *E. coli* by the Hoechst researchers.

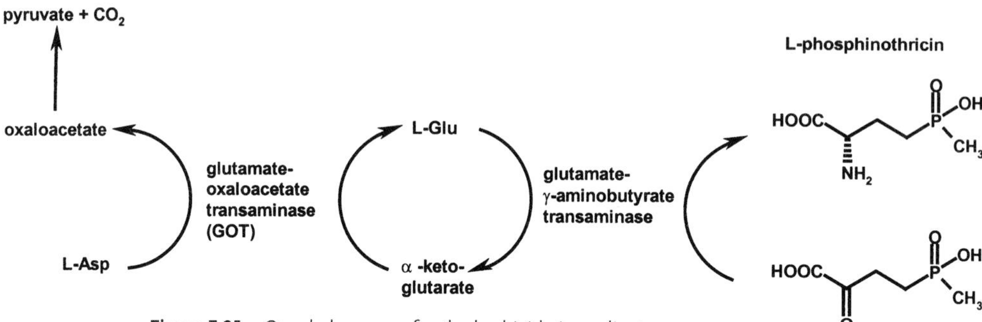

Figure 7.31 Coupled process for the herbicide ingredient L-phosphinothricin with transaminases (Drauz, 2002).

(S)-MOIPA: building block of (S)-metolachlor

(*S*)-metolachlor (Dual Magnum®, Ciba-Geigy/Syngenta, Basel, Switzerland) is a broad-spectrum herbicide; (*S*)-methoxyisopropylamine [(*S*)-MOIPA] is its key building block. One promising route to (*S*)-MOIPA is transamination of methoxyacetone, a nearly unactivated ketone, with the help of (*S*)-transaminase. Although transamination proceeded with a nearly complete yield and > 99% enantioselectivity, this reaction offers a (fairly rare) case where organometallic catalysis presented a better processing alternative. The enantioselectivity of 80% e.e. which could be achieved with the Ir/josiphos 4 (substituted ferrocene) catalyst is sufficient for a herbicide

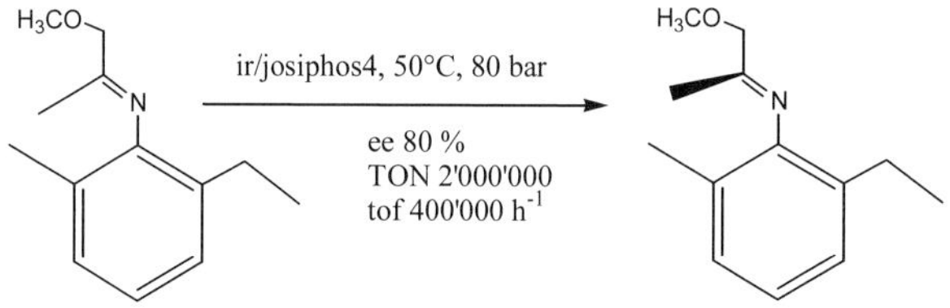

H_3CO ... N — ir/josiphos4, 50°C, 80 bar; ee 80 %; TON 2'000'000; tof 400'000 h^{-1} → H_3CO ... N

Figure 7.32 Chiral reduction to (*S*)-metolachlor with organometallic catalyst.

with an innocuous unwanted enantiomer (eutomer). With respect to the remaining performance data, the chemically catalyzed process performed admirably with a turnover frequency (tof) of 100 s^{-1} and a TTN of 2 000 000 (Blaser, 2001) (Figure 7.32).

7.5 Enzymes for Large-Scale Pharma Intermediates

7.5.1 Penicillin G (or V) Amidase (PGA, PVA): β-Lactam Precursors, Semi-synthetic β-Lactams

β-Lactams, i.e., penicillins and cephalosporins, represent the most important class of antibiotics. Penicillins consist of a common core, 6-aminopenicillanic acid (6-APA) and different side chains. Penicillin G (pen G), with a phenyl acetate side chain, and penicillin V (pen V), with a phenoxyacetate side chain, are fermented from the fungus *Penicillium chrysogenum*; all the others are produced from 6-APA, which nowadays is produced mostly from pen G via penicillin G amidase (pen G amidase, pen G acylase, PGA, E.C. 3.5.1.11; Figure 7.33). Cephalosporins feature 7-aminocephalosporanic acid (7-ACA) or its deacetyl-form, 7-aminodesacetylcephalosporanic acid (7-ADCA) as their common core; cephalosporin C (Ceph C) is obtained through fermentation, and all the others are derived from 7-A(D)CA.

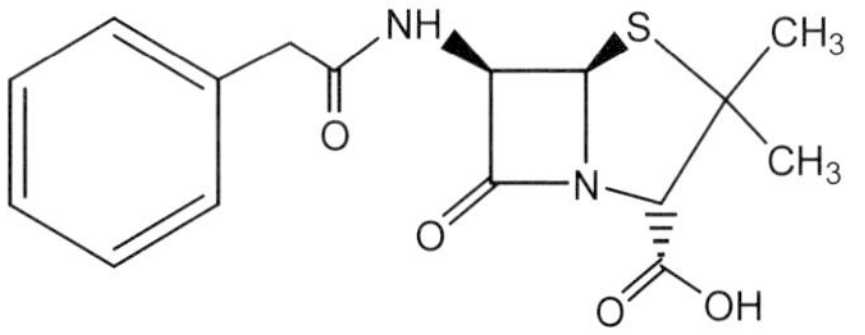

Penicillin G

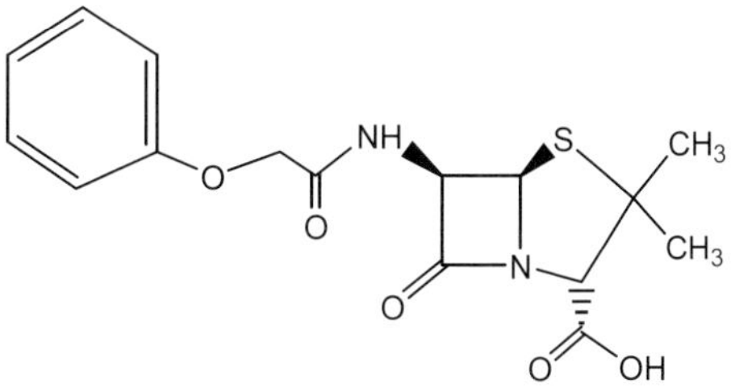

Penicillin V

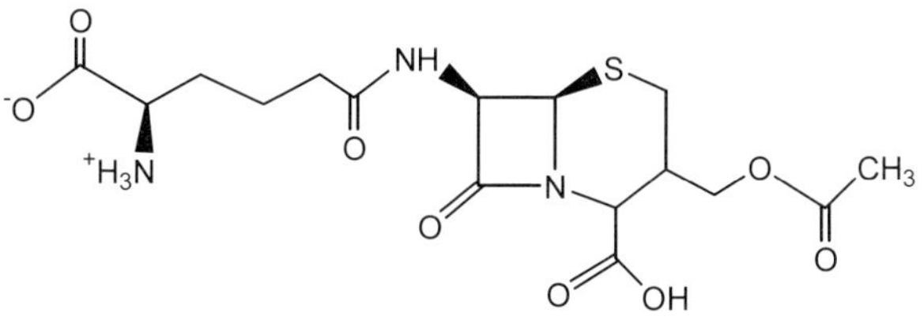

Cephalosporin C

Figure 7.33 Semi-synthetic penicillins.

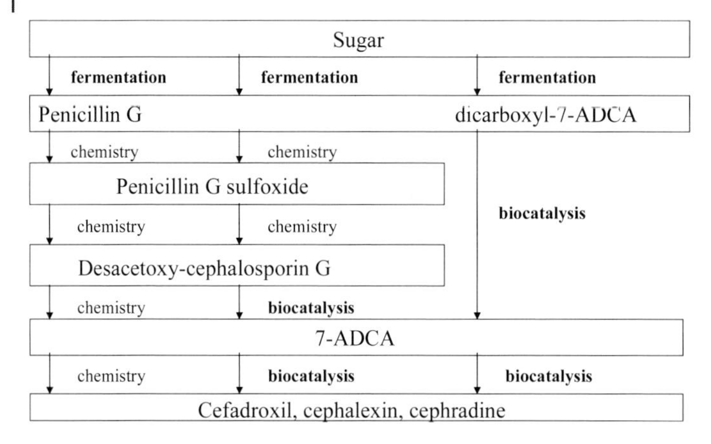

Figure 7.34 Conventional and biotechnological routes to β-lactam antibiotics.

Possibly owing to the long-term and stable outlook, boosted by a newly surging need for novel antibiotics (see Chapter 13), an example of the successful application of biocatalysis in the manufacture of already existing target compounds is found in the case of β-lactam antibiotics. An overview of the routes to 6-aminopenicillanic acid (6-APA), 7-aminodesacetoxycephalosporanic acid (7-ADCA), semi-synthetic penicillins (SSPs), and semi-synthetic cephalosporins (SSCs) (Figure 7.34) reveals that several breakthroughs have been achieved by applying fermentation and biocatalysis (Van de Sandt, 2000).

With the aid of genetic engineering it is now possible to construct an effective pathway towards the 7-ADCA structure within the producing organism (Van de Sandt, 2000). The enzyme pen G acylase catalyzes several steps in many of the current processes (De Vroom, 1999) (Figure 7.35). With PGA-catalyzed reactions it is now possible to directly remove side chains from and add them to the β-lactam nucleus (Van de Sandt, 2000); in the case of addition of side chains, this is a long sought-after goal. Without the enzyme, several protecting steps are required to perform these conversions. A schematic of the new process is provided in Figure 7.36.

As expected, the kinetically controlled PGA-catalyzed synthesis of ampicillin, via acylation of 6-aminopenicillanic acid with D-phenylglycine amide, is accompanied by formation of the hydrolysis product D-phenylglycine. Recycling of D-phenylglycine via D-phenylglycine methyl ester HCl salt on a laboratory scale, resulting in a process using both amide and ester as donors, i.e., mixed-donor coupling, led to a more economic process with respect to materials use and waste formation (van Langen, 2001).

In part, all the breakthroughs mentioned depended on process modifications in enzyme isolation, in purification, and in immobilization as well as in handling precipitating product in an immobilized enzyme reactor system. These new processes involving PGA in several central roles result in shorter and more cost-effective routes to semi-synthetic β-lactam antibiotics.

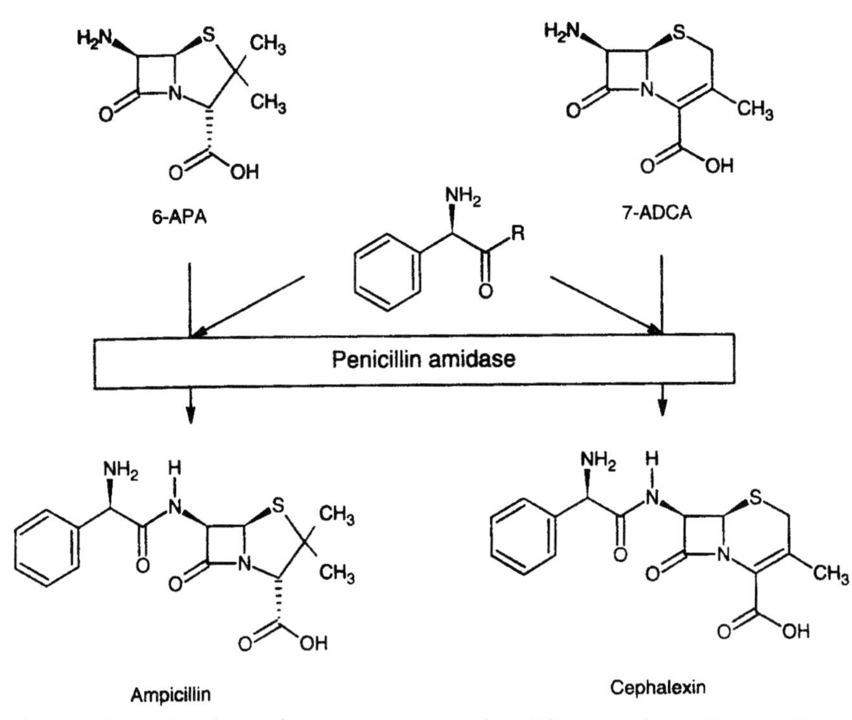

Figure 7.35 Pen G acylase in the process to semi-synthetic β-lactam antibiotics (Drauz, 2002).

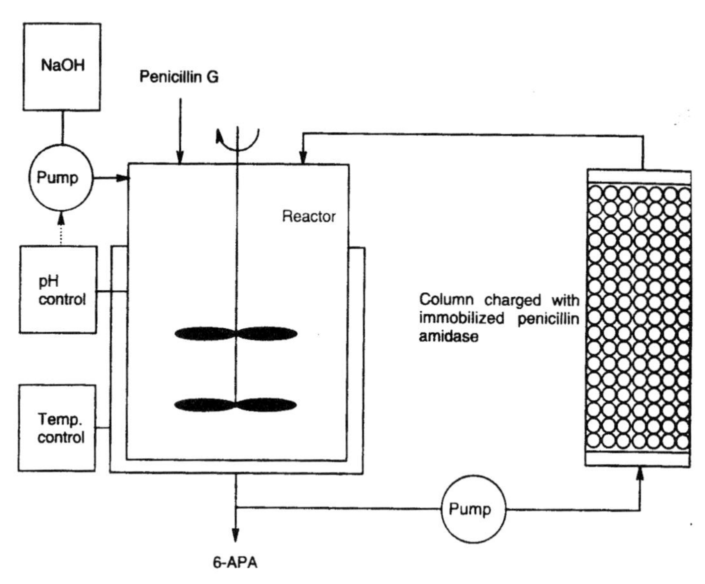

Figure 7.36 Schematic of the new process to semi-synthetic antibiotics with pen G acylase (Drauz, 2002).

7.5.2
Ephedrine

Pyruvate decarboxylase (PDC, E.C. 4.1.1.1) accepts other substrates besides pyruvate, its natural reactant. As early as 1921, Neuberg and Hirsch described the reaction of yeast with benzaldehyde and pyruvate to phenylacetylcarbinol (2-keto-3-hydroxypropylbenzene) (Neuberg, 1921) in a carboligase side reaction which yields ephedrine after reaction with methylamine and catalytic hydrogenation (Figure 7.37).

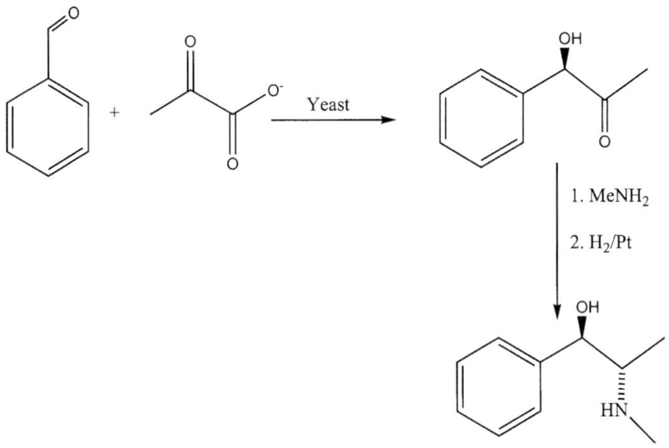

Yeast

1. $MeNH_2$

2. H_2/Pt

Figure 7.37 Process to ephedrine by carboligation of benzaldehyde and pyruvate.

The reaction was already being used in the 1930s by Knoll (Ludwigshafen, Germany) (later BASF, since 2001 Abbott) (German Patent DE 546459, 1932) to produce ephedrine via (*R*)-phenylacetylcarbinol, (*R*)-PAC. Recently, efforts have been renewed to obtain an isolated enzyme process or a whole-cell process based on recombinant DNA technology. It was found that pyruvate could be substituted by the less expensive acetaldehyde, which did not produce CO_2 as a side product.

The suitability of three pyruvate decarboxylases from *Saccharomyces cerevisiae*, from *Zymomonas mobilis*, and the PDC W392M mutant of the latter was tested in the biotransformation with acetaldehyde and benzaldehyde as substrates. The mutant enzyme was the most active and most stable one. Initial rate data at various concentrations of each substrate revealed both substrate and product inhibition and were fitted into a kinetic model; consequently, a continuous reactor system operating as a CSTR and semi-batch feeding of equimolar concentrations of each substrate were chosen (circumventing substrate inhibition). With each substrate at 50 mM, the continuously operated enzyme membrane reactor gave (*R*)-PAC with a space–time yield of 81 g $(L \cdot d)^{-1}$ (Goetz, 2001). As expected from the model, the yield was easily increased by cascading of enzyme–membrane reactors and thus by moving towards a plug-flow configuration (see Chapter 5). The process configuration allows the synthesis of (*R*)-PAC from inexpensive substrates in aqueous reaction without the by-product spectrum of the fermentation process.

Inlet Figure Enzyme membrane reactor pilot plant at Degussa's research center (Wolfgang, Germany).

Suggested Further Reading

K. FABER, *Biotransformations in Organic Chemistry*, Springer-Verlag, Berlin,4th edition, **2000**.

K. DRAUZ and H. WALDMANN (eds.), *Enzyme Catalysis in Organic Synthesis*, Wiley-VCH, Weinheim, 2nd edition, **2002**.

A. LIESE and M. V. FILHO, Production of fine chemicals using biocatalysis, *Curr. Opin. Biotechnol.* **1999**, *10*, 595–603.

A. LIESE, K. SEELBACH, and C. WANDREY, *Industrial Biotransformations: A Collection of Processes*, Wiley-VCH, Weinheim, **2000**.

A. SCHMID, J. S. DORDICK, B. HAUER, A. KIENER, M. WUBBOLTS, and B. WITHOLT, Industrial biocatalysis today and tomorrow, *Nature*, **2001**, *409*, 258–268.

A. ZAKS, Industrial biocatalysis, *Curr. Opin. Chem. Biol.* **2001**, *5*, 130–136.

References

M. B. Ansorge and M.-R. Kula, Production of recombinant L-leucine dehydrogenase from *Bacillus cereus* in pilot scale using the runaway replication system E. coli[pIET98], *Biotechnol. Bioeng.* **2000a**, *68*, 557–562.

M. B. Ansorge and M.-R. Kula, Investigating expression systems for the stable large-scale production of recombinant L-leucine-dehydrogenase from *Bacillus cereus* in *Escherichia coli*, *Appl. Microbiol. Biotechnol.* **2000b**, *53*, 668–673.

P. C. Babbitt, M. S. Hasson, J. E. Wedekind, D. R. Palmer, W. C. Barrett, G. H. Reed, I. Rayment, D. Ringe, G. L. Kenyon, and J. A. Gerlt, The enolase superfamily: a general strategy for enzyme-catalyzed abstraction of the alpha-protons of carboxylic acids, *Biochemistry* **1996**, *35*, 16489–16501.

F. Balkenhohl, K. Ditrich, B. Hauer, and W. Ladner, Optically active amines via lipase-catalyzed methoxyacetylation, *J. Prakt. Chem./Chem. Ztg.* **1997**, *339*, 381–384.

K. Bartsch, R. Schneider, and A. Schulz, Stereospecific production of the herbicide phosphinothricin (glufosinate): purification of aspartate transaminase from *Bacillus stearothermophilus*, cloning of the corresponding gene, *aspC*, and application in a coupled transaminase process, *Appl. Environ. Microbiol.* **1996**, *62*, 3794–3799.

W. Becker, H. Freund, and E. Pfeil, Stereospecific synthesis of D-hydroxy nitriles and optically active ethanolamines, *Angew. Chem. Int. Ed. Engl.* **1965**, *4*, 1079; *Angew. Chem.* **1965**, *77*, 1139.

H. U. Blaser, F. Spindler, and M. Studer, Enantioselective catalysis in fine chemicals production, *Appl. Catalysis, A: General* **2001**, *221*, 119–143.

A. S. Bommarius, K. Drauz, U. Groeger, and C. Wandrey, Membrane bioreactors for the production of enantiomerically pure α-amino acids, in: *Chirality in Industry*, A. N. Collins, G. N. Sheldrake, and J. Crosby (eds.), Wiley, London, **1992**, Chapter 20, pp. 371–397.

A. S. Bommarius, K. Drauz, W. Hummel, M.-R. Kula, and C. Wandrey, Some new developments in reductive amination with cofactor regeneration, *Biocatalysis* **1994**, *10*, 37–47.

A. S. Bommarius, M. Schwarm, K. Stingl, M. Kottenhahn, K. Huthmacher, and K. Drauz, Synthesis and use of enantiomerically pure *tert.*-leucine, *Tetrahedron Asymm.* **1995**, *6*, 2851–2888.

S. Borman, Chemical engineering focuses increasingly on the biological, *Chem. Eng. News* **1993**, January 11, 26–36.

A. J. Brazier, M. D. Lilly, and A. B. Herbert, Toluene *cis*-glycol synthesis by *Pseudomonas putida*; kinetic data for reactor evaluation, *Enzyme Microb. Technol.* **1990**, *12*, 90–94.

J. Brussee, W. T. Loos, C. G. Kruse, and A. van der Gen, Synthesis of optically active silyl protected cyanohydrins, *Tetrahedron* **1990**, *46*, 979–986.

H. A. J. Carless, J. R. Billinge, and O. Z. Oak, Photochemical routes from arenes to inositol intermediates: the photooxidation of substituted *cis*-cyclohexa-3,5-diene-1,2-diols, *Tetrahedron Lett.* **1989**, *30*, 3113–3116.

F. Cecere, G. Galli, G. Della Penna, and B. Rappuoli, D-Carbamylamino acids and corresponding D-amino acids, German Patent DE 2422737, **1976**.

S. Chauhan, S. Wu, S. Blumerman, R. D. Fallon, J. E. Gavagan, R. DiCosimo, and M. S. Payne, Purification, cloning, sequencing and over-expression in *Escherichia coli* of a regioselective aliphatic nitrilase from *Acidovorax facilis* 72W, *Appl. Microbiol. Biotechnol.* **2003**, *61*, 118–122.

B. Cooper, W. Ladner, B. Hauer, and H. Siegel, Microbial manufacture of 2-(4-hydroxyphenoxy)propionic acid, PCT Int. Patent Appl. WO 9011362, **1990** (to BASF).

D. Cotoras and F. Wagner, Stereospecific hydrolysis of 5-monosubstituted hydantoins, *Eur. Congr. Biotechnol., 3rd* **1984**, *1*, 351–6.

R. Crossley, *Chirality and the Biological Activity of Drugs*, CRC Press, Boca Raton, **1995**, Chapter 1.

W. Crueger and A. Crueger, *Biotechnology: A Textbook of Industrial Microbiology*, Science Tech Publishers, Madison/WI, **1989.**

S. P. Crump, J. S. Heier, and J. David Rozzell, The production of amino acids by transamination, in: *Biocatalysis*, D. A. Abramowicz (ed.), Van Nostrand Reinhold, New York, **1990**, pp. 115–133.

S. P. Crump and J. David Rozzell, Biocatalytic production of amino acids by transamination, in: *Biocatalytic Production of Amino Acids and Derivatives: New Developments and Process Considerations*, J. D. Rozzell and F. Wagner (eds.), Hanser Publishers, Munich, **1992**, pp. 43–58.

E. De Vroom, The central role of penicillin acylase in antibiotics production, *Chimica Oggi* **1999**, *17*, 65–68.

D. DiTullio, D. Anderson, C. S. Chen, and C. J. Sih, L-carnitine via enzyme-catalyzed oxidative kinetic resolution, *Bioorg. Med. Chem.* **1994**, *2*, 415–420.

K. Drauz and H. Waldmann (eds.), *Enzyme Catalysis in Organic Synthesis*, Wiley-VCH, Weinheim, 2nd edition, **2002**.

A. L. L. Duchateau, M. G. Hillemans, C. H. M. Schepers, P. E. F. Ketelaar, H. F. M. Hermes, and J. Kamphuis, Determination of biocatalyst consumption in an aminopeptidase process using automated sample preparation and high-performance liquid chromatography, *J. Chromatogr.* **1991**, *566*, 493–498.

F. Effenberger, T. Ziegler, and S. Foerster, Enzyme-catalyzed cyanohydrin preparation in organic solvents, *Angew. Chem.* **1987**, *99*, 491–492.

F. Effenberger, B. Hoersch, S. Foerster, and T. Ziegler, Enzyme-catalyzed reactions. 5. Enzyme-catalyzed synthesis of (*S*)-cyanohydrins and subsequent hydrolysis to (*S*)-α-hydroxy-carboxylic acids, *Tetrahedron Lett.* **1990**, 31, 1249–1252.

F. Effenberger, B. Hoersch, F. Weingart, T. Ziegler, and S. Kuehner, Enzyme catalyzed reactions. 9. Enzyme-catalyzed synthesis of (*R*)-ketone cyanohydrins and their hydrolysis to (*R*)-α-hydroxy-α-methyl carboxylic acids, *Tetrahedron Letters* **1991**, *32*, 2605–2608.

F. Effenberger, S. Forster, and H. Wajant, Hydroxynitrile lyases in stereoselective catalysis, *Curr. Opin. Biotechnol.* **2000**, *11*, 532–539.

A. Fischer, A. S. Bommarius, K. Drauz, and C. Wandrey, A novel approach to enzymatic peptide synthesis using highly solubilizing Na-protecting groups of amino acids, *Biocatalysis* **1994**, *8*, 289–307.

I. G. Fotheringham, N. Grinter, D. P. Pantaleone, R. F. Senkpeil, and P. P. Taylor, Engineering of a novel biochemical pathway for the biosynthesis of L-2-aminobutyric acid in *Escherichia coli* K12, *Bioorgan. Med. Chem*, **1999**, *7*, 2209–2213.

D. T. Gibson, J. R. Koch, and R. E. Kallio, Oxidative degradation of aromatic hydrocarbons by microorganisms. I. Enzymic formation of catechol from benzene, *Biochemistry* **1968**, *7*, 2653–2662.

G. Goetz, P. Iwan, B. Hauer, M. Breuer, and M. Pohl, Continuous production of (*R*)-phenylacetylcarbinol in an enzyme–membrane reactor using a potent mutant of pyruvate decarboxylase from *Zymomonas mobilis*, *Biotechnol. Bioeng.* **2001**, 74, 317–325.

H. Griengl, H. Schwab, and M. Fechter, The synthesis of chiral cyanohydrins by oxynitrilases, *Trends Biotechnol.* **2000**, *18*, 252–256.

C. Gross, C. Syldatk, and F. Wagner, Screening method for microorganisms producing L-amino acids from DL-5-monosubstituted hydantoins, *Biotechnol. Technol.* **1987**, *1*, 85–90.

J. Haberland, A. Kriegesmann, E. Wolfram, W. Hummel, and A. Liese, Diastereoselective synthesis of optically active (2*R*,5*R*)-hexanediol, *Appl Microbiol Biotechnol.* **2002**,*58*, 595–599.

E. C. Hann, A. Eisenberg, S. K. Fager, N. E. Perkins, F. G. Gallagher, S. M. Cooper, J. E. Gavagan, B. Stieglitz, S. M. Hennessey, and R. DiCosimo, 5-Cyanovaleramide production using immobilized *Pseudomonas chlororaphis* B23, *Bioorgan. Med. Chem.* **1999**, *7*, 2239–2245.

E. C. Hann, A. E. Sigmund, S. M. Hennessey, J. E. Gavagan, D. R. Short, A. Ben-Bassat, S. Chauhan, R. D. Fallon, M. S. Payne, and R. DiCosimo, Optimization of an immobilized-cell biocatalyst for production of 4-cyanopentanoic acid, *Org. Proc. Res. Dev.* **2002**, *6*, 492–496.

R. L. Hanson, J. Singh, T. P. Kissik, R. N. Patel, L. J. Szarka, and R. H. Mueller, Synthesis of L-β-hydroxyvaline from α-keto-β-hydroxyisovalerate using leucine dehydrogenase from *Bacillus* species, *Bioorg. Chem.* **1990**, *18*, 116–130.

R. L. Hanson, M. D. Schwinden, A. Banerjee, D. B. Brzozowski, B.-C. Chen, B. P. Patel, C. G. McNamee,

G. A. Kodersha, D. R. Kronenthal, R. N. Patel, and L. J. Szarka, Enzymatic synthesis of L-6-hydroxynorleucine, *Bioorgan. Med. Chem.* **1999**, *7*, 2247–2252.

H. F. M. Hermes, T. Sonke, P. J. H. Peters, J. A. M. van Balken, J. Kamphuis, L. Dijkhuizen, and E. M. Meijer, Purification and characterization of an L-aminopeptidase from *Pseudomonas putida* ATCC 12633, *Appl. Environm. Microbiol.* **1993**, *59*, 4330–4334.

W. Hibino, I. Onishi, S. Abe, and K. Yokozeki, Production of D-amino acid, Japanese Patent Appl. JP 09280073, **1999** (to Ajinomoto).

B. L. Hirschbein and G. M. Whitesides, Laboratory-scale enzymic/chemical syntheses of D- and L-β-chlorolactic acid and D- and L-potassium glycidate, *J. Am. Chem. Soc.* **1982**, *104*, 4458–4460.

F. W. J. M. M. Hoeks, H. Kulla, and H. P. Meyer, Continuous cell-recycle process for L-carnitine production: performance, engineering and downstream processing aspects compared with discontinuous processes, *J. Biotechnol.* **1992**, *22*, 117–27.

F. W. J. M. M. Hoeks, J. Muehle, L. Boehlen, and I. Psenicka, Process integration aspects for the production of fine chemicals illustrated with the biotransformation of γ-butyrobetaine into L-carnitine, *Chem. Eng. J.* **1996**, *61*, 53–61.

G. A. Homandberg, J. A. Mattis, and M. Laskowski Jr., Synthesis of peptide bonds by proteinases. Addition of organic cosolvents shifts peptide bond equilibriums toward synthesis, *Biochemistry* **1978**, *17*, 5220–5227.

T. Hudlicky, H. Luna, G. Barbieri, and L. D. Kwart, Enantioselective synthesis through microbial oxidation of arenes. 1. Efficient preparation of terpene and prostanoid synthons, *J. Am. Chem. Soc.* **1988**, *110*, 4735–4741.

T. Hudlicky, H. Luna, J. D. Price, and F. Rulin, An enantiodivergent approach to D- and L-erythrose via microbial oxidation of chlorobenzene, *Tetrahedron Lett.* **1989**, *30*, 4053–4054.

T. Hudlicky, H. Luna, J. D. Price, and F. Rulin, Microbial oxidation of chloroaromatics in the enantiodivergent synthesis of pyrrolizidine alkaloids: trihydroxyheliotridanes, *J. Org. Chem.* **1990**, *55*, 4683–4687.

T. Hudlicky, Design constraints in practical syntheses of complex molecules: current status, case studies with carbohydrates and alkaloids, and future perspectives, *Chem. Rev.* **1996**, *96*, 3–30.

W. Hummel, H. Schütte, and M.-R. Kula, Leucine dehydrogenase from *Bacillus sphaericus*. Optimized production conditions and an efficient method for 1st large scale purification, *Eur. J. Appl. Micobiol. Biotechnol.* **1981**, *12*, 22–27.

W. Hummel, H. Schütte, E. Schmidt, and M.-R. Kula, Isolation and characterization of acetamidocinnamate acylase, a new enzyme suitable for production of L-phenylalanine, *Appl. Microbiol. Biotechnol.* **1987**, *27*, 283–291.

W. Hummel, Large-scale applications of NAD(P)-dependent oxidoreductases: recent developments, *Trends Biotechnol.* **1999**, *17*, 487–492.

K. Isobe and Y. Hirose, Production of D-amino acid, Japanese Patent Appl. JP 09286147, **1999** (to Amano).

M. Kataoka, K. Yamamoto, H. Kawabata, M. Wada, K. Kita, H. Yanase, and S. Shimizu, Stereoselective reduction of of ethyl 4-chloro-3-oxobutanoate by *Escherichia coli* transformant cells coexpressing the aldehyde reductase and glucose dehydrogenase genes, *Appl. Microbiol. Biotechnol.* **1999**, *51*, 486–490.

M. Kato, H. Kitagawa, and T. Miyoshi, Microbial racemization of optically active 5-substituted hydantoins, Japanese Patent JP 62122591, **1987**.

A. Kiener, Biosynthesis of functionalized aromatic N-heterocycles, *CHEMTECH* **1995**, *25*, 31–35.

N. Kizaki, Y. Yasohara, J. Hasegawa, M. Wada, M. Kataoka, and S. Shimizu, Synthesis of optically pure iso ethyl(S)-4-chloro-3-hydroxybutanoate by Escherichia coli transformant cells coexpressing the carbonyl reductase and glucose dehydrogenase genes, *Appl. Microbiol. Biotechnol.* **2001**, *55*, 590–595.

Knoll, German Patent DE 546459, **1932**.

U. Kragl, D. Vasic-Racki, and C. Wandrey, Continuous processes with soluble enzymes, *Chem. Ing. Tech.* **1992**, *64*, 499–509.

G. Krix, A. S. Bommarius, K. Drauz, M. Kottenhahn, M. Schwarm, and M.-R. Kula, Enzymatic reduction of α-keto acids leading to L-amino acids or D-hydroxy acids, *J. Biotechnol.* **1997**, *53*, 29–39.

M.-R. KULA and C. WANDREY, Continuous enzymic transformation in an enzyme–membrane-reactor with simultaneous NADH regeneration, *Meth. Enzymol.* **1987**, *136*, 9–21.

M.-R. KULA, U. NIEDERMEYER, and I. M. STUERTZ, The production of (*S*)-cyanohydrins, EP 350908, **1990** (to Degussa).

V. S. LAMZIN, A. E. ALESHIN, B. V. STROKOPYTOV, M. G. YUKHNEVICH, V. O. POPOV, E. H. HARUTYUNYAN, and K. S. WILSON, Crystal structure of NAD-dependent formate dehydrogenase, *Eur. J. Biochem.* **1992**, *206*, 441–521.

W. LEUCHTENBERGER, M. KARRENBAUER, and U. PLOECKER, Scale-up of an enzyme membrane reactor process for the manufacture of L-enantiomeric compounds, *Ann. N.Y. Acad. Sci. USA* **1984**, *434*, 78–86.

S. V. LEY, Stereoselective synthesis of inositol phosphates, *Pure Appl. Chem.* **1990**, *62*, 2031–2034.

A. LIESE and M. V. FILHO, Production of fine chemicals using biocatalysis, *Curr. Opin. Biotechnol.* **1999**, *10*, 595–603.

G. W. MATCHAM and A. R. ST. G. BOWEN, Biocatalysis for chiral intermediates: meeting commercial and technical challenges, *Chimica Oggi* **1996**, *(6)*, 20–24.

T. MATSUDA, T. HARADA, N. NAKAJIMA, and K. NAKAMURA, Mechanism for improving stereoselectivity for asymmetric reduction using acetone powder of microorganism, *Tetrahedron Lett.* **2000**, *41*, 4135–4138.

A. MATSUYAMA and S. TOKUYAMA, Methods for racemizing *N*-acylamino acids and producing optically active amino acids, Eur. Patent Appl. EP 1130108, **2001**

O. MAY, M. SIEMANN, M. PIETZSCH, M. KIESS, R. MATTES, and C. SYLDATK, Substrate-dependent enantioselectivity of a novel hydantoinase from *Arthrobacter aurescens* DSM 3745: purification and characterization as new member of cyclic amidases, *J. Biotechnol.* **1998**, *61*, 1–13.

O. MAY, P. NGUYEN, and F. H. ARNOLD, Inverting enantioselectivity by directed evolution of hydantoinase for improved production of L-methionine, *Nature Biotechnol.* **2000**, *18*, 317–320.

O. MAY, S. VERSECK, A. S. BOMMARIUS, and K. DRAUZ, Development of dynamic kinetic resolution processes for biocatalytic production of natural and non-natural L-amino acids, *Org. Proc. Res. Dev.* **2002**, *6*, 452–457.

R. H. MAZUR, J. M. SCHLATTER, and A. H. GOLDKAMP, Structure–taste relationships of some dipeptides, *J. Am. Chem. Soc.* **1969**, *91*, 2684–2691.

H. MISONO, K. SUGIHARA, Y. KUWAMOTO, S. NAGATA, and S. NAGASAKI, Leucine dehydrogenase from *Corynebacterium pseudodiphtheriticum*: purification and characterization, *Agric. Biol. Chem.* **1990**, *54*, 1491–1498.

S. NAGATA, K. TANISAWA, N. ESAKI, Y. SAKAMOTO, T. OHSHIMA, H. TANAKA, and K. SODA, Gene cloning and sequence determination of leucine dehydrogenase from *Bacillus stearothermophilus* and structural comparison with other $NAD(P)^+$-dependent dehydrogenases, *Biochemistry* **1988**, *27*, 9056–9062.

S. NAGATA, H. MISONO, S. NAGASAKI, N. ESAKI, H. TANAKA, and K. SODA, Gene cloning, purification, and characterization of the highly thermostable leucine dehydrogenase of *Bacillus* sp., *J. Ferment. Biotechnol.* **1990**, *69*, 199–203.

C. NEUBERG and J. HIRSCH, An enzyme which brings about union into carbon chains (carboligase), *Biochem. Z.* **1921**, *115*, 282–310.

C. NEUBERG and J. HIRSCH, A process to ephedrine, German patent DE 546 459, **1932**.

U. NIEDERMEYER, U. KRAGL, M.-R. KULA, C. WANDREY, K. MAKRYALEAS, and K. DRAUZ, Enzymic preparation of optically active cyanohydrins, EP 326063, **1989** (to Degussa).

U. NIEDERMEYER and M.-R. KULA, Enzyme-catalyzed synthesis of (*S*)-cyanohydrins, *Angew Chem.* **1990**, *102*, 423–425.

K. NIEFIND, J. MUELLER, B. RIEBEL, W. HUMMEL, and D. SCHOMBURG, The crystal structure of *R*-specific alcohol dehydrogenase from *Lactobacillus brevis* suggests the structural basis of its metal dependency, *J. Mol. Biol.* **2003**, *327*, 317–328.

Y. NISHIDA, K. NAKAMICHI, K. NABE, and T. TOSA, Enzymic production of L-tryptophan from DL-5-indolylmethylhydantoin by *Flavobacterium* sp., *Enzyme Microb. Technol.* **1987**, *9*, 721–725.

M. NORTH, Catalytic asymmetric cyanohydrin synthesis, *Synlett* **1993**, 807–820.

T. C. NUGENT and T. HUDLICKY, Chemoenzymatic synthesis of all four stereoisomers of sphingosine from chlorobenzene: glycosphingolipid precursors(1)(a), *J. Org. Chem.* **1998**, *63*, 510–520.

V. I. OGNYANOV, V. K. DATCHEVA, and K. S. KYLER, Preparation of chiral cyanohydrins by an oxynitrilase-mediated transcyanation, *J. Am. Chem. Soc.* **1991**, *113*, 6992–6996.

T. OHSHIMA, S. NAGATA, and K. SODA, Purification and Characterization of thermostable leucine dehydrogenase from *Bacillus stearothermophilus*, *Arch. Microbiol.* **1985a**, *141*, 407–411.

T. OHSHIMA, C. WANDREY, M. SUGIURA, and K. SODA, Screening of thermostable leucine and alanine dehydrogenases in thermophilic *Bacillus strains*, *Biotechnol. Lett.* **1985b**, *7 (12)*, 871–876.

T. OHSHIMA and K. SODA, Thermostable amino acid dehydrogenases: applications and gene cloning, *Trends Biotechnol.* **1989a**, *7*, 210–214.

T. OHSHIMA and K. SODA, Biotechnological aspects of amino acid dehydrogenases, *Int. Ind. Biotechnol.* **1989b**, *9*, 5–11.

T. OHSHIMA and K. SODA, Biochemistry and biotechnology of amino acid dehydrogenases, *Adv. Biochem. Eng./Biotechnol.* **1990**, *42*, 187–209.

M. OKA, Y.-S. YANG, S. NAGATA, N. ESAKI, H. TANAKA and K. SODA, *Biotechnol. Appl. Biochem.* **1989**, *11*, 307–311.

K. OYAMA, K. KIHARA, and Y. NONAKA, On the mechanism of the action of thermolysin: kinetic study of the thermolysin-catalyzed condensation reaction of *N*-benzyloxycarbonyl-L-aspartic acid with L-phenylalanine methyl ester, *J. Chem. Soc. Perkin Trans. 2* **1981a**, 356–360.

K. OYAMA, S. NISHIMURA, Y. NONAKA, K. KIHARA, and T. HASHIMOTO, Synthesis of an aspartame precursor by immobilized thermolysin in an organic solvent, *J. Org. Chem.* **1981b**, *46*, 5241–5242.

K. OYAMA, S. IRINO, T. HARADA, and N. HAGI, Enzymic production of aspartame, *Ann. N. Y. Acad. Sci. (Enzyme Eng. 7)* **1984**, *434*, 95–98.

K. OYAMA and K. KIHARA, A new horizon for enzyme technology, *CHEMTECH* **1984**, *14*, 100–105.

K. OYAMA, S. IRINO, and N. HAGI, Production of aspartame by immobilized thermoase, *Meth. Enzymol.* **1987**, *136*, 503–516.

D. R. PALMER, J. B. GARRET, V. SHARMA, R. MEGANATHAN, P. C. BABBITT, and J. A. GERLT, Unexpected divergence of enzyme function and sequence: *N*-acyl-amino acid racemase is *o*-succinylbenzoate synthase, *Biochemistry* **1999**, *38*, 4252–4258.

C. C. PFEIFFER, Optical isomerism and pharmacological action, a generalization, *Science* **1956**, *124*, 29–31.

M. PIETZSCH, C. SYLDATK, and F. WAGNER, Isolation and characterization of a new, non pyridoxal-5′-phosphate dependent hydantoin racemase, *DECHEMA Biotechnol. Conf.* **1990**, *4*, 259–262.

B. RIEBEL, Biochemical and molecular biological characterization of novel microbial NAD(P)-dependent alcohol dehydrogenases, PhD thesis, University of Düsseldorf, Düsseldorf, Germany, **1997**.

S. RODRIGUEZ, K. T. SCHROEDER, M. M. KAYSER, and J. D. STEWART, Asymmetric synthesis of I-hydroxy esters and α-alkyl-β-hydroxy esters by recombinant *Escherichia coli* expressing enzymes from baker's yeast, *J. Org. Chem.* **2000**, *65*, 2586–2587.

S. RODRIGUEZ, M. M. KAYSER, and J. D. STEWART, Highly stereoselective reagents for β-keto ester reductions by genetic engineering of baker's yeast, *J. Am. Chem. Soc.* **2001**, *123*, 1547–1555.

J. D. ROZZELL, Production of amino acids by transamination, US Patent 4518692, **1985**.

J. D. ROZZELL, Immobilized aminotransferases for amino acid production, *Meth. Enzymol.* **1987**, *136*, 479–497.

H. E. SCHOEMAKER, W. H. J. BSOESTERN, B. KAPTEIN, H. F. M. HERMES, T. SONKE, Q. B. BROXTERMAN, W. J. J. VAN DEN TWEEL, and J. KAMPHUIS, Chemo-enzymic synthesis of amino acids and derivatives, *Pure Appl. Chem.* **1992**, *64*, 1171–1175.

E. SCHOFFERS, A. GOLEBIOWSKI, and C. R. JOHNSON, Enantioselective synthesis through enzymic asymmetrization, *Tetrahedron* **1996**, *52*, 3769–3826.

H. SCHÜTTE, J. FLOSSDORF, H. SAHM, and M.-R. KULA, Purification and properties of formaldehyde dehydrogenase and formate dehydrogenase from *Candida boidinii*, *Eur. J. Biochem.* **1976**, *62*, 151–160.

H. Schütte, W. Hummel, H. Tsai, and M.-R. Kula, L-Leucine dehydrogenase from *Bacillus cereus, Appl. Microbiol. Biotechnol.* **1985**, *22*, 306–317.

K. Seelbach, B. Riebel, W. Hummel, M.-R. Kula, V. I. Tishkov, A. M. Egorov, C. Wandrey, and U. Kragl, A novel, efficient regenerating method of NADPH using a new formate dehydrogenase, *Tetrahedron Lett.* **1996**, *37*, 1377–1380.

R. A. Sheldon, Consider the environmental quotient, *CHEMTECH* **1994**, *(3)*, 38–47.

W.-R. Shieh, A. S. Gopalin, and C. J. Sih, Stereochemical control of yeast reductions. 5. Characterization of the oxidoreductases involved in the reduction of β-keto esters, *J. Am. Chem. Soc.* **1985**, *107*, 2993–2994.

W.-R. Shieh and C. J. Sih, Stereochemical control of yeast reductions. 6. Diastereoselectivity of 2-alkyl-3-oxobutanoate oxidoreductases, *Tetrahedron Asymm.* **1993**, *4*, 1259–1269.

J.-S. Shin and B.-G. Kim, Kinetic modeling of ω-transamination for enzymic kinetic resolution of α-methylbenzylamine, *Biotechnol. Bioeng.* **1998**, *60*, 534–540.

J.-S. Shin and B.-G. Kim, Asymmetric synthesis of chiral amines with ω-transaminase, *Biotechnol. Bioeng.* **1999**, *65*, 206–211.

H. Slusarczyk, Stabilization of NAD-dependent formate dehydrogenase from *Candida boidinii* by site-directed mutagenesis, Ph.D. thesis, Heinrich-Heine University, Düsseldorf, Germany, **1997**.

H. Slusarczyk, S. Felber, M.-R. Kula, and M. Pohl, Stabilization of NAD-dependent formate dehydrogenase from *Candida boidinii* by site-directed mutagenesis of cysteine residues, *Eur. J. Biochem.* **2000**, *267*, 1280–1289.

H. Slusarczyk, S. Felber, M.-R. Kula, and M. Pohl, Variants of the formate dehydrogenase of *Candida boidinii* with improved catalytic activity and stability for industrial use, Eur. Patent Appl. EP 1295937, **2003** (to Degussa).

Solvay Duphar N. V., Eur. Patent Appl. 203214, **1991**.

N. O. V. Sonntag, The reactions of aliphatic acid chlorides, *Chem. Rev.* **1953**, *52*, 237–416.

J. D. Stewart, Dehydrogenases and transaminases in asymmetric synthesis, *Curr. Opin. Chem. Biol.* **2001**, *5*, 120–129.

S. C. Stinson, Chiral drugs, *Chem. Eng. News* **1994**, Sept. 19, 38–72.

D. I. Stirling, The use of aminotransferases for the production of chiral amino acids and amines, in *Chirality in Industry*, A. N. Collins, G. N. Sheldrake, and J. Crosby (eds.), Wiley, New York, **1992**, pp. 209–222.

W. F. H. Sybesma, A. J. J. Straathof, J. A. Jongejan, J. T. Pronk, and J. J. Heijnen, Reductions of 3-oxo esters by baker's yeast: current status, *Biocatal. Biotrans.* **1998**, *16*, 95–134.

C. Syldatk, O. May, J. Altenbuchner, R. Mattes, and M. Siemann, Microbial hydantoinases – industrial enzymes from the origin of life?, *Appl. Microbiol. Biotechnol.* **1999**, *51*, 293–309.

T. Takahashi and K. Hatano, Microbial acylamino acid racemase, production and use in L- and D-amino acid manufacture, Eur. Patent EP 304021, **1989.**

S. M. Thomas, R. DiCosimo, and V. Nagarajan, Biocatalysis: applications and potentials for the chemical industry, *Trends Biotechnol.* **2002**, *20*, 238–242.

S. Tokuyama, H. Miya, K. Hatano, and T. Takahashi, Purification and properties of a novel enzyme, *N*-acylamino acid racemase, from *Streptomyces atratus* Y-53, *Appl. Microbiol. Biotechnol.* **1994**, *40*, 835–840.

S. Tokuyama and K. Hatano, Purification and properties of thermostable *N*-acylamino acid racemase from *Amycolatopsis* sp. TS-1-60, *Appl. Microbiol. Biotechnol.* **1995a**, *42*, 853–859.

S. Tokuyama and K. Hatano, Cloning, DNA sequencing and heterologous expression of the gene for thermostable *N*-acylamino acid racemase from *Amycolatopsis* sp. TS-1-60 in *Escherichia coli, Appl. Microbiol. Biotechnol.* **1995b**, *42*, 884–889.

S. Tokuyama and K. Hatano, Overexpression of the gene for *N*-acylamino acid racemase from *Amycolatopsis* sp. TS-1-60 in *Escherichia coli* and continuous production of optically active methionine in a bioreactor, *Appl. Microbiol. Biotechnol.* **1996**, *44*, 774–777.

S. Tokuyama, D-Aminoacylase, Eur. Patent Appl. EP 0950706, **1999** (to Daicel Chemical Industries).

E. J. A. X. van de Sandt, and E. de Vroom, Innovations in cephalosporin and penicillin production: painting the antibiotics industry green, *Chimica Oggi* **2000**, *18*, 72–75.

W. J. J. van de Tweel, J. A. M. De Bont, M. J. A. W. Vorage, E. H. Marsman, J. Tramper, and J. Koppejan, Continuous production of *cis*-1,2-dihydroxycyclohexa-3,5-diene (*cis*-benzeneglycol) from benzene by a mutant of a benzene-degrading *Pseudomonas* sp., *Enzyme Microb. Technol.* **1988**, *10*, 134–142.

W. J. J. van de Tweel, T. J. G. M. van Dooren, P. H. De Jonge, B. Kaptein, A. L. L. Duchateau, and J. Kamphuis, *Ochrobactrum anthropi* NCIMB 40321: a new biocatalyst with broad-spectrum L-specific amidase activity, *Appl. Microbiol. Biotechnol.* **1993**, *39*, 296–300.

L. M. van Langen, E. de Vroom, F. van Rantwijk, and R. A. Sheldon, Enzymatic coupling using a mixture of side chain donors affords a greener process for ampicillin, *Green Chem.* **2001**, *3*, 316–319.

S. Verseck, A. S. Bommarius, and M.-R. Kula, Screening, overexpression and characterization of an *N*-acylamino acid racemase from *Amycolatopsis orientalis* subsp. *lurida*, *Appl. Microbiol. Biotechnol.* **2001**, *55*, 354–361.

F. Wagner, B. Hantke, T. Wagner, K. Drauz, and A. Bommarius, New microorganism, its use, and process for the manufacture of L-α-amino acids, US Patent 5714355, **1996**.

T. Wagner, B. Hantke, and F. Wagner, Production of L-methionine from D,L-5-(2-ethylthioethyl) hydantoin by resting cells of a new mutant strain of *Arthrobacter* species DSM 7330, *J. Biotechnol.* **1996**, *46*, 63–68.

C. Wandrey and B. Bossow, Continuous cofactor regeneration. Utilization of polymer bound NAD(H) for the production of optically active acids, *Biotekhnologiya Biotekhnika* **1986a**, *3*, 8–13.

C. Wandrey, in: *Enzymes as Catalysts in Organic Synthesis* (ed.: M. Schneider), D. Reidel, Dordrecht, **1986b**, pp. 263–284.

G. M. Whited, J. C. Downie, T. Hudlicky, S. P. Fearnley, T. C. Dudding, H. F. Olivo, D. Parker, Oxidation of 2-methoxynaphthalene by toluene, naphthalene and biphenyl dioxygenases: structure and absolute stereochemistry of metabolites, *Bioorg. Med. Chem.* **1994**, *2*, 727–734.

M. Whittaker, C. D. Floyd, P. Brown and A. J. H. Gearing, Design and therapeutic application of matrix metalloprotease inhibitors, *Chem. Rev.* **1999**, *99*, 2735–2776.

R. Wichmann, C. Wandrey, A. F. Bückmann, and M.-R. Kula, Continuous enzymatic transformation in an enzyme membrane reactor with simultaneous NADH regeneration, *Biotechnol. Bioeng.* **1981**, *23*, 2789–2802.

M. Wolberg, W. Hummel, and M. Mueller, Biocatalytic reduction of beta,delta-diketo esters: a highly stereoselective approach to all four stereoisomers of a chlorinated beta,delta-dihydroxy hexanoate, *Chemistry* **2001**, *7*, 4562–4571.

C.-H. Wong and G. M. Whitesides, Enzyme-catalyzed organic synthesis: NAD(P)H cofactor regeneration by using glucose-6-phosphate and the glucose-5-phosphate dehydrogenase from *Leuconostoc mesenteroides*, *J. Am. Chem. Soc.* **1981**, *103*, 4890–4899.

J. M. Woodley, A. J. Brazier, and M. D. Lilly, Lewis cell studies to determine reactor design data for two-liquid-phase bacterial and enzymic reactions, *Biotechnol. Bioeng.* **1991**, *37*, 133–140.

H. Yamada and T. Nagasawa, Production of useful amides by enzymatic hydration of nitriles, *Ann. N. Y. Acad. Sci. (Enzyme Engineering 10)* **1990**, *613*, 142–154.

Th. P. Zimmermann, K.-T. Robins, J. Werlen, and F. W. J. M. M. Hoeks, Biotransformation in the production of L-carnitine, in: *Chirality in Industry II*, A. N. Collins, G. N. Sheldrake, and J. Crosby (eds.), Wiley, Chichester, UK, **1997**, pp. 287–305.

8
Biotechnological Processing Steps for Enzyme Manufacture

Summary

Every biotechnological process can be divided into three parts: (i) medium preparation, (ii) reaction, and (iii) downstream processing. Fermentations are more complex than enzyme reactions because of: (i) the requirements to work under sterile conditions; (ii) the autocatalytic nature of fermentation which necessitates a starting amount of cells, the *inoculum;* (iii) the need to sparge air or oxygen into the fermenter, and (iv) the need for other elements such as C, N, P, S, and others. After a *lag phase,* cells enter an *exponential growth phase* before settling into the *stationary phase.* Nutrients are required for three processes: cell growth; maintenance; and product formation.

After full-scale production, the product mixture has to be purified by *downstream processing* steps. In the case of intracellular products, which constitute the majority of enzymes produced, *downstream processing* begins with centrifugation and washing of the cells, followed, in case of intracellular proteins, by cell homogenization. *Primary purification* continues either by another centrifugation at much higher g-forces, aqueous two-phase extraction, or precipitation from a high salt environment. A dialysis step usually concludes the primary purification protocol. Should a higher degree of purity be required, *final purification* steps are added, such as chromatography (ion exchange or affinity chromatography), dialysis steps, and storage after drying or lyophilization, possibly with excipients.

During downstream processing, a trade-off has to be managed between *purity, yield,* and *concentration level*:

- *Purity* is measured most conveniently by specific activity [U (mg protein)$^{-1}$]. The ratio of the specific activity to the initial specific activity in the crude extract is called the *purification factor.*
- *Yield* is expressed as a percentage of the amount of protein (in International Units, I.U.) at the processing step in question compared to the amount of target protein contained in the crude extract after homogenization (crude lysate).
- Operating at a high *concentration level* is important because complexity, duration, and cost of downstream processing depend linearly on the concentration difference between initial and final purity levels.

Purification protocols should always aim for brevity, to minimize complexity and cost. The goal of the protein purification procedure, besides a pure protein, is a *purification table* listing all the operations undertaken, with results on overall yield, specific activity, and purification factor. An assay for both protein function and protein concentration is required at every step.

The flow rate Q of a centrifuge depends on properties of the fermentation broth (ρ_p, ρ_m, d_p, η), centrifuge design (R_o, R_i, n, θ) and the operating parameter ω. Protein release during cell wall disruption often obeys a first-order law. Protein mixtures can be fractionated by precipitation with salt [kosmotropes such as $(NH_4)_2SO_4$] or organic solvent (low ε). The flux of water J_w through a membrane at low membrane resistance is proportional to transmembrane pressure ΔP.

Chromatography (elution or frontal mode), often the final purification step for high purity, is a separation process that depends on partitioning between a flowing fluid (the *mobile* phase) and a solid adsorbent (the *stationary* phase).

8.1
Introduction to Protein Isolation and Purification

While plants and animal organs are often good sources of specific enzymes, such as α-chymotrypsin from bovine liver or peroxidase from soybeans, the vast majority of enzymes are obtained from microorganisms, mostly bacteria but also yeast or fungi. To obtain enzymes from microorganisms in known quantities and quality, at any time of year independently of nature's growth and harvest cycle, the appropriate production processes have to be developed.

Every biotechnological process can be divided into three parts: (i) medium preparation; (ii) reaction; and (iii) downstream processing (Figure 8.1). Enzymatic processes are often comparatively straightforward to conduct owing to simple medium preparation and downstream processing steps as there are only a few components in the product mixture. Fermentations, on the other hand, are more complex: medium components have to be sterilized before use and in many cases an initial growth cycle on a small scale has to be run. The final broth from this initial stage then serves as the inoculum for the full-scale production run.

After fermentation, the product mixture has to be purified by downstream processing steps. In the case of intracellular products, which constitute the majority of enzymes produced, *downstream processing* begins with centrifugation and washing of the cells, followed by cell homogenization, with either a ball mill or a French press. The cell homogenization step is skipped in the case of extracellular enzymes, such as proteases for laundry detergents. *Primary purification* continues through either another centrifugation step at much higher g-forces, aqueous two-phase extraction, or precipitation from a high-salt environment. A dialysis step usually concludes the primary purification protocol. Enzymes are stored at that step either in dry form after precipitation or, on an industrial scale, after spray drying, or with excipients (stabilizers) such as sugars or polyols (glycerol) in soluble form.

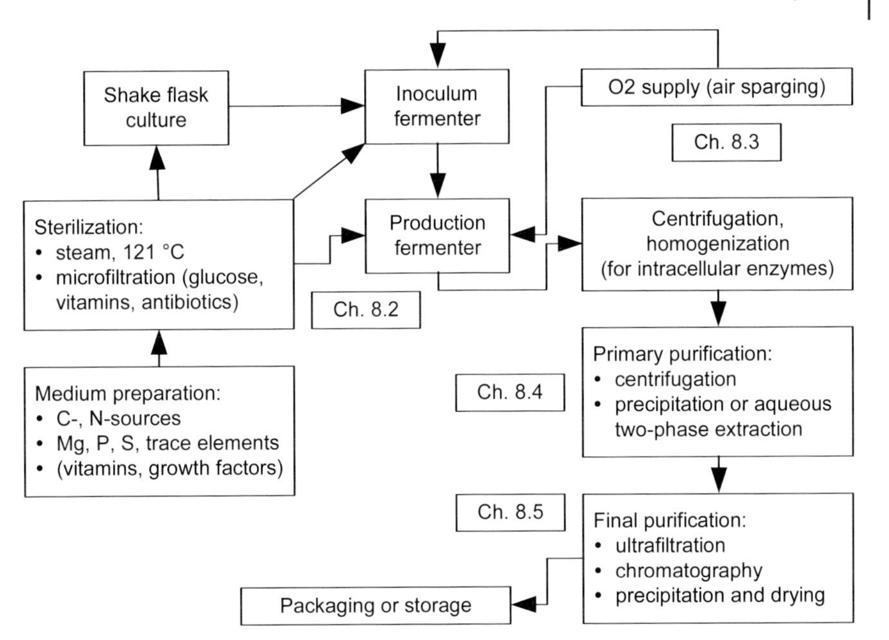

Figure 8.1 The steps of a biotechnological process.

Should a higher degree of purity be required, final purification steps are added to the procedure. Final purification typically includes various forms of chromatography such as ion-exchange and, commonly as the last step, affinity chromatography. Dialysis steps to change the salt level are added as required. A pure enzyme is either stored with excipients in liquid form or is lyophilized, i.e., dried by sublimation of water.

This chapter is organized around the different processing steps of fermentation, primary and final purification, as illustrated in Figure 8.1.

During downstream processing, a trade-off has to be managed between *purity*, *yield*, and *concentration level*:

- *Purity* is measured most conveniently by comparing specific activity [U [(mg protein)$^{-1}$] at the processing stage in question to the specific activity of the pure protein. If the specific activity of the pure protein is not known, an estimate of the concentration of the target protein from gel electrophoresis can be substituted. The ratio of the specific activity at the processing step in question to the initial specific activity in the crude extract is called the *purification factor*.

- *Yield* is expressed as a percentage of the amount of protein (in International Units, I.U.) at the processing step in question compared to the amount of target protein contained in the crude extract after homogenization. Overexpression of the target protein during fermentation results in a much higher amount of target protein so that a high yield might not be as necessary as for a less efficient fermentation.

- Operating at high *concentration level* is important because complexity and time, and thus also cost of downstream processing, are linearly dependent on the concentration difference between the initial and final purity levels. Often the final stage is homogeneity or near-homogeneity of the target protein, so that the initial concentration of target protein in the crude extract predetermines the complexity of the downstream processing operation. This is also the rationale for aiming for overexpression of target protein during fermentation.

To obtain meaningful numbers, both an assay for protein function as well as an assay for protein concentration (see Chapter 9, Section 9.3.1) are required. Purification protocols should always aim for brevity, in order to minimize complexity and cost. Any additional step to the protocol, with rare exceptions, results in a yield loss but usually helps purity. The incompatibility of high yield combined simultaneously with high purity has to be emphasized again here. Some steps, such as ultrafiltration, serve the sole purpose of enhancing concentration while adding little to enhance yield. At the end of a protein purification, a *purification table* should be set up (see Section 8.6 for examples) listing all the operations undertaken and their effect on the overall yield, specific activity, and purification factor, the last two being measures of purity.

8.2 Basics of Fermentation

While enzymes can be prepared from cells grown in either batch or continuous culture, we will cover only batch fermentation here as, even for pilot-scale biocatalysis runs, the amounts of enzyme obtainable in a batch fermentation are sufficient.

In a fermenter with growing cells, three processes occur simultaneously: (i) cell growth; (ii) cell maintenance; and (iii) product formation. On some occasions, the product is the biomass resulting from cell growth, as in the case of yeast for the brewing of beer. In most cases, the product is sought; it can be a protein, for either biocatalytic or for therapeutic purposes, or a low-molecular-weight substance ranging from inexpensive compounds such as ethanol or lysine for animal feed to expensive ones such as vitamin B_{12} or pharmaceuticals such as lovastatin or macrolide antibiotics. Figure 1.1 in Chapter 1 illustrated the continuum in modes of product formation, from fermentation via precursor fermentation to biotransformation, which can be performed in growing, resting, or dead cells.

There are four characteristic features that differentiate a fermentation from most other reactions: (i) the requirement to work under sterile conditions; (ii) the autocatalytic nature of fermentation which necessitates a starting amount of cells, the *inoculum;* (iii) the need to sparge air or oxygen into the fermenter; and (iv) the need for other elements such as C, N, P, S, and others discussed below.

8.2.1 Medium Requirements

To grow and to generate product, a cell needs a source of essential elements such as carbon, nitrogen, oxygen, sulfur, phosphorus, magnesium, and a variety of trace elements such as metals. In the most advantageous cases of well-researched fermentations and simple metabolisms, a *simple medium* is sufficient. Such a simple medium contains well-defined compounds containing carbon, nitrogen, sulfur, phosphorus, and magnesium. The carbon source (C-source) can be a sugar (most commonly glucose but also sucrose), an alcohol such as methanol, a hydrocarbon (such as octane), or an organic acid such as acetic acid. The nitrogen source (N-source) can be an ammonium salt or a nitrate. Magnesium sulfate is added to provide the needs for both magnesium and sulfur; phosphorus is added as a potassium salt as many phosphorus salts are insoluble in water. A special case is the source for oxygen, which in the vast majority of cases is air, and sometimes is pure oxygen. Hydrogen might be considered as the other special case, as no specific source for hydrogen is required in a fermentation: the C- or N-source takes care of the need for hydrogen.

If the needs of the cellular organism are more complex and/or are not very well known, the fermentation requires a *complex medium*. The C-source, which simultaneously also serves as an N-, S-, and P-source, for such a complex medium can be yeast extract, tryptone, corn steep liquor, or molasses. Requirements for the other elements can be supplemented by the addition of the respective salts. Oxygen is again an exception and has to be supplied in the form of air or, much less often, as pure oxygen.

Cells do not need C-, N-, O-, and other sources just for cell mass and product formation but also to satisfy energy needs. Except in rare cases, the C-source simultaneously serves as the energy source; the product of such a use as an energy source is CO_2. As can already be justified from the different heats of combustion, C-sources feature vastly different values as energy sources, depending on their initial oxygen content: hydrocarbons have considerably higher values than acetic acid as energy sources (Table 8.1).

Table 8.1 Heats of combustion, ΔG_{com}, for different C-sources.

	C-source							
	Pentane	Benz-aldehyde	Acetone	Ethanol	Sucrose	Glucose	Acetic acid	Oxalic acid
Formula	C_5H_{12}(l)	C_7H_6O(l)	C_3H_6O(l)	C_2H_6O(s)	$C_{12}H_{22}O_{11}$	$C_6H_{12}O_6$	$C_2H_4O_2$(l)	$C_2H_2O_4$(s)
ΔG_{com} [kJ mol^{-1}]	–3509.0	–3525.1	–1789.9	–1366.8	–5644.9	–2805.0	– 874.2	–251.1
MW [g mol^{-1}]	72.15	106.12	58.08	46.07	342.11	180.06	60.05	90.04
Oxygen [%]	0	15.1	27.5	34.7	51.1	53.3	53.3	71.1
ΔG_{com} [kJ g^{-1}]	–48.63	–33.22	–30.82	–29.67	–16.50	–15.58	– 14.56	–2.79

Source: Duran (1995), Table B.8.

8.2.2 Sterilization

Sterility means total absence of contaminating organisms. Unsterile conditions lead to competition between the target organism and the contaminant and thus often to outgrowth of the target organism, but to spoilage of the fermentation in any case. Sterility can be achieved by microfiltration, or by chemical or heat treatment. Heat treatment is the common method to achieve sterility, but sensitive compounds such as vitamins cannot be subjected to it and must be sterilized by microfiltration separately from the rest of the fermentation broth. Heat treatment is achieved by using steam, most often at 121 °C, for a given period of time. As cell death is an exponential process over time, we can define a criterion $\nabla = \ln(n_0/n)$ (n is the number of contaminant cells); after each ∇, only 10% of the original contaminant organisms are left. Killing of contaminants occurs during heating of the fermentation medium to 121 °C, during the hold time, and during the cooling phase. Equation (8.1) therefore holds.

$$\nabla_{total} = \nabla_{heating} + \nabla_{holding} + \nabla_{cooling} \tag{8.1}$$

If sterility is to be achieved in all except one out of 100 cultures, so the probability level N of contamination is 1/100 (a good number for laboratory fermentations), and we assume an initial concentration of 10^3 mL^{-1} (potable water quality) to 10^6 mL^{-1} (visible contamination starts) and a fermentation volume of 10 L, the ∇_{total} to be achieved is given by Eq. (8.2a) or (8.2b).

$$\text{lower limit} \qquad \nabla_{total} = \ln(10^3\ \text{mL}^{-1} \times 10^4\ \text{mL}/10^{-2}) = \ln 10^9 = 20.7 \tag{8.2a}$$

or

$$\text{upper limit} \qquad \nabla_{total} = \ln(10^6\ \text{mL}^{-1} \times 10^4\ \text{mL}/10^{-2}) = \ln 10^{12} = 27.6 \tag{8.2b}$$

8.2.3 Phases of a Fermentation

A batch fermentation can be divided into four phases: after *inoculation*, there often is a *lag phase*, where apparently nothing happens until the fermentation goes into its exponential growth phase. In that phase, often just called the *exponential phase*, cells divide into two in successive generations of one, two, four (= 2^2), eight (= 2^3), and so on, cells. The time between doublings, the *doubling time* $t_{1/2}$, is characteristic for each organism and conditions; fast-growing bacteria such as *E. coli* need about 20 min, yeast and fungi typically about an hour, and mammalian cells often more than a day. After the limiting nutrient (usually the C-source) is consumed or after a certain cell density is achieved, the fermentation enters the *stationary phase* where cell density stays constant. After more time, cell lysis sets in or cells consume each other to supply maintenance; the fermentation now is in its *death phase*.

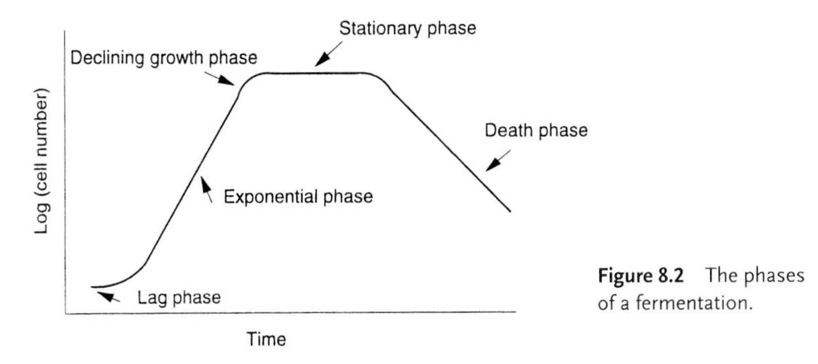

Figure 8.2 The phases of a fermentation.

Harvesting of cells or product occurs best during at the end of the logarithmic phase or during the stationary phase.

8.2.4
Modeling of a Fermentation

As cells divide into two and grow exponentially, the rate of change of cell number dN/dt is proportional to the number of cells already existing (Eq. (8.3), where μ is the *specific growth rate* [h^{-1}] [even though it is a rate *constant* (!), not a rate]).

$$dN/dt = \mu \cdot N \tag{8.3}$$

As cells in a fermenter cannot be counted easily, it is more convenient to employ cell density X, expressed as dry cell weight (dcw) per volume [$g\ L^{-1}$], as the dependent variable [Eq. (8.4)], which yields Eqs. (8.5a) and (8.5b) upon integration with the inoculum.

$$dX/dt = \mu \cdot X \tag{8.4}$$

$$X = X_0 \exp(\mu \cdot t) \quad \text{which is equivalent to} \quad \ln(X/X_0) = \mu \cdot t \tag{8.5a,b}$$

With incorporation of the lag time t_{lag} at the starting point of the exponential phase, Eqs. (8.6a) and (8.6b) are obtained.

$$X = X_0 \exp\{\mu \cdot (t - t_{lag})\} \text{ or likewise} \quad \ln(X/X_0) = \mu \cdot (t - t_{lag}) \tag{8.6a,b}$$

The stationary phase can be represented by X = const. whereas the death phase can be modeled according to Eq. (8.7) with k_d as the *death rate constant* [time]$^{-1}$ or upon integration based on X_{max} as the cell density after the stationary phase [Eqs. (8.8a) and (8.8b)].

$$dX/dt = -k_d \cdot X \tag{8.7}$$

$$X = X_{max} \exp(-k_d \cdot t) \text{ which again is equivalent to } \ln(X/X_{max}) = -k_d \cdot t \tag{8.8a,b}$$

8.2.5
Growth Models

The equations describing increase in cell density [Eqs. (8.3)–(8.8)] so far do not contain any information about the nature and concentration of any substrate such as the C-source. As the specific growth rate μ tends to depend on quality and amount of substrate, however, we require a growth model which provides the function $\mu = \mu([S])$. The most widely used growth model is the Monod model (Monod, 1950) which assumes that only one substrate limits cell growth and proliferation. The corresponding equation [Eq. (8.9), in which μ_{max} is the *maximum specific growth rate* [h^{-1}]] reads very similarly to the Michaelis–Menten equation.

$$\mu = \mu_{max} \cdot [S]/(K_s + [S]) \tag{8.9}$$

Just as the Michaelis–Menten equation, the Monod equation features a saturation curve. At $[S] \ll K_s$, Eq. (8.9) simplifies to first-order kinetics with respect to substrate: $\mu = (\mu_{max}/K_s) \cdot [S]$. When $[S] \gg K_s$, we obtain the expression for the zeroth-order limit: $\mu = \mu_{max}$. Typical values for μ_{max} [h^{-1}] are: animal cells, 0.03–0.1; fungi, 0.1–0.34; yeasts, 0.34–0.6; bacteria, 0.7–3.0. The growth rate μ does not depend just on substrate concentration, of course, but also on other variables such as temperature. Every organism features its optimum growth temperature, which can range from 10 to 20 °C for psychrophiles (cold-loving organisms), through 25–40 °C for mesophiles and 50–70 °C for moderate thermophiles, to 80–100 °C for extreme thermophiles. In Chapter 3, we have discussed these and similar situations for extreme pH values and salt contents.

8.2.6
Fed-Batch Culture

In many enzyme fermentations, the limiting component, usually the C-source, has to be added semi-continuously to keep its concentration at a predetermined, usually low, value. This measure makes it possible either to influence selectivity between different pathways or to uncouple predominantly cell growth during the first phase of the fermentation from predominantly product (i.e., enzyme) formation in the later stages of the fermentation cycle. Often, protein formation is induced by adding an inducer (see Chapter 4). During the fed-batch phase, the broth volume increases. Either the broth is harvested when the maximum volume is reached, or broth is withdrawn from time to time. The product is present in high concentrations.

When analyzing a fed-batch fermentation, we cannot assume constant volume as in the batch fermenter, so the balances for cell density, substrate, product, and volume read according to Eqs. (8.10)–(8.13).

$$\text{Cell density:} \quad d(XV)/dt = \mu XV \tag{8.10}$$

$$\text{Substrate:} \quad d(SV)/dt = FS_f - (1/Y_{X/S}) \cdot \mu \cdot XV - (1/Y_{P/S}) \cdot q_p \cdot XV \tag{8.11}$$

Product: $d(PV)/dt = q_p \cdot XV$ (8.12)

Volume: $dV/dt = F$ (8.13)

In Eq. (8.11), S_f is the substrate feed rate, and FS_f consequently is the molar flow into the fermenter (there is no flow out of the fermenter). $Y_{X/S}$ and $Y_{P/S}$ are the two empirical yield coefficients of cell X or product P on substrate S [g (g substrate)$^{-1}$] and q_p is the *specific production formation rate* [g product (g cells)$^{-1}$ h^{-1}] ($q_p \cdot X = r_p$, the product formation rate [g product h^{-1}]).

As $d(AV)/dt = A \cdot dV/dt + V \cdot dA/dt$ in Eqs. (8.10)–(8.12) and with Eq. (8.13), the equations can be rewritten as Eqs. (8.14)–(8.16).

Cell density: $dX/dt = \mu \cdot X - (F/V) \cdot X$ (8.14)

Substrate: $dS/dt = (F/V)(S_f - S) - (1/Y_{X/S}) \cdot \mu X$ (8.15)

Product: $dP/dt = q_p \cdot X - (F/V) \cdot P$ (8.16)

Combining cell and substrate mass balance [Eqs. (8.14) and (8.15) respectively] and integrating yields Eq. (8.17), which, with the feed concentration satisfying the condition stipulated by Eq. (8.18), transforms into Eq. (8.19).

$$X(t) + Y_{X/S} \cdot S(t) = Y_{X/S} \cdot S_f - \{Y_{X/S} \cdot S_f \cdot (Y_{X/S} \cdot S(0) + X(0))\} \exp(-Ft/V) \quad (8.17)$$

$$Y_{X/S} \cdot (S_f - S(0)) = X(0) \quad (8.18)$$

$$X(t) + Y_{X/S} \cdot S(t) = Y_{X/S} \cdot S_f \quad (8.19)$$

The most important special case of fed-batch fermentation increases F exponentially so as to keep F/V constant. Then, Eq. (8.13) can be rewritten as Eq. (8.20), or integrated to give Eqs. (8.21) and (8.22).

$$dV/dt = F = (F/V) \cdot V \quad (8.20)$$

$$V = V_0 \cdot \exp\{(F/V) \cdot t\} \quad (8.21)$$

$$F = (F/V) \cdot V_0 \cdot \exp\{(F/V) \cdot t\} \quad (8.22)$$

Importantly, both X and S can be kept at constant values under such conditions during fed-batch fermentations. This can be demonstrated by inserting Eq. (8.20) into Eq. (8.14) to yield Eq. (8.23).

$$dX/dt = (\mu - (F/V)) \cdot X \quad (8.23)$$

As, however, the growth rate μ in a fed-batch fermenter equals the dilution rate F/V [h^{-1}], $dX/dt = 0$ and thus X = const. $\neq f(t)$. If X = const. is now inserted into

Eq. (8.19), S = constant $\neq f(t)$ is obtained, as X, $Y_{X/S}$, and S_f are all constant. (The dilution rate is the inverse of the fermenter residence time [h] and indicates the number of fermenter volumes per unit time.)

8.3 Fermentation and its Main Challenge: Transfer of Oxygen

Owing to the low concentration of soluble oxygen in aqueous media (8 mM or 32 mg L^{-1} at 25 °C) oxygen is often the limiting component in the fermentation. Therefore, the amount of oxygen required to be transferred into a fermentation culture for the desired cell growth and product formation has to be determined.

8.3.1 Determination of Required Oxygen Demand of the Cells

With the assumption that the supply of O_2 is limiting in the fermentation, Eq. (8.24) follows, where

$Y_{x/o}$ = yield coefficient of biomass on O_2 [g cells (g $O_2)^{-1}$]
q_{O_2} = specific respiration rate (oxygen transfer rate) [mM O_2(g cells)$^{-1}$ h^{-1})]
k' = conversion factor [mM O_2 (g O_2^{-1}] = 1000/32 = 31.3 mM O_2 (g $O_2)^{-1}$

$$q_{O_2} = k' \cdot \mu / Y_{x/o} \tag{8.24}$$

With X = cell density [g cells L^{-1}], this gives Eq. (8.25).

$$N_{O_2} = q_{O_2} \cdot X = k' \cdot \mu \cdot X / Y_{x/o} \ [\text{mM } O_2 \ (\text{L h})\text{–}1] \tag{8.25}$$

It should be noted that this equation only considers cell growth but not maintenance or product formation. Empirically, oxygen uptake is connected with heat production according to Eq. (8.26) (Cooney, 1969, 1971).

$$Q \ [\text{kcal (L h)}^{-1}] = 0.12 \cdot N_{O_2} \ [\text{mM } O_2/(\text{L h})^{-1}] \tag{8.26}$$

As a first pass or estimate, the yield coefficient $Y_{x/o}$ can be determined from Eq. (8.27, which is empirical (Mateles, 1971).

$$1/Y_{x/o} = 18 \left[\frac{2\,C + H/2 - O}{Y_{x/s} \cdot M} + \frac{O'}{1600} - \frac{C'}{600} + \frac{N'}{933} - \frac{H'}{200} \right] \tag{8.27}$$

where M is the molar mass of the substrate and C, H, and N are the numbers of respective atoms in the substrate molecule. O', C', N', and H' are the percentages of the respective element of the composition of the microorganism (as dry mass) ("molecular formula" or elemental formula). As mentioned above, $Y_{x/s}$ is the yield coefficient of biomass on substrate [g cells (g substrate)$^{-1}$]. Table 8.2 lists elemental formulae of a few microorganisms.

Table 8.2 Elemental formulae of some microorganisms.

Organism	*Elemental formula*
Escherichia coli	$CH_{1.77}O_{0.49}N_{0.24}$
Klebsiella aerogenes	$CH_{1.73}O_{0.43}N_{0.24}$
Pseudomonas C12B	$CH_{2.00}O_{0.52}N_{0.23}$
Saccharomyces cerevisiae	$CH_{1.81}O_{0.51}N_{0.17}$
Candida utilis	$CH_{1.87}O_{0.56}N_{0.20}$

Source: Roels, 1980.

As an average, the composition $CH_{1.8}O_{0.5}N_{0.25}$ often suffices. The average elemental composition of bacteria is different from that of yeasts or fungi, of which the approximate compositions are listed in Table 8.3:

Table 8.3 Elemental compositions of different cell types [%].

Element	*C*	*H*	*O*	*N*	*P*	*Ash*
Bacteria	48	6	32	11	1	1
Yeasts	44	6.7		8.6		8
Average	46	6.5	32.5	9	1	5

Source: Roels, 1983, p. 31.

8.3.2 Calculation of Oxygen Transport in the Fermenter Solution

Oxygen is supplied to a fermentation almost exclusively in the form of air which is sparged into the fermenter from the bottom in gaseous form. Thus, the oxygen supply problem translates into a gas–liquid oxygen transfer problem. The corresponding oxygen transfer equation is Eq. (8.28).

$$\mathrm{OTR} = k_L \cdot a \cdot ([O_2]_{equ} - [O_2]) \tag{8.28}$$

OTR = oxygen transfer rate [mol (L h)$^{-1}$]
k_L = mass transfer coefficient [m h^{-1}]
a = specific gas exchange area [m² m^{-3}]
$k_L \cdot a$ = volume-based oxygen mass transfer coefficient [time^{-1}]
$[O_2]_{equ}$ = O_2-equilibrium concentration [mM] (8 mM in H_2O at 25°C)
$[O_2]$ = oxygen concentration [mM]

If a fermentation is limited by oxygen in the fermenter, Eq. (8.29a) and hence Eq. (8.29b) hold.

$$N_{O_2} = \mathrm{OTR} \tag{8.29a}$$

$$k_L \cdot a \cdot ([O_2]_{equ} - [O_2]) = k' \cdot \mu \cdot X/Y_{x/o} \tag{8.29b}$$

The left-hand side of Eq. (8.29b) contains only quantities which are dependent on the physical quantities of the system: k_L depends on the surface properties between the liquid and gaseous phases, such as surface tension; a depends on the degree of partitioning of gas bubbles; $[O_2]_{equ}$ is constant at constant temperature and constant pressure.

The right-hand side only contains quantities which are dependent on the biological parameters of the system: the growth rate μ primarily depends on cell type, $Y_{x/o}$ depends on cell type (molecular composition) and C-source, and cell density X is determined by the status of the fermentation. The amount of air required in a fermentation is given by the quantity vvm [Eq. (8.30)] and is usually of the order of 0.1–2.

$$\mathrm{vvm} = \frac{\text{volume of air}}{\text{volume of fermenter liquid} \times \text{minutes}} \tag{8.30}$$

According to Henry's law ($p_{O_2} = H \cdot [O_2]$), an increase in pressure favors an increased oxygen concentration in the liquid phase; however, the increased partial pressure of CO_2 has to be taken into account.

8.3.3
Determination of k_L, a, and k_La

Empirically, it has been found that the mass transfer coefficient k_L is proportional to the square root of the impeller speed N [Eq. (8.31)] (Richards, 1961).

$$k_L = k_1 \cdot N^{0.5} \tag{8.31}$$

k_1 is a constant oreover, k_L can be obtained from an equation of dimensionless quantities, Eq. (8.32).

$$\mathrm{Sh} = k_2\, \mathrm{Gr}^a\, \mathrm{Sc}^b \tag{8.32}$$

Sh = Sherwood number = convection/diffusion = k_Ld/D
Gr = Grashof number = free convection/viscous forces = $d^3 Drg/\nu$
Sc = Schmidt number = momentum transfer/mass transfer = ν/D

It should be noted that $\nu = \eta/\rho$. For small Grashof numbers, both exponents a and b are about 1/3. In larger reactors and for hard spheres, $b \sim 0.5$. A simple empirical correlation for the specific surface a is given by Eq. (8.33) (Calderbank, 1959, 1961).

$$a = k_3(P/V)^{0.4}v_s^{0.5} \quad (8.33)$$

P/V is the specific power input into the fermenter via the impeller (P = power [W], V = volume [m³]), and v_s is the superficial velocity of the gas bubbles [m s^{-1}]. With Eq. (8.31), Eq. (8.34) is obtained for the product $k_L a$.

$$k_L a = k_4(P/V)^{0.4}v_s^{0.5}N^{0.5} \quad (8.34)$$

8.3.2.1 Methods of Measurement of the Product $k_L a$

Direct method with the Clark electrode

Via polarography, through constant renewal of the mercury surface, the reaction in Eq. (8.35) can be followed at a positive potential of +1.4 V.

$$O_2 + 2\,H_2O + 4\,e^- \rightarrow 4\,OH^- \quad (8.35)$$

From the half-potential the quantity $[O_2]$ can be inferred.

Sulfite method

The sulfite method is a chemical method which is based on the reaction between sodium sulfite and oxygen [Eq. (8.36)].

$$Na_2SO_3 + \tfrac{1}{2}O_2 \xrightarrow{Cu^{2+},\,Co^{+2}} Na_2SO_4 \quad (8.36)$$

The reaction is catalyzed by divalent heavy metal ions, especially by Cu^{2+} and Co^{2+}. It follows zeroth order, so it is independent of the concentration of the reactants Na_2SO_3 and O_2(!) but only depends on the rate of mass transfer. The remaining sulfite is oxidized to sulfate by the addition of iodine and the iodide generated is back-titrated with thiosulfate [Eqs. (8.37), (8.38)].

$$SO_3^{2-} + I_2 + 2\,OH^- \rightarrow SO_4^{2-} + 2\,I^- + H_2O \quad (8.37)$$

$$2\,I^- + S_2O_3^{2-} + H_2O \rightarrow I_2 + S_2O_4^{2-} + 2\,OH^- \quad (8.38)$$

Usually, iodine is not detected as its usual brown form but as a blue inclusion complex (cage complex) with starch. The sulfite method, besides being labor-intensive, has the disadvantage of yielding accurate data in cell-free systems only, as living cells interfere with the processes in Eqs. (8.36)–(8.38) necessary for the analysis.

Dynamic method

The dynamic method is applicable in both batch and continuous fermenters. During the experiment the oxygen concentration is constantly observed. The oxygen balance for the case of a continuous fermenter is given by Eq. (8.39).

$$d[O_2]/dt = (F/V)\cdot([O_2]_{equ} - [O_2]) + k_L\cdot a\cdot([O_2]_{equ} - [O_2]) - q_{O_2}\cdot X \quad (8.39)$$

As (F/V) with about 0.1 h^{-1} is much smaller than $k_L \cdot a$ with about 100 h^{-1} even for a continuous fermenter, the term $(F/V) \cdot ([O_2]_{equ} - [O_2])$ can be neglected and Eq. (8.39) simplifies to Eq. (8.40).

$$d[O_2]/dt = k_L \cdot a \cdot ([O_2]_{equ} - [O_2]) - q_{O_2} \cdot X \tag{8.40}$$

If the oxygen supply is interrupted and (i) surface aeration is unimportant and (ii) the bubbles disengage quickly (which they do, owing to low oxygen solubility in solution and large surface tension differences) $d[O_2]/dt = 0$, but the cells continue to consume oxygen for a while as if the oxygen supply were still on. During this phase, the oxygen concentration decreases linearly [Eq. (8.41)] until the oxygen supply is resumed.

$$k_L \cdot a \cdot ([O_2]_{equ} - [O_2]) = q_{O_2} \cdot X \tag{8.41}$$

Then, the oxygen concentration increases asymptotically to the new steady-state value $[O_2]_{t=\infty}$ ("$C_{O_2}^{\infty}$", Figure 8.3). The product $k_L a$ can then be calculated by rearranging Eqs. (8.40) and (8.41) into Eq. (8.42).

$$k_L a = r \cdot X/(c_{equ} - c) \tag{8.42}$$

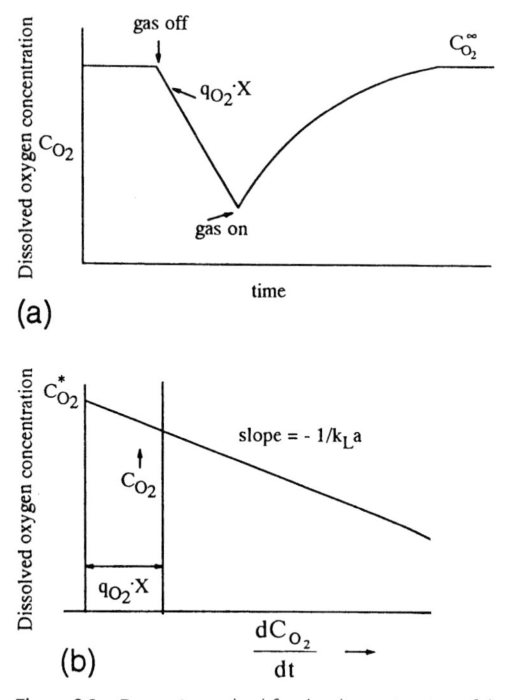

Figure 8.3 Dynamic method for the determination of the oxygen concentration.

8.4
Downstream Processing: Crude Purification of Proteins

8.4.1
Separation (Centrifugation)

After fermentation, the biomass is commonly harvested by subjecting the fermentation broth to a centrifugation process to separate insolubles, notably cells. On a large scale, a continuous centrifugation apparatus called a separator is typically employed. Centrifugation functions on the basis of density differences between two phases, typically a liquid and a solid. As the density difference $\Delta\rho$ between cells and fermentation broth can be assumed to be around 0.05 g cm^{-3}, it would take an area of tens of thousands of square meters, or several soccer (or football) fields, to separate the equivalent amount of cells in a day under gravity. The required centrifugal force and area for other measures can be calculated by the following procedure. If we set up the mass balance on a small particle, we obtain Eq. (8.43), in which the first and second term can be expressed for the situation in a centrifugal field by Eqs. (8.44) and (8.45), where ρ_p is the particle density and ρ_m the medium density, whereas the term for the drag force can be described by Stokes law at low Re ($Re \leq 1$) [Eq. (8.46)].

$$F_{gravity} - F_{buoancy} - F_{drag} = m \cdot dv/dt \quad (8.43)$$

$$F_{gravity} = m \cdot a = m \cdot \omega^2 \cdot r \quad (8.44)$$

$$F_{buoancy} = (m/\rho_p) \cdot \rho_m \cdot a = (\rho_m/\rho_p) \cdot m \cdot \omega^2 \cdot r \quad (8.45)$$

$$F_{drag} = 6\pi \cdot v \cdot \eta \cdot r_p = 3\pi \cdot v \cdot \eta \cdot d_p \quad (8.46)$$

The resulting mass balance [Eqs. (8.44)–(8.46)] yields Eq. (8.47), so that after canceling the mass, the settling velocity is given by Eq. (8.48).

$$m \cdot \omega^2 \cdot r - (\rho_m/\rho_p) \cdot m \cdot \omega^2 \cdot r - 3\pi \cdot v \cdot \eta \cdot d_p = m \cdot dv/dt \quad (8.47)$$

$$dv/dt = [1 - (\rho_m/\rho_p)] \cdot \omega^2 \cdot r - 18\pi \cdot v \cdot \eta/(\rho_p \cdot d_p^2) \quad (8.48)$$

Thus at steady-state ($dv/dt = 0$) the settling velocity during centrifugation v_c in the Stokes regime ($Re \leq 1$) can be written as Eq. (8.49).

$$v_c = (\rho_p - \rho_m) \cdot d_p^2 \cdot \omega^2 \cdot r/18\eta \quad (8.49)$$

The corresponding equation for sedimentation, meaning settling of particles under just gravity is Eq. (8.50).

$$v_s = (\rho_p - \rho_m) \cdot d_p^2 \cdot g/18\eta \quad (8.50)$$

The ratio $\zeta = \omega^2 \cdot r/g = v_c/v_s$ is called the acceleration ratio. To relate the settling velocity to parameters important for the operation and scale-up of centrifuges, we first define the equivalent settling area Σ [length²] which determines the effectiveness of the centrifuge and describes the equivalent area needed for a settler operating under the influence of just gravity. The capacity of the centrifuge is described by the liquid flow rate Q [volume/time] [Eq. (8.51)], so that Σ is given by Eq. (8.52).

$$Q = v_s \cdot A \cdot \zeta \quad \text{or} \quad Q = v_s \cdot \Sigma \quad \text{or} \quad Q = v_c \cdot (g/\omega^2 \cdot r) \cdot \Sigma \tag{8.51}$$

$$\Sigma = A \cdot \zeta \tag{8.52}$$

For a continuous disk-bowl centrifuge (Figure 8.4) with its liquid passing between the outer (R_o) and inner (R_i) radius, the appropriate definition for Q is Eq. (8.53) with the z-direction defined as the axial dimension.

$$Q = \pi\,(R_o^2 - R_i^2) \cdot dz/dt \tag{8.53}$$

The velocity of flow in the r-direction is defined in Eq. (8.54).

$$dr/dt = v = v_g\,(\omega^2 \cdot r/g) \tag{8.54}$$

Particles travel from $x = y = 0$ (Figure 8.4) to $x = (R_o - R_i)/\sin\theta$ and $y = (R_o - R_i)\cos\theta$ which, for a disk-bowl centrifuge, yields Eq. (8.55) upon integration with respect to r from (R_i) to (R_o).

$$Q = v_c\,(g/\omega^2 \cdot r) \cdot (2\pi \cdot \omega^2/g) \cdot \{(R_o^3 - R_i^3)\cot\theta/3\} \cdot n \tag{8.55}$$

Thus, Eq. (8.56) applies for a disk-bowl centrifuge, and the flow rate, using Eq. (49) and canceling the term $g \cdot \omega^2 \cdot r$, is given by Eq. (8.57).

$$\Sigma = (2\pi/3g) \cdot n \cdot \omega^2 \cdot ((R_o^3 - R_i^3)\cot\theta) \tag{8.56}$$

$$Q = \underbrace{[\pi/27]}_{\text{geometry}} \cdot \underbrace{[(\rho_p - \rho_m) \cdot d_p^2/\eta]}_{\text{fermentation broth}} \cdot \underbrace{[n \cdot \omega^2 \cdot (R_o^3 - R_i^3)\cot\theta]}_{\text{centrifuge design}} \tag{8.57}$$

Equation (8.57) elucidates the point that the flow rate Q in a separator depends on a geometric factor ($\pi/27$) as well as on both the properties of the fermentation medium to be separated, ρ_p, ρ_m, d_p, and η (the terms in the second brackets) and on the properties of the apparatus, n, ω, R_o, R_i, and θ (the terms in the third brackets). Again, without a density difference $\Delta\rho = \rho_p - \rho_m$, no centrifugation will occur. For a given fermentation broth, the only parameter under the operator's control is ω, as n, θ, R_o, and R_i depend on the centrifuge design and cannot be varied by the operator.

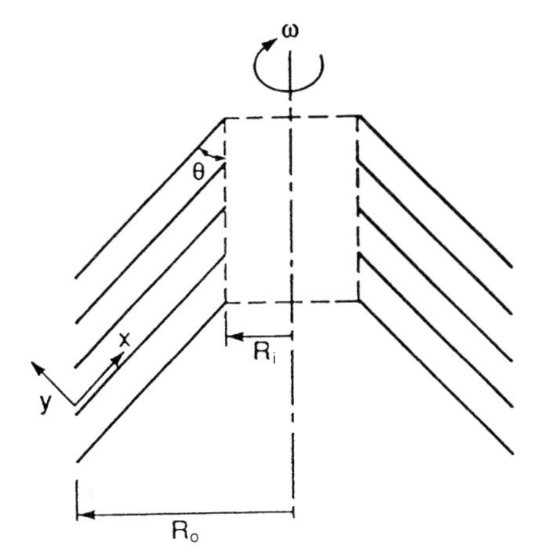

Figure 8.4 Model of a disk centrifuge (from Blanch, 1997).

8.4.2 Homogenization

After centrifugation and thus concentration of the fermentation broth, the cell walls have to be broken, as the majority of products, especially proteins, are intracellular products that cannot pass the cell wall. This breakage process is called homogenization and can be achieved mechanically, chemically, and enzymatically. Whereas on a laboratory scale homogenization with the help of solvents such as toluene, or enzymes such as lysozyme, or detergents such as octylglucosides, is often preferable to mechanical methods, on a large scale mechanical methods are the only viable techniques. The two most prevalent homogenizer types are the ball mill and the French press. In the former, the cells are subjected to shear forces created by balls rolling against each other; in the latter, the cells are forced through a narrow channel with subsequent sudden widening in which the cell walls are ruptured owing to differences in stresses due to the Bernoulli law.

In a simple model of homogenization in a French press, the process of the disruption of the cell wall is modeled by a first-order law. Then, the amount of protein released, R, varies with the number of passes of broth through the homogenizer, N [Eq. (8.58)], where R_{max} is the maximum amount of protein obtainable through homogenization and k' is a constant.

$$dR/dN = k' (R_{max} - R) \tag{8.58}$$

Separation of variables and integration yields Eq. (8.59).

$$\log (R_{max}/(R_{max} - R)) = k' \cdot N = k\,(T, \text{cell wall, physiology}) \cdot N \cdot P\alpha \tag{8.59}$$

The rate constant k' at typical operating pressures of 500–1000 bar is found to be proportional to the pressure raised to a power α dependent on the organism ($\alpha = 2.9$ in *Saccharomyces cerevisiae* and 2.2 in *Escherichia coli* (Hetherington, 1971); $\alpha = 2.5$

is a useful average value). The protein release constant k depends in complex manner on a number of factors such as temperature, cell wall composition, and physiology (which depends on the state of fermentation; see Section 8.2). While disruption is more rapid at higher temperatures, homogenizers have to be pre-cooled and cooled between passes because of a temperature rise owing to adiabatic compression (about 12°C at 500 bar). For sturdy cell walls of yeast (*S. cerevisiae*), a pressure of 400–600 MPa was necessary for disruption by decompression, whereas for weak Gram-negative bacteria, such as *E. coli*, pressures of 200 MPa commonly suffice (Edebo, 1969, 1973).

Data of release rates for seven enzymes from baker's yeast with a high-pressure homogenizer revealed that differences in release rates agreed with reported locations in the cell but are not sufficient to fractionate the enzymes; release rates did not seem to depend much on operating pressure, temperature, or initial cell concentration (Follows, 1971). Both the dependence of release rates on the location within the cell and the description of release by a first-order law has been confirmed in both yeast and *E. coli* with several disruption techniques such as sonication, high-pressure homogenization, and hydrodynamic cavitation (Balasundaram, 2001).

Ball mills are best suited for breaking up strong cell walls such as those encountered in Gram-positive bacteria, but scale-up is limited beyond the small pilot scale. French presses are the apparatus of choice for large-scale homogenization of cells.

8.4.3 Precipitation

After homogenization, it is often advantageous to perform a salt precipitation, most commonly with ammonium sulfate. The purpose of this precipitation is the separation of cell debris and nucleic acids rather than the purification of the target protein from impurities. Whereas the purification factor of ammonium sulfate precipitation is usually around only 1.5 to 2, the separation of non-proteineous impurities and stabilization of the target protein in ammonium sulfate usually provide sufficient benefit to include this step in any purification protocol.

The basis for characterizing fractionated precipitation of proteins is the Cohn–Edsall equation (Cohn, 1943) [Eq. (8.60), where S is the solubility of the protein, β is the solubility in salt-free solution, K_s is the salting-out constant, and I represents the ionic strength].

$$\log S = \beta - K_s \cdot I \tag{8.60}$$

The typical situation of a multi-component mixture is depicted in Figure 8.5, which demonstrates that fibrinogen is the least soluble protein across all ionic strengths; therefore, it can be precipitated quite easily at any ionic strength I. Pseudoglobulin is more soluble than hemoglobin at low ionic strengths ($\log K_{S,\text{pseudoglobulin}} < \log K_{S,\text{hemoglobin}}$), so the ionic strength has to be optimized and should be at least about 6 M ammonium sulfate in the above case. Usually, especially in the case of a

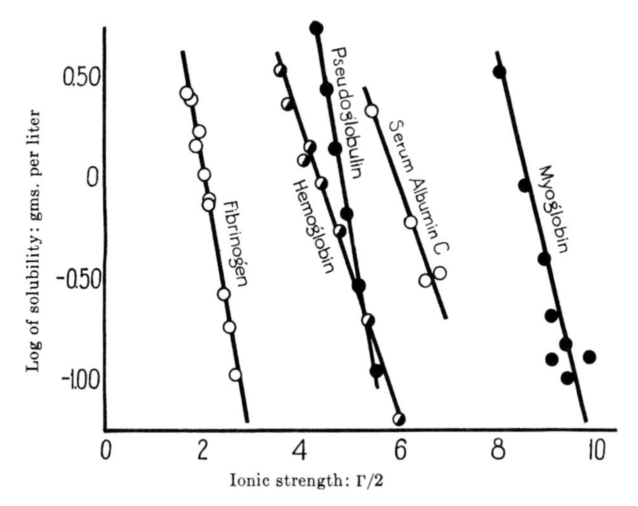

Figure 8.5 Solubility curves of several proteins (Cohn, 1943).

novel protein, the solubility–I curve is not known, so the optimum salt level, expressed as % saturation of ammonium sulfate, has to be found by dialyzing a concentrated ammonium sulfate solution against a low ionic strength solution or, even more easily, by adding small portions of solid ammonium sulfate to the protein solution, centrifuging off the pellet, and analyzing the activity distribution between the pellet and the supernatant.

Whether the goal is purification of a large-scale batch of protein for biocatalytic purposes or purification of an analytical amount to homogeneity for biochemical characterization of the protein, there is going to be a conflict between yield and purity. Despite the high salt levels required, ammonium sulfate precipitation remains one of the most effective methods for initial purification of proteins after fermentations, especially if an overexpressed protein has to be isolated; even a purification factor of 2 increases purity significantly. Typical yields for this step should be expected to fall between 60 and 80%.

The dependence of the solubility of proteins on the nature and concentration of salts had already been found by Hofmeister (1888) and by Setschenow (1889) in the 19th century. Hofmeister, in fact, based the series that now bears his name (and which we have already encountered in Chapter 3, Section 3.4; Figure 8.6) on the ease of precipitation of hen egg-white lysozyme.

⟵ Increasing chaotropic effect (destabilizing effect)	
Cations	Al^{3+}, Ca^{2+}, Mg^{2+}, Li^{+}, Na^{+}, K^{+}, NH_4^{+}, $(CH)_3)_4N^4$
Anions	SCN^-, I^-, ClO_4^-, Br^-, Cl^-, $HCOO^-$, CH_3COO^-, SO_4^{2-}, HPO_4^{2-}, citrate^{3-}
Increasing kosmotropic effect (stabilization) ⟶	

Figure 8.6 Hofmeister series of kosmotropic and chaotropic ions.

Setschenow (1889) gave the first quantitative description of protein solubility as a function of salt concentration, as in Eq. (8.61), where [E] is the solubility of the enzyme, [S] the salt concentration, and K_s the salting-out constant.

$$\log \{[E]/[E]_0\} = -K_s \cdot [S] \tag{8.61}$$

It is apparent from Figure 8.6 that the use of ammonium sulfate for protein precipitation is rationalized with the salt's kosmotropic properties; in addition, $(NH_4)_2SO_4$ is highly soluble in aqueous solution.

Other agents besides ammonium sulfate or other salts can be employed to effect precipitation. As an example, for acid-stable enzymes such as are often encountered in proteins from *Lactobacilli*, responsible for spoiling milk, low pH conditions often lead to precipitation of unwanted proteins while the target protein survives (for an example, see Riebel, 2002). Likewise, heat (typically 1 h at 50–60°C) can be employed to separate heat-labile proteins from more stable ones.

8.4.3.1 Precipitation in Water-Miscible Organic Solvents

Addition of a water-miscible organic solvent tends to lower the dielectric constant ε, which in turn enhances charge–dipole and dipole–dipole interactions, which again in turn reduces protein solubility and induces protein precipitation. The empirical correlation of Eq. (8.62) has been found very useful; [S] is protein solubility, $[S]_0$ is solubility at infinite ε, and K is a constant ($K' < 0$) at constant I and T but does depend on the charge and shape of the protein.

$$\log [S]/[S]_0 = K \cdot \varepsilon^2 \tag{8.62}$$

8.4.3.2 Building Quantitative Models for the Hofmeister Series and Cohn–Edsall and Setschenow Equations

While the Hofmeister series and the Cohn–Edsall and Setschenow equations are useful tools for the estimation of protein stability and precipitation behavior, their usefulness is limited because of the lack of a quantitative relationship to molecular or solution properties. The goal of past and current efforts is to quantify the Hofmeister series and to predict the constants K_s and β (or K_s and $\log [E]_0$) in the Cohn–Edsall or Setschenow equations, respectively. Some of the most relevant efforts focus on:

- the balance between electrostatic and hydrophobic interactions on a protein molecule (Melander, 1977);
- intermolecular potentials (Coen, 1995), the construction of protein–aqueous phase diagrams (Coen, 1997), or their calculation from statistical mechanics (Curtis, 2001);
- correlating protein–protein interactions with the free energy for protein surface desolvation (Curtis, 2002); or
- interactions of chaotropes and kosmotropes with proteins through layers of hydrations (Washabaugh, 1986; Collins, 1995, 1997).

One approach (Melander, 1977) treats the protein as a dipole and applies the Kirkwood equation (Kirkwood, 1943), which attributes the influence of ionic strength to both electrostatic and hydrophobic forces [Eqs. (8.63) or (8.64)].

$$\Delta G = \Delta G_0 - \Delta G_{\text{electrost.}} - \Delta G_{\text{hydrophobic}} \tag{8.63}$$

$$\Delta G = \Delta G_0 - B \cdot I^{0.5}/(1 + C \cdot I^{0.5}) - (\Lambda - \Omega \cdot \sigma) \cdot I \tag{8.64}$$

B, *C*, and *D* are constants, the electrostatic term is expressed by $\Lambda = D\mu/RT$ in which μ is the dipole moment of the protein, and the hydrophobic term is expressed by $\Omega = [N\Phi + 4.8N^{1/3}V^{2/3}(\kappa^e - 1)]/RT$, where *N* is Avogadro's number, Φ is the non-polar surface area of the protein, *V* the molar volume of the solvent, and σ is the surface tension increment, i.e., the difference between the surface tensions with and without salt; κ^e is a correction factor for the surface tension to take account of the curvature of the protein surface at molecular dimensions.

At high salt concentrations the second term, the Debye term, drops out. The remaining equation, $\Delta G = \Delta G_0 - (\Lambda - \Omega \cdot \sigma) \cdot I$, bears a close resemblance to the Cohn–Edsall [Eq. (8.60)] and Setschenow [Eq. (8.61)] for the solubility of proteins. ΔG_0 can be identified with β or $\log [E]_0$ whereas K_s can be modeled by Eq. (8.65).

$$K_s = (\Lambda - \Omega \cdot \sigma) \cdot I \tag{8.65}$$

Good qualitative agreement of Eq. (8.65) with experimental data has been found (Melander, 1977), especially regarding the prediction of the transition from salting-in ($K_s > 0$) to salting-out behavior ($K_s < 0$) at $\Lambda \approx \Omega \cdot \sigma$.

The second attempt describes the influence of salt as an interaction of chaotropes and kosmotropes with layers of hydration on the surface of the respective ion. This interaction can be captured by measuring relative viscosity as a function of salt concentration or ionic strength (Jones, 1929) [Eq. (8.60)].

Here η_0 is the viscosity in salt-free medium and *A* and *B* are constants; at high salt concentrations, the second term becomes irrelevant. The constant *B*, which is the second virial coefficient signifying ion–solvent interactions, is termed the Jones–Dole coefficient after the inventors (Jones, 1929). Chaotropes have a coefficient *B* which is less than zero, whereas kosmotropes are characterized by $B > 0$.

$$\eta/\eta_0 = 1 + A \cdot c^{0.5} + B \cdot c \tag{8.66}$$

8.4.4 Aqueous Two-Phase Extraction

Upon addition of certain polymers such as polyethylene glycol (PEG) and dextran or salt to water, a phase boundary forms even though the system consists of only one solvent, water. When a mixture of biomolecules such as a fermentation broth or a solution of lysed cells is added to such a system, each type of biomolecule partitions uniquely between the two phases, achieving separation (Kula, 1979, 1990;

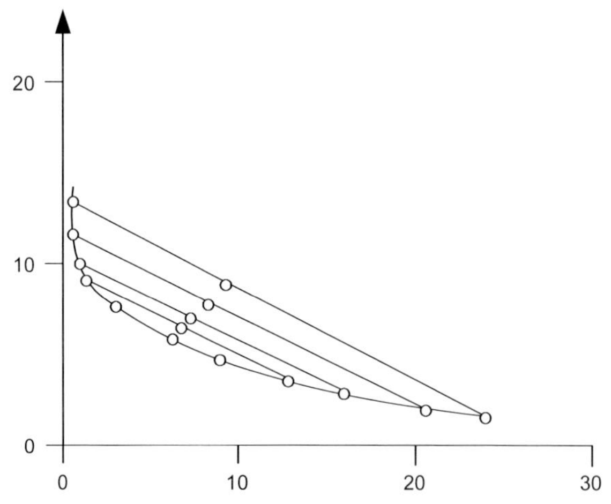

Figure 8.7 Phase diagram of an aqueous two-phase system: *x*-axis: PEG concentration [%]; *y*-axis: salt concentration [%]. The parallel tie-lines in the two-phase region indicate the phase split of a system with a composition in the two-phase region into two phases with compositions at the intersections with the phase boundary line.

Hatti-Kaul, 2001). A low degree of complexity and ease of scale-up in combination offer a potentially attractive method for initial purification of a protein mixture. In principle, only a contactor and a mixer–settler are required.

Formation of two phases upon addition of polymer(s) and salt leads to both one-phase and two-phase regions in the phase diagram (Figure 8.7), separated by the phase boundary. A mixture of biomolecules introduced into this medium will separate into two phases along the connodals, the tie lines connecting the boundary in the two-phase area, until the phase boundary is reached. The phase ratio of top and bottom phases can be discerned by comparing the length of the tie lines from the point designating the composition introduced into the system to the phase boundary. The smaller the area under the phase boundary, the more efficient is the separation. To conduct a two-phase extraction, knowledge of the phase diagram is very helpful.

The quality of separation depends on several factors, among them the partition coefficients of biomolecules between the phases, phase behavior, temperature, and the type and concentration of both salt and polymer. The partition coefficient of species X between the top and bottom phases is defined by Eq. (8.67).

$$K_P = [X]_{top}/[X]_{bottom} \tag{8.67}$$

The partition coefficient K_P varies exponentially with the molecular mass of the soluble protein [Eq. (68)], where T = absolute temperature.

$$K_P = \exp\,(\text{const} \cdot M/T) \tag{8.68}$$

Addition of kosmotropic salts such as $(NH_4)_2SO_4$ or KH_2PO_4 to the system enhances K_P, as the protein is salted out of the bottom phase.

Typically, in a system such as PEG/dextran or PEG/phosphate, cell debris of a lysed and homogenized fermentation broth will partition to the bottom (dextran or salt) phase whereas the target protein partitions go to the top (PEG) phase. As the partition coefficient often is not much different from 1 (typically between 1 and 5) and PEG would disturb further processing, a second two-phase extraction with the PEG phase, typically with a much lower salt content to salt-in the protein, is necessary to recover the target protein.

The major theories developed to predict phase separation and biomolecule partitioning in aqueous two-phase systems are mostly extensions of the widely known polymer solution theories of Flory and Huggins and the osmotic virial expansion. For modeling of aqueous two-phase extractions systems, see Abbott (1990, 1992a–c, 1993).

Other applications besides purification of biomolecules are affinity partitioning (Labrou, 1994), extractive bioconversions (Andersson, 1990; Kaul, 1991; Zijlstra, 1998), liquid–liquid partition chromatography, and analytical assays. Although industrial applications of aqueous two-phase systems have not gained widespread use to date, practical and economic feasibility has been proved (Tjerneld, 1990; Cunha, 2002).

8.5 Downstream Processing: Concentration and Purification of Proteins

8.5.1 Dialysis (Ultrafiltration) (adapted in part from Blanch, 1997)

For dilute solutions at a homogeneous, isotropic membrane the transmembrane fluxes of water and solute in the steady state is given by Eqs. (8.69) and (8.70).

$$J_w \approx L_p\,(\Delta P - \sigma \cdot \Delta\pi) \tag{8.69}$$

$$J_s \approx P_s\,\Delta c_s + J_w\,(1-\sigma)\;\hat{c}_s \tag{8.70}$$

Here $\Delta\pi = RT \cdot c_{osm}$, L_p = permeability for solute (cm s^{-1}), ΔP = hydrostatic pressure difference, $\Delta\pi$ = osmotic pressure, σ = reflection coefficient, Δc_s = solute concentration difference, and $\hat{c}_s$ = average solute concentration in upstream solution. Membranes with a reflection coefficient $\sigma \to 0$ are permeable to all components whereas a membrane with $\sigma \to 1$ rejects all solutes.

From mass conservation, we can infer Eq.s (8.71) and (8.72).

$$J_s = c_{downstream} \cdot J_{total} \tag{8.71}$$

$$J_s = (1-R) \cdot J_{total} \cdot c_{upstream} \tag{8.72}$$

R is the rejection coefficient, i.e., the fraction of solute in upstream solution rejected by the membrane [Eq. (8.73)].

$$R = (c_{\text{upstream}} - c_{\text{downstream}})/c_{\text{upstream}} \tag{8.73}$$

If the flux is perpendicular to the membrane surface, solute tends to accumulate at the membrane surface up to the solute solubility limit. This phenomenon, called concentration polarization, often results in solute leakage and in reduced driving force owing to increased osmotic pressure.

Integration of the steady-state mass balance over solute on the polarization layer from Eq. (8.69) results in Eq. (8.74).

$$J_{\text{w}} = (D_{\text{s}}/\delta) \ln [(c_{\text{osm}} - c_{\text{downstream}})/(c_{\text{upstream}} - c_{\text{downstream}})] \tag{8.74}$$

If $R = 1$, then $c_{\text{downstream}} = 0$, and Eq. (8.74) becomes Eq. (8.75).

$$J_{\text{w}} = (D_{\text{s}}/\delta) \ln (c_{\text{osm}}/c_{\text{upstream}}) \tag{8.75}$$

With k_{s} = mass transfer coefficient; this can most easily be determined through correlations as in Eq. (8.76), with $\text{Re} = d_n \cdot u \cdot \rho/\mu$ (d_n = equivalent channel diameter).

$$k_{\text{s}} \cdot d_{\text{n}}/D_{\text{s}} = \text{Sh} = A \cdot \text{Re}^{\text{a}} \cdot \text{Sc}^{1/3} \tag{8.76}$$

Under laminar conditions ($\text{Re} < 2100$), the constants can be calculated to be $A = 0.33$ and $a = 0.5$, whereas under turbulent conditions ($\text{Re} > 10^4$), calculations yield $A = 0.023$ and $a = 0.8$.

For the crossflow microfiltration of cells, values of $A = 1.0$ and $a = 1.3$ were determined experimentally. More generally, mass transfer to a stationary layer at a surface in a stirred cyclindrical vessel can be described by Eq. (8.77), with r as the cell radius.

$$k_{\text{s}} \cdot r/D_{\text{s}} = \text{Sh} = A \cdot (\omega \cdot r^2/\nu)^{\text{a}} \cdot \text{Sc}^{1/3} \tag{8.77}$$

In the laminar regime ($8000 < (\omega \cdot r^2/\nu) < 32000$), the constants were found to be $A = 0.255$ and $a = 0.55$, whereas under turbulent conditions, $\omega \cdot r^2/\nu > 32000$, $A = 0.0443$, and $a = 0.75$ (Colton, 1969).

Concentration polarization J_{w} can be modeled as another resistance in series, as shown in Eq. (8.78), wherein $(1/L_{\text{p}})$ is the membrane resistance and R_{p} the resistance of the polarized layer. If it is small, Eq. (8.78) reveals that J_{w} is proportional to the transmembrane pressure ΔP, as $(1/L_{\text{p}})$ is a constant.

$$J_{\text{w}} = \Delta P/[(1/L_{\text{p}}) + R_{\text{p}}] \tag{8.78}$$

8.5.2 Chromatography

After the initial purification steps of salt precipitation of an aqueous two-phase extraction, an enzyme is often already pure enough for preparative use. If a high degree of purity or even homogeneity is required, such as for biochemical investigations of a novel enzyme, fine purification steps have to be added. The most common such step is chromatography.

Chromatography is any separation process which depends on partitioning between a flowing fluid and a solid adsorbent. The *stationary* phase is a packed column of adsorbent particles (charged resin, dextran, agarose, silica). The *mobile* phase is any flowing solvent or carrier stream (aqueous solution, in HPLC: H_2O/acetonitrile). Two kinds of chromatography can be distinguished:

- *elution chromatography*, commonly understood as "chromatography"; a pulse is injected at one end of the column and eluted at the other. Different solutes exit the column at different retention times. Typical examples of this mode are gas chromatography (GC), high-pressure liquid chromatography (HPLC), or FPLC (fast-performance liquid chromatography).

- *frontal chromatography*, also called "fixed-bed adsorption"; no resolution of multicomponent mixture into constituents occurs during that operation. During the multi-step process, the column is contacted first with feed until breakthrough. Then, a wash step removes unspecifically bound material, and finally elution is achieved with buffer of a different pH or I.

8.5.2.1 Theory of Chromatography

The effectiveness of a chromatographic separation step is characterized by differential migration of different solutes and by clean, compact, and separated elution bands. Owing to axial dispersion and differential pathlengths of solutes caused by column packing, all the elution bands are spread out to some degree. Whereas peak spreading is unimportant for solutes with widely differing retention times, it markedly influences separation results for closely related molecular structures. Peak spreading can be analyzed with the concept of theoretical plates: the column is considered to consist of a number of segments of height h, called the height equivalent of theoretical plate, HETP, where h is of the same order of magnitude as the diameter of the packing material. The smaller h, the less spread a peak is at the exit of the column and thus the better is the column performance. The HETP concept, while empirical and thus not well founded theoretically, nevertheless is very helpful in describing column performance. If the solute peak appears at the end of the column in Gaussian form, the number of theoretical plates n can be calculated from Eq. (8.79), with V_e = elution volume and w = peak width at the base line of the chromatogram. The higher n, the better the column performance.

$$n = 16\,(V_e/w)^2 \tag{8.79}$$

Height h and number n of theoretical plates are related to the length l of the column by a simple equation, Eq. (8.80).

$$l = n \cdot h \tag{8.80}$$

Empirically, the plate height h has been found to depend on the linear liquid velocity u through the column in the manner described by Eq. (8.81), as found by van Deemter (1961).

$$h = A/u + B \cdot u + C \tag{8.81}$$

A, B, and C are constants. Importantly, the *van Deemter equation*, Eq. (8.81), states that there is a point of minimum plate height h_{min}, at $u = (A/B)^{0.5}$, whereas h increases at both *higher* and *lower* velocity through the column.

The key performance parameter in a chromatography column is its *resolution* R_n, a measure of the overlap of different peaks. It is defined by Eq. (8.82), where V_{e2} and V_{e1} are elution volumes for peaks 2 and 1, as manifested by their distances on the chromatogram, and w_1 and w_2 are their corresponding peak widths at the baseline.

$$R_n = 2(V_{e2} - V_{e1})/(w_1 + w_2) \tag{8.82}$$

If both widths are approximately the same, Eq. (8.82) can be written as Eq. (8.83) and, with substitution for w from Eq. (8.79), as Eq. (8.84).

$$R_n = (V_{e2} - V_{e1})/w \tag{8.83}$$

$$R_n = \frac{1}{4}\sqrt{n\frac{V_{e2} - V_{e1}}{V_{e2}}} \tag{8.84}$$

The last term, $[(V_{e2} - V_{e1})/V_{e2}]$, can be expressed [Eq. (8.85)] in terms of the *capacity factor* k, where $k = [(V_e - V_o)/V_o]$, and in turn V_o is the void volume and thus also the minimum volume possible in the chromatogram, and of the *relative retention* or *selectivity* δ, where $\delta = k_2/k_1$.

$$R_n = \frac{1}{4}\sqrt{\frac{n(\delta - 1)}{\delta\frac{k_2}{k_2 + 1}}} \tag{8.85}$$

Finally, with Eq. (8.80) Eq, (8.85) becomes Eq. (8.86).

$$R_n = \frac{1}{4}\sqrt{\frac{\frac{l}{h}(\delta - 1)}{\delta\frac{k_2}{k_2 + 1}}} \tag{8.86}$$

As is apparent from Eq. (8.86), resolution improves with increasing column length l and decreasing h as well as with both increasing δ and k_2. Resolution suffers both at similar retention times [$\delta \rightarrow 1$, $(\delta - 1)/\delta \rightarrow 0$] and at capacity factors close to the void volume ($k_i \rightarrow 0$).

8.5.2.2 Different Types of Chromatography

Ion exchange chromatography

One of two most useful chromatography steps is ion exchange chromatography. Depending on whether the protein's isoelectric point (the pI value, the pH value at which the protein is globally neutral) is above or below the operating pH, either cationic or, more often, anionic ion exchange chromatography is employed (anionic chromatography refers to the charge of exchanged ions, not the charge of the material on the column). The typical anionic column material Q-Sepharose (Q = quaternary ammonium ion, $-NCH_3^+$) binds proteins above their pI; variations of Q-Sepharose, such as Q-Sepharose FF (FF = fast flow), Mono Q, or Resource Q have been developed. Ion exchange chromatography columns, both anionic (Q-Sepharose) and cationic ion exchange columns such as S-Sepharose (S = sulfonic acid group, $-SO_3^-$) columns have to be eluted with an *increasing* salt gradient, typically 0 to 1000 mM NaCl.

Hydrophobic interaction chromatography

The other important chromatography technique besides ion exchange chromatography is hydrophobic interaction chromatography (hic); the common types are phenyl, octyl, and butyl columns. In contrast to ion exchange chromatography columns, hic columns are eluted with a salt gradient of *decreasing* molarity, typically from 1000 to 0 mM $(NH_4)_2SO_4$.

Affinity chromatography

If the degree of purity achieved is still insufficient, further chromatographic steps have to be included. As the protein mixture at this point is already quite pure, so that only a few proteins tend to remain in the mixture, the next chromatography step has to be very specific. The method of choice is affinity chromatography, in which a specific group in the target protein is picked to interact with the column material. The principle is discussed with the two examples of his-tag and NAD^+ affinity chromatography:

- *His tag.* The ability of the imidazole moiety of the histidine residue to bind divalent metal ions such as nickel, iron, and cobalt can be used to purify histidine-containing proteins on a column which has such divalent metals bound to it. To streamline the purification procedure of any desired protein, such histidines are deliberately added to a protein of choice by cloning a four- to sixfold CAG repeat sequence into the expression construct upstream of the gene of interest so as to express an N-terminal or C-terminal tetra- or hexa-His tag. From experience, especially in *E. coli*, C-terminal tagging often yields superior results to N-termi-

nal tagging; in each case, results have to be established empirically. The crude protein extract is then poured onto an immobilized metal affinity column (IMAC), consisting of a basic bead material, a chelator such as dialkylaminotetraacetic acid and either nickel or cobalt ions. Cobalt ions (such as used in talon resin) have been found to be more specific than nickel ions (and, of course, less oxidation-sensitive than Fe^{II} ions).
As is common to all affinity chromatography protocols, only the desired protein is supposed to bind to the column. The target protein then has to be eluted off the column, in this case with 0.5 M imidazole which displaces the imidazole rings of the histidine residues of the target protein. Non-specific binding is often observed, which results in several intermittent washing steps being required as well as optimization of the protocol for each protein.

- *NAD^+ affinity tag.* By offering an NAD^+-laden column, or more often a 2-ADP-Sepharose column, to an NAD^+-dependent enzyme such as a dehydrogenase, such enzymes are bound while all others are passing through the column. Subsequent elution with 0.5 mM NAD^+ after a washing step to remove unspecifically bound protein results in a relatively pure fraction of protein which should contain only NAD^+-dependent proteins.

After chromatography, a protein solution can be prepared for storage either by drying or by stabilization in solution with an excipient. We will now discuss both options.

8.5.3 Drying: Spray Drying, Lyophilization, Stabilization for Storage

During both spray drying and lyophilization, water is withdrawn from the protein formulation, during the former with the help of heat and during the latter with the help of low pressure at low temperature.

Spray drying

Spray drying, employed preferentially for inexpensive enzymes on a mass scale, evaporates water by spraying the protein solution through a nozzle at high temperature, utilizing the Bernoulli effect. As contact times are short (on the order of less than 1 s), the enzyme is not deactivated. Not much modeling has been performed on this operation.

Lyophilization

Lyophilization employs sub-zero temperatures in combination with very low pressure to withdraw water from the protein solution; typical values would be –80 °C and 1–3 mbar. In these conditions, as the water phase diagram reveals, water sublimes, leaving a fluffy porous enzyme powder. As typical run times are on the order of a day, lyophilization is not the method of choice for large-scale enzyme drying operations. On a laboratory scale, however, lyophilization is a very effective method.

Stabilization for storage: excipients

A very useful method for storage of biocatalysts is the addition of glycerol so that water and glycerol are present in a ratio of about 1 : 1. Glycerol, like many other polyols, stabilizes protein structures (Gekko, 1981). Additionally, a 1 : 1 water/glycerol mixture does not freeze when stored at –20 °C, a most common freezer temperature in laboratories and industrial plants. Thus, storage at –20 °C does not cause biocatalyst deactivation owing to incorporation of water crystals into the protein structure. Alternative stabilizers to glycerol for storage of proteins are sucrose (Lee, 1981) as well as other sugars or sugar alcohols (Arakawa, 1982a). One very useful description of the cause of the effects of polyols and other additives on proteins is preferential interaction, either preferential binding of the cosolvent or its preferential exclusion (preferential hydration) (Arakawa, 1982b; Timasheff, 2002).

8.6 Examples of Biocatalyst Purification

In the following examples, typical purification protocols are covered and explained.

8.6.1 Example 1: Alcohol Dehydrogenase [(R)-ADH from L. brevis (Riebel, 1997)]

The purification table (Table 8.4) can be regarded as typical for a novel biocatalyst with unknown properties.

Table 8.4 Purification table for (*R*)-ADH from *L. brevis*.

Step	*Specific activity [U mg^{-1}]*	*Yield [%]*	*Total activity [U]*	*Purification factor [fold]*
Crude extract	30	100	143	1
Phenyl Sepharose	64	51	70	2
Octyl Sepharose	86	23	36	3
Mono Q	211	17	24	7
2′,5′-ADP-Sepharose	306	9.6	14	10

Mono-Q: Q-Sepharose (Pharmacia);
2′,5′-ADP-Sepharose: affinity column to pick up NADP-dependent enzymes.
Source: Riebel (1997).

From this table, we learn that

- the overall yield of the purification (to homogeneity) is less than one-tenth, 9.6%;
- the purification factor (to homogeneity) is 10 so the enzyme had been expressed to about 10% of cell protein in the crude extract,
- the two most critical steps are the two Sepharose steps at the beginning because the relative losses are highest.

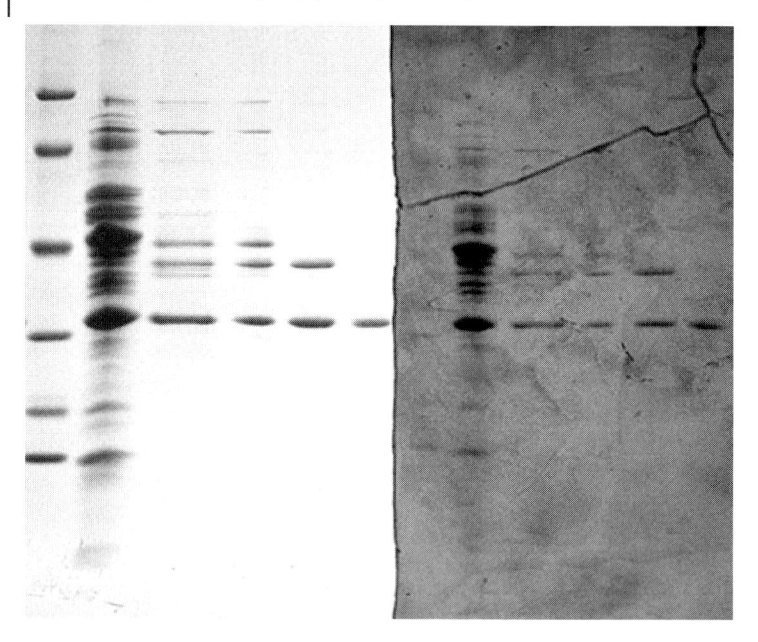

Figure 8.8 SDS-PAGE of purification of (*R*)-ADH from *L. brevis* (Riebel, 1997). Left: Coomassie stain; right: silver stain. The lanes correspond to the following steps (concentrations [μg]): lanes 1, 7: marker (1); 2, 8: crude extract (30); 3, 9: phenyl (20); 4, 10: octyl (18); 5, 11: Mono-Q (10); 6, 12: 2′,5′-ADP (5).

Figure 8.8 depicts the corresponding step-by-step protein gel (SDS-PAGE) of the (*R*)-ADH purification process.

In the present case, both the Coomassie and the more sensitive silver nitrate stain yield the same results. After the last step, the target protein, (*R*)-ADH, can be regarded as homogeneous.

8.6.2
Example 2: L-Amino Acid Oxidase from *Rhodococcus opacus* (Geueke 2002a,b)

The second example describes another redox enzyme, L-amino acid oxidase, which oxidizes L-amino acids to α-keto acids. Table 8.5 lists the results in a purification table, and Figure 8.9 describes the SDS-PAGE results.

The purification table reveals that the first operation, anionic chromatography over a Macro Q column, either activates the enzyme or binds and removes inhibitors to yield 131%; yields in excess of 100% are not uncommon in initial purifications steps. The purification factor of 144 after the third step indicates that the enzyme was not purified from an overexpressed gene but rather obtained from a non-recombinant strain. The SDS-PAGE diagram (Figure 8.9) reveals. The specific activity after the final step (4.6 U mg^{-1}) is rather low.

Table 8.5 Purification table for L-amino acid oxidase from *Rhodococcus opacus*.

	Total activity [U]	*Total protein [mg]*	*Volume [mL]*	*Specific activity [U mg^{-1}]*	*Purification factor [fold]*	*Yield [%]*
Crude extract	9.6	262	28	0.032	1	100
Anionic exchange, Macro Q	12.6	17	110	0.74	23	131
Phenyl Sepharose	8.2	2.9	35	2.8	88	88
Hydroxyapatite	6.5	1.4	26	4.6	144	68

Source: Geueke (2002).

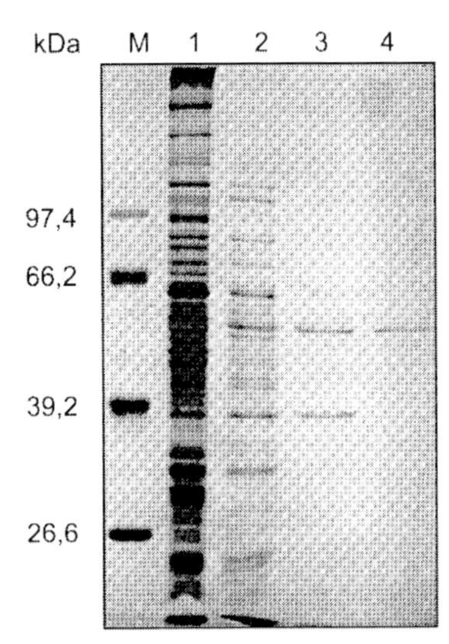

Figure 8.9 SDS-PAGE diagram of L-amino acid oxidase from *Rhodococcus opacus*.

Table 8.6 Purification table for xylose isomerase from *Thermoanaerobium* strain JW/SL-YS 489.

Step	*Total protein [mg]*	*Total activity [U]*	*Specific activity [U mg^{-1}]*	*Yield [%]*	*Purification factor [fold]*
Cell extract	11 760	1744	0.148	100	1.0
$(NH_4)_2SO_4$ fractionation	4328	1794	0.41	103	2.8
DEAE Sepharose	1478	1408	0.95	81	6.4
Superose 12	284	1041	3.66	60	24.7
Phenyl Sepharose	193	808	4.18	46	28.2

Source: Liu (1996).

8.6.3
Example 3: Xylose Isomerase from *Thermoanaerobium* Strain JW/SL-YS 489

The third example concerns an industrially relevant enzyme, glucose isomerase (see Chapters 7, 10, and 19). Table 8.6 reproduces the purification data, and Figure 8.10 shows the SDS PAGE and IEF (isoelectric focussing) diagrams.

The corresponding picture of SDS PAGE:

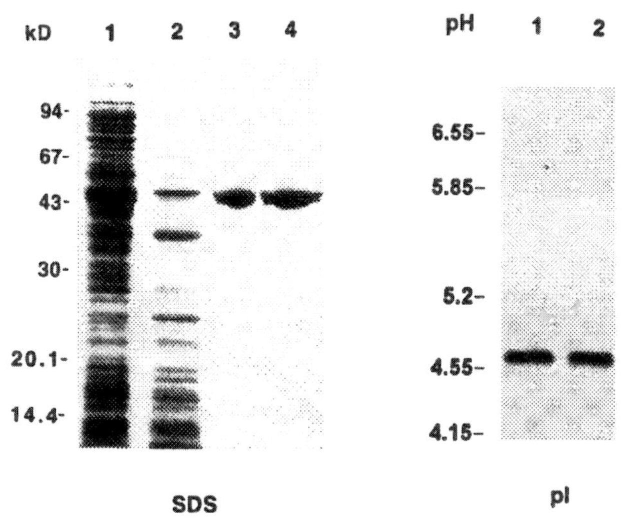

Figure 8.10 SDS-PAGE for the determination of subunit molecular mass and IEF for the isoelectric point of xylose isomerase from *Thermoanaerobium*. Source: Liu (1996).

Both the purification table and the electropherograms make it clear that:

- ammonium sulfate fractionation does not lead to a yield loss;
- no single operation seems to be slated for optimization as every step improves purification in a similar fashion; and
- the specific activity after purification with 4.2 U mg^{-1} is rather low.

Suggested Further Reading

H. W. Blanch and D. S. Clark, *Biochemical Engineering*, Marcel Dekker, New York, 1997.

R. K. Scopes, *Protein Purification: Principles and Practice*, Springer Verlag, 3rd edition, 1994.

M. L. Shuler and F. Kargi, *Bioprocess Engineering*, Prentice Hall, Upper Saddle River, NJ 07458.

References

N. L. Abbott, D. Blankschtein, and T. A. Hatton, On protein partitioning in two-phase aqueous polymer systems, *Bioseparation* **1990**, *1*, 191–225.

N. L. Abbott, D. Blankschtein, and T. A. Hatton, Protein partitioning in two-phase aqueous polymer systems. 2. On the free energy of mixing globular colloids and flexible polymers, *Macromolecules* **1992a**, *25*, 3917–3931.

N. L. Abbott, D. Blankschtein, and T. A. Hatton, Protein partitioning in two-phase aqueous polymer systems. 3. A neutron scattering investigation of the polymer solution structure and protein–polymer interactions, *Macromolecules* **1992b**, *25*, 3932–3941. N. L. Abbott, D. Blankschtein, and T. A. Hatton, Protein partitioning in two-phase aqueous polymer systems. 4. Proteins in solutions of entangled polymers, *Macromolecules* **1992c**, *25*, 5192–5200.

N. L. Abbott, D. Blankschtein, and T. A. Hatton, Protein partitioning in two-phase aqueous polymer systems. 5. Decoupling of the effects of protein concentration, salt type, and polymer molecular weight, *Macromolecules* **1993**, *26*, 825–828.

E. Andersson and B. Hahn-Haegerdal, Bioconversions in aqueous two-phase systems, *Enzyme Microb. Technol.* **1990**, *12*, 242–254.

T. Arakawa and S. N. Timasheff, Stabilization of protein structure by sugars, *Biochemistry* **1982a**, *21*, 6536–6544.

T. Arakawa, S. N. Timasheff, Preferential interactions of proteins with salts in concentrated solutions, *Biochemistry* **1982b**, *21*, 6545–6552.

B. Balasundaram and A. B. Pandit, Significance of location of enzymes on their release during microbial cell disruption, *Biotechnol. Bioeng.* **2001**, *75*, 607–614.

H. W. Blanch and D. S. Clark, *Biochemical Engineering*, Marcel Dekker, New York, **1997**.

P. H. Calderbank, Physical rate processes in industrial fermentation. II. Mass transfer coefficients in gas–liquid contacting with and without mechanical agitation, *Trans. Inst. Chem. Engrs.* **1959**, *37*, 173–185.

P. H. Calderbank and M. B. Moo-Young, Power characteristics of agitators for the mixing of Newtonian and non-Newtonian fluids, *Trans. Inst. Chem. Engrs.* **1961**, *39*, 337–347.

C. J. Coen, H. W. Blanch, and J. M. Prausnitz, Salting out of aqueous proteins: phase equilibria and intermolecular potentials, *AIChE J.* **1995**, 41, 996–1004.

C. J. Coen, J. M. Prausnitz and H. W. Blanch, Protein salting-out: phase equilibria in two-protein systems, *Biotechnol. Bioeng.* **1997**, *53*, 567–574.

E. J. Cohn and E. T. Edsall, *Amino Acids and Peptides*, Reinhold, New York, **1943**.

K. D. Collins, Sticky ions in biological systems, *Proc. Natl. Acad. Sci.* **1995**, *92*, 5553–5557.

K. D. Collins, Charge density-dependent strength of hydration and biological structure, *Biophys. J.* **1997**, *72*, 65–76.

C. K. Colton, PhD thesis, Dept. of Chemical Engineering, MIT, **1969**.

C. L. Cooney, D. I. C. Wang, and R. I. Mateles, Measurement of heat evolution and correlation with oxygen consumption during microbial growth, *Biotechnol. Bioeng.* **1969**, *11*, 269–281.

C. L. Cooney and F. A. Acevedo, *Biotechnol. Bioeng.* **1971**, *19*, 1449–1462.

T. Cunha and R. Aires-Barros, Large-scale extraction of proteins, *Mol. Biotechnol.* **2002**, *20*, 29–40.

R. A. Curtis, H. W. Blanch, and J. M. Prausnitz, Calculation of phase diagrams for aqueous protein solutions, *J. Phys. Chem. B* **2001**, *105*, 2445–2452.

R. A. Curtis, J. Ulrich, A. Montaser, J. M. Prausnitz, and H. W. Blanch, Protein–protein interactions in concentrated electrolyte solutions: Hofmeister-series effects, *Biotechnol. Bioeng.* **2002**, *79*, 367–380.

P. M. Duran, *Bioprocess Engineering Principles*, Academic Press, London, **1995**.

L. Edebo and K. E. Magnusson, Disintegration of cells and protein recovery, *Pure Appl. Chem.* **1973**, *36*, 325–338.

L. Edebo, Disintegration of cells, in: *Fermentation Advances*, D. Perlman (ed.), Academic Press, New York, **1969**, pp. 249–271.

M. Follows, P. J. Hetherington, P. Dunnill, and M. D. Lilly, Release of enzymes from bakers' yeast by disruption in an industrial homogenizer, *Biotechnol. Bioeng.* **1971**, *13*, 549–560.

K. Gekko and S. N. Timasheff, Thermodynamic and kinetic examination of protein stabilization by glycerol, *Biochemistry* **1981**, *20*, 4677–4686.

B. Geueke, PhD thesis, University of Düsseldorf, Düsseldorf, Germany, **2002a**.

B. Geueke and W. Hummel, A new bacterial L-amino acid oxidase with a broad substrate specificity: purification and characterization, *Enzyme Microb. Technol.* **2002b**, *31*, 77–87.

R. Hatti-Kaul, Aqueous two-phase systems: a general overview, *Mol. Biotechnol.* **2001**, 19, 269–277.

P. J. Hetherington, M. Follows, P. Dunnill, and M. D. Lilly, Release of protein from baker's yeast (*Saccharomyces cerevisiae*) by disruption in an industrial homogenizer, *Trans. Inst. Chem. Eng.* **1971**, *49*, 142–148.

F. Hofmeister, Zur Lehre von der Wirkung der Salze. Zweite Mitteilung, *Arch. Exp. Pathol. Pharmakol.* **1888**, *24*, 247–260.

G. Jones and M. Dole, The viscosity of aqueous solutions of strong electrolytes with special reference to barium chloride, *J. Am. Chem. Soc.* **1929**, *51*, 2950–2964.

R. Kaul and B. Mattiasson, Extractive bioconversions in aqueous two-phase systems, *Bioproc. Technol.* **1991**, *11*, 173–188.

J. G. Kirkwood, in: *Proteins, Amino Acids and Peptides*, E. J. Cohn and E. T. Edsall (eds.), Reinhold, New York, **1943**, pp. 276–303.

M.-R. Kula, Extraction and purification of enzymes using aqueous two-phase systems, *Appl. Biochem. Bioeng.* **1979**, *2*, 71–95.

M.-R. Kula, Trends and future prospects of aqueous two-phase extraction, *Bioseparation* **1990**, *1*, 181–189.

N. Labrou and Y. D. Clonis, The affinity technology in downstream processing, *J. Biotechnol.* **1994**, 36, 95–119.

J. C. Lee and S. N. Timasheff, The stabilization of proteins by sucrose, *J. Biol. Chem.* **1981**, *256*, 7193–7201.

S.-Y. Liu, J. Wiegel, and F. C. Gherardini, Purification and cloning of a thermostable xylose (glucose) isomerase with an acidic pH optimum from *Thermoanaerobacterium* strain JW/SL-YS 489, *J. Bacteriol.* **1996**, *178*, 5938–5945.

R. I. Mateles, Calculation of the oxygen required for cell production, *Biotechnol. Bioeng.* **1971**, *19*, 581–582.

W. Melander and C. Horvath, Salt effects on hydrophobic interactions in precipitation and chromatography of proteins: an interpretation of the lyotropic series *Arch. Biochem. Biophys.* **1977**, *183*, 200–215.

J. Monod, Continuous culture technique. Theory and applications, *Ann. Inst. Pasteur Paris* **1950**, *79*, 390–410.

J. W. Richards, Studies in aeration and agitation, *Progr. Ind. Microbiol.* **1961**, *3*, 143–172.

B. Riebel, Biochemical and molecular biological characterization of novel microbial NAD(P)-dependent alcohol dehydrogenases, PhD thesis, University of Düsseldorf, Düsseldorf, Germany, **1997**.

B. R. Riebel, P. R. Gibbs, W. B. Wellborn, and A. S. Bommarius, Cofactor regeneration of NAD^+ from NADH: novel water-forming NADH oxidases, *Adv. Synth. Catal.* **2002**, *344*, 1156–1169.

J. A. Roels, Application of macroscopic principles to microbial metabolism, *Biotechnol. Bioeng.* **1980**, *22*, 2457–2514.

J. A. Roels, *Energetics and Kinetics in Biotechnology*, Elsevier, Amsterdam, **1983**.

J. Setschenow, Über die Konstitution der Salzlösungen auf Grund ihres Verhaltens zu Kohlensäure, *Z. Physik. Chem.* **1889**, *4*, 117–125.

S. N. Timasheff, Protein hydration, thermodynamic binding, and preferential hydration, *Biochemistry* **2002**, *41*, 13473–13482.

F. Tjerneld and G. Johansson, Aqueous two-phase systems for biotechnical use, *Bioseparation* **1990**, *1*, 255–263.

J. J. van Deemter, Mixing and contacting in gas–solid fluidized beds, *Chem. Eng. Sci.* **1961**,*13*, 143–154.

M. W. Washabaugh and K. D. Collins, The systematic characterization by aqueous column chromatography of solutes which affect protein stability, *J. Biol. Chem.* **1986**, *261*, 12477–12485.

G. M. Zijlstra, C. D. De Gooijer, and J. Tramper, Extractive bioconversions in aqueous two-phase systems, *Curr. Opin. Biotechnol.* **1998**, *9*, 171–176.

9
Methods for the Investigation of Proteins

Summary

The postulation of a model of the mechanism of an enzyme reaction encompasses investigation of the elementary steps. The presence of a rate-limiting step is of the greatest importance both for the understanding of the mechanism itself and for the kinetic expression resulting from such a mechanism. Enzyme mechanisms can be investigated by tools such as (i) determination of product distribution in case the Hammond postulate can be applied, (ii) stationary kinetics methods, (iii) non-stationary methods for the trapping of intermediates, (iv) determination of the influence of Brønsted and Hammett effects, (v) kinetic isotope effects, and finally, (vi) use of transition-state analogs.

The distribution ratio between two products (such as enantiomers) is indicative of their rate of formation if the energy barrier between the two substrate isomers or conformers is small (Curtin–Hammett principle). Acceleration of a reaction by addition of buffer at a constant pH value suggests general acid–base catalysis and proton transfer in the rate-determining step. Even though no stable molecule can mimic an unstable transition-state structure in all details, the transition-state analog concept can provide the basis for an effective strategy for inhibitor design and insight into structural and mechanistic details of catalysis.

Protein concentration can be quantified by assays such as the Bradford assay, which is a relative measure against bovine serum albumin (BSA) as a standard. A gene only codes for a protein monomer; however, proteins are able to assemble in a quaternary structure into homo- or heterooligomers, such as dimers, tetramers, hexamers, or even octamers. The native size of a (possibly oligomeric) protein can be determined by mass spectrometry or by size exclusion chromatography (SEC). The molecular mass of the monomeric state can be derived from gel electrophoresis (SDS-PAGE), in which migration is determined not by intrinsic electrical charge of the polypeptide but by molecular weight.

The active site of serine proteases is characterized by a catalytic triad of serine, histidine, and aspartate. The mechanism of lipase action can be broken down into (i) adsorption of the lipase to the interface, responsible for the observed interfacial activation; (ii) binding of substrate to enzyme; (iii) chemical reaction; and (iv) release of product(s).

Zinc proteases, such as carboxypeptidases A and B or thermolysin, contain a tightly bound zinc which is bound to two imidazole residues (His), a carboxylate (Glu), and a molecule of H_2O. Zinc interacts with the substrate carbonyl oxygen and increases its polarization, thereby displacing water.

ADH features another catalytic triad, Ser-Tyr-Lys. Whereas the liver ADH kinetic mechanism is highly ordered, coenzyme associating first and dissociating last, the yeast ADH mechanism is largely random. In both cases, the actual chemical reaction is a hydride transfer. In the oxidation of secondary alcohols by *Drosophila* ADH (DADH), the release of NADH from the enzyme–NADH complex is the rate-limiting step, so v_{max} is independent of the chemical nature of the alcohol. With primary alcohols, as v_{max} is much lower and depends on the nature of alcohol, Theorell–Chance kinetics are not observed and the rate-limiting step is the chemical interconversion from alcohol to aldehyde.

9.1 Relevance of Enzyme Mechanism

The postulation of a model of the mechanism of an enzyme reaction encompasses consideration of the elementary steps. In an enzyme reaction, there are always several steps which make up a mechanism, including at least one binding step and a catalytic one. The Michaelis–Menten equation [Eq. (9.1)] discussed in Chapters 2 and 5 is the kinetic manifestation of a simple mechanism encompassing binding and reaction:

$$E + S \Leftrightarrow ES \rightarrow E + P \tag{9.1}$$

Whereas in the simple scheme in Eq. (9.1) the rate-limiting step is easy to discern (it must be product formation, ES → E + P, or in rare cases (see Section 2.4) enzyme–substrate association, E + S ⇔ ES; there are no other options), more complicated schemes benefit from knowing the *rate-limiting* or *rate-determining* step. Most mechanisms feature a rate-limiting step as only one, or possibly two, steps in a complex mechanism tend to be the slowest. Recognition of the rate-limiting step is of the greatest importance both for the understanding of the mechanism itself and for the kinetic expression resulting from such a mechanism. In a mechanism of consecutive steps, the rate-determining step determines the timescale of all other steps, owing to the steady-state principle. In a multi-step reaction, all steps after the rate-determining step exert no influence on the overall kinetics, with the assumption that no intermediate is more stable than the substrate or product (Murdoch, 1981).

9.2 Experimental Methods for the Investigation of an Enzyme Mechanism

Included in these methods are (i) determination of product distribution, (ii) steady-state kinetics, (iii) non-stationary methods for the trapping of intermediates, (iv) determination of the influence of Brønsted and Hammett effects, (v) kinetic isotope effects, and finally (vi) use of transition-state analogs.

9.2.1 Distribution of Products (Curtin–Hammett Principle)

According to the Curtin–Hammett principle, the ratio of the distribution between two products depends only on their rates of formation (Figure 9.1). If the two products are conformers, the energy barrier between the two states tends to be small and the ratio between the two products is an indicator of the free enthalpies of the transition states. The Curtin–Hammett principle can be applied when two conformations, tautomers, or isomers of a starting material are in rapid equilibrium compared to the rate of a reaction. In this case, the product ratio provides no information about which conformation, tautomer, or isomer was present in the starting material.

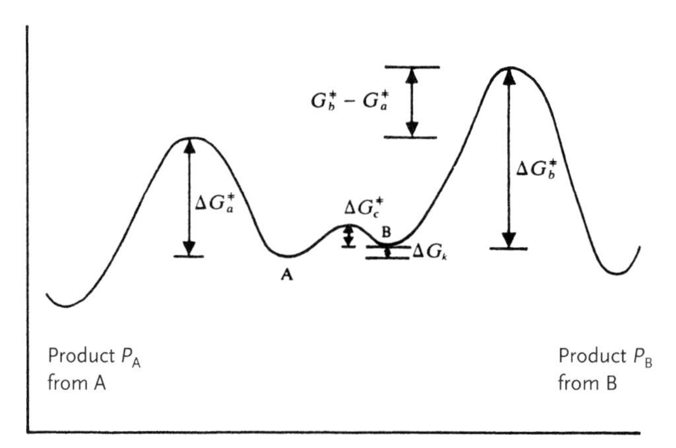

$$\text{rate at which product } P_A \text{ is formed} = \frac{dP_A}{dt} = k_a[A] = k_a K_k[B]$$

$$\text{rate at which product } P_B \text{ is formed} = \frac{dP_B}{dt} = k_b[B]$$

$$\text{product ratio} = \frac{dP_A/dt}{dP_B/dt} = \frac{k_a K_k[B]}{k_b[B]} = \frac{k_a K_k}{k_b}$$

$$\text{from the transition-state theory: } k_r = \frac{\kappa k T}{h} e^{-\Delta G^*/RT} \quad \text{and} \quad K_k = e^{-(-\Delta G_k)/RT}$$

$$\text{product ratio} = \frac{(\kappa k T/h)\, e^{-\Delta G_a^*/RT}\, e^{+\Delta G_k/RT}}{(\kappa k T/h)\, e^{-\Delta G^b*/RT}} = e^{(-\Delta G_a^* + \Delta G_b^* + \Delta G_k)/RT}$$

Figure 9.1 Curtin–Hammett principle.

The general equation relating the product ratio at infinite time, P_∞, to the rate constants (k_1, k_{-1} = rates of equilibration of conformers X and Y, k_2 = rate of formation of product A from X, k_3 = rate of formation of B from Y) for the conformationally mobile system described by Eq. (9.2) is given by Eq. (9.3), with $K = k_1/k_{-1}$ (Zefirov, 1977).

$$\mathrm{A} \xleftarrow{k_2} \mathrm{X} \underset{k_{-1}}{\overset{k_1}{\Leftrightarrow}} \mathrm{Y} \xrightarrow{k_3} \mathrm{B} \tag{9.2}$$

$$P_\infty = K \cdot k_3 \cdot (k_1 + k_{-1} + k_2)/\{k_2 \cdot (k_1 + k_{-1} + k_3)\} \tag{9.3}$$

The limiting cases of Eq. (9.3) are represented by

- the Curtin–Hammett principle if $k_1/k_{-1} \gg k_2, k_3$, and thus the product distribution ratio [A]/[B] is determined by the respective heights of the transition states to A and B; or by
- *conformational-equilibrium control*, if $k_1, k_{-1} \ll k_2, k_3$, and thus the ratio of the products is controlled by the ratio of the population of the starting states.

As a consequence of the Curtin–Hammett principle, one of the most useful means of investigating a mechanism is (not altogether unexpectedly) to measure the distribution of product(s). One prerequisite for measuring product distribution is a proper assay.

9.2.2 Stationary Methods of Enzyme Kinetics

Most aspects of enzyme kinetics and its quantification have been covered in Chapter 5. In this section, we will discuss several methods of obtaining the necessary data, with which the kinetic and mechanistic studies covered in Chapter 5 and in this chapter are performed.

All methods for following an enzymatic reaction can be classified into two groups: *on-line* methods, such as UV–VIS spectrophotometry, fluorimetry, or polarimetry, and *off-line* methods, such as gas chromatography (GC) or high-performance liquid chromatography (HPLC). On-line methods feature the advantage of producing many data points, which are often even taken continuously, for rate calculations, whereas off-line methods provide only a few data points, or even just one, from which a rate is calculated. On the other hand, chromatographic methods in particular directly identify concentrations of the key component or even of several key components, thus rendering the rate measurement more meaningful. The simultaneous availability of concentration data for substrate(s) and product(s) is a particular advantage as this allows calculation of the *mass balance;* if the mass balance closes, i.e., if the sum of all the material found matches the value expected from the stoichiometry, the measurement can usually be trusted. The importance of this point cannot be overemphasized!

Despite the remarks above, in a well-defined reaction system on-line analysis methods are very valuable. The most important on-line methods are, in this order, UV–VIS spectrophotometry, fluorimetry, and polarimetry. All three methods are based on the validity of equivalent laws linking the measured optical parameter with concentration:

- in spectrophotometry and fluorimetry, the Lambert–Beer law, Eq. (9.4a) or (in the form of the time derivative) Eq. (9.4b),

$$E = \varepsilon \cdot d \cdot c \tag{9.4a}$$

$$\Delta E/\Delta t = \varepsilon \cdot d \cdot \Delta c/\Delta t \tag{9.4b}$$

- in polarimetry, the Biot law [correspondingly, Eqs. (9.5a) and (9.5b)], with the specific optical rotation at 20 °C and the Na d-line at 589 nm, $[\alpha]_d^{20}$ [(M cm degrees)$^{-1}$] in polarimetry taking the position of the extinction coefficient ε [(M cm)$^{-1}$] in spectrophotometry or fluorimetry.

$$\alpha = [\alpha]_d^{20} \cdot d \cdot c \tag{9.5a}$$

$$\Delta\alpha/\Delta t = [\alpha]_d^{20} \cdot d \cdot \Delta c/\Delta t \tag{9.5b}$$

Before applying either the Lambert–Beer or the Biot law uncritically, one should check the range of linearity. As a rule of thumb, extinctions E above 1.5, but definitely those above 2.0, are to be discarded because linearity no longer holds.

A short list of particularly important assays and their extinction coefficients ε or specific rotations $[\alpha]_d^{20}$ is given in Table 9.1.

The limit of stationary kinetics is reached when the reaction proceeds faster than data taking is possible with the above-mentioned equipment, on timescales faster than about 10 s.

Table 9.1 Important optically-based assays and their coefficients.

Assay	*Wavelength [nm]*	ε *[M cm^{-1}]*	$[\alpha]_d^{20}$ *[(M cm deg)$^{-1}$]*
Reduction of NAD^+ to NADH	340	6 220	–
Hydrolysis to *p*-nitrophenolate (pH 7.4)	400	17 500	–
Hydrolysis of *o*-nitrophenylgalactoside (ONPG) (pH 7.3 at 37 °C)	410	3 500	–
Peroxide reaction with ABTS	405	36 800	–
Isomerization of glucose to fructose	589	–	39.3
Invertase reaction to glucose/fructose	589	–	–27.2

ABTS = 2,2′-azino-bis(3-ethylbenzthiazoline-6-sulfonic acid)

9.2.3 Linear Free Enthalpy Relationships (LFERs): Brønsted and Hammett Effects

A reaction which is accelerated by the addition of a Brønsted acid is catalyzed by general acid–base catalysis. If a buffer is added at a constant pH value the reaction is accelerated with rising amounts of buffer (which leaves the acid or the base to be determined as the catalyst). Catalysis of a reaction by buffer suggests proton transfer in the rate-determining step but does not identify such a step. With general acid–base catalysis a connection between acid strength and acceleration of the reaction is observed. With the assumption that the mechanism and the rate-limiting step do not change, the logarithm of the rate constant, log k, and the pK_a yield a straight line according to Eq. (9.6), where α is the Brønsted α value; in base-catalyzed reactions, the parameter β is used [Eq. (9.7)].

$$\log k = A - \alpha \cdot \mathrm{p}K_a \tag{9.6}$$

$$\log k = A + \beta \cdot \mathrm{p}K_a \tag{9.7}$$

The coefficients α and β are bounded between 0 and 1. In Table 9.2, the acceleration factors are listed for different α and β values and for different pH values (Fersht, 1999).

In similar fashion to the Brønsted equation, the Hammett equation can be used: in this case, the rate constants of two or more substituent groups are plotted against the pKa values [Eq. (9.8)].

$$(\mathrm{p}K_a)_X = (\mathrm{p}K_a)_0 - \sigma_X \tag{9.8}$$

Eq. (8) holds if X is a *meta* or *para* substituent of a benzoic acid and either the rate constants or the equilibrium constants with such a benzene ring system are investigated. For systems other than the benzoic acid system, Eq. (9.9a) or (9.9b) holds, where ρ is the *reaction parameter*, which describes the strength of influence of substituents on a reaction.

Table 9.2 Influence of Brønsted values on acceleration factors of reactions (Fersht, 1999).

	(Rate in 1 M solution of base)/ (rate in water)			***(Rate in 1 M solution of acid)/ (rate in water)***		
β	pK_a = 5	pK_a = 7	α	pK_a = 5	pK_a = 7	pK_a = 9
0	1	1	0	1	1	1
0.3	2.9	8.6	0.3	31	8.6	2.9
0.5	44	427	0.5	4.3×10^3	427	44
0.7	951	2.4×10^4	0.7	6×10^5	2.4×10^4	951
0.85	9.7×10^3	4.9×10^5	0.85	2.5×10^7	4.9×10^5	9.7×10^3
1	1×10^5	1×10^7	1	1×10^9	1×10^7	1×10^5

$$(pK_a)_X = (pK_a)_0 - \rho \cdot \sigma_X \quad (9.9a)$$

$$\log k_X = \log k_0 + \rho \cdot \sigma_X \quad (9.9b)$$

Brønsted α and β values provide evidence on whether the transition state is similar to the substrate or the product:

- The reaction of basic nucleophiles with esters of activated leaving groups shows a β value of +0.1 to 0.2 for the variation of the pK_a of the nucleophile and –0.1 to – 0.2 for the variation of the p*K*a of the leaving group. The transition state involves only very little bond formation or breakage and thus resembles the starting material.
- In contrast, a β value of +1.5 is measured for the change of pK_a of the nucleophile during the attack of a tertiary amine on basic alcohols (–1.5 when the pK_a of the alcohol is varied). In this case, the transition is close in structure to the products.

The lower the pK_a, the more efficient are the acid catalysts; conversely, the higher the base pK_a, the more efficient are the base catalysts. This does not hold for biologically relevant buffer solutions at pH 7.0.

Efficient catalysts at pH 7.0 are shown to possess pK_a values around 7 because a higher percentage of the acid is unionized and a higher percentage of the base is protonated. Therefore, the acid–base pair is more reactive, which can be measured as lower Brønsted values. This explains the widespread use of histidine (pK_a = 6.5) as an acid–base catalyst in enzymes.

9.2.4 Kinetic Isotope Effects

If the rate-determining step involves the formation or breakage of a bond, isotopic labeling will change the rate constant; in the case of replacement of an isotope by a heavier one, such as replacement of H by D, the reaction will be decelerated (Andres, 1997). Smaller changes are expected in the replacement of ^{14}N by ^{15}N or ^{16}O by ^{18}O. The magnitude of the isotope effect gives a clue about the relevance of bond formation or breakage of the respective bond. Owing to the different zero-point energies, a C–H bond is broken seven times (range between 2 and 14 times) more easily than a C–D bond. In the case of replacement of H by T, $(k_H/k_T) = (k_H/k_D)^{1.44}$ holds at room temperature.

9.2.5 Non-stationary Methods of Enzyme Kinetics: Titration of Active Sites

9.2.5.1 Determination of Concentration of Active Sites

For α-chymotrypsin, the procedure of active-site titration for the calculation of active enzyme concentration and thus of the catalytic constant k_{cat} is long established. The original active-site titration experiment on α-CT by Hartley and Kilby (Hartley, 1954) was performed with ethyl *p*-nitrobenzoate (Figure 9.2).

Meanwhile, reagents (and methods; see Hsia, 1996) have been improved. A short series of titration experiments at different concentrations of electrophile should be conducted to determine the percentage of active enzyme (without nucleophile, so as not to start the reaction). The calculation of the true [E] (= $[E]_{true}$) is performed with [E] after weighing in, with the factor f from the active-site titration experiment from Eq. (9.10).

$$[E]_{true} = [E]_{weigh\text{-}in} \times \text{protein content [\%]} \times f \qquad (9.10)$$

With $[E]_{true}$ the "true" k_{cat} can be determined according to Eq. (9.11) (n = number of active sites per subunit of enzyme).

$$k_{cat} = v_{max}/([E]_{true} \cdot n) \qquad (9.11)$$

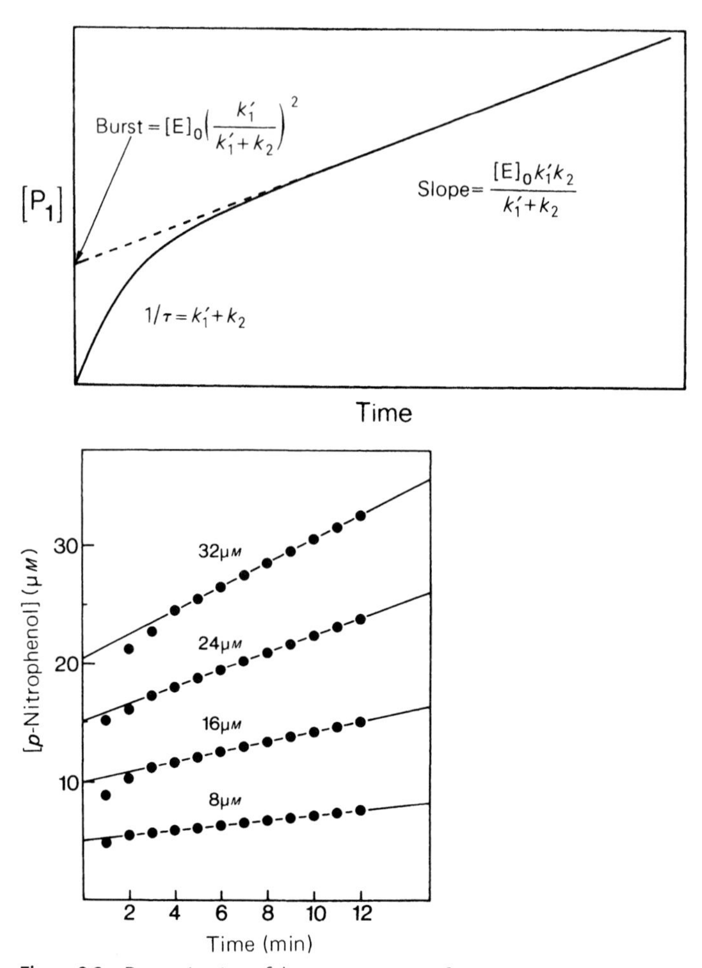

Figure 9.2 Determination of the concentration of active sites (Fersht, 1999).

9.2.6
Utility of the Elucidation of Mechanism: Transition-State Analog Inhibitors

According to Hammond's postulate, the structure of transition states resembles either the starting material or product, depending on which is closer in energy. Thus for exothermic reactions the transition state resembles the starting material, whereas for endothermic reactions it resembles the product. As no stable molecule can mimic the geometric and electronic characteristics of an unstable transition-state structure in all the details, the full power of transition-state stabilization (discussed in Chapter 2) cannot be attained. Applied appropriately, however, the transition-state analog concept provides the conceptual basis for an effective strategy for inhibitor design, and insight into structural and mechanistic details of catalysis (for reviews, see Wolfenden, 1991; Mader, 1997; Schramm, 1999):

- orally active renin inhibitors, useful as antihypertensives, moved during their successive generations from the concept of "transition-state" inhibitors to drugs that are stable to protease cleavage (Weidmann, 1991);
- determination of the X-ray crystal structures of transition-state (TS) analogs bound to lipases helped to elucidate enzyme–substrate interactions; such elucidation accelerated the change from empirical approaches of using lipases in organic synthesis to more rational design of substrates or inhibitors using molecular modeling and site-directed mutagenesis (Kazlauskas, 1991).

What are the criteria for regarding a compound as a TS analog? The observation that the binding affinity of an inhibitor is greater than that of a substrate, i.e., $K_I < K_M$, is insufficient as many potent inhibitors bind differently to an enzyme than the substrate; examples are methotraxate, inhibiting dihydrofolate reductase (DHFR) $K_I = 0.15$ pM (Werkheiser, 1961), and sulfonyl urea herbicides, inhibiting acetolactate synthase (ALS) at picomolar levels.

A more stringent criterion emanates from the thermodynamic cycle already encountered in Chapter 2 (Figure 2.2). For an inhibitor, which necessarily is an imperfect mimic of the transition state, and thus for which clearly $K_I \neq K_T$, the relationship $k_{cat} = k_{uncat}\,(K_S/K_T)$ [the Kurz equation; Chapter 2, Eq. (2.8)], has to be modified to Eq. (9.12), reflecting the proportionality of K_I and K_T, and setting $K_S \approx K_M$.

$$K_I = c \cdot K_T = c \cdot K_M \cdot (k_{uncat}/k_{cat}) \tag{9.12}$$

The greatest utility in characterizing transition-state analogy is derived from rearranging Eq. (9.12) in logarithmic form as Eq. (9.13).

$$\log K_I = \log (K_M/k_{cat}) + \log (c \cdot k_{uncat}) \tag{9.13}$$

Equs. (9.12) and (9.13) point out that alterations in structure or environment should produce parallel effects on enzymatic rate enhancement and the affinity of a transition-state analog relative to the substrate (Frick, 1989). If matching structural alterations are made in a series of inhibitors and substrates, such that the alteration does not affect the rate of the uncatalyzed reaction, then a linear relationship with a slope of 1 will be obtained from plotting log K_I against log (K_M/k_{cat}), if the inhibitors are transition-state analogs (Westerik, 1972: Thompson, 1973). Provided that there is no change in the rate-limiting step, this relationship holds regardless of the magnitude of the inhibition constants, since the proportionality c from Eq. (9.13), along with the unmeasurable k_{uncat}, becomes part of the invariant log $(c \cdot k_{uncat})$.

This approach has been applied extensively to peptidases, as the substrate or inhibitor structure can be varied most liberally. Correlations between inhibitor K_I and substrate K_M/k_{cat} have been demonstrated for the phosphorus-containing peptide inhibitors of thermolysin (Figure 9.3) (Bartlett, 1983; Morgan, 1991), carboxypeptidase A (Hanson, 1989), and pepsin (Bartlett, 1996). Each series of inhibitors shows a slope close to unity in the log K_I vs. log (K_M/k_{cat}) plot; importantly, the structural variations at the P2′ residue should not affect k_{uncat} significantly.

Importantly, the value of the slope does not represent a measure of how well the inhibitor approximates to the transition state; that criterion, c or rather log $(c \cdot k_{uncat})$, according to Eq. (9.13) is only reflected in the intercept. Instead, a slope of unity in

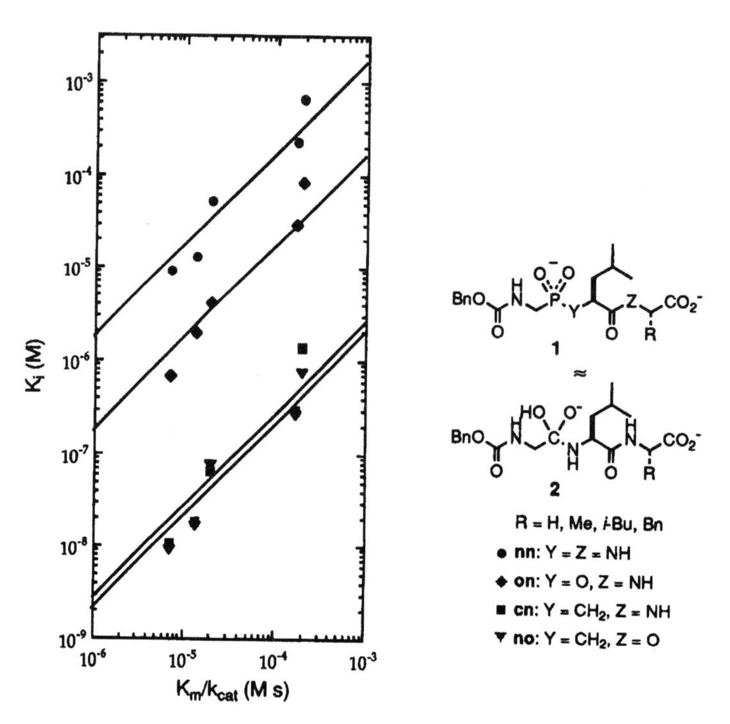

Figure 9.3 Comparison of K_I values for phosphonate inhibitors of thermolysin with K_M/k_{cat} values for the corresponding substrates (Bartlett, 1983; Morgan, 1991). Diagonal lines correspond to slopes of unity.

the correlation of Eq. (9.13) links a structural alteration changing the binding energy of the transition state to the same effect in the inhibitors, an intuitive definition of transition-state analogy.

The conclusion that can be drawn from each of these correlations is that the P2′ residues of the inhibitors, and only those, interact with the protein active site in the same manner as in the transition state. The tightness of the correlation may indicate the degree to which the binding geometries remain constant.

9.3 Methods of Enzyme Determination

9.3.1 Quantification of Protein

Protein quantification in a solution is an extremely important problem; no specific activity can be reported without employing both an activity assay and a protein quantification assay. The outcome of a protein quantification assay is a measure of total protein in a sample solution, routinely reported in mg mL^{-1} or mg L^{-1}. Compared to an ideal assay, which should be accurate, very sensitive, reliable, fast, and easy to perform, all commonly employed assays for total protein suffer several shortcomings.

The most frequently used protein assay is based on a method after Bradford (Bradford, 1976), which combines a fast and easily performed procedure with reliable results. However, the Bradford assay has sensitivity limitations and its accuracy depends on comparison of the protein to be analyzed with a standard curve using a protein of known concentration, commonly bovine serum albumin (BSA). Many commercially available protein assays such as those from Pierce or BioRad rely on the Bradford method. The assay is based on the immediate absorbance shift from 465 nm (brownish-green) to 595 nm (blue) that occurs when the dye Coomassie Brilliant Blue G-250 binds to proteins in an acidic solution. Coomassie dye-based assays are known for their non-linear response over a wide range of protein concentrations, requiring comparison with a standard. The dye is assumed to bind to protein via an electrostatic attraction of the dye's sulfonic groups, principally to arginine, histidine, and lysine residues. It also binds weakly to the aromatic amino acids, tyrosine, tryptophan, and phenylalanine via van der Waals forces and hydrophobic interactions.

Other commonly used ways to determine protein concentrations are the Lowry assay (Legler, 1985) and the BCA (bicinchoninic acid) method (Smith, 1985). Both methods, however, suffer from definite limitations in terms of sensitivity, dynamic range and – in the Lowry assay – compatibility with reducing agents. Recently, labeling of proteins with fluorescers and subsequent detection of the label via fluorescence has proven to be a convenient way of quantifying total protein. As an example, the NanoOrange® protein quantitation kit from Molecular Probes (Eugene/OR, USA; Leiden, The Netherlands) is based on the binding of a dye reagent to detergent-covered and hydrophobic regions of proteins: only the bound dye fluoresces. This assay shows high selectivity towards proteins with little interference by sample contaminants. In addition, there is little variability between protein types and both bound and unbound reagents are stable for hours.

9.3.2 Isoelectric Point Determination

The *isoelectric point*, or pI, is the pH at which the overall net charge on the protein is zero. It serves as a characteristic for every protein; proteins can be precipitated most easily at the pI, and similarly, charge interactions can be assessed when both the pI and pH of interest are known. The pI is determined by forming a pH gradient when an electric field is applied across a solution containing the protein. The variously charged ampholyte species migrate in solution until they reach their respective isoelectric points and thus form a pH gradient. Simultaneously, the target protein molecules also migrate until the reach their respective isoelectric points. To avoid pH drifts towards the cathode, and thus a plateau with a lack of conductivity in the middle of the gel with ensuing gradient inhomogeneity and inconsistent results, an immobilized pH gradient is created with immobilized acrylamide derivatives together with buffering groups, the immobilines (Pharmacia). The more precise an immobilized pH gradient, the more (≥ 2) different immobilines are required. If the immobiline is a base, the process can be viewed as an acid–base titration, following the Henderson–Hasselbach equation [Eq. (9.14)].

$$\mathrm{pH} = \mathrm{p}K_\mathrm{B} + \log\{(C_\mathrm{Base} - C_\mathrm{Acid})/C_\mathrm{Acid}\} \tag{9.14}$$

The gradient is established through linear mixing of two different acrylamide–immobiline solutions using a gradient mixer. Nowadays, isoelectric points can be predicted with computer programs based on the target protein's amino acid sequence (see Chapter 14). Such predicted pI values should always be verified experimentally, as neighboring amino acids within the tertiary structure influence charge interactions in proteins and cause variations of predicted from actual pIs.

9.3.3 Molecular Mass Determination of Protein Monomer: SDS-PAGE

The easiest method for estimating the molecular weight of a protein is to derive the amino acid sequence and thus the mass from the corresponding size of the gene. However, a corresponding gene can only provide the exact size of a protein monomer. Monomers of fully folded proteins with a tertiary structure are often able to associate further to oligomers such as dimers, tetramers, hexamers, or octamers, or with an even higher degree of association (rarely, trimers or pentamers are formed). Protein oligomers can be classified as either homo- or heterooligomers, the latter requiring several different genes. For the determination of its native oligomeric structure, the protein can be submitted either to mass spectrometry, discussed in Section 9.3.5 and in Chapter 15, Section 15.2.4, or to size exclusion chromatography, described in Section 9.3.4. The monomeric molecular weight state can be inferred from SDS-PAGE.

Electrophoresis causes the migration of charged molecules in solution along the x-axis in response to an electric field. The migration rate or velocity v can be obtained from a balance of electrical force and Stokes drag and expressed by Eq. (9.15).

$$v = dx/dt = \nu E/(3\pi \cdot \eta \cdot d_p) = z \cdot e \cdot E/(3\pi \cdot \eta \cdot d_p) \tag{9.15}$$

It thus depends on the field strength E, the diameter d_p (and shape) of the migrating molecule, the net charge ν or the number of charges per molecule z, the elementary charge e (1.6×10^{-19} C), and the viscosity η of the surrounding medium. The electrophoretic mobility m is just the velocity divided by the field strength [Eq. (9.16)].

$$m = v/E = (e/3\pi \cdot \eta) \cdot (z/d_p) \tag{9.16}$$

Thus, electrophoresis commonly measures size/charge ratios of migrating molecules (or, better, charge/size ratios (z/d_p)). The Stokes diameter of a protein [nm] is related to the molar mass M_r [g mol^{-1}] by Eq. (9.17).

$$M_r = (f_0/f)^3 \times 433 \times d_p^{\ 3} \tag{9.17}$$

The frictional ratio $(f_0/f)^3$ for globular proteins equals 1.23 for the range 45 kDa $< M_r <$ 100 kDa, 1.28 for the range 100 kDa $< M_r <$ 500 kDa, and 1.43 between 0.5 MDa and 1 MDa (Felgenhauer, 1974). By employing sodium dodecyl sulfate (SDS), *s*odium *d*odecyl *s*ulphate *p*oly*a*crylamide *g*el *e*lectrophoresis (or SDS-PAGE) screens out charge effects and measures only size differences strongly correlating to molecular weight. SDS is an anionic detergent which denatures proteins by surrounding the polypeptide backbone. SDS binds to proteins in a specific mass ratio of 1.4 : 1 and thus confers a negative charge onto the polypeptide in proportion to its length. After SDS treatment, each protein features a negative charge proportional to its size. As disulfide bridges in proteins disturb the process of unfolding for separation by size, they have to be reduced by the addition of 2-mercaptoethanol (ME) or dithiothreitol (DTT). Migration in denaturing SDS-PAGE separations is determined by the molecular weight, not by the intrinsic electrical charge, of the polypeptide.

The logarithm of the molecular weight of an SDS-denatured polypeptide scales linearly with the polypeptide's R_f value [Eq. (9.18)]. The R_f value is calculated as the ratio of the distance migrated by the molecule to that migrated by a low-molecular-weight marker dye-front.

$$R_f = a \cdot \log M_r + b \tag{9.18}$$

Separated proteins can be compared to markers, a mixture of proteins with known molecular weights, in adjacent lanes in the same gel. A simple procedure of determining relative molecular weight (MW) by electrophoresis is to plot a standard curve of distance migrated vs. $\log_{10}$ MW for known samples, and to read off the log M_r of the sample after measuring the distance migrated on the same gel.

Since the denaturing and reducing conditions destroy quaternary and tertiary structures of proteins, SDS-PAGE is unable to determine the molecular weight of a protein oligomer, such as the quaternary structure of a protein, but can only determine the mass of a monomer, or more specifically of any single, uninterrupted polypeptide chain. For the determination of the mass of oligomeric proteins other techniques such as size exclusion chromatography or mass spectrometry, both of which are described below, have to be used.

9.3.4 Mass of an Oligomeric Protein: Size Exclusion Chromatography (SEC)

The principle of size exclusion chromatography (SEC) or gel permeation chromatography (GPC) relies on molecule separation depending on size, caused by the molecular sieving effect of the pore structure of gel beads. The gel beads in turn form a packed bed (Figure 9.4). Pore size distribution within the gel depends on the degree of crosslinking of the matrix: the smaller a molecule, the easier its diffusion into the pores and thus the greater its delay in passing through the column. Larger molecules can be separated from smaller molecules provided their size is of the same order of magnitude as the pore size. As charge interactions in the pores also affect salts and buffer molecules, gel filtration is a suitable method for a buffer exchange.

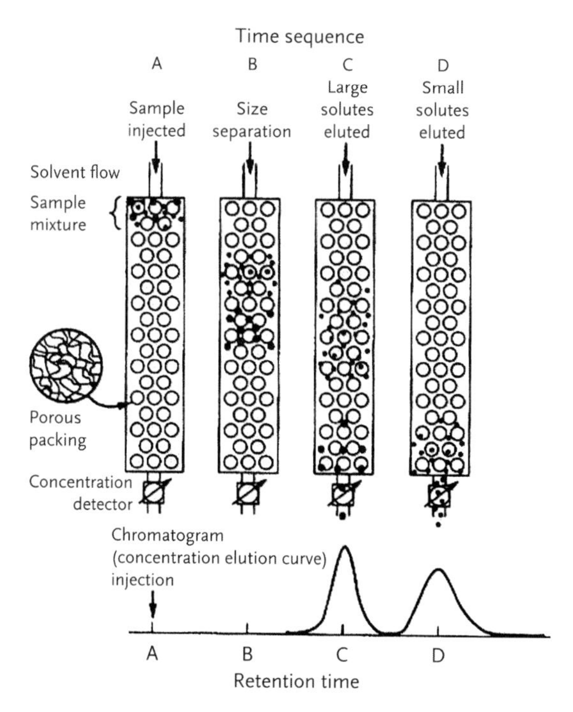

Figure 9.4 The principle of gel filtration.

The accuracy of SEC is determined by the precision of packing of the column: the more uniform the particles and the more pressure applied during packing, the higher is the accuracy of the columns. Prepacked columns with different size exclusion ranges are available and are preferred over self-packed columns. The characteristic quantity to be measured is the relative capacity factor k_i (see Chapter 8, Section 8.5.2), defined by Eq. (9.19), with V_0 representing the void volume, the minimum volume or the volume a molecule travels through the column without entering the pores [determined with a protein of very high mass, such as Dextran Blue (MW: 2 MDa)], and V_t the volume of the gel filtration column.

$$k_i = (V - V_0)/(V_t - V_0) \tag{9.19}$$

SEC columns have to be standardized with proteins of known molecular weight. The elution time or elution volume is used to determine the k_i values, which are plotted in a semilogarithmic graph of k_i against the logarithm of the known molecular weights. Regression yields a straight line which is used to evaluate molecular weights for unknown proteins from their k_i values.

9.3.5
Mass Determination: Mass Spectrometry (MS) (after Kellner, Lottspeich, Meyer)

Mass spectrometry (MS) is a very sensitive method of determining the mass of native proteins, as well as their purity. The two most common techniques are ESI-MS (electron spray ionization–MS) and MALDI-TOF-MS (matrix-assisted laser desorption ionization–time-of-flight MS) with resolutions down to ±1 to 10 au for a protein of maximally 100 kDa.

Traditional mass spectrometry is useful for analyzing low-molecular-weight compounds but is of little use for underivatized higher-mass molecules such as proteins. In the case of proteins, a polar, non-volatile biopolymer has to be converted into intact, isolated, ionized molecules in the gas phase. ESI-MS and MALDI-TOF-MS use different physical approaches for the conversion. ESI forms ions directly from small, charged liquid droplets. Sample preparation has to provide the liquid droplets. For ESI measurements, the samples are dissolved in a suitable solvent such as a mixture of methanol or acetonitrile with water. Using the ESI source (a metal capillary tip at elevated voltage relative to a counter electrode) a spray of fine, highly charged droplets is created under atmospheric pressure. The mass analyzer separates and records ions based on their mass-to-charge-ratio (m/z). This method is regarded as the softest mass analysis.

MALDI-TOF uses short, intense laser pulses to induce the formation of intact gaseous ions from proteins adsorbed onto a special matrix. The matrix serves three distinct functions: adsorption of energy from the laser light, isolation of the single biomolecules from each other, and ionization of the single biomolecules. Using a TOF analyzer the mass-to-charge-ratio (m/z) of an ion is determined by measuring its flight time. Proteins in the range from a few to a few hundred kilodaltons can be analyzed.

9.3.6
Determination of Amino Acid Sequence by Tryptic Degradation, or Acid, Chemical or Enzymatic Digestion

Proteins can be degraded for further analysis to determine amino acid composition or sequence. The amino acid composition can be determined through total acid hydrolysis; the resulting amino acids can be analyzed using ninhydrin coupling and comparison to known amino acid standards. For protein sequencing, the enzyme can be digested into peptide structures with enzymatic or chemical methods. Enzymatically, specific proteases can be used such as trypsin or endoproteases such as Lys-C or Glu-C proteases cleaving the peptide bond C-terminally behind a lysine or glutamic acid. The resulting peptides can be separated through reversed-phase HPLC and analyzed in a protein sequencer using Edman degradation. Chemically, the most prevalent method is cyanogen bromide (CNBr) cleavage, specifically cleaving peptides C-terminally from methionine. The peptides can be separated by SDS-PAGE and blotted onto a PVDF membrane. These blot pieces can be cut out and again analyzed through Edman degradation.

9.4
Enzymatic Mechanisms: General Acid–Base Catalysis

General acid–base catalysis is often the controlling factor in many mechanisms and acts via highly efficient and sometimes intricate proton transfers. Whereas log *K* versus pH profiles for conventional acid–base catalyzed chemical processes pass through a minimum around pH 7.0, this pH value for enzyme reactions is often the maximum. In enzymes, the transition metal ion Zn^{2+} usually displays the classic role of a Lewis acid, however, metal-free examples such as lysozyme are known too. Good examples of acid–base catalysis are the mechanisms of carbonic anhydrase II and both heme- and vanadium-containing haloperoxidase.

9.4.1
Carbonic Anhydrase II

Carbonic anhydrase II, present in human red blood cells (RBCs), catalyzes the reversible hydration of CO_2. It is one of the most efficient enzymes and only diffusion-limited in its turnover numbers. The catalytic Zn^{II} is ligated by three histidine residues and OH; this $ZnOH^+$ structure renders the zinc center an efficient nucleophile which is able to attack the CO_2 molecule and capture it in an adjacent hydrophobic pocket. The catalytic mechanism is shown in Figure 9.5.

Zn^{II} enhances the acidity of the coordinated water and therefore vastly enhances the concentration of coordinated OH^- at pH 7.5. It coordinates both OH^- and CO_2 in close proximity; its sufficient lability towards ligand exchange and ability to change coordination numbers from 4 to 5 lead to fast overall turnover (step 2 in Figure 9.5). Water then attacks the five-coordinated Zn complex and HCO_3^- is released,

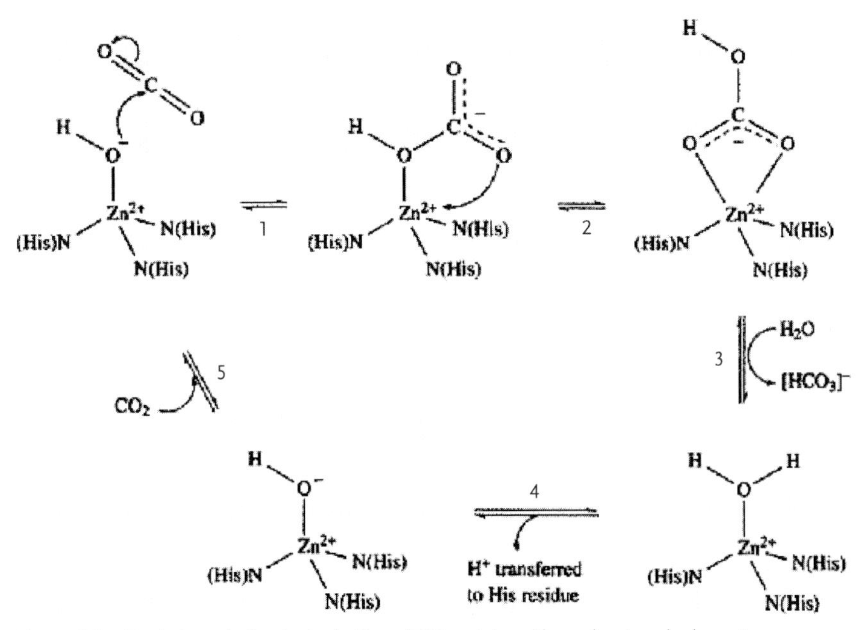

Figure 9.5 Catalytic cycle for the hydration of CO_2 catalyzed by carbonic anhydrase II. Taken from http://chemlearn.chem.indiana.edu/C430/C430L16.pdf.

leaving a water molecule bound to Zn^{II}. Proton transfer to a coordinated histidine residue restores the catalytic $ZnOH^+$ structure.

Even though studies of the pH dependence of carbonic anhydrase catalysis well before the advent of structural work suggested the involvement of the zinc–hydroxyl ($ZnOH^+$) species as a nucleophile attacking CO_2, the relative importance of basicity versus nucleophilicity was not clear for quite a while. Whereas a metal hydroxo complex can be up to ~10^{10} times weaker as a base towards the proton than free OH^-, as a nucleophile towards CO_2 it might be only one or two orders of magnitude weaker: the complex $[Co(NH_3)_5OH]^{2+}$ is derived from an aqua species with a pK_a of 6.5 while OH^- is derived from water (pK_a 15.7), suggesting that the coordinated OH^- is $10^{9.2}$ times weaker as a base than OH^-. However, the experimental rate of attack of coordinated OH^- on CO_2 is only 40 times slower than that of free OH^-, so that a 1 M solution of $[Co(NH_3)_5OH]^{2+}$ at pH 7.5 would have a nucleophilic power equivalent to a 0.025 M solution of NaOH. As zinc aqua complexes are generally significantly weaker acids than their Co^{III} analogs, it is reasonable to expect that $[Zn(OH)]^+$ might be an even better nucleophile than $[Co(OH)]^{2+}$ and hence a very effective component of an enzyme.

A simple calculation reveals that the picture cannot be quite as simple. Carbonic anhydrase has an exceptionally high overall rate of reaction, its turnover number k_{cat} is ~$5 \times 10^5\ s^{-1}$; consequently, the rate constants of individual steps must be greater than this number. The acid dissociation of a Zn^{II} aqua species seems to be inconsistent with this requirement. The dissociation constant K_a can be written as the ratio of forward k_f and backward k_b rate constants [Eq. (9.20)].

$$K_a = k_f/k_b = 10^{-7} \tag{9.20}$$

Therefore $k_f = 10^{-7}k_b$, and with $k_{b,max} = 10^{10}$ M^{-1} s^{-1} (the diffusional limit), Eq. (21) follows, showing an observed rate constant at least 1000-fold slower than expected.

$$k_{f,max} = (10^{-7}\ \text{M})(10^{10}\ \text{M}^{-1}\ \text{s}^{-1}) = 10^3\ \text{s}^{-1} \tag{9.21}$$

The mechanism can only be retained if a base very much better than water is present at the enzyme active site to deprotonate the $Zn(OH_2)^{2+}$ species. In fact, the structure determination reveals a histidine residue with its side chain positioned approximately halfway down the ~15 Å-deep cleft in the protein structure within which the Zn site is located. This arrangement could act as a proton shuttle between the $Zn(OH_2)^{2+}$ and external solvent water, possibly via another two water molecules also found within the cleft. As a consequence, the enhancement of ligand acidity by Zn^{II} is more important in the kinetic than the equilibrium sense (taken from http://www.chem.uwa.edu.au/enrolled_students/BIC_sect4/sect4.2.html).

9.4.2 Vanadium Haloperoxidase

Chloroperoxidase (CPO; E.C. 1.11.1.10) catalyzes the halogenation of a number of aliphatic substrates in the reaction of Eq. (9.22) (AH = substrate; X = Cl, Br, I, but not F).

$$AH + X^- + H^+ + H_2O_2 \rightarrow AX + 2H_2O \tag{9.22}$$

A number of structurally different enzymes have been characterized: the heme-thiolate CPO from the marine fungus *Caldariomyces fumago* (*Cf*CPO), metal-free bacterial CPOs (CPO-L), and vanadium-containing CPOs (VCPOs) from marine algae.

Vanadium chloroperoxidases are enzymes that are abundant in marine algae, and they halogenate activated C–H bonds employing H_2O_2 and Cl^- at pH 5–8. As is known from X-ray structures of the native and peroxide-bound form of VCPO from *Curvularia inequalis* (Messerschmidt, 1996, 1997), the active site is very different from CPO: VO_3^- is bound via several hydrogen bridges to various amino acids, resulting in a trigonal bipyramidal coordination of V^{5+} to four non-protein oxygen donors (O_2^-, OH^-, or OH_2). The only covalent attachment to the protein is an apical $V–N_{His}$ bond. Another nearby histidine is believed to act as an acid–base catalyst (His-404). No enzyme mimic has yet been able to reproduce the unique binding features found in native VCPO, so the catalytic cycle could not be elucidated from such mimics.

VCPOs catalyze the two-electron oxidation of halides (X^- = Cl^-, Br^-, I^-) using activated peroxide through a Lewis acid-promoted mechanism (in contrast to possible redox cycling at the V center). Peroxide is bound to V in η^2-fashion after release of the apical oxygen (Littlechild, 2002); this release is catalyzed through the

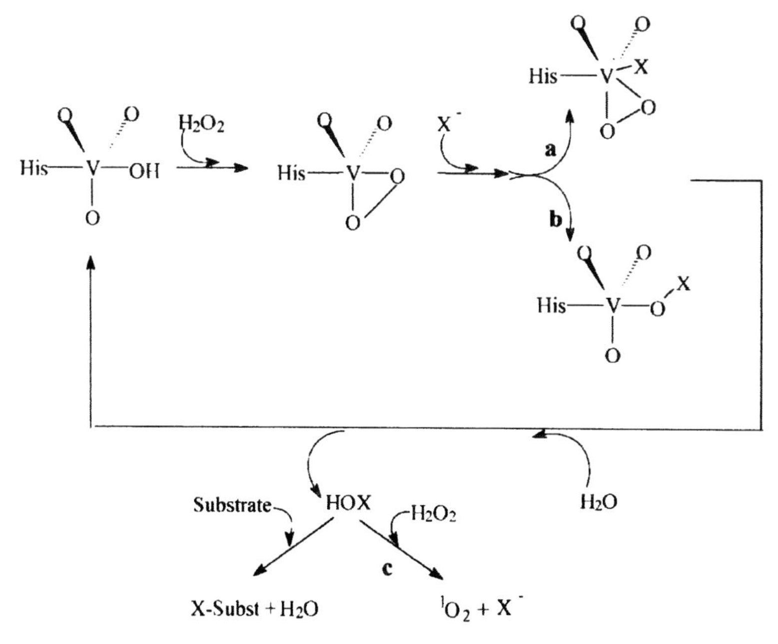

Figure 9.6 Proposed mechanism of vanadium-dependent haloperoxidases.

acid–base catalyst function of His-404, which polarizes the V-bound hydroxyl group and facilitates its nucleophilic attack on the approaching peroxide. Thus the peroxide is singly deprotonated and water exits from the V-coordinated sphere, leaving an empty coordination site for the peroxide. Currently, pathway b in Figure 9.6 is favored because the crystal structure confirms that V never changes its redox or coordination state. The peroxo complex of vanadium binds the halide and accepts its two electrons. The OX^- (X^- = halide) deprotonates a surrounding water molecule and quits the active site as HOX. Kinetic studies have demonstrated release of singlet oxygen in a secondary reaction as shown in Figure 9.6, route c.

9.5 Nucleophilic Catalysis

9.5.1 Serine Proteases

Proteases belong to one of four broad families:

- serine proteases, such as chymotrypsin or trypsin, in which catalysis involves formation of an acyl-enzyme with the acyl group on the serine;
- cysteine proteases, such as papain, with a somewhat similar mechanism involving an acyl-cysteine;

- acid proteases, such as pepsin, with an acid pH optimum, and two aspartate residues involved in the mechanism; and
- metalloproteases, such as carboxypeptidase, which use a metal ion, usually zinc.

There are other hydrolytic enzymes, such as lipases (see below) and alkaline phosphatase, with a mechanism closely related to that of the serine proteases or glyceraldehydephosphate dehydrogenase (GAP-DH) containing a cysteine in the active site.

Serine proteases are characterized by a catalytic triad of Ser, His, and Asp. They work in pairs of Ser-His and His-Asp. Replacement of Asp by a second His revealed the auxiliary nature of aspartate, probably in orienting the histidine with respect to the catalytic serine. All of the catalytic steps are performed by the dyad serine–histidine.

The serine proteases are divided into at least two genetic families: the mammalian serine proteases, such as trypsin, chymotrypsin, elastase, the enzymes of the blood clotting system, and many other proteases with specific roles in control of systems; and the bacterial proteases called subtilisins (first to be isolated from *Bacillus amyloliquefaciens*), which are genetically unrelated to the mammalian enzymes but independently evolved the same mechanism (evolutionary convergence).

They also feature the same catalytic triad and the same system of hydrogen bonds for binding the carbonyl oxygen and the acetamido group of the substrate (Table 9.3). The bacterial enzymes are useful for site-directed mutagenesis studies because they are easily expressed in bacterial cells (for secretion) and because they do not require the activation reactions typical of the mammalian enzymes.

While members of the same family feature similarity of tertiary structure, their sequence is less similar. Trypsin has been converted to a chymotrypsin-like protease upon altering (i) the S1 binding pocket residues, Q192M and I138T; (ii) an adjacent residue Y172W; and (iii) the two chymotrypsin surface loops 185–188 and 221–225 (which actually do not contact the substrate) to the analogous trypsin loops (Hedstrom, 1992, 1994a,b; Kurth, 1997). This mutant enzyme is improved to a protease with 2–15% of the activity of chymotrypsin. [For conversion of trypsin to elastase, see Hung (1998); for a review see Hedstrom, 2002.]

Table 9.3 Catalytic triad nomenclature in different enzymes.

Enzyme	*E.C. number*	*Ser/Cys*	*His*	*Asp/Glu*
Acetylcholine esterase	3.1.1.7	Ser200	440	Glu327
Trypsin	3.4.21.4	Ser195	57	Asp189
α-Chymotrypsin	3.4.21.1	Ser195	57	Asp102
Subtilisin Carlsberg	3.4.21.62	Ser221	64	Asp32
Papain	3.4.22.2	Cys25	159	

The serine proteases act by forming and hydrolyzing an ester on a serine residue. This was initially established using the nerve gas diisopropyl fluorophosphate, which inactivates serine proteases as well as acetylcholinesterase. It is a very potent inhibitor (it essentially binds in a 1 : 1 stoichiometry and thus can be used to titrate the active sites) and is extremely toxic in even low amounts. Careful acid or enzymatic hydrolysis (see Section 9.3.6.) of the inactivated enzyme yielded *O*-phosphoserine, and the serine was identified as residue 195 in the sequence. Chymotrypsin acts on the compound cinnamoylimidazole, producing an acyl intermediate called cinnamoyl-enzyme which hydrolyzes slowly. This fact was exploited in an active-site titration (see Section 9.2.5.). Cinnamoyl-CT features a spectrum similar to that of the model compound *O*-cinnamoylserine, on denaturation of the enzyme in urea the spectrum was identical to that of *O*-acetylserine. Serine proteases act on both esters and amides.

The kinetic mechanism is an acyl-enzyme mechanism (Figure 9.7): the substrate binds non-covalently, the serine displaces the alcohol or amine part of the substrate to form an acyl-enzyme, and water then displaces the serine to yield the acid product and free enzyme.

For esters the deacylation step (k_3) is rate-limiting; for amides the acylation step (k_2) is much slower and rate-limiting. Note that an ester and an amide of the same acid yield the same acyl-enzyme and have the same rate constant for deacylation; but the acyl-enzyme accumulates only when the substrate is an ester or acylimidazole. This was shown originally with *p*-nitrophenyl acetate as a substrate of chymotrypsin – the acetyl-enzyme could be isolated at acid pH – and confirmed with more specific substrates such as acetyltryptophan *p*-nitrophenyl ester: one equivalent of *p*-nitrophenol is released at a rate (k_2) much faster than that of the overall catalytic reaction (essentially k_3). Trypsin and chymotrypsin work well on esters and amides of single amino acids, as long as the α-amino group is blocked, but have a higher k_{cat} for peptides with several residues to the amino side of the residue where cleavage occurs (referred to as the P_2, P_3, P_4 residues, which bind to the corresponding sites S_2, S_3, S_4). The v_{max} of the reaction depends on the basic form of a group with $pK_a = 7$ and the acidic form of a group with $pK_a = 9$. The latter is the α-amino group of Ile16, which forms an ion-pair with the carboxyl side chain of Asp194 in the active site and causes changes in the position of the peptide chain. The pK_a 7 group was identified as a histidine by using tosylphenylalanine chloromethyl ketone as His57 contact in chymotrypsin. This does not act by nucleophilic attack – cinnamoyl-chymotrypsin does not have the spectrum of cinnamoylimidazole – but by general base catalysis (the reaction is slowed in D_2O). It is generally considered that the histidine pulls the proton off Ser195 as it attacks the substrate carbonyl, facilitating its nucleophilic attack, and then puts it on the leaving oxygen or nitrogen. In the enzyme without substrate bound, it is far enough from the serine so that only a very weak H-bond could be formed, but the serine moves closer to it as substrate binds.

X-ray crystallography of chymotrypsin found an unexpected aspartic acid residue, Asp102, buried in the protein and next to the other side of His57. The group of three residues, Asp102, His57, and Ser195, has been referred to as a catalytic triad or a charge-relay system; it is seen also in subtilisin, and something similar is seen in short-chain alcohol dehydrogenases (see Section 9.6). Considerable argument ensued as to whether the pK_a = 7 was that of His57, with Asp102 ionized at all reasonable pH values (above 2) and H-bonded to a proton on His57, or whether the pK_a was actually that of Asp102, picking up the proton from the inside of His57, which would then pick up a proton from the medium and remain uncharged. This was settled in the most direct way, by neutron diffraction studies. Neutron diffraction studies locate hydrogen atoms, which are unable to be seen in X-ray crystallography because of the low electron density of these atoms. The scattering of neutrons by the hydrogen nucleus gives enough density to compare the diffraction pattern to that obtained by X-ray crystallography; thus the proton was found to be located on the histidine instead of the aspartate. Other uses of neutron diffraction studies are exchange of protons by deuterons from solvent, and protein–water interactions when clarification of enzyme mechanisms is needed. X-ray crystallography of chymotrypsin with various inhibitors bound has contributed greatly to understanding the mechanism.

First, the substrate and enzyme associate to form a non-covalent complex. Then, nucleophilic attack of the side chain oxygen of Ser195 (hydroxyl group) on the carbonyl carbon of the scissile bond (bond to be cleaved) occurs, and a tetrahedral

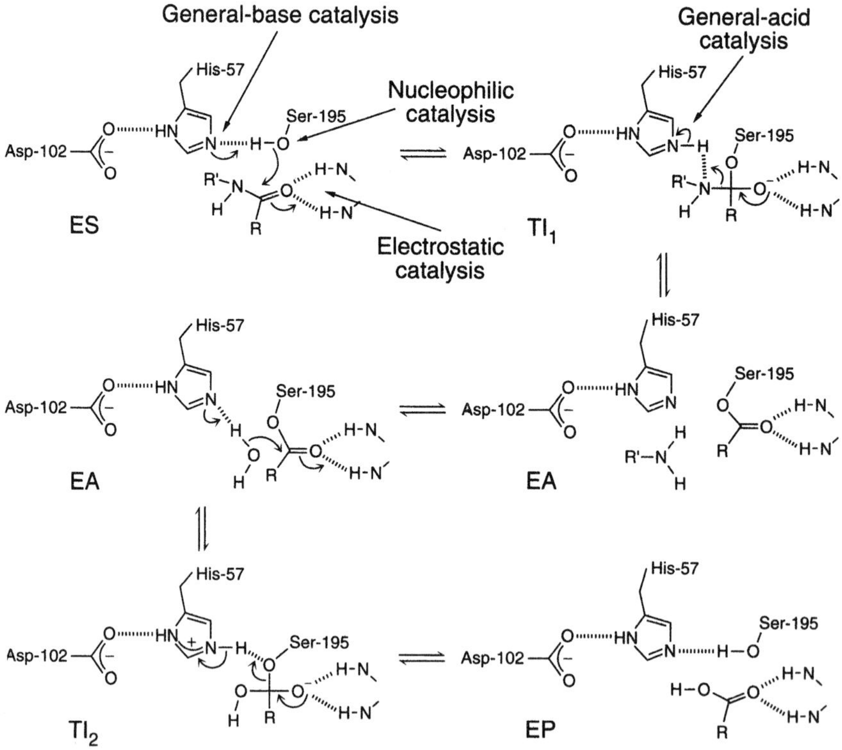

Figure 9.7 Reaction mechanism of chymotrypsin (Fersht, 1999).

intermediate is formed. During this intermediate His57 accepts a proton from the hydroxyl group of Ser195. The intermediate collapses to give the acyl-enzyme formed by the remaining peptide and the serine, and releases the first product, an amine. A nucleophilic attack by a water molecule on the acyl intermediate leads to the second tetrahedral intermediate. The intermediate is stabilized by amides of Ser195 and Gly193. In the last step the acyl-intermediate is decomposed and the second amine is released. Asp102 is not involved in the cleavage of peptide bonds but ionizes His57 so that it is able to accept a proton from serine.

A 10^6 enhancement in rate implies about a 21 kJ/mol difference in activation free energy between catalyzed and uncatalyzed reaction. This difference in free energy can be related to a difference in binding of the substrate compared to the transition state.

Evidence for the tetrahedral intermediate includes a Hammett ρ constant of +2.1 for the deacylation reaction of substituted benzoyl-chymotrypsins and the formation of tetrahedral complexes with many inhibitors, such as boronates, sulfonyl fluorides, peptide aldehydes, and peptidyl trifluoromethyl ketones. In these last the chemical shift of the imidazole proton is 18.9 ppm, indicating a good low-barrier H-bond, and the pK_a of the imidazolium is 12.1, indicating that it is stabilized by 7.3 kcal mol^{-1} compared to substrate-free chymotrypsin. The imidazole in effect is a much stronger base, facilitating proton removal from the serine.

(Taken and modified from:
http://aesop.rutgers.edu/~dbm/serineproteases.html.)

9.5.2
Cysteine in Nucleophilic Attack

Cysteine proteases are a class of enzymes that have been widely studied over the years. The overall principles of substrate recognition, catalysis, and inhibition are now reasonably well documented. This enzyme class includes the plant proteases such as papain, actinidin, and bromelain, and several mammalian lysosomal cathepsins. By far the majority of the literature reports dealing with cysteine proteases describe results obtained with the enzyme papain, because it is considered to be the archetype of this enzyme class.

As in serine proteinases, catalysis proceeds through formation of a covalent intermediate and involves a cysteine and a histidine residue. The essential Cys25 and His159 (papain numbering) play the same role as Ser195 and His57 respectively. The nucleophile is a thiolate ion rather than a hydroxyl group. The thiolate ion is stabilized through the formation of an ion-pair with a neighboring imidazolium group, His159.

Probably the aspect of primary importance for the catalytic activity of cysteine proteases is the high nucleophilicity of the active-site thiol group. It is now generally accepted that the active form of papain and of cysteine proteases in general consists of a thiolate–imidazolium ion-pair, built from Cys25 and His159.

9.5.3
Lipase, Another Catalytic Triad Mechanism

The mechanism of lipase action can be broken down into the following steps:
1. adsorption of the lipase to the interface, which is called *interfacial activation*, and is the big difference between esterases and lipases;
2. binding of substrate to enzyme;
3. chemical reaction;
4. release of product(s).

ad 1) *Interfacial activation*. The crystal structures of lipases from human pancreas and *Candida rugosa* lipase (CRL) in complexes with transition-state inhibitors reveal conformational changes compared to the crystal structures of the native lipases. A lid covering the active site is displayed, creating an open, accessible active site and a large, hydrophobic lipid-binding site. In CRL, the conformational change results in an increased hydrophobic surface area of about 800 $\AA^2$ (see Inlet Figure, at the end of the chapter) The structure determination of unique closed and open conformations supports the enzyme hypothesis with a two-state model: a stable, closed (inactive) conformation in water and a stable, open (active) conformation at the water/lipid interface or in organic solvents. Comparison of the crystal structures of a fungal lipase and a human pancreatic lipase, which evidently have a divergent evolutionary history, has revealed the structural and chemical basis of the interfacial activation. These studies reveal that:

- Both enzymes have the Asp-His-Ser catalytic side chains, the same as in the serine proteases. The active atoms of this catalytic triad have essentially identical stereochemistry in the serine proteases and in these lipases. The amino acids themselves, however, have quite different conformations and orientations.
- In both enzymes the catalytic groups are buried and inaccessible to the surrounding solvent. Burial in these two lipases is brought about by a small stretch of helix (the lid) which is situated over the active site.
- In both enzymes this helical lid presents non-polar side chains over the catalytic group, and polar side chains to the enzyme surface. Although the "lids" are very similar in construction for the two enzymes, they belong to very different parts of the polypeptide chain.
- Although the amino acid sequences have no identity (except at the active serine), the two enzymes show a similar architectural framework consisting of a central five-stranded parallel β-sheet structure. The catalytic groups are positioned on this β-sheet structure in a strikingly similar way, though there are also some significant differences.

ad 2) *Substrate binding and catalytic triad*. The catalytic triad reacts with the carbonyl group. Since the carbonyl group must be very near the active site, the acyl chain must also be near the surface of the enzyme. It is the interaction of the carbonyl group and the acyl chain with the enzyme that allows the substrate to bind.

The location of the acyl chain is of primary importance in the binding process because of its size. Due to the movement of lid during interfacial activation, a hydrophobic trench is created between the lid and enzyme surface. The trench size is ideal to accommodate the acyl chain. Interactions between the non-polar residues of the trench and the non-polar acyl chain stabilize the coupling. It has been postulated that the configuration of the trench is responsible for substrate specificity. This hypothesis seems plausible since lipases usually discriminate against certain acyl chain lengths, degrees of unsaturation, and location of double bonds in the chain. Any of these factors could affect the interaction between the acyl chain and the trench.

The carbonyl group binds to the enzyme near the active site. A pocket is created at the active site that is of the proper size to accommodate the carbonyl group. The carbonyl is also attracted to the pocket, owing to hydrogen bonding with the residues near the catalytic triad. When the carbonyl group is in position near the active site, the chemical reaction can occur.

ad 3) *Chemical reaction.* The chemical reaction is due to the action of the catalytic triad, which acts as a charge-relay system. The carboxylate group on aspartic acid is hydrogen-bonded to histidine, and the nitrogen on histidine is hydrogen-bonded to the alcohol on serine. The first step in the reaction is to make the serine alcohol a better nucleophile. This task is performed by the histidine, which completely pulls the proton off the serine alcohol, forming an oxyanion. The serine oxyanion then attacks the substrate's carbonyl carbon, forming a tetrahedral intermediate **1** (Figure 9.8). The created oxyanion is stabilized by nearby amino acids which hydrogen-bond to the oxyanion. Next, the electrons on the oxyanion are pushed back to the carbonyl carbon, and the proton currently on the histidine is transferred to the diglyceride, which is subsequently released.

ad 4) *Product release.* The serine ester formed must react with water to complete the hydrolysis. When a water molecule nears the active site, it must be converted to a better nucleophile, just as was done with the serine alcohol previously (acyl enzyme complex, Figure 9.8). The histidine nitrogen removes hydrogen from the water molecule, forming a hydroxide anion. The hydroxide attacks the carbonyl carbon, the intermediate oxyanion is stabilized by hydrogen bonds (tetrahedral intermediate **2**, Figure 9.8), the electrons are pushed back to the carbonyl carbon, and the free fatty acid is formed. The serine oxygen then reclaims the hydrogen situated on the histidine to re-establish the hydrogen-bonding network. The aspartic acid serves to pull positive charge from the histidine during the times it is fully protonated (Figure 9.8).

(This section was obtained through the URL
http://www.chemnet.ee/~imre/lipases.html.)

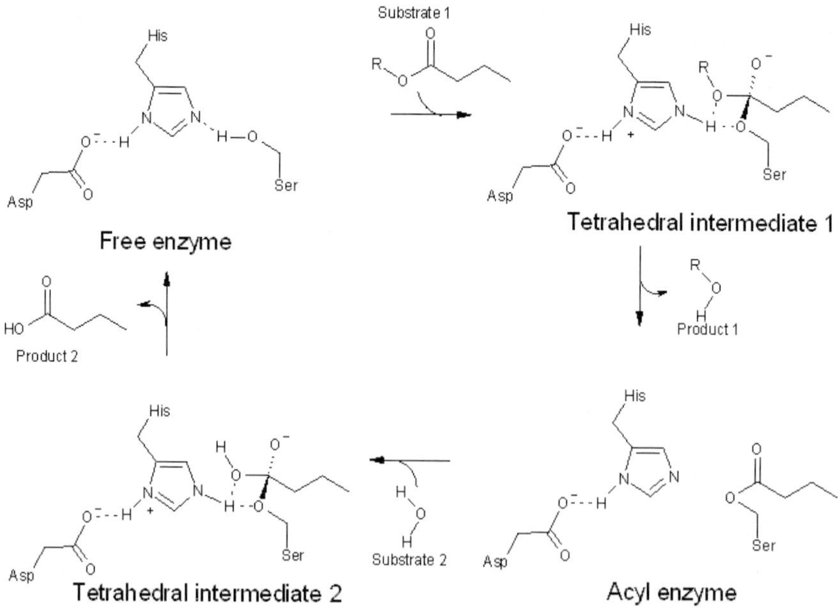

Figure 9.8 Schematic reaction mechanism of a lipase-catalyzed hydrolysis of butyric acid ester.

9.5.4
Metalloproteases

Zinc proteases are metalloenzymes containing tightly bound zinc; examples are carboxypeptidases A and B, collagenase, and thermolysin. The zinc atom is bound to the imidazole moiety of two histidines and the carboxylate of Glu; the fourth ligand is a molecule of H_2O.

In carboxypeptidase A, substrate binding occurs by induced fit: Arg145 and COO^- of Glu270 move approximately 2 Å, thereby effectively closing the active site. The C-terminus of the substrate forms a salt link with the Arg145 side chain. The hydroxyl group of Tyr248 moves approximately 12 Å (a quarter of the protein's diameter) and closes the active site. The hydroxyl group is now positioned close to the amino group of the scissile bond and forms a hydrogen bond. Water is excluded from the active site except for a single H_2O bound to the zinc ion. There is an entropy increase due to the removal of the aromatic side chain of Tyr248 from water, which lowers the activation barrier.

Possible Mechanism

The peptide carbonyl oxygen interacts with zinc, withdrawing electrons from carbon. Water is displaced from zinc. Glu270 accepts a proton from water – it acts as a general base – and the hydroxyl anion of water acts as a nucleophile to attack the carbonyl carbon of the peptide bond and forms a negatively charged tetrahedral intermediate. Glu270 acts as a general acid and donates a proton to the departing amine. Stabilization of the transition state (a tetrahedral intermediate) is performed by a negative charge on O interacting with Zn^{2+} and by the side chain of Arg127.

9.6 Electrophilic catalysis

9.6.1 Utilization of Metal Ions: ADH, a Different Catalytic Triad

Alcohol dehydrogenases can be classified in three different groups:

- medium-chain dehydrogenases (about 350 amino acids), represented by horse-liver ADH and yeast ADH (YADH);
- short-chain alcohol dehydrogenases (about 250 amino acids), represented by the *Drosophila* ADH (DADH) as well as the co-author's own *R*-specific ADH from *L. brevis* (Riebel, 1996); and
- long-chain or iron-activated alcohol dehydrogenases (600–750 amino acids) using pyrroloquinoline quinone (PQQ) as a cofactor. Iron-activated ADHs include 1,2-propanediol dehydrogenase of *E. coli*, ADH II of the bacterium *Zymomonas mobilis*, which ferments glucose and sucrose to ethanol by the Entner–Doudoroff pathway, and ADH IV of yeast.

9.6.1.1 Catalytic Mechanism of Horse Liver Alcohol Dehydrogenase, a Medium-Chain Dehydrogenase

In the case of all pyridine nucleotide-dependent dehydrogenases, the reaction mechanism is characterized by two substrates interacting with the enzymes and two products dissociating from the catalyst molecule after reaction. Whereas the liver ADH kinetic mechanism is highly ordered, the coenzyme associating first and dissociating last, the yeast ADH mechanism is largely random, at least as observed under ordinary laboratory conditions. The actual chemical reaction is a hydride transfer, formally transferring a negatively charged hydrogen atom, though possibly atom and charge are transferred separately. In YADH this chemical step is largely rate-limiting, especially with aromatic alcohols on which substituent effects on the benzene ring can be examined with LFERs (see Section 9.2.4).

The original mechanism of liver ADH by Theorell and Chance in 1951 (Theorell, 1951) postulated that the reaction proceeds in highly ordered fashion, the cofactor NAD(H) binding first, followed by substrate, then hydride transfer occurring, and product being released before coenzyme. No significant concentration of central complex (ES_1S_2) was found, however, with alcohol association, hydride transfer, and aldehyde dissociation occurring as one concerted step to the point where aldehyde and alcohol are mutually competitive product inhibitors. This so-called "Theorell–Chance mechanism" represents a compulsory ordered pathway where the ternary complexes are kinetically not significant (Figure 9.9). In the case of primary alkyl alcohols the apparent rate-limiting step is coenzyme dissociation at

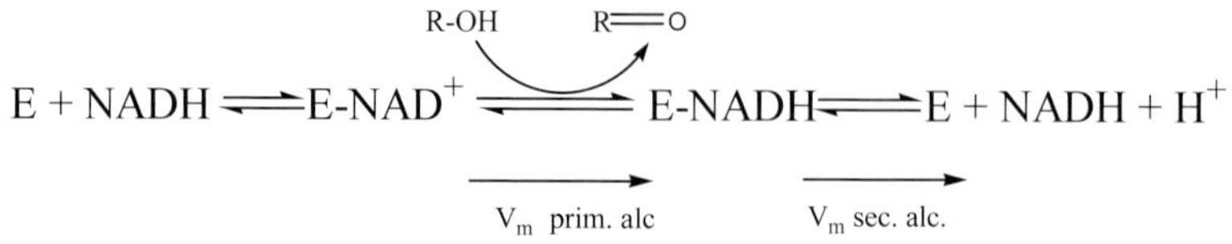

$$E + NADH \rightleftharpoons E\text{-}NAD^+ \xrightleftharpoons[]{R\text{-}OH \rightarrow R{=}O} E\text{-}NADH \rightleftharpoons E + NADH + H^+$$

$\longrightarrow$ V_m prim. alc $\qquad$ $\longrightarrow$ V_m sec. alc.

Figure 9.9 Kinetic mechanism according to Theorell–Chance.

$k_{cat} = 3\ s^{-1} = 4.5\ \mu mol\ min^{-1}\ (mg\ protein)^{-1}$ whereas the rate-limiting step is aldehyde dissociation for aromatic alcohols and hydride transfer for secondary alcohols. Substrate inhibition is observed for primary alcohols owing to formation of a ternary E–NADH.alcohol complex which dissociates NADH more slowly than the binary E–NADH complex; with the secondary alcohol cyclohexanol, NADH dissociates faster from the ternary than from the binary complex so that substrate activation occurs at high cyclohexanol concentration.

Like all NAD(P)-dependent enzymes HL-ADH features a conserved folding pattern in the cofactor binding domain, the Rossmann fold (which is different from the catalytic domain). The catalytic zinc atom in the enzyme center is ligated by Cys46, Cys174, and His67, whereas the structural zinc is located in a loop at the edge of the structure. Zinc-chelating compounds such as o-phenanthroline bind at the active site, competing with both substrate and coenzyme. Coenzyme stretches in an open conformation from the coenzyme binding domain down towards the catalytic zinc. The substrate binds in a long narrow pocket, with the catalytic zinc at the bottom.

Two types of inhibitors, pyrazoles and imidazoles (with E–NAD^+) and isobutyramide (with E–NADH), form tight ternary complexes with E–coenzyme, allowing single turnover to be observed (through photometry at 290 nm or fluorescence caused by NADH) and thus titration of the active sites (see Section 9.2.3.). Pyrazole and isobutyramide are kinetically competitive with ethanol and acetaldehyde, respectively. If the reaction E + NADH + aldehyde is run in the presence of a high concentration of pyrazole, the complex E–NAD^+ formed by dissociation of alcohol immediately binds pyrazole for a single turnover only. Under favorable conditions, a single NADH oxidation can be observed by stopped-flow techniques to find a k_{cat} of about 150 s^{-1} and a deuterium isotope effect $k_D \approx 4$ as expected (see Section 9.2.5).

Overall reaction and chemical mechanism

Alcohol oxidation requires release of a proton, which formally comes from the alcohol. In other dehydrogenases such as lactate dehydrogenase, proton release occurs simultaneously with hydride transfer. In liver ADH proton release can be demonstrated, by reaction of the proton with an indicator such as thymol blue or phenol red in stopped-flow spectrophotometry, to be faster than hydride transfer, 270 vs. 150 s^{-1}, and unaffected by use of deuterated substrate, so it occurs before hydride transfer. Binding of the NAD^+ nicotinamide ring is accompanied by a conformational change of ADH bringing the catalytic zinc about 0.1 nm closer to the

NAD^+ binding domain. NAD^+ is then in a position to receive the actual or incipient hydride from the zinc-bound substrate as the aldehyde or ketone product forms. In the reverse reaction, Zn^{2+} polarizes the carbonyl bond and promotes hydride ion transfer from NADH to the carbonyl group. The proton relay system in the apo-enzyme (NAD^+ absent) involves a chain of the hydrogen-bonded groups Zn–OH_2, Ser48, H_2O, and His52. When both OH^- and ROH are presumed to be bound to Zn^{2+} the OH^- is stabilized by a hydrogen bond to Ser48, which is hydrogen-bonded to His51. Hydride transfer to NAD^+ from the CH_2 group of ROH, which lost its proton to OH^-, then completes the oxidation to aldehyde. In most dehydrogenases, hydride transfer occurs enantiospecifically: in HL-ADH, and H^- transfers from the pro-*R* position of the substrate to C4 of the NAD^+. The two hydrogen atoms at the 4-position in NADH are not equivalent; in HL-ADH, the intermediate carboxamide formed in the transition state would collide with the catalytic zinc and residues 178 and 203 upon attempted binding of the pro-*S* hydride.

(Modified from http://aesop.rutgers.edu/~dbm/AllcoholDeh.html.)

9.6.1.2 Catalytic Reaction Mechanism of Drosophila ADH, a Short-Chain Dehydrogenase

When trying to determine the 3D structure of binary and ternary complexes of *Drosophila* alcohol dehydrogenase (DADH), researchers' initial attempts to soak apo-form crystals with the oxidized coenzyme (NAD^+) failed. The crystals cracked after several hours and became unusable. This suggested that the coenzyme, upon binding to the enzyme, induced a conformational change that seriously affected the crystal packing. The same phenomenon prevented solution of the HL-ADH 3D structure for many years.

A possible mechanism for DADH is shown schematically in Figure 9.10.

(This section was taken in part from the URL http://www.csb.ki.se/users/xray/jordi/thesis/thesis.html.)

Substrate and stereo-specificity

The bifurcated and hydrophobic active site of DADH favors oxidation of secondary compared to primary alcohols. In HL-ADH the alcohol binding region is a funnel-like single cavity resulting in higher affinity towards primary than secondary alcohols. The non-polar character of the DADH active site decreases the specificity for multi-substituted alcohols such as polyols or carbohydrates that contain hydroxyl groups or other polar groups. Owing to their rigidity and size, binding of large or rigid molecules such as prostaglandins or steroids seems to be inhibited. Conversely, the active-site cavity seems too large to allow for productive binding of small substrates such as methanol. Higher affinity towards (*R*)-enantiomeric secondary alcohols in DADH can be explained by the different size and disposition of the R1 and R2 sub-cavities, whereas HL-ADH favors (*S*)-enantiomeric secondary alcohols. Branching at the C3 atom in 3-methyl-2-butanol causes a decrease in the binding strength in the (*R*) enantiomer but not the (*S*) form, indicating that the R2 sub-cavity is slightly wider than R1 and that it can accept an extra methyl group without affecting the binding of the alcohol.

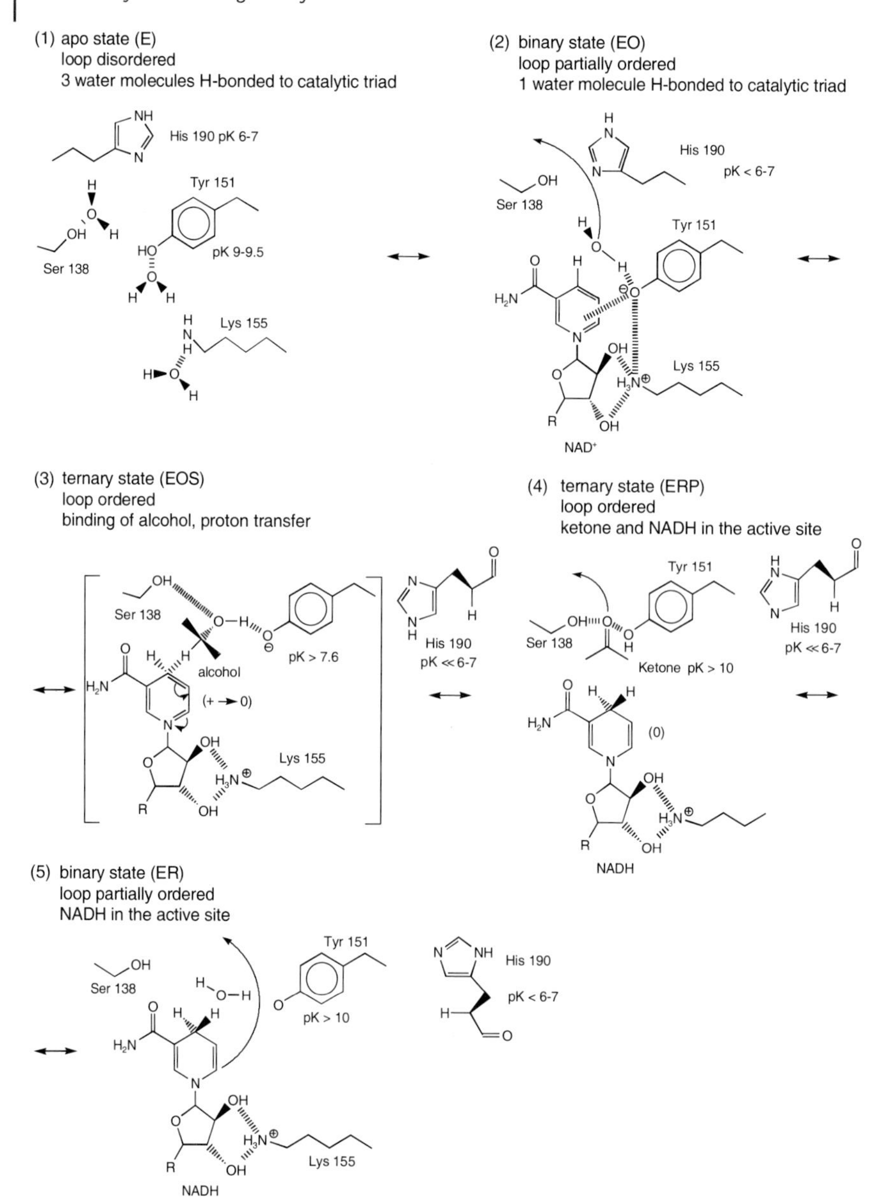

Figure 9.10 Catalytic mechanism of DADH, involving Tyr151 as proton-abstracting group.

Possible roles of the catalytic triad: Ser138, Tyr151, Lys155

The Theorell–Chance mechanism describes the predominating pathway for the conversion of a wide range of primary and also some secondary alcohols by HL-ADH. The same kind of mechanism is also valid for the reaction of DADH with secondary alcohols. As mentioned in the previous section, ternary complexes are kinetically insignificant for the compulsory ordered Theorell–Chance mechanism.

In the oxidation of secondary alcohols by DADH, the coenzyme is the leading substrate, the release of NADH from the enzyme–NADH complex is the rate-limiting step, and the maximum velocity v_{max} is independent of the chemical nature of the alcohol. In the case of primary alcohols, as v_{max} is much lower and depends on the nature of the alcohol, Theorell–Chance kinetics (Figure 9.9) are not observed and the rate-limiting step is the chemical interconversion from alcohol to aldehyde. With all this biochemical information it is possible to delineate a catalytic reaction mechanism that is in agreement with the crystal structures and the steps of alcohol oxidation observed in the kinetic analysis of the DADH reaction.

Formation/dissociation of the E–NAD^+ complex

As with HL-ADH, the enzyme undergoes conformational changes upon binding of coenzyme which, together with the formation of 11 hydrogen bonds with the enzyme, can be related to the observed protection against unfolding. The entrance to the active site, probably the region around residues 186–191, has to be flexible for the substrate to bind. Formation of this complex is kinetically regulated by a group with pK_a 8.5–9.5 assigned to the residue Tyr151. At pH > pK_a, the deprotonated form of this tyrosine interacts favorably with the charged nicotinamide ring of the NAD^+ molecule. The presence of the charged side chain of Lys155 might influence the pK_a of Tyr151. Whereas a fully solvated tyrosine amino acid has a pK_a near 10, a positive charge at this distance would stabilize a charged tyrosine by about –1 kcal mol^{-1} and thus lower the pK_a. The hydrophobic environment of the active site cavity exerts the opposite effect, resulting in a final pK_a of about 9 for this amino acid.

The release of NAD^+ from the E–NAD^+ complex is controlled by only a single deprotonated group with a pK_a of 7.6 in *D. melanogaster* ADH, again assigned to Tyr151. The pK_a change from 9.0 in the apo form to 7.6 in the E–NAD^+-complex can be attributed to the influence of the positively charged nicotinamide ring in the E–NAD^+ complex. As in the case with Lys155, the presence of a positive charge in the neighborhood of Tyr151 stabilizes its deprotonated state and lowers its pK_a to 7.6.

Interconversion of the ternary complex

There seems to be no proton transfer to the solvent during interconversion of the ternary complex, which is in agreement with a closed catalytic cavity with no connection to the bulk solvent. The active-site cavity supports proper orientation of the substrate, its alkyl chains are positioned in the sub-cavities R1 and R2, and the hydroxyl group of the alcohol forms a hydrogen bond on one side to the Ser138 and on the other side to the deprotonated hydroxyl group of Tyr151. This step is rate-limiting for the oxidation of primary alcohols. The reason might be unspecific binding of a single-chain alcohol in a bifurcated active site, which can result in an increased $\Delta G^{\neq}$ and thus a decrease in the rate constant. The unprotonated tyrosinate-151 can act as a general base, resulting in a proton transfer from the alcohol to Tyr151. Substitution of Tyr151 by Phe, Glu, Gln, or His results in inactive enzyme; with Cys, 0.25% activity is retained. The task of Lys155 is (i) to bind the OH groups of the coenzyme nicotinamide ribose moiety, and (ii) to lower the pK_a of Tyr151. The importance of Lys155 has also been studied by mutagenesis: Lys155Ile is inactive, Lys155Arg is 2.2% active. A hydride is then transferred from the alcohol C1 atom to the C4 position of the oxidized coenzyme either simultaneously or in two steps, and the ketone/aldehyde is formed. Ser138 is assumed to have the role of stabilizing the transient state during catalysis. Its importance is such that its substitution by Ala or Cys in DADH totally inactivates the enzyme.

Dissociation/formation of the E–NADH·product complex

Upon conversion from alcohol to ketone/aldehyde the active site contains a reduced NADH molecule, which, in conjunction with the hydrophobic nature of the active-site cavity, is believed to raise the pK_a of Tyr151 to above 10. Evidence is provided by studies demonstrating pH-independent binding of acetaldehyde to the E–NADH complex so that the pK_a value for the amino acid responsible for interaction with acetaldehyde is higher than 10.

An analogous mechanism

A second short-chain alcohol dehydrogenase, the *R*-specific ADH from *L. brevis*, seems to follow the same mechanism, as revealed by structural analysis of crystals soaked in cofactor and substrate solution. The catalytic cycle is initiated by deprotonation of Tyr155, transferring it to the carbonyl of the substrate. The hydroxyl group of Ser142 stabilizes the negative charge of Tyr155 in the transition state and fixes the substrate in position so that the hydride transfer can occur. Ser142 is essential and mutation at this position results in loss of activity. The excessive distance between Lys159 and Tyr155 precludes any direct interaction; rather, Lys159 seems to be more important in stabilizing the position of the NADP-ribose during catalysis. In the second step, hydride transfer occurs from the pro-*S* hydrogen of the nicotinamide ring onto the carbonyl of the substrate, while a water molecule protonates Tyr155. After these two steps, so far there is no evidence whether this mechanism is sequential or simultaneous (Mueller, 2000).

9.6.2
Formation of a Schiff Base, Part I: Acetoacetate Decarboxylase, Aldolase

Aldolases such as fructose-1,6-bisphosphate aldolase (FBP-aldolase), a crucial enzyme in glycolysis, catalyze the formation of carbon–carbon bonds, a critical process for the synthesis of complex biological molecules. FBP-aldolase catalyzes the reversible condensation of dihydroxyacetone phosphate (DHAP) and glyceraldehyde-3-phosphate (G3P) to form fructose-1,6-bisphosphate. There are two classes of aldolases: the first, such as the mammalian FBP-aldolase, uses an active-site lysine to form a Schiff base, whereas the second class features an active-site zinc ion to perform the same reaction. Acetoacetate decarboxylase, an example of the second class, catalyzes the decarboxylation of β-keto acids. A lysine residue is required for good activity of the enzyme; the ε-amine of lysine activates the substrate carbonyl group by forming a Schiff base.

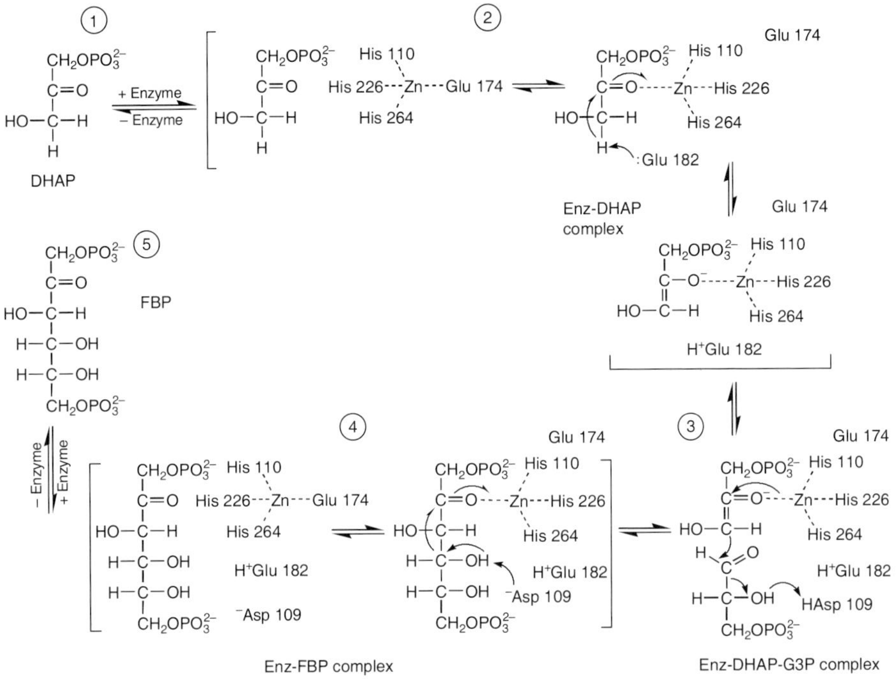

Figure 9.11 Mechanism of action of the class II aldolase from *E.coli*.

Figure 9.11 demonstrates the Schiff base formation reaction for class II aldolases, which are mainly found in prokaryotes; *E.coli* aldolase uses a zinc ion to ionize the carbonyl group of DHAP (Alan Berry, University of Leeds); it is now Figure 9.12.

9.6.3
Formation of a Schiff Base with Pyridoxal Phosphate (PLP): Alanine Racemase, Amino Acid Transferase

Vitamin B_6 enzymes play important roles in amino acid metabolism, and catalyze a wide range of amino acid transformations (transamination, decarboxylation, deamination, racemization, and aldol cleavages) by labilizing one of three bonds at a carbon atom of an amino acid substrate. Aspartate aminotransferase (AAT) is one of the best-studied vitamin B_6-dependent enzymes. Importantly, one enzyme usually catalyzes one reaction with its substrate specificity. The vitamin B_6-dependent enzymes are classified into fold types I–IV, indicating that the enzymes evolved from four common ancestors, and the versatility of the enzymes has been attained by divergent and convergent molecular evolution. Systematic single and multiple replacement studies have been applied to *Escherichia coli* AAT, and the role of active-site residues or the catalytic mechanism has been proposed from X-ray crystallographic and enzymological studies.

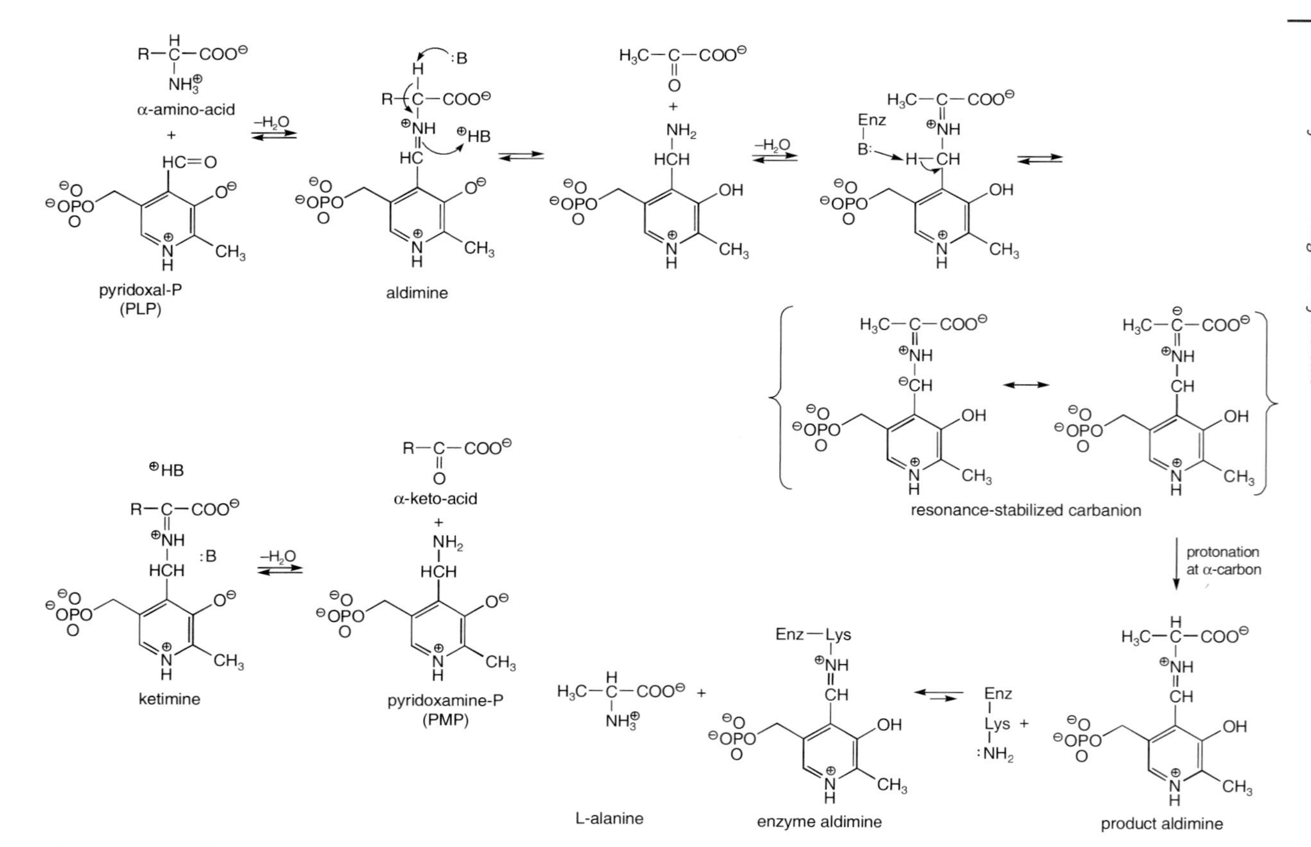

Figure 9.12 Schiff base mechanism utilizing PLP, demonstrated on amino acid transferase.

Alanine racemase, as another PLP-dependent enzyme, is a bacterial enzyme used to create D-alanine from L-alanine for incorporation into the bacterial cell wall. Its role is to act as an electron sink to stabilize carbanionic intermediates generated in enzymatic catalysis.

The general reaction mechanism of AAT is shown in Figure 9.12.

9.6.4 Utilization of Thiamine Pyrophosphate (TPP): Transketolase

Transketolase (E.C. 2.2.1.1) catalyzes the interconversion of sugars by transferring a two-carbon ketol unit from a ketose donor substrate to an aldose acceptor substrate (Figure 9.13). The transfer is accomplished by the formation of a covalent intermediate between the ketol moiety and the thiazole ring of a thiamine diphosphate cofactor. The structure, cofactor binding sites, subunit interactions, and catalytic mechanism of transketolase have been studied extensively and are well understood. Transketolase catalyzes, in a rate-limiting manner for the pathway, several reactions in the non-oxidative portion of the pentose phosphate pathway. By doing so, transketolase, along with transaldolase, provides a reversible link between the

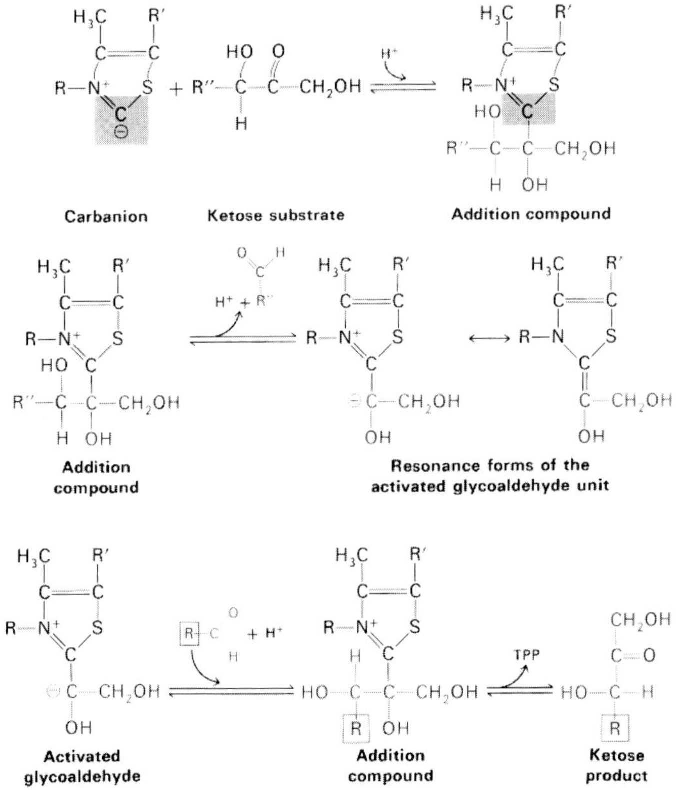

Figure 9.13 Catalytic mechanism of transketolase (taken from Stryer, 1995).

oxidative portion of the pathway and glycolysis. The thiamine pyrophosphate (TPP) cofactor molecule is covalently bound to the enzyme in its activated form. Upon reaction, the thiazolium ring adds to the carbonyl to perform a C2 addition.

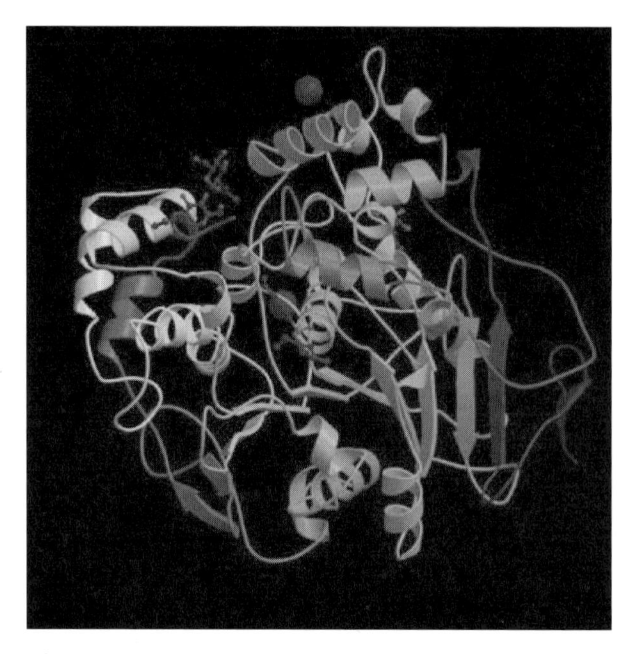

Inlet Figure Crystal structure of *Candida rugosa* lipase (CRL, E.C. 3.1.1.3), complexed with (1*S*)-menthyl hexyl phosphonate (Cygler, 1994).

Suggested Further Reading

Alan Fersht, *Structure and Mechanism in Protein Science*, Freeman, New York, **1999**.

Robert A. Copeland, *Enzymes – A Practical Introduction to Structure, Mechanism, and Data Analysis*, Wiley-VCH, New York, **1996**.

William P. Jencks, *Mechanisms in Chemistry and Enzymology*, Dover, New York, **1975**.

R. Kellner, F.Lottspeich, and H. E. Meyer, *Microcharacterization of Proteins*, John Wiley; 2nd edition, **1999**.

References

J. Andres, V. Moliner, V. S. Safont, and L. R. Domingo, Transition structures characterization and kinetic isotope effects. Understanding chemical reactivity in enzymes, *Recent Res. Devel. Phys. Chem.* **1997**, *1*, 99–116.

P. A. Bartlett and C. K. Marlowe, Phosphonamidates as transition-state analogue inhibitors of thermolysin, *Biochemistry* **1983**, *22*, 4618–4624.

P. A. Bartlett and M. A. Giangiordano, Transition state analogy of phosphonic acid peptide inhibitors of pepsin, *J. Org. Chem.* **1996**, *61*, 3433–3438.

M. M. Bradford, A rapid and sensitive method for the quantitation of microgram quantities of protein utilizing the principle of protein–dye binding, *Anal. Biochem.* **1976**, *72*, 248–254.

M. Cygler, P. Grochulski, R. J. Kazlauskas, J. D. Schrag, F. Bouthillier, B. Rubin, A. N. Serreqi, and A. K. Gupta, A structural basis for the chiral preferences of lipases, *J. Am. Chem. Soc.* **1994**, *116*, 3180–3186.

K. Felgenhauer, Evaluation of molecular size by gel electrophoretic techniques, *Hoppe-Seyler's Z. Physiol. Chem.*, **1974**, *355*, 1281–1290.

A. Fersht, *Structure and Mechanism in Protein Science*, Freeman, New York, **1999**.

J. Frick and R. Wolfenden, in: *Design of Enzyme Inhibitors as Drugs*, M. Sandler, J. H. Smith (eds), Oxford University Press, New York, **1989**, 19.

J. E. Hanson, A. P. Kaplan, and P. A. Bartlett, Phosphonate analogs of carboxypeptidase A substrates are potent transition-state analog inhibitors, *Biochemistry*, **1989**, *28*, 6294–305.

B. S. Hartley and B. A. Kilby, Reaction of p-nitrophenyl esters with chymotrypsin and insulin, *Biochem. J.* **1954**, *56*, 288–297.

L. Hedstrom, L. Szilagyi, and W. J. Rutter, Converting trypsin to chymotrypsin: the role of surface loops, *Science* **1992**, *255*, 1249–53.

L. Hedstrom, J. J. Perona, and W. J. Rutter, Converting Trypsin to Chymotrypsin: Residue 172 Is a Substrate Specificity Determinant, *Biochemistry*, **1994a**, *33*, 8757–63.

L. Hedstrom, S. Farr-Jones, C. A. Kettner, W. J. Rutter, Converting Trypsin to Chymotrypsin: Ground-State Binding Does Not Determine Substrate Specificity, *Biochemistry* **1994b**, *33*, 8764–9.

L. Hedstrom, Serine Protease Mechanism and Specificity, *Chem. Rev.* **2002**, *102*, 4501–4523.

C. Y. Hsia, G. Gantshaw, C. Paech and C. J. Murray, Active-Site Titration of Serine Proteases Using a Fluoride Ion Selective Electrode and Sulfonyl Fluoride Inhibitors, *Anal. Biochem.* **1996**, *242*, 221–227.

S.-H. Hung, L. Hedstrom, Converting trypsin to elastase: substitution of the S1 site and adjacent loops reconstitutes esterase specificity but not amidase activity, *Prot. Eng.* **1998**, *11*, 669–673.

R. J. Kazlauskas, Elucidating structure–mechanism relationships in lipases: prospects for predicting and engineering catalytic properties, *Trends Biotechnol.* **1994**, *12*, 464–72.

R. Kellner, F. Lottspeich, and H. E. Meyer, *Microcharacterization of Proteins*, John Wiley, 2nd edition, **1999**.

T. Kurth, D. Ullmann, H.-D. Jakubke, and L. Hedstrom, Converting trypsin to chymotrypsin: Structural determinants of S1' specificity, *Biochemistry* **1997**, *36*, 10098–10104.

G. Legler, C. M. Mueller-Platz, M. Mentges-Hettkamp, G. Pflieger, and E. Juelich, On the chemical basis of the Lowry protein determination, *Anal. Biochem.* **1985**, *150*, 278–87.

J. Littlechild, E. Garcia-Rodriguez, A. Dalby, and M. Isupov, Structural and functional comparisons between vanadium haloperoxidase and acid phosphatase enzymes, *J. Mol. Recognit.* **2002**, *15*, 291–296.

M. M. Mader and P. A. Bartlett, Binding Energy and catalysis: The Implications for Transition-State Analogs and Catalytic Antibodies, *Chem. Rev.* **1997**, *97*, 1281–1301.

A. Messerschmidt and R. Wever, X-ray structure of a vanadium-containing enzyme: chloroperoxidase from the fungus *Curvularia inaequalis*, *Proc. Natl. Acad. Sci. USA* **1996**, *93*, 392–96.

A. Messerschmidt, X. Prade, and R. Wever, Implications for the catalytic mechanism of the vanadium-containing enzyme chloroperoxidase from the fungus Curvularia inaequalis by X-ray structures of the native and peroxide form, *Biol. Chem.* **1997**, *378*, 309–315.

B. P. Morgan, J. M. Scholtz, M. D. Ballinger, I. D. Zipkin, and P. A. Bartlett, Differential binding energy: a detailed evaluation of the influence of hydrogen-bonding and hydrophobic groups on the inhibition of thermolysin by phosphorus-containing inhibitors, *J. Amer. Chem. Soc.* **1991**, *113*, 297–307.

J. Mueller, X-Ray Crystal structure analysis of the R-specific alcohol dehydrogenase from Lactobacillus brevis with 0.99 Å resolution, PhD thesis, University of Cologne, Cologne, Germany, **2000**.

J. R. Murdoch, What is the rate-limiting step of a multistep reaction?, *J.Chem.Ed.* **1981**, *58*, 32–6.

D. B. Northrop and F. B. Simpson, Beyond Enzyme Kinetics: Direct Determination of Mechanisms by Stopped-Flow Mass Spectrometry, *Bioorg. Med. Chem.* **1997**, *4*, 641–644.

B. Riebel, Biochemical and molecular biological characterization of novel microbial NAD(P)-dependent alcohol dehydrogenases, PhD thesis, University of Düsseldorf, Düsseldorf, Germany, **1997**.

V. L. Schramm, Enzymatic transition-state analysis and transition-state analogs, *Meth. Enzymol.* **1999**, *308*, 301–355.

P. K. Smith, R. I. Krohn, G. T. Hermanson, A. K. Mallia, F. H. Gartner, M. D. Provenzano, E. K. Fujimoto, N. M. Goeke, B. J. Olson, and D. C. Klenk, Measurement of protein using bicinchoninic acid, *Anal. Biochem.* **1985**, *150*, 76–85.

L. Stryer, *Biochemistry*, W. H. Freeman, New York, 4th edition, **1995**.

H. Theorell and B. Chance, Liver alcohol dehydrogenase. II. Kinetics of the compound of horse-liver alcohol dehydrogenase and reduced diphosphopyridine nucleotide, *Acta Chem. Scand.* **1951**, *5*, 1127–1144.

R. C. Thompson, Use of peptide aldehydes to generate transition-state analogs of elastase, *Biochemistry* **1973**, *12*, 47–51.

B. Weidmann, Renin inhibitors. from transition state analogs and peptide mimetics to blood pressure lowering drugs, *Chimia* **1991**, *45*, 367–376.

W. C. Werkheiser, Specific binding of 4-amino folic acid analogs by folic acid reductase, *J. Biol. Chem.* **1961** *236*, 888–893.

J. O. Westerik and R. Wolfenden, Aldehydes as inhibitors of papain, *J. Biol. Chem.* **1972**, *247*, 8195–8197.

R. Wolfenden and A. Radzicka, Transition-state analogs, *Curr. Opin. Struct. Biol.* **1991**, *1*, 780–787.

N. S. Zefirov, Stereochemical studies. XII. General equation of relationship between products ratio and conformational equilibrium, *Tetrahedron* **1977**, *33*, 2719–2722.

10
Protein Engineering

Summary

Protein engineering results in the targeted mutation of one or more specific nucleotides in a gene, which then codes for an altered protein compared to the wild type. Targeted exchange of specific amino acids against others, termed "site-directed mutagenesis", requires the gene of the targeted protein as well as knowledge of its three-dimensional structure, and encompasses the following steps:

1. purification of target enzyme to homogeneity and analysis of the amino acid sequence;
2. determination and isolation of the corresponding gene, and recombinant expression of protein (steps 1 and 2 are often interchanged);
3. crystallization of the homogeneous protein, followed by X-ray analysis of the protein crystals;
4. solution of the 3D structure; for activity studies, elucidation of the enzyme's mechanism;
5. picking of target amino acids for improvement through site-directed mutagenesis;
6. site-directed point mutations on the DNA level, cloning, and recombinant expression of the mutated gene;
7. test of mutein for changes in characteristic parameters; if unsuccessful, repeat steps 5–7.

Primers can be exploited to introduce a mutation into the target DNA by changing one or two base pairs within a codon but still binding to the target DNA and creating a small loop during the first annealing process. Polymerization will proceed from these primers and create an altered sequence.

Successful applications of protein engineering include the following examples:

- Thermal inactivation (high fructose corn syrup) of tetrameric glucose isomerase (GI), important in the manufacture of HFCS, is caused in high glucose concentrations (≈ 3 M) by chemically driven glycation of lysine-253, involved in dimer–dimer contact. Exchange of lysine-253 for arginine by site-directed mutagenesis, creating the K253R mutant, resulted in a three-fold reduced deactivation constant k_d and a beneficial pH shift from 6.0 to 6.5.

- The serine protease subtilisin is used in bulk amounts in laundry detergents for its broad specificity and high alkaline stability. The oxidative bleaching force of hydrogen peroxide, another laundry detergent component, deactivates native subtilisin by oxidizing Met222, immediately adjacent to the active-site Ser221, to sulfoxide or sulfone. Site-specific mutagenesis of Met222 into alanine (M222A) or serine (M222S) resulted in remarkably stabilized enzyme, expressed as residual activity in 2 M H_2O_2 solution after 1 h.

- Structurally and sequentially related NAD(H)-dependent lactate and malate dehydrogenase, LDH and MDH, have been revealed to contain potential amino acid targets for site-specific mutagenesis that will change the substrate binding pocket without affecting their identical catalytic mechanisms. Exchange of a crucial amino acid within the substrate binding pocket (Q102R) resulted in a highly specific catalyst for the new substrate oxaloacetate.

10.1
Introduction: Elements of Protein Engineering

A protein consisting of 100 amino acids allows 20^{100} possibilities of different protein sequences. Nature is unable to test and fulfill all the possible protein sequences and the subsequent structures using the existing 20 amino acids. Especially for events which occur only improbably or rarely in nature, there is a good chance of improving or creating new structures which are more fitting than the ones nature has provided. There are two possible strategies for addressing a problem for which a new enzyme or organism is needed. First, protein engineering results in rearrangement or mutation of specific protein sequences on the DNA level that can lead to faster and more efficient proteins. Secondly, changes can be introduced through directed evolution (see Chapter 11), which mimics Nature and improves the target for the desired problem. To achieve directed mutagenesis of certain amino acids on proteins, several steps are necessary to obtain enzymatic material that is useful for mutagenesis:

1. purification of the target enzyme to homogeneity (see Chapter 8);
2. analysis of the primary structure (amino acid sequence);
3. determination and isolation of the corresponding gene (steps 2 and 3 are reversed, if the protein is searched for on the DNA level) (see Chapter 4);
4. crystallization of the homogeneous protein (this can be done parallel to steps 2 and 3, but more often requires step 3 to achieve large amounts of pure recombinant protein),
5. X-ray analysis of the protein crystals (sometimes NMR of the protein in solution is sufficient, depending on the size of the protein);

6. determination of the 3D structure (if there are no similar structures available then heavy metal derivation is necessary);
7. analysis of target amino acids to achieve improvement of desired characteristics (for changes in activity, knowledge of the mechanism is crucial);
8. point mutations on the DNA level at the target sites;
9. cloning and expression of the mutated gene (see Chapter 4);
10. testing the mutein for changes in characteristics;
11. if positive, isolation and purification of the new enzyme;
12. repetition of steps 7–10 if success was not achieved the first time.

At the point when protein engineering is considered in a project, steps 1–3 usually have been done already (although this might be time-consuming), the characteristics of the protein in question are known, and certain problems regarding catalysis, specificity, or stability have arisen. There are two possible choices for obtaining fast results using protein engineering. Crystallization of the recombinant over-expressed protein should start as soon as possible, as protein crystallization is an empirical method with timelines impossible to predict. In the meantime, bio-informatics tools can be used to look for similar enzymes, where it is to be hoped that X-ray structures are already known or, at least, that important amino acids have been determined to be crucial for certain characteristics. Then the waiting time for the crystals can be reduced by starting point mutations on crucial amino acids derived from the related enzymes.

Classically, protein engineering was limited to be a definition of artifically changing protein characteristics by *specific* amino acid changes. Nowadays, with more methods developed that belong in the field of directed evolution, crossovers between these two fields occur and some methods such as domain swapping are also regarded as protein engineering (Ostermeier, 2000). This chapter is limited to examples that used the specific amino acid change approach, and refers to Chapter 11 for examples involving directed evolution.

10.2 Methods of Protein Engineering

When protein engineering started to become a powerful tool to achieve new or enhanced characteristics, only limited methods were available.

The key to changing an amino acid lies within the development of PCR and the exploitation of the ability to use primers which match parts of the target DNA.

Usually primers have the exact DNA sequence corresponding to both strands of the target DNA, being able to amplify the DNA between them as a double-stranded DNA molecule (see Chapter 4 for more details). Primers can be exploited to introduce a mutation into the target DNA by changing one or two base pairs within a codon (consisting of triplet base pairs coding for one amino acid). These primers can still bind to the target DNA and will create a small loop during the first annealing process. Polymerization will start from these primers and thus create a changed sequence.

Usually introduction of a point mutation is desired to occur in the middle of the target gene and not at the N- or C-terminal end, for which the primers are normally designed. To ensure that both DNA strands will carry the mutation, a fusion PCR can be performed.

10.2.1 Fusion PCR

Figure 10.1 shows the two successive PCR steps of an early method to introduce a point mutation into a gene. First, sense and antisense primers within the point mutation region are designed (corresponding to the 5′ primer at the top left of the first diagram and the 3′ primer on the lower right). Using the mutation primer and the existing N- and C-terminal primers, two gene fragments are created, each of them carrying the mutation on both strands. In the second diagram, those fragment are fused together to create the whole gene again, but introducing the point mutation at the target site. Fusion of the two fragments was achieved by running five PCR cycles without any primers; annealing occurred only on the overlapping mutation region, and polymerization started in both directions from the annealed mutated regions. After five cycles, the N- and C-terminal primers were added.

Fusion PCR was a successful but not very efficient method because it involved two successive PCR reactions, which can introduce errors into the gene. Amplification of the fusion product can be difficult when the overlapping region of the mutation features a low melting temperature.

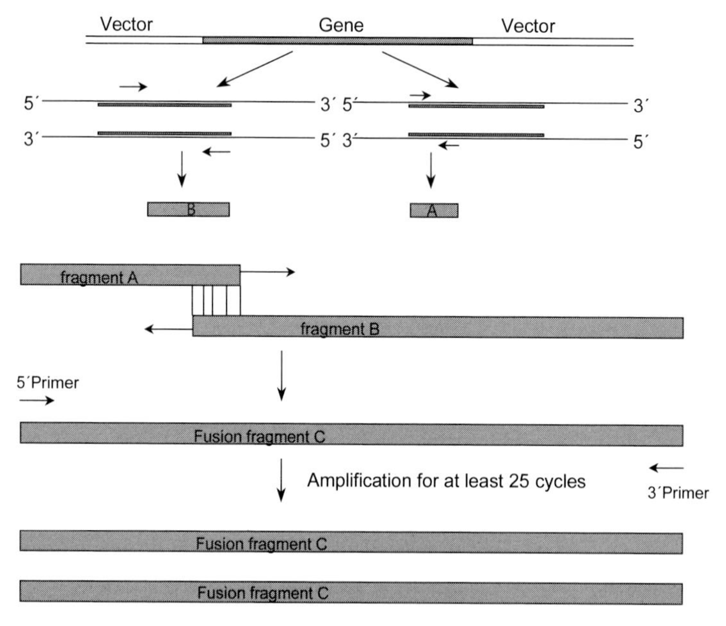

Figure 10.1 Fusion PCR, an early method to introduce point mutations in site-directed mutagenesis.

10.2.2 Kunkel Method

A different method was developed by Kunkel (1985), who exploited the fact that the base uracil (which occurs in RNA, but not in DNA) cannot be translated in the host (such as *E. coli*) when introduced recombinantly (Figure 10.2). Single-stranded viral DNA from phages grown in an *E. coli* strain deficient in uracil glycosylase (*ung*$^-$) and dUTPase (*dut*$^-$) is used as the template for site-specific mutagenesis. The template has several uracil residues incorporated instead of thymine, and would be biologically inactive when transformed into a regular *E. coli* strain, because the enzyme uracil glycosylase would remove the uracil bases, leaving abrasive sites in a single-strand DNA, which are lethal.

Using this DNA in vitro for standard site-specific mutagenesis with oligonucleotides carrying the base exchange and the M13 phage system for replication will create a second single-strand DNA incorporating the mutation and devoid of uracil. The advantage lies in avoiding the use of uracil for the mutational strand, thereby giving selective advantage to the mutated strand over the parental template strand. After transfection of the viral phage DNA into *E. coli* hosts containing the *ung* and *dut* genes, the mutated strand is favored over the parental strand, later being destroyed through the action of uracil glycosylase. Therefore, expression of the desired change, present in the newly synthesized non-uracil-containing, covalently closed, circular complementary strand, is thus strongly favored.

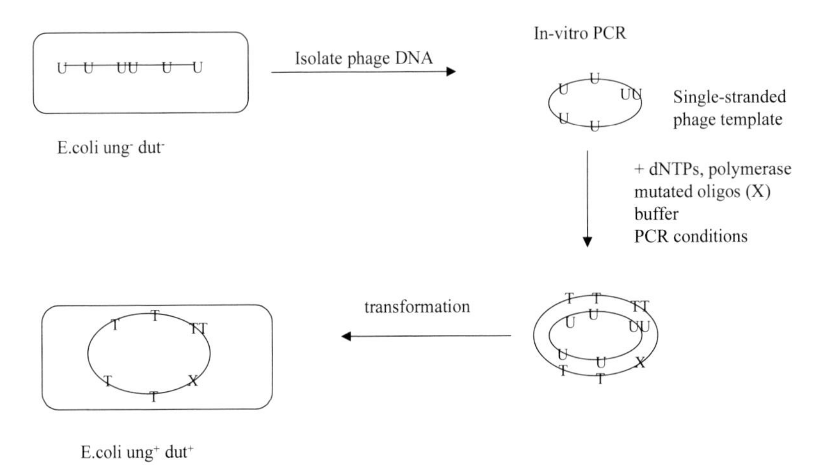

Figure 10.2 Schematic description of the site-specific mutagenesis method of Kunkel (1985). The first *E. coli* host is *ung*$^-$ *dut*$^-$ deficient, creating uracil containing templates through phage replication. The single-stranded phage molecule is used in PCR together with mutated oligos; the mutation site is marked as X. The now double-stranded molecule is transformed into a new *E. coli* host, which is now *ung*$^+$ *dut*$^+$. The parental strand is degraded, leaving the new mutated strand as the only recombinant DNA source.

Successful results were achieved by Niggemann and Steipe (Niggemann, 2000) using this method to explore protein stability after changing the structural motif. It is also a good example of how methods in directed evolution are overlapping in protein engineering, since the authors have used shuffling to create some of their molecules.

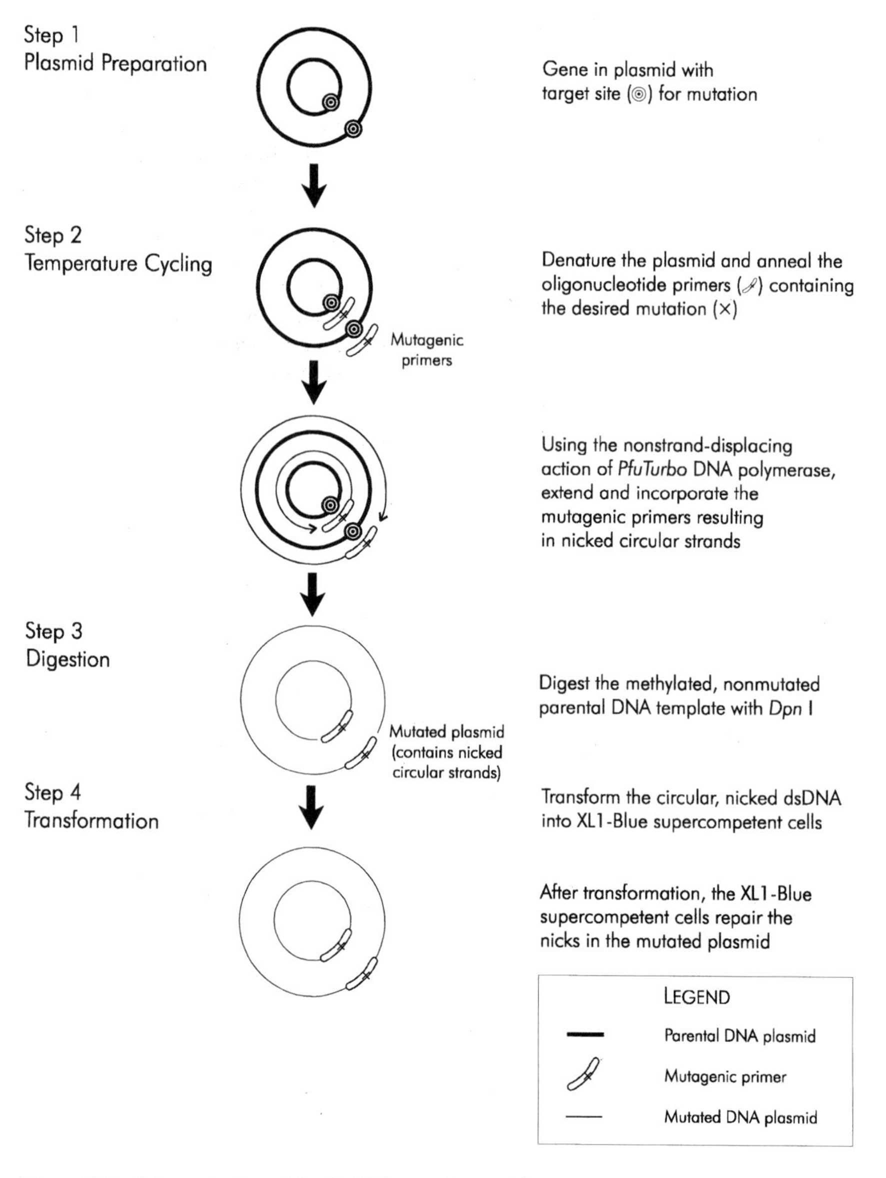

Figure 10.3 Schematic view of the QuikChange site-specific mutagenesis kit from Stratagene as described in the text.

10.2.3
Site-Specific Mutagenesis Using the QuikChange Kit from Stratagene

The technique offered by Stratagene (La Jolla/CA, USA) exploits the fact that newly, in-vitro synthesized DNA is not methylated (Figure 10.3). Using the parental plasmid carrying the target gene of interest and mutated oligonucleotide primers, DNA replication of the plasmid is performed with Pfu DNA polymerase during temperature cycling (a procedure similar to PCR). Upon incorporation of the primers, a mutated plasmid containing staggered nicks is generated. The product is then treated with the restriction enzyme DpnI, which digests the parental, methylated plasmid DNA. The remaining nicked vector DNA containing the mutated gene is transformed into an *E. coli* host. The small amount of starting DNA template required to perform QuikChange mutagenesis, the high fidelity of the *PfuTurbo* DNA polymerase employed, and the low cycle number all contribute to the high mutation efficiency and decreased potential for random mutations.

An extension to this existing kit is the QuikChange Multi, which enables multiple mutations at different sites. It uses only one mutated oligonucleotide primer, in contrast to the kit described above, but follows the same procedure using the restriction enzyme DpnI. The single-stranded plasmid is converted into a double strand in vivo after transformation into an *E. coli* host.

Features of the Stratagene kits are summarized in Table 10.1.

Table 10.1 Site-directed mutagenesis kits.

	QuikChange®	*QuikChange® XL*	*QuikChange® Multi*
Feature			
Phosphorylated primers	no	no	yes
No. of primers per site	2	2	1
Plasmid size	< 8 kb	> 8 kb	< 8 kb
Specialized plasmid	no	no	no
Competent cells	XL1-Blue	XL10-Gold®	XL10-Gold®
Protocol length	1 day	1 day	1 day
Mutation efficiency	> 80%	> 80%	> 80% (1–2 sites) > 50% (3–4 sites)
Application			
Insertions	yes	yes	NR
Deletions	yes	yes	NR
Point mutations	yes	yes	yes
Domain swapping	yes	yes	no
No. of sites mutated per reaction	1	1	up to 5
Engineered mutant collections	no	no	yes
Randomize amino acid substitutions at key residues	no	no	yes

Taken from the Stratagene catalogue, 2002.

10.2.4
Combined Chain Reaction (CCR)

Another method for site-directed mutagenesis is described by Bi and Stambrook (Bi, 1998) using the combination of PCR and ligase chain reaction (LCR); therefore the new method is termed CCR (combined chain reaction). A CCR can introduce double mutations in one round of cycling, comparably to the Stratagene method, but can also be used for introducing deletions. It seems to be a very efficient method for introducing single-point mutations, showing 100% efficiency in 23 tested clones for a single mutation (Bi, 1998). The method uses a thermostable polymerase which lacks 5′–3′ exonuclease activity (such as native Pfu polymerase). Each CCR cycle (Figure 10.4) consists of four steps: (i) denaturation of template DNA; (ii) annealing of primers to single-stranded templates; (iii) elongation of primers by the Pfu polymerase (following the PCR protocol); (and iv) ligation of the 3′-OH and 5′-phosphate ends of the extended primers using a thermostable ligase.

The method employs four primers for the reaction. Two primers anneal at the 5′ and 3′ ends of the target, which offers advantages because these primers can be complementary to the commonly used T7 or T3 promoter sequences used in many standard vectors and thus be independent of the gene of interest. The other two primers are carrying the mutation and are located within the gene. Melting temperatures of the mutation-specific primers should be higher than those of the flanking primers to achieve successful mutagenesis. Cycle processing results in amplification of double-strand mutated, ligated target DNA.

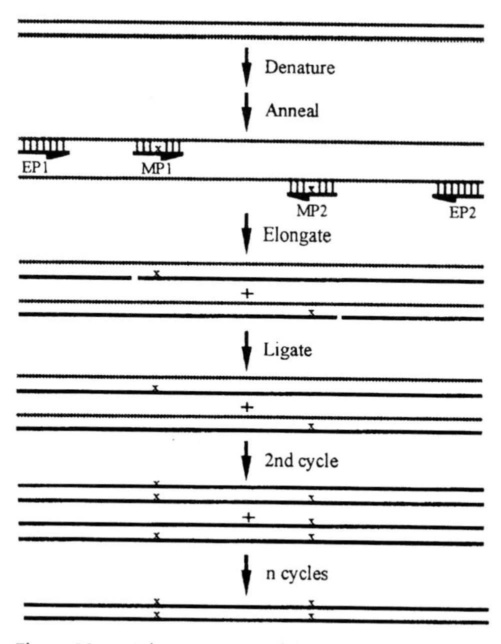

Figure 10.4 Schematic view of the CCR method (Bi, 1998).

10.3 Glucose (Xylose) Isomerase (GI) and Glycoamylase: Enhancement of Thermostability

10.3.1 Enhancement of Thermostability in Glucose Isomerase (GI)

Industrial-scale isomerization of glucose to fructose, catalyzed by glucose isomerase (GI), features process conditions involving glucose syrup initial concentrations of 3 M (≈ 50% dry wt.), pH 6.5, and temperatures around 60–75 °C. The enzyme is immobilized together with small amounts of Mg^{2+}; care has to be taken to control the Ca^{2+} level as Ca^{2+} tends to inhibit the enzyme. Thermodynamic equilibrium favors higher fructose/glucose ratios with increasing temperature, so enzymes are immobilized to be stabilized for temperatures from 55 °C to 75 °C, the effective upper limit of GI thermostability. These conditions yield 45% fructose/55% glucose syrups; as 55% fructose syrups are desired for soft drinks and food to match the sweetness of sucrose at the same bulk, they have to be concentrated chromatographically. This expensive step could be omitted if the enzyme displayed higher temperature stability at 90–95 °C, where theoretically 55% fructose syrups could be achieved directly.

GI is a tetramer and thermal inactivation proceeds via the tetrameric apoenzyme. The GI tetramer of *Actinoplanes missouriensis* features a lysine at position 253, which is involved in dimer–dimer contact. Teams at Gist Brocades (Delft, The Netherlands) and Plant Genetic Systems (Gent, Belgium) found this lysine to be partly responsible for deactivation because high glucose concentrations caused a chemically driven glycation of Lys253. By exchanging lysine for arginine by site-directed mutagenesis at this position, creating the K253R mutant, a three times lower deactivation constant k_d was achieved as well as a beneficial pH shift from 6.0 to 6.5 (Quax, 1991). It is interesting to note that R252 is naturally present in *Thermus thermophilus* xylose isomerase which displays higher thermostability than the *Actinoplanes* enzyme (Hartley, 2000).

Hartley (2000) et al. review the list of GIs used in processes: in general, the apoenzymes are far less stable than the metal-containing holoenzymes. More tightly bound Co and Mn ions provide more stability than Mg ions. Thermostability of Mg class 1 enzymes appears to be in the order *T. thermophilus* > *Actinoplanes* > *Arthobacter S. olivochromogenes* > *S. rubiginosus*. At high temperatures, the thermal inactivation constant k_d is much higher than metal binding k_{on} and release constants k_{off}, so that the apoenzyme is denatured faster than it can rebind the metal ion, which is regarded as the Achilles' heel of GIs. Owing to its industrial importance, a large amount of protein engineering has been applied to GI. Mutations designed to increase thermostability are listed in Tables 10.2 and 10.3 (Hartley, 2000).

Active-site mutations had no effect on thermostability (H53K, H100F, W136A, E185Q, D244A, D256L, or K294L). Most of these mutants involved in metal ion binding fail to express, which suggests the necessity for the metal ion to be present during the folding process.

Table 10.2 Point mutations in the active site of GI (Hartley, 2000).

Residue	*Structural role*	*Mutants*	*Activity (%wt) (k_{cat}/K_m)*	*Thermostability*
W15	shields C1 and C2	A, L, F	1%, 1%, 11%	
		F, A, R	*30%, 10%, 9%	
A24	near M2	K	52%	
F25	shields C1	K	inactive	
H33c	H1'...H1 insert	F	115%	
H53	ring-opening	K	–	–
		A, N, Q	4%, 3%, 4%	
		S, F	3%, inactive	
		N	3%, 45%	
		Q, N, E, D	10%, 14%, 5%, 9%	~100%
		E, D	*5%, 9%	~100%
D56	H-bond H53	N	37%	
		N, H	60%, 34%	
W87	[Art M] Contact C6 G	F, Y	G/X; 15, 9	
		F, Y	60%, 28%	
		M, L, V, A, K	44%, 52%, 52%, 56%, 28%	
W87/V134	contacts C6 glucose	F/T, F/S	G/X ratio; 25, 52	
F93	contacts F25, W136	S, K	4%, 0.8%	
H100	B3–H3 loop	F	92%	~100%
W136	shields C1, C2	A	–	–
		E; F	inactive; 2%, 4%	
		H	< 0.3%	
V134	contacts C6 G	T, S, A	G/X ratio; 3, 6, 2	
T141	contacts C6 G	S	G/X ratio; 2	
E180	binds M1	A, S	inactive	
		Q	< 0.1%	
		D	0.6 %, 2%	
K182	hydride transfer	S, Q, R	inactive	
		M	inactive	
		Q	inactive	
E185	salt bridge K182	D	25%	
		Q	10%, 45%	Mn^{+}; Mg^{-}
E216	binds M1	D, Q	inactive	
H219	binds M2	N, E, S	< 0.1%	
		Q	1%, 3%	
D244	gbinds M1	A, L	–, –	–, –
D244	binds M1	N	0.1%, 0.3%	
D254	binds M2	A, N	< 1 %	
		E	–	–
D254/D256	M2 ligands	E/E	1.6%	–
D256	binds M2	A	–	–
		S, K; E	inactive; 2%	
		N	0.5%, 10%	
D292	binds M1	A, E, N	inactive	
K294	salt bridge D256	L	–	–
		Q, R	3.6%, 61%	

Table 10.3 Point mutations within GI to increase thermostability (Hartley, 2000).

Residue	Structural role	Mutants	Activity (%wt) (k_{cat}/K_m)	Thermostability
E203/Q204	A-A* salt bridge	D/R	100%	100%
L199/A200	A-A* S-S bridge	C199/C200	100%	100%
A223	A-A* S-S bridge	C	100%	100%
Y253	A-B* S-S bridge	C	reduced 15%	*Mg –10 °C
			oxidized 10%	*Apo –15 °C
W15	shields C1, C2	A	?	< 1%
		F	?	4%
		R	100%	157%
F25	shields C1, C2	H	40%	71%
W87	(*arthrobacter* M87)	F	?	191%
	contacts C6 glucose	M	?	166%
		A	?	143%
F93	contacts F25, W136	K	?	< 1%
W136	shields C1, C2	H	?	< 1%
K252	S.b. A-A*, A-B*	R	100%	130%
K294	S.b. D256, D292	R	100%	78%
A73	H2 helix	S	100%	76%
G70/G74	H2 helix	S/T	100%	95%
A73/G70/G74	H2 helix		100%	114%
K252	S.b. A-A*, A-B*	R	93%	130%
		Q	69%	< 1%
K309	H8 surface variable	R	~100 %	150%
		Q		80%
K319	H9	R	~100%	150%
K323	H9 only in Ami XI	R	~100%	100%
		Q		90%
K309/319/323	triple, H8 and H9	R/R/R	100%	240%

* 'Melting points' for 50% inactivation after 10 min at pH 8.

Class II xylose isomerases exhibit more variety in their thermostability. Comparing two members of this class, enzymes from *Thermoanaerobacterium thermosulfurigens* (TTGI) and *Thermotoga neapolitana* (TNGI) (Sriprapundh, 2000) with 75% mutual identity at the protein level, TNGI is much more thermostable than TTGI. TNGI has additional prolines and fewer asparagine and glutamine residues. Corresponding positions were exchanged by site-directed mutagenesis in TTGI. Only one mutation, Q58P, stabilized the TTGI enzyme, shifting the denaturation temperature from 84 °C (wild type) to 90 °C (mutant Q58P). Further mutations were introduced to shift catalytic efficiency towards glucose instead of xylose, the natural substrate. A V185T mutation in TNGI turned out to be an efficient glucose isomerase with high thermostability: with an activity level at 97 °C of 45 U mg^{-1} towards glucose, it is the most active class II enzyme so far.

10.3.2
Resolving the Reaction Mechanism of Glucose Isomerase (GI): Diffusion-Limited Glucose Isomerase?

Detailed information about this enzyme led to the assumption that the mechanism should be similar to the ones resolved for triose phosphate isomerase (TIM) and mandelate racemase (MR). TIM converts dihydroxyacetone phosphate (DHAP) to glutaraldehyde phosphate (GAP) through isomerization of a single H atom. The reaction follows an enediol intermediate, but the contributing role of the histidine was only resolved after analyzing MR with its 3D structure which is nearly identical to TIM. Mandelate racemase, however, utilizes metal ions in its catalytic site to accomplish the racemization, another isomerization of an H atom (push–pull mechanism) (Bialy, 1993). Resolving the structure of MR using a competitive inhibitor in the active site demonstrated that glutamic acid provides the push (it shows a protonated oxygen near the hydrogen to be transferred) and lysine and histidine in close proximity create a more basic environment for the pull mechanism. This is a good example of how enzymes with different features can exhibit the same mechanism. GI should have had the same mechanism, given its structural and mechanistic similarities to MR and TIM with its classification as an isomerase like TIM and its catalytic metal ion like MR. Surprisingly, however, no enediol intermediate was found, so that the H atom must be transferred differently.

By trying to convert the GI mechanism into a push–pull mechanism as seen in TIM, one could improve reaction kinetics towards the catalytically perfect (optimal k_{cat}/K_M) characteristics of TIM. A commercially important enzyme could thus be catalytically improved by several orders of magnitude. Farber (1989) provided a detailed structure of a GI from *Streptomyces olivochromogenes* (all GI tertiary structures so far are superimposable) which suggested that despite the structural and mechanistic similarities to TIM and MR, the mechanism pointed to a hydride shift involving His219. Further insight into the mechanism (Collyer, 1990) revealed that the mechanism involves eight steps under the participation of both Mg atoms present. One crucial step is the ring opening of the pyranose form, catalyzed by a charge-relay system analogous to chymotrypsin using Glu180, Glu216, Asp244, and Asp292, so that the substrate can be oriented to transfer a C1 hydroxyl to O5 (Hartley, 2000). The second crucial step is isomerization of the substrate via a symmetric C1/C2 anionic transition state induced by translocation of one of the Mg^{2+} metal ions adjacent to each other. Shielding of the ensuing hydride ion situated midway from the solvent is provided by a hydrophobic cluster of an adjacent subunit and is essential for the transition state. This seems to be the rate-limiting step (Rangarajan, 1992). All these experiments were carried out with class I GIs, which are the most important ones used in HFCS production.

HFCS is such an important industrial product that other enzymes involved in its manufacture were also subjected to mutagenesis to increase yield and stability so as to cut production costs subsequently. Glycoamylase (GA) is the second of the three-enzyme series amylase–glycoamylase–glucose isomerase required to convert starch to glucose and fructose. Glycoamylase converts starch dextrin to glucose. The by-products, such as isomaltose, have to be removed chromatographically, which is energy- and cost-intensive. Protein engineering focuses on increasing the glucose yield (affecting enzyme selectivity) and enzyme thermostability.

Recent results on GAs from industrially relevant strains *Aspergillus awamori* and *Aspergillus niger*, through multiple mutations, achieved higher enzyme selectivity in the form of an increased glucose yield, from 96% to 97.5%, theoretically reducing the by-product formation by half. The challenge of increased thermostability was addressed in two ways: the first, strengthening weak bonds, did not turn out to be favorable whereas the second, stiffening α-helices by substituting glycine residues, adding Pro residues, and introducing new disulfide bonds, led to increased thermostability after a series of mutations, including G137A, S436P, N20C, and A27C as single and combined double and triple mutations. The best combination yielded a fourfold increase in thermostability with a 15% increase in activity (Reilly, 1999).

10.4 Enhancement of Stability of Proteases against Oxidation and Thermal Deactivation

10.4.1 Enhancement of Oxidation Stability of Subtilisin

Extracellular proteases are of commercial value and find multiple applications in various industrial sectors. A good number of bacterial alkaline proteases are commercially available, such as *Subtilisin Carlsberg*, subtilisin BPN' and Savinase, with their major application as detergent enzymes.

Fermentation by microorganisms accounts for a two-thirds share of commercial protease production in the world (Kumar, 1999). Depending on their pH optimum and active-site characteristics, microbial proteases are classified as aspartyl proteases (E.C. 3.4.23, acidic), metalloproteases (E.C. 3.4.24, neutral), cysteine or sulfhydryl proteases (E.C. 3.4.22, alkaline), or serine proteases (E.C. 3.4.21, also alkaline) (Kalisz 1988; Rao, 1998). Their commercial uses are listed in Table 10.4.

The serine protease subtilisin cleaves a broad variety of peptide bonds with mainly aromatic or hydrophobic residues such as tyrosine, phenylalanine, methionine, or leucine. The enzyme is used in bulk amounts in laundry detergents because of its broad specificity and high alkaline stability. Besides proteases, laundry detergents contain hydrogen peroxide to support oxidative bleaching of the soiled parts of garments. The oxidative force of hydrogen peroxide deactivates native subtilisin, rendering it a useful target for protein engineering. The cause of deactivation has been found in the oxidation of Met222 to sulfoxide or even sulfone [Eqs. (10.1) and (10.2)]. Met222 is immediately adjacent to the active site (an anion hole on Ser 221) and therefore influences the proteolytic activity of subtilisin. Using *cassette mutagenesis,* scientists from Genentech removed deactivation through site-specific mutagenesis of Met222 into alanine (M222A) or serine (M222S) (Oxender, 1987).

Table 10.4 Commercially utilized proteases.

Supplier	*Product trade name*	*Microbial source*	*Applications*
Novo Nordisk, Denmark	Alcalase	*Bacillus licheniformis*	detergent, silk degumming
	Savinase	*Bacillus* sp.	detergent, textiles
	Esperase	*B. lentus*	detergent, food, silk degumming
	Biofeed pro	*B. licheniformis*	feed
	Durazym	*Bacillus* sp.	detergent
	Novozyme 471 MP	n.s.	photographic gelatin hydrolysis
	Novozyme 243	*B. licheniformis*	denture cleaners
	Nue	*Bacillus* sp.	leather
Genencor Int., USA	Purafact	*B. lentus*	detergent
	Primatan	bacterial source	leather
Gist-Brocades, The Netherlands	Subtilisin	*B. alcalophilus*	detergent
	Maxacal	*Bacillus* sp.	detergent
	Maxatase	*Bacillus* sp.	detergent
Solvay Enzymes, Germany	Opticlean	*B. alcalophilus*	detergent
	Optimase	*B. licheniformis*	detergent
	Maxaperm	protein engineering variant of *Bacillus* sp.	detergent
	HT proteolytic	*B. subtilis*	alcohol, baking, brewing. feed, food, leather
Amano Pharmaceuticals, Japan	Proleather	*Bacillus* sp.	food
	Collagenase	*Clostridium* sp.	technical
	Amano protease S	*Bacillus* sp.	food
Enzyme Development, USA	*Enzeco* alkaline protease	*B. licheniformis*	industrial
	Enzeno alkaline protease L-FG	*B. licheniformis*	food
	Enzeco high alkaline protease	*Bacillus* sp.	industrial
Nagase Biochemicals, Japan	Bioprase concentrate	*B. subtilis*	cosmetics, pharmaceuticals
	Ps. protease	*Pseudomonas aeruginosa*	research
	Ps. elastase	*P. aeroginosa*	research
	Bioprase	*B. subtilis*	detergent, cleaning
	Bioprase SP-10	*B. subitilis*	food
Ghodo Shusei, Japan	Godo-Bap	*B. licheniformis*	detergent, food
Rohm, Germany	Corolase 7089	*B. subtilis*	food
Wuxi Synder Bioproducts, China	Wuxi	*Bacillus* sp.	detergent
Advance Biochemicals, India	Protosol	*Bacillus* sp.	detergent

Source: Gupta (2002). n.s. = not specified

Stability against hydrogen peroxide, expressed as residual activity in 2 M H_2O_2 solution after 1 h, was remarkably improved.

$$-S- + H_2O_2 \longrightarrow -S(=O)- + H_2O \quad (10.1)$$

$$-S(=O)- + H_2O_2 \longrightarrow -S(=O)_2- + H_2O \quad (10.2)$$

10.4.2
Thermostability of Subtilisin

In addition to stabilizing the enzyme against oxidation, enhanced thermostability is also favored in detergents in the laundry business. Using subtilisin BPN′ six individual amino acid substitutions (N218S, G169A, Y217K, M50F, Q206C, N76D) at separate positions in the tertiary structure were found to increase the stability of the enzyme at elevated temperatures (65 °C) and extreme alkalinity (pH 12) (Pantoliano, 1989). Under these denaturing conditions, the rate of deactivation of the sixfold variant was 300 times slower than that of the wild-type subtilisin BPN′. An additional disulfide bond linkage through site-directed mutagenesis between residues Cys61 and Cys98 of subtilisin E resulted in increased thermostability (Takagi, 1990). A single amino acid change, Ser236Cys, in subtilisin E resulted in the forming of an intermolecular disulfide bridge between two subtilisin E molecules and also in enhanced thermostability at 60 °C (Yang, 2000).

10.5
Creating New Enzymes with Protein Engineering

10.5.1
Redesign of a Lactate Dehydrogenase

Lactate dehydrogenase (LDH, E.C. 1.1.1.27) catalyzes the reduction of pyruvate to lactate using NADH as a cofactor:

$$CH_3COCOO^- + NADH \xrightarrow{LDH} CH_3CHOHCOO^- + NAD$$

$$\begin{array}{l} O{=}C\text{-}COO^- + NADH \\ \quad | \\ \quad H_2C\text{-}COO^- \end{array} \xrightarrow{MDH} \begin{array}{l} OH\text{–}C\text{-}COO^- + NAD \\ \quad | \\ \quad H_2C\text{-}COO^- \end{array}$$

Lactate dehydrogenase and malate dehydrogenase (MDH) are structurally related enzymes, both catalyzing redox interconversion of keto and hydroxy acids. Nevertheless, each enzyme is highly specific for its substrate, as revealed by their k_{cat}/K_M values. Comparing the two substrates for each enzyme, the values differ by seven and three orders of magnitude, respectively, for MDH and LDH. Comparison of the primary and tertiary structures of LDHs and MDHs revealed that the potential amino acid targets for site-specific mutagenesis change the substrate binding pocket without affecting the catalytic mechanism, which is identical for the two enzymes.

Using the NAD-dependent lactate dehydrogenase from *Bacillus stearothermophilus*, the Holbrook group (Bristol, UK) exchanged three crucial amino acids within the substrate binding pocket (D197N, T246G, and Q102R); each exchange resulted in a substrate specificity shift from pyruvate towards malate but only one led to a highly specific catalyst for the new substrate (Wilks, 1988).

The first exchange, D197N, decreased the negative charge within the substrate pocket, thus allowing oxaloacetate with its two negative charges (instead of one for lactate) to be bound. The catalytic efficiency was reduced 32-fold against pyruvate and 1.3-fold against oxaloacetate. While selectivity was shifted towards oxaloacetate the change did not produce an effective MDH. The second exchange, T246G, enlarged the substrate pocket to accommodate the larger oxaloacetate in the active site. In a more pronounced shift than above, this modification resulted in a 3200-fold decrease in k_{cat}/K_M for pyruvate but a 1.7-fold increase for oxaloacetate. The greatly reduced K_M value for oxaloacetate indicated that threonine-246 prevents oxaloacetate binding in native LDH.

To finally achieve total reversion of substrate acceptance, computer graphic modeling was applied and revealed that substrate and coenzyme binding induces a rearrangement of protein structure, causing a loop (residues 98–110) to close over the active site. Within that loop the third mutation was chosen, Q102R, as arginine is able to form an ion pair with the 3-carboxyl group of oxaloacetate and is always conserved in known MDHs. The resulting truly effective MDH featured an 8400-fold preference for oxaloacetate over pyruvate and a k_{cat}/K_M value as high for the new MDH as for the native LDH. This is because the rearrangement during substrate binding, and not the hydride ion transfer, is the rate-limiting step. This outstanding work demonstrated not only the usefulness of crystal structures and computer modeling for site-specific mutagenesis, but also the need for thermodynamic and kinetic evaluation, as knowledge about the rate-limiting step within the reaction mechanism opened the route to success.

Likewise, on LDH, Eszes et al. (Eszes, 1996) investigated the influence of a single residue change on substrate inhibition. Excess concentrations of keto acid substrates such as pyruvate inhibit most native hydroxy acid dehydrogenases by forming an enzyme–NAD–ketoacid complex, thus reducing their efficiency as a biocatalyst. Sequence comparisons revealed one unique enzyme from the malarial parasite *Plasmodium falsiparium*, which did not display the mentioned substrate inhibition. It features an amino acid exchange at position 163 from a serine to a leucine, S163L. This serine bonds to a hydroxyl group of the nicotinamide via a water molecule, as derived from LDH crystal structures. Likewise, after exchange of the corresponding serine in human LDH, no substrate inhibition could be detected. The reason for the lack of substrate inhibition seems to be a significantly decreased prior binding of coenzyme to the protein which subsequently also weakens the enzyme–keto acid binding process. The same results were achieved using the LDH from *Lactobacillus delbrueckii*, exchanging H205Q, thus losing an ionic contribution to cofactor binding via the pyrophosphate group (Bernard, 1997).

10.5.2
Synthetic Peroxidases

A more ancestral method of protein engineering before site-directed mutagenesis was employed to design a semi-synthetic peroxidase from subtilisin by chemically modifying the active-site serine to a selenocysteine (Häring, 1999). Incorporating selenium converts the active-site serine-221 into an oxidized version of selenocysteine (Figure 10.5). Thereby the enzyme catalytic mechanism is changed from that of a protease into that of a peroxidase, catalyzing the selective reduction of hydroperoxides in the presence of thiols. Kinetic resolution of racemic hydroperoxides resulted in valuable chiral hydroperoxides, which are of special interest for reactions such as the Sharpless epoxidation of allylic alcohols to epoxy alcohols. X-ray studies revealed that overall structure and substrate binding sites of the native subtilisin and the new seleno-subtilisin are identical. The catalytic efficiency of the selenocysteine peroxidase is on the same order of magnitude as that of the native horseradish peroxidase.

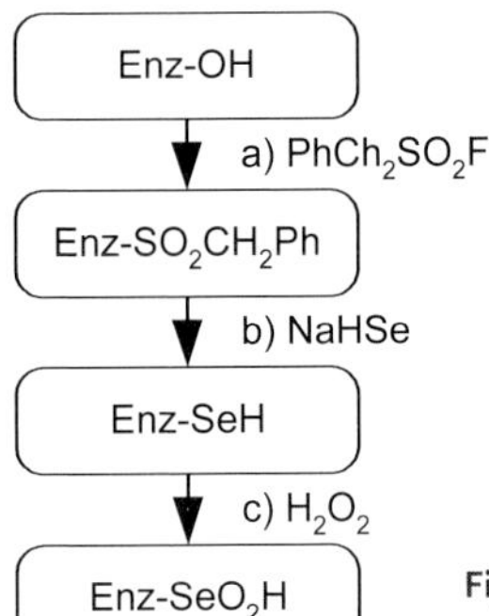

Figure 10.5 Schematic of the chemical approach towards new enzymes using incorporation of selenium.

A similiar approach was performed by van de Velde (1999), using incorporation of vanadate into an acid phosphatase (phytase) to create a semi-synthetic peroxidase similar to the heme-dependent chloroperoxidase. The latter is a useful enzyme for the asymmetric epoxidation of olefins, but less stable due to oxidation of the porphyrin ring and difficult to express outside the native fungal host. The authors exploited the structural similarity of active sites from vanadate-dependent haloperoxidases and acid phosphatases and have shown the useful application as an enantioselective catalyst for the synthesis of chiral sulfoxides (van de Velde, 1999).

10.6 Dehydrogenases, Changing Cofactor Specificity

Dehydrogenases are very valuable enzymes in biocatalysis, although one of their challenges is the use of an additional cofactor, such as NAD(P)(H). There are several advantages to dealing only with dehydrogenases that are dependent on NAD(H) but not on NADP(H):

- the cost of NAD^+ is about one order of magnitude less than for $NADP^+$, independently of scale;
- NAD^+ is more stable than $NADP^+$, especially in the reduced form and at similar pH values (Chenault, 1987, and references therein).

A possible strategy to obtain economical reductases is to redesign NADP(H)-dependent reductases to accept NAD(H). This strategy was demonstrated in the 1990s on the examples of glutathione (Perham, 1991) and isocitrate dehydrogenase (Yaoi, 1996), albeit with limited success with respect to the activity level of the mutants. It turns out that the reverse change, from NAD(H)- to NADP(H)-specificity, is often more successful:

- Lactate dehydrogenase (LDH) from *Bacillus stearothermophilus*, in its I51K:D52S double mutant, exhibited a 56-fold increased specificity to NADPH over the wild-type LDH in a reaction mixture containing 15% methanol. Furthermore, the NADPH turnover number of this mutant was increased almost fourfold compared with wild-type LDH (Holmberg, 1999).
- The best k_{cat}/K_M value for $NADP^+$ shown for the triple mutant D203A–I204R–D210R leucine dehydrogenase (LeuDH) from *Thermoactinomyces intermedius* was 2% of the corresponding value of the wild-type enzyme. The mutant LeuDH preferred $NADP^+$ and features a specific activity of 19 µmol mg^{-1} min^{-1} (Galkin, 1997).

One successful example regarding a specificity change from NADP(H) to NAD(H) was achieved by Riebel and Hummel at the Research Center Jülich (Jülich, Germany); they succeeded in affecting a specificity change on the (*R*)-ADH of *L. brevis* (Hummel, 1999; Riebel, 2002). NAD(P)-specificity is conferred by a single residue in position 37: an Asp or Glu shifts the specificity to NAD, whereas a small uncharged residue such as Gly next to an Arg leads to preference for NADP over NAD. However, while the specificity now favors NAD by a factor of 300, the rate went down by a factor of 4. The properties of the wild-type enzyme and two mutants are listed in Table 10.5.

Mutein 2 was created using site-specific mutagenesis after evaluation of the primary structure. Crystals of the recombinant enzyme and the subsequent determination of the 3D structure were available later (Niefind, 2000) and resulted in the introduction of the single mutation of G37D. Figure 10.6 shows the superimposition of an NAD-dependent 3α,20β-hydroxysteroid dehydrogenase with the NADP-

Table 10.5 Comparison of the properties of wild-type and mutant (*R*)-alcohol dehydrogenase from *L. brevis*.

Property	*Wild type*	*Mutein 2 (R38L/K48M/A9G)*	*Mutein G37D*
pH opt., reduction	6.5	6.5	5.5
pH opt., oxidation	8.0	6.5	6.5
pH stability, 24 h	4.5–9.0 (70%)	5.5–8.5 (70%)	6.5–8.5 (80%)
Temp. opt. [°C]	55	50	40
Temp. stab.30 °C	150 h	16.5 h	148 h
Temp. stab.42 °C	7.15 h	0.19 h	257 h
K_M NAD [mM]	2.94	0.77	0.89
K_M NADP [mM]	0.24	0.11	14.04
V_{max} NAD [nmol mL^{-1} s^{-1}]	467	439	236
V_{max} NADP [nmol mL^{-1} s^{-1}]	1420	623	402
k_{cat} NAD [s^{-1}]	21.4	33.11	34.57
k_{cat} NADP [s^{-1}]	65.2	46.98	58.88
k_{cat}/K_M NAD [s^{-1} mM^{-1}]	7.3	43	38.84
k_{cat}/K_M NADP [s^{-1} mM^{-1}]	270	427	4
NAD/NADP	0.03 : 1	0.1 : 1	9 : 1

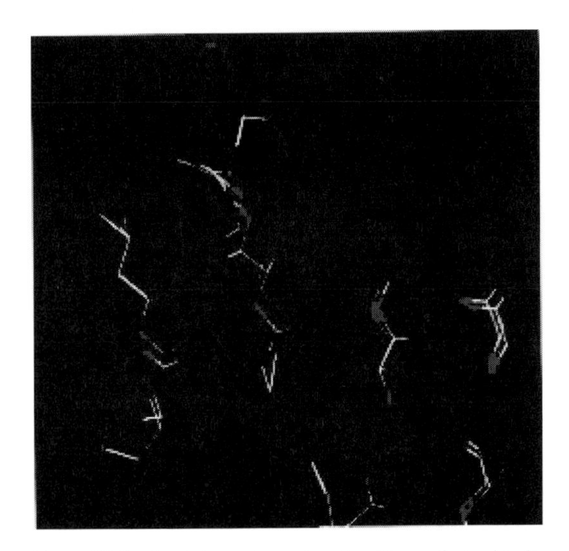

Figure 10.6 Crystal structure of parts of the cofactor binding pocket of the (*R*)-alcohol dehydrogenase from *L. brevis*.

dependent alcohol dehydrogenase from *L. brevis*. As is clearly seen, the NAD-dependent enzyme has a glutamic acid at the same position of G37 in the ADH, which blocks the NADP-pocket and favors the smaller NAD. Moreover, the additional negative charge of glutamic acid favors NADH over the negatively charged NADPH.

Comparing several dehydrogenases at the amino acid sequence level, it is evident that there is always a small amino acid such as glycine next to the cofactor binding site in NADP-dependent enzymes, which has to be exchanged for a negatively charged amino acid, such as occurs regularly in NAD-dependent dehydrogenases, to change the substrate specificity from $NAD(P)^+$ to NAD^+-dependent dehydrogenases.

A similar approach resulting in a cofactor exchange from NAD to NADP was conducted using the D-lactate dehydrogenase from *Lactobacillus delbrueckii*. Here, a single exchange from D175 to A resulted in the NADP-dependent enzyme (Bernard, 1995). The D175A mutant displays a 40-fold preference for NADP over NAD. As above, the crucial residue is located 17 amino acids from the GXGXXG consensus sequence. Strikingly, this represents the exact opposite to the example above, exchanging a negatively charged amino acid for a small neutral residue within the same distance, truly emphasizing the importance of this residue for discrimination between the two coenzymes.

Another example of cofactor exchange was performed using a mammalian cytochrome P450, changing the specificity from NADP to NAD, which is discussed in detail Section 10.7.

10.7 Oxygenases

Oxygenases are an important class of enzymes because of their ability to perform difficult chemistry with high enantioselectivity. They are regarded as a priority target for engineering because of their poor stability and often low catalytic rates. They comprise several protein families, introducing either one (monooxygenases) or two (dioxygenases) oxygen atoms into their substrates. Sometimes special peroxidases, which exhibit selective monooxygenase activity, are included in this group; an example is the chloroperoxidase from *Caldariomyces fumago*. A good review on work with oxygenases is given by Cirino (2002). Some examples of valuable catalyzed reactions are given in Figure 10.7.

Hydroxylation and epoxidation reactions performed by monooxygenases are of particular interest, the cytochrome P450 monooxygenases representing a superfamily that catalyzes such reactions.

Cytochrome P450 BM-3 enzyme from *Bacillus megaterium* hydroxylates long-chain fatty acids and contains the reductase subunit within the same enzyme. It is expressed in soluble form in *E. coli* and represents a good target for protein engineering. The residue F87 plays an important role in determining regioselectivity in the substrates. Mutation of this amino acid to alanine, F87A, not only shifts

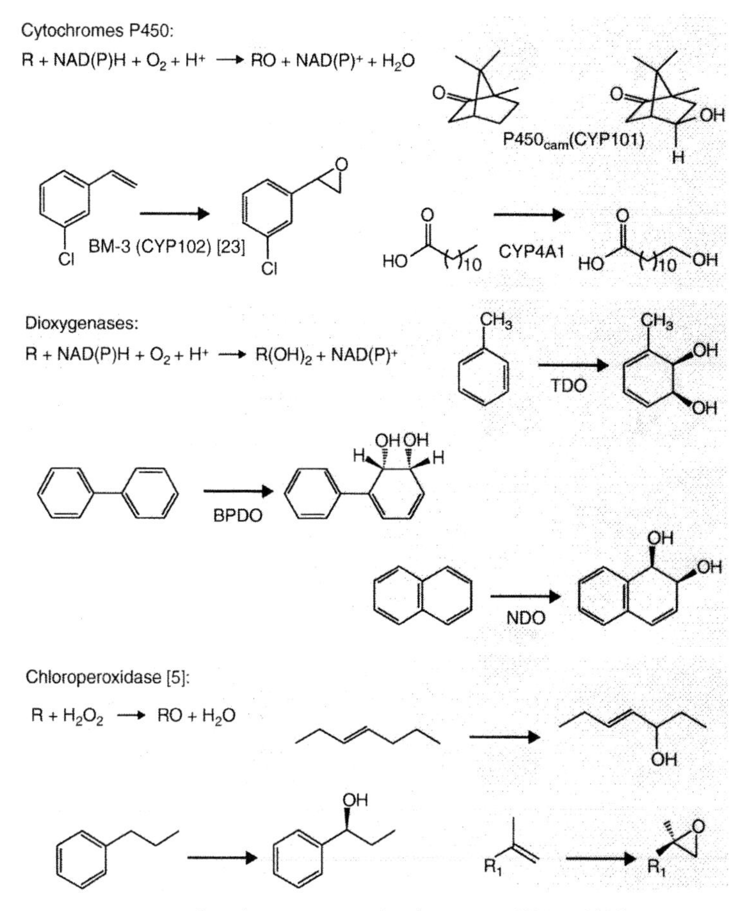

Figure 10.7 Examples of oxygenase-catalyzed reactions (Cirino, 2002).

hydroxylation towards the ω-position but is also able to accept electrons directly from a chemical source, thereby rendering the cofactor NADPH unneccessary, which is often a significant cost factor (Schwaneberg, 2000). The chemical source cobalt(iii) sepulchrate could be regenerated using zinc dust. The mutation F87G changed stereoselectivity upon epoxidation of 3-chlorostyrene to (*R*)-(+)-3-chlorostyrene oxide from –61% e.e. (wild type) to 94.6% e.e. for the mutant (Li, 2001). This is a good example to show that the task of changing stereoselectivity can be done but does depend on the substrate, not just on the enzyme.

Another cytochrome P450 system was used as a target for site-specific mutagenesis to change cofactor specificity. As mentioned in Section 10.5, a single mutation was sufficient to change the cofactor preference from NADP(H) to NAD(H). Human NADPH cytochrome P450 reductase, the redox partner for the monooxygenase, was altered at W676A to create an NAD-dependent enzyme that has a 1000-fold higher specificity for NADH over NADPH. In this mutant, NADPH was a potent inhibitor of cytochrome reduction. The K_M value for NADH was lowered from 47.6 mM to 0.32 mM and the k_{cat}/K_M value was raised from 0.06 $(\mu M\ min)^{-1}$ to 11.8 $(\mu M\ min)^{-1}$ (Döhr, 2001).

Excellent examples of protein engineering work on dioxygenases are given in Section 10.8, changing the regio- and stereoselectivity on Rieske non-heme iron oxygenases.

10.8
Change of Enantioselectivity with Site-Specific Mutagenesis

While enzymes have been widely accepted as useful catalysts, especially for kinetic resolution of racemates and asymmetrical reaction with prochiral or meso compounds, only one of the two product enantiomers is accessible for the vast majority of reactions. In contrast to chemical catalysts, where a catalyst with the opposite enantioselectivity can often be designed just as easily as the one for the original situation, enzymes are composed of L-amino acid building blocks only and thus do not have a mirror-image counterpart enzyme molecule. In most cases, enzymes with the opposite enantioselectivity have not been found. Utilizing site-specific mutagenesis or directed evolution (Chapter 11) to create the opposite stereoselectivity in a given enzyme with interesting substrate specificity would thus be highly relevant. Here we present two successful examples of a change of enantioselectivity; in both cases, templates for the enantioselectivities in both directions were available.

Lipase

Lipase AH (Amano, Najoya, Japan) from *Pseudomonas* sp. and Lipase PS (Amano, Najoya, Japan) from *Pseudomonas cepacia* in particular have been demonstrated to be useful for production, especially for enantiomerically pure secondary alcohols in organic media (Hirose, 1995) (Chapter 12, Section 12.6). Despite almost identical amino acid sequences in the two enzymes (their proteins differ only in 16 amino acid residues), these two lipases exhibit opposite enantioselectivity. As no crystal structure of these enzymes was available, all experiments were based on knowledge of the primary sequence.

The authors prepared 16 single mutants of lipase PS covering all the 16 amino acid positions differing between lipases PS and AH. A check of the enantioselectivity revealed that three mutants had slightly decreased selectivities towards the (*R*)-enantiomer of the target compound, nifedipine. Combining these three single mutants to yield double and triple mutants, however, it was found that the triple mutant F221L/V266L/L287I, called FVL, displayed 100% selectivity towards the (*S*)-enantiomer. Moreover, in the case of lipase PS, but not AH, the results were usually found to depend heavily on the organic solvent used. The triple mutant, however, was found not to yield solvent-dependent results, indicating true conversion of lipase PS to the desired lipase AH. Only three amino acid exchanges are necessary to accomplish a changeover of enantioselectivity.

Naphthalene dioxygenase (NDO)

The other example uses naphthalene dioxygenase (NDO) as a template. NDO from *Pseudomonas* sp. is a multi-component enzyme that catalyzes dioxygenation of naphthalene using an iron–sulfur flavoprotein reductase and a Rieske ferredoxin to shuttle electrons from NADH to the oxygenase component. NDO is an a_3b_3 hexamer with a solved crystal structure (Kauppi, 1998). The a subunit shows 84% amino acid identity with 2-nitrotoluene-2,3-dioxygenase from *Pseudomonas* sp. (2NTDO), which degrades 2-nitrotoluene.

The authors used both NDO and 2NTDO as templates; the former produces enantiomerically pure (+)-naphthalene *cis*-1,2-dihydrodiol from naphthalene whereas the latter forms a mixture of 70% (+)/30% (–). The 3D structure of NDO was analyzed to determine crucial amino acids in the substrate pocket which should influence substrate specificity. Of 17 amino acids lining the substrate pocket, five residues inside the pocket were picked (A206, L253, V260, H295, F352) to be changed to the corresponding amino acids in the 2NTDO enzyme. Among the 14 mutants created, which contained between one and five mutations, the mutations F352I and A206I were found to be crucial for enantioselectivity, decreasing the e.e. value of the favored (+)-enantiomer from 90% to 76% and 57%, respectively (Yu, 2001). Reactions with different substrates corroborated the result. The double mutant A206I/F352I increased the enantioselectivity for naphthalene *cis*-1,2-dihydrodiol towards the (–)-enantiomer from 6% to 30%. A triple mutant adding the mutation H295I yielded the best result with 37% e.e. in favor of the (–)-enantiomer, identical to the e.e. value for 2NTDO. As a result, exchange of again just three amino acids accomplished the changeover of enantioselectivity from that of NDO to that observed in the similar 2NTDO.

10.9 Techniques Bridging Different Protein Engineering Techniques

10.9.1 Chemically Modified Mutants, a Marriage of Chemical Modification and Protein Engineering

Wild-type subtilisin from *Bacillus lentus* (SBL) shows a marked preference for substrates with large hydrophobic residues in the P1-position, such as in the standard subtilisin substrate, succinyl-AAPF-*p*-nitroanilide (A: alanine, P: proline, F: phenylalanine). To confer more relaxed P1-specificity, a strategy of chemical modification in combination with site-directed mutagenesis was applied to create chemically modified mutants (CMMs) (DeSantis, 1999). The completely specific thiol-modifying reaction of methanethiosulfonate can be exploited to introduce positive or negative charges, one or several (≤ 3), in place of thiol residues of cysteine. Thus, site-directed mutagenesis of a chosen site to cysteine with subsequent chemical modification allows a broad range of new protein surface environments to be designed. SBL is an ideal protein for such an approach as it does not contain any natural cysteines.

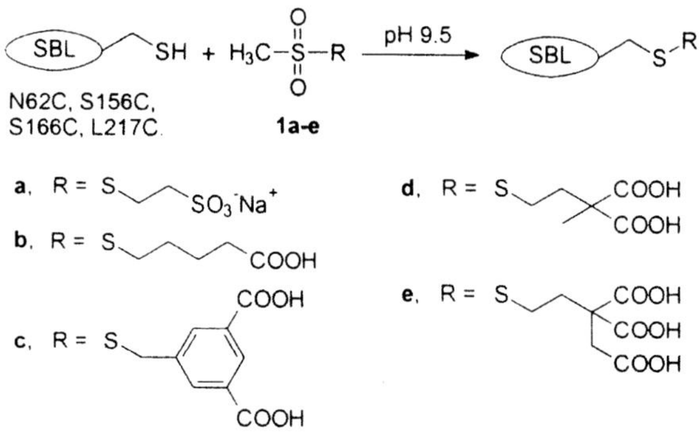

Figure 10.8 Methanethiosulfonates for CMMs (Davis, 1999).

To create an S1-pocket providing a better fit for small P1 side chains, a large cyclohexyl group was introduced by the CMM approach at position S166C with the aim of partially filling up the S1-pocket. As expected, the S166C-S-CH_2-*c*-C_6H_{11} CMM thus created showed twofold improvement in k_{cat}/K_M with the succinyl-AAPA-*p*-nitroanilide substrate (A: alanine). A series of mono-, di-, and trinegatively charged CMMs were generated from S166C; all showed improved k_{cat}/K_Ms with the positively charged P1-residue containing substrate, succinyl-AAPR-*p*-nitroanilide (R: arginine). The specificity ratios k_{cat}/K_M improved almost linearly with the number of negative charges on the S166C-R side chain, up to a ninefold increase for the succinyl-AAPA-*p*-nitroanilide substrate. Conversely, the positively charged S166C-S-$CH_2CH_2NH_3^+$ CMM generated showed a 19-fold improvement in k_{cat}/K_M for the succinyl-AAPE–*p*-nitroanilide substrate (E: glutamic acid).

A series of mono-, di- and triacidic acid methanethiosulfonates were synthesized and used to modify cysteine mutants of SBL at positions 62 in the S2 site, 156 and 166 in the S1 site and 217 in the S1′ site (Davis, 1999) (Figure 10.8). Investigation of the kinetic parameters of the CMMs at pH 8.6, ensuring complete ionization of all the unnatural amino acid side chains introduced, revealed an increased K_M value and decreased k_{cat} with every unit of negative charge. The reason is possibly interference with oxyanion stabilization of the transition state of the catalyzed reactions. Surprisingly, the biggest decrease is only 11-fold, indicating that SBL tolerates high levels of charge at single sites without drastic attenuation of catalytic efficiency.

10.9.2
Expansion of Substrate Specificity with Protein Engineering and Directed Evolution

To expand the substrate specificity and stereoselectivity of the aldolase DERA (2-deoxyribose-5-phosphate aldolase, E.C. 4.1.2.4), both site-specific mutagenesis and random mutagenesis have been investigated (DeSantis, 2003). The goal was to extend substrate specificity to the unnatural non-phosphorylated substrate, D-2-deoxyribose.

Five amino acid residues were picked for mutation, based on the 3D crystal structure of DERA complexed with its natural substrate.

The S238D variant was successful in three different ways:

- it exhibited a 2.5-fold improvement over the wild-type enzyme in the retroaldol reaction of 2-deoxyribose'
- it reacts with 2-methyl-substituted aldehydes and non-phosphorylated 2-deoxyribose as substrates in an aldol reaction with inversion in enantioselectivity; the product was used in the asymmetric synthesis of a deoxypyranose as a new effective synthon for the total synthesis of epothilones, anti-cancer compounds (Figure 10.9); and
- it accepts 3-azidopropynaldehyde as a substrate, unlike the wild-type enzyme, in a sequential asymmetric aldol reaction to form a deoxyazidoethylpyranose, which is a precursor to the corresponding lactone and the cholesterol-lowering agent Lipitor (Chapter 13, Section 13.3.2) (Figure 10.9).

Lastly, with the goal of further testing enhancement of substrate specificity and reaction rate on a larger scale with respect to the number of samples, a novel engineered *E. coli* strain has been developed, termed SELECT (Δace, adhC, DE3), which requires acetaldehyde as C-source for growth and maintenance. The scheme constitutes one of the first cases where in-vivo selection has identified unnatural enzyme specificity.

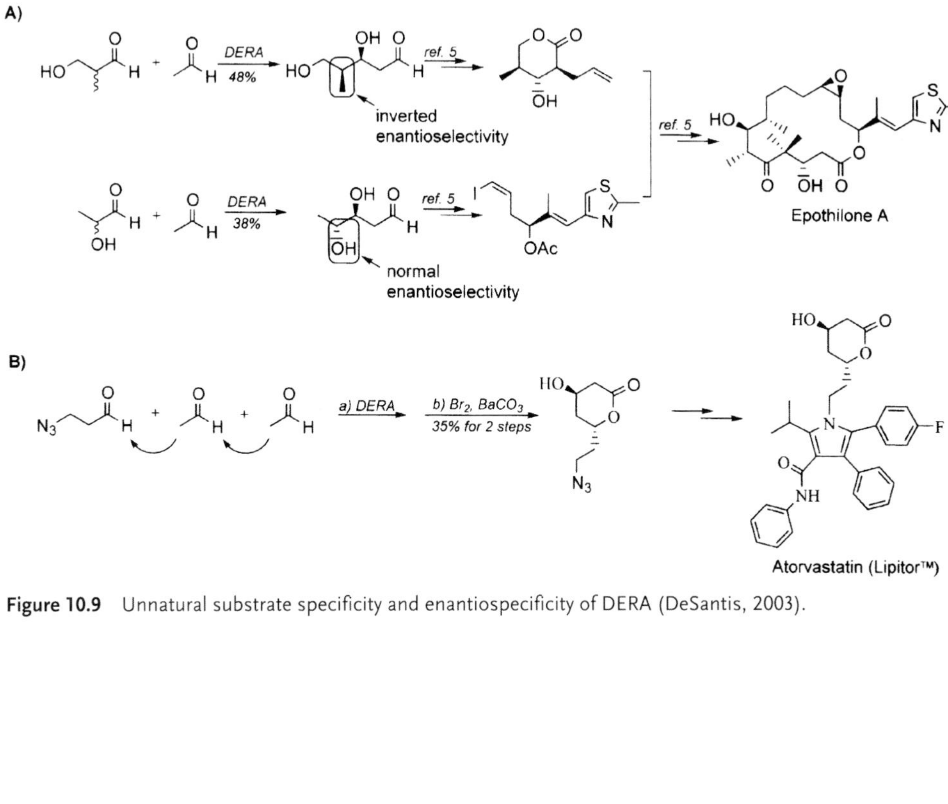

Figure 10.9 Unnatural substrate specificity and enantiospecificity of DERA (DeSantis, 2003).

Suggested Further Reading

J. L. Cleland and C. S. Craik, *Protein Engineering: Principles and Practice*, Wiley–Liss, New York, **1996**.

D. L. Oxender and C. F. Fox (eds.), *Protein Engineering*, Alan R. Liss, New York, **1987**.

References

N. Bernard, K. Johnson, J. J. Holbrook, and J. Delcour, D175 discriminates between NADH and NADPH in the coenzyme binding site of *Lactobacillus delbrueckii* subsp. *bulgaricus* D-lactate dehydrogenase, *Biochem.Biophys.Res. Commun.* **1995**, *208*, 895–900.

N. Bernard, K. Johnson, J. L. Gelpi, J. A. Alvarez, T. Ferain, D. Garmyn, P. Hols, A. Cortes, A. R. Clarke, J. J. Holbrook, and J. Delcour, D-2-Hydroxy-4-methylvalerate dehydrogenase from *Lactobacillus delbrueckii* subsp. *bulgaricus II.* Mutagenic analysis of catalytically important residues, *Eur. J. Biochem.* **1997**, *244*, 213–219.

W. Bi and P. J. Stambrook, Site-directed mutagenesis by combined chain reaction, *Anal. Biochem.* **1998**, *256*, 137–140.

H. Bialy, Protein engineering succeeds, *Bio/Technology* **1993**, *11*, 278–279.

H. K. Chenault and G. M. Whitesides, Regeneration of nicotinamide cofactors for use in organic synthesis, *Appl Biochem Biotechnol.* **1987**, *14*, 147–197.

P. C. Cirino and F. H. Arnold, Protein engineering of oxygenases for biocatalysts, *Curr. Opin. Chem. Biol.* **2002**, *6*, 130–135.

C. A. Collyer, K. Henrick, and D. M. Blow, Mechanism for aldose–ketose interconversion by D-xylose isomerase involving ring opening followed by a 1,2-hydride shift, *J.Mol.Biol.* **1990**, *212*, 211–235.

B. G. Davis, X. Shang, G. DeSantis, R. R. Bott, and J. B. Jones, The controlled introduction of multiple negative charge at single amino acid sites in subtilisin *Bacillus lentus, Bioorg. Med. Chem.* **1999**, *7*, 2293–2301.

G. DeSantis, X. Shang, and J. B. Jones, Toward tailoring the specificity of the S1 pocket of subtilisin *B. lentus*: chemical modification of mutant enzymes as a strategy for removing specificity limitations, *Biochemistry* **1999**, *38*, 13391–13397.

G. DeSantis, J. Liu, D. P. Clark, A. Heine, I. A. Wilson, and C. H. Wong, Structure-based mutagenesis approaches toward expanding the substrate specificity of D-2-deoxyribose-5-phosphate aldolase, *Bioorg. Med. Chem.* **2003**, *11*, 43–52.

O. Döhr, M. J. I. Paine, T. Friedberg, G. C. K. Roberts, and C. R. Wolf, Engineering of a functional human NADH-dependent cytochrome P450 system, *Proc. Natl. Acad. Sci. USA* **2001**, *98*, 81–86.

C. M. Eszes, R. B. Sessions, A. R. Clarke, K. M. Moreton, and J. J. Holbrook, Removal of substrate inhibition in a lactate dehydrogenase from human muscle by a single residue, *FEBS Lett.* **1996**, *399*, 193–197.

G. K. Farber, A. Glasfield, G. Tiraby, D. Ringe, and G. A. Petsko, Crystallographic studies of the mechanism of xylose isomerase, *Biochemistry* **1989**, 28, 7297–7299.

A. Galkin, L. Kulakova, T. Ohshima, N. Esaki, and K. Soda, Construction of a new leucine dehydrogenase with preferred specificity for $NADP^+$ by site-directed mutagenesis of the strictly NAD^+-specific enzyme, *Protein Eng.* **1997**, *10(6)*, 687–690.

R. Gupta, Q. K. Beg, and P. Lorenz, Bacterial alkaline proteases: molecular approaches and industrial applications, *Appl Microbiol Biotechnol.* **2002**, *59*, 15–32.

D. Häring and P. Schreier, Chemical engineering of enzymes: altered catalytic activity, predictable selectivity and exceptional stability of the semisynthetic peroxidase seleno-subtilisin, *Naturwissenschaften* **1999**, *86*, 307–312.

B. S. Hartley, N. Hanlon, R. J. Jackson, and M. Rangarajan, Glucose isomerase: insights into protein engineering for increased thermostability, *Biochim. Biophys. Acta* **2000**, *1543*: 294–335.

Y. Hirose, K. Kariya, Y. Nakanishi, Y. Kurono, and K. Achiwa, Inversion of enantioselectivity in hydrolysis of 1,4-dihydropyridines by point mutation of lipase PS, *Tetrahedron Lett.* **1995**, *36*, 1063–1066.

N. Holmberg, U. Ryde, and L.Bulow, Redesign of the coenzyme specificity in L-lactate dehydrogenase from *Bacillus stearothermophilus* using site-directed mutagenesis and media engineering, *Protein Eng.* **1999**, *12*, 851–856.

W. Hummel and B. Riebel, Dehydrogenase mutants with improved NAD-dependence, their manufacture, with recombinant microorganisms, and their use for chiral hydroxy compound preparation, PCT Int. Appl. WO 9947684 A2, **1999**.

H. M. Kalisz, Microbial proteinases, *Adv Biochem Eng Biotechnol.* **1988**, *36*, 1–65.

B. Kauppi, K. Lee, E. Carredano, R. E. Parales, D. T. Gibson, H. Eklund, and S. Ramaswamy, Structure of an aromatic ring-hydroxyl-dioxygenase–naphthalene-1,2-dioxygenase, *Structure* **1998**, *6*, 571–575.

C. G. Kumar and H. Tagaki, Microbial alkaline proteases: from a bioindustrial viewpoint, *Biotechnol. Adv.* **1999**, 17, 561–594.

T. A. Kunkel, Rapid and efficient site-specific mutagenesis without phenotypic selection, *Proc. Natl. Acad. Sci. USA* **1985**, *82*, 488–492.

Q. S. Li, J. Ogawa, R. D. Schmid, S. Shimizu, Residue size at position 87 of cytochrome P450BM-3 determines its stereoselectivity in propylbenzene and 3-chlorostyrene oxidation, *FEBS Lett.* **2001**, *508*, 249–252.

K. Niefind, B. Riebel, J. Mueller, W. Hummel, and D. Schomburg, Crystallization and preliminary characterization of crystals of *R*-alcohol dehydrogenase from *Lactobacillus brevis, Acta Crystallogr D Biol Crystallogr.* **2000**, *56*, 1696–1698.

K. Niefind, J. Mueller, B. Riebel, W. Hummel, and D. Schornburg, The crystal structure of *R*-specific alcohol dehydrogenase from Lactobacillus brevis suggests the structural basis of its metal dependency, *J. Mol. Biol.* **2003**, *327*, 317–328.

M. Niggemann and B. Steipe, Exploring local and non-local interactions for protein stability by structural motif engineering, *J. Mol. Biol.* **2000**, *296*, 181–195.

M. Ostermeier and S. J. Benkovic, Evolution of protein function by domain swapping, *Adv. Protein Chem.* **2000**, *55*, 29–77.

D. L. Oxender and C. F. Fox (eds.), *Protein Engineering*, Alan R. Liss, New York, **1987**, Chapter 25, pp. 279–287.

M. W. Pantoliano, M. Whitlow, J. F. Wood, S. W. Dodd, K. D. Hardman, M. L. Rollence and P. N. Bryan, Large increases in general stability for subtilisin BPN′ through incremental changes in the free energy of unfolding, *Biochemistry* **1989**, *28*, 7205–7213.

R. N. Perham, N. S. Scrutton, and A. Berry, New enzymes for old: redesigning the coenzyme and substrate specificities of glutathione reductase, *Bioessays* **1991**, *13*, 515–525.

W. J. Quax, N. T. Mrabet, R. G. M. Luiten, P. W. Schuurhuizen, P. Stanssens, and I. Lasters, Enhancing the thermostability of glucose isomerase by protein engineering, *Bio/Technology* **1991**, *9*, 738–742.

M. Rangarajan and B. S. Hartley, Mechanism of D-fructose isomerization by Arthrobacter D-xylose isomerase, *Biochem. J.* **1992**, *283*, 223–233.

M. B. Rao, A. M. Tanksale, M. S. Ghatge, and V. V. Deshpande, Molecular and biotechnological aspects of microbial proteases, *Microbiol. Mol. Biol. Rev.* **1998**, *62*, 597–635.

P. J. Reilly, Protein engineering of glycoamylase to improve industrial performance – a review, *Starch* **1999**, *51*, 269–274.

B. Riebel, W. Hummel, and A. Bommarius, Dehydrogenase mutants capable of using NAD as coenzyme and their preparation and use for chiral hydroxy compound preparation, Eur. Pat. Appl. EP 1176203 A1, **2002**.

U. Schwaneberg, C. Schmidt-Dannert, J. Schmitt, and R. D. Schmid, A continuous spectrophotometric assay for P450 BM-3, a fatty acid hydroxylating enzyme, and its mutant F87A, *Anal Biochem.* **1999**, *269*, 359–366.

D. Sriprapundh, C. Vielle, and J. G. Zeikus, Molecular determinants of xylose isomerase thermal stability and activity: analysis of thermoenzymes by site-directed mutagenesis, *Prot. Eng.* **2000**, *13*, 259–265.

H. Takagi, T. Takahashi, H. Momose, M. Inouye, Y. Maeda, H. Matsuzawa, and T. Ohta, Enhancement of the thermostability of subtilisin E by introduction of a disulfide bond engineered on the basis of structural comparison with a thermophilic serine protease, *J. Biol. Chem.* **1990**, *265*, 6874–6878.

F. van de Velde, L. Könemann, F. van Rantwijk, and R. A. Sheldon, The rational design of semisynthetic peroxidases, *Biotechnol. Bioeng.* **1999**, *67*, 87–96.

H. M. Wilks, K. W. Hart, R. Feeney, C. R. Dunn, H. Muirhead., W. N. Chia, D. A. Barstow, T. Atkinson, A. R. Clarke, and J. J. Holbrook, A specific, highly active malate deyhdrogenase by redesign of a lactate dehydrogenase framework, *Science* **1988**, *242*, 1541–1544.

Y. H. Yang, Y. J. Wu, L. Jiang, L. Q. Zhu, and S. L. Yang, A mutant subtilisin E with enhanced thermostability, *World J. Microbiol. Biotechnol.* **2000**, *16*, 249–251.

T. Yaoi, K. Miyazaki, T. Oshima, Y. Komukai, and M. Go, Conversion of the coenzyme specificity of isocitrate dehydrogenase by module replacement, *J. Biochem. (Tokyo)* **1996**, *119*, 1014–1018.

C.-L. Yu, R. E. Parales, and D. T. Gibson, Multiple mutations at the active site of naphthalene dioxygenase affect regio-selectivity and enantioselectivity, *J. Ind. Microbiol. Biotechnol.* **2001**, *27*, 94–103.

11
Applications of Recombinant DNA Technology: Directed Evolution

Summary

A protein with an average size of about 250 amino acids can form 20^{250} linear combinations, and thus cover 4×10^{287} points in sequence space. Even if only a very small fraction, 1 in 10^{10}, can form a 3D structure, more protein structures would exist than atoms in the universe. As there are only about 10 000 protein structures, nature so far could only probe a small fraction of the possible structures, which opens up opportunities to achieve success in improving enzymes in the test tube.

A breakthrough in recombinant DNA technology and protein engineering was achieved by recognizing that the process of natural selection can be harnessed to evolve effective enzymes in artificial circumstances. In this framework of "directed evolution", the processes of natural evolution for selecting proteins with the desired properties are accelerated in a test tube. The starting point is an enzyme with a measurable desired activity which still has to be improved.

An experimental cycle of directed evolution consists of the following steps:

1. A library of parent DNA sequences encoding for the desired protein is chosen. Sequence diversity is created or increased through a *mutagenesis* step, either by introduction of random point mutations through error-prone PCR or by recombination of DNA fragments such as DNA shuffling or RACHITT.
2. The DNA sequences created are ligated into an expression vector, they are transformed into host cells, and the protein is expressed. Mutagenesis and fragmentation render all but a few sequences inactive.
3. A *screening* or *selection* procedure is employed next to isolate the few transformants containing the encoded sequences for active enzymes or functional proteins.
4. These selected sequences are then *amplified* and the cycle of mutagenesis, screening, and amplification is repeated until proteins with the desired property or function are found.

The realization that one could screen or select beneficial point mutations or recombinations and discard deleterious ones, thus mimicking sexual recombination, spawned the field of directed evolution.

Branched-chain aminotransferases (BCATs) evolved from aspartate aminotransferases (AATs) showed a record 10^5- to 2×10^6-fold improvement in catalytic efficiency (k_{cat}/K_M). Not only were the 13–17 amino acid substitutions concentrated in the most active mutants, but all but one mutated amino acid residues are located far from the active site. With directed evolution, enantioselectivities can be improved on enantiounspecific enzymes (from $E = 1.1$ to 25.8) and even inverted to yield the opposite enantiomer in comparison to the wild type (40% D- to both 90% D- and 20% L-).

Work on gaining antibiotic resistance demonstrated the power of multiple gene-shuffling over single gene-shuffling, especially if the maximum possible sequence space has to be covered. Combining different directed evolution methods yields improved results. Initially beneficial mutations in early generations might be detrimental for performance in subsequent generations, so all winners should be investigated independently. A combined approach of rational pathway assembly and directed evolution techniques opens the perspective for discovery and production of compounds that are either rare and hard to access in nature or even entirely non-natural, in simple laboratory organisms.

11.1 Background of Evolvability of Proteins

11.1.1 Purpose of Directed Evolution

Natural selection has created a large variety of enzymes superbly adapted to catalyze an array of chemical reactions. However, there are still too few enzymes available for catalyzing reactions of interest and many of these have sub-optimal properties for processing conditions. Even in the case of processing conditions amenable to enzymes and cofactors, a process designed around naturally occurring enzymes is likely to be suboptimal (Burton, 2002). Therefore, tailoring the enzyme to the processing conditions would be preferred (Figure 11.1). Recent advances in random mutagenesis and directed evolution offer the possibility of achieving this goal, even in the absence of detailed structural information on the protein.

Custom tailoring of enzyme characteristics has been a long-sought goal, especially since the advent of recombinant DNA and site-directed mutagenesis methods (Ling, 1997; Duncan, 1997). While the discovery of the structure of DNA and the molecular basis of heredity enhanced understanding of the phenomena of mutation and evolution, the process of inducing mutations and selecting improved variants was still a matter of chance using mutagenic agents such as NTG, UV light, and X-rays. The advent of recombinant DNA in 1979 and site-directed mutagenesis methods have enabled the direct control and amplification of the mutations of the enzyme of interest. This endeavor, known as "protein engineering" and rational design, discussed in Chapter 10, has had several notable successes (Graycar, 1992). However, research quickly showed that the understanding of the

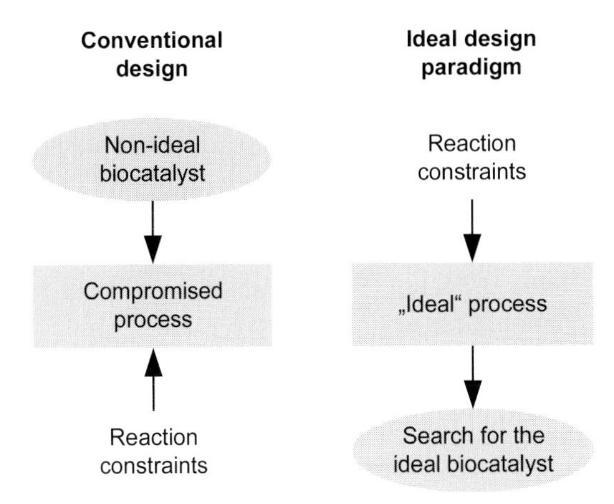

Figure 11.1 Paradigm shift in design of biocatalysis reactions (Burton, 2002).

structure–function relationship in enzymes was insufficient to engineer enzymes with desired characteristics easily. In addition, there still does not exist any detailed structural information for many enzymes of interest in biocatalysis.

More recently, in the 1990s, a breakthrough was achieved by recognizing that the process of natural selection can be harnessed to evolve effective enzymes in artificial circumstances. This framework of *"directed evolution"* was pioneered by Stemmer (Stemmer, 1994a) and Arnold (You, 1996). In directed evolution, the processes of natural evolution are accelerated in a test tube for selecting proteins with the desired properties. Whereas other researchers had previously shown that PCR was capable of producing chimeric DNA molecules (Paabo, 1990), it has only been fairly recently that this property has been exploited to evolve new enzymes. Remarkable success in directed evolution has been reported, ranging from industrial enzymes with substantially improved activities and thermal stabilities to vaccines and pharmaceuticals. These accomplishments mark the onset of enormous possibilities for future uses of directed evolution in basic research, for understanding protein function and for creating new biocatalysts.

11.1.2 Evolution and Probability

Considering the availability of the 20 proteinogenic amino acids, a protein of an average size of about 250 amino acids can form 20^{250} possibilities of linear combinations, and thus cover 20^{250} or 4×10^{287} points in *sequence space*. Even if only a very small fraction, such as 1 in 10^{10} can form a three-dimensional structure, there would exist more protein 3D structures than atoms in the universe. As this clearly is not the case, given the just over 10 000 protein structures (Chapter 14, Section 14.3), nature so far could only probe a small fraction of the possible structures, which opens up opportunities to achieve success in improving enzymes in the test tube. The starting point is an enzyme with a measurable desired activity which still has to be improved.

As there are 19 possibilities for the introduction of proteinogenic amino acids besides the one contained in the wild-type protein, the number of possible variants of a protein obtained by introducing M substitutions over the length of the protein sequence of N amino acids is given by Eq. (11.1).

$$\# = 19^M \, [N!/\{(N-M)!M!\}] \tag{11.1}$$

The number of variants as a function of the number of mutated amino acids M and the length of the protein sequence N is given in Table 11.1.

Table 11.1 Number of possible variants as a function of the length of protein sequence length N.

No. of substitutions M	*N = 10*	*N = 100*	*N = 200*	*N = 300*
0	1	1	1	1
1	190	1 900	3 800	5 700
2	16245	1 786 950	7 183 900	16 190 850
3	823 080	1 109 100 300	9 008 610 600	30 557 530 900
4	27 367 410	511 017 963 225	8 429 807 368 950	43 109 036 717 100
5	623 976 948	186 419 352 984 480	6 278 520 528 393 960	48×10^{15}

It is useful to calculate the chances of effecting changes in the amino acid sequence of a protein by mutation (Steipe, 2000). In a protein 250 amino acids long and with 19 alternatives for changes from the wild type, the chance of a specific change in the amino acid sequence are calculated in Eq. (11.2).

$$p = \frac{1}{\text{sequence length}} \cdot \frac{1}{\text{amino acid alternatives}} = \frac{1}{250} \cdot \frac{1}{19} = 2.1 \times 10^{-4} \tag{11.2}$$

As change has to be effected via the DNA level, however, we have to calculate the probability of such a change through mutagenesis at the DNA level. With the assumption of random nucleotide changes in the protein's corresponding DNA sequence (which in practice is certainly not always met), the probability of specific changes at the nucleotide level effecting a specific change in the amino acid sequence depends on the number of nucleotides that have to be changed [Eq. (11.3)].

$$p = \left(\frac{1}{\text{gene length}} \cdot \frac{1}{\text{nucleotide alternatives}}\right)^{\text{number of changes}} = \left(\frac{1}{250 \times 3} \cdot \frac{1}{3}\right)^{d} \tag{11.3}$$

A specific amino acid sequence change might require one, two, or all three nucleotides of a coding triplet to be exchanged ($d = 1$, 2, or 3).

- *Case 1:*
  $d = 1$, which occurs in 40% of all mutations (such as Tyr(TAC) to Phe (TTC):
  $p_1 = \{1/(250 \times 3) \cdot (1/3)\}^1 = 4.4 \times 10^{-4}$

- *Case 2:*
  $d = 2$, which occurs in 53% of all mutations (such as Tyr(TAC) to Trp (TGG):
  $p_2 = \{1/(250 \times 3) \cdot (1/3)\}^2 = 2.0 \times 10^{-7}$

- *Case 3:*
  $d = 3$, which occurs in 7% of all mutations (such as Tyr(TAC) to Met(ATG):
  $p_3 = \{1/(250 \times 3) \cdot (1/3)\}^3 = 8.8 \times 10^{-11}$

To exhaustively screen a library to be sure to include a (statistically) certain change from one specific amino acid in a certain position to another specific one in the same position, a library size of $1/p_3 = (1/8.8 \times 10^{-11}) = 1.14 \times 10^{10}$ is required! Given the possible library sizes of up to 10^{15} or less (see Section 11.3 below), only the distance of one or two amino acids can be covered with certainty. In other words, variants that require more than one or two amino acid changes, or variants that require cooperative changes in amino acid sequence, stand a low chance of being detected quickly, even with a perfect assay.

11.1.3
Evolution: Conservation of Essential Components of Structure

Although there are thousands of different protein functions and millions of protein sequences, the number of basic 3D structural folds is believed not to exceed about 30 (Brandon, 1999). Nature uses the same fold over and over again: about 10% of all enzymes have the α/β-barrel as an essential structural feature (Figure 11.2a). α-helices on the outside and β-sheets on the inside form an n-fold (usually 8-fold) symmetry with a hole in the middle that extends deep into the center of the enzyme structure.

The α/β-barrel proteins do feature a highly conserved barrel structure, but neither the amino acid sequence nor the function of these proteins is constant over the timescale of evolution. As is demonstrated in Figure 11.2a as well, many different enzymes can result from one ancestral protein. This evolutionary force is also at work nowadays, as is evidenced by the fairly sudden development of a diffusion-controlled phosphotriesterase which can only have evolved after the development and deployment of chemical weapons and pesticides, respectively, in the last 50 years. Conversely, an enzyme with a defined function such as glyceraldehyde phosphate dehydrogenase (GAPDH) can have very different properties, apart from its function (Figure 11.2b); the melting point T_m of GAPDH varies from 50 °C for the American lobster to 98 °C for an archaea from a deep-sea vent. The essential structural features of GAPDH such as the α/β-barrel are conserved but not the detailed amino acid sequence, resulting in different physical properties.

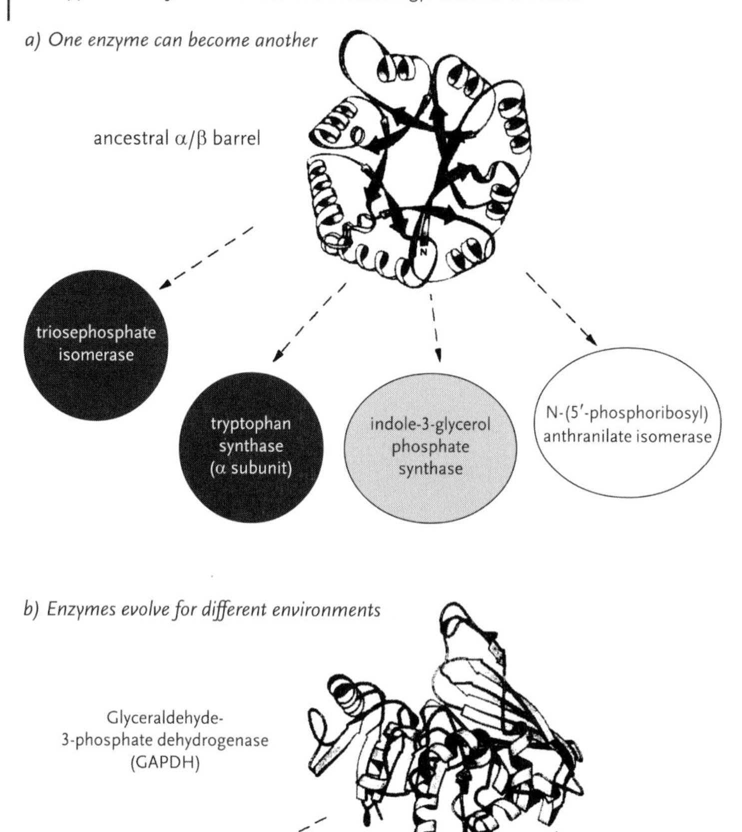

Figure 11.2 Connection of structure, catalytic function, and physical properties of enzymes (Arnold, 1996).

11.2
Process steps in Directed Evolution: Creating Diversity and Checking for Hits

A breakthrough was achieved by recognizing that the process of natural selection can be harnessed to evolve effective enzymes in artificial circumstances. In directed evolution the processes of natural evolution are accelerated in a test tube in order to select proteins with the desired properties. The realization that one could screen or select beneficial point mutations or recombinations and discard deleterious ones, thus mimicking sexual recombination, spawned the field of directed evolution.

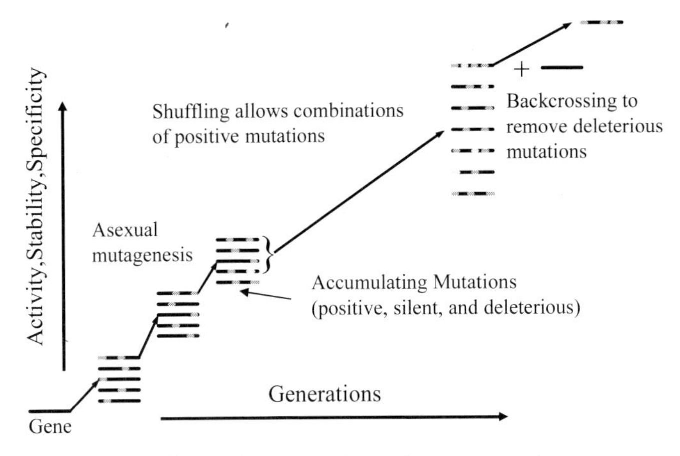

Figure 11.3 Evolution of enzymes with random mutagenesis and DNA shuffling (adapted from Arnold, 1996).

An experimental cycle of directed evolution consists of the following steps:

1. A library of parent DNA sequences encoding for the desired protein is chosen. Sequence diversity is created or increased through a *mutagenesis* step, either by introduction of random point mutations through error-prone PCR or by recombination of DNA fragments such as DNA shuffling or RACHITT.
2. The created DNA sequences are ligated into an expression vector, they are transformed into host cells, and the protein is expressed. Mutagenesis and fragmentation render all but a few sequences inactive.
3. A *screening or selection* procedure is employed next to isolate the few transformants containing the encoded sequences for active enzymes or functional proteins.
4. These selected sequences are then *amplified* and the cycle of mutagenesis, screening, and amplification is repeated until proteins with the desired property or function are found.

Therefore tailoring the enzyme to the processing conditions would be preferred; recent advances in random mutagenesis and directed evolution, notable features conceptually illustrated in Figure 11.3, offer the possibility of achieving this goal in the absence of detailed structural information on the protein.

11.2.1 Creation of Diversity in a DNA Library

In contrast to site-directed mutagenesis, which is covered in the context of protein engineering in Chapter 10, alteration of the DNA of a gene in the situation discussed in this chapter is desired to occur as *random mutagenesis*. Random mutagenesis is performed by methods that do not control the location at which mutagenesis occurs, but it relies on methods that statistically yield a certain number of mutations along a gene.

- In the simplest fashion, random mutagenesis can occur when, during amplification of the gene by PCR, the inevitable *point mutation* errors during replication not only are not corrected but are actually enhanced through the use of manganese ions instead of magnesium ions as activators for DNA polymerase. Thus, this method is termed "error-prone PCR" (You, 1996). Through control of Mn^{2+} levels, the error rate can be controlled at levels between mutations 0 and 10%, i.e., up to 10 residues per 100 amino acids in the corresponding protein (Vartanian, 1996). However, only very few mutations can be introduced per gene, or the mutant will be completely inactive. The resulting mutant variant is therefore still almost identical to the original protein – in other words, error-prone PCR can only cover a small sequence space around the original protein.

- A much wider sequence space, corresponding to a much more altered gene, than with error-prone PCR can be covered with *recombination* of DNA. For recombination, the DNA has to be cleaved in several places and then combined to the original length in a different sequence than before. With a single parental gene, however, this approach results in a gene so altered as to not be likely to result in any active protein after transcription and translation. For recombination to be successful, i.e., to result in active protein, several parental genes (usually rather more than two) are required which in addition have to possess a certain level of *sequence identity*. For this reason, this approach often is termed "sexual recombination". Recombination methods can cover much larger sequence space than methods introducing point mutations such as error-prone PCR. However, the fraction of inactive clones invariably rises with greater coverage of sequence space.

In recent years, to achieve the goals of directed evolution, new DNA recombination techniques have been introduced that have several advantages over the commonly practiced error-prone PCR and DNA shuffling techniques (Table 11.2) (Kurtzman, 2001). The two main shortcomings of DNA shuffling to address are the high incidence of non-shuffled background clones recovered as parental clones, and the requirement for significant sequence identity, usually not less than ~90%. Unlike early DNA recombination techniques such as family shuffling (Stemmer, 1994) and the staggered extension process, "StEP" (Zhao, 1998), which utilize double-stranded DNA to provide templates for shuffling, techniques using single-stranded DNA have been developed. Single-stranded family shuffling (Kikuchi, 2000) and **r**ecombined **e**xtension on **t**runcated **t**emplates, (RETT: S. H. Lee, K. H. Jung, J. Y. Jeon, and Y. C. Shin, unpublished work, 2001), reduce background of unshuffled clones, **d**egenerate **olig**onucleotide **s**huffling (DOGS: Gibbs, 2001), RACHITT (Coco, 2001), and L-shuffling (Sonet, 2001) increase crossover frequency and lower the required sequence identity (~70% for RACHITT vs. 90% for family shuffling).

Of all the methods for introduction of random mutations into genes, pathways, and organisms, only those have been discussed in detail here that are frequently used in conjunction with directed evolution protocols. Several approaches exist which represent in essence a hybrid between random point mutation and recombination; these approaches are covered in Section 11.4 and listed in Table 11.2.

Table 11.2 Selected DNA mutagenesis and recombination methods (Kurtzman, 2001).

Method	*Features and problems*	*Reference*
Mutator strain	deficient DNA repair, stability of host strain is problematic	Greener, 1996
Mutator plasmid	mutation rate of host is enhanced; curing strain of plasmid stabilizes the mutation	Selifonova, 2001
Random elongation mutagenesis	random peptides fused to the C-terminus	Matsuura, 1999
Error-prone PCR	alteration of PCR amplification protocol introduces errors; low yield of product and some amino acid substitutions are inaccessible; libraries contain increasing fraction of inactive clones with higher mutational loads	You, 1996
Mutazyme	operates under optimum PCR conditions and has a different mutational spectrum from standard polymerases	Stratagene
Family shuffling	creates diverse, highly functional libraries; closely related genes are required; requires separation of small DNA fragments and libraries contain large percentage of unshuffled clones	Stemmer, 1994
Staggered Extension Process (StEP)	comparable diversity to family shuffling but with no fragment purification required; same problems as family shuffling	Zhao, 1998
Combinatorial Library Enhanced by Recombination in Yeast (CLERY)	similar diversity to family shuffling; limited to protein screening in yeast	Abecassis, 2000
Single-stranded DNA shuffling	higher proportion of shuffled clones than family shuffling	Kikuchi, 2000
Random Chimeragenesis on Transient Templates (RACHITT)	all generate library members are shuffled; more crossovers than possible with PCR methods	Coco, 2001
L-shuffling	similar to RACHITT but lower diversity library due to use of larger fragments	Proteus Corp.
Recombined Extension on Truncated Templates (RETT)	single-stranded version of StEP; higher frequency of shuffled clones and control of regions of gene where shuffling occurs	Lee, 2001
Sequence Homology-Independent Protein Recombination (SHIPREC)	no requirement for sequence homology; requires size separation of DNA prior to screening; only one crossover per gene	Sieber, 2001
Incremental Truncation for the Creation of Hybrid Enzymes (ITCHY; THIO-ITCHY)	no requirement for sequence homology; operationally easier to implement than SHIPREC; only one crossover	Ostermeier, 1999b

Table 11.2 (continued)

Method	*Features and problems*	*Reference*
Combination of THIO-ITCHY and family shuffling (SCRATCHY)	more diverse family created than either method alone; useful for generating shuffling candidates where higher sequence homology is required than available genes	Lutz, 2001
Exon shuffling	no homology required; high percentage of functional clones; limited to intron-containing genes; diversity is proportional to the number of exons	Kolkman, 2001
Gene Site Saturation Mutagenesis (GSSM)	All single amino acid substitutions explored. Technically out of reach for most researchers.	US Patent 6171820
Tunable gene reassembly	no homology required from starting genes; technically out of reach for most researchers	US Patent 5965408
Degenerate Oligonucleotide gene-shuffling (DOGS)	easy to implement via PCR; tunable levels of recombination; reduced unshuffled background	Gibbs, 2001

11.2.2
Testing for Positive Hits: Screening or Selection

At least as important as the mutation or recombination method for generating mutants is the assay for their testing. All assays fall into the categories of either screening or selection [for recent reviews, see Soumillion (2001) and Wahler, (2001)]. By linking the growth rate of an organism and its chances for successful reproduction as a condition for survival a *selection pressure* is exerted. Selection strategies result in a qualitative "yes" or "no" measure for each library member (i.e., bind/eluted or live/die for antibiotic resistance) (Table 11.5, below). Through selection of a fast-growing organism even large populations of organisms (> 10^9 for bacteria) can be screened. Many of the possible properties of an enzyme, however, cannot be linked to a selection scheme, so every mutated variant has to be subjected to screening in a separate test for the changed and improved property. This fact limits the size of a screening to about 10^4–10^6 variants. As, according to Table 11.1, there are only 5700 variants with one mutated amino acid (for a protein with 300 amino acids) and already 16.2 million with two changed amino acids and 30.6 billion with three, one will find only those with few changes in amino acid sequence in a screening of proteins from randomly mutated genes. Screens (Table 11.3) are very versatile but have to be adapted to the specific question being investigated; careful choice of the screening method is absolutely crucial for success as "you get what you screen for" (Arnold, 1998). Screens are capable of giving more quantitative measures of individual library members but are limited by throughput (Table 11.4, below) since each candidate must be tested individually. In cases where a selection or screening strategy is not available, directed evolution is incapable of improving the situation, since the improvement of library candidates cannot be assessed.

11.3
Experimental Protocols for Directed Evolution

11.3.1
Creating Diversity: Mutagenesis Methods

In general, the published mutagenesis methods fall into two categories (Kurtzman, 2001):

- *DNA replication under low-fidelity conditions,* typically error-prone PCR (Arkin, 1992: Zaccolo, 1999): the method is simple and convenient, but the conservative nature of the genetic code limits the available substitutions to only five to seven amino acid substitutions per codon and the incorporation of stop codons limits the acceptable mutation rate to around two amino acid substitutions per gene (Smith, 1993; Arnold, 2000);

- *gene synthesis*: given the falling cost of synthesized oligonucleotides, synthetic methods have become more accessible to the average researcher (Meldrum, 2000a,b). These synthetic methods are not limited by stop codons and either can be operated with no bias at any given position or, alternatively, they can be directed just towards a subset of amino acids at any given position (Ness, 2002; Coco, 2002).

11.3.2
Creating Diversity: Recombination Methods

Extensive attention has been given to "DNA shuffling" techniques since the first article was published showing that improved variants could be obtained from libraries produced by such methods (Stemmer, 1994a). These methods are analogous to the crossover operator in genetic algorithms (Spears, 1998) but, as with mutagenesis methods, can be grouped into general classes based on the techniques employed. The most common shuffling techniques utilize PCR reaction conditions to encourage crossover between different starting genes (Stemmer, 1994b; Zhao, 1998). Methods for producing the starting fragments or genes for these shuffling reactions vary with the techniques but can involve restriction fragments, synthetic oligonucleotides, or full-length ssDNA genes with rapid PCR cycles. While the PCR reaction is generally considered straightforward, modeling of these PCR reactions has shown them to be limited to about four crossovers per gene (Moore, 2000, 2001). Template-based hybridization methods such as RACHITT (Coco, 2001) have been shown to achieve a much higher crossover ratio per gene but the requirement for ssDNA and modified template (uracil-labeled) has limited the application of the technique to large industrial customers to date. The recent introduction of an improved synthetic template method may increase its adoption. Other methods based on gene truncation (Ostermeier, 1999; Lutz, 2000) and ligation of gene fragments also exist but the advantage of these techniques over other methods has not been demonstrated. Again, the practitioner has many choices in the parameters of the protocol as well as the protocols to use for shuffling. These decisions will be supported through simulation, the results of which will be filtered and interpreted in terms of population diversity and likelihood of activity being displayed.

11.3.2.1 DNA Shuffling

A typical protocol for DNA shuffling (Stemmer, 1994b; Figure 11.4) is employed in conjunction with a selection assay, such as resistance against antibiotics (see Section 11.5.1). The starting DNA molecule is fragmented physically, most often by sonication, or digested enzymatically by DNAse I to yield fragments typically in the 50–100 bp range, which are gel-purified and then subjected to 40–60 rounds of PCR but without flanking primers. By self-priming of annealed fragments, the DNA chain grows. The extended fragments dissociate during the next PCR cycle and grow further by self-priming after annealing, often to a different location than the previous cycle. Full-length products can be recovered by conducting a few conventional PCR cycles (with primers) and gel purification of the amplified variant and cloning into the target vector.

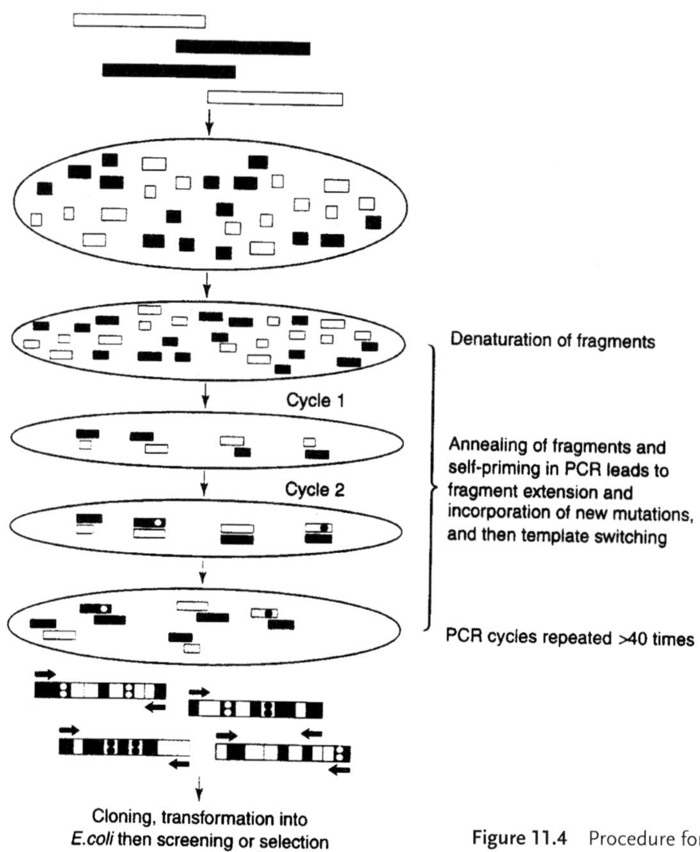

Figure 11.4 Procedure for DNA shuffling (McPherson, 2000).

11.3.2.2 Staggered Extension Process (StEP)

The staggered extension process (Zhao, 1998; Figure 11.5) is a recombination method which often starts from a few useful variants obtained through error-prone PCR with subsequent screening, and originated from the question of whether the combination of beneficial mutations itself is beneficial or not. The variants can be subjected to recombination by conducting PCR with usually high-fidelity polymerases in standard conditions but with very short extension times. Still, the polymerase causes some extension upon the temperature rise from the annealing to the denaturation temperature. The extended DNA molecules dissociate and, upon annealing, possibly attach to a different template, thus causing "template switching". The process continues until full-length DNA molecules are generated which can now result from one or several different templates and thus can carry zero, one, two, or more mutations. The library, often still small at 100 000 clones or less, is then screened for the desired property. The closer two mutations are, the lower the probability of recombination; the minimum recombination distance during even short extension has been estimated at 34 nucleotides.

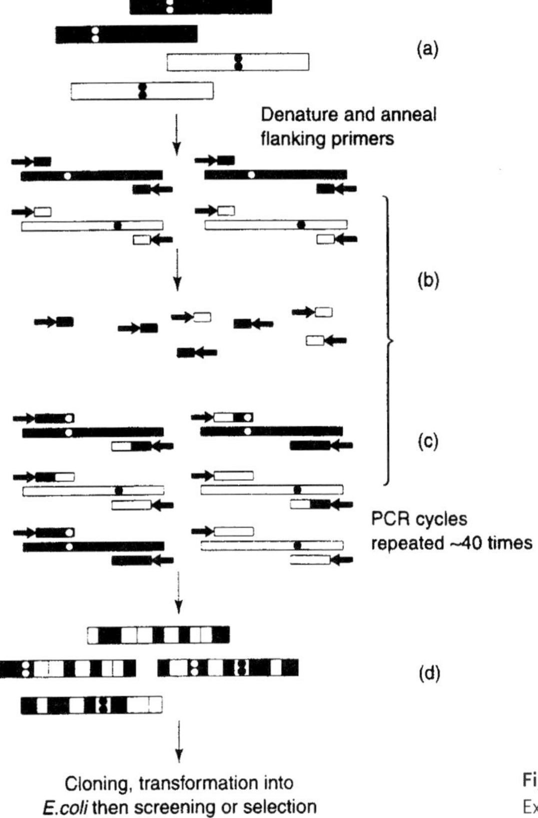

Figure 11.5 The Staggered Extension Process (StEP) procedure (McPherson, 2000).

11.3.2.3 RACHITT (Random Chimeragenesis on Transient Templates)

While there are a variety of directed evolution protocols to be found in the literature (Kurtzman, 2001), the application of the RACHITT method (Coco, 2001) depicted in Figure 11.6, possibly coupled with mutagenesis techniques, should afford the most rapid adaptation of enzymes to the chosen property to be optimized. The major advantage of RACHITT is the generation of highly diverse libraries devoid of any unshuffled background. Thus, all of the candidates screened will be novel enzymes, maximizing the results gained in screening efforts since screening of unshuffled clones provides no useful information. Other methods can have 90–99% unshuffled clones (Kikuchi, 2000), increasing the screening effort 10- to 100-fold to gain the same information. In addition, the lower sequence identity requirement between shuffling partners for RACHITT is also a significant advantage over other methods since closely related shuffling partners are not always available. Therefore, lowering the sequence identity threshold required increases the probability that a good shuffling candidate can be obtained. In those cases where no shuffling partners are available, DNA shuffling has still been successfully applied to shuffling partners generated with random mutagenesis methods on the target gene (Stemmer, 1994b).

The reduced background of unshuffled clones is a significant advantage with these techniques, with over 75% shuffled clones in most cases and 100% with RACHITT. In addition, the higher crossover frequency observed with RACHITT, and the average of 14 per gene compared to one to four for other PCR-based methods, leads to a considerable increase in library diversity (Pelletier, 2001).

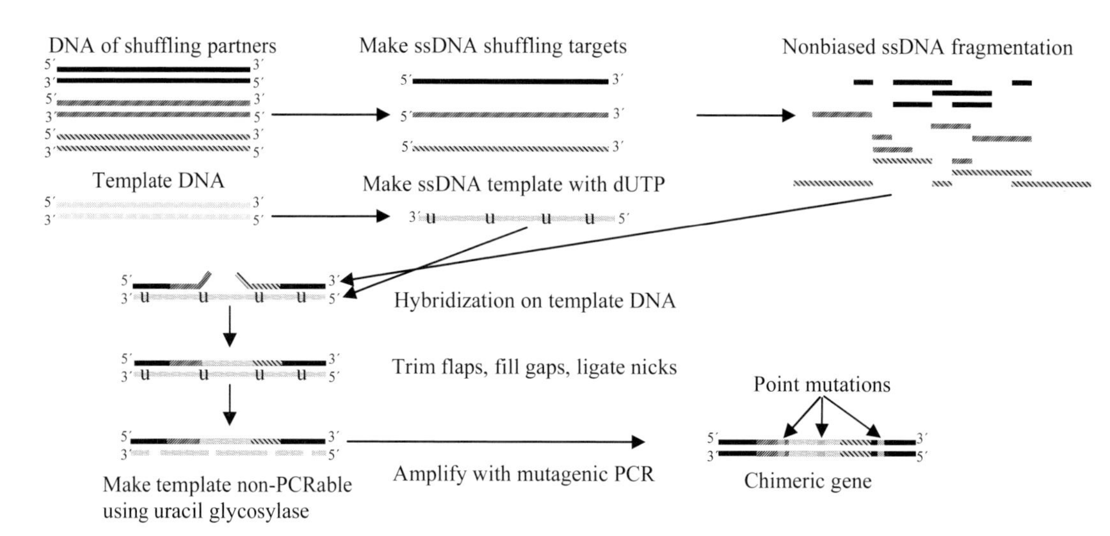

Figure 11.6 Random Chimeragenesis on Transient Templates (RACHITT) with mutagenesis (Gibbs, private communication).

11.3.3
Checking for Hits: Screening Assays

If screening is applied in directed evolution it is of extreme importance: the community often cites the phrase "you get what you screen for". An efficient assay is the starting point for any directed evolution effort since the ability to determine winners accurately is assumed in all directed evolution schemes (Zhao, 1997). However, in addition to the selective effect the screen has on the progression of individual genes, the quality of measurement (Zhang, 1999) and the throughput capability can also significantly impact the information obtained from each round of evolution (Gonzalez, 1999; Taylor, 1999). For a screening assay, an inexpensive substrate with no background, a high dynamic range, and very low variability are always preferred (Zhang, 1999); however, such an assay is not always available for the reaction of interest. Understanding the impact of the assay may encourage the researcher to invest in assay improvement before starting the directed evolution.

During the directed evolution of an esterase from *Pseudomonas fluorescens* (PFE) to resolve a sterically hindered 3-hydroxy ethyl ester, enzyme libraries were assayed on minimal media agar plates supplemented with the pH indicators Neutral Red and Crystal Violet (Bornscheuer, 1999). Active esterase variants were identified by formation of a red color caused by a pH decrease owing to release of acid product. Use of the glycerol ester, releasing glycerol upon hydrolysis, complemented the screening strategy as glycerol conferred a growth advantage on minimal medium for the cells containing active variants.

High-throughput screening is essential to evaluate the many candidates from gene-shuffling and mutagenesis experiments sufficiently well to have a reasonable probability of finding an improved enzyme (Reetz, 1999; Cohen, 2001; Fox, 2000). Table 11.4 compares some of the higher-throughput screening capabilities. Absorbance or fluorescence for common sample loads of up to 10^3–10^4 per day can be monitored by a microplate reader, which is common in many laboratories today. More sophisticated techniques, such as GC, HPLC, or mass spectrometry, have a much lower output (in the hundreds per day) but for some substrates of

Table 11.3 Representative screening assays.

Method type	*Endpoint*	*Kinetic profile*
Chromogenic/fluorimetric	TLC colored substrate/product product staining	chromogenic/fluorimetric substrates/products pH indicators
Coupled reagents	QUEST ELISA	FRET substrates
Analytical instrumentation	HPLC MS CE GC	Homogeneous cat.–ELISA IR thermography GFP folding reporter

Table 11.4 Comparison of screening capabilities.

Method	*Throughput library sizes [samples/day]*
Microplate reader	10^3–10^4
Colony imaging	10^5–10^6
Flow cytometry	10^6–10^9
Solid phase capture	10^6–10^9

interest, with assays such as those based on chirality, readily available higher-throughput solutions currently do not exist (Reetz, 2001). Even two-phase systems have been investigated (Khmelnitsky, 1999).

11.3.4
Checking for Hits: Selection Procedures

The fundamental steps of evolution (mutation, selection, and amplification) are exploited in directed evolution in the laboratory to create and characterize protein catalysts on a human timescale. In-vivo genetic selection strategies enable the exhaustive analysis of protein libraries with 10^{10} different members, and even larger ensembles can be studied with in-vitro methods. Evolutionary approaches can consequently yield statistically meaningful insight into the complex and often subtle interactions that influence protein folding, structure, and catalytic mechanism. Such methods are also being used increasingly as an adjunct to design, thus providing access to novel proteins with tailored catalytic activities and selectivities (Table 11.5). For a recent review, see Taylor (2001).

Table 11.5 Representative selection assays.

Method	*Selection pressure*	*Features*	*Reference*
Antibiotic	ability to degrade or modify substrate	commonly used molecular biology tool	Chakraborty, 1995
Substrate-selected growth	ability to degrade compound	degradation of substrate provides elements (C, N, S, etc.) required for growth	Hammann, 1998
Auxotrophic complementation	restored metabolic function	Strong positive selection	Andersen, 1984
Phage display	binding	display of peptides/proteins of virus surface can be segregated on the basis of binding affinity	Bach, 2001
CAT folding reporter	antibiotic resistance	correctly folded proteins impart antibiotic resistance to organism via a chloramphenicol acetyl-transferease fusion tag	Maxwell, 1999

11.3.5
Additional Techniques of Directed Evolution

Several approaches exist which represent a mix between the two techniques of error-prone PCR and DNA recombination described in Section 11.2.1 above, for example:

- Not all methods of directed evolution or DNA recombination require a high level of sequence identity. A series of approaches utilizes *incremental truncation* of genes, gene fragments, or even gene libraries (Ostermeier, 1999a). To achieve incremental truncation, exonuclease III is applied for various periods of time to create a library of all possible single base-pair deletions of a given piece of DNA. This technique is totally independent of DNA sequence identity.
- Exhaustive replacement of all the nucleotides in a gene by all the other three nucleic acids has been achieved by the Diversa company (San Diego/CA, USA) (Short, 2001). This approach, termed gene-site saturation mutagenesis (GSSM™), results in total coverage of sequence space with a distance of one amino acid change with respect to the parental protein.

11.4
Successful Examples of the Application of Directed Evolution

11.4.1
Application of Error-prone PCR: Activation of Subtilisin in DMF

Subtilisin is sufficiently stable in high concentrations of DMF but catalytically not very active. It was expected that even a subtilisin molecule mutated in one or two positions would not be much more active, so that it would be necessary to measure even small improvements in activity. By error-prone PCR, a library of DNA mutated in a controlled manner was generated and in every generation a single variant was picked as the parent for the next generation. The mutated genes were ligated back into the plasmids and the *E. coli* bacteria were transformed with these plasmids. A relatively crude assay screens for protein which features an enhanced activity in DMF: subtilisin is excreted into the medium, so that a halo is generated around the colony on an agar plate by diffusion of a reaction front of casein (which is hydrolyzed). Active colonies generate larger halos.

In the first generation a multiple-fold activity could already be generated through just three amino acid substitutions, a fourth amino acid could be exchanged by a known site-directed mutagenesis, and a 40-fold activity could be achieved in 60% DMF (Figure 11.7). Another six generations with screening of hundreds of colonies in each generation yielded an increase by a factor of more than 500 in comparison with the wild type; altogether, only 10 000 colonies were checked. The 12 substitutions of the best mutated enzyme were all located on the surface of the protein and randomly distributed across the gene. However, this simple approach

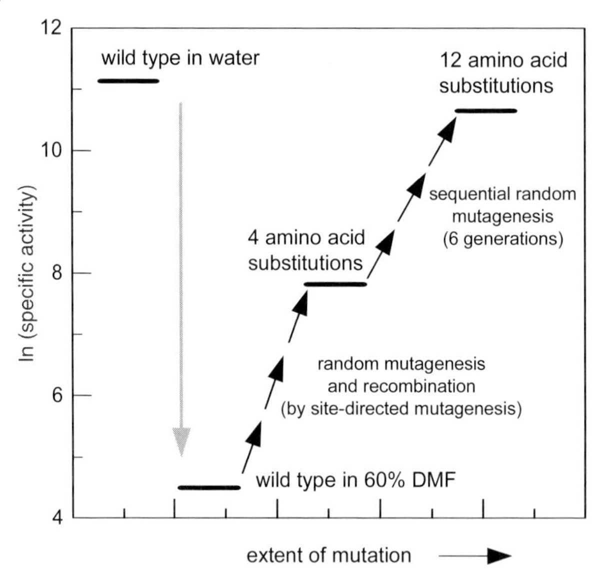

Figure 11.7 Activation of subtilisin in DMF (Arnold, 1996; You, 1996).

to a new subtilisin does not guarantee access to the best enzyme, as the enzyme was evolved little by little, generation after generation.

11.4.2
Application of DNA Shuffling: Recombination of p-Nitrobenzyl Esterase Genes

Eli Lilly (Indianapolis/IN, USA) needed a *p*-nitrobenzyl esterase to hydrolyze the *p*-nitrobenzyl ester of a β-lactam compound in DMF. The purely organic route with zinc salts and large amounts of organic solvent was unattractive, so an enzyme was sought which could hydrolyze the ester. One with rather low activity was found which served as a starting point for directed evolution. As nature never had to hydrolyze an antibiotic in DMF, this approach seemed very promising.

The wild-type esterase is not secreted and does not have any easily measurable assay reaction. To avoid the labor-intensive HPLC screening of thousands of colonies a *p*-nitrophenyl esterase was sought at first, which would yield *p*-nitrophenol after hydrolysis which could be measured colorimetically in a 96-well plate. After four generations of random PCR mutagenesis and screening of about 1000 colonies each, a 15-fold enhanced esterase activity in 15% DMF was found. After the fourth generation, 64 clones were collected which were investigated for correlation between *p*-nitrophenyl and *p*-nitrobenzyl esterase activity (Figure 11.8).

A perfect correlation would be indicated if all the points lay on a 45° line; the actual results point towards sufficient correlation. The five best mutants in the broken-line circle were selected to be subjected to DNA recombination, the “sexual PCR” (Stemmer, 1994a). The method is depicted in Figure 11.9. Genes are pooled in the test tube and split with DNAse at random positions. After electrophoresis

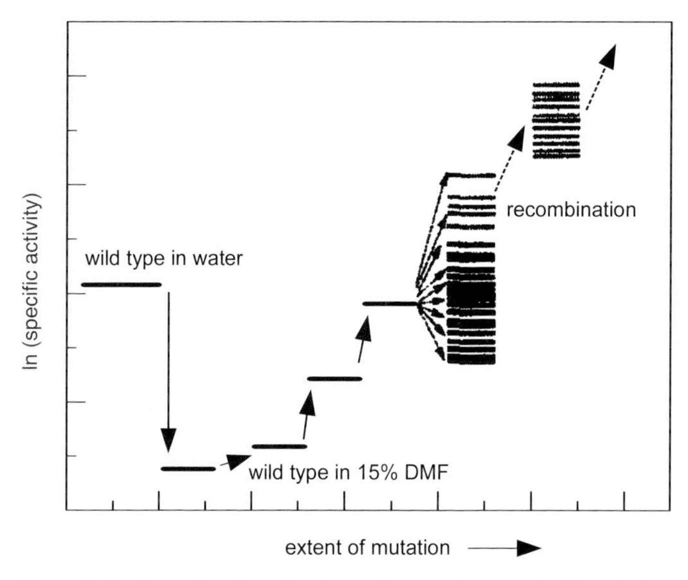

Figure 11.8 Recombination of mixtures of genes ("sex in a test tube") (Arnold, 1996).

the 200–300 kbp DNA pieces are extracted from the gel. Through application of PCR the whole gene can be reassembled in the test tube, and a new gene library is generated. The recombined genes are ligated back into the plasmid as usual, *E. coli* cells are transformed with the plasmids, and the genes are expressed. Screening of only 400 colonies yielded eight clones with at least a 20-fold activity enhancement with respect to the parent generation.

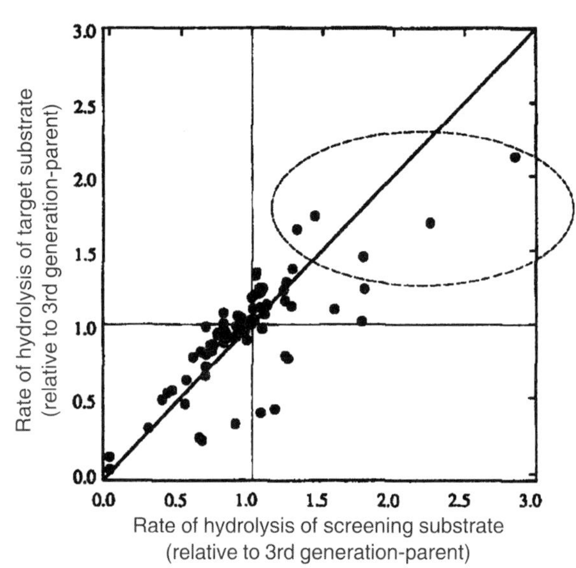

Figure 11.9 Correlation between screening (*p*-nitrophenyl ester) and target substrate (*p*-nitrobenzyl ester) and enhancement profile in DMF (Arnold, 1996).

11.4.3
Enhancement of Thermostability: p-Nitrophenyl Esterase

With the example of *p*-nitrophenyl esterase the Arnold group demonstrated that even the thermostability of an enzyme is amenable to directed evolution (Giver, 1998). As a test of thermostability the activity in standard conditions (pH 7.5, 0.25 mM *p*-nitrophenyl acetate) at 30 °C (A_i) was compared with the activity in the standard assay at 30 °C (A_r) but after heating to T_m for 10 min (with subsequent cooling on ice and incubation for 30 min at room temperature). The results are shown in Figure 11.10. This work demonstrates that enhanced activity and enhanced stability need not be mutually exclusive. However, the improvements along one of the two axes sometimes tend to be minor, i.e., by a factor between 3 and 5.

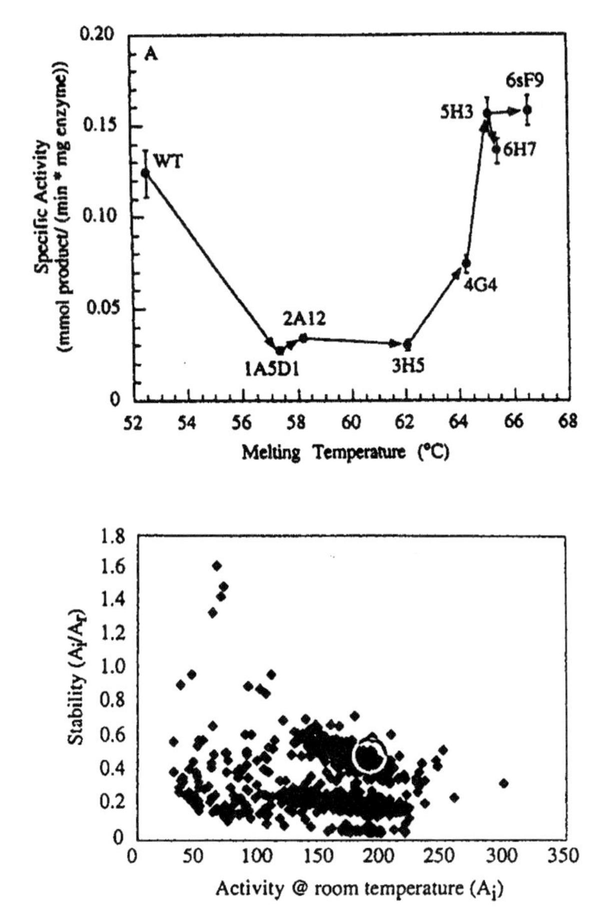

Figure 11.10
a) Thermostability profile of the esterase variants (Giver, 1996);
b) enhancement of activity and stability (Giver, 1998).

11.4.4
Selection instead of Screening: Creation of a Monomeric Chorismate Mutase

Chorismate mutase (CM) catalyzes conversion of chorismate into prephenate in a Claisen rearrangement; prephenate is a precursor in the biosynthesis of both L-phenylalanine (L-Phe) and L-tyrosine (L-Ttyr). CM occurs as a dimer; for structural studies, the monomer was needed (MacBeath, 1998). The assay for CM activity was based on growth of the colonies in the absence of L-Phe and L-Tyr and thus was based on selection, not screening. One resulting mutant was found to be a monomer and to contain a somewhat polar Ala-Arg-Trp-Pro-Trp-Ala sequence.

11.4.5
Improvement of Enantioselectivity: Pseudomonas aeruginosa Lipase

For many cases of chiral chemoenzymatic synthesis, it is difficult or impossible to identify an enzyme that possesses the desired enantioselectivity. Therefore, there is a strong need to develop enzymes which display high enantioselectivity. A wild-type bacterial lipase from *Pseudomonas aeruginosa* hydrolyzed a racemic model substrate, 2-methyldecanoic acid *p*-nitrophenyl ester, nearly enantiounspecifically ($E = 1.1$) (Liebeton, 2000). Error-prone PCR, adjusted to yield a mutation rate of 1–2 base substitutions per gene, was applied during six generations with 1000–7000 clones screened per round. The best variant gave an E value of 13.5 (Figure 11.11).

Next, saturation mutagenesis was performed at all residues found in optimum variants from previous generations. Only the second generation variant S155L improved to S155F; all other optimum variants gave no improvement. However, saturation mutagenesis in position 155 on the optimum variant after the third

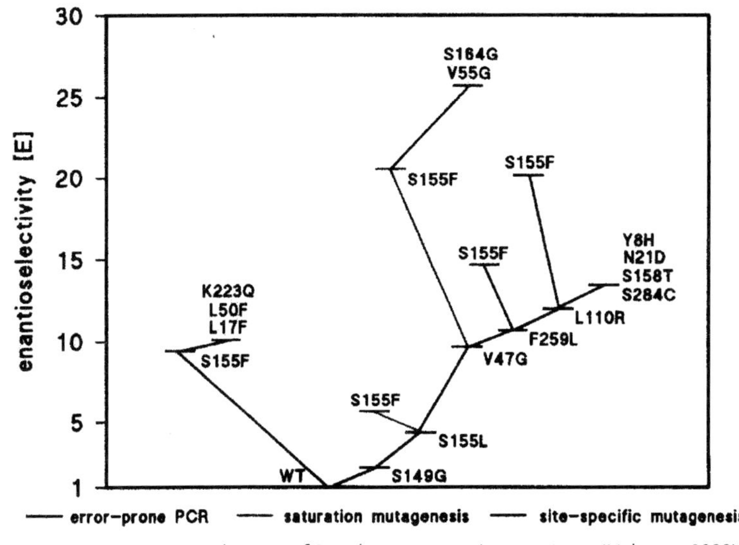

Figure 11.11 Enantioselectivity of *Pseudomonas aeruginosa* variants (Liebeton, 2000).

generation resulted the *E* value improving from 9.7 to 20.5, with similar but not quite such strong results on the optimum variants of the fourth and fifth generation. Another round of error-prone PCR resulted in a variant with an *E* value of 25.8 and a total of five substitutions (V47G, V55G, S149G, S155F, and S164G; Figure 11.11). The 3D structure of *Pseudomonas aeruginosa* lipase solved afterwards elucidates the probable reason for the improvement: the flexibility of distinct loops of the enzyme is markedly enhanced (except residue 155, all other replacements are glycines). The following lessons can be learned from this work:

- combination of directed evolution methods, such as error-prone PCR and saturation mutagenesis, is beneficial;
- initially beneficial mutations in early generations might be detrimental for performance in subsequent generations; an example is the negative effect of the mutation F259L in the fourth generation on the effect of mutation S155F. Therefore, it pays to investigate winners of every generation *independently*.

Subsequent directed evolution work on *Pseudomonas aeruginosa* demonstrates that protein sequence space with respect to enantioselectivity is best explored by a three-step procedure (Reetz, 2001): (i) generation of mutants by error-prone PCR at a high-mutation rate; (ii) identification of hot regions and spots in the enzyme by error-prone PCR and substantiation by simplified combinatorial multiple-cassette mutagenesis; (iii) extension of the process of combinatorial multiple-cassette mutagenesis to cover a defined region of protein sequence space.

The topic of improving enantioselectivity, especially in lipases and esterases, has been reviewed recently (Bornscheuer, 2002).

11.4.6 Inversion of Enantioselectivity: Hydantoinase

One decisive reason for the lack of large-scale application of the hydantoinase/carbamoylase process to enantiomerically pure L-amino acids (Chapter 7, Section 7.2.3.3) is the insufficient enantiospecificity of the hydantoinase, whose specificity varies with the 5′-substituent on the substrate. To improve the space–time-yield of this process to acceptable levels, an L-selective hydantoinase was required. As screening did not provide better hydantoinases, inversion of enantioselectivity of the hydantoinase was attempted by directed evolution (May, 2000). Indeed, inversion of enantioselectivity can be achieved rapidly by this approach. Mutants differing in only two effective amino acid substitutions showed remarkable differences in enantioselectivity: One mutant (Variant 1CF3) produced the D-enantiomer at an enantiomeric excess of 90% whereas an L-selective mutant (Variant Q2H4) produced the opposite enantiomer with 20% e.e. at 40% conversion. Although the L-selectivity of the designed enzyme was not impressive and leaves room for further improvements, the productivity of the process could be dramatically improved. These improvements have been confirmed at the cubic meter (m^3) scale using a simple batch reactor concept coupled to a continuous centrifuge for cell separation.

11.4.7
Redesign of an Enzyme's Active Site: KDPG Aldolase

The carbon–carbon forming ability of aldolases has been limited in part by their narrow substrate utilization. Site-directed mutagenesis of various enzymes to alter their specificity has most often not produced the desired effect. Directed evolution approaches have furnished novel activities through multiple mutations of residues involved in recognition; in no instance has a key catalytic residue been altered while activity is retained. Random mutagenesis resulted in a double mutant of *E. coli* 2-keto-3-deoxy-6-phosphogluconate (KDPG) aldolase with reduced but measurable enzyme activity and a synthetically useful substrate profile (Wymer, 2001). The new aldolase differs from all other existing ones with respect to the location of its active site in relation to its secondary structure and still displays enantiofacial discrimination during aldol addition. Modification of substrate specificity is achieved by altering the position of the active site lysine from one β-strand to a neighboring strand rather than by modification of the substrate recognition site. Determination of the 3D crystal structure of the wild type and the double mutant demonstrated how catalytic competency is maintained despite spatial reorganization of the active site with respect to substrate. It is possible to perturb the active site residues themselves as well as surrounding loops to alter specificity.

11.5
Comparison of Directed Evolution Techniques

11.5.1
Comparison of Error-Prone PCR and DNA Shuffling: Increased Resistance against Antibiotics

Crameri et al. (Crameri, 1998) investigated and compared the power of error-prone PCR and DNA shuffling by testing both procedures against the following problem: four cephalosporinase C genes from *Enterobacter*, *Yersinia*, *Citrobacter*, and *Klebsiella* were subjected to either error-prone PCR or DNA shuffling. The assay was activity against the β-lactam antibiotic Moxalactam.

Colonies were grown in presence of Moxalactam, so activity enhancement also served as a selection criterion: successful cephalosporinase mutants would cleave the β-lactam antibiotic and allow the cells of the respective colony to grow. The procedure was tested over four generations, and about 50 000 colonies per cycle were tested.

While the best mutant from the library of point mutations after single gene-shuffling yielded an eightfold activity increase, the best mutants in the library of chimeras from multiple gene-shuffling showed enhancement of 270- to 540-fold. The best mutant turned out to be a chimera of all four original organisms; the homology was 27% to *Citrobacter*, 37% to *Enterobacter*, 47% to *Klebsiella* and 50% to *Yersinia*. Even more remarkably, the best mutant from multiple gene-shuffling featured a total of 33 amino acid mutations and seven crossovers. Such a result would have been impossible to achieve with single gene-shuffling alone.

This work demonstrates the power of multiple gene-shuffling over single gene-shuffling, especially if maximum possible sequence space has to be covered.

11.5.2 Protein Engineering in Comparison with Directed Evolution: Aminotransferases

Aminotransferase (AAT), the enzyme catalyzing the reversible transformation of aspartate and glutamate into the respective oxo acids, has been studied most among the vitamin B_6-dependent enzymes. An X-ray crystal structure is now known for the aspartic–glutamic aminotransferase from *E. coli* (Smith, 1986). Active-site residues have been identified, laying the groundwork for further detailed mechanistic studies and modification of the enzyme by specific mutagenesis. Several groups have been successful at changing the relative activity of aminotransferase towards different groups of substrates or even different reactions through protein engineering and directed evolution.

Multiple active-site site-specific mutations of AAT led to an increase in β-decarboxylase activity with the double mutant Y225R/R386A (1380-fold) (Graber, 1995). Coupled with a transaminase activity that was decreased by a factor of 500 in the single mutant R292K (Vacca, 1997), the authors found a combined 20 000-fold decrease in the rate of transamination in the triple mutant Y225R/R292K/R386A (Graber, 1999). In fact, the triple mutant catalyzed β-decarboxylation eightfold faster than transamination, a change of ratio from the wild-type enzyme by a factor of 25 million. The observed changes in substrate specificity were rarely additive, however, because triple mutants containing R292X, i.e., mutations to amino acids other than lysine, were mostly completely inactive towards β-decarboxylation even though they contained the double mutant Y225R/R386A eliciting β-decarboxylase activity.

Previously, AAT had been transformed into an L-tyrosine aminotransferase (TAT) by site-specific mutation of up to six amino acid residues lining the active site of wild-type AAT. The hextuple AAT-mutant achieved kinetic data towards the transamination of aromatic substrates such as L-phenylalanine within an order of magnitude of wild-type TAT (Onuffer, 1995).

11.5.2.1 Directed Evolution of Aminotransferases

Meanwhile, directed evolution methods that combine mutagenesis of genes with high-throughput screening of functional gene products have developed rapidly. Recently, efforts to evolve aminotransferases with improved activity on new keto acid substrates have been initiated with encouraging results (Rozzell, 1999). Using directed evolution, the substrate specificity of AAT has been changed to one favoring β-branched amino acids and their respective oxo acids, effectively converting AAT into a branched-chain aminotransferase (BCAT). By employing an *E. coli* auxotroph deficient in the branched-chain aminotransferase (BCAT) gene, *ilvE*, the authors set up a stringent selection system which provided a powerful advantage for cell growth to the mutated AAT systems (Yano, 1998; Oue, 1999).

The resulting aminotransferases evolved had 13 (Yano, 1998) and 17 (Oue, 1999) amino acid substitutions and showed 10^5-fold and 2×10^6-fold improvement in catalytic efficiency (k_{cat}/K_M), respectively, towards the unnatural substrate valine, and between 10- and 100-fold decrease towards the natural substrate L-aspartate, compared to the wild type. A high degree of conserved amino acid residues was found in the most active mutants. Interestingly, only one mutated amino acid residue in each case is located at a distance from the substrate that would allow interactions, the remainder were mutated far away from the active site. This work demonstrates that 10^6-fold shifts in substrate specificity can be achieved when employing directed evolution methods, that combinatorial or evolutionary methods are probably superior to rational design methods when changing substrate specificity, and most importantly, that remote residues and their interactions with the active-site environment are important determinants of enzyme activity and specificity. Such remote residues act cumulatively, possibly by remodeling the active site, by altering the subunit interfaces, or by shifting different enzyme domains.

11.5.3
Directed Evolution of a Pathway: Carotenoids

The carotenoid biosynthetic pathway leads to products such as vitamin A and its precursors carotene, responsible for the color of carrots or salmon, and lycopene, the color agent in tomatoes (Figure 11.12). Pathway products are sought for a variety of applications in food, medicine, and materials. Genes in the pathway from different organisms were combined and heterologously expressed in *E. coli*, a non-

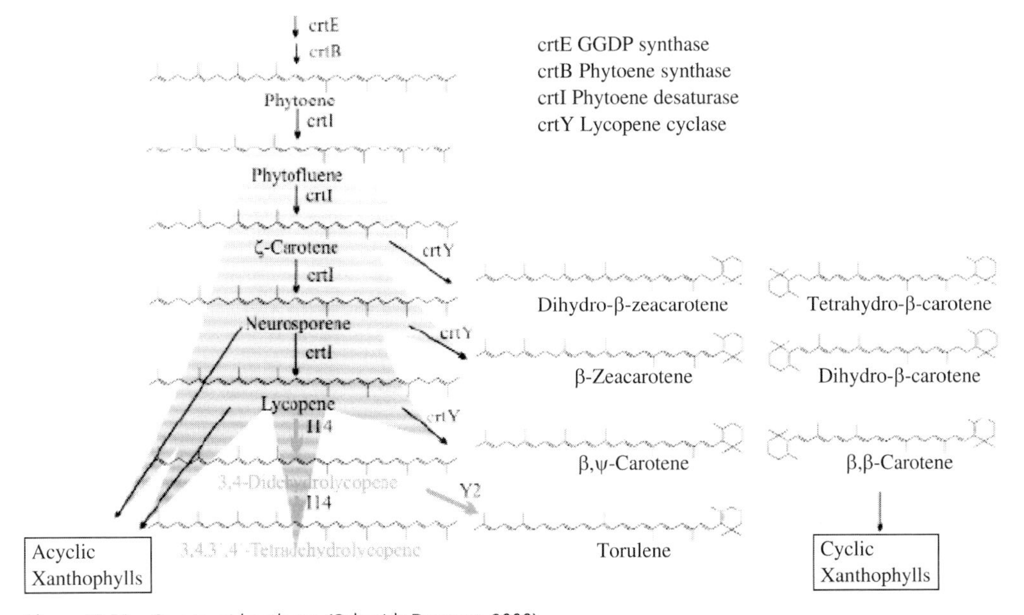

Figure 11.12 Carotenoid pathway (Schmidt-Dannert, 2000).

carotenoic organism, as well as being modified by directed evolution techniques. As an example, one phytoene desaturase variant obtained by recombination introduced six rather than four double bonds into phytoene and thus favored production of the fully conjugated carotenoid, 3,4,3′,4′-tetradehydrolycopene (Schmidt-Dannert, 2000). This new pathway was extended with a second library of shuffled lycopene cyclases to produce a variety of colored products (see the Inlet Figure at the end of this chapter), such as cyclic torulene, not previously observed in *E. coli*. This work paves the road to high-value added products of the carotenoid pathway, including those not found in nature (de Boer, 2003). Further extension of these C_{30} pathway compounds to C*40* carotenoid enzymes resulted in violet-colored phillipsisaxanthin, catalyzed by spheroidene monooxygenase, and novel torudene derivatives with the help of carotene hydroxylase, desaturase, glucosylase, and ketolase (Lee, 2003). Such results illustrate the utility of the concept of using a central metabolic pathway combined with catalytically promiscuous downstream enzymes to generate structurally novel compounds.

A combined approach of rational pathway assembly and directed evolution techniques opens the perspective for discovery and production of compounds that are either rare and hard to access in nature or even entirely non-natural, in simple laboratory organisms.

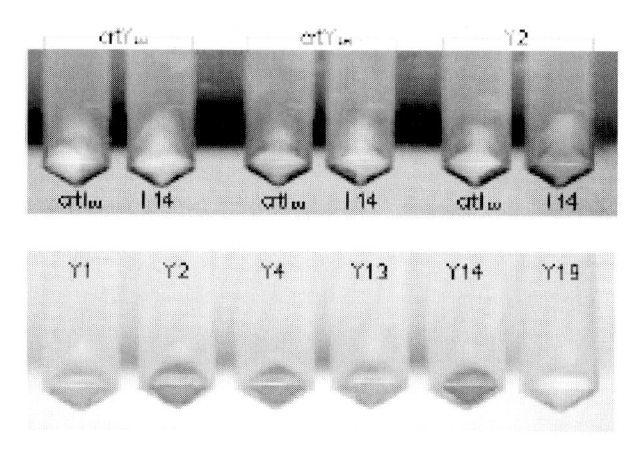

Inlet Figure Cell pellets with carotenoid pathway products (Schmidt-Dannert, 2000). JM 109 transformants expressing wild-type and mutant cyclases.

Suggested Further Reading

F. H. Arnold, Combinatorial and computational challenges for biocatalyst design, *Nature* **2001**, *409*, 253–257.

K. A. Powell, S. W. Ramer, S. B. Del Cardayre, W. P. C. Stemmer, M. B. Tobin, P. F. Longchamp, amd G. W. Huisman, Directed evolution and biocatalysis, *Angew. Chem. Int. Ed.* **2001**, *40*, 3948–3959.

Website

Frances Arnold: http://www.che.caltech.edu/groups/fha/Enzyme/directed.html

Review articles

F. A. Arnold, Directed evolution: creating biocatalysts for the future, *Chem. Eng. Sci.* **1996**, *51*, 5091–5102.
F. H. Arnold and J. C. Moore, Optimizing industrial enzymes by directed evolution, *Adv. Biochem. Eng. Biotechnol.* **1997**, *58*, 1–14.
D. N. Bolon, C. A. Voigt, and S. L. Mayo, De novo design of biocatalysts, *Curr. Opin. Chem. Biol.* **2002**, *6*, 125–129.
U. T. Bornscheuer, Directed evolution of enzymes for biocatalytic applications, *Biocatal. Biotransf.* **2001**, *19*, 85–97.
U. T. Bornscheuer and M. Pohl, Improved biocatalysts by directed evolution and rational protein design, *Curr. Opin. Chem. Biol.* **2001**, *5*, 137–143.
N. Cohen, S. Abramov, Y. Dror, and A. Freeman, In vitro enzyme evolution: the screening challenge of isolating the one in a million, *Trends Biotechnol.* **2001**, *19*, 507–510.
C. Schmidt-Dannert and F. H. Arnold, Directed evolution of industrial enzymes, *Trends Biotechnol.* **1999**, *17*, 135–137.
S. V. Taylor, P. Kast, and D. Hilvert, Investigating and engineering enzymes by genetic selection, *Angew. Chem, Int. Ed.* **2001**, *40*, 3310–3335.
P. L. Wintrode and F. H. Arnold, Temperature adaptation of enzymes: lessons from laboratory evolution, *Adv, Protein Chem.* **2000**, *55*, 161–225.

References

V. Abecassis, D. Pompon, and G. Truan, High efficiency family shuffling based on multi-step PCR and in vivo DNA recombination in yeast: statistical and functional analysis of a combinatorial library between human cytochrome P450 1A1 and 1A2, *Nucleic Acids Res.* **2000**, *28*, E88.
K. Andersen and M. Wilke-Douglas, Construction and use of a gene bank of Alcaligenes eutrophus in the analysis of ribulose bisphosphate carboxylase genes, *J. Bacteriol.* **1984**, *159*, 973–978.
A. P. Arkin and D. C. Youvan, Optimizing nucleotide mixtures to encode specific subsets of amino acids for semi-random mutagenesis, *Biotechnology (NY)* **1992**, *10*, 297–300.
F. A. Arnold, Directed evolution: creating biocatalysts for the future, *Chem. Eng. Sci.* **1996**, *51*, 5091–5102.
F. H. Arnold, Enzyme engineering reaches the boiling point, *Proc. Natl. Acad. Sci. USA* **1998a**, *95*, 2035–2036.
F. H. Arnold, Design by directed evolution, *Acc. Chem. Res.* **1998b**, *31*, 125–131.
F. H. Arnold, Advances in protein chemistry. Introduction. *Adv. Protein Chem.* **2000**, *55*, ix–xi.
H. Bach, Y. Mazor, S. Shaky, A. Shoham-Lev, Y. Berdichevsky, D. L. Gutnick, and I. Benhar, *Escherichia coli* maltose-binding protein as a molecular chaperone for recombinant intracellular cytoplasmic single-chain antibodies, *J. Mol. Biol.* **2001**, *312*, 79–93.
U. T. Bornscheuer, J. Altenbuchner, and H. H. Meyer, Directed evolution of an esterase: screening of enzyme libraries based on pH-indicators and a growth assay, *Bioorgan. Med. Chem.* **1999**, *7*, 2169–2173.

U. T. Bornscheuer, Methods to increase enantioselectivity of lipases and esterases, *Curr. Opin. Biotechnol.* **2002**, *13*, 543–547.

C. Brandon and J. Tooze, *Introduction to Protein Structure*, Garland, New York, 2nd edition, **1999**.

S. G. Burton, D. A. Cowan, and J. M. Woodley, The search for the ideal biocatalyst, *Nature Biotechnol.* **2002**, *20*, 37–45.

D. Chakraborty, B. Mondal, S. C. Pal, and S. K. Sen, Characterisation and identification of broad spectrum antibiotic producing *Streptomyces hygroscopicus* D 1.5, *Hindustan Antibiot. Bull.* **1995**, *37*, 37–43.

W. M. Coco, W. E. Levinson, M. J. Crist, H. J. Hektor, A. Darzins, P. T. Pienkos, C. H. Squires, and D. J. Monticello, DNA shuffling method for generating highly recombined genes and evolved enzymes, *Nature Biotechnol.* **2001**, *19*, 354–359.

W. M. Coco, L. P. Encell, W. E. Levinson, M. J. Crist, A. K. Loomis, L. L. Licato, J. J. Arensdorf, N. Sica, P. T. Pienkos, and D. J. Monticello, Growth factor engineering by degenerate homoduplex gene family recombination, *Nature Biotechnol.* **2002**, *20*, 1246–1250.

A. Crameri, S.-A. Raillard, E. Bermudez, and W. P. C. Stemmer, DNA shuffling of a family of genes from diverse species accelerates directed evolution, *Nature* **1998**, *391*, 288–291.

A. L. de Boer and C. Schmidt-Dannert, Recent efforts in engineering microbial cells to produce new chemical compounds, *Curr. Opin. Chem. Biol.* **2003**, *7*, 273–278.

J. H. Duncan, A. H. B. Gaskin, N. A. Turner, and E. N. Vulfson, Directed evolution of an industrially important enzyme, *Biochem. Soc. Trans.* **1997**, *25*, 15 S.

S. J. Fox, M. A. Yund, and S.-F. Jones, Assay innovations vital to improving HTS, *Drug Discov. Dev.* **2000**, 40–43.

M. D. Gibbs, K. M. Nevalainen, and P. L. Bergquist, Degenerate oligonucleotide gene shuffling (DOGS): a method for enhancing the frequency of recombination with family shuffling, *Gene* **2001**, *271*, 13–20.

L. Giver, A. Gershenson, P.-O. Freskgard, and F. H. Arnold, Directed evolution of a thermostable esterase, *Proc. Natl. Acad. Sci. USA* **1998**, *95*, 12809–12813.

J. E. Gonzalez and P. A. Negulescu, Intracellular detection assays for high-throughput screening, *Curr. Opin. Biotechnol.* **1998**, *9*, 624–631.

R. Graber, P. Kasper, V. N. Malashkevich, E. Sandmeier, P. Berger, H. Gehring, J. N. Jansonius, and P. Christen, Changing the reaction specificity of a pyridoxal-5′-phosphate-dependent enzyme, *Eur. J. Biochem.* **1995**, *232*, 686–690.

R. Graber, P. Kasper, V. N. Malashkevich, P. Strop, H. Gehring, J. N. Jansonius, and P. Christen, Conversion of aspartate aminotransferase into an L-aspartate β-decarboxylase by a triple active-site mutation, *J. Biol. Chem.* **1999**, *274*, 31203–31208.

T. P. Graycar, R. R. Bott, R. M. Caldwell, J. L. Dauberman, P. J. Lad, S. D. Power, I. H. Sagar, R. A. Silva, G. L. Weiss and L. R. Woodhouse, Altering the proteolytic activity of subtilisin through protein engineering, *Ann. N. Y. Acad. Sci.* **1992**, *672*, 71–79.

A. Greener, M. Callahan, and B. Jerpseth, An efficient random mutagenesis technique using an *E. coli* mutator strain, *Methods Mol. Biol.* **1996**, *57*, 375–385.

R. Hammannn and H. J. Kutzner, Key enzymes for the degradation of benzoate, *m*- and *p*-hydroxybenzoate by some members of the order *Actinomycetales*, *J. Basic Microbiol.* **1998**, *38*, 207–220.

Y. L. Khmelnitsky and J. O. Rich, Biocatalysis in nonaqueous solvents, *Curr. Opin. Chem. Biol.* **1999**, *3*, 47–53.

M. Kikuchi, K. Ohnishi, and S. Harayama, An effective family shuffling method using single-stranded DNA, *Gene* **2000**, *243*, 133–137.

J. A. Kolkman and W. P. Stemmer, Directed evolution of proteins by exon shuffling, *Nature Biotechnol.* **2001**, *19*, 423–428.

A. L. Kurtzman, S. Govindarajan, K. Vahle, J. T. Jones, V. Heinrichs, and P. A. Patten, Advances in directed protein evolution by recursive genetic recombination: applications to therapeutic proteins, *Curr. Opin. Biotechnol.* **2001**, *12*, 361–370.

P. C. Lee, A. Z. R. Momen, B. N. Mijts, and C. Schmidt-Dannert, Biosynthesis of structurally novel carotenoids in *Escherichia coli*, *Chem. Biol.* **2003**, *10*, 453–462.

K. Liebeton, A. Zonta, K. Schimossek, M. Nardini, D. Lang, B. W. Dijkstra, M. T. Reetz, and K.-E. Jaeger, Directed evolution of an enantioselective lipase, *Chem. Biol.* **2000**, *7*, 709–718.

M. M. Ling and B. H. Robinson, Approaches to DNA mutagenesis: an overview, *Anal. Biochem.* **1997**, *254*, 157–178.

S. Lutz and S. J. Benkovic, Homology-independent protein engineering, *Curr. Opin. Biotechnol.* **2000**, *11*, 319–324.

S. Lutz, M. Ostermeier, G. L. Moore, C. D. Maranas, and S. J. Benkovic, Creating multiple-crossover DNA libraries independent of sequence identity, *Proc. Natl. Acad. Sci. USA* **2001**, *98*, 11248–11253.

G. MacBeath, P. Kast, and D. Hilvert, Redesigning enzyme topology by directed evolution, *Science* **1998**, *279*, 1958–1961.

T. Matsuura, K. Miyai, S. Trakulnaleamsai, T. Yomo, Y. Shima, S. Miki, K. Yamamoto, and I. Urabe, Evolutionary molecular engineering by random elongation mutagenesis, *Nature Biotechnol.* **1999**, *17*, 58–61.

K. L. Maxwell, A. K. Mittermaier, J. D. Forman-Kay, and A. R. Davidson, A simple in vivo assay for increased protein solubility, *Protein Sci.* **1999**, *8*, 1908–1911.

O. May, P. T. Nguyen, and F. H. Arnold, Inverting enantioselectivity by directed evolution of hydantoinase for improved production of L-methionine, *Nature Biotechnol.* **2000**, *18*, 317–320.

M. J. McPherson and S. G. Møller, *PCR*, Bios Scientific Publishers Ltd. London, **2000**.

D. Meldrum, Automation for genomics, part one: preparation for sequencing, *Genome Res.* **2000a**, *10*, 1081–1092.

D. Meldrum, Automation for genomics, part two: sequencers, microarrays, and future trends, *Genome Res.* **2000b**. *10*, 1288–1303.

G. L. Moore and C. D. Maranas, Modeling DNA mutation and recombination for directed evolution experiments, *J. Theor. Biol.* **2000**, *205*, 483–503.

G. L. Moore, C. D. Maranas, S. Lutz, and S. J. Benkovic, Predicting crossover generation in DNA shuffling, *Proc. Natl. Acad. Sci. USA* **2001**, *98*, 3226–3231.

J. E. Ness, S. Kim, A. Gottman, R. Pak, A. Krebber, T. V. Borchert, S. Govindarajan, E. C. Mundorff, and J. Minshull, Synthetic shuffling expands functional protein diversity by allowing amino acids to recombine independently, *Nature Biotechnol.* **2002**, *20*, 1251–1255.

J. J. Onuffer and J. F. Kirsch, Redesign of the substrate specificity of *Escherichia coli* aspartate aminotransferase to that of *Escherichia coli* tyrosine aminotransferase by homology modeling and site-directed mutagenesis, *Protein Sci.* **1995**, *4*, 1750–1757.

M. Ostermeier, A. E. Nixon, and S. J. Benkovic, Incremental truncation as a strategy in the engineering of novel biocatalysts, *Bioorgan. Med. Chem.* **1999a**, *7*, 2139–2144.

M. Ostermeier, A. E. Nixon, J. H. Shim, and S. J. Benkovic, Combinatorial protein engineering by incremental truncation, *Proc. Natl. Acad. Sci. USA* **1999b**, *96*, 3562–3567.

S. Oue, A. Okamoto, T. Yano, and H. Kagamiyama, Redesigning the substrate specificity of an enzyme by cumulative effects of the mutations of non-active site residues, *J. Biol. Chem.* **1999**, *274*, 2344–2349.

S. Paabo, D. M. Irwin, and A. C. Wilson, DNA damage promotes jumping between templates during enzymatic amplification, *J. Biol. Chem.* **1990**, *265*, 4718–4721.

J. N. Pelletier, A RACHITT for our toolbox, *Nature Biotechnol.* **2001**, *19*, 314–315.

M. T. Reetz, M. H. Becker, H.-W. Klein, and D. A. Stöckigt, Method for high-throughput screening of enantioselective catalysts, *Angew. Chem. Int. Ed.* **1999**, *38*, 1758–1761.

M. T. Reetz, Combinatorial and evolution-based methods in the creation of enantioselective catalysts, *Angew. Chem. Int. Ed.* **2001a**, *40*, 284–310.

M. T. Reetz, S. Wilensek, D. Zha, and K.-E. Jaeger, Directed evolution of an enantioselective enzyme through combinatorial multiple-cassette mutagenesis, *Angew. Chem. Intl Ed.* **2001b**, *40*, 3589–3591.

J. D. Rozzell, Methods for producing amino acids by transamination, U. S. Patent Application 09/334821, **1999**.

C. Schmidt-Dannert, D. Umeno, and F. H. Arnold, Molecular breeding of carotenoid biosynthetic pathways, *Nature Biotechnol.* **2000**, *18*, 750–753.

O. Selifonova, F. Valle, and V. Schellenberger, Rapid evolution of novel traits in microorganisms, *Appl. Environ. Microbiol.* **2001**, *67*, 3645–3649.

J. M. Short, Saturation Mutagenesis in Directed Evolution, US patent 6,171,820, **2001**.

V. Sieber, C. A. Martinez, and F. H. Arnold, Libraries of hybrid proteins from distantly related sequences, *Nat. Biotechnol.* **2001**, *19*, 456–460.

D. L. Smith, D. Ringe, W. L. Finlayson, and J. F. Kirsch, Preliminary X-ray data for aspartate aminotransferase from *Escherichia coli*, *J. Mol. Biol.* **1986**, *191*, 301–302.

K. D. Smith, A. Valenzuela, J. L. Vigna, K. Aalbers, and C. T. Lutz, Unwanted mutations in PCR mutagenesis: avoiding the predictable, *PCR Methods Appl.* **1993**, *2*, 53–57.

J.-M. Sonet, Screening for novel biocatalysts from biodiversity, *Bioforum Int.* **2001**, *5*, 232–234.

J. K. Song and J. S. Rhee, Simultaneous enhancement of thermostability and catalytic activity of phospholipase A1 by evolutionary molecular engineering, *Appl. Env. Microbiol.* **2000**, *66*, 890–894.

P. Soumillion and J. Fastrez, Novel concepts for selection of catalytic activity, *Curr. Opin. Biotechnol.* **2001**, *12*, 387–394.

W. M. Spears, *Evolutionary Algorithms: The Role of Mutation and Recombination*, Springer-Verlag, New York, **1998**.

B. Steipe, Evolutionary approaches to protein engineering, *Curr. Top. Microbiol. Immunol.* **1999**, *243*, 55–86.

W. P. C. Stemmer, Rapid evolution of a protein in vitro by DNA shuffling, *Nature* **1994a**, *370*, 389–391.

W. P. Stemmer, DNA shuffling by random fragmentation and reassembly: *in vitro* recombination for molecular evolution, *Proc. Natl. Acad. Sci. USA* **1994b**, *91*, 10747–10751.

S. J. Taylor, R. C. Brown, P. A. Keene, and I. N. Taylor, Novel screening methods – the key to cloning commercially successful biocatalysts, *Bioorg. Med. Chem.* **1999** *7*, 2163–2168.

S. V. Taylor, P. Kast, and D. Hilvert, Investigating and engineering enzymes by genetic selection, *Angew. Chem. Int. Ed.* **2001**, *40*, 3310–3335.

R. A. Vacca, S. Giannattasio, R. Graber, E. Sandmeier, E. Marra, and P. Christen, Active-site Arg→Lys substitutions alter reaction and substrate specificity of aspartate aminotransferase, *J. Biol. Chem.* **1997**, *272*, 21932–21937.

J. P. Vartanian, M. Henry, and S. Wain-Hobson, Hypermutagenic PCR involving all four transitions and a sizeable proportion of transversions, *Nucleic Acids Research* **1996**, *24*, 2627–2631.

D. Wahler and J. L. Reymond, Novel methods for biocatalyst screening, *Curr. Opin. Chem. Biol.* **2001**, *5*, 152–158.

N. J. Wymer, L. V. Buchanan, D. Henderson, N. Mehta, C. H. Botting, L. Pocivavsek, C. A. Fierke, E. J. Toone, and J. H. Naismith, Directed evolution of a new catalytic site in 2-keto-3-deoxy-6-phosphogluconate aldolase from *Escherichia coli*, *Structure* **2001**, *9*, 1–10.

T. Yano, S. Oue, and H. Kagamiyama, Directed evolution of an aspartate aminotransferase with new substrate specificities, *Proc. Natl. Acad. Sci. USA* **1998**, *95*, 5511–5515.

L. You and F. H. Arnold, Directed evolution of subtilisin E in *Bacillus subtilis* to enhance total activity in aqueous dimethylformamide, *Protein Eng.* **1996**, *9*, 77–83.

M. Zaccolo and E. Gherardi, The effect of high-frequency random mutagenesis on *in vitro* protein evolution: a study on TEM-1 beta-lactamase, *J. Mol. Biol.* **1999**, *285*, 775–783.

J. H. Zhang, T. D. Chung, and K. R. Oldenburg, A simple statistical parameter for use in evaluation and validation of high throughput screening assays, *J. Biomol. Screen.* **1999**, *4*, 67–73.

H. Zhao and F. H. Arnold, Combinatorial protein design: strategies for screening protein libraries, Curr. Opin. Struct. Biol. **1997**, *7*, 480–485.

H. M. Zhao, L. Giver, Z. X. Shao, J. A. Affholter, and F. H. Arnold, Molecular evolution by staggered extension process (StEP) in vitro recombination, *Nature Biotechnol.* **1998**, *16*, 258–261.

12
Biocatalysis in Non-conventional Media

Summary

Contrary to expectations that enzymes are only active in aqueous solution, activity in almost anhydrous organic solvents was already demonstrated in the 1930s and rediscovered in 1977. It was not water-miscible hydrophilic solvents such as methanol or acetone that proved to be the best reaction media, but hydrophobic water-immiscible solvents such as toluene or cyclohexane. Supposedly, the cause is the partitioning of water between the enzyme surface and the bulk phase of the organic solvent. As comparably hydrophilic solvents such as methanol or acetone can take up basically infinite amounts of water, they strip the remaining water molecules off the enzyme surface. As a consequence, the enzyme is no longer active because it requires a small but measurable amount of water for developing its activity.

The advantages of organic media for enzyme reactions are: (i) an enhancement of the solubility of reactants; (ii) a shift of equilibria in organic media; (iii) easier separation of organic solvents than water; (iv) enhanced stability of enzymes in organic solvents; and (v) altered selectivity, including substrate specificity, enantioselectivity, prochiral selectivity, regioselectivity, and chemoselectivity, of enzymes in organic solvents. In water (with no organic phase present) the yield of the ester is about 0.01%, whereas in a biphasic water–water-immiscible organic solvent system consisting of porous glass impregnated with aqueous buffer solution it is practically 100%. Upon dehydration, many enzymes become extremely thermostable, up to and beyond 100 °C.

Most categories of enzymes have been found to be active in organic solvents, not just lipases but also proteases, dehydrogenases, peroxidases, and several others. Enzymatic reactions in organic solvents follow saturation kinetics; mechanisms and active sites have been found to be the same as in aqueous solution. Kinetic constants, however, do not correlate with a few simple solvent parameters such as hydrophobicity, dielectric constant, or dipole moment; instead, case-by-case correlations are found.

While enzymes in organic solvents often display activities that are orders of magnitude of lower than in water, there are several activation mechanisms which each yield about an order of magnitude of improvement: (i) sufficient hydration in organic solvent (between 1% v/v and saturation); (ii) lyophilization at the pH of maximum activity in water; (iii) addition of lyoprotectants such as polyols; (iv) addition of hydrophobic binding pocket protectors such as phenols or anilines; and (v) lyophilization in strong salts such as KCl (if present at > 90% w/w).

Whereas "protein engineering" covers knowledge of improving the biocatalyst, the complementary "medium engineering" optimizes the reaction medium, resulting in improved prochiral, regio-, and enantioselectivity. Many such changes have been observed, demonstrating that not only the catalyst but also the medium exerts a decisive influence on the outcome of an enzymatic reaction.

12.1 Enzymes in Organic Solvents

Conventional wisdom says that enzymes are active in water and in water only, just as in nature; according to these ideas, enzymes immediately deactivate in organic solvents. This picture is wrong for several reasons:

- in nature, enzymes are not active in water but in the proximity of cell membranes, so a double layer of lipid molecules or a micelle would be a more suitable environment to emulate than an aqueous bulk phase;
- enzymes do not necessarily deactivate in organic solvents – other phenomena such as inaccessibility owing to insolubility or denaturation during lyophilization can account for lack of activity.

Despite these arguments and classic papers demonstrating activity of enzymes in partially or wholly organic solvents (Sym, 1936), results from the Lomonosov University in Moscow, Russia, at the end of the 1970s came somewhat as a surprise: Klibanov et al. demonstrated esterification of *N*-acetyl-L-tryptophan with ethanol in chloroform (Klibanov, 1977). As porous glass was impregnated with aqueous buffer solution, Klibanov et al. termed this system a biphasic water–water-immiscible organic solvent system. They considered the system to consist of two phases, the organic phase and the aqueous phase on the glass bead, noted the shift of equilibria for water-forming reactions in nearly anhydrous media and provided equations modeling the situation. In water (no organic phase) the yield of the ester is about 0.01%, whereas in the biphasic system it is practically 100%.

However, it was not water-miscible hydrophilic solvents such as methanol or acetone that proved to be the best reaction media but, on the contrary, hydrophobic water-immiscible solvents such as toluene or cyclohexane. Supposedly, the cause is the partitioning of water between the enzyme surface and the bulk phase of the organic solvent. As comparably hydrophilic solvents such as methanol or acetone can take up basically infinite amounts of water, they strip the remaining water molecules off the enzyme surface. As a consequence, the enzyme is no longer active because it requires a small but measurable amount of water for developing its activity.

What are the advantages of organic solvents as media for biocatalytic reactions?

1. The main advantage is an enhanced solubility of many, mostly hydrophobic, reactants insoluble or only sparingly soluble in water. Even for reactants with a moderate solubility in water, that solubility can often limit the attainable reaction rate.
2. In nearly anhydrous media, equilibria of hydrolytic reactions can be shifted towards condensation. In this fashion, reactions are possible in organic media whose equilibrium constants favor the reactants to such a degree that they alone are present in water.
3. As the enzyme in most cases is not dissolved in hydrophobic media but only suspended, it can be separated relatively easily after the reaction is finished. As the enzyme is not immobilized, connected disadvantages such as low activity yields or mass transfer limitations are avoided.
4. Enzymes are sometimes more stable in anhydrous organic solvents than in water. Lysozyme in water at 100 °C was demonstrated to be 50% deactivated after 30 s (pH 8) or after 100 min (pH 4), but after 140 h in cyclohexane or even after 200 h as a dry powder (Zaks, 1986a).
5. Owing to the influence of the medium, enzymes should feature different substrate specificities and selectivities than water. This would allow for tuning of these parameters under the control of the operator.
6. Organic solvents can be recycled more easily than water with its known high enthalpy of vaporization.

12.2 Evidence for the Perceived Advantages of Biocatalysts in Organic Media

12.2.1 Advantage 1: Enhancement of Solubility of Reactants

The rates of asymmetric sulfoxidation of thioanisole in nearly anhydrous (99.7%) isopropyl alcohol and methanol catalyzed by horseradish peroxidase (HRP) were determined to be tens to hundreds of times faster than in water under otherwise identical conditions (Dai, 2000). Similar effects were observed with other hemoproteins. This dramatic activation is due to a much higher substrate solubility in organic solvents than in water and occurs even though the intrinsic reactivity of HRP in isopropyl alcohol and in methanol is hundreds of times lower than in water. In addition, the rates of spontaneous oxidation of the model prochiral substrate thioanisole in several organic solvents was observed to be some 100- to 1000-fold slower than in water. This renders peroxidase-catalyzed asymmetric sulfoxidations synthetically attractive.

12.2.2
Advantage 2: Shift of Equilibria in Organic Media

12.2.2.1 Biphasic Reactors

Multi-phase reactors are of interest in biocatalytic reactions if one or several components of the reaction are insoluble or insufficiently soluble in aqueous phases but if an aqueous phase has to be kept, if only for the biocatalyst. However, a two-phase system can be utilized advantageously for the shifting of an equilibrium; this is demonstrated below. We analyze the simple reaction A ⇔ B in an organic–aqueous two-phase system with the assumption that reactant A and product B partition between the two phases. The partition coefficients P_w and P_{org} are defined by Eq. (12.1).

$$P_X = [X]_{org}/[X]_w \tag{12.1}$$

In both phases, equilibrium exists between A and B [Eq. (12.2)].

$$K_w = [B]_w/[A]_w \quad \text{and} \quad K_{org} = [B]_{org}/[A]_{org} \tag{12.2}$$

A biphasic equilibirum constant also can be defined as in Eq. (12.3).

$$K_{biphas} = [B]_{tot}/[A]_{tot} \tag{12.3}$$

$[X]_{tot}$ can be linked to $[X]_{org}$ and $[X]_w$ via a mass balance [Eq. (12.4)].

$$[X]_{tot}V_{tot} = [X]_w V_w + [X]_{org} V_{org} \tag{12.4}$$

In addition, the sum $V_{tot} = V_w + V_{org}$ is valid, so insertion of Eq. (12.4) into Eq. (12.3) yields Eq. (12.5).

$$K_{biphas} = \{[B]_w V_w + [B]_{org} V_{org}\}/\{[A]_w V_w + [A]_{org} V_{org}\} \tag{12.5}$$

Upon division by $[A]_w V_w$ and introduction of the phase ratio α ($\alpha = V_{org}/V_w$) we finally obtain Eq.:

$$K_{biphas} = K_w \cdot \{(1 + \alpha P_B)/(1 + \alpha P_A)\} \tag{12.6}$$

It can be seen from this equation that the effective equilibrium constant $K_{bisphas}$ is a function of the phase ratio α. Within limits, the equilibrium can be shifted in the desired direction by controlling and shifting α. If the reaction is of the form A + B ⇔ C + D, Eq. (12.7) is equivalent to Eq. (12.6).

$$K_{biphas} = K_w \cdot \{(1 + \alpha P_C)(1 + \alpha P_D)/(1 + \alpha P_A)(1 + \alpha P_B)\} \tag{12.7}$$

Figure 12.1 further elucidates the connection between K_{biphas} and α:

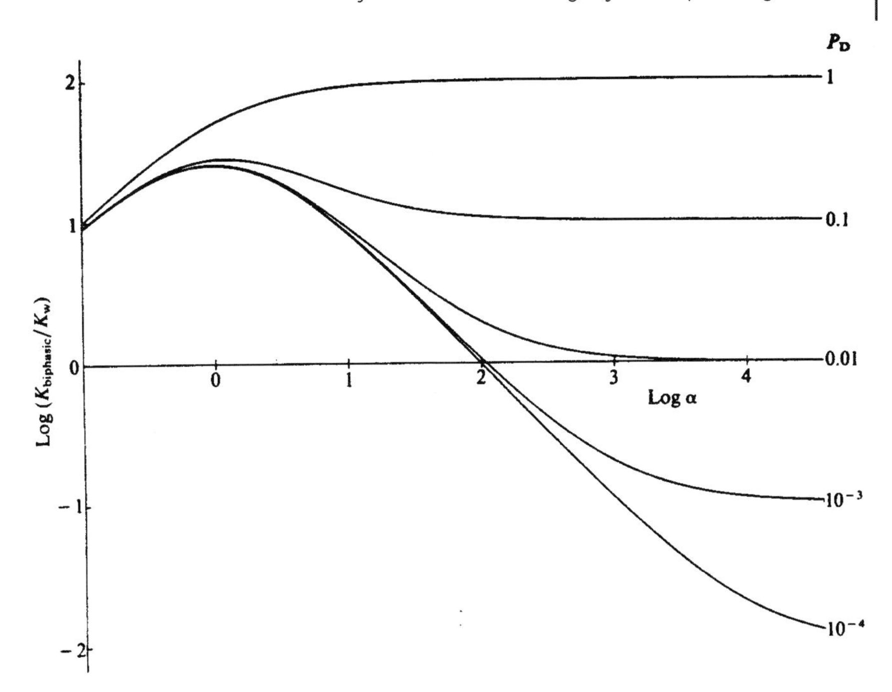

Figure 12.1 Dependence of the effective equilibrium constant K_{biphas} on the phase ratio α (from Chaplin, 1990).

12.2.3 Advantage 3: Easier Separation

In comparison to the high enthalpy of evaporation of water (57.36 kJ mol^{-1}), the respective values are much lower for organic solvents, which are thus much easier to separate than water. Table 12.1 lists some boiling points and enthalpies of evaporation for common organic solvents.

Table 12.1 Boiling points and enthalpies of evaporation for common organic solvents.

	Solvent										
	MTBE	Acetone	CH_3OH	THF	Hexane	DIPE	EA	C_2H_5OH	C_6H_{12}	Toluene	DMF
B.p. [°C]	55.2	56.2	65.0	67.0	69.0	69.0	77.1	78.5	80.7	110.6	149
ΔH_{vap} [kJ mol^{-1}]	29.70	32.00	39.26	28.76	31.93	32.56	34.75	40.50	32.79	39.23	60.45

MTBE: methyl *t*-butyl ether; THF: tetrahydrofuran; EA: ethyl acetate; C_6H_{12}: cyclohexane; DIPE: diisopropyl ether; DMF: dimethylformamide.

12.2.4
Advantage 4: Enhanced Stability of Enzymes in Organic Solvents

Porcine pancreatic lipase catalyzes the transesterification reaction between tributyrin and various primary and secondary alcohols in a 99% organic medium (Zaks, 1984). Upon further dehydration, the enzyme becomes extremely thermostable. Not only can the dry lipase withstand heating at 100 °C for many hours, but it exhibits a high catalytic activity at that temperature. Reduction in water content also alters the substrate specificity of the lipase: in contrast to its wet counterpart, the dry enzyme does not react with bulky tertiary alcohols.

During transesterification of amino acid esters with alcohols in the presence of chymotrypsin and subtilisin, both thermal and storage stabilities of chymotrypsin in non-aqueous solvents are observed to be greatly enhanced compared to water (Zaks, 1998).

12.2.5
Advantage 5: Altered Selectivity of Enzymes in Organic Solvents

One of the most profound influences a solvent system can have on a reaction is a change of selectivity. Enzymes in organic solvents have been discovered in many cases to feature altered selectivity, including substrate specificity, enantioselectivity, prochiral selectivity, regioselectivity, and chemoselectivity (Wescott, 1994; Hirose, 1995).

12.3
State of Knowledge of Functioning of Enzymes in Solvents

12.3.1
Range of Enzymes, Reactions, and Solvents

Not only do enzymes work in anhydrous organic media, but in this unnatural milieu they acquire remarkable properties such as enhanced stability, altered substrate and enantiomeric specificities, molecular memory, and the ability to catalyze unusual reactions (Klibanov, 1989). Regarding the latter point, hydrolases, such as lipases, catalyze not only transesterifications in organic media but also other types of reactions, including esterification, aminolysis, thiotransesterification, and oximolysis. As all of these reactions compete with hydrolyses, which tend to dominate in aqueous media, some of them proceed to an appreciable extent only in non-aqueous solvents.

Soon after the initial discovery, it became apparent that neither the source of the enzyme, nor the type of enzyme, nor the type of solvent seem to constrain the use of organic solvents (Zaks, 1986a). Various types of enzymes, such as lipases, proteases (chymotrypsin, subtilisin), oxidoreductases (alcohol dehydrogenase, oxidases, and peroxidases), and others, react in organic solvents. A selection of enzymes

Table 12.2 A selection of enzymes active in organic solvents.

Class	E.C. number	Enzyme	Reaction	Reference
Oxido-reductases	1.1.1.1	yeast ADH	ethanol oxidation	Deetz, 1988
	1.1.1.1	horse-liver ADH	alcohol oxidation	Zaks, 1988b
	1.1.3.13	alcohol oxidase	ethanol oxidation	Zaks, 1988b
	1.11.1.	HR peroxidase	sulfoxidation	Dai, 2000
Hydrolases	3.1.1.3	PP lipase	transesterification	Zaks, 1984
	3.2.1.17	lysozyme		Zaks, 1986a
	3.4.21.1	α-chymotrypsin	transesterification	Zaks, 1988a
	3.4.21.62	subtilisin	transesterification	many
	3.4.21.4	trypsin	transesterification	van Unen, 2001
	3.4.24.27	thermolysin	dipeptide synthesis	Bedell, 1998
	3.5.1.5	urease	hydrolysis of urea	Ghatorae, 1994
	3.5.1.11	Pen G acylase	resolution of β-amino esters	Roche, 1999
Lyases	4.1.2.10	(*R*)-oxynitrilase	HCN addition to C=O	Wajant, 1996
	4.2.1.84	nitrile hydratase	H_2O addition to nitriles	Thomas, 2002

HR peroxidase: horseradish peroxidase; PP lipase: porcine pancreatic lipase.

found to be active is listed in Table 12.2. Lipases from three different sources, porcine pancreatic, yeast, and mold, act vigorously as catalysts in a number of nearly anhydrous organic solvents. The range of solvents encompasses the whole spectrum from hydrophilic ones such as methanol, isopropanol, and acetonitrile via THF to the hydrophobic ones such as dialkyl ethers, toluene, hexane, cyclohexane, and heptane.

12.3.2 The Importance of Water in Enzyme Reactions in Organic Solvents

12.3.2.1 Exchange of Water Molecules between Enzyme Surface and Bulk Organic Solvent

Early after the (re)discovery of enzyme activity in organic solvents, the higher level of activity in very hydrophobic solvents, such as hydrocarbons, in comparison with rather hydrophilic solvents, such as alcohols, was noted. Soon the hypothesis was raised that water molecules partition between the enzyme surface and the bulk organic solvent, and they are "stripped off" the enzyme surface in dry organic solvents. This idea explained why hydrophilic solvents, which strip water off an enzyme surface more readily than hydrophobic ones, deactivate enzymes much more readily than the latter. The hypothesis was proven correct by exchange of water on the enzyme surface by tritiated water, T_2O: suspension of the tritiated enzyme in the organic solvent, filtration of the enzyme, and measurement of the amount of tritiated water in the solvent by scintillation counting demonstrated that several different enzymes (α-chymotrypsin, Subtilisin Carlsberg, and horseradish peroxidase) yielded water molecules immediately after suspension in the solvent. Depending on the solubility of water in the solvent, different amounts of T_2O were desorbed, from 0.4% in hexane to 62% in methanol (Gorman, 1992).

The amount of water required by chymotrypsin and subtilisin for catalysis in organic solvents is several hundred molecules per protein molecule, less than required to form a monolayer on the surface. While subtilisin and α-chymotrypsin act as catalysts in a variety of dry organic solvents, the vastly different catalytic activities in these organic solvents are partly due to stripping of the essential water from the enzyme by more hydrophilic solvents and partly due to the solvent directly affecting the enzymatic process.

Enzymes as different as yeast alcohol oxidase, mushroom polyphenol oxidase, and horse-liver alcohol dehydrogenase demonstrated vastly increased enzymatic activity in several different solvents upon an increase in the water content, which always remained below the solubility limit (Zaks, 1988b). While much less water was required for maximal activity in hydrophobic than in hydrophilic solvents, relative saturation seems to be most relevant to determining the level of catalytic activity. Correspondingly, miscibility of a solvent with water is not a decisive criterion: upon transition from a monophasic to a biphasic solvent system, no significant change in activity level was observed (Narayan, 1993). Therefore, the level of water essential for activity depends more on the solvent than on the enzyme.

12.3.2.2 **Relevance of Water Activity**

While water level itself seems to be of no significance for enzyme activity in organic solvents, the relative amount of water with respect to total water solubility seems to be very influential. Correspondingly, instead of using concentration c_w or mole fraction x_w of water as a measure of the amount of water, *water activity* a_w should be preferred. [Activity, linked to concentration by the activity coefficient γ_w, $a_w = \gamma_w \cdot c_w$, written here for the case of water, is the thermodynamically rigorous way to consider amounts per volume in thermodynamic formulae such as mass action laws.] At the solubility limit in any solvent, water activity is always equal to unity, regardless of the mole fraction. The transesterification of vinyl butyrate with *n*-octanol with *Pseudomonas cepacia* lipase yielded increasing K_M values with rising water activity; however, v_{max} values demonstrated a maximum at water activity levels a_w between 0.11 and 0.38 (Bovara, 1993).

Activity maxima at certain a_w values have also been found with many other enzymes, such as optima at $a_w = 0.55$ for *Mucor miehei* lipase-catalyzed reaction in several solvents of different polarity from hexane to 3-pentanone (Valivety, 1992a). In comparing five different lipases, considerable variations not just of the optima of a_w but also the dependence of activity on a_w itself were found (Valivety, 1992b). Moreover, in polar solvents, correlations of activity with a_w often cannot be found at all, probably owing to solvent effects beyond those controlling hydration of the enzyme (Bell, 1997).

Summarizing the findings above, control of water activity a_w during enzyme reactions in organic solvents is extremely important, as water activity exerts a crucial influence, and enzyme reactivity crucially depends on it.

12.3.3
Physical Organic Chemistry of Enzymes in Organic Solvents

12.3.3.1 Active Site and Mechanism

In organic solvents, enzymes react at the same active site as in water; covalent modification of the active center of the enzymes by a site-specific reagent renders them catalytically inactive in organic solvents (Zaks, 1988a). Upon replacement of water with octane as the reaction medium, the specificity of chymotrypsin towards competitive inhibitors reverses.

Early results with enzymes in organic solvents demonstrated that enzymes work best in organic solvents if there are lyophilized from aqueous solution at or near the optimum pH value in water (seeSection 12.4, below). Any specific pH effect of organic solvents, however, can be dismissed: it was found that the p*K*a of amino, carboxylic, and phenolic compounds does not differ by more than 0.3 units in the aqueous and lyophilized states (Costantino, 1997).

Enzymatic transesterifications in organic solvents follow Michaelis–Menten kinetics, and in quite a few cases, the values of v/K_M roughly correlate with the solvent's hydrophobicity (Zaks, 1988a). The dependence of the catalytic activity of the enzyme in organic media on the pH of the aqueous solution from which it was recovered is bell-shaped, with the maximum coinciding with the pH optimum of the enzymatic activity in water.

Hammett analysis and finding of corresponding LFERs (see Chapter 9, Section 9.2.3) suggests that the mechanism of an enzyme reaction in organic solvents is the same as in aqueous solution (Kanerva, 1989).

12.3.3.2 Flexibility of Enzymes in Organic Solvents

From experiments on (i) replacement of water with other hydrogen bond forming additives and (ii) titration of enzyme amino groups in an organic medium, as well as the literature data on dehydrated enzymes, it is concluded that the water required by enzymes in non-aqueous solvents provides them with sufficient conformational flexibility for catalysis (Zaks, 1988b).

The activity of lyophilized HL-ADH in several solvents increased by orders of magnitude upon addition of small amounts of water to solvents of dielectric constant ε from 1.9 to 36 (Guinn, 1991). Enzyme flexibility, as measured by electron spin resonance (ESR) spectroscopy with a spin label at the active site, did not depend on water content but on ε of the pure solvent: HL-ADH turns less flexible with decreasing ε.

Correlating positively with the hydrophobicity of the solvent, different fractions of inactivated active centers were measured in different solvents with solid-state NMR spectroscopy (^{13}C-cross-polarization/magic angle spinning (MAS) NMR) (Burke, 1992). Just as with tritiated water (see above), immediate desorption of water molecules from the protein surface was observed after addition to the organic solvent.

In some instances, a relationship between enzyme enantioselectivity and flexibility has been found: while there are some positive examples in the literature

(Broos, 1995), other investigations either found a slightly negative correlation or did not find any correlation at all (Rariy, 2000).

12.3.3.3 **Polarity and Hydrophobicity of Transition State and Binding Site**

As enzyme active sites commonly have to be totally anhydrous to allow chemical catalysis to occur, substrates have to be desolvated when transferring from solvent to active site. The effect of six solvents ranging from water to hexane on the standard test reaction in organic solvents, the transesterification of *N*-acetyl-L-Phe-OEt with *n*-PrOH, with the help of Subtilisin Carlsberg, has been interpreted by taking into account substrate desolvation, as measured by relative substrate solubility (Kim, 2000). Analysis of a thermodynamic cycle, analogous to the one in Figure 2.2 (Chapter 2), between ground (E + S) and transition state ($ES^{\neq}$) in the solvent and in acetone yields Eq. (12.8).

$$\Delta\Delta G^{\neq}{}_{obs} = \Delta\Delta G^{\neq}{}_{intr} + \Delta G_s = \Delta G^{\neq}{}_{ES} - \Delta G_E \tag{12.8}$$

Equation (12.8) states that the apparent and intrinsic differential activation energies $\Delta\Delta G^{\neq}{}_{obs}$ and $\Delta\Delta G^{\neq}{}_{intr}$ differ by the differential Gibbs free solvation energy ΔG_s, with ΔG_s relating the saturation solubilities of *N*-Ac-L-Phe-OEt in a solvent to acetone as the standard state [Eq. (12.9)], so that, with $\Delta G = \Delta H - T \cdot \Delta S$, Eq. (12.10) follows.

$$\Delta G_s = -\ RT \ln ([S]_{solvent}/[S]_{acetone}) \tag{12.9}$$

$$\Delta\Delta G^{\neq}{}_{obs} = \Delta\Delta H^{\neq}{}_{intr} - T \cdot \Delta\Delta S^{\neq}{}_{intr} - RT \ln ([S]_{solvent}/[S]_{acetone}) \tag{12.10}$$

While $\Delta\Delta G^{\neq}{}_{intr}$ failed to yield a straight-line correlation, an LFER (see Chapter 9, Section 9.2.3), with log ε (manifesting solvent polarity) and $\Delta\Delta G^{\neq}{}_{obs}$ in fact did feature such a correlation. The results indicate that substrate desolvation is important and that the intrinsic activation energy of subtilisin catalysis is lowest in polar organic solvents, which may be due to transition-state stabilization of the enzyme's polar transition state for transesterification. Furthermore, the results strongly point towards the same transition state in water and organic solvents across a wider range of solvent polarity ε.

The dramatic activation of serine proteases in non-aqueous media owing to prior lyophilization in the presence of KCl could be due in principle to the relaxation of potential substrate diffusional limitations in clumps of heterogeneous, undissolved enzyme in organic solvent. This hypothesis was checked with a classic experiment for diffusional limitation: Subtilisin Carlsberg was lyophilized with KCl and phosphate buffer in varying proportions ranging from 1 to 99% (w/w) enzyme in the final lyophilized solids (Bedell, 1998). The active enzyme content of a given biocatalyst preparation was controlled by mixing native subtilisin with subtilisin preinactivated with the serine protease inhibitor PMSF and lyophilizing the enzyme mixture in the presence of different fractions of KCl and phosphate buffer. Initial catalytic rates, normalized per weight of total enzyme in the catalyst material, were measured as a function of active enzyme for biocatalyst preparations containing different ratios of active to inactive enzyme.

Plots of initial reaction rates as a function of percentage active subtilisin in the biocatalyst were found to be linear for all biocatalyst preparations. Thus, enzyme activation, as high as 3750-fold in hexane for the transesterification of *N*-Ac-L-Phe-OEt with *n*-PrOH, is a manifestation of intrinsic enzyme activation and not of relaxation of diffusional limitations resulting from diluted enzyme preparations. Thus, the above-mentioned hypothesis was refuted. As similar results were found for the metalloprotease thermolysin, activation due to lyophilization in the presence of KCl may be a general phenomenon.

12.3.4
Correlation of Enzyme Performance with Solvent Parameters

Given the multitude of potential organic solvents for enzyme reactions, design rules or at least heuristics were sought early on for picking organic solvents without trying a large number of them experimentally Organic solvents can be described and classified according to several criteria, such as volatility (melting and boiling points, T_m and T_{boil}), size (molar mass, M [g mol^{-1}]), accommodation of charge (dielectric constant ε), hydrophobicity (water/octanol partition coefficient, log P), distribution of charge (dipole moment μ), or polarizability (polarization constant α); for data, see Table 12.3. The water/octanol partition coefficient log P of a component A is defined by Eq. (12.11), where A is in dilute solution below the solubility limit.

$$\log P = \log \{[A]_{water}/[A]_{n\text{-octanol}}\} \quad (12.11)$$

n-Octanol is chosen as a reference because it mimics biological membranes with a hydrophobic tail and a hydrophilic head capable of hydrogen bonding.

Table 12.3 Some properties of common organic solvents.

	Solvent										
	MTBE	Acetone	CH_3OH	THF	Hexane	DIPE	EA	C_2H_5OH	C_6H_{12}	Toluene	DMF
B.p. [°C]	55.2	56.2	65.0	67.0	69.0	69.0	77.1	78.5	80.7	110.6	153
M [g/mol]	88.15	58.08	32.04	72.10	86.17	102.17	88.10	46.07	84.16	92.13	73.09
ε [–]	4.5	20.7	32.63	7.58	1.89	3.23	6.02	24.3	2.02	2.38	36.7
log P	1.15	– 0.23	– 0.76	0.49	3.5	1.9	0.68	– 0.24	3.2	2.5	– 1.0
μ [D]	1.23	2.70	1.71	1.74	0.0	1.24	1.83	1.74	0.0	0.30	3.24

MTBE: methyl *t*-butyl ether; THF: tetrahydrofuran; EA: ethyl acetate;
C_6H_{12}: cyclohexane; DIPE: diisopropyl ether; DMF: dimethylformamide.

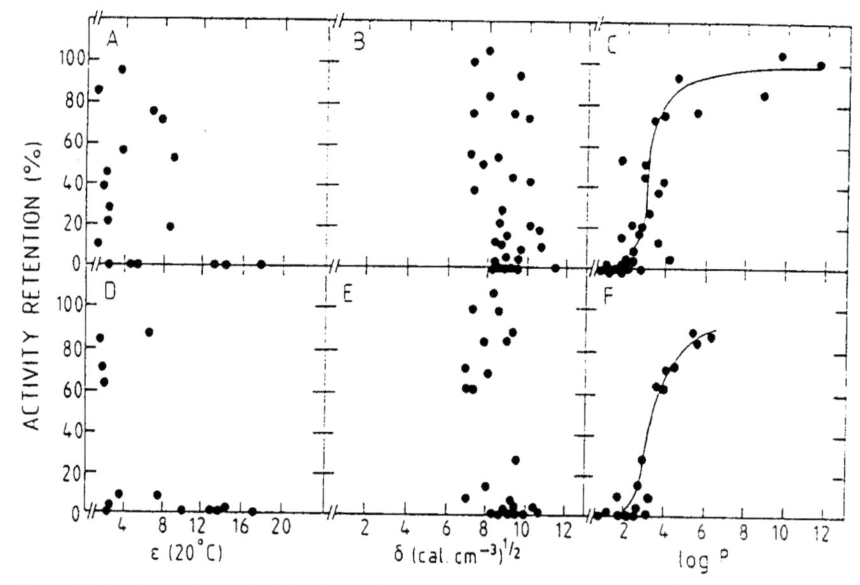

Figure 12.2 Correlation of activity data with solvent parameters (Laane, 1987).

12.3.4.1 Control through Variation of Hydrophobocity: log P Concept

In an influential early investigation, correlation of biocatalytic activity data of aerobic and anaerobic whole-cell biocatalysis with solvent properties resulted in the strongest correlation for the partition coefficient log *P*, whereas both Hildebrand's solubility parameter δ and the dielectric constant ε showed either a weak correlation with activity data or none at all (Laane, 1985, 1987) (Figure 12.2).

This simple picture, unfortunately, did not hold for all enzyme reactions in organic solvents.

12.3.4.2 Correlation of Enantioselectivity with Solvent Polarity and Hydrophobicity

The enantioselectivity of the transesterification of vinyl butyrate with *sec*-phenylethanol was influenced greatly by the solvent; however, not log *P* (hydrophobicity) but both the dielectric constant ε and the dipole moment μ of the solvent correlated best with the *E* values (Fitzpatrick, 1992). Table 12.4 demonstrates the correlation of μ with *E*.

Table 12.4 Correlation of the enantioselectivity *E* of subtilisin in the transtesterification of racemic 1-phenylethanol in various organic solvents, and the dielectric constants ε of the solvents (Fritzpatrick, 1992).

Solvent	ε_r	E	Solvent	ε_r	E
Dioxane	2.2	61	Dimethylformamide	36.7	9
Benzene	2.3	54	Nitromethane	35.9	5
Triethylamine	2.4	48	Acetonitrile	35.9	3
Tetrahydrofuran	7.6	40	Methylacetamide	191.3	3
pyridine	12.9	31			

Table 12.5 Effects of the hydrophobicity (log *P*) of solvents on the prochiral selectivity (*S*) of *Pseudomonas sp. lipase* in the monohydrolysis of 2-(1-naphthoylamino)trimethylene dibutyrate (Terradas, 1993).

Solvent	*log P*	*S*
Acetonitrile	–0.33	> 30
Nitrobenzene	1.8	> 30
Acetone	–0.23	18
Cyclohexanone	0.96	18
Butanone	0.29	16
2-Pentanone	0.80	16
Chloroform	2.0	9.9
Tetrahydrofuran	0.49	9.1
2-hexanone	1.3	8.8
Dioxane	–1.1	5.4
t-Butyl acetate	1.7	5.3
t-Butyl alcohol	0.80	4.9
t-Amyl alcohol	1.4	4.8
Triethylamine	1.6	4.7
Toluene	2.5	3.5
Benzene	2.0	3.2
Carbon tetrachloride	3.0	2.6

In a third example, the prochiral selectivity results of the *Pseudomonas* sp. lipase-catalyzed monohydrolysis of 2-(1-naphthoylamino)trimethylene dibutyrate correlate inversely with hydrophobicity, as expressed by log *P* (Table 12.5; Terradas, 1993).

The three examples provided in this section demonstrate that each case has to be considered individually. None of the solvent parameters has been found to correlate reliably with enzyme activity or selectivity data, except in individual cases; predictions at this point are not possible.

12.4
Optimal Handling of Enzymes in Organic Solvents

While different enzyme reactions yield different results in different solvents, thus rendering comparison difficult, a procedure akin to a "standard protocol" for conducting enzyme reactions in organic solvents has been developed. It calls for

- lyophilization at optimum pH in aqueous solution;
- meeting the requirement for water for enzyme function; and
- testing the enzyme with a standard reaction under standard conditions.

In the case of subtilisin, the standard reaction is the transesterification of *N*-acetyl-L-phenylalanine ethyl ester with *n*-propanol [Eq. (12.12)].

$$N\text{-Ac-L-Phe-OEt} + n\text{-PrOH} \rightarrow N\text{-Ac-L-Phe-OPr} + \text{EtOH} \qquad (12.12)$$

More data have been obtained with this reaction on enzymes in organic solvents than with any other.

12.4.1
Enzyme Memory in Organic Solvents

Repeatedly, researchers have found the phenomenon that enzyme molecules seem to "remember" the aqueous conditions from which they were prepared (Klibanov, 1995). As mentioned above, early results with enzymes in organic solvents demonstrated that enzymes work best in organic solvents if there are lyophilized from aqueous solution at or near the optimum pH value in water. No specific pH effect of organic solvents on enzyme reactions was found: the pK_a of amino, carboxylic, and phenolic compounds does not differ by more than 0.3 units in the aqueous and lyophilized state (Costantino, 1997).

Catalytic activities of α-chymotrypsin and Subtilisin Carlsberg in various hydrous organic solvents were measured as a function of how the enzyme suspension had been prepared (Ke, 1998). Direct suspension of the lyophilized enzyme in the solvent containing 1% water was compared with precipitation of the same enzyme from its aqueous solution by a 100-fold dilution with anhydrous solvent. The reaction rate in a given non-aqueous enzymatic system was found to depend on the nature of both enzyme and solvent, but to depend strongly on the mode of enzyme preparation.

The common mode of suspending lyophilized enzymes in organic solvents containing very little water results in just a low specific activity. Much higher specific activity can be achieved if the enzymes are dissolved in water first and then diluted with anhydrous organic solvent to the same water content (Dai, 1999); the lower the water content of the medium, the greater the discrepancy.

The mechanism of this phenomenon, termed *"lyophilization-induced inactivation"*, was found to arise from reversible denaturation of the oxidases on lyophilization: because of its conformational rigidity, the denatured enzyme exhibits very limited activity when directly suspended in largely non-aqueous media but renatures and thus yields much higher activity if first redissolved in water. Lyophilization-induced inactivation could be minimized, at least for the case of four oxidases employed, by at least two strategies, both involving the addition of excipients to the aqueous enzyme solution before lyophilization:

- addition of lyoprotectants such as polyols or polyethylene glycol to preserve the overall enzyme structure during dehydration;
- addition of phenolic and aniline substrates that presumably bind to the hydrophobic pocket of the enzyme active site.

Not only was the effect of these excipients found to extend activity by up to orders of magnitude but the effect of the two excipient groups was found to be additive, affording up to complete protection against lyophilization-induced inactivation when representatives of the two are present together.

12.4.2
Low Activity in Organic Solvents Compared to Water

One important issue in the discussion of the biotechnological opportunities afforded by non-aqueous enzymology and its impact on biocatalysis is the often drastically diminished enzymatic activity in organic solvents compared to that in water. The loss of comparative specific activity can be as high as 10^3–10^5. Several recent studies have addressed the issue and have made great strides towards both elucidating causes of the phenomenon of activity loss and designing strategies that systematically enhance activity by multiple orders of magnitude (Klibanov, 1997). The goal ultimately is to bring the activity level of enzymes in organic solvents to aqueous-like level.

Dipeptide synthesis in acetonitrile is found to be enhanced 425-fold in the α-chymotrypsin-catalyzed reaction between the 2-chloroethyl ester of *N*-acetyl-L-phenylalanine and L-phenylalaninamide upon lyophilization of the enzyme in the presence of 50 equivalents of 18-crown-6 (van Unen, 1998). Acceleration is observed in different solvents and for various peptide precursors.

The enhancement of enzyme activity upon addition of 18-crown-6 to the organic solvent can be reconciled with a mechanism in which macrocyclic interactions of 18-crown-6 with the enzyme play an important role (van Unen, 2002). Macrocyclic interactions (e.g., complexation with lysine ammonium groups of the enzyme) can lead to a reduced formation of inter- and intramolecular salt bridges and, consequently, to lowering of the kinetic conformational barriers, enabling the enzyme to refold into thermodynamically stable, catalytically (more) active conformations. This assumption is supported by the observation that the crown-ether-enhanced enzyme activity is retained after removal of the crown by washing with a dry organic solvent. Much stronger crown ether activation is observed when 18-crown-6 is added prior to lyophilization, and this can be explained by a combination of two effects: the before-mentioned macrocyclic complexation effect, and a less specific, non-macrocyclic, lyoprotecting effect. The magnitude of the total crown ether effect depends on the polarity and thermodynamic water activity of the solvent, the activation being highest in dry and apolar media, where kinetic conformational barriers are highest. By determination of the specific activity of crown-ether-lyophilized enzyme as a function of the enzyme concentration, the macrocyclic crown ether (linearly dependent on the enzyme concentration) and the non-macrocyclic lyoprotection effect (not dependent on the enzyme concentration) could be separated. These measurements reveal that the contribution of the non-macrocyclic effect is significantly larger than the macrocyclic refolding effect.

Immobilization in a sol–gel matrix accelerated the propanolysis of *N*-acetyl-L-phenylalanine ethyl ester in cyclohexane for several serine proteases compared to the non-immobilized lyophilized enzymes: 31-fold for Subtilisin Carlsberg, 43-fold for α-chymotrypsin, and 437-fold for trypsin (van Unen, 2001). The activity yield upon immobilization was 90% (α-chymotrypsin). The rate enhancement effect of immobilization on the enzyme activity is highest in hydrophobic solvents.

Table 12.6 Approximate activation factors for enzymes in organic solvents for various strategies.

Action	*Approx. activation factor*
Sufficient hydration in organic solvent (≥ 1%, < saturation)	10
Lyophilization at pH of maximum activity in water	10
Lyoprotectants such as polyols	10
Hydrophobic binding pocket protectors (phenols, anilines)	10
Strong salts such as KCl (if present at > 90% w/w)	10
total:	10^5

The activation mechanisms, with the corresponding activation ratios by which the maximum attainable rate constants of enzymes in organic solvents can be enhanced, are summarized in Table 12.6.

Activation by a factor of 10^5 closes most of the gaps between specific activity levels in water and organic solvents. The state of affairs regarding the preparation of highly active enzyme formulations for use in non-aqueous media is summarized by Lee and Dordick (Lee, 2002). Improved mechanistic understanding of enzyme function and activation in dehydrated environments will lead to the development of a broad array of techniques for generating more active, stable, and enantioselective and regioselective tailored enzymes for synthetically relevant transformations. This, in turn, should result in an increase in the opportunities for enzymatic processes to be developed on a commercial scale.

12.4.3 Enhancement of Selectivity of Enzymes in Organic Solvents

Enzymatic enantioselectivity in organic solvents can be markedly enhanced by temporarily enlarging the substrate via salt formation (Ke, 1999). In addition to its size, the stereochemistry of the counterion can greatly affect the enantioselectivity enhancement (Shin, 2000). In the *Pseudomonas cepacia* lipase-catalyzed propanolysis of phenylalanine methyl ester (Phe-OMe) in anhydrous acetonitrile, the *E* value of 5.8 doubled when the Phe-OMe/(*S*)-mandelate salt was used as a substrate instead of the free ester, and rose sevenfold with (*R*)-mandelic acid as a Brønsted–Lewis acid. Similar effects were observed with other bulky, but not with petite, counterions. The greatest enhancement was afforded by 10-camphorsulfonic acid: the *E* value increased to 18 ± 2 for a salt with its *R*-enantiomer and jumped to 53 ± 4 for the *S*. These effects, also observed in other solvents, were explained by means of structure-based molecular modeling of the lipase-bound transition states of the substrate enantiomers and their diastereomeric salts.

The relatively meager stereoselectivity of horseradish peroxidase in the sulfoxidation of thioanisole in 99.8% (v/v) methanol is vastly improved when the enzyme forms a complex with benzohydroxamic acid (Das, 2002). The generality of the observed "ligand-induced stereoselectivity enhancement" is demonstrated with other hydrophobic hydroxamic acids, as well as with additional thioether substrates, and rationalized by means of molecular dynamics simulations and energy minimization.

12.5
Novel Reaction Media for Biocatalytic Transformations

12.5.1
Substrate as Solvent (Neat Substrates): Acrylamide from Acrylonitrile with Nitrile Hydratase

One of the best examples for discussing biotransformations in neat solvents is the enzymatic hydrolysis of acrylonitrile, a solvent, to acrylamide, covered in Chapter 7, Section 7.1.1.1. For several applications of acrylamide, such as polymerization to polyacrylamide, very pure monomer is required, essentially free from anions and metals, which is difficult to obtain through conventional routes. In Hideaki Yamada's group (Kyoto University, Kyoto, Japan), an enzymatic process based on a nitrile hydratase was developed which is currently run on a commercial scale at around 30 000–40 000 tpy with resting cells of third-generation biocatalyst from *Rhodococcus rhodochrous* J1 (Chapter 7, Figure 7.1).

Characteristic of this process are its extremely high concentration level and space–time-yield (Figure 7.2). Solid nitrile substrates render precipitating product up to a solid medium at high degrees of conversion; liquid nitriles can be run as neat substrate. Figure 12.3 illustrates the connection between substrate concentration up to 15 M (!) and achievable degree of conversion for the example of the transformation of 3-cyanopyridine to nicotinamide discussed in Chapter 7, Section 7.1.1.2.

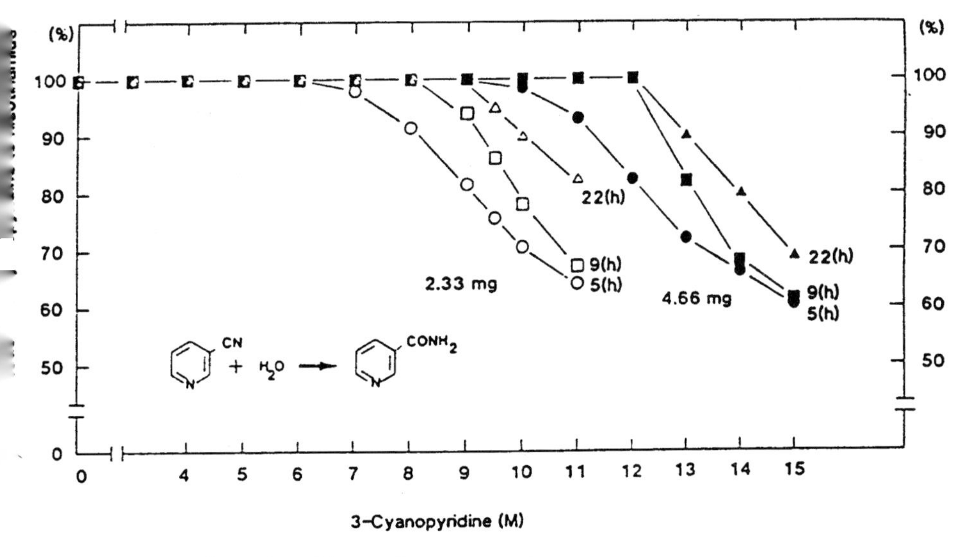

Figure 12.3 Relationship of degree of conversion with substrate concentration for the reaction of cyanopyridine to form nicotinamide (Oyama, 1984).

12.5.2
Supercritical Solvents

Supercritical solvents combine extremely low viscosity as well as diffusivity of gases with the high solubility of compounds in solution. In addition, products can be separated by simply depressurizing the reaction vessel and boiling off the solvent. After having been tested for chemical reactions, supercritical solvents were also tested for enzymatic reactions in around 1985. Owing to the expected dependence of the reaction parameters on temperature and pressure, supercritical solvents should allow the investigation of that influence. Supercritical CO_2 ($SCCO_2$), especially, with its accessible critical parameters ($T_{crit} = 31°C$, $p_{crit} = 50$ bar) offers access to this technique.

However, results demonstrated that $SCCO_2$ does not show good results for all reactions. It was found for the synthesis of acrylates by transesterification of methylacrylate with lipase from *Candida rugosa* that owing to its hydrophilic character $SCCO_2$ was inferior to more hydrophobic supercritical solvents such as propane, ethane, or even $SCSF_6$ with respect to reactivity (Kamat, 1992).

A comparison between $SCCO_2$ and *n*-hexane for the esterification of oleic acid showed that in $SCCO_2$ inhibition by ethanol was less important but that v_{max} in *n*-hexane was about an order of magnitude higher. Enzyme stability was comparable and satisfactory in both cases (Marty, 1990).

For a review of the field of biocatalysis in supercritical solvents, see Mesiano (1999).

12.5.3
Ionic Liquids

Ionic liquids are composed of organic cations with a compatible anion; examples are [BMIM][PF_6] and [MMIM][$MeSO_4$] (Figure 12.4). Unlike conventional organic solvents, ionic liquids possess no vapor pressure but good solvent properties, and a wide range of thermal stability over a wide range of liquid state (≤ 300°C); also, they can be used to form two-phase systems with many solvents (Kragl, 2002).

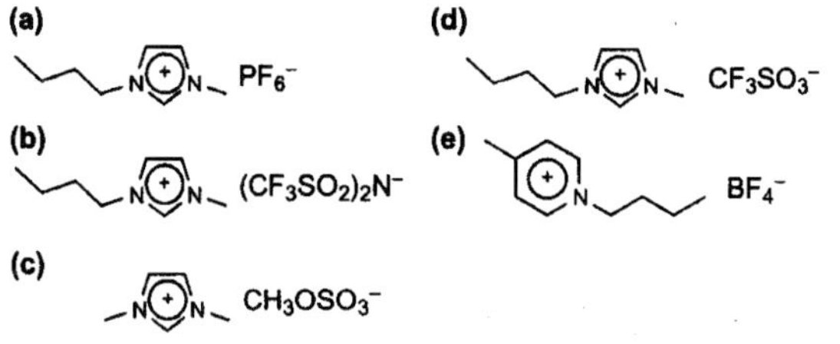

Figure 12.4 Some ionic liquids (Kragl, 2002):
a) 1-butyl-3-methylimidazolium hexafluorophosphate, [BMIM][PF_6];
b) 1-butyl-3-methylimidazolium bis((trifluoromethyl)sulfonyl)amide, [BMIM][$CF_3SO_2]_2N$;
c) 1-methyl-3-methylimidazolium methylsulfate, [MMIM][$MeSO_4$];
d) 1-butyl-3-methylimidazolium triflate, [BMIM][CF_3SO_3];
e) 1-butyl-4-methylpyridium tetrafluoroborate, [MBPy][BF_4].

Given these properties, they can extend the application of solvent engineering to biocatalytic reactions. Instead of just serving as a replacement for organic solvents, they often lead to improved process performance. The first report on thermolysin-catalyzed Z-aspartame synthesis from Z-L-aspartate and L-phenylalanine methyl ester · HCl in 1-butyl-3-methylimidazolium hexafluorophosphate [BMIM][PF_6] achieved an initial rate (1.2 nmol (min^{-1} (mg protein)$^{-1}$) comparable to rates in organic solvents, and remarkable enzyme stability obviating immobilization (Erbeldinger, 2000). Since then, a range of enzymatic reactions and whole-cell processes have achieved remarkable yields (Kaftzik, 2002; Eckstein, 2002a), (enantio) selectivity (Park, 2001; Eckstein, 2002b), or enzyme stability (Lozano, 2001; Kaftzik, 2002).

12.5.4 Emulsions [Manufacture of Phosphatidylglycerol (PG)]

Fatty acids are important raw materials which are gained from naturally occurring triglycerides. However, thermal processes to obtain fatty acids result in discoloration, so alternative processes have been sought.

Lipases (E.C. 3.1.1.3.) catalyze the hydrolysis of lipids at an oil/water interface. In a membrane reactor, the enzymes were immobilized both on the side of the water phase of a hydrophobic membrane as well as on the side of the organic phase of a hydrophilic membrane. In both cases, no other means for stabilization of the emulsion at the membrane were required. The synthesis reaction to *n*-butyl oleate was achieved with lipase from *Mucor miehei*, which had been immobilized at the wall of a hollow fiber module. The degree of conversion reached 88%, but the substrate butanol decomposed the membrane before the enzyme was deactivated.

Phosphatidylglycerol (PG), which is useful as a lung surfactant for the treatment of respiratory illnesses, can be obtained from phosphatidylcholine (PC) in a transphosphatidylation reaction [Eq. (12.13)] catalyzed by phospholipase D (PLD, E.C. 3.1.4.4.), with the hydrolysis to phosphatidic acid (PA) as a side reaction [Eq. (12.14)].

$$\text{PC} + \text{glycerol} \rightarrow \text{PG} + \text{choline} \quad (12.13)$$

$$\text{PC} + \text{H}_2\text{O} \rightarrow \text{PA} + \text{choline} \quad (12.14)$$

Commonly, the reaction is conducted in an emulsion stabilized by surfactants and containing the substrate (PC). In a microporous membrane reactor, PLD stability could be increased sevenfold by the addition of ethers to the chamber side. Additionally, no surfactants were required as the product could be separated in simple fashion; 20% product was formed, compared with 4% in the simple emulsion system.

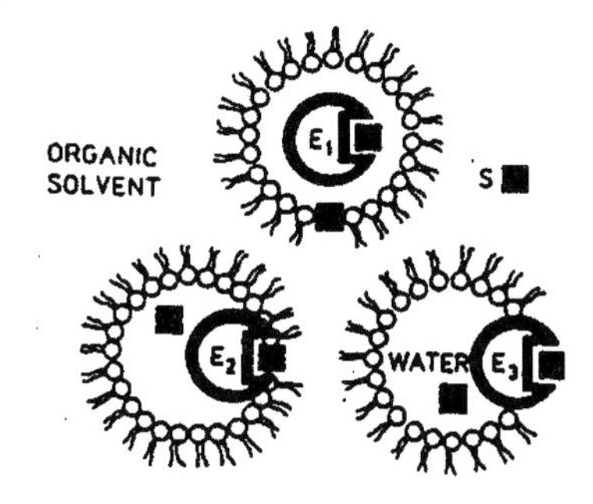

Figure 12.5 Microemulsion droplets and possible average location of enzyme molecules (Martinek, 1986).

12.5.5
Microemulsions

If the objective is to keep the enzyme active and stable in an aqueous phase but otherwise to use as much organic phase as possible, microemulsions are an option as a reaction medium. In contrast to ordinary emulsions they are thermodynamically stable and, at a particle diameter of 1–20 nm, accommodate most often only one enzyme molecule (Figure 12.5). The microemulsion droplets communicate rapidly and exchange their contents through elastic collisions. The boundary between microemulsions and reversed micelles is not clearly delineated, and the two notions are often used interchangeably. Enzyme of almost all classes and structures have been solubilized in microemulsion systems and used for reactions (Shield, 1986).

Activity and stability are often comparable to values in aqueous media. Many substrates which cannot be made to react in water or in pure organic solvents such as hexane owing to lack of solubility can be brought to reaction in microemulsions. Whereas enzyme structure and mechanism do not seem to change upon transition from water to the microemulsion phase (Bommarius, 1995), partitioning effects often are very important. Besides an enhanced or diminished concentration of substrates in the vicinity of microemulsion droplets and thus of enzyme molecules, the effective pH values in the water pool of the droplets can be shifted in the presence of charged surfactants. Frequently, observed acceleration or deceleration effects on enzyme reactions can be explained with such partitioning effects (Jobe, 1989).

12.5.6
Liquid Crystals

As described above, microemulsions feature the disadvantage that the reaction product can be separated only with difficulty from other components of the system. This disadvantage can be compensated by employing highly viscous to solid liquid-crystalline phases which consist of the same components as microemulsions (oil, surfactant, water). Many enzymes display activity in several phases of a three-component mixture (Figure 12.6) (Martinek, 1986).

At the beginning of the investigation a surfactant/water/oil phase diagram has to be prepared in case it has not been done already. Subsequently, one can generate

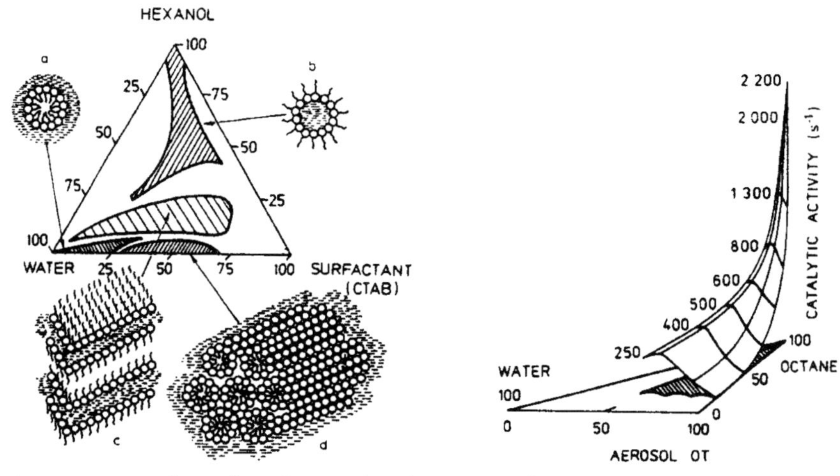

Figure 12.6 Liquid-crystalline phases and catalytic activity of peroxidase in ternary water/surfactant/organic solvent phases (Martinek, 1986). Phases: a = normal micelles; b = reversed micelles; c = lamellar aggregates; d = hexagonal aggregates.

the reaction medium by mixing corresponding amounts of components according to the phase diagram; in most cases, several phases will form. In cubic inverse phases, which are mechanically very stable, a series of enzymatic reactions have been conducted (Klyachko, 1986). Often, stability of enzymes has been found to be significantly higher in such liquid-crystalline phases than in pure isotropic continuous phases and media.

12.5.7
Ice–Water Mixtures

Even more than 30 years ago, it had been found that hydroxylaminolysis of amino acid esters with the help of trypsin catalysis is accelerated in frozen aqueous phases at –23°C in comparison to liquid aqueous phases at 1°C (Grant, 1966). For kinetically controlled peptide synthesis with different serine or cysteine proteases, strongly enhanced yields of di- and oligopeptides have been observed (Schuster, 1990) (Table 12.7; Figure 12.7). The strength of nucleophiles seems to be much higher in ice/water phases than in the purely liquid phase, as even unprotected dipeptides render good yields and even unprotected amino acid render at least moderate yields (Table 12.7; Figure 12.7).

The reverse action of a trypsin-free elastase isolated from porcine pancreas was studied in frozen aqueous systems and was found to catalyze peptide bond formation more effectively than in solution at room temperature (Haensler, 1998). The acceptance of free amino acids as nucleophilic amino components indicates a changed specificity of the endoprotease in frozen reaction mixtures. In elastase-catalyzed formation of Ser-, Ile-, and Val-X-bonds in frozen aqueous reaction mix-

Table 12.7 Yields of α-chymotrypsin-catalyzed peptide synthesis in water and ice using maleoyl-L-tyrosine methyl ester as acyl donor and various peptides, amino acid amides, and amino acids as nucleophiles* (Schuster, 1990).

Amino component of the reaction	***Peptide yield [%]***	
	25°C	–25°C
H-Gly-Ala-OH	5.8	94.7
H-Gly-Gly-OH	2.6	95.4
H-Gly-Gly-Gly-OH	5.1	91.2
H-D-Leu-NH_2	9.9	73.0
H-L-Leu-NH_2	79.1	91.8
H-Arg-OH	< 2	32.7
H-Lys-OH	< 2	43.8
H-β-Ala-Gly-OH	< 2	78.8

* [amino component] = 50 mM (25 mM as free base); [Maleoyl-L-Tyr-OMe] = 2 mM; [α-chymotrypsin] = 0.15 mM at 25°C and 0.3 μM at –25°C.

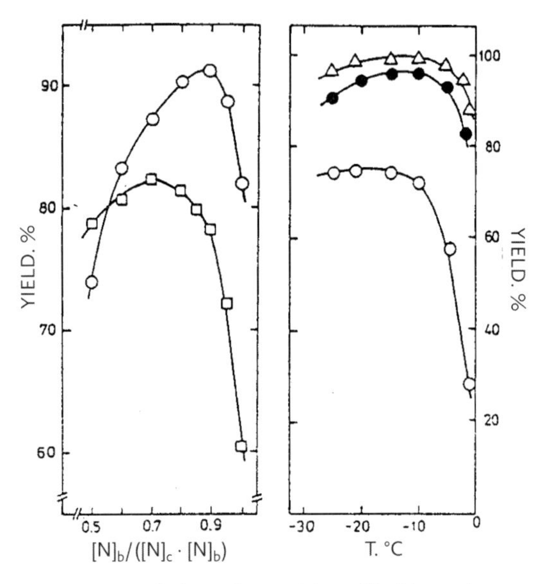

Figure 12.7 Left: dependence of the yield of the α-chymotrypsin-catalyzed synthesis of the peptides maleoyl-Tyr-D-Leu-NH_2 (o) and maleoyl-Tyr-β-Ala-Gly-OH (□) in ice upon the relative amount of nucleophile free base (for conditions, see Table 12.7). Right: influence of temperature on the yield of the α-chymotrypsin-catalyzed peptide synthesis in ice using maleoyl-L-Tyr methyl ester as acyl donor and H-D-Leu-NH_2 (o), H-Gly-Gly-Gly-OH (•), and H-Ala-Ala-OH (△) as amino components (for conditions, see Table 12.7) (Schuster, 1990).

tures, peptide yields obtained depended on the P1 amino acid and the acyl donor chain length.

The highest yields have been found at temperatures between –20 and –10°C and at pH values where more than 90% of the nucleophile was present as free base. Beyond a tenfold excess of nucleophile over electrophile, no further increase in yield could be observed. In a typical case, for the α-chymotrypsin (E.C. 3.4.21.1)-catalyzed reaction of maleyl-Phe-OMe with H-Leu-NH_2 investigated under a range of reaction conditions, a variety of results were demonstrated (Gerisch, 1994):

- the peptide yield is independent of the method of shock-freezing;
- the optimal reaction temperature is 248–263 K (–25 to –10 °C), and lower temperatures result in clearly retarded reactions;
- addition of DMSO leads to decreasing peptide yields; and
- peptide bond formation is catalyzed by the active enzyme, since unspecific protein surface catalysis gave no peptide yields at all.

The best explanation of the good results for peptide syntheses in ice–water mixtures are based on the freeze–concentration-model, which just provides for a volume-reducing function for the ice while the liquid aqueous part is still the only relevant phase for the reaction. All observed enhancements of reaction rate would then have to be attributed to an increase in effective concentration. ^{1}H-NMR relaxation time measurements have been used to determine the amount of unfrozen water in partially frozen systems, thus quantifying the extent of the "freeze–concentration effect" (Ullmann, 1997). Comparative studies in ice and at room temperature verify the importance of freeze–concentration which, however, is not sufficient for a complete understanding of the observed effects.

12.5.8
High-Density Eutectic Suspensions

Enzymatic catalysis in heterogeneous eutectic mixtures, i.e., at comparable mole fractions of each substrate (in the case of peptide synthesis, electrophile and nucleophile) and water, offers advantages over conventional reactions conducted in organic solvents. Such benefits include the avoidance of bulk solvents and the attainment of greatly improved productivities. A wide range of proteases have been shown to retain their catalytic activity in eutectic mixtures of substrates, and have been used to synthesize various model and bioactive oligopeptides (Gill, 1994). In similar fashion, the results of an initial study of enzymatic catalysis in metastable supersaturated solutions of carbohydrates show that such solutions, formed in the presence of small amounts of water and alcohol as plasticizers, are sufficiently stable under ambient conditions to enable enzymatic transformations of substrates (Millqvist-Fureby, 1998). A partial phase diagram for a system consisting of glucose, water, and polyethylene glycol was constructed to identify the regions which are most suitable for biotransformations (Figure 12.8). It was confirmed that the glass transition in this system occurred below the reaction temperature at any given

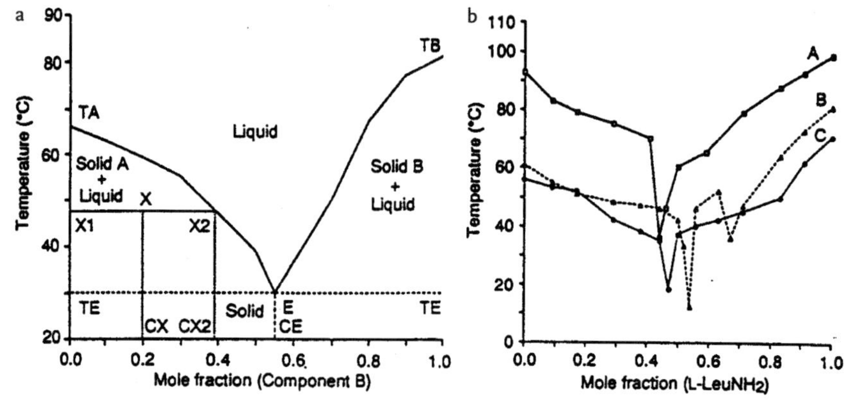

Figure 12.8 Phase diagram for a binary eutectic system (Gill, 1994).
a) E: eutectic point; TA: melting point of component A;
TB: melting point of component B; CE: eutectic composition.
b) Phase diagram of the system *N*-Z-L-Tyr-OEt + L-Leu-NH_2.
Curve A: binary mixture of substrates in the absence of adjuvant;
curve B: ternary mixture with 3% (w/w) water and 6% (w/w) ethanol;
curve C: ternary phase diagram containing 10% (w/w) triethylene glycol dimethyl ether.

composition of the constituent components. Several glycosidases were found to be catalytically active in this medium and the activity of β-glucosidase from almond was determined at several compositions of the reaction mixture and related to the corresponding regions of the phase diagram. The synthetic utility of the system was illustrated by glucosylation of several α,ω-alkyldiols, short-chain polyethylene glycols, and hydroxyalkyl and glyceryl monoacrylates.

The influence of eutectic media on the kinetics and productivity of biocatalysts has yet to be fully elucidated. Syntheses in eutectic suspensions have been scaled up to the pilot scale in a rotating drum reactor. The bioactive peptide *N*α-Cbz-L-Lys(*N*ε-Cbz)-Gly-L-Asp(OAll)-L-Glu(OAll)OEt was synthesized via a sequential N-to-C strategy in a heterogeneous solid–liquid mixture of the substrates in the presence of chymopapain and subtilisin as well as 16–20% (w/w) water and ethanol (Gill, 2002). At substrate concentrations of around 1 M, yields of 67–74% per step at product concentrations of 0.36, 0.49, and 0.48 kg kg^{-1} were achieved. The corresponding space–time yields were between 0.30 and 0.64 kg $(kg\ d)^{-1}$ and biocatalyst reuse provided productivities of 166–312 kg product $(kg\ enzyme)^{-1}$.

12.5.9 High-Density Salt Suspensions

A measure as simple as adding certain inorganic salts to aqueous enzyme solutions prior to lyophilization can result in dramatic activation of the dried powder in organic media relative to enzyme lyophilized without added salt (Ru, 1999).

Upon salt-induced lyophilization of Subtilisin Carlsberg, the optimum specificity constant k_{cat}/K_M for transesterification in hexane exceeded the value for salt-free enzyme 20 000-fold, substantially more than the previously reported enhancement of 3750-fold (Khmelnitsky, 1994). As for pure enzyme, salt-activated enzyme seems to exhibit the greatest activity when lyophilized from a solution of pH equal to the pH for optimal activity in water. The activation ratio and active-site content in the presence of 98% KCl (1% phosphate buffer and 1% enzyme) are highly sensitive to the lyophilization time and water content of the sample, over a range of up to 13-fold for Subtilisin Carlsberg and 11-fold for a lipase from *Mucor javanicus*.

The activity and water content were found to correlate directly with the kosmotropicity of the activating salt (kosmotropic salts bind water molecules strongly relative to the strength of water–water interactions in bulk solution) (Ru, 2001). Combinations of kosmotropic salts with known lyoprotectants such as polyethylene glycol (PEG) and sugars did not yield an appreciably more active catalyst. However, the combination of the kosmotropic sodium acetate with the strongly buffering sodium carbonate activated the enzyme more than the individual additives alone. Enzyme activity was enhanced further by the addition of small amounts of water to the organic solvent. Substrate selectivity experiments suggest that a mechanism other than selective partitioning of substrate into the enzyme–salt matrix is responsible for salt-induced activation of enzymes in organic solvents. Under optimal conditions, the enzyme activity in hexane was improved over 27 000-fold relative to the salt-free enzyme, reaching a catalytic efficiency that was within one order of magnitude of k_{cat}/K_M for hydrolysis of the same substrate in aqueous buffer.

12.5.10
Solid-to-Solid Syntheses

When one is using proteases in a direct reversal of their normal hydrolytic function, the equilibrium position is very important in limiting the attainable yield in equilibrium-controlled enzymatic peptide synthesis. If both reactants and products are largely undissolved in the reaction medium as suspended solids, thermodynamic analysis of such a system shows the reaction will proceed until at least one reactant has dissolved completely, towards either products or reactants ("switch-like" behavior). In case of a favorable equilibrium for synthesis, the yield is maximized in the solvent of least solubility for the starting materials (Halling, 1995). Thermolysin-catalyzed reactions of X-Phe-OH (X = formyl, Ac, Z) with Leu-NH_2 yielded X-Phe-Leu-NH_2 with equilibrium yields > 90% over a range of solvents. Some predictions, such as a linear decrease in yield with the reciprocal of the initial reactant concentrations, could be verified (Halling, 1995).

This approach enables high peptide yields in equilibrium-controlled peptide synthesis in high-density aqueous media with an equimolar supply of substrates. Scale-up to molar amounts verified the concepts as well as demonstrate the synthetic utility of this approach: Z-His-Phe-NH_2 and Z-Asp-Phe-OMe, precursors for cyclo-[-His-Phe-] and the low-calorie sweetener Aspartame, respectively, were synthesized in preparative yields of 84–88% (Eichhorn, 1997). For a review of the field of peptide synthesis in unusual media, see Jakubke (1996).

In the thermolysin-catalyzed solid-to-solid dipeptide synthesis of equimolar amounts of Z-Gln-OH and H-Leu-NH_2 as model substrates, the water content was varied from 0 to 600 mL water (mol substrate)$^{-1}$ and enzyme concentration in the range 0.5–10 g (mol substrate)$^{-1}$ to achieve 80% yield and initial rates of 5–20 mmol (s kg)$^{-1}$ (Erbeldinger, 1998). When the water content is decreased from the 1.6-molal lowest substrate concentration, the initial rate increases tenfold to a pronounced optimum at 40 mL water (mol substrate)$^{-1}$ and falls to much lower values in a system with no added water, and to zero in a rigorously dried system. The behavior at a higher water content was demonstrated through variation of the enzyme content to be caused by mass transfer limitations; at low water levels, the effects reflect the stimulation of the enzymatic activity by water. Preheating of the substrates or ultrasonic treatment had no significant effect on the system.

During melting point experiments, the formation of a third compound, the salt of the two substrates, was discovered and coincided with a very strong dependence of kinetics on the exact substrate ratio: at 60% Leu-NH_2 and 40% Z-Gln the rate was twice as high as at an equimolar substrate ratio (Erbeldinger, 1999). The reaction rate of both Z-Gln-Leu-NH_2 and Z-Asp-Phe-OMe formation is strongly dependent on the addition of basic salts to the reaction system, pointing to the importance of acid–base effects in solid-to-solid systems (Erbeldinger, 2001). If either $KHCO_3$ or K_2CO_3 is raised to 2.25-fold the equimolar amount of the Phe-OMe · HCl starting material, the rate increases 20-fold and remains at that level with further addition of $KHCO_3$, but it drops sharply when further K_2CO_3 is added. Beyond the addition of 1.5 equivalents of basic salt, the final yield of the reaction is affected negatively. These effects can be rationalized using a model estimating the pH of these systems, taking into account the possible formation of up to ten different solid phases. Together with influences of different mixing and water distribution methods, the pH model seems to explain most of the experimental results.

Speeding up of enzymatic conversions of substrate in aqueous suspensions is often attempted by raising the temperature or adding organic solvents to promote dissolution of the substrate. Results of an α-chymotrypsin-catalyzed hydrolysis of dimethylbenzyl methyl malonate demonstrated that, upon addition of organic cosolvents, longer process times were actually required, even though the substrate solubility increased severalfold as expected; however, on raising the temperature from 25 to 37 °C, the substrate solubility, the substrate dissolution rate, and the enzymatic reaction rate all increased, leading to shorter process times (Wolff, 1999a). A simple relationship for the prediction of the overall process time was established by evaluating the time constants for three sub-processes involved: substrate dissolution, enzymatic conversion, and enzyme deactivation [Eqs. (12.15)–(12.17)].

$$\text{Substrate dissolution:} \quad \tau_{dis} = 1/(k_L a)_0 \tag{12.15}$$

$$\text{Enzymatic conversion:} \quad \tau_{enz} = K_M/(k_{cat} \cdot [E]_0) \tag{12.16}$$

$$\text{Enzyme deactivation:} \quad \tau_{deact} = 1/(k_d \cdot [E]_0) \tag{12.17}$$

With the assumptions of second-order deactivation and first-order reaction, the regimes are characterized as follows: ($[A]_{solid,0}$ = initial concentration of solid substrate, $[A]_{lim}$ = substrate solubility):

- In the *deactivation-limited regime,* if $\tau_{deact} \approx [A]_{lim} \cdot \tau_{enz}/(0.4 \cdot [A]_{solid,0})$, the process time is totally dominated by deactivation. If $[A]_{solid,0} \geq [A]_{lim} \cdot \tau_{enz}/(0.4 \cdot \tau_{deact})$ the enzyme is deactivated before the reaction is complete.

- In the *dissolution-limited regime,* if deactivation is insignificant, $\tau_{deact} \gg [A]_{lim} \cdot \tau_{enz}/(0.4 \cdot [A]_{solid,0})$, overall product formation rate r_P and the overall process time t_{pro} are given by Eqs. (12.18).

$$r_P = k_L a \cdot [A]_{lim} \quad \text{and} \quad t_{pro} \approx 2\tau_{dis}\,([A]_{solid,0}/[A]_{lim}) \tag{12.18}$$

- In the *reaction-limited regime,* $\tau_{deact} \gg [A]_{lim} \cdot \tau_{enz}/(0.4 \cdot [A]_{solid,0})$ and $\tau_{enz} \gg 2 \cdot \tau_{dis}$, and r_P and t_{pro} are then given by Eq. (12.19).

$$r_P = (k_{cat}/K_M) \cdot [E]_0 \cdot [A]_{lim} \quad \text{and} \quad t_{pro} \approx \tau_{enz}\,([A]_{solid,0}/[A]_{lim}) \tag{12.19}$$

  In this case, the concentration of dissolved substrate equals $[A]_{lim}$.

The subprocesses, such as dissolution/crystallization of substrates and products, enzymatic synthesis of the product(s), undesired enzymatic hydrolysis of substrates and/or products, and deactivation, often influence the pH value and are influenced in turn by it as the reactants are weak electrolytes.

Investigating the kinetically controlled synthesis of the β-lactam antibiotic amoxicillin from 6-aminopenicillanic acid and D-*p*-hydroxyphenylglycine methyl ester in a solid suspension system in which the reaction nevertheless occurred in the liquid phase, Diender et al. found that the pH value and dissolved concentrations took a very different course at different initial substrate amounts (Diender, 2000). These results were described reasonably well by the model based on mass and charge balances, pH-dependent solubilities of the reactants, and enzyme kinetics.

Whereas enzymatic kinetic resolutions in a solid substrate suspension are usually treated as conventional kinetic resolutions of a fully dissolved substrate, a modeling study has demonstrated that there are important differences leading to much better performance in a solid suspension case (Wolff, 1999b). In the suspension processes the liquid-phase concentration of substrate enantiomer that should be converted can be kept close to the maximum value, i.e., the solubility, when process conditions are properly chosen, whereas in a conventional process this concentration gradually decreases. Calculations show that this leads to a productivity that is about sixfold higher in the suspension processes. Also, for enzymes with a low enantioselectivity, a severalfold increase in yield of remaining enantiopure substrate is predicted, compared to the conventional kinetic resolution of dissolved enantiomers. Other potential advantages of using suspension reactions are that the initial substrate concentration may be higher [up to 25% (w/w)] and that the desired remaining substrate may be recovered by simply filtering off the solid crystals.

12.6
Solvent as a Parameter for Reaction Optimization ("Medium Engineering")

The term "medium engineering" was coined as a complement to protein engineering (Wescott, 1994). Whereas the latter changes the catalyst, the former changes the reaction medium with resulting changes in prochiral, regio-, and enantioselectivity. Many such changes have been observed in the literature (Wescott, 1994; Carrea, 1995, 2000), however, only a few examples are given here to demonstrate that not only the catalyst but also the medium exerts a decisive influence on the outcome of an enzymatic reaction.

12.6.1
Change of Substrate Specificity with Change of ReactionM: Specificity of Serine Proteases

When switching from water to an organic solvent, or switching between organic solvents, the substrate specificity can change. In the example of the standard reaction, transesterification of *N*-acetyl-L-phenylalanine ethyl ester with *n*-propanol by Subtilisin Carlsberg, which has been mentioned several times in this chapter already, the relative specificity between the rather hydrophobic phenylalanine compound and its more hydrophilic analog *N*-acetyl-L-serine ethyl ester varies with the solvent (Table 12.8) (Wescott, 1993).

Partitioning of the two substrates between the bulk organic solvent and the binding pocket of the enzyme active site can act as an explanation. It is known that any binding pocket of enzymes can be identified with a value for the hydrophobicity. A hydrophobic substrate in water partitions readily into the binding pocket, but

Table 12.8 Substrate specificity S $(k_{cat}/K_M)_{Ser}/(k_{cat}/K_M)_{Phe}$, in the transesterification of N-Ac-L-amino acid ethyl ester with *n*-propanol, catalyzed by Subtilisin Carlsberg in various anhydrous organic solvents (Wescott, 1993).

Solvent	*S*	*Solvent*	*S*
Dichloromethane	8.2	*tert*-Butyl methyl ether	2.5
Chloroform	5.5	Octane	2.5
Toluene	4.8	Isopropyl acetate	2.2
Benzene	4.4	Acetonitrile	1.7
N,N-Dimethylformamide	4.3	Dioxane	1.2
tert-Butyl acetate	3.7	Acetone	1.1
N-Methylacetamide	3.4	Pyridine	0.53
Diethylether	3.2	*tert*-Amyl alcohol	0.27
Carbon tetrachloride	3.2	*tert*-Butyl alcohol	0.19
Ethyl acetate	2.6	*tert*-Butylamine	0.12

the more hydrophobic the solvent becomes the worse is the partitioning that results. From Table 12.8 it can be discerned, however, that this monocausal explanation does not suffice, as specificity *S* does not follow hydrophobicity very well.

12.6.2
Change of Regioselectivity by Organic Solvent Medium

Ortho-substituted phenolic diesters, such as bis-*p*-(1,4-butyroyl)-2-octylhydroquinone (Figure 12.9), can be hydrolyzed in the more hindered 1-position or in the less-hindered 4-position. The regioselectivity of hydrolysis to the monoester, as expressed by the ratio of k_{cat}/K_M values, depends approximately linearly on log *P* over the range from acetonitrile (log *P*: –0.33) to cyclohexane (log *P*: 3.2), and is thus controlled by the hydrophobicity of the solvent (Rubio, 1991).

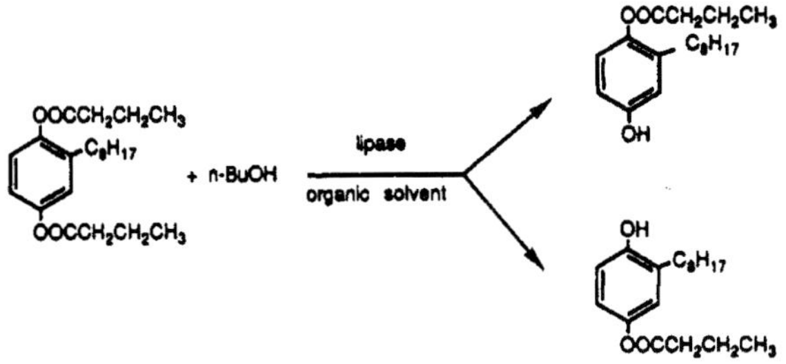

Figure 12.9 Regioselectivity of the hydrolysis of hydroquinone bis-esters (Rubio, 1991).

In another case, 4-alkylphenols can be transformed with the help of the flavoenzyme vanillyl alcohol oxidase (VAO) to either (*R*)-1-(4′-hydroxyphenyl) alcohols or to 1-(4′-hydroxyphenyl)alkenes. Both products pass through a common intermediate, the *p*-quinone methide, which then either is attacked by water or rearranges. The product spectrum can be controlled by medium engineering: in organic solvents such as acetonitrile and toluene, more *cis*-alkene but not *trans*-alkene, and less alcohol, are produced (van den Heuvel, 2001). A similar shift in *cis*/*trans*-alkene was achieved by the addition of monovalent anions that bind specifically close to the active site.

12.6.3
Solvent Control of Enantiospecificity of Nifedipines

The probably most striking example so far of solvent influence on enzyme selectivity has been found by Yoshihiko Hirose and co-workers at the Amano Corp. in Nagoya, Japan, studying nifedipines, which provide a befitting conclusion of this chapter and a transition to the next on use of enzyme catalysis in the pharma industry. Nifedipines, structurally 1-substituted dihydropyridine mono- or diesters, are used in cardiovascular therapy, where they are termed calcium antagonists (Goldmann, 1991); the most prominent representative is Adalat® from Bayer (Leverkusen, Germany). While the simplest achiral dihydropyridines can be syn-

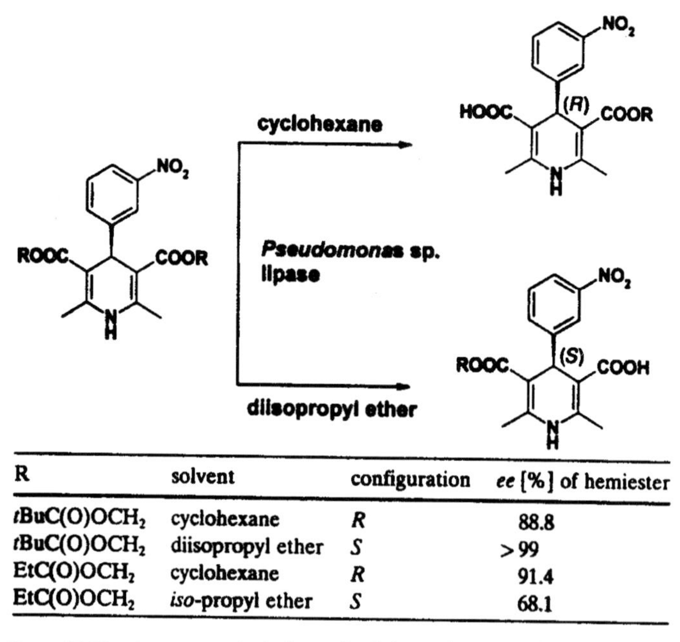

R	solvent	configuration	*ee* [%] of hemiester
*t*BuC(O)OCH$_2$	cyclohexane	*R*	88.8
*t*BuC(O)OCH$_2$	diisopropyl ether	*S*	>99
EtC(O)OCH$_2$	cyclohexane	*R*	91.4
EtC(O)OCH$_2$	*iso*-propyl ether	*S*	68.1

Figure 12.10 Asymmetric hydrolysis of nifedipine diesters; influence of solvent on stereochemical preferences (Hirose, 1992; Carrea, 2000).

thesized by simple mixing and heating of benzaldehydes, ammonia, and alkyl acetoacetate, the asymmetric – and thus difficult to synthesize – dihydropyridines are therapeutically the most valuable today.

The Amano researchers found not just drastically different enantiomeric excesses in the *Pseudomonas* sp. (Amano AH) lipase-catalyzed hydrolysis of methyleneoxypropionyl or -pivaloyl diesters, but a near-total reversal of specificity. Whereas in cyclohexane the different esters yielded half-esters with 88.8–91.4% e.e. *R*-specificity for a triple mutant ("FVL") of Amano PS lipase, the same transformation with the same enzyme in diisopropyl ether (DIPE) yielded between 68.1 and > 99% e.e. of the *S*-product (Figure 12.10) (Hirose, 1992, 1995).

Suggested Further Reading

G. Carrea, G. Ottolina, and S. Riva, Role of solvents in the control of enzyme selectivity in organic media, *Trends Biotechnol.* **1995**, *13*, 63–70.

G. Carrea and S. Riva, Properties and synthetic applications of enzymes in organic solvents, *Angew. Chem. Int. Ed.* **2000**, *39*, 2226–2254.

A. M. Klibanov, Improving enzymes by using them in organic solvents, *Nature* **2001**, *409*, 241–246.

References

B. A. Bedell, V. V. Mozhaev, D. S. Clark, and J. S. Dordick, Testing for diffusion limitations in salt-activated enzyme catalysts operating in organic solvents, *Biotechnol. Bioeng.* **1998**, *58*, 654–657.

G. Bell, A. E. M. Janssen, and P. J. Halling, Water activity fails to predict critical hydration level for enzyme activity in polar organic solvents: interconversion of water concentrations and activities, *Enzyme Microb. Technol.* **1997**, *20*, 471–477.

A. S. Bommarius, D. I. C. Wang, and T. A. Hatton, Xanthine oxidase reactivity in reversed micellar systems: a contribution to the prediction of enzymatic activity in organized media, *J. Am. Chem. Soc.* **1995**, *117*, 4515–4523.

R. Bovara, G. Carrea, G. Ottolina, and S. Riva, Effects of water activity on V_{max} and K_M of lipase catalyzed transesterification in organic media, *Biotechnol. Lett.* **1993**, *15*, 937–942.

J. Broos, A. J. W. G. Visser, J. F. J. Engbersen, W. Verboom, A. van Hoek, and D. N. Reinhoudt, Flexibility of enzymes suspended in organic solvents probed by time-resolved fluorescence anisotropy. Evidence that enzyme activity and enantioselecticity are directly related to enzyme flexibility, *J. Am. Chem. Soc.* **1995**, *117*, 12657–12663.

P. A. Burke, R. G. Griffin, and A. M. Klibanov, Solid-state NMR assessment of enzyme active center structure under nonaqueous conditions, *J. Biol. Chem.* **1992**, *267*, 20057–20064.

G. Carrea, G. Ottolina, and S. Riva, Role of solvents in the control of enzyme selectivity in organic media, *Trends Biotechnol.* **1995**, *13*, 63–70.

G. Carrea and S. Riva, Properties and synthetic applications of enzymes in organic solvents, *Angew. Chem. Int. Ed.* **2000**, *39*, 2226–2254.

M. F. Chaplin and C. Bucke, *Enzyme Technology*, Cambridge University Press, Cambridge, UK, **1990**, p. 229.

H. R. Costantino, K. Griebenow, R. Langer, and A. M. Klibanov, On the pH memory of lyophilized compounds containing protein functional groups, *Biotechnol. Bioeng.* **1997**, *53*, 345–348.

L. Dai and A. M. Klibanov, Striking activation of oxidative enzymes suspended in nonaqueous media, *Proc. Natl. Acad. Sci.* **1999**, *USA 96*, 9475–9478.

L. Dai and A. M. Klibanov, Peroxidase-catalyzed asymmetric sulfoxidation in organic solvents versus in water, *Biotechnol. Bioeng.* **2000**, *70*, 353–357.

P. K. Das, J. M. Caaveiro, S. Luque, and A. M. Klibanov, Binding of hydrophobic hydroxamic acids enhances peroxidase's stereoselectivity in nonaqueous sulfoxidations, *J. Am. Chem. Soc.* **2002**, *124*, 782–787.

J. S. Deetz and J. D. Rozzell, Enzymic catalysis by alcohol dehydrogenases in organic solvents, *Ann. N. Y. Acad. Sci.* **1988**, *542* (*Enzyme Eng.* 9), 230–234.

M. B. Diender, A. J. Straathof, T. van der Does, M. Zomerdijk, and J. J. Heijnen, Course of pH during the formation of amoxicillin by a suspension-to-suspension reaction, *Enzyme Microb. Technol.* **2000**, *27*, 576–582.

M. Eckstein, M. Sesing, U. Kragl, and P. Adlercreutz, At low water activity α-chymotrypsin is more active in an ionic liquid than in non-ionic organic solvents, *Biotechnol. Lett.* **2002a**, *24*, 867–872.

M. Eckstein, P. Wasserscheid, and U. Kragl, Enhanced enantioselectivity of lipase from *Pseudomonas* sp. at high temperatures and fixed water activity in the ionic liquid, 1-butyl-3-methylimidazolium bis[(trifluoromethyl)sulfonyl]amide, *Biotechnol. Lett.* **2002b**, *24*, 763–767.

U. Eichhorn, A. S. Bommarius, K. Drauz, and H.-D. Jakubke, Synthesis of dipeptides by suspension-to-suspension conversion via thermolysin catalysis: from analytical to preparative scale, *J. Pept. Sci.* **1997**, *3*, 245–251.

M. Erbeldinger, X. Ni, and P. J. Halling, Effect of water and enzyme concentration on thermolysin-catalyzed solid-to-solid peptide synthesis, *Biotechnol. Bioeng.* **1998**, *59*, 68–72.

M. Erbeldinger, X. Ni, and P. J. Halling, Kinetics of enzymatic solid-to-solid peptide synthe-sis: intersubstrate compound, substrate ratio, and mixing effects, *Biotechnol. Bioeng.* **1999**, *63*, 316–321.

M. Erbeldinger, A. J. Mesiano, and A. J. Russell, Enzymatic catalysis of formation of Z-aspartame in ionic liquid – an alternative to enzymatic catalysis in organic solvents, *Biotechnol. Prog.* **2000**, *16*, 1129–1131.

M. Erbeldinger, X. Ni, and P. J. Halling, Kinetics of enzymatic solid-to-solid peptide synthesis: synthesis of Z-aspartame and control of acid–base conditions by using inorganic salts, *Biotechnol. Bioeng.* **2001**, *72*, 69–76.

P. A. Fitzpatrick and A. M. Klibanov, How can the solvent affect enzyme enantioselectivity? *J. Am. Chem. Soc.* **1991**, *113*, 3166–3171.

P. A. Fitzpatrick, D. Ringe, and A. M. Klibanov, Computer-assisted modeling of subtilisin enantioselectivity in organic solvents, *Biotechnol. Bioeng.* **1992**, *40*, 735–742.

S. Gerisch, G. Ullmann, K. Stubenrauch, and H.-D. Jakubke, Enzymatic peptide synthesis in frozen aqueous systems: influence of modified reaction conditions on the peptide yield, *Biol. Chem. Hoppe Seyler,* **1994**, *375*, 825–828.

A. S. Ghatorae, G. Bell, and P. J. Halling, Inactivation of enzymes by organic solvents: new technique with well-defined interfacial area, *Biotechnol. Bioeng.* **1994**, *43*, 331–336.

I. Gill and E. Vulfson, Enzymic catalysis in heterogeneous eutectic mixtures of substrates, *Trends Biotechnol.* **1994**, *12*, 118–122.

I. Gill and R. Valivety, Pilot-scale enzymatic synthesis of bioactive oligopeptides in eutectic-based media, *Org. Proc. Res. Dev.* **2002**, *6*, 684–691.

S. Goldmann and J. Stoltefuss, 1,4-Dihydropyridines: effect of chirality and conformation on the calcium-antagonistic and -agonistic effects, *Angew. Chem.* **1991**, *103*, 1587–605; *Angew. Chem. Int. Ed. Engl.* **1991**, *30*, 1559–1578.

L. A. S. Gorman and J. S. Dordick, Organic solvents strip water off enzymes, *Biotechnol. Bioeng.* **1992**, *39*, 392–397.

N. H. Grant and H. E. Alburn, Acceleration of enzyme reactions in ice, *Nature* **1966**, *212*, 194.

R. M. Guinn, P. S. Skerker, P. Kavanaugh, and D. S. Clark, Activity and flexibility of alcohol dehydrogenase in organic solvents, *Biotechnol. Bioeng.* **1991**, 37, 303–308.

M. Haensler, N. Wehofsky, S. Gerisch, J. D. Wissmann, and H.-D. Jakubke, Reverse catalysis of elastase from porcine pancreas in frozen aqueous systems, *Biol. Chem. Hoppe Seyler* **1998**, *379*, 71–74.

P. J. Halling, U. Eichhorn, P. Kuhl, and H. D. Jakubke, Thermodynamics of solid-to-solid conver-sion and application to enzymic peptide synthesis, *Enzyme Microb. Technol.* **1995**, *17*, 601–606.

Y. Hirose, K. Kariya, I. Sasaki, Y. Kurono, H. Ebiike, and K. Achiwa, Drastic solvent effect on lipase-catalyzed enantioselective hydrolysis of prochiral 1,4-dihydropyridines, *Tetrahedron Lett.* **1992**, *33*, 7157–7160.

Y. Hirose, K. Kariya, Y. Nakanishi, Y. Kurono, and K. Achiwa, Inversion of enantioselectivity in hydrolysis of 1,4-dihydropyridines by point mutation of lipase PS, *Tetrahedon Lett.* **1995**,*36*,1063–1066.

H. D. Jakubke, U. Eichhorn, M. Hansler, and D. Ullmann, Non-conventional enzyme catalysis: application of proteases and zymogens in biotransformations, *Biol. Chem. Hoppe Seyler* **1996**, *377*, 455–464.

D. Jobe, in: *Reactions in Compartmentalized Liquids*, W. Knoche and R. Schomäcker (eds.), Springer Verlag, Heidelberg, **1989**, pp. 39–51.

N. Kaftzik, P. Wasserscheid, and U. Kragl, Use of ionic liquids to increase the yield and enzyme stability in the β-galactosidase catalyzed synthesis of *N*-acetyllactosamine, *Org. Proc. Res. Dev.* **2002**, *6*, 553–557.

S. Kamat, J. Barrera, E. J. Beckman, and A. J. Russell, Biocatalytic synthesis of acrylates in organic solvents and supercritical fluids: I. Optimization of enzyme environment, *Biotechnol. Bioeng.* **1992a**, *40*, 158–166.

S. Kamat, E. J. Beckman, and A. J. Russell, Role of diffusion in nonaqueous enzymology. 1. Theory, *Enz. Microb. Technol.* **1992b**, *14*, 265–271.

L. T. Kanerva and A. M. Klibanov, Hammett analysis of enzyme action in organic solvents, *J. Am. Chem. Soc.* **1989**, *111*, 6864–6865.

T. Ke and A. M. Klibanov, **1998**, On enzymatic activity in organic solvents as a function of enzyme history, *Biotechnol. Bioeng.* *57*, 746–750.

T. Ke and A. M. Klibanov, Markedly enhancing enzymic enantioselectivity in organic solvents by forming substrate salts, *J. Am. Chem. Soc.* **1999**, *121*, 3334–3340.

J. Kim, D. S. Clark, and J. S. Dordick, **2000**, Intrinsic effects of solvent polarity on enzymic activation energies, *Biotechnol. Bioeng.* *67*, 112–116.

A. M. Klibanov, **1989**, Enzymatic catalysis in anhydrous organic solvents, *Trends Biochem. Sci.* *14*, 141–144.

A. M. Klibanov, **1995**, Enzyme memory. What is remembered and why?, *Nature*, *374*, 596.

A. M. Klibanov, **1997**, Why are enzymes less active in organic solvents than in water? *Trends Biotechnol.* *15*, 97–101.

A. M. Klibanov, G. P. Samokhin, K. Martinek, and I. V. Berezin, **1977**, A new approach to preparative enzymatic synthesis, *Biotechnol. Bioeng.* *19*, 1351–1361, reprinted in *Biotechnol. Bioeng.* **2000**, *67*, 737–747.

N. L. Klyachko, A. V. Levashov, A. V. Pshezhetskii, N. G. Bogdanova, I. V. Berezin, and K. Martinek, Catalysis by enzymes entrapped into hydrated surfactant aggregates having lamellar or cylindrical (hexagonal) or ball-shaped (cubic) structure in organic solvents, *Eur. J. Biochem.* **1986**, *161*, 149–154.

U. Kragl, M. Eckstein, and N. Kaftzik, **2002**, Enzyme catalysis in ionic liquids, *Curr. Opin. Biotechnol.* *13*, 565–571.

C. Laane, S. Boeren, and K. Vos, On optimizing organic solvents in multi-liquid-phase biocatalysis, *Trends Biotechnol.* **1985**, *3*, 251–252.

C. Laane, S. Boeren, K. Vos, and C. Veeger, Rules for optimization of biocatalysis in organic solvents, *Biotechnol. Bioeng.* **1987**, *30*, 81–87.

M. Y. Lee and J. S. Dordick, **2002**, Enzyme activation for nonaqueous media, *Curr. Opin. Biotechnol.* *13*, 376–384.

P. Lozano, T. De Diego, J.-P. Guegan, M. Vaultier, and J. L. Iborra, Stabilization of α-chymotrypsin by ionic liquids in transesterification reactions, *Biotechnol. Bioeng.* **2001**, *75*, 563–569.

K. Martinek, A. V. Levashov, N. Klyachko, Yu. L. Khmel'nitskii, and I. V. Berezin, Micellar enzymology, *Eur. J. Biochem.* **1986**, *155*, 453–468.

A. Marty, W. Chulalaksananukul, J. S. Condoret, R. M. Willemot, and G. Durand, Comparison of lipase-catalyzed esterification in supercritical carbon dioxide and in *n*-hexane, *Biotechnol. Lett.* **1990**, *12*, 11–16.

A. J. Mesiano, E. J. Beckman, and A. J. Russell, **1999**, Supercritical biocatalysis, *Chem. Rev.* *99*, 623–634.

A. Millqvist-Fureby, I. S. Gill, and E. N. Vulfson, **1998**, Enzymatic transformations in supersaturated substrate solutions: I. A general study with glycosidases, *Biotechnol. Bioeng.* *60*, 190–196.

V. S. Narayan and A. M. Klibanov, Are water-immiscibility and apolarity of the solvent relevant to enzyme efficiency? *Biotechnol. Bioeng.* **1993**, *41*, 390–393.

K. Oyama, S. Irino, T. Harada, and N. Hagi, Enzymic production of aspartame, *Ann. N. Y. Acad. Sci. (Enzyme Eng. 7)* **1984**, *434*, 95–98.

S. Park and R. J. Kazlauskas, Improved preparation and use of room-temperature ionic liquids in lipase-catalyzed enantio- and regioselective acylations, *J. Org. Chem.* **2001**, *66*, 8395–8401.

R. V. Rariy and A. M. Klibanov, On the relationship between enzymatic enantioselectivity in organic solvents and enzyme flexibility, *Biocat. Biotransf.* **2000**, *18*, 401–407.

D. Roche, K. Prasad, and O. Repic, Enantioselective acylation of β-aminoesters using penicillin G acylase in organic solvents, *Tetrahedron Lett.* **1999**, *40*, 3665–3668.

M. T. Ru, J. S. Dordick, J. A. Reimer, and D. S. Clark, **1999**, Optimizing the salt-induced activation of enzymes in organic solvents: effects of lyophilization time and water content, *Biotechnol. Bioeng.* *63*, 233–141.

M. T. Ru, K. C. Wu, J. P. Lindsay, J. S. Dordick, J. A. Reimer, and D. S. Clark, **2001**, Towards more active biocatalysts in organic media: increasing the activity of salt-activated enzymes, *Biotechnol. Bioeng.* *75*, 187–196.

E. Rubio, A. Fernandez-Mayorales, and A. M. Klibanov, Effect of the solvent on enzyme regioselectivity, *J. Am. Chem. Soc.* **1991**, *113*, 695–696.

M. Schuster, A. Aaviksaar, and H.-D. Jakubke, Enzyme-catalyzed peptide synthesis in ice, *Tetrahedron* **1990**, *24*, 8093–8102.

J. W. Shield, H. D. Ferguson, A. S. Bommarius, and T. A. Hatton, Enzymes in reversed micelles as catalysts for organic-phase reactions, *Ind. Eng. Chem. Fundam.* **1986**, *25*, 603–612.

J. S. Shin, S. Luque, and A. M. Klibanov, **2000**, Improving lipase enantioselectivity in organic solvents by forming substrate salts with chiral agents, *Biotechnol. Bioeng.* *69*, 577–583.

E. A. Sym, The action of esterase in the presence of organic solvents, *Biochem. J.* **1936**, *30*, 609–617.

F. Terradas, M. Teston-Henry, P. A. Fitzpatrick, and A. M. Klibanov, Marked dependence of enzyme prochiral selectivity on the solvent, *J. Am. Chem. Soc.* **1993**, *115*, 390–396. S. M. Thomas, R. Dicosino, and V. Najarajan, Biocatalysis: applications and potentials for chemical industry, *Trends Biotechnol.* **2002**, *20*, 238–242.

G. Ullmann, M. Haensler, W. Gruender, M. Wagner, H. J. Hofmann, and H.-D. Jakubke, Influence of freeze–concentration effect on proteinase-catalysed peptide synthesis in frozen aqueous systems, *Biochim. Biophys. Acta* **1997**, *1338*, 253–258.

R. H. Valivety, P. J. Halling, and A. R. Macrae, Reaction rate with suspended lipase catalyst shows similar dependence on water activity in different organic solvents, *Biochim. Biophys. Acta* **1992a**, *1118*, 218–222.

R. H. Valivety, P. J. Halling, A. D. Peilow, and A. R. Macrae, Lipases from different sources vary widely in dependence of catalytic activity on water activity, *Biochim. Biophys. Acta* **1992b**, *1122*, 143–146.

R. H. H. van den Heuvel, J. Partridge, C. Laane, P. J. Halling, and W. J. H. van Berkel, Tuning of the product spectrum of vanillyl-alcohol oxidase by medium engineering, *FEBS Lett.* **2001**, *503*, 213–216.

D. J. van Unen, J. F. Engbersen, and D. N. Reinhoudt, **1998**, Large acceleration of alpha-chymotrypsin-catalyzed dipeptide formation by 18-crown-6 in organic solvents, *Biotechnol. Bioeng.* *59*, 553–556.

D. J. van Unen, J. F. Engbersen, and D. N. Reinhoudt, **2001**, Sol–gel immobilization of serine proteases for application in organic solvents, *Biotechnol. Bioeng.* *75*, 154–158.

D. J. van Unen, J. F. Engbersen, and D. N. Reinhoudt, **2002**, Why do crown ethers activate enzymes in organic solvents? *Biotechnol. Bioeng.* *77*, 248–255.

H. Wajant, S. Forster, A. Sprauer, F. Effenberger, and K. Pfizenmaier, Enantioselective synthesis of aliphatic (*S*)-cyanohydrins in organic solvents using hydroxynitrile lyase from *Manihot esculenta*, *Ann. N. Y. Acad. Sci.* **1996**, *799*, 771–776.

C. R. Wescott and A. M. Klibanov, Solvent variation inverts substrate specificity of an enzyme, *J. Am. Chem. Soc.* **1993**, *115*, 1629–1631.

C. R. Wescott and A. M. Klibanov, The solvent dependence of enzyme specificity, *Biochim. Biophys. Acta* **1994**, *1206*, 1–9.

A. Wolff, L. Zhu, Y. W. Wong, A. J. Straathof, J. A. Jongejan, and J. J. Heijnen, Understanding the influence of temperature change and cosolvent addition on conversion rate of enzymatic suspension reactions based on regime analysis, *Biotechnol. Bioeng.* **1999a**, *62*, 125–134.

A. Wolff, V. van Asperen, A. J. Straathof, and J. J. Heijnen, Potential of enzymatic kinetic resolution using solid substrates suspension: improved yield, productivity, substrate concentration, and recovery, *Biotechnol. Prog.* **1999b**, *15*, 216–227.

A. Zaks and A. M. Klibanov, Enzymatic catalysis in organic media at 100 degrees C, *Science* **1984**, *224*, 1249–1251.

A. Zaks and A. M. Klibanov, Enzyme-catalyzed processes in organic solvents, *Proc. Natl. Acad. Sci. USA* **1986a**, *82*, 3192–3196.

A. Zaks and A. M. Klibanov, Substrate specificity of enzymes in organic solvents vs. water is reversed, *J. Am. Chem. Soc.* **1986b**, *108*, 2767–2768.

A. Zaks and A. M. Klibanov, Enzymatic catalysis in nonaqueous solvents, *J. Biol. Chem.* **1988a**, *263*, 3194–3201.

A. Zaks and A. M. Klibanov, The effect of water on enzyme action in organic media, *J. Biol. Chem.* **1988b**, *263*, 8017–8021.

13
Pharmaceutical Applications of Biocatalysis

Summary

Two of the most promising perspectives in the search for novel pharmacologically active structures are (i) studies of the inhibition of enzymes, and (ii) studies of the modification or blockage of a receptor. The two main steps in the development of pharmacologically active compounds are *lead structure development* and *lead structure optimization*. A *"lead structure"* is a template compound used for further optimization with promising but not optimal properties. Every pharmacologically active structure has to feature sufficient performance in three dimensions: (i) *inhibition of the target enzyme* (low inhibition constant K_I); (ii) *bioavailability*, the fraction of active compound available to organism or tissue; and (iii) *stability* of the compound in the body (long half-life $\tau_{1/2}$).

Since the 1960s, registration, following a rigorous testing procedure and review by a panel, is a crucial step before any new drug can be brought onto the market. The process required by the Food and Drug Administration (FDA) of the United States is the generally accepted model all over the world. The research and development process for a new drug consists of the steps of (i) preclinical research; (ii) clinical phase I; (iii) clinical phase II; (iv) clinical phase III; (v) registration period; and (vi) commercialization. At every step, a fraction of the candidates drop out owing to insufficient performance, i.e., insufficient efficacy in treating the desired condition and/or excessive side effects.

Besides acid stability and resistance against β-lactamases, the ideal antibiotic should feature broad-spectrum utility against various bacteria and effectiveness against bacterial problem strains which cannot be combated by naturally occurring antibiotics or which develop resistance, such as the problem strain in hospitals, *Staphylococcus aureus*.

Currently, there are five approved HIV protease inhibitors available which are mostly used in combination therapy with other drugs, such as reverse transcriptase inhibitors (NTIs). Most HIV protease inhibitors are still synthesized chemically. Trends apparent during the development of HIV protease inhibitors are: (i) the labile amide bond of peptides is replaced by isosteres such as hydroxyethylamine; (ii) the original size of lead structures is reduced from seven to four residues; (iii) bioavailability, i.e., water solubility, is enhanced; and (iv) the design of small inhibitors (MW < 600) with small clearance rates out of the body and a convenient sequence of processing steps is sought.

Whereas several anti-cholestemic drugs are produced wholly by fermentation, the side chain of several others is accessible through enzymatic synthesis. (3*R*,5*S*)-Dihydroxyhexanoate, a key intermediate of fluvastatin, is accessible by reduction of the diketo acid, either by bioreduction with whole cells (85% yield, 97% e.e.), or cell extracts (72% yield, 98.5% e.e.), or, for *syn*-(3*R*,5*S*)-dihydroxy-6-Cl-hexanoate, regioselective and (*R*)-specific reduction with ADH to yield (5*S*)-6-chloro-3-ketohexanoate.

Several examples, such as Trimegestone (menopausal diseases), Trusopt (anti-glaucoma), or Montelukast (anti-asthma), are described of the use of whole cells, mostly yeast, for the reduction of keto functions.

13.1 Enzyme Inhibition for the Fight against Disease

13.1.1 Introduction

Two of the most promising perspectives in the search for novel pharmacologically active structures are (i) studies of the inhibition of enzymes; and (ii) studies of the modification or blockage of a receptor. Although receptors are a very important topic, they cannot be covered here. Table 13.1 lists a selection of important enzymes which cause pharmacological effects through selective inhibition.

Every pharmacologically active structure has to feature sufficient performance in three dimensions:

- sufficient *inhibition of the target enzyme*, expressed as a sufficiently low inhibition constant K_I;
- *bioavailability*, determined as the percentage of active substance available to the organism or tissue; and
- *stability* of the compound in the body, characterized by a sufficiently long half-life $\tau_{1/2}$.

Table 13.1 Pharmacologically important inhibition of enzymes.

Enzyme	*Inhibitor*	*K_I value [M]*	*Indication*
Renin	pepstatin	4.7×10^{-11}	high blood pressure
ACE	captopril	1.7×10^{-9}	high blood pressure
ACE	enalapril	1.3×10^{-9}	high blood pressure
HMG-CoA reductase	mevastatin	1.0×10^{-9}	cholesterol level
HIV protease	amprenavir	6.3×10^{-10}	AIDS
HIV protease	saquinavir	1.0×10^{-8}	AIDS
HIV protease	ritonavir	1.5×10^{-11}	AIDS

ACE: angiotensin-converting enzyme; HIV: human immunodeficiency virus; HMG-CoA reductase: 3-hydroxy-3-methyl-glutaryl-coenzyme A reductase.

Insufficient performance along even one dimension leads to discontinuation of this compound as a pharmaceutical. For a long time, the relevance of the bioavailability and stability was underestimated; a K_I value in the nanomolar range was regarded as sufficient for development as an active compound. It is important to note that, beyond fulfilling the three criteria of inhibition of target enzyme, bioavailability, and stability, any pharmaceutical under development has to pass muster in in-vivo situations, not just in in-vitro tests.

An exemplary optimization of the three above-mentioned criteria is embodied by the development of VX-478 (Figure 13.1), an anti-AIDS drug by Vertex, a company in Cambridge/MA, USA (Borman, 1995). To improve bioavailability, a particular weakness of competing structures, the molar masses of all approved pharmaceutical agents were plotted in a frequency diagram. A maximum of the distribution was found at 350 Da, with a tail ranging to 1000 Da. The Vertex scientists then decreased the molecular mass of their compound in a targeted fashion, such as by introduction of a 4-amino substituent of the arylsulfonic acid moiety to increase bioavailability, while attempting to simultaneously keep the favorable inhibitory properties with respect to HIV protease inhibition. VX-478 finally features a molecular mass of 506.

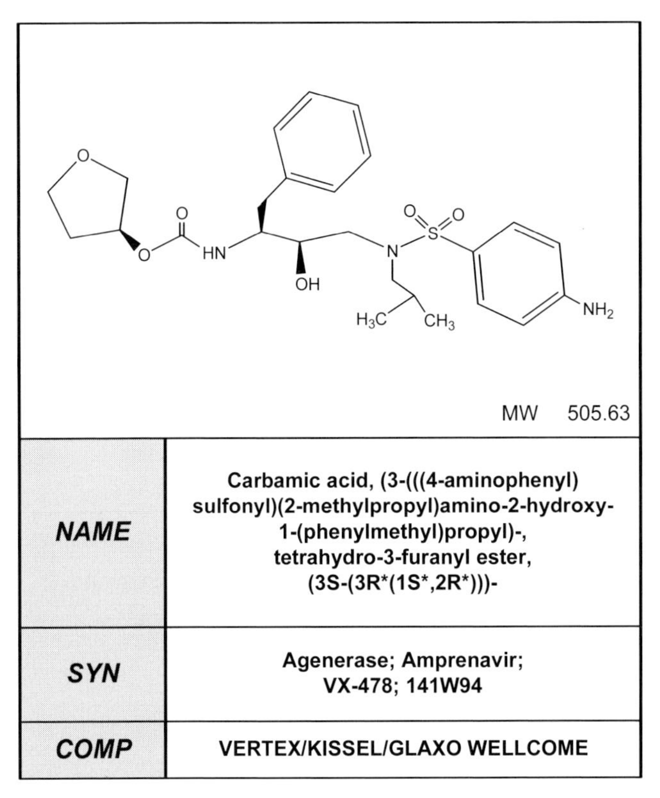

Figure 13.1 Structure of VX-478 (Amprenavir, Agenerase®, Vertex/Kissei/GlaxoSmithKline).

13.1.2
Procedure for the Development of Pharmacologically Active Compounds

Figure 13.2 features the currently practiced paradigm of structure-based drug design (Dixon, 1992). The key step, besides finding an inhibitory compound, is the elucidation of the 3D structure of the complex between the binding pocket and inhibitor.

The two main steps in the development of pharmacologically active compounds are *lead structure development* and *lead structure optimization*. A *"lead structure"* is a compound with promising but not optimal properties which acts as a basis for further optimization, often on the basis of varying side chains. Whereas the techniques of computational chemistry and molecular modeling are applied routinely and with great success in the optimization of pharmacological structures, success has been much less pronounced in the development of lead structures. Traditionally, lead structures are obtained from random testing of molecules from natural sources or from ensembles of synthetic compounds (combinatorial search) or from modification of naturally occurring ligands.

Even if the structure of the binding site on the enzyme or the structure of the receptor is known, finding of lead structures remains a difficult challenge. In general, one proceeds as follows:

1. A database of molecules is analyzed for fit with the receptor or binding pocket and optimized for maximization of steric interactions. During this procedure, the geometric fit of the target structure is optimized concurrently.
2. Subsequently, or often even concurrently, other interactions are drawn upon for super(juxta)position between the binding site and the molecule to be bound.

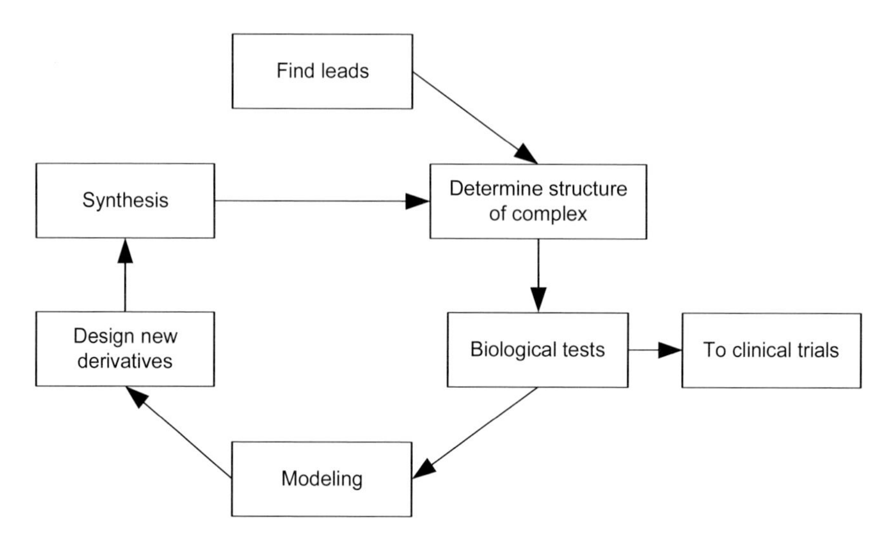

Figure 13.2 Paradigm of current drug development (cycle of steps in the process for structure-based drug design).

Electrostatic interactions as well as van der Waals forces can be visualized and optimized with the computer nowadays.

3. In yet another way, one attempts to let molecules "grow" in the binding pocket, either atom by atom or through utilization of prescribed choices of alternative substituents for existing groups on the molecule. At the point of generating such alternative substitutions, questions regarding chemical stability of the structure generated, a convenient route of synthesis, or the presence of realistic conformations should be taken into account.

13.1.3
Process for the Registration of New Drugs

New drugs cannot just be brought onto the market once the development is done. From today's perspective, the fact that up to the 1960s new pharmaceuticals were essentially tested only on the researchers themselves, or their co-workers and friends, before introduction into the market seems like a scenario from a different world. Today, registration, with a rigorous testing procedure and review by a panel, is a crucial step before any commercial activity can occur. As the registration process required by the Food and Drug Administration (FDA) of the United States is the generally accepted model all over the world, it is described in the following paragraphs. Other countries have similar procedures.

The research and development process to a new drug consists of the following steps:

1. preclinical research
2. clinical phase I
3. clinical phase II
4. clinical phase III
5. registration period
6. commercialization

At every step, a fraction of the drug candidates drops out owing to insufficient performance in the tests required, especially during the clinical phases. Insufficient performance usually means either insufficient efficacy of the drug candidate in treating the desired condition and/or excessive side effects. If the drug candidate survives the clinical phase III, the FDA is usually petitioned to grant approval for the new drug. Even at that stage, there is a distinct chance of failure, usually for the same reasons as above. In some cases, insufficiency of data is the cause for rejection. Table 13.2 illustrates the volume, risk of failure, timeline, and costs associated with each stage of the process.

From Figure 13.3 and Table 13.2, we can discern that the development of new drugs is a situation of statistics with long odds. The average number of compounds necessary to be tested at the preclinical stage per compound making it to the market has exceeded 10 000. In addition to the resources required for the synthesis and characterization of such numbers of compounds, the clinical phases are often

Table 13.2 The development pipeline for novel drugs: probabilities and time commitments.

Stage	*Probability of success*			*Time spent at each step*	
	Bph, individual	Bph, cumulative	NCE, cumulative	Bph [years]	NCE [years]
Preclinical	0.57	0.4	0.11	2.3	2.4
Clin. phase I	0.88	0.71	0.25	1.8	
Clin. phase II	0.86	0.8		2.2	
Clin. phase III	0.93	0.93	0.66	2.0	
Pre-registration	1.00	1.00		2.0	
Registration	1.00	1.00		1.6 ($\Sigma = 11.9$)	

Bph: biopharmaceuticals; NCE: new chemical entities.
Source: Struck (1994).

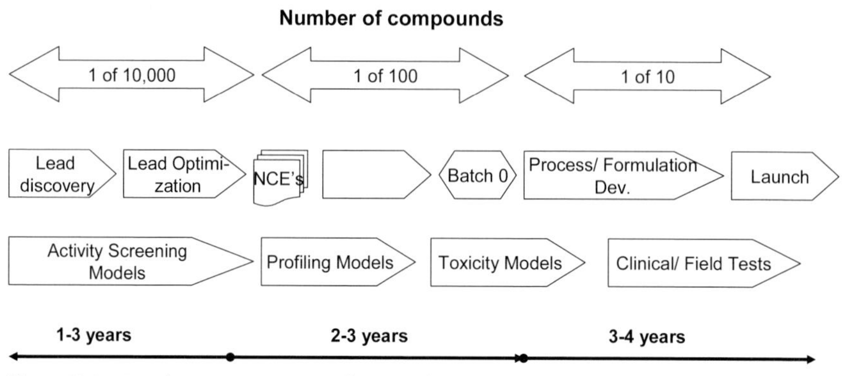

Figure 13.3 Development process in pharma (Blaser, 2001).

very expensive, up to $ 400 million per successful drug candidate reaching the market. In all, the average expenses per typical successful drug candidate are between $ 800 million and more than $ 1 billion. Such numbers are behind the recent consolidation of the pharmaceutical industry, as only large entities can afford such costs and can assemble a reasonably diversified portfolio of drug candidates expected to contain at least some winners.

Another way to control costs and reap higher profits, besides improving the statistics for a novel pharmaceutical hit by mergers and consolidations, lies in accelerating research and development through the six phases delineated in Table 13.2. Figure 13.4 illustrates the consequences of a faster development pace by comparing the traditional timeline with the novel, accelerated pace.

While expenditure owing to expensive clinical trials grow more quickly for the accelerated pace, introduction to the market is reached more quickly (the goal is eight years in comparison to 12 years traditionally). Importantly, however, cumulative revenues between market introduction in, let us say, year 8 and patent expiration in year 18 to 20 tend to be considerably higher, as (i) more years under patent protection are available, and (ii) the initial market penetration and build-up phase is not as significant.

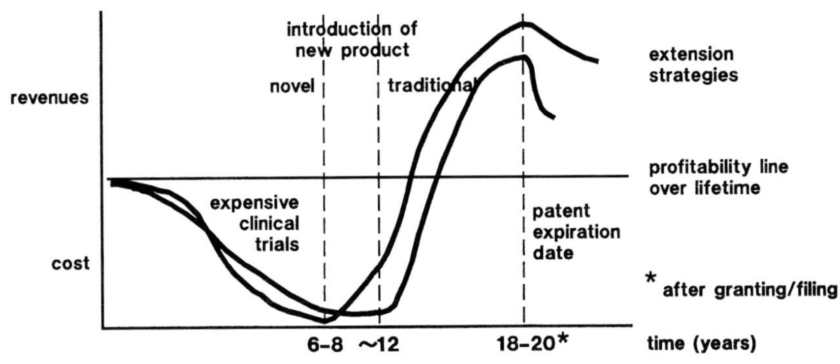

Consequences for developers of (biocatalytic) processes

- **well-defined set of skills**
- **short reaction time**
- **early commitment to process and customer**

Figure 13.4 Development cycle and costs in pharma (Bommarius, 1998).

13.1.4 Chiral versus Non-chiral Drugs

The vast majority of these potential drugs contain one or several chiral centers (Stinson, 1994). As the wrong enantiomer can cause harmful side effects, very high enantiomeric purity of therapeutics is essential (Crossley, 1995). The observation that chirality can play a major role in the toxicity and specificity of therapeutic agents was first made in 1956 by Carl Pfeiffer (Pfeiffer, 1956) and is commonly referred to as Pfeiffer's rule, which states "...the greater the difference between the pharmacological activity of the D and L isomers the greater is the specificity of the active isomer for the response of the tissue under test." In retrospect, it is not surprising that, given the chiral nature of most biological components such as amino acids, sugars, or nucleotides, different enantiomers would produce different pharmacological effects.

The realization that enantiomeric purity played a critical role in the specificity and toxicity of pharmacological agents has prompted the Food and Drug Administration to increase its regulatory oversight of enantiomeric purity of approved pharmaceuticals. Recent FDA policy (Docket No. 97D-0448) requires that any drug component comprising over 1% of the total composition be tested with separate toxicology studies. Thus, a racemic drug candidate would require separate trials for each enantiomer. While enantiomeric purity is not mandated by the FDA, the huge cost of clinical trials virtually ensures that pharmaceutical companies will develop enantiomerically pure drug candidates, provided the enantiomers do not readily interconvert.

13.2
Enzyme Cascades and Biology of Diseases

13.2.1
β-Lactam Antibiotics

Antibiotics are substances for defense against non-endogenous bacteria. Antibiotics are typically antibacterial drugs, interfering with some structure or process that is essential to bacterial growth or survival without harm to the eukaryotic host harboring the infecting bacteria. Their discovery by Alexander Fleming in 1928 allowed for the first time a causal therapy of infectious diseases, which constituted a major cause of death at the time. In the 1980s and 1990s the fight against most infectious agents seemed to have been won, so that in many drug companies the research and development capacities (and in most cases also the capabilities) were reduced or at least not increased. However, in all parts of the world bacterial agents are on the offensive again (Begley, 1994). The reasons are either lack of hygiene, including lack of measures against transmitters such as rodents, or lack of overall health status, such as in the case of patients with severely compromised immune systems, or finally, the progressive development of resistance of many bacteria against β-lactam antibiotics. We live in an era when antibiotic resistance has spread at an alarming rate and when dire predictions concerning the lack of effective antibacterial drugs occur with increasing frequency. The situation can very well be labeled a race between progress in medicine versus development of resistance in bacteria. In this context it is especially appropriate to understand the nature, origin, mode of action, and cause of effectiveness of antibiotics and how the development of antibiotic-resistant "superbugs" can be slowed down (Walsh, 2000).

After the discovery of penicillin in 1929 it took until 1941 until the difficulties were overcome through the use of column chromatography and freeze-drying to isolate the acid-, base- and heat-sensitive molecule. Between 1941 and 1945, five naturally occurring penicillins could be isolated which all demonstrated similar antibacterial properties (Nayler, 1991). Owing to its superior suitability for large-scale production, penicillin G (Pen G) with the phenylacetyl group as side chain was picked as the standard for clinical development (Rolinson, 1988). It was found that, although *Penicillium chrysogenum* transformed many acyl compounds to penicillins, addition of phenylacetic acid effectively suppressed the formation of all other penicillins besides the desired pen G, so that the yield of pen G increased drastically.

At the Biochemie Kundl company in Kundl, Austria, it was found in 1954 that the penicillin with the phenoxy group in the side chain (Pen V) is much more stable against acid, so oral administration became possible. Besides pen G and pen V, penicillins with yet different side chains were isolated. Table 13.3 lists naturally occurring and synthetic species. The majority of antibiotics in use today (about 80%) belongs to the category of β-lactam antibiotics, mainly the *penicillins* as well as the related *cephalosporins*, but also higher-generation antibiotics developed more recently such as *carbapenems* or *monobactams* (Figure 13.5).

Table 13.3 Naturally occurring and synthetic penicillins (Nayler, 1991).

Name	*N-Acyl side chain (R)*	*Name of side chain*
Benzylpenicillin, pen G	Ph-CH_2-CO-	phenylacetic acid
p-Hydroxybenzylpenicillin, pen X	*p*-HO-Ph-CH_2-CO-	*p*-hydroxyphenylacetic acid
Phenoxymethylpenicillin, pen V	Ph-O-CH_2-CO-	phenoxyacetic acid
Heptylpenicillin, pen K	H_3C-$(CH_2)_6$-CO-	octanoic acid
2-Pentenylpenicillin, pen F	H_3CCH_2CH=$CHCH_2$-CO-	β,γ-hexanoic acid
Amylpenicillin, Pen-dihydro-F	H_3C-$(CH_2)_4$-CO-	hexanoic acid
Isopenicillin, pen M	HOOC-CH(NH_2)$(CH_2)_3$-CO-	L-α-aminoadipic acid
Penicillin N, cephalosporin M	HOOC-CH(NH_2)$(CH_2)_3$-CO-	D-α-aminoadipic acid

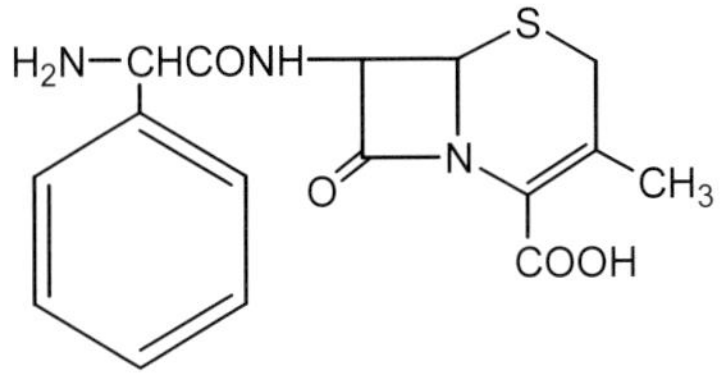

Figure 13.5 Structures of the most important β-lactam antibiotics.

As the β-lactam ring is sensitive to acid- or base-catalyzed hydrolysis as well as to lactamases, owing to torsion within the ring, the moiety poses a challenge for synthesis and process development. The biosynthetic pathway to the simplest naturally occurring penicillin, penicillin N, was elucidated in the 1970s. The role of phenylacetic acid as an indispensable component of pen G, however, had been recognized much earlier. The three amino acids L-α-aminoadipic acid, L-cysteine, and L-valine are joined to form the tripeptide δ-(L-α-*a*minoadipyl)-L-*c*ysteinyl-D-*v*aline (ACV), catalyzed by ACV synthase with ATP consumption. Isopenicillin N synthetase (IPNS), which requires Fe^{2+}, O_2, and ascorbic acid as cofactors, then converts the two-step ring closure to isopenicillin N. The L-α-aminoadipyl side chain is epimerized in part to the D-α-aminoadipyl compound, penicillin N, catalyzed by epimerases. With the help of desacetoxycephalosporin C synthase (DAOC synthase, also called expandase, depending on the cofactors α-ketoglutarate, Fe^{2+}, O_2, and ascorbic acid) cephalosporin C (Ceph C) is synthesized.

β-Lactam antibiotics operate through inhibition of murein synthesis in the cell membrane of growing Gram-positive bacteria. The mechanism of action is among the milder ones against bacteria. Other antibiotics affect bacterial metabolism much more strongly: gramicidin increases cell membrane permeability, streptomycin and erythromycin impede ribosomal function of protein biosynthesis, and rifamycin even inhibits RNA biosynthesis. The standard penicillin, pen G, does not even remotely feature all the desired properties of an antibiotic for therapy (Table 13.4).

Table 13.4 Properties of natural and semisynthetic penicillins (Diekmann, 1991).

Compound	***Residue R***	***Acid stability***	***β-Lactamase stability***	***Broad spectrum***	***Fights problem strains***
Naturally occurring					
Benzylpenicillin, pen G	Ph-CH_2-CO-	–	–	–	–
Phenoxymethylpenicillin, Pen V	Ph-O-CH_2-CO-	+	–	–	–
Semisynthetic					
Ampicillin	Ph-CH(NH_2)-CO-	+	–	+	–
Carbenicillin	Ph-CH(COOH)-CO-	–	–	+	+

Every inexpensive high-value-added therapy aims to act through resorption in the intestine via oral administration or through resorption of nasal muci via nasal administration (inhalation). One condition for oral administration, not fulfilled by pen G, is sufficient stability against the strongly acidic milieu of the stomach (pH 1–2). The subsequently developed phenoxy analog pen V owes its increased acid stability to the electron-withdrawing property of the phenoxy group, which withdraws electron density from the side chain carbonyl group so that it can no longer attack the β-lactam ring. The stability of penicillins to acid can be correlated with the p*K* value of the side chain acid (Doyle, 1961).

Another increasingly important problem is resistance of some bacteria against certain antibiotics. Resistance is frequently based on the ability of β-lactamases (previously: penicillinases, E.C. 3.5.2.6.) contained in corresponding strains to hydrolyze the amide bond of the β-lactam ring of penicillins (or cephalosporins). Hydrolysis of the β-lactam ring renders β-lactam antibiotics ineffective. Development of inhibitors against β-lactamases in the 1970s broadened the utility of penicillins significantly: clavulanic acid, itself a β-lactam, was the first successful example. Besides acid stability and resistance against β-lactamases, the ideal antibiotic should feature broad-spectrum utility against various bacteria and effectiveness against bacterial problem strains which cannot be combated by naturally occurring antibiotics or which develop resistance, such as the problem strain in hospitals, *Staphylococcus aureus.*

13.2.2
Inhibition of Cholesterol Biosynthesis (in part after Suckling, 1990)

Enhanced cholesterol levels in blood are known for many years to constitute one of the principal risk factors for atherosclerosis and coronary heart disease, the main cause of death in industrialized countries. Several modes of treatment to reduce cholesterol in the blood have been proposed. Cationic resins capture some of the bile acids which are biosynthesized from cholesterol. Such bile acid sequesters yield 15–30% lower serum cholesterol values. At least 50% of total cholesterol in the body is formed through de-novo synthesis so that inhibition of cholesterol biosynthesis seems to be an option for therapy. However, ubiquinones and isopentyladenosine, a component of transfer-RNA, are formed through the same biosynthetic route. In addition, cholesterol serves as a precursor for the synthesis of other steroids such as estrogen, testosterone, and progesterone, so that questions with respect to the feasibility of inhibition of such a fundamentally important biosynthetic route remain. The pathway is depicted in Figure 13.6.

Triparanol, the first inhibitor candidate in the clinic, interferes late in the metabolic pathway and was withdrawn quickly owing to side effects. In contrast, the

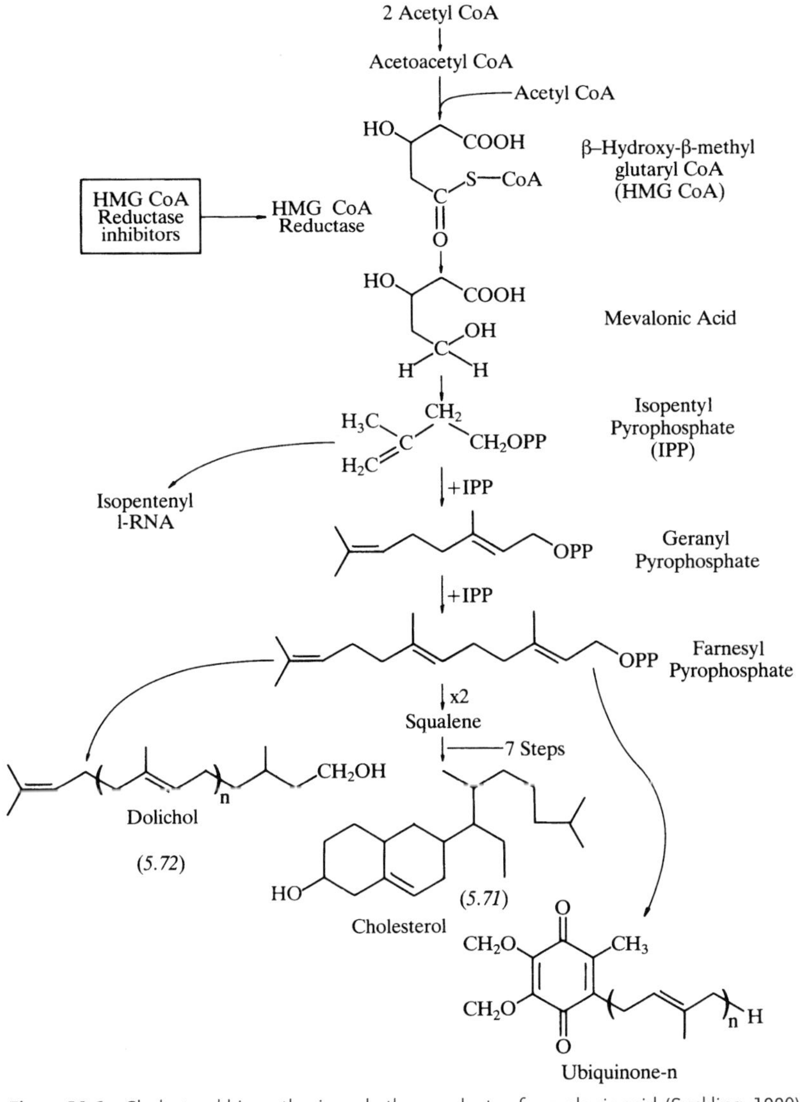

Figure 13.6 Cholesterol biosynthesis and other products of mevalonic acid (Suckling, 1990).

rate-limiting enzyme in the pathway, 3-hydroxy-3-methylglutaryl-coenzyme A reductase (HMG-CoA reductase), as the key regulated enzyme in the pathway (through feedback regulation) is an eminently suitable therapeutic target for the regulation of cholesterol biosynthesis. For this reason, HMG-CoA reductase rapidly became the target for the development for anti-cholestemics.

After screening 8000 microorganisms, three novel inhibitors of HMG-CoA reductases were found at Sankyo (Tokyo, Japan), Mevastatin among them. The open form inhibits HMG-CoA reductase with a K_I value of 1 nM and influences dramatically the biosynthesis of cholesterol, just like the later developed analogs Lovastatin and Simvastatin (Figure 13.15, below).

Vertebrates normally transport cholesterol in the blood as esterified compounds with long-chain fatty acids in the hydrophobic core of plasma lipoproteins (low-density lipoprotein–cholesterol complexes, LDL–C). After endocytosis of these complexes through the cell membrane the cholesterol esters are cleaved again and superfluous cholesterol excreted via incorporation into bile acids. Owing to hereditary hypercholesterolemia, a condition with a defective LDL–C receptor gene, blood plasma cholesterol levels are often extremely high, but can be controlled at a sustainable level by the above-mentioned inhibitors. After broad-based clinical studies over several years, Mevastatin was withdrawn because of toxic side effects, Lovastatin, however, developed by Merck, Sharpe, and Dohme (MSD, today Merck and Co.), is still a very successful pharmaceutical today and can be counted as one of the main breakthroughs for the demonstration of the effectiveness of drugs in clinical studies.

All these potent anti-cholestemics have been obtained without much knowledge about the mechanism and the structure of the active site of the key enzyme HMG-CoA reductase. Although the enzyme from some sources has been purified to homogeneity (Kleinsek, 1981) only a few details are known, such as the two reactions catalyzed by the enzyme (Figure 13.7). Structural similarity between the dihydroxy moiety of the inhibitors and mevalonic acid is regarded as the reason for the former's effectiveness. Especially interesting seems to be the fact that Mevastatin inhibits the enzyme through binding at two different sites (Nakamura, 1985), the hydroxymethylglutaryl domain as well as the hydrophobic region; the 10^8-fold improved binding of Mevastatin in comparison with 3,5-dihydroxyvalerate demonstrates impressively the role of the hydrophobic region of Mevastatin.

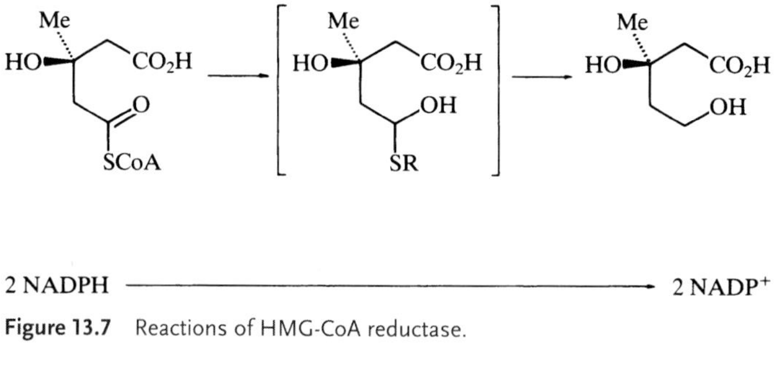

Me
HO — CO₂H
O
SCoA
→
[Me
HO — CO₂H
OH
SR]
→
Me
HO — CO₂H
OH

2 NADPH → 2 NADP⁺

Figure 13.7 Reactions of HMG-CoA reductase.

13.2.3
Pulmonary Emphysema, Osteoarthritis: Human Leucocyte Elastase (HLE)

Emphysema is accompanied by destruction of elastin, a structural protein in the lung; the disease occurs frequently in smokers. In the healthy body, the effect of elastase, (human leucocyte elastase, HLE), a rather unspecific protease disassembling many structural proteins, is controlled by α1-protease inhibitor (α1-PI). However, in cases of genetic deficiency of α1-PI, or excess supply of HLE owing to inflammation, or destruction of α1-PI, possibly through cigarette smoke, the finely tuned balance between HLE and α1-PI is destroyed. Another therapeutic concept besides supply of synthetic recombinant α1-PI is the inhibition of HLE with inhibitors of low molecular weight. Whereas the ICI Pharmaceuticals Group attempted to mimic the tetrahedral adduct of the serine protease through transition state analogs, another group at Merck (MSD) planned to construct inhibitors based on the similarity of HLE to other serine proteases such as PGA. Starting from the structure of elastin and of α1-PI as well as initial results by James Powers (Georgia Tech, Atlanta/GA, USA) towards a peptide substrate model (Figure 13.8), the ICI group managed to prepare peptide aldehydes with very good selectivity and po-

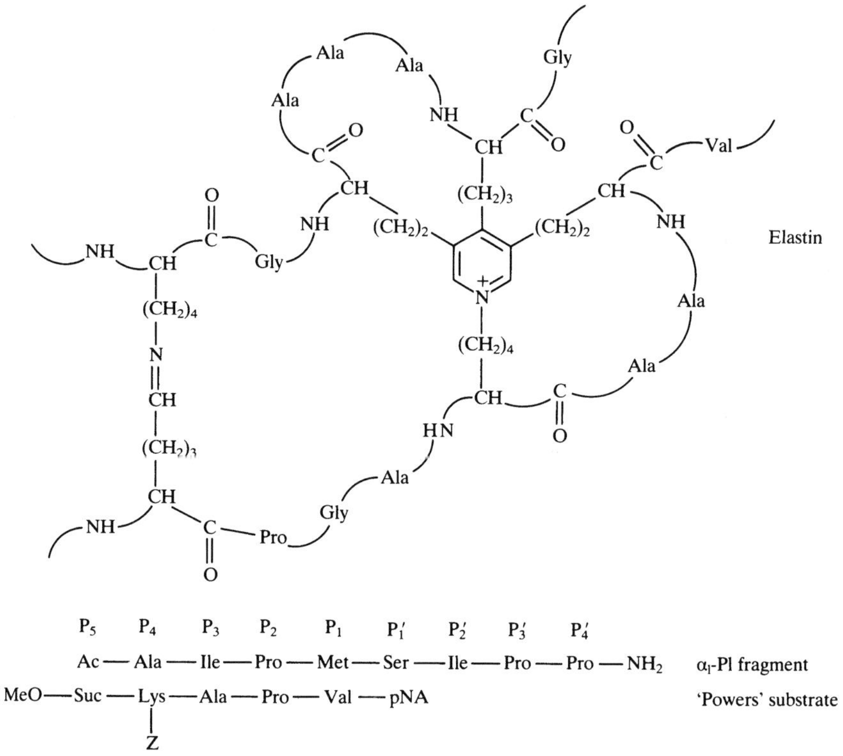

Elastin

P_5	P_4	P_3	P_2	P_1	P_1'	P_2'	P_3'	P_4'		
Ac—	Ala—	Ile—	Pro—	Met—	Ser—	Ile—	Pro—	Pro—	NH_2	α_1-PI fragment

MeO—Suc—Lys(Z)—Ala—Pro—Val—pNA 'Powers' substrate

Figure 13.8 Elastin, α1-PI fragment and the "Powers" substrate (Suckling, 1990).

P5	P4	P3	P2	P1	K_i(nM)
				Z—VAL—CF_3	13,000
			Z—PRO—	VAL—CF_3	1,800
		Z—VAL—	PRO—	VAL—CF_3	1.6
	Z—LYS(Z)—	VAL—	PRO—	VAL—CF_3	<0.1
				MeOSuc—VAL—CF_3	53,000
			MeOSuc—PRO—	VAL—CF_3	3,200
		MeOSuc—VAL—	PRO—	VAL—CF_3	13
MeOSuc—	LYS(Z)—	VAL—	PRO—	VAL—CF_3	<0.3

Figure 13.9 Effect of peptide length on inhibition constant K_i of trifluoromethyl ketones (Suckling, 1990).

tency against HLE. The concept of peptide aldehydes was abandoned when they were found to feature low metabolic stability and low shelf-life.

The mechanism-based trifluoromethyl ketones (TFMKs) turned out to be just as selective and potent as peptide aldehydes. The trifluoromethyl group stabilizes the oxyanion of the tetrahedral complex so that it imitates the state in the actual initermediate well and it enhances the electrophilicity of the carbonyl carbon, as evidenced by the enhanced hydration constant of trifluoromethyl ketone versus acetone (35 to 1). Owing to the rate-limiting conformational change of the enzyme–inhibitor complex EI, TFMKs are competitive, slow-binding inhibitors (see Chapter 5, Section 5.3.1). Due to the extended binding pocket the dependence of inhibition on the length of the peptide residue on the TFMK was not surprising (Figure 13.9).

Based on the necessary balance of acidity and water/lipid solubility at the locus of action, i.e., the lung, compounds with N-terminal acylsulfonamide residues turned out to be the most suitable to combine, not only in vitro but also in vivo, high activity with sustained action (Figure 13.10).

After the Merck group had recognized that significant identity exists between β-lactamases and vertebrate serine proteases and that the benzyl ester of clavulanic acid inhibits HLE, the cephalosporin sulfone structural element (Figure 13.11) was discovered after a large screening process as a lead structure. Bacterial enzymes are inhibited by the $C_7\beta$ configuration, in contrast to vertebrate enzymes which are inhibited by the $C_7\alpha$ configuration; for this reason, β-lactam antibiotics do not act as HLE inhibitors, and vice versa.

Owing to the endopeptidase character of HLE, free clavulanic acids are not good inhibitors. Interpretation of the mechanism of inhibition of HLE by cephalosporin

A was supported by a 3D X-ray structure of the inhibitor at the closely related porcine elastase (Navia, 1987) (Figure 13.12). After initially reversible inhibition (Figure 13.12, top two rows), time-dependent irreversible inhibition through covalent bonding of the dihydrothiazine ring of the inhibitor to the His-57 residue of the active site of the enzyme increasingly takes over (Figure 13.12, bottom row). Peptidyl chloromethyl ketones, known unspecific inhibitors, act in similar fashion.

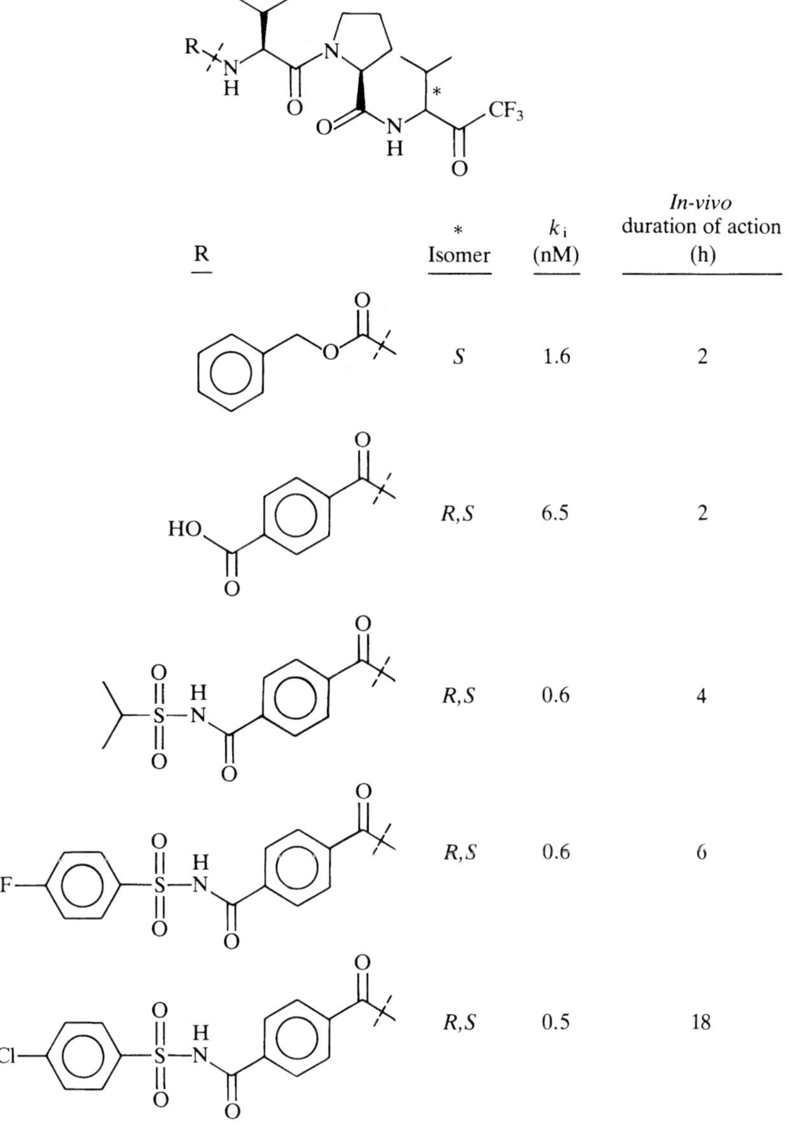

Figure 13.10 TFMK inhibitors with sustained action in the hamster model of emphysema (Suckling, 1990).

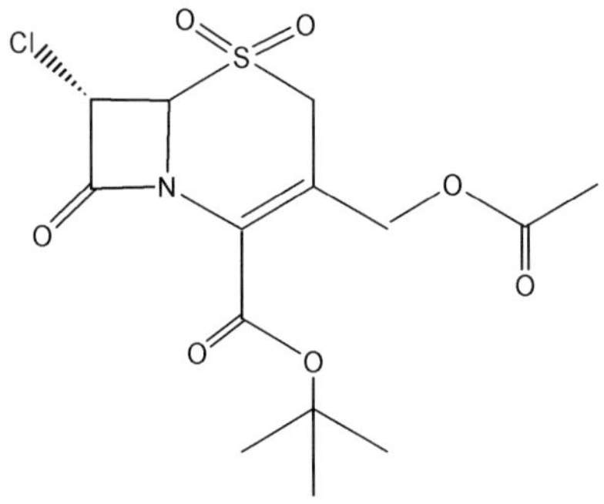

Figure 13.11 Lead structure for HLE inhibition: cephalosporin sulfone.

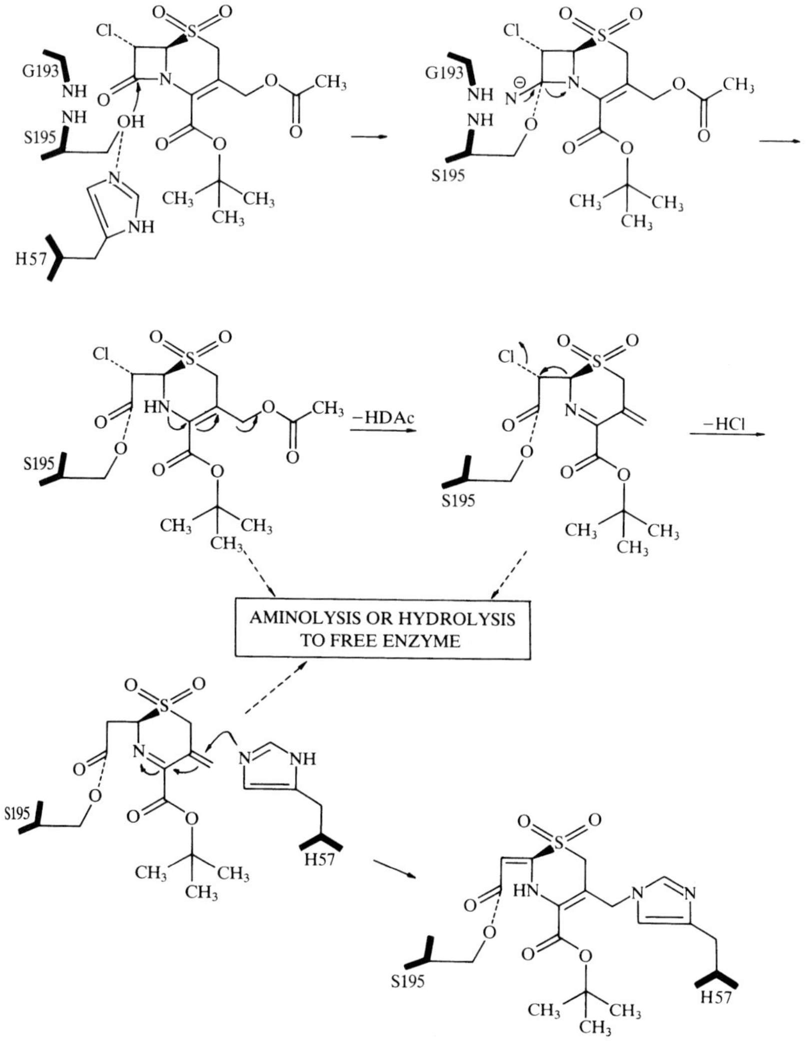

Figure 13.12 Proposed mechanism of inhibition of elastase by β-lactams (Suckling, 1990).

13.2.4 AIDS: Reverse Transcriptase and HIV Protease Inhibitors

The history of combating the human immunodeficiency virus (HIV) can be viewed as a model of today's development of pharmaceuticals (Stone, 1995). Since the middle of the 1980s a total of 20 anti-AIDS pharmaceuticals based on 16 distinct chemical entities have been introduced into the market; the first was AZT by Glaxo in 1987. A causal therapy, however, has not been developed yet. Most advantageous effects of today's actives are temporary because HIV is characterized by an extreme mutability and can cause resistance phenomena through rapid mutation within months or sometimes weeks against most approved preparations or those in clinical phases. The criterion for efficacy is the reduction of the number, or even total elimination, of viruses in the blood combined with an enhanced count of CD4-T-lymphocytes.

The development of high-resolution 3D structures of proteins or protein–inhibitor complexes, however, was crucial in rapidly developing medicinals, although more so in the optimization phase of lead structures than in the discovery phase. The majority of preparations in clinical phases have been found through conventional screening methods and interfere in the inhibition of either reverse transcriptase or HIV protease. The cycle of replication is shown in Figure 13.13. New programs target control and inhibition of all three HIV enzymes, such as reverse transcriptase, protease, and integrase, as well as regulator proteins such as Tat, Rev, and Nef receptors.

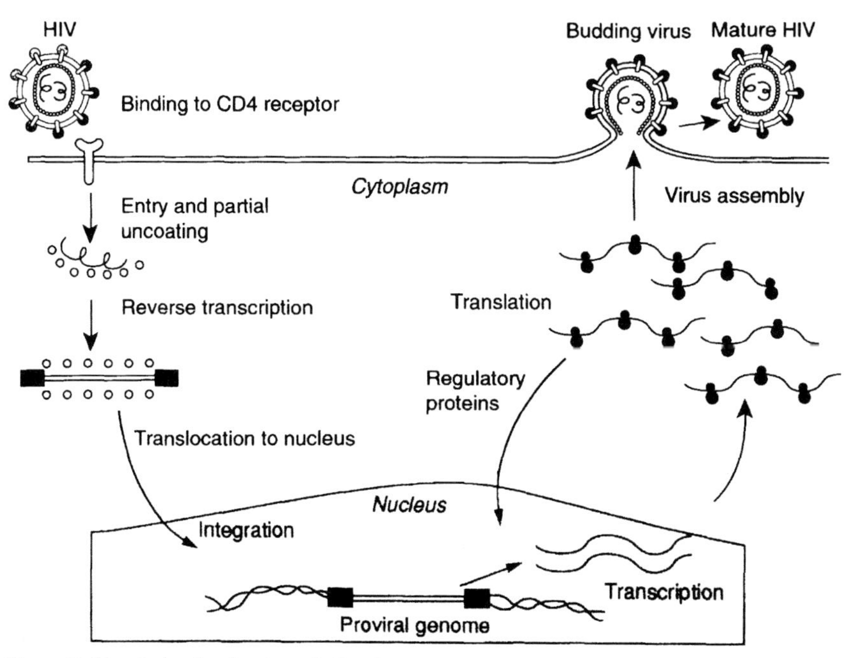

Figure 13.13 Cycle of replication of HIV (Gait, 1995).

Experience teaches us that combinations of active compounds offer the most promising perspective – for example, the combination of AZT with 3TC of BioChemPharma (Laval, Quebec, Canada). Currently (2003), there are seven approved entities of nucleoside analogs (NRTIs) and three approved non-nucleosides (NNRTIs) that inhibit reverse transcriptase (RT) by mimicking the structure of DNA building blocks and thus the copy process of RNA into DNA by reverse transcriptase. In addition, five HIV protease inhibitors and one viral fusion inhibitor have been approved.

HIV protease inhibitors are transition-state analogs (Chapter 9, Section 9.2.6): HIV protease binds them much more tightly than the natural substrate because the substrate must be distorted to assume its transition-state configuration. Thus, HIV protease inhibitors are competitive enzyme inhibitors and thereby prevent the maturation and infectivity of the viral particle. They offer advantages over RTIs with respect to efficacy, safety, and occurrence of resistance. However, just like RTIs, HIV protease inhibitors are most often offered in combination with therapies based on another mode of action. Structures and data for the five approved HIV protease inhibitors are shown in Figure 13.14 and in Table 13.5.

Invirase® (saquinavir, Ro-318959), the product of Roche (Basel, Switzerland), demonstrated only limited efficacy even at doses of up to 7200 mg d^{-1}. In addition, owing to the difficult and cumbersome synthesis of its active ingredient, Roche was able only in 1995 to offer sufficient product. Although clinical phase III had begun in January 1994, Roche announced only in June of 1995 that 4000 patients had been selected by coin toss for supply of the drug. During the clinical studies the following combinations were tested: (i) saquinavir, AZT, and ddC (Roche's RT inhibitor); (ii) AZT and ddC; and finally (iii) ddC and saquinavir. Combination (i), the donation of all three drugs, resulted in the highest count of CD4-T-lymphocytes. In phase III, Invirase is offered three times a day at 600 mg each dose. Saquinavir is active against both HIV-1 and HIV-2 protease with EC_{50} values in the range of 1–30 nM (Ohta, 1997c; Roberts, 1990). Low bioavailability (~4%) due to poor absorption and extensive first-pass metabolism limits the use of saquinavir

Table 13.5 Currently approved HIV protease inhibitors.

Name	*Trade name*	*Internal name*	*Company*	*Marketing partners*	*Approval [US, EU]*	*K_i [nM]*	*Bioavail. [%]*	*$\tau_{1/2}$ [h]*	*MW [g/mol]*
Saquinavir	Invirase	Ro-318959	Roche	–	12/95, 07/96, 09/97 (J)	10	4		670.9
Indinavir	Crixivan	MK639; L-735,524	Merck	–	03/96, 07/96	0.5	23	1.5	613.8
Ritonavir	Norvir	ABT-538	Abbott	–	03/96, 08/96	0.015	78	1.2	721.0
Nelfinavir	Viracept	AG-1343	Agouron	Pfizer, J. Tobacco	03/97, 01/98	2.0	≈50	4.0	663.9
Amprenavir	Agenerase	VX-478	Vertex	Kissei, GSK	04/99, 06/00	0.6	70	8	505.6

GSK: Glaxo SmithKline; J. Tobacco: Japan Tobacco.

Sources: saquinavir: Ohta (1997c); indinavir: Vacca (1994), Ohta (1997b); ritonavir: Kempf (1995), Ohta (1997a); nelfinavir: Kaldor (1997), Pai (1999).

Structure		Developer	clinical trials
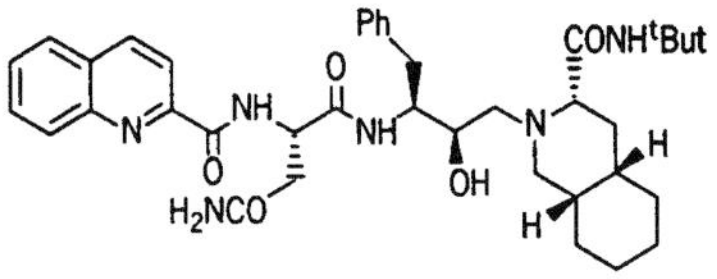	Ro31-8959	Roche	Phase III
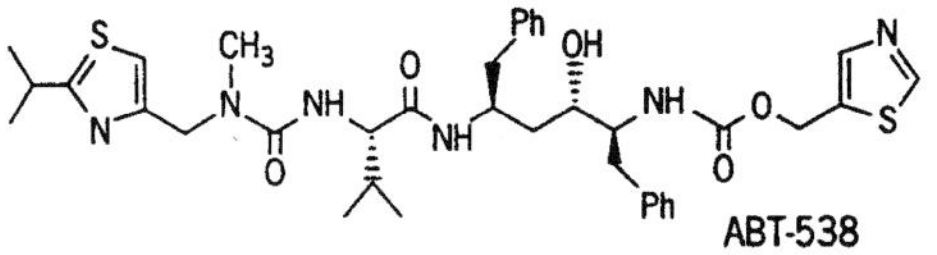	ABT-538	Abbott	Phase II/III
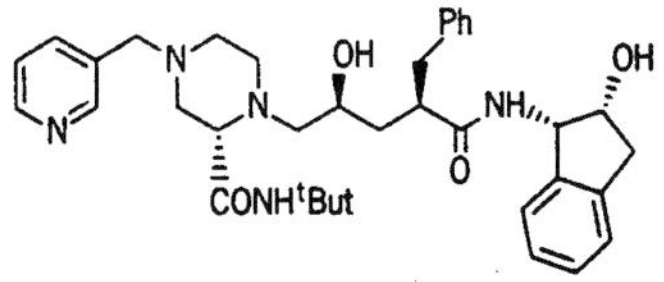	MK-639 (L735524)	Merck	Phase II
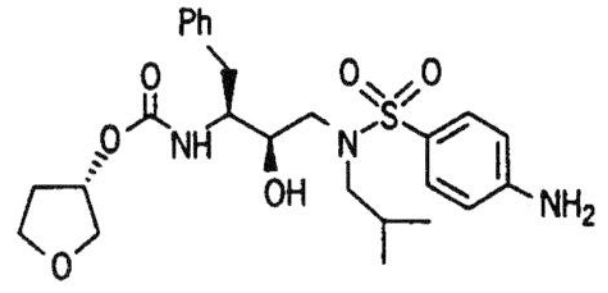	VX-478	Vertex-Wellcome	Phase I
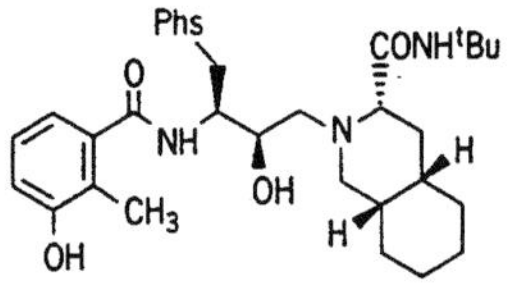	AG1343	Agouron	Phase I
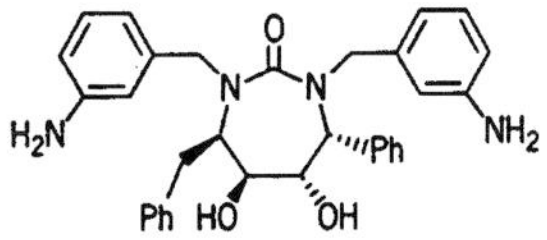	XM412 (DMP 450)	Dupont-Merck	Phase I (terminated)
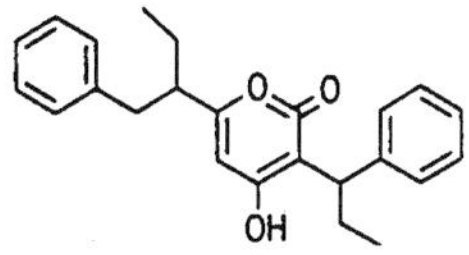	U96988	Upjohn	Phase I

Figure 13.14 Structures of currently (June, 2003) approved HIV protease inhibitors (Gait, 1995).

in potent triple combination therapies. A new soft gel formulation which increases bioavailability threefold has recently been approved.

Crixivan® from Merck (Rahway/NJ, USA) showed a higher level of CD4-T cell count but also a faster occurrence of resistance. After an initially 100-fold decreased viral count only 20% of the patients showed values that had decreased below the baseline after 24 weeks. Liver toxicity limits the daily doses to three times 600 mg, too little for total repression of viral replication. Crixivan is difficult to produce, as is Invirase, and Merck limited access to its program to only 1400 patients. At the end of 1995, Crixivan was approved by the FDA in record time after only a few weeks of testing; this was certainly due at least in part to political pressure by groups of activists. Pharmacological data on Crixivan are IC_{50} = 0.5 nM, CIC_{95} = 25–50 nM, oral bioavailability = 23% and 70% for rat and dog respectively, half life $\tau_{1/2}$ = 1–2 h.

Norvir® (ritonavir, ABT-538) by Abbott (Kalamazoo/MI, USA) in phase II demonstrated a 100-fold reduction of viral count, more than Crixivan and AZT. The threefold enhancement of CD4-T cell count after 12 weeks is also an improvement over AZT. However, even ABT-538 develops resistances, albeit more slowly than Crixivan. Hepatic side effects limit doses to 600 mg twice daily. Ritonavir has high oral bioavailability, about 78% in rats, good solubility (5.3 g L^{-1} (pH 7.4) to 6.9 g L^{-1} (pH 4.0)), and a plasma half-life of 1.2 h (Kempf, 1995). The molecule is difficult to produce: in phase II, the overall yield in production was 2% (!); in addition, the product was cherry-red. The combination of ritonavir and saquinavir has also been reported to dramatically increase saquinavir plasma concentrations (by as much as 50-fold). Ritonavir is believed to act by inhibiting cytochrome P450 (CYP 3A4), the enzyme responsible for saquinavir first-pass metabolism.

Viracept® (nelfinavir, AG-1343) of Agouron (La Jolla/CA, USA) in collaboration with Japan Tobacco (Tokyo, Japan) began only in July 1995 with phase II investigations. Dosages in phase II are set between 770 and 1030 mg daily. The viral count decreased between 15- and 75-fold in comparison with the base case.

In July 1995, Vertex (Cambridge/MA, USA) presented encouraging preclinical and phase I results for Agenerase® (amprenavir,VX-478): the compound demonstrated no toxicity for single doses up to 1200 mg but 70–78% bioavailability in tablet form and a half-life of 8 h in blood plasma, in combination with a 25-fold decrease of viral count in comparison with the base case. Even versus the most resistant HIV strain, VX-478 is only 15-fold less sensitive than against an unmutated virus.

Amprenavir came from a program of rational drug design to be small, orally bioavailable, and potent and is licensed by Glaxo Wellcome (today Glaxo SmithKline) and Kissei Pharmaceuticals (Yoshino, Japan) from Vertex Pharmaceuticals. The sulfonamide is synergistic with AZT, ddI, and other nucleoside analogs of Glaxo Wellcome and is active against a variety of HIV strains including AZT-resistant mutants. X-ray structure analysis of the protease–amprenavir complex reveals a very tight fit, with 397 $Å^2$ of solvent-accessible surface area being excluded upon complex formation, and hydrogen bonds to the active-site resident water molecule and to the catalytic aspartate D25.

HIV-1 genetic resistance to protease inhibitors occurs via specific mutations. Genotypic analysis of the HIV protease gene from isolates selected in vitro indicated that Gly48Val and Leu90Met mutants had reduced susceptibility to saquinavir (Ohta, 1997). Indinavir and ritonavir resistance maps to residue 82, whereas for amprenavir the key mutation is at residue 50 (I50V) and confers a threefold decline in viral sensitivity to amprenavir. Two additional mutations at residues 46 and 47 follow development of mutation at position 50, resulting in a 20-fold decline of sensitivity in the triple mutant. Viral mutations causing resistance to amprenavir leave sensitivity to other inhibitors intact; therefore, amprenavir should be used advantageously in combination with other protease inhibitors.

Several trends can be discerned in the area of HIV protease inhibitors:
- the labile amide bond of peptides is replaced by isosteres such as hydroxyethylamine;
- the original size of lead structures, about seven amino acid residues, is reduced to four residues;
- bioavailability, and thus water solubility, are increasingly enhanced;
- increasingly, the design of small inhibitors (MW < 600) with small clearance rates out of the body and a convenient sequence of processing steps is sought.

13.3 Pharmaceutical Applications of Biocatalysis

This section deals with applications of biocatalysis within chemoenzymatic synthesis routes. Although the amounts of product produced, even in the case of success, will never be large compared to applications such as are described in Chapter 7, biocatalysis nevertheless is an essential tool within such syntheses because it acts as an enabling technology.

13.3.1 Antiinfectives (see also Chapter 7, Section 7.5.1)

13.3.1.1 Cilastatin

A key step in the synthesis of the β-lactamase inhibitor cilastatin (Bayer, Leverkusen, Germany) is the preparation of (*S*)-2,2-dimethylcyclopropane carboxamide. The chemically synthesized corresponding nitrile, 1-cyano-2,2-dimethylcyclopropane, is hydrolyzed by a highly active but enantiounspecific nitrile hydratase to the racemic carboxamides. An amidase from *Comomonas acidovorans* overexpressed in *E. coli* selectively hydrolyzes the undesired (*R*)-isomer to the acid. The remaining (*S*) enantiomer is obtained with > 99% e.e. and 48% conversion (the resulting *E* value thus exceeds 100). The (*R*)-acid is recycled by chemical amidation with thionyl chloride and ammonia. The process has been developed by Lonza (Visp, Switzerland) and runs on a 15 m^3 scale (Rasor, 2001).

13.3.2 Anticholesterol Drugs

Figure 13.15 depicts the five leading anti-hypercholemic HMG-CoA reductase inhibitors.

Whereas several anti-cholestemic drugs such as lovastatin and pravastatin are produced wholly by fermentation, the side chain of several others such as fluvastatin

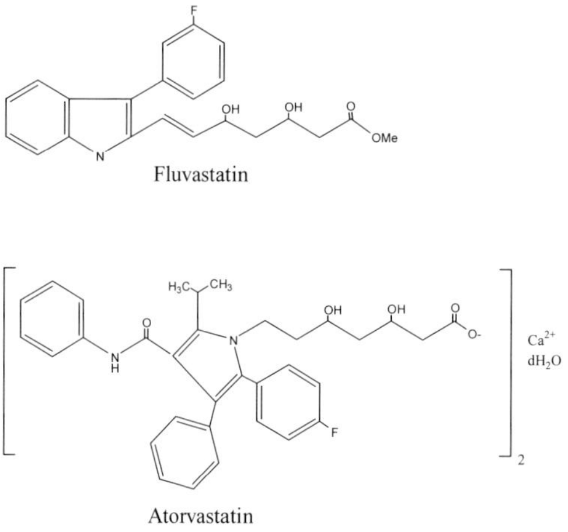

Fluvastatin

Atorvastatin

Figure 13.15 Structures of HMG-CoA reductase inhibitors.

(Lescol®, Novartis) and atorvastatin (Lipitor®, Pfizer) are accessible through enzymatic synthesis.

The side chain of such statins, the (3*R*,5*S*)-dihydroxyhexanoates, is a key intermediate and are accessible by reduction of the diketo acids. One way is bioreduction with whole cells (Patel, 1993); with glycerol-grown suspensions of *Acinetobacter calcoaceticus* SC 13876, yields of 85% and 97% e.e. were achieved for the benzyloxy-(3*R*,5*S*)-derivative. If cell extracts were used and NAD^+, glucose, and glucose dehydrogenase supplied, both intermediate monohydroxy (in the 3- and 5-position) compounds were produced before the desired dihydroxyhexanoate, and the overall reaction resulted in 72% yield and 98.5% e.e. (Patel, 2001).

Another way to statin side chains, via the intermediate *syn*-(3*R*,5*S*)-6-chlorohexanoate, employs regioselective and (*R*)-specific reduction with alcohol dehydrogenase (ADH) from *Lactobacillus brevis* to yield the intermediate (5*S*)-6-chloro-3-ketohexanoate from the 3,5-diketo acid (Wolberg, 2001) (Figure 13.16). Further reduction of (5*S*)-6-chloro-3-ketohexanoate to *syn*-(3*R*,5*S*)-6-chlorohexanoate is afforded chemically with $NaBH_4/B(OMe)Et_2$.

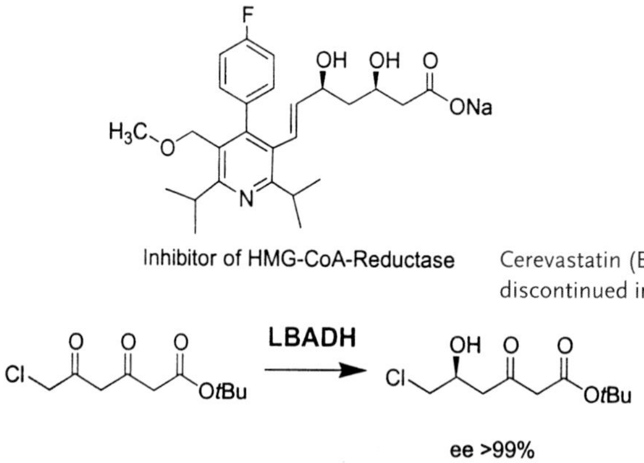

Inhibitor of HMG-CoA-Reductase

Cerevastatin (Baycol®, Bayer; discontinued in 2001)

LBADH

ee >99%

Figure 13.16 Reduction of a statin intermediate with ADH.

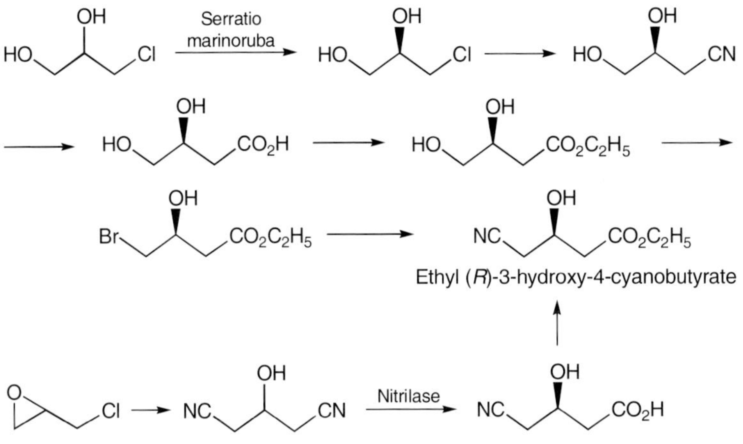

Figure 13.17 Novel synthesis route to Lipitor as developed by Diversa (Rouhi, 2002).

In another development, the statin side chain en route to Atorvastatin (Lipitor®, Pfizer) is synthesized via the key intermediate alkyl 3-hydroxy-4-cyanobutyrate (Figure 13.17). Instead of the currently practiced six-step route, a much more concise three-step route starts from epichlorohydrin via C1 chain length enhancement by both nucleophilic substitution of chloride and nucleophilic ring opening of the epoxide with cyanide to yield symmetric dicyanoisopropanol. Nitrilase action desymmetrizes the dinitrile intermediate with the creation of a chiral center in C3 to yield (*R*)-3-hydroxy-4-cyanobutyrate, which is esterified to the key intermediate ethyl (*R*)-3-hydroxy-4-cyanobutyrate.

13.3.2.1 **Cholesterol Absorption Inhibitors**

A concept different than blockage of cholesterol synthesis by inhibition of HMG-CoA reductase focuses on the inhibition of cholesterol uptake by absorption into the intestinal wall. As cholesterol uptake inhibitors are often not very water-soluble, they have to be modified for intake and transport to the desired organ or for studies of metabolic profiles. The first step of the metabolism of many drugs is their glucuronidation. The glucuronide of the novel cholesterol absorption inhibitor SCH 58325 (Figure 13.18) of Schering–Plough (Parsippany/NJ, USA) was synthesized on a 200 mg scale in one step via bovine liver microsome glucuronyl-transferase-catalyzed coupling of the glucuronyl moiety of UDP-glucuronic acid with the phenolic hydroxyl of SCH 58235 (Zaks, 1998). Glucuronide yield initially was limited by UDP-glucuronic acid hydrolysis from impurities present in the microsomal transferase preparation. Partial purification of the transferase, increase of enzyme concentration, and optimization of reaction conditions resulted in 95% conversion and 88% isolated product yield. Phenolic hydroxy derivatives of SCH 58235 could be glucuronidated as well with glucuronyl transferases derived from bovine- and dog-liver microsomes (Reiss, 1999), followed by iodination with ^{125}I to generate the labeled glucuronides needed for metabolic studies.

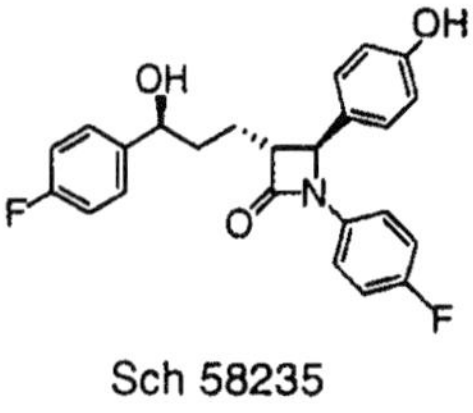

Sch 58235

Figure 13.18 Cholesterol inhibitor SCH 58235 of Schering–Plough (Zaks, 1998).

13.3.3
Anti-AIDS Drugs

13.3.3.1 Abacavir Intermediate

It has been mentioned repeatedly that a high substrate concentration correlates with a high space–time-yield and thus with an efficient process. One example is the process to an intermediate of the reverse transcriptase inhibitor and anti-HIV agent Abacavir (Glaxo Wellcome) (Figure 13.19a). In a first-generation process, the racemic γ-lactam [(–)-2-azabicyclo[2.2.1]hept-5-en-3-one] substrate was enantioselectively hydrolyzed with the help of an alkaline protease (Savinase) in a water–THF mixture; the substrate concentration was reported as an impressive 100 g L^{-1}. Chirotech (Exeter, UK), now part of Dow Chemical, screened (from organisms *Pseudomonas solonacearum* NCIMB 40249 and *Rhodococcus* NCIMB 40213) and cloned a selective γ-lactamase to react with the unprotected lactam substrate (Figure 13.19b) to yield the chiral lactam and the corresponding β-amino acid. At substrate concentrations of 500 g L^{-1}, the process is performed with enantioselectivities (*E* values) of > 400. The overall yield, of course, is still limited to 50%.

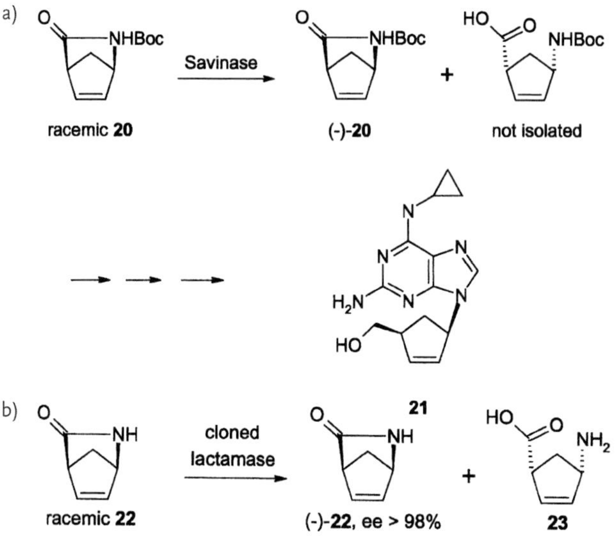

a) racemic **20** —Savinase→ (-)-**20** + not isolated → → → **21**

b) racemic **22** —cloned lactamase→ (-)-**22**, ee > 98% + **23**

Figure 13.19
a) Hydrolysis of a γ-lactam en route to Abacavir;
b) improved hydrolysis of the unprotected γ-lactam (Rasor, 2001).

13.3.3.2 **Lobucavir Intermediate**

Lobucavir (BMS-180194) is a guanine nucleoside analog for treatment of herpes and HBV (Ireland, 1997), a valine derivative (BMS-233866) acts as a prodrug and was developed alongside. The synthesis of a regioselective mono-derivatized product with L-valine ester, chemically difficult if not impossible, was achieved enzymatically. Monosubstitued lobucavir was achieved either by regioselective hydrolysis of the disubstituted N-protected or unprotected di-L-valine lobucavir in 90% toluene with lipase M (*N*-Cbz compound) or CCL (unprotected di-L-valyl compound) or by regioselective aminoacylation of lobucavir in acetone/DMF (2 : 1) with PEPTICLEC® BL Subtilisin and *N*-Cbz- and *N*-Boc-L-valine *p*-nitrophenyl ester (61% yield) or with *Pseudomonas cepacia* lipase (PCL) (84% yield), the route ultimately taken for development (Patel, 2001; Hanson, 2000a).

13.3.3.3 **cis-Aminoindanol: Building Block for Indinavir (Crixivan®)**

(–)-*cis*-(1*S*,2*R*)-1-Aminoindan-2-ol is a key chiral precursor to Crixivan (Indinavir), Merck's HIV protease inhibitor (see Section 13.2.4. above). Currently, the precursor is synthesized chemically at the multi-ton level from indene by enantioselective epoxidation to indane oxide with NaOCl in 90% yield and 87% e.e. via Jacobsen asymmetric epoxidation with a manganese salen complex followed by two subsequent chemical steps (Hughes, 1997).

An alternative biotransformation route to enantiomerically pure 1,2-indanediol has been worked out starting from indene (Buckland, 1999). Cultures of *Pseudomonas putida* (DSM 6899 and ATCC 55687) and *Rhodococcus* sp. strains B264-1 and I24 oxidized indene to mixtures of *cis*- and *trans*-indanediols and related metabolites (Figure 13.20), which is consistent with monooxygenase and dioxygenase ac-

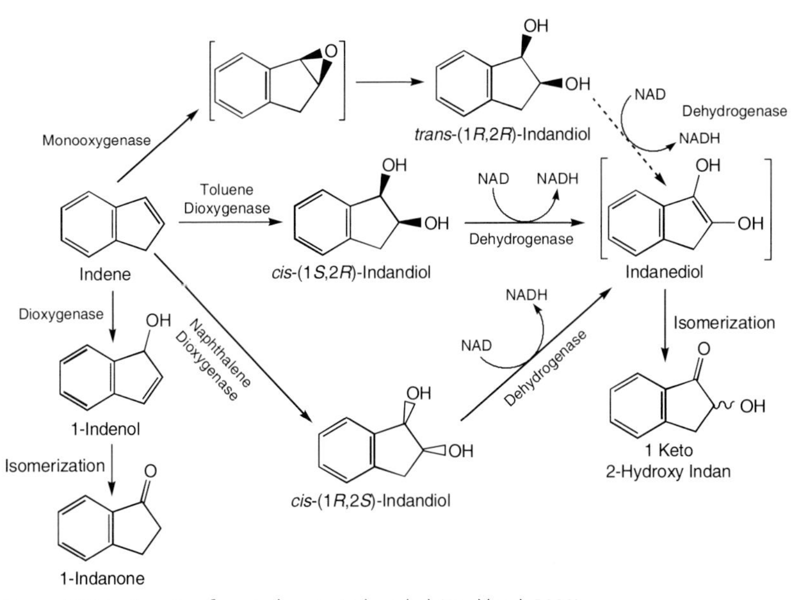

Figure 13.20 Reaction from indene to indanediol (Buckland, 1999).

tivity. *Ps. putida* resolves racemic mixtures of *cis*-1,2-indanediol by oxidizing the undesired (1*R*,2*S*)-enantiomer to yield the desired *cis*-(1*S*,2*R*)-indanediol with > 98% e.e. Cloning of toluene dioxygenase (TOD) confirmed that dioxygenase activity leads to formation of *cis*-indanediol whereas toluene monooxygenase results in *trans*-indanediol; *cis*-indanediol is resolved to *cis*-(1*S*,2*R*)-indanediol with the help of dihydrodiol dehydrogenase.

To reduce undesirable 1-indenol and 1-indanone formation by TOD during indene bioconversion, the recombinant TDO expressed in *E. coli* was evolved using the error-prone PCR method (Chapter 11, Section 11.3.1) (Zhang, 2000). High-throughput fluorimetric and spectrophotometric assays were developed for rapid screening of the mutant libraries in a 96-well format. The mutant with the highest reduced 1-indenol concentration and the highest *cis*-(1*S*,2*R*)-indanediol e.e. value was carried forward to each subsequent round of mutagenesis. After three rounds of mutagenesis and screening, mutant 1C4-3G was identified to have a threefold reduction in 1-indenol formation over the wild type (20% vs, 60% of total products) and a 40% increase in product (*cis*-indanediol) yield.

13.3.4 High Blood Pressure Treatment

13.3.4.1 Biotransformations towards Omapatrilat

[4*S*-(4*I*,7*I*,10*aJ*)]-1-Octahydro-5-oxo-4-{[(phenylmethoxy)carbonyl]amino}-7H-pyrido-[2,1-*b*][1,3]thiazepine-7-carboxylic acid methyl ester (BMS-199541-01; Figure 13.21) is a key chiral intermediate for the synthesis of Omapatrilat (BMS-186716, Vanlev®), a novel dual-action neutral endopeptidase/ACE vasopeptidase inhibitor. Bristol–Myers–Squibb (BMS, New Brunswick/NJ, USA) developed it with great hopes as a novel therapeutic concept, in addition to constituting a novel compound. Ultimately, in March of 2002, however, Omapatrilat was abandoned owing too little efficacy compared to standards. Nevertheless, as the synthetic approaches involve interesting biotransformations, they deserve coverage.

L-Lysine ε-aminotransferase reactions

A novel L-lysine ε-aminotransferase, obtained from *Sphingomonas paucimobilis* SC 16113 by using a selective enrichment culture technique, catalyzed the oxidation of the ε-amino group of lysine in the dipeptide dimer N^2-({[*N*-(phenylmethoxy) carbonyl]-L-homocysteinyl}-L-lysine) 1,1-disulfide (BMS-201391-01) to produce BMS-199541-01 (Figure 13.21) (Patel, 2000). The aminotransferase reaction required α-ketoglutarate as amino acceptor (Chapter 7, Section 7.2.3.3); the resulting L-glutamate is recycled back to α-ketoglutarate by glutamate oxidase from *Streptomyces noursei* SC 6007. Both L-lysine ε-aminotransferase and glutamate oxidase went through a development cycle consisting of the development of fermentation processes, purification to homogeneity, biochemical characterization of the enzymes (both are dimers of subunit size of 40 and 60 kDa, respectively), partial protein sequencing (the transferase only) as well as finding, cloning, and overexpressing their genes, in *E. coli* and *Streptomyces lividans*, respectively.

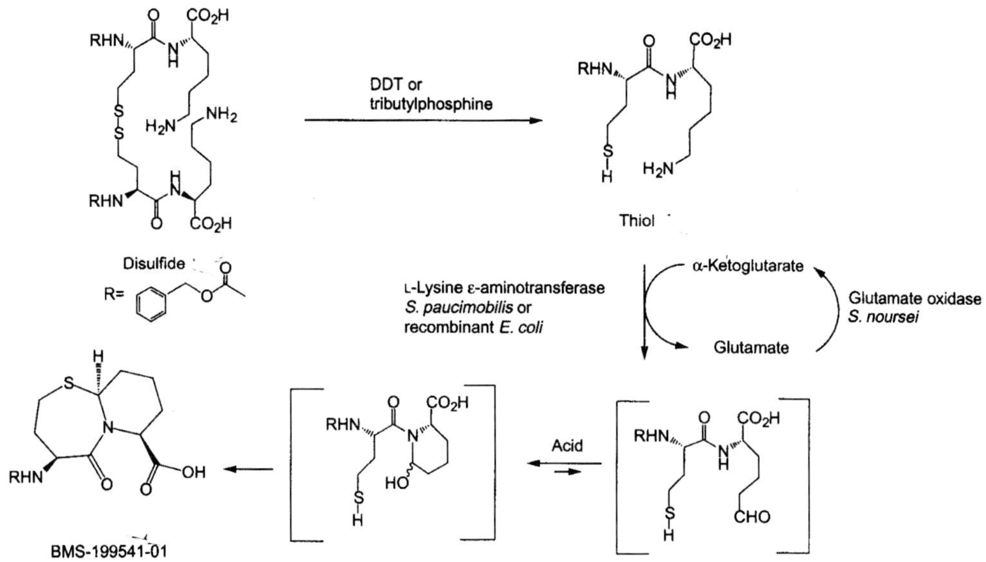

Figure 13.21 L-Lysine ε-aminotransferase reactions to Omapatrilate precursors (Patel, 2000).

In the biotransformation process with L-lysine ε-aminotransferase in recombinant *E. coli* cells, a reaction yield of 65–70% mol/mol was obtained for conversion of BMS-201391-01 (Cbz protecting group) to BMS-199541-01 depending upon the reaction conditions used in the process. Similar molar yields were obtained with phenylacetyl- or phenoxyacetyl-protected analogs of BMS-201391-01. Even Nα-protected L-Met/L-Homocys-L-Lys dipeptides served as substrates, as did *N*-ε-Cbz- or Boc-protected L-lysines.

Enzymatic synthesis of (S)-allysine ethylene acetal

Allysine ethylene acetal [(*S*)-2-amino-5-(1,3-dioxolan-2-yl)pentanoic acid], a precursor to Omapatrilat, was obtained by reductive amination of the corresponding keto acid employing phenylalanine dehydrogenase (PheDH) from *Thermoactinomyces intermedius* ATCC 33205 (Hanson, 2000b). Other amino acid dehydrogenases were less effective. The NAD^+ produced from NADH during the reaction was regenerated via formate oxidation to CO_2 with formate dehydrogenase (FDH) (Chapter 7, Section 7.2.2.4). PheDH was produced by growth of *T. intermedius* ATCC 33205 or recombinantly expressed in *E. coli* or *Pichia pastoris*.

Three generations of processes were tested:

- Employing heat-dried *T. intermedius* and heat-dried *Candida boidinii* SC13822 as sources of PheDH and FDH, respectively, reaction yields averaged 84% mol/mol and the e.e. was > 98%; however, production of *T. intermedius* could not be scaled up.

- Replacing *T. intermedius* with *E. coli* as a source for PheDH, 197 kg of (*S*)-allysine ethylene acetal product was produced in three batches with an average yield of 91% mol/mol and the e.e. was > 98%.
- In a third-generation process, heat-dried methanol-grown *P. pastoris* expressing endogenous FDH and recombinant *Thermoactinomyces* PheDH were used to produce 15 kg (*S*)-allysine ethylene acetal with a yield of 97% mol/mol and > 98% e.e. Both 5-(1,3-dioxolan-2-yl)-2-oxopentanoic acid and lithium-6,6-dimethoxy-2-oxohexanoate served as substrates for the enzymatic reductive amination.

Enzymatic synthesis of L-6-hydroxynorleucine

L-6-Hydroxynorleucine, a different key chiral intermediate used for synthesis of the vasopeptidase inhibitor Omapatrilat (Vanlev®), was prepared in 89% yield and > 99% optical purity by reductive amination of 2-keto-6-hydroxyhexanoic acid using glutamate dehydrogenase from beef liver (Hanson, 1999) (Figure 13.22). In an alternative process, racemic 6-hydroxynorleucine produced by hydrolysis of 5-(4-hydroxybutyl)hydantoin was treated with D-amino acid oxidase to prepare a mixture containing 2-keto-6-hydroxyhexanoic acid and L-6-hydroxynorleucine followed by the reductive amination procedure to convert the mixture entirely to L-6-hydroxynorleucine, with yields of 91–97% and optical purities of > 99%.

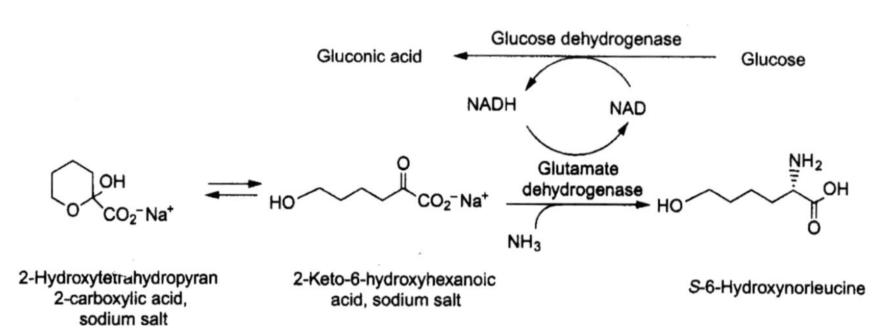

Figure 13.22 L-6-Hydroxynorleucine, intermediate to Omapatrilat (Patel, 2001).

13.3.4.2 Lipase Reactions to Intermediates for Cardiovascular Therapy

(R)-Glycidyl butyrate: an intermediate to *β*-blockers

(*R*)-Glycidyl butyrate, an intermediate to β-blocker circulatory drugs, is obtained in moderate enantiomeric purity ($E = 18$) by treatment of the racemic ester with porcine pancreatic lipase (PPL) (Figure 13.23; Ladner, 1984).

The degree of conversion has to be considerably higher than 50% to obtain the desired product in sufficiently high enantiomeric purity. The incomplete enantioselectivity of PPL ($E = 18$) was countered by continuation of the reaction to 67% conversion. The productivity in a multi-phase reactor (Figure 13.24), in which the racemic ester was circulated in the lumen, was 17.6 g $(h\ m^2)^{-1}$ or 28.4 mmol (h mg enzyme)$^{-1}$. The main problem of the multi-phase reactor is the low catalyst effectiveness factor, which is normally found to lie between 30 and 50%.

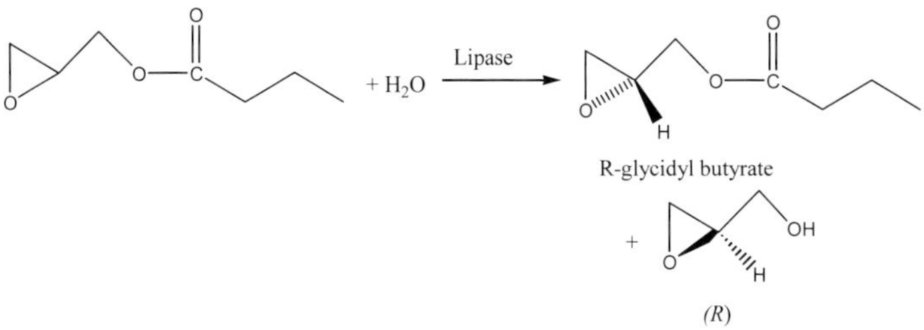

+ H_2O —Lipase→

R-glycidyl butyrate

+ (R)

Figure 13.23 Lipase hydrolysis to (*R*)-glycidyl butyrate (Ladner, 1984).

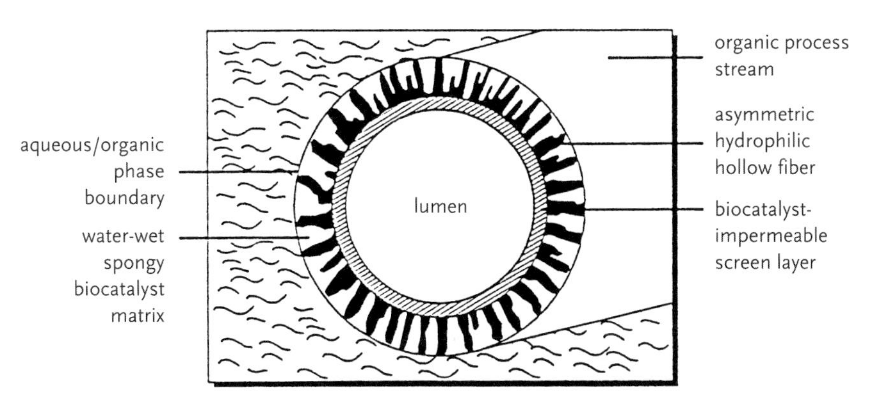

Figure 13.24 Cut through a hollow-fiber membrane of a multi-phase reactor.

Figure 13.25 displays the route to the product as well as competing routes: asymmetric epoxidation à la Sharpless, and separation of racemates with lipases.

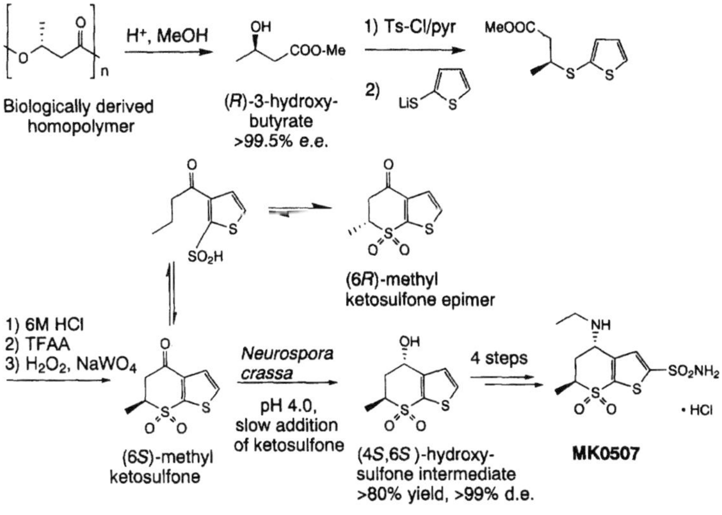

H^+, MeOH

1) Ts-Cl/pyr
2) LiS

Biologically derived homopolymer

(*R*)-3-hydroxy-butyrate >99.5% *e.e.*

(6*R*)-methyl ketosulfone epimer

1) 6M HCl
2) TFAA
3) H_2O_2, $NaWO_4$

Neurospora crassa

pH 4.0, slow addition of ketosulfone

4 steps

(6*S*)-methyl ketosulfone

(4*S*,6*S*)-hydroxy-sulfone intermediate >80% yield, >99% d.e.

MK0507

Figure 13.25 Production route to β-blockers (Sheldon, 1993).

Nifedipines: dihydropyridines for calcium antagonists

At the end of the previous chapter, we discussed the medium dependence of the lipase-catalyzed synthesis of nifedipines, dihydropyridines with an aromatic substituent in the 4-position and often asymmetric ester derivatives in the 3- and 5-positions, which are active as calcium antagonists in cardiovascular therapy. At the Amano Company in Nagoya, Japan, lipase-catalyzed hydrolyses of methyleneoxypropionyl or -pivaloyl diesters with Amano PS (*Pseudomonas* sp.) lipase were found to yield varying enantiomeric excesses depending on the solvent: in cyclohexane the different esters yielded half-esters with 88.8–91.4% e.e. (*R*)-specificity for a triple mutant ("FVL") of Amano PS lipase, whereas the same transformation with the same enzyme in diisopropyl ether (DIPE) yielded between 68.1 and > 99% e.e. of the (*S*)-product (Chapter 12, Figure 12.10) (Hirose, 1992, 1995).

Besides the substituents on the two carboxyl groups in the 3- and 5-positions, the aromatic group in the 4-position can be varied to achieve greater efficacy. Testing on guinea-pig ileal smooth muscle against nifedipine as a standard, replacement of the *o*-nitrophenyl group by a 1-methyl-2-methylsulfonyl-5-imidazolyl substituent resulted in higher activity in some cases (Shafiee, 1998a).

13.4 Applications of Specific Biocatalytic Reactions in Pharma

13.4.1 Reduction of Keto Compounds with Whole Cells

13.4.1.1 Trimegestone

Trimegestone (Figure 13.26) is a new progestomimetic compound developed for the treatment of postmenopausal diseases (Roussel Uclaf, Romainville, France). Its nine-step industrial-scale synthesis starts from a 3-ketal, a 17-keto norsteroid, and involves (i) hydrocyanation of the 17-keto group; (ii) alkylation of the protected cyanohydrin with EtMgBr to give a 17α-hydroxy 20-ketone; (iii) stereospecific 17α-methylation of the corresponding 17,20-enolate; and (iv) oxidation of the 20,21-enolate with air and ketal deprotection to give 3,20,21-triketone. The key step of the synthesis is the chemo-, regio-, and almost stereospecific reduction of the triketone with baker's yeast in water to the desired 21(*S*)-alcohol trimegestone (d.e. = 99%) (Crocq, 1997). The transformation demonstrated for the first time the feasibility of industrial-scale whole-cell yeast reductions of keto compounds.

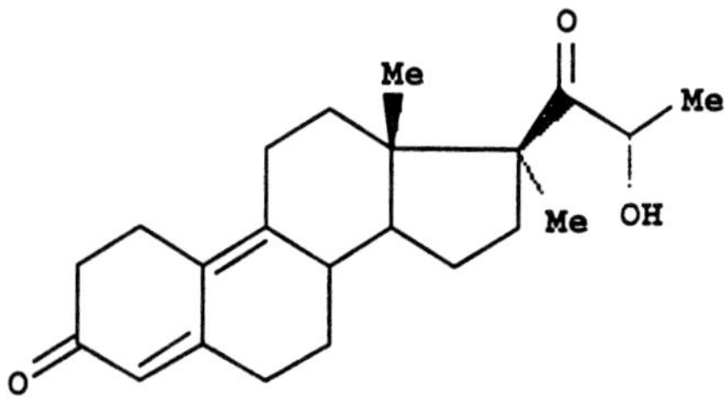

Figure 13.26 Trimegestone, an anti-postmenopausal agent (Crocq, 1997).

Trusopt

Trusopt® (MK-0507) (Figure 13.27) is a carbonic anhydrase inhibitor developed at and marketed by Merck and Co. (Rahway/NJ, USA) for the treatment of glaucoma in the eye. A key step is the reduction of a keto function in the presence of both a sulfone opposite the keto group in the six-membered ring and another sulfur function in the annelated thiophene ring, the (6*S*)-methylketosulfone. Reduction of the keto group in the (6*S*)-ketosulfide intermediate (sulfide instead of sulfone) by $LiAlH_4$ reduction led to the predominant (95%) formation of the unwanted *cis*-isomer; epimerization with sulfuric acid at 0 °C was incomplete so a biocatalytic route was sought. Microbial screening on the more water-soluble (6*S*)-methylketosulfone yielded several microorganisms capable of reducing the precursor to the desired (4*S*,6*S*)-hydroxysulfone intermediate, among them the fungus *Neurospora crassa* (IMI 19419). A particular challenge not encountered in the chemical process, however, had to be overcome as the (6*S*)-methylketosulfone, but not the reduced (4*S*,6*S*)-hydroxy-sulfone, tends to epimerize to the (6*R*)-compound via a ring-opening reaction at pH values above 5. As a consequence, the fungal growth medium was adjusted to pH 4 before ketosulfone addition. The low pH and slow addition of the substrate helped to achieve a high *trans*/*cis* ratio (99.8% d.e.) for the hydroxysulfone as well as almost complete conversion and good isolated yield (80%).

Incidentally, the raw material for the six-membered ring of Trusopt® is (*R*)-3-hydroxybutyrate, which in turn is obtained from biologically derived poly-(*R*)-3-hydroxybutyrate.

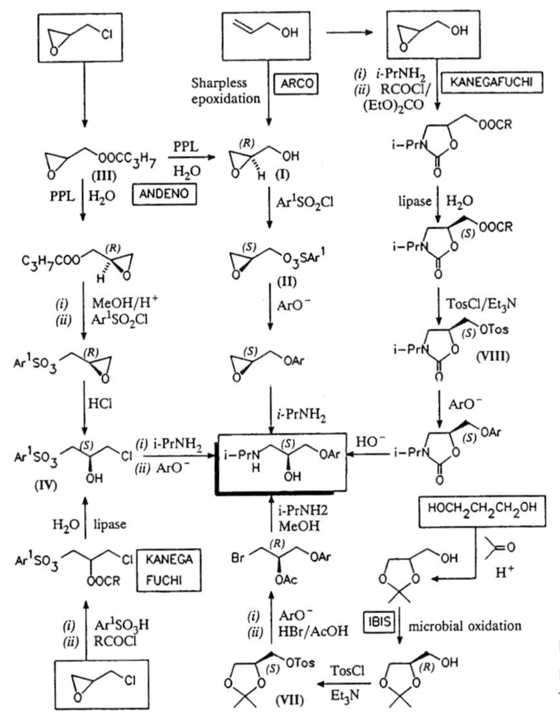

Figure 13.27 Biotransformation route to the anti-glaucoma agent Trusopt® (Zaks, 1997).

13.4.1.2 **Reduction of Precursor to Carbonic Anhydrase Inhibitor L-685393**

An analog of the (6*S*)-methylketosulfone towards Trusopt®, 5,6 dihydro-(6*S*)-propyl-4*H*-thieno[2,3*b*]thiopyran-4-one 7,7-dioxide ("ketosulfone", Figure 13.28), the precursor to the carbonic anhydrase inhibitor L-683393 (Merck), could be reduced to the *trans*-hydroxysulfone 5,6 dihydro-(4*S*)-hydroxy-(6*S*)-propyl-4*H*-thieno[2,3*b*] thiopyran 7,7-dioxide by whole cells of the yeast *Rhodotorula rubra* MY 2169 from the Merck collection (Lorraine, 1996). Low water-solubility limited the optimum substrate concentration to 2 g L^{-1}, as organic solvents added to increase solubility resulted in lower rates, which at 0.04 g (g dcw)$^{-1}$ h^{-1} were not high to start with. Diastereomeric excess ranged from 89 to 94% d.e., and decreased with increasing conversion.

Figure 13.28 Whole-cell reduction of a precursor to anti-glaucoma agent L-683393 (Lorraine, 1996).

13.4.1.3 **Montelukast**

One of the critical steps in the synthesis of the anti-asthmatic, anti-rhinitis (seasonal allergic hay fever) agent Montelukast (Singulair®, Merck, Rahway/NJ, USA), an antagonist towards the leukotriene LTD_4 receptor, is the reduction of a hydrophobic ketone (keto ester M) to the (*S*)-alcohol (hydroxy ester S, Figure 13.29). Although an established chemical route exists employing β-chlorodiisopinocampheylborane (King, 1993), alternative economical routes were sought. A culture of *Microbacterium campoquemadoensis* converted the keto ester M substrate to the hydroxy ester with > 95% e.e. at a maximum average rate of 15.62 μg (g cells)$^{-1}$$h^{-1}$ (Shafiee, 1998b). The ketoreductase was isolated, purified, and catalyzed keto ester M with a specific activity of 16 ng (mg min)$^{-1}$. The enzyme was found to be NADPH-dependent and, in immobilized form, stable to DMSO and hexane (in a two-phase system) but labile to other solvents. A volumetric productivity of 0.04 g (L d)$^{-1}$ was achieved, at a substrate concentration of 0.5 g L^{-1}.

13.4.1.4 **LY300164**

LY300164, an orally administered benzodiazepine developed by Eli Lilly (Indianapolis/IN, USA), is efficaceous against amylotropic lateral sclerosis (ALS, Lou Gehrig's disease). In an early key step, a 3,4-disubstituted phenylacetone is reduced to the (*S*)-alcohol by yeast whole cells from *Zygosacchamyces rouxii* (ATCC 14462) in > 99.9% e.e. and 96% yield. The subsequent steps encompass a series of chemical steps (Figure 13.30). Volumetric productivity is optimized in remarkable fashion: the ketone substrate is introduced adsorbed on an XAD-7 resin to keep the concentration in solution (2 g L^{-1}) below the toxicity limit (6 g L^{-1}). Product alcohol likewise is adsorbed so that a loading of 80 g L^{-1} of substrate or product is achieved.

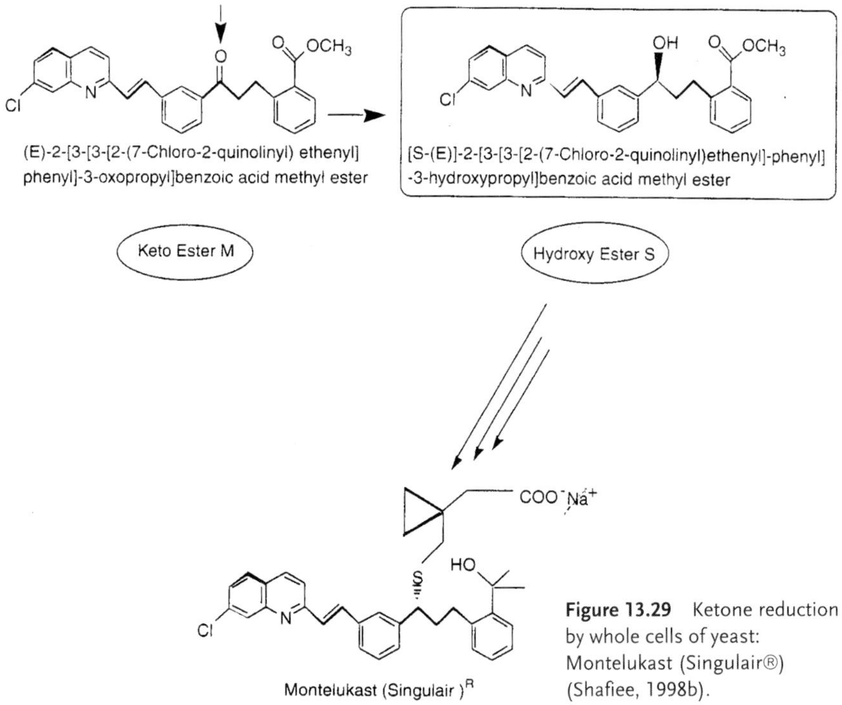

(E)-2-[3-[3-[2-(7-Chloro-2-quinolinyl) ethenyl] phenyl]-3-oxopropyl]benzoic acid methyl ester

[S-(E)]-2-[3-[3-[2-(7-Chloro-2-quinolinyl)ethenyl]-phenyl] -3-hydroxypropyl]benzoic acid methyl ester

Keto Ester M

Hydroxy Ester S

Montelukast (Singulair)[R]

Figure 13.29 Ketone reduction by whole cells of yeast: Montelukast (Singulair®) (Shafiee, 1998b).

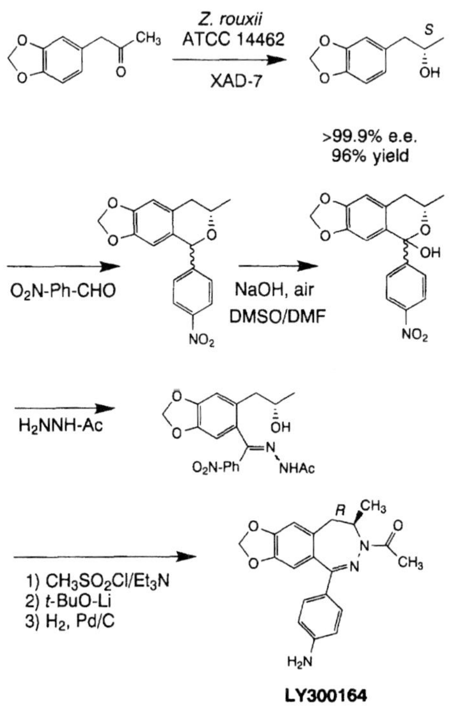

Z. rouxii ATCC 14462

XAD-7

>99.9% e.e.
96% yield

O_2N-Ph-CHO

NaOH, air
DMSO/DMF

H_2NNH-Ac

1) CH_3SO_2Cl/Et_3N
2) *t*-BuO-Li
3) H_2, Pd/C

LY300164

Figure 13.30 Synthesis route to Eli Lilly's LY300164 (Zaks, 1997).

13.4.2 Applications of Pen G Acylase in Pharma

13.4.2.1 Loracarbef®

In Chapter 7, Section 7.5.1, we covered the use of pen G acylase (PGA) for the catalysis of condensation reactions in the synthesis of β-lactam antibiotics. The same principle is applied in the commercial route to Loracarbef® from Eli Lilly (Indianapolis/IN, USA) (Figure 13.31). Instead of 6-APA, the nucleophile is *cis*-3-aminoazetidinone, importantly substituted in the 4-position by a 2′-furylethyl group; the electrophile is methyl phenoxyacetate. Pen G acylase from *E. coli* does not hydrolyze the phenoxyacetyl ester of the pen V side chain but condenses the ester with the nucleophile to the (2*R*,3*S*)-β-lactam intermediate with 44% conversion and 97% e.e. (Briggs, 1994).

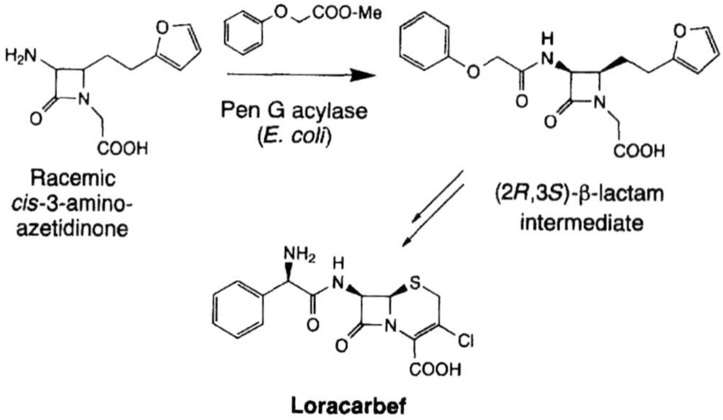

Figure 13.31 Synthesis route to Loracarbef® (Zaks, 1997).

13.4.2.2 Xemilofibran

The β-amino acid (*S*)-ethyl-3-amino-4-pentynoate is a chiral synthon used in the synthesis of xemilofiban hydrochloride, an anti-platelet agent. (*R*)- and (*S*)-enantiomers of ethyl 3-amino-4-pentynoate were obtained in enantiomerically pure form by employing penicillin acylase (Pen G acylase, PGA, Chapter 7, Section 7.5.1) (Figure 13.32) (Topgi, 1999). Both the acylation and deacylation modes were ap-

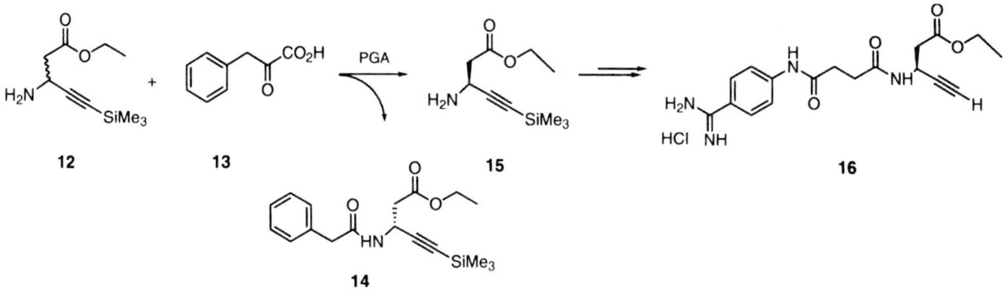

Figure 13.32 Synthesis of xemilofibran (Zaks, 2001).

plied successfully. In the acylation, phenylacetic acid was used as an acylating agent. PGA activity can be controlled by maintaining an appropriate pH of the reaction medium.

13.4.3 Applications of Lipases and Esterases in Pharma

13.4.3.1 LTD_4 Antagonist MK-0571

Besides Montelukast (see above), Merck (Rahway/NJ, USA) has developed another LTD_4 receptor antagonist, MK-0571 (Figure 13.33). The two thioethyl groups protruding from the central aromatic ring are almost equivalent, one being a carboxyl group, the other dimethylamide. In a remarkable reaction, the bis-β-thiopropionic acid methyl ester compound is hydrolyzed to the half-ester by *Pseudomonas* sp. lipase at pH 7 and 40 °C (5% Triton X-100 surfactant has been added to aid solubility) (Hughes, 1989). The prochiral center is four carbon atoms away from the hydrolyzed ester group!

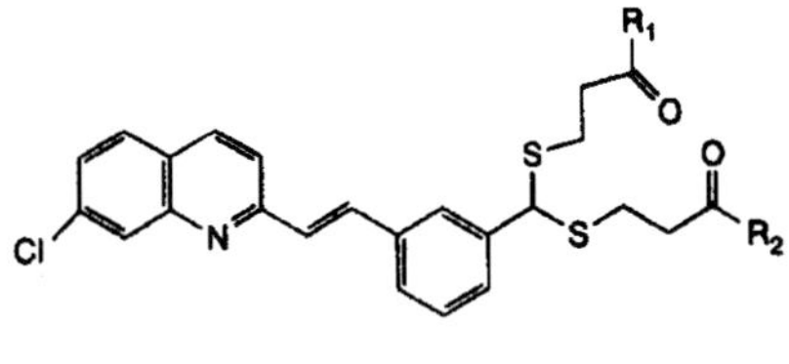

MK0571

R_1 = OH
R_2 = NMe2

Figure 13.33 LTD_4 receptor antagonist MK-0571 (Zaks, 1997).

13.4.3.2 Tetrahydrolipstatin

Tetrahydrolipstatin is a lipase inhibitor developed and marketed by Roche (Basel, Switzerland) as an anti-obesity drug. With the incidence of obesity rising rapidly in the industrialized nations, having reached 33% of all adults in the United States and more than a quarter of all French schoolchildren, this problem will rapidly cease to remain one of lifestyle and enter the arena of medical costs associated with the diseases stemming from obesity.

In the synthesis route, the 3,5-diketohexadecanoate ester is substituted with a hexyl group in the 2-position and reduced to a hydroxy group in the 5-position, resulting in lactonization (Wirz, 1996). After hydrogenation, the remaining 3-hydroxy group is esterified in hexane at 67 °C with vinyl butyrate in the presence of lipase PL or PS to generate the (2*S*,3*S*,5*R*)-lactone in 43% yield and $>$ 99% e.e. After several further chemical steps, tetrahydrolipstatin is obtained.

Suggested Further Reading

B. C. Buckland, D. K. Robinson, and M. Chartrain, Biocatalyis for pharmaceuticals – status and prospects for a key technology, *Metabolic Eng.* **2000**, *2*, 42–48.

R. N. Patel (ed.), *Stereoselective Biocatalysis*, Marcel Dekker, Amsterdam, **1999.**

R. N. Patel, Biocatalytic synthesis of intermediates for the synthesis of chiral drug substances, *Curr. Opin. Biotechnol.* **2001**, *12*, 584–604.

J. P. Rasor and E. Voss, Enzyme-catalyzed processes in pharmaceutical industry, *Appl. Catalysis A: General* **2001**, *221*, 145–158.

A. Zaks and D. R. Dodds, Application of biocatalysis and biotransformations to the synthesis of pharmaceuticals, *Drug Discovery Today* **1997**, *2*, 513–531.

References

S. Begley, The end of antibiotics, *Newsweek* **1994**, March 28, 39–43.

H. U. Blaser, F. Spindler, and M. Studer, Enantioselective catalysis in fine chemicals production, *Appl. Catal. A: General*, **2001**, *221*, 119–143.

A. S. Bommarius, M. Schwarm, and K. Drauz, Biocatalysis to amino acid-based chiral pharmaceuticals – examples and perspectives, *J. Mol. Cat B: Enzymatic* **1998**, *5*, 1–11.

S. Borman, Clinical trials to begin on designed AIDS drug, *Chem Eng. News* **1995**, *73(7)*, 39–40.

B. S. Briggs, I. G. Wright, M. J. Zmijewski, J. N. Levy, and M. Stukus, Side chain selectivity and kinetics of penicillin G amidase in acylating a *cis*-racemic β-lactam intermediate in the synthesis of Loracarbef, *New J. Chem.* **1994**, *18*, 425–434.

B. C. Buckland, S. W. Drew, N. C. Connors, M. M. Chartrain, C. Lee, P. M. Salmon, K. Gbewonyo, W. Zhou, P. Gailliot, R. Singhvi, R. C. Olewinski Jr., W. J. Sun, J. Reddy, J. Zhang, B. A. Jackey, C. Taylor, K. E. Goklen, B. Junker, and R. L. Greasham, Microbial conversion of indene to indanediol: a key intermediate in the synthesis of CRIXIVAN, *Metab. Eng.* **1999**, *1*, 63–74.

V. Crocq, C. Masson, J. Winter, C. Richard, G. Lemaitre, J. Lenay, M. Vivat, J. Buendia, and D. Prat, Synthesis of trimegestone: the first industrial application of bakers' yeast mediated reduction of a ketone, *Org. Proc. Res. Dev.* **1997**, *1*, 2–13.

R. Crossley, *Chirality and the Biological Activity of Drugs*, CRC Press, Boca Raton, **1995**, Chapter 1.

H. Diekmann and H. Metz, *Basics and Practice of Biotechnology* (in German), Gustav Fischer Verlag, Stuttgart, **1991**, p. 226.

J. S. Dixon, Computer-aided drug design: getting the best results, *TIBTECH* **1992**, *10*, 357–363.

F. P. Doyle, J. H. C. Nayler, H. Smith, and E. R. Stove, Some novel acid-stable penicillins, *Nature* **1961**, *191*, 1091–1092.

M. J. Gait and J. Kern, Progress in anti-HIV structure-based drug design, *Trends Biotechnol.* **1995**, *13*, 430–438.

R. L. Hanson, M. D. Schwinden, A. Banerjee, D. B. Brzozowski, B.-C. Chen, B. P. Patel, C. G. McNamee, G. A. Kodersha, D. R. Kronenthal, R. N. Patel, and L. J. Szarka, Enzymatic synthesis of L-6-hydroxynorleucine, *Bioorgan. Med. Chem.* **1999**, *7*, 2247–2252.

R. L. Hanson, Z. Shi, D. B. Brzozowski, A. Banerjee, T. P. Kissick, J. Singh, A. J. Pullockaran, J. T. North, J. Fan, J. Howell, S. C. Durand, M. A. Montana, D. R. Kronenthal, R. H. Mueller, and R. N. Patel, Regioselective aminoacylation of lobucavir to give an intermediate for lobucavir prodrug, *Bioorg. Med. Chem.* **2000a**, *8*, 2681–2687.

R. L. Hanson, J. M. Howell, T. L. LaPorte, M. J. Donovan, D. L. Cazzulino, V. Zannella, M. A. Montana, V. B. Nanduri, S. R. Schwarz,

R. F. Eiring, S. C. Durand, J. M. Wasylyk, W. L. Parker, M. S. Liu, F. J. Okuniewicz, B.-C. Chen, J. C. Harris, K. J. Natalie, K. Ramig, S. Swaminathan, V. W. Rosso, S. K. Pack, B. T. Lotz, P. J. Bernot, A. Rusowicz, D. A. Lust, K. S. Tse, J. J. Venit, L. J. Szarka, and R. N. Patel, Synthesis of allysine ethylene acetal using phenylalanine dehydrogenase from *Thermoactinomyces intermedius*, *Enzyme Microb. Technol.* **2000b**, *26*, 348–358.

Y. Hirose, K. Kariya, I. Sasaki, Y. Kurono, H. Ebiike, and K. Achiwa, Drastic solvent effect on lipase-catalyzed enantioselective hydrolysis of prochiral 1,4-dihydropyridines, *Tetrahedron Lett.* **1992**, *33*, 7157–7160.

Y. Hirose, K. Kariya, Y. Nakanishi, Y. Kurono, and K. Achiwa, Inversion of enantioselectivity in hydrolysis of 1,4-dihydropyridines by point mutation of lipase PS, *Tetrahedron Lett.* **1995**, *36*, 1063–1066.

D. L. Hughes, J. J. Bergan, J. S. Amato, P. J. Reider, and E. J. J. Grabowski, Synthesis of chiral dithio-acetals: a chemoenzymic synthesis of a novel LTD_4 antagonist, *J. Org. Chem.* **1989**, *54*, 1787–1788.

D. L. Hughes, J. J. Bergan, J. S. Amato, M. Bhupathy, J. L. Leazer, J. M. McNamara, D. R. Sidler, P. J. Reider, and E. J. J. Grabowski, Lipase-catalyzed asymmetric hydrolysis of esters having remote chiral/prochiral centers, *J. Org. Chem.* **1990**, *55*, 6252–6259.

D. L. Hughes, G. B. Smith, J. Liu, G. C. Dezeny, C. H. Senanayake, R. D. Larsen, T. R. Verhoeven, and P. J. Reider, Mechanistic study of the Jacobsen asymmetric epoxidation of indene, *J. Org. Chem.* **1997**, *62*, 2222–2229.

International Conference on Harmonisation; Guidance on Q&A Specifications: Test Procedures and Acceptance Criteria for New Drug Substances and New Drug Products, *Chemical Substances. Federal Register* **2000**, December 29, *65(251)*, Notices Food and Drug Administration [Docket No. 97D-0448].

C. Ireland, P. A. Leeson, and J. Castaner, Lobucavir: antiviral, *Drugs Future* **1997**, *22*, 359–370.

S. W. Kaldor, V. J. Kalish, J. F. Davies II, B. V. Shetty, J. E. Fritz, K. Appelt, J. A. Burgess, K. M. Campanale, N. Y. Chirgadze, D. K. Clawson, B. A. Dressman, S. D. Hatch, D. A. Khalil, M. B. Kosa, P. P. Lubbehusen, M. A. Muesing, A. K. Patick, S. H. Reich, K. S. Su, and J. H. Tatlock, Viracept (nelfinavir mesylate, AG1343): a potent, orally bioavailable inhibitor of HIV-1 protease, *J. Med. Chem.* **1997**, *40*, 3979–3985.

D. J. Kempf, K. C. Marsh, J. F. Denissen, E. McDonald, S. Vasavanonda, C. A. Flentge, B. E. Green, L. Fino, C. H. Park, X.-P. Kong, N. E. Wideburg, A. Saldivar, L. Ruiz, W. M. Kati, H. L. Sham, T. Robins, K. D. Stewart, A. Hsu, J. J. Plattner, J. M. Leonard, and D. W. Norbeck, ABT-538 is a potent inhibitor of human immunodeficiency virus protease and has high oral bioavailability in humans, *Proc. Natl. Acad. Sci. USA* **1995**, *92*, 2484–2488.

A. O. King, E. G. Corely, R. K. Anderson, R. D. Larsen, T. R. Verhoeven, P. J. Reider, Y. B. Xiang, M. Belley, Y. Leblanc, M. Labelle, P. Prasit, and R. J. Zamboni, An efficient synthesis of LTD_4-antagonist L-699,392, *J. Org. Chem.* **1993**, *58*, 3731–3735.

D. A. Kleinsek, R. E. Dugan, T. A. Baker, and J. W. Porter, 3-Hydroxy-3-methylglutaryl-CoA reductase from rat liver, *Methods Enzymol.* **1981**, *71*, 462–479.

W. Ladner and G. M. Whitesides, Lipase-catalyzed hydrolysis as a route to esters of chiral epoxy alcohols, *J. Am. Chem. Soc.* **1984**, *106*, 7250–7251.

A. Liese and M. V. Filho, Production of fine chemicals using biocatalysis, *Curr. Opin. Biotechnol.* **1999**, *10*, 595–603.

K. Lorraine, S. King, R. Greasham, and M. Chartrain, Asymmetric bioreduction of a ketosulfone to the corresponding *trans*-hydroxysulfone by the yeast *Rhodotorula rubra* MY 2169, *Enzyme Microb. Technol.* **1996**, *19*, 250–255.

C. E. Nakamura and R. H. Abeles, Mode of interaction of β-hydroxy-β-methylglutaryl coenzyme A reductase with strong binding inhibitors: compactin and related compounds, *Biochemistry* **1985**, *24*, 1364–1376.

M. A. Navia, J. P. Springer, T. Y. Lin, H. R. Williams, R. A. Firestone, J. M. Pisano, J. B. Doherty, P. E. Finke, and K. Hoogsteen, Crystallographic study of a β-lactam inhibitor complex with elastase at 1.84 Å resolution, *Nature* **1987**, *327*, 79–82.

J. H. C. Nayler, Early discoveries in the penicillin series, *Trends Biochem. Sci.* **1991**, *16*, 195–197, 234–237.

Y. Ohta and I. Shinkai, Ritonavir, *Bioorg. Med. Chem.* **1997a**, *5*, 461–462.

Y. Ohta and I. Shinkai, Indinavir, *Bioorg. Med. Chem.* **1997b**, *5*, 463–464.

Y. Ohta and I. Shinkai, Saquinavir, *Bioorg. Med. Chem.* **1997c**, *5*, 465–66.

V. B. Pai and M. C. Nahata, Nelfinavir mesylate: a protease inhibitor, *Ann. Pharmacother.* **1999**, *33*, 325–339.

R. N. Patel, A. Banerjee, C. G. McNamee, D. B. Brzozowski, R. L. Hanson, and L. J. Szarka, Enantioselective microbial reduction of 3,5-dioxo-6-(benzyloxy) hexanoic acid, ethyl ester, *Enzyme Microb. Technol.* **1993**, *15*, 1014–1021.

R. N. Patel, A. Banerjee, V. B. Nanduri, S. L. Goldberg, R. M. Johnston, R. L. Hanson, C. G. McNamee, D. B.Brzozowski, T. P. Tully, R. Y. Ko, T. L. LaPorte, D. L. Cazzulino, S. Swaminathan, C.-K. Chen, L. W. Parker, and J. J. Venit, Biocatalytic preparation of a chiral synthon for a vasopeptidase inhibitor: enzymatic conversion of N^2-[*N*-phenylmethoxy)carbonyl] L-homocysteinyl]- L -lysine (1,1′)-disulfide to [4S-(4*I*,7*I*,10*aJ*)]-1-octahydro-5-oxo-4-[phenylmethoxy)carbonyl]amino]-7*H*-pyrido-[2,1-*b*][1,3]thiazepine-7-carboxylic acid methyl ester by a novel L -lysine ε-aminotransferase, *Enzyme Microb. Technol.* **2000**, *27*, 376–389.

R. N. Patel, Biocatalytic synthesis of intermediates for the synthesis of chiral drug substances, *Curr. Opin. Biotechnol.* **2001**, *12*, 584–604.

C. C. Pfeiffer, Optical isomerism and pharmacological action, a generalization, *Science* **1956**,*124*, 29–31.

J. P. Rasor and E. Voss, Enzyme-catalyzed processes in pharmaceutical industry, *Appl. Catal. A: General* **2001**, *221*, 145–158.

P. Reiss, D. A. Burnett, and A. Zaks, An enzymatic synthesis of glucuronides of azetidinone-based cholesterol absorption inhibitors, *Bioorg. Med. Chem.* **1999**, *7*, 2199–2202.

N. A. Roberts, J. A. Martin, D. Kinchington, A. V. Broadhurst, C. J. Craig, I. B. Duncan, S. A. Galpin, B. K. Handa, and J. Kay, Rational design of peptide-based HIV proteinase inhibitors, *Science* **1990**, *248*, 358–361.

G. N. Rolinson, The influence of 6-aminopenicillanic acid on antibiotic development, *J. Antimicrob. Chemother.* **1988**, *22*, 5–14.

A. M. Rouhi, Biocatalysis buzz, *Chem. Eng. News* **2002**, *80*, *7*, 86–87.

A. Shafiee, A. R. Dehpour, F. Hadizadeh, and M. Azimi, Syntheses and calcium channel antagonist activity of nifedipine analogues with methylsulfonylimidazolyl substituent, *Pharm. Acta Helv.* **1998a**, *73*, 75–79.

A. Shafiee, H. Motamedi, and A. King, Purification, characterization, and immobilization of an NADPH-dependent enzyme involved in the chiral specific reduction of the keto ester M, an intermediate in the synthesis of an anti-asthma drug, Montelukast, from *Microbacterium campoquemadoensis* (MB5614), *Appl. Microbiol. Biotechnol.* **1998b**, *49*, 709–717.

R. A. Sheldon, *Chirotechnology: Industrial Synthesis of Optically Active Compounds*, Marcel Dekker, Amsterdam, **1993**.

S. C. Stinson, Chiral drugs, *Chem. Eng. News* **1994**, Sept. 19, 38–72.

D. Stone, HIV protease inhibitors head to market, *Biotechnology* **1995**, *13*, 940–941.

M. M. Struck, Biopharmaceutical R&D success rates and development times. A new analysis provides benchmarks for the future, *Bio/Technology* **1994**, *12*, 674–677.

C. J. Suckling (ed.), *Enzyme Chemistry – Impact and Applications*, Chapman and Hall, **1990**.

R. S. Topgi, J. S. Ng, B. Landis, P.Wang, and J. R. Behling, Use of enzyme penicillin acylase in selective amidation/amide hydrolysis to resolve ethyl 3-amino-4-pentynoate isomers, *Bioorgan. Med. Chem.* **1999**, *7*, 2221–2229.

J. P. Vacca, B. D. Dorsey, W. A. Schleif, R. B. Levin, S. L. McDaniel, P. L. Darke, J. Zugay, J. C. Quintero, O. M. Blahy, E. Roth, V. V. Sardana, A. J. Schlabach, P. I. Graham, J. H. Condra, L. Gotlib, M. K. Holloway, J. Lin, I.-W. Chen, K. Vastag, D. Ostovic, P. S. Anderson, E. A. Emini, and J. R. Huff, L-735,524: an orally bioavailable HIV-1 protease inhibitor, *Proc. Natl. Acad. Sci. USA* **1994**, *91*, 4096–4100.

C. Walsh, Molecular mechanisms that confer antibacterial drug resistance, *Nature* **2000**, *406*, 775–781.

B. Wirz, T. Weisbrod, and H. Estermann, Enzymic reactions in process research – the importance of parameter optimization and work-up, *Chimica Oggi* **1996**, *14*, 37–41.

M. Wolberg, W. Hummel, and M. Mueller, **2001**, Biocatalytic reduction of beta,delta-diketo esters: a highly stereoselective approach to all four stereoisomers of a chlorinated beta,delta-dihydroxy hexanoate, *Chem. Eur. J.* *7*, 4562–4571.

A. Zaks and D. R. Dodds, Application of biocatalysis and biotransformations to the synthesis of pharmaceuticals, *Drug Discovery Today* **1997**, *2*, 513–531.

A. Zaks, and D. R. Dodds, Enzymic glucuronidation of a novel cholesterol absorption inhibitor, SCH 58235, *Appl. Biochem. Biotechnol.* **1998**, *73*, 205–214.

A. Zaks, Industrial biocatalysis, *Curr. Opin. Chem. Biol.* **2001**, *5*, 130–136.

N. Zhang, B. C. Stewart, J. C. Moore, R. L. Greasham, D. K. Robinson, B. C. Buckland, and C. Lee, Directed evolution of toluene dioxygenase from *Pseudomonas putida* for improved selectivity towards *cis*-(1S,2*R*)-indanediol during indane bioconversion, *Metab. Eng.* **2000**, *2*, 339–348.

14
Bioinformatics

Summary

Whereas the conventional way to investigate genes, proteins, and biochemical function used to be from function to protein to gene, the massive amounts of available data from genome studies suggest the opposite path today. Conventionally, the first attribute known about an enzyme was its function, which was screened for in microbial cultures or tissue samples. The protein was then purified to homogeneity and subjected to biochemical studies, including investigation of its catalytic profile. Proteolytic digestion led to the primary sequence, from which the gene sequence was deduced. Establishment of a 3D structure was attempted from the crystallized homogeneous protein to find residues for change by site-directed mutagenesis. Nowadays, an annotated function is verified (or falsified) by searching the gene with suitable DNA primers in genomic DNA, amplifying the gene, sequencing it for a control, and cloning it into an expression system to obtain protein, which is then checked for function. This process is much less time-consuming and increasingly relies on standard protocols.

Bioinformatics is the field of science in which biology, computer science, and information technology merge into a single discipline. The ultimate goal of the field is to enable the discovery of new biological insights as well as to create a global perspective from which unifying principles in biology can be discerned. The rationale for applying computational approaches to problems in biology includes an explosive growth in the amount of biological information which necessitates the use of computers for information cataloging and retrieval, as well as the desire to obtain a more global perspective in experimental design. Bioinformatics involves the development of tools that can compare vast amounts of sequence data and reduce them to a human scale.

An overview of databases yields the number of publicly accessible genomes: 112 finished microbial and two eukaryotic genomes with over 300 overall in progress (as of April 2003). The number of enzymes with E.C. nomenclature is 4159 according to the ExPASy website and 3225 according to Brenda databases; the number of accessible 3D structures totals 20 946, 18 872 of which are proteins.

14.1
Starting Point: from Consequence (Function) to Sequence

14.1.1
Conventional Path: from Function to Sequence

Conventionally, the first attribute known about an enzyme used to be its function, usually in a crude extract. This property was screened for in microbial cultures or in tissue samples. The crude extract was then purified to homogeneity and the protein subjected to biochemical studies to learn of its pH and *T* profiles, its pI and subunit composition, catalytically important residues, and other properties. Proteolytic digestion of the protein with subsequent Edman degradation led to the primary sequence, but no information on the secondary structures such as α-helices and β-sheets or the folding in three dimensions of the polypeptide chain. The primary sequence could have been used to deduct the gene sequence but, with the degeneration of the code, several possibilities for certain amino acids occur, which makes prediction of the gene sequence a risk.

An attempt was then made to obtain the homogeneous protein as a crystal for subsequent X-ray crystallography. After the 3D structure had been discerned, often the most laborious part of the whole enzyme investigation, the structure revealed the 3D coordination of the amino acids in folding.

14.1.2
Novel Path: from Sequence to Consequence (Function)

In recent years, a different path has been established which follows almost the reverse direction of the conventional route, from sequence to consequence, but stands a higher chance of success with fewer resources, including less time:

From genome sequences, a plethora of DNA sequence information is, often publicly, accessible in databases. Pieces of genome sequence have often been annotated, based on sequence similarity to other genes with similar function, to possess a certain function. From the gene sequence, the corresponding protein sequence of a putative enzyme can of course be readily derived so that at this stage the target of the search for a new enzyme is defined.

To obtain the desired gene, either the genomic DNA itself can be ordered from a culture collection or the cell line or bacterial strain is obtained and the genomic DNA isolated from a grown culture and the gene found in the genomic DNA with suitable primers (Chapter 4). If only EST tags are available, these sequences can be used as gene hybridization probes. The target protein corresponding to the isolated gene is then commonly overexpressed in a suitable expression system, purified sufficiently and subjected to a series of assays for function. An alternative way is to create the knockout cell without the gene of interest and investigate the knockout for its phenotype (a set of functional properties).

One potential problem of the novel path is the process of annotation of function. Owing to the magnitude of the sequences waiting to be annotated the process is often performed by computer programs searching for little more than sequence similarity combined with some very interesting fingerprints, such as for consensus sequence. Consequently, machine annotation of a gene is not fully reliable and certainly is not proof of function. Only an assay on an isolated and appropriately purified protein constitutes proof of function.

The conventional and current routes of characterization of function in a protein are depicted in Figure 14.1.

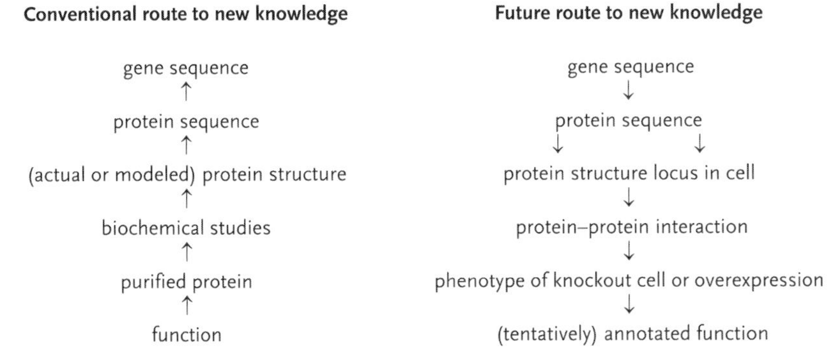

Figure 14.1 Conventional and novel ways of finding function on a protein.

14.2 Bioinformatics: What is it, Why do we Need it, and Why Now? (NCBI Homepage)

14.2.1 What is Bioinformatics?

Bioinformatics is the field of science in which biology, computer science, and information technology merge into a single discipline. The ultimate goal of the field is to enable the discovery of new biological insights as well as to create a global perspective from which unifying principles in biology can be discerned. There are three important sub-disciplines within bioinformatics:

- the development of new algorithms and statistics with which to assess relationships among members of large data sets;
- the analysis and interpretation of various types of data including nucleotide and amino acid sequences, protein domains, and protein structures;
- the development and implementation of tools that enable efficient access and management of different types of information.

Data-mining: a novel way to do science

Data-mining is the process by which testable hypotheses are generated regarding the function or structure of a gene or protein of interest by identifying similar sequences in better characterized organisms. For example, new insight into the molecular basis of a disease may come from investigating the function of homologs of the disease gene in model organisms. Equally exciting is the potential for uncovering phylogenetic relationships and evolutionary patterns.

In a different context the catchword "genome mining" occurs; this refers to the possibility of searching the available genomes for suitable or interesting ORFs (open reading frames), which can be used as a basis of cloning.

14.2.2 Why do we Need Bioinformatics?

Why deal with bioinformatics?

It is nearly impossible to function as a biologically inclined scientist today without some basic skills as a bioinformatist. The informational potential in genomics (Chapter 15) is enormous, but understanding and extracting the information take skills in bioinformatics.

Why use bioinformatics?

The rationale for applying computational approaches to problems in biology includes:

- an explosive growth in the amount of biological information which necessitates the use of computers for information cataloging and retrieval;
- the desire for a more global perspective in experimental design. As we move from the one scientist–one gene/protein/disease paradigm of the past to a consideration of whole organisms, we gain opportunities for new insights into metabolic pathways as well as health and disease.

Why do we need bioinformatics?

The magnitude of the flood of sequence information is a problem – how can it be handled? There is also the problem of making sense of the information. Bioinformatics involves the development of tools that can compare vast amounts of sequence data and reduce them to a human scale.

14.2.3 Why Bioinformatics Now?

The last 20 or 30 years have witnessed the coincidence of two of the most powerful trends in technology development in human history: the simultaneous exponential growth of the capacity of microchips and the capacity for DNA sequencing. These two forces have created an environment which enables the collection and analysis of large amounts of DNA sequences for relevant information regarding enzyme function and thus also biocatalysis.

Ever since the early 1960s, progress in microelectronics has been such that about every 18 months the density of electronic circuits per unit area has doubled. In parallel, processing power per unit of money doubled every 1.5–2 years, or, alternatively, the price of a constant amount of processing power fell by half. This phenomenon is called Moore's Law, after Gordon Moore, one of the founders of Intel, who first observed this behavior.

In 1970, Maxam and Gilbert and, in parallel, Sanger and Coulson, invented their method of DNA sequencing, of which, in principle, the latter is still in use today. In those days, sequencing took several months. In 1982, the first complete DNA sequence was performed by Sanger et al. on the phage Lambda with about 48 502 nucleotides. Progress has been largely exponential, just as with microchip performance. In recent years, progress again accelerated markedly once again, however: in 1995, the sequencing of the first genome was completed, that of *Haemophilus influenzae* KW20 provided through TIGR with 1 830 137 bp; other genomes followed, with 112 bacterial genomes alone by the time of writing. Finally, the much publicized completion of the human genome with 3 billion base pairs, on which work was started in 1990 by Celera Pharmaceuticals led by Craig Venter, is now imminent.

Figure 14.2 depicts the exponential increase in capacity of processors and DNA sequences as well as of DNA sequencing (Yin, 1999; M. Waterman, personal communication).

The current situation in bioinformatics is characterized by an avalanche of DNA sequences from the human genome project and similar programs and, consequently, an *exponential* increase in DNA sequences but only a *linear* increase in protein 3D structures. While multitudes of putative genes have been annotated, up to 90% of all known DNA sequences have no assigned, i.e., experimentally proven, function. From this situation arise the need for interpretation of DNA sequences by information technology, and moreover, analysis of functional genomics and proteomics (see Chapter 15).

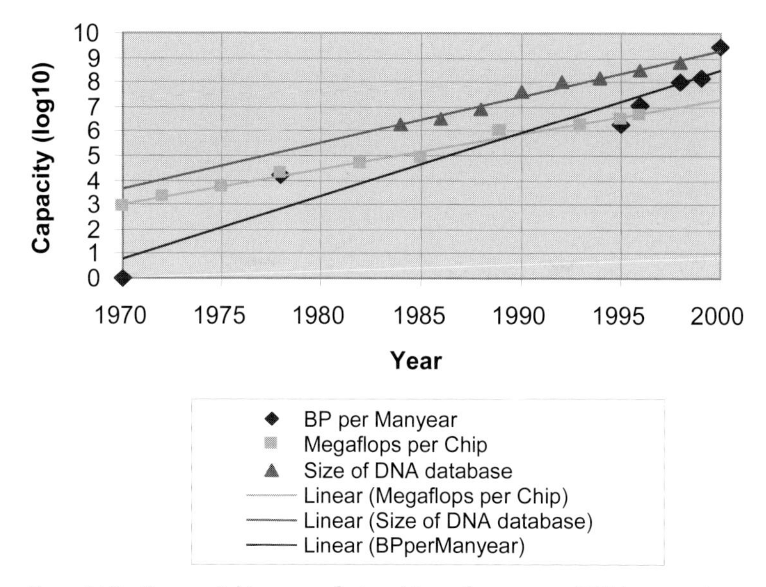

Figure 14.2 Exponential increase of microchip performance and DNA sequencing.

14.3
Tools of Bioinformatics: Databases, Alignments, Structural Mapping

14.3.1
Available Databases

An overview of several databases yields, among other items, the following information:

Number of publically accessible genomes: 112 finished microbial and 2 eucaryotic with over 300 overall in progress (April 2003)

- *Size and point in time of finished sequencing of genomes*:
  1995: prokaryote (*Haemophilus influenzae*), 1.8 Mbp,
  1996: eukaryote (*Saccharomyces cerevisiae*), 12 Mbp,
  1999: insect (*Drosophila melanogaster*), 130 Mbp (not completely annotated yet),
  2003: human (*Homo sapiens*), 3000 Mbp (not completely annotated yet).

- *Number of enzymes with E.C. nomenclature*:
  4159 according to http://us.expasy.org/enzyme/(April 28, 2003),
  3225 according to http://www.brenda.uni-koeln.de/(September 21, 2000)

- *Number of accessible protein 3D structures (PDB current holdings)*:
  Total: 20 946 total structures, **18 872** protein structures (May 13, 2003)

A concise discussion of the most important databases follows.

14.3.2
Protein Data Bank (PDB)

PDB is one of the oldest protein data bases, founded in 1971. It has three locations, Rutgers University in New Jersey, San Diego Supercomputer Center (SDSC) at the University of California, and the National Institute of Standards and Technology (NIST) in Gaithersburg, Maryland. The PDB is a source for protein characterization and structure as well. The PDB archive contains macromolecular structure data on proteins, nucleic acids, protein–nucleic acid complexes, and viruses. Approximately 50–100 new structures are deposited each week, which are annotated and released upon the depositor's specifications. PDB data are freely available worldwide. PDB formats, annotates, validates, and releases dozens of complicated structure files each week; some of them take only a couple of hours, others take weeks to process. Data processing is the main task of people at the PDB and validation is the most time-consuming part (Smith-Schmidt, 2002).

A variety of information associated with each structure is available, including sequence details, atomic coordinates, crystallization conditions, 3D structure neighbors computed using various methods, derived geometric data, structure factors, 3D images, and various links to other resources. Information on structures can be retrieved from the main PDB web site at http://www.pdb.org. Structure files can also be obtained through the main FTP site at ftp://ftp.rcsb.org. Searches with keywords can retrieve single or multiple structures.

14.3.3 Protein Explorer

Protein Explorer is free software for visualizing the 3D structures of protein, DNA, and RNA macromolecules, and their interactions and binding of ligands, inhibitors, and drugs found at the URL: http://molvis.sdsc.edu/protexpl/frntdoor.htm.

Protein Explorer makes it possible to see the relationships of 3D molecular structure to function. The image can be simplified by hiding everything except the region of interest. A variety of one-click renderings and color schemes help to visualize the backbone, secondary structure, distributions of hydrophobic versus hydrophilic residues, non-covalent bonding interactions, salt bridges, amino acid or nucleotide sequences, sequence-to-structure mappings and locations of residues of interest, and patterns of evolution and conservation.

Protein Explorer is a knowledge base with introductory information on many topics about protein structure, such as the origins and limitations of 3D protein structure data, specific oligomers vs. crystal contacts, hydrogen bonds, cation–pI interactions, etc. Protein Explorer is more user-friendly than the comparable software RASMOL, also available on servers such as ExPASy.

14.3.4 ExPASy Server: Roche Applied Science Biochemical Pathways

The ExPASy server (www.expasy.ch) is one of the most useful servers, where almost any bioinformatic tool can be found, together with useful links to other websites such as NCBI or EBI. The several access databases are descriptive, easy to follow, and up to date. Protein data bank searches with SwissProt or Trembl, as well as sequence alignments using either SimAlign (for two sequences) or ClustalW (for more than two protein sequences) can be started from ExPASy, to name just a few of the possibilities available. Access is also given to the Roche Applied Science Biochemical pathways where either keyword searches for particular enzymes or for metabolites can be performed, or entire metabolic pathways or sections thereof can be visualized. Proteomics evaluation is also available on ExPASy, which features free 2D-PAGE software called Melanie.

14.3.5 GenBank

One point to remember: when exploring either genomes or GenBank nucleotide entries, searchers should be aware that, especially in GenBank, there exist multiple entries for the same gene and this confuses links between the gene of interest, the GenBank ID and the nucleotide sequence. Bacterial genomes especially are critical, where all genes are grouped under a single entry. GenBank distributes IDs according to the date of entry, so there can be several entries for one gene, incomplete and complete ones, depending on when the nucleotide sequence was entered. This causes redundancy in the current GenBank.

The easiest access route when exploring GenBank is with the NCBI server, choosing either nucleotide or protein or genome under the search button. Searches can be done with GenBank ID numbers or keywords, but keywords sometimes give confusing results. The most relevant parts in a Genbank entry are the source of the gene (organism, genome) and the features link, which shows the nucleotide range and what kind of gene region we are looking at. In genomes this can range from promoter regions, to ribosome binding sites (RBS), to genes with actual or annotated functions and repeat structures. When a gene of interest is found the CDS (coding sequence) should be used and then transformed into the FASTA format for further use.

14.3.6 SwissProt

SwissProt is, next to GenBank, the most important server for biologists, but it has a different philosophy behind it. In GenBank all entries are submitted by the authors themselves and underlie their authority; in SwissProt all entries are screened by one person and his group, which makes entry acceptance slower (this speed has been increased by creating an accession buffer called trembl) but accuracy and efficacy is higher, and redundancy does not normally occur.

There are many ways of accessing SwissProt, through either ExPASy or the SRS link at EBI, or directly through their own website www.swissprot.com, or through any genomic database. Since SwissProt is for exploring proteins, there is much information behind well characterized proteins, accessible through the cross-reference section. Examples are the corresponding nucleotide database embl; PIR (a protein information source, which gives information about amino acid modification); Swiss-D-Page, which contains links to proteomics or profiles/domains and contains a series of fields that deal with the classification of proteins in families or superfamilies of related sequences based on their sharing of domains.

Another interesting part to explore is the feature section, which displays characteristics of the protein displayed, such as signal peptides, domains, transmembrane regions, or repeated sequence segments.

14.3.7 Information on an Enzyme: the Example of dehydrogenases

From the viewpoint of trying to find information on a particular enzyme, Figure 14.3 elucidates the possibilities with leucine dehydrogenase as an example.

Bioinformatics can help to find two kinds of information about an enzyme: (i) its sequence, and (ii) structure information.

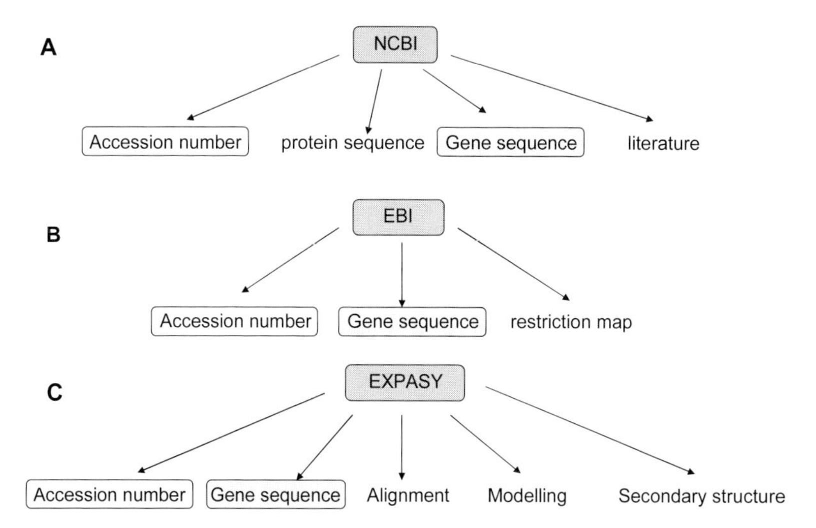

Figure 14.3 Information about an enzyme (e.g., leucine dehydrogenase, LeuDH) in the databases.

14.3.7.1 Sequence Information

DNA sequences can be obtained readily from databases such as Genbank or EMBL. In addition, a restriction map and also a suggestion for the design of primers can be obtained to help in cloning a gene into a desired target plasmid or host (see Chapter 4). Comparison between putative and actual sequences yields valuable information about possible mutations.

Protein (amino acid) sequences are available from databases such as SwissProt/Trembl or Protein Data Bank (PDB). Most useful is the ability of such databases to perform alignments, the comparison between different sequences. Simple alignment compares two sequences, multiple alignment more than two.

As an example, Table 14.1 depicts the results of a multiple alignment of different sequences of amino acid dehydrogenases (from Bommarius, 2000).

Table 14.1 Multiple alignment of six amino acid dehydrogenase sequences: identities [%] of protein sequences in amino acid dehydrogenases (from Bommarius, 2000).*

Protein	*LeuDH, B. cereus*	*LeuDH, B. sphaer*	*PheDH, Rh. rhodoc*	*PheDH, Th. interm*	*GluDH, C. symbios*
LeuDH, *B. stearotherm*	82.5	79.9	32.0	45.6	12.6
LeuDH, *B. cereus*	–	76.9	31.5	44.54	14.0
LeuDH, *B. sphaericus*		–	31.7	41.8	12.4
PheDH, *Rh. rhodocrous*			–	26.4	14.2
PheDH, *T. intermedius*				–	

* Data gathering was through BLAST search in SwissProt (Peitsch, 1995, 1996).

When comparing sequence identities from Table 14.1 it is readily apparent that GluDH from *Clostridium symbiosus* is not at all related to the other dehydrogenases, while the three LeuDHs with ≥ 77% sequence identity are very similar enzymes and most probably are evolutionarily related.

14.3.7.2 **Structural Information**

Given an amino acid sequence, the secondary structure can be discerned with a program called psipred, which is available on the ExPASy site. Also, prediction of transmembrane location is possible nowadays. The procedure is illustrated in Figure 14.4 for formate dehydrogenase from *Pseudomonas* sp. 101 as an example.

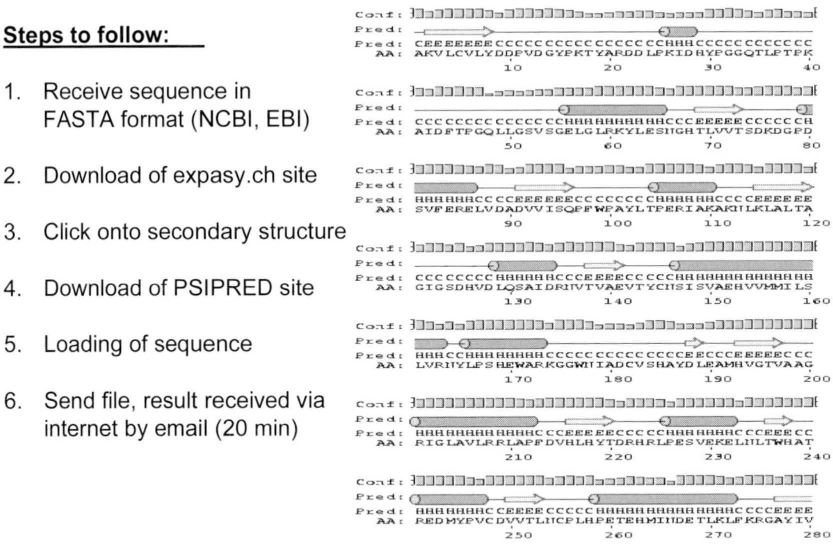

Steps to follow:

1. Receive sequence in FASTA format (NCBI, EBI)
2. Download of expasy.ch site
3. Click onto secondary structure
4. Download of PSIPRED site
5. Loading of sequence
6. Send file, result received via internet by email (20 min)

```
Conf:
Pred:
Pred: CEEEEEEEECCCCCCCCCCCCCCCCCCHHHCCCCCCCCCCC
  AA: AKVLCVLYDDPVDGYPKTYARDDLPKIDHYPGGQTLPTPK
               10        20        30        40
Conf:
Pred:
Pred: CCCCCCCCCCCCCCCHHHHHHHHHHCCCEEEEECCCCCCH
  AA: AIDFTPGQLLGSVSGELGLRKYLESNGHTLVVTSDKDGPD
               50        60        70        80
Conf:
Pred:
Pred: HHHHHHCCCCEEEEEECCCCCCCCHHHHHHCCCCEEEEEE
  AA: SVFERELVDADVVISQPFWPAYLTPERIAKAKNLKLALTA
               90       100       110       120
Conf:
Pred:
Pred: CCCCCCCCHHHHHHCCCEEEECCCCCHHHHHHHHHHHHHH
  AA: GIGSDHVDLQSAIDRNVTVAEVTYCNSISVAEHVVMMILS
              130       140       150       160
Conf:
Pred:
Pred: HHHCCHHHHHHHHCCCCCCCCCCCCCCEECCCEEEEECCC
  AA: LVRNYLPSHEWARKGGWNIADCVSHAYDLEAMHVGTVAAG
              170       180       190       200
Conf:
Pred:
Pred: HHHHHHHHHHHHCCCEEEEEECCCCCHHHHHHHCCCEEECC
  AA: RIGLAVLRRLAPFDVHLHYTDRHRLPESVEKELNLTWHAT
              210       220       230       240
Conf:
Pred:
Pred: HHHHHHHCCEEEECCCCCHHHHHHHHHHHHHHCCCCEEEEE
  AA: REDMYPVCDVVTLNCPLHPETEHMINDETLKLFKRGAYIV
              250       260       270       280
```

Figure 14.4 Prediction of secondary structure, with formate dehydrogenase as an example.

14.4 Applied Bioinformatics Tools, with Examples

14.4.1 BLAST

BLAST is the most successful bioinformatics tool ever, and makes it possible to search for similar sequences within a database, when submitting a protein or DNA sequence. BLAST stands for Basic Local Alignment and Search Tool. When exploring protein sequences, one can search either for similar proteins within the database using blastp or for new genes similar to the protein sequence at hand using tblastn, which scans DNA sequences being translated into their six possible reading frames (three on each strand) for similarity to the given sequence.

As an example, a protein sequence of a bacterial NADH oxidase, in this case the protein sequence of a water-forming NADH oxidase from *Streptococcus mutans*, is submitted to a blastp search:

The results (Figure 14.5) give a graphic display where the query sequence is similar to other sequences, with the scores in different colors (not shown), which is then deciphered into the following alignments, called the hit list. They are listed in descending order, starting with the highest similarity. The accession numbers (underlined) can be accessed directly.

```
Sequences producing significant alignments:                        (bits) Value

gi|547996|sp|P37062|NAPE ENTFA  NADH PEROXIDASE (NPXASE)             159   3e-39
gi|547994|sp|P37061|NAOX ENTFA  NADH OXIDASE (NOXASE)                156   3e-38
gi|2500132|sp|P75389|NAOX MYCPN  Probable NADH oxidase (NOXASE)      152   7e-37
gi|2500131|sp|Q49408|NAOX MYCGE  PROBABLE NADH OXIDASE (NOXASE)      137   2e-32
gi|2500133|sp|Q58065|NAOX METJA  Putative NADH oxidase (NOXASE)      127   2e-29
gi|1171661|sp|P42435|NASD BACSU  Nitrite reductase [NAD(P)H]         102   8e-22
gi|1723803|sp|P37596|YGBD ECOLI  Hypothetical protein ygbD            93   5e-19
gi|1171659|sp|P42433|NASB BACSU  Assimilatory nitrate reduct...       91   2e-18
gi|1174665|sp|P43494|THCD RHOER  RHODOCOXIN REDUCTASE                 76   4e-14
gi|1173329|sp|P42454|RURE ACICA  Rubredoxin-NAD(+) reductase          75   7e-14
gi|1346195|sp|P48639|GSHR BURCE  GLUTATHIONE REDUCTASE (GR) ...       75   1e-13
gi|417867|sp|P33009|TERA PSESP  TERPREDOXIN REDUCTASE                 74   2e-13
gi|1170040|sp|P42770|GSHC ARATH  Glutathione reductase, chlo...       74   3e-13
gi|2500123|sp|Q54465|MERA SHEPU  Mercuric reductase (Hg(II) ...       73   4e-13
gi|1706442|sp|P31052|DLD2 PSEPU  DIHYDROLIPOAMIDE DEHYDROGEN...       72   8e-13
gi|1346197|sp|P48642|GSHR ORYSA  GLUTATHIONE REDUCTASE, CYTO...       72   9e-13
gi|585562|sp|Q06458|NIRB KLEPN  NITRITE REDUCTASE [NAD(P)H] ...       72   9e-13
gi|126995|sp|P08663|MERA STAAU  Mercuric reductase (Hg(II) r...       72   1e-12
gi|1170041|sp|P43783|GSHR HAEIN  Glutathione reductase (GR) ...       70   2e-12
gi|118676|sp|P14218|DLDH PSEFL  DIHYDROLIPOAMIDE DEHYDROGENA...       70   4e-12
gi|2500121|sp|P94702|MERA ENTAG  Mercuric reductase (Hg(II) ...       70   4e-12
gi|115557|sp|P16640|CAMA PSEPU  PUTIDAREDOXIN REDUCTASE               70   4e-12
gi|2500115|sp|Q43154|GSHC SPIOL  GLUTATHIONE REDUCTASE, CHLO...       70   4e-12
gi|118677|sp|P09063|DLD1 PSEPU  DIHYDROLIPOAMIDE DEHYDROGENA...       69   6e-12
gi|1171719|sp|P08201|NIRB ECOLI  Nitrite reductase [NAD(P)H]...       69   7e-12
gi|118672|sp|P21880|DLD1 BACSU  Dihydrolipoamide dehydrogena...       69   7e-12
gi|8928150|sp|Q60151|GSHR STRTR  Glutathione reductase (GR) ...       69   9e-12
gi|2500120|sp|P94188|MERA ALCSP  Mercuric reductase (Hg(II) ...       68   1e-11
gi|461933|sp|Q04829|DLDH HALVO  DIHYDROLIPOAMIDE DEHYDROGENASE        68   1e-11
gi|2500119|sp|Q52109|MERA ACICA  Mercuric reductase (Hg(II) ...       67   2e-11
gi|126988|sp|P00392|MERA PSEAE  Mercuric reductase (Hg(II) r...       67   2e-11
gi|2500122|sp|Q51772|MERA PSEFL  Mercuric reductase (Hg(II) ...       67   3e-11
gi|121676|sp|P27456|GSHC PEA  GLUTATHIONE REDUCTASE, CHLOROP...       67   4e-11
gi|8928125|sp|O15770|GSHR PLAF7  Glutathione reductase (GR) ...       66   4e-11
gi|1346194|sp|P48641|GSHR ARATH  GLUTATHIONE REDUCTASE, CYTO...       66   5e-11
gi|731029|sp|P39051|TYTR TRYBB  TRYPANOTHIONE REDUCTASE (TR)...       66   5e-11
gi|3123310|sp|O10499|YDGE SCHPO  PUTATIVE FLAVOPROTEIN C26F1...       65   7e-11
gi|121677|sp|P23189|GSHR PSEAE  Glutathione reductase (GR) (...       64   2e-10
gi|2500116|sp|Q43621|GSHR PEA  GLUTATHIONE REDUCTASE, CYTOSO...       64   2e-10
gi|14194687|sp|Q9PJI3|DLDH CHLMU  Dihydrolipoamide dehydroge...       64   2e-10
gi|118671|sp|P11959|DLD1 BACST  DIHYDROLIPOAMIDE DEHYDROGENA...       64   3e-10
gi|1708059|sp|P80461|GSHC TOBAC  GLUTATHIONE REDUCTASE, CHLO...       64   3e-10
```

Figure 14.5 Results of a blastp search (in BLAST) of a water-forming NADH oxidase from *Streptococcus mutans*.

```
gi|6014977|sp|Q59822|DLDH_STAAU  DIHYDROLIPOAMIDE DEHYDROGEN...   64   3e-10
gi|126994|sp|P08332|MERA_SHIFL  MERCURIC REDUCTASE (HG(II) R...   64   3e-10
gi|7531102|sp|Q9Z773|DLDH_CHLPN  DIHYDROLIPOAMIDE DEHYDROGEN...   63   3e-10
gi|126996|sp|P17239|MERA_THIFE  Mercuric reductase (Hg(II) r...   63   3e-10
gi|1706443|sp|P52992|DLDH_ALCEU  DIHYDROLIPOAMIDE DEHYDROGEN...   63   4e-10
gi|1346193|sp|P48638|GSHR_ANASP  GLUTATHIONE REDUCTASE (GR) ...   63   5e-10
gi|8928151|sp|Q94655|GSHR_PLAFK  Glutathione reductase (GR) ...   63   5e-10
gi|118670|sp|P18925|DLDH_AZOVI  DIHYDROLIPOAMIDE DEHYDROGENA...   63   5e-10
gi|121674|sp|P06715|GSHR_ECOLI  Glutathione reductase (GR) (...   62   1e-09
gi|1168642|sp|Q07946|BEDA_PSEPU  Benzene 1,2-dioxygenase sys...   61   1e-09
gi|136620|sp|P28593|TYTR_TRYCR  TRYPANOTHIONE REDUCTASE (TR)...   61   2e-09
gi|1708060|sp|P41921|GSHR_YEAST  GLUTATHIONE REDUCTASE (GR) ...   60   2e-09
gi|416906|sp|Q04933|DLDH_TRYBB  DIHYDROLIPOAMIDE DEHYDROGENASE    60   3e-09
gi|7531097|sp|O34324|DLD3_BACSU  Dihydrolipoamide dehydrogen...   60   3e-09
gi|731027|sp|P39040|TYTR_CRIFA  TRYPANOTHIONE REDUCTASE (TR)...   60   3e-09
gi|1346192|sp|P48640|GSHC_SOYBN  GLUTATHIONE REDUCTASE, CHLO...   60   3e-09
gi|126987|sp|P16171|MERA_BACSR  Mercuric reductase (Hg(II) r...   60   4e-09
gi|6166121|sp|P90597|DLDH_TRYCR  DIHYDROLIPOAMIDE DEHYDROGENASE   60   4e-09
gi|399391|sp|P31046|DLD3_PSEPU  DIHYDROLIPOAMIDE DEHYDROGENA...   58   1e-08
gi|118675|sp|P09623|DLDH_PIG  DIHYDROLIPOAMIDE DEHYDROGENASE...   58   2e-08
gi|6014973|sp|O08749|DLDH_MOUSE  DIHYDROLIPOAMIDE DEHYDROGEN...   58   2e-08
gi|6014978|sp|O18480|DLHD_MANSE  DIHYDROLIPOAMIDE DEHYDROGEN...   58   2e-08
gi|118674|sp|P09622|DLDH_HUMAN  DIHYDROLIPOAMIDE DEHYDROGENA...   57   2e-08
gi|136619|sp|P13110|TYTR_TRYCO  TRYPANOTHIONE REDUCTASE (TR)...   57   3e-08
gi|13431764|sp|O95831|PCD8_HUMAN  PROGRAMED CELL DEATH PROTE...   57   4e-08
gi|1706444|sp|P49819|DLDH_CANFA  DIHYDROLIPOAMIDE DEHYDROGEN...   57   4e-08
gi|113644|sp|P17052|RURE_PSEOL  Rubredoxin-NAD(+) reductase       57   4e-08
gi|14916975|sp|P31023|DLDH_PEA  Dihydrolipoamide dehydrogena...   57   4e-08
gi|135972|sp|P13452|TODA_PSEPU  TOLUENE 1,2-DIOXYGENASE SYST...   57   4e-08
gi|7531099|sp|O84561|DLDH_CHLTR  Dihydrolipoamide dehydrogen...   56   5e-08
gi|6014975|sp|P95596|DLDH_RHOCA  Dihydrolipoamide dehydrogen...   55   7e-08
gi|5902749|sp|O06465|AHPF_XANCH  ALKYL HYDROPEROXIDE REDUCTA...   55   8e-08
gi|13431757|sp|Q9JM53|PCD8_RAT  Programed cell death protein...   55   1e-07
gi|12643685|sp|P78965|GSHR_SCHPO  GLUTATHIONE REDUCTASE (GR)...   55   1e-07
gi|5902748|sp|O82864|AHPF_PSEPU  ALKYL HYDROPEROXIDE REDUCTA...   55   1e-07
gi|128349|sp|P22944|NIR_EMENI  NITRITE REDUCTASE [NAD(P)H]        55   2e-07
gi|13431781|sp|Q9Z0X1|PCD8_MOUSE  Programed cell death prote...   54   2e-07
gi|11132261|sp|P57303|DLDH_BUCAI  Dihydrolipoamide dehydroge...   54   2e-07
gi|15213976|sp|Q50068|DLDH_MYCLE  Dihydrolipoamide dehydroge...   54   3e-07
gi|14917007|sp|P47791|GSHR_MOUSE  Glutathione reductase, mit...   54   3e-07
gi|13124714|sp|O00087|DLDH_SCHPO  DIHYDROLIPOAMIDE DEHYDROGE...   54   3e-07
gi|11182440|sp|P50529|STHA_VIBCH  Soluble pyridine nucleotid...   53   5e-07
gi|11182439|sp|P27306|STHA_ECOLI  Soluble pyridine nucleotid...   53   5e-07
gi|1169352|sp|P43784|DLDH_HAEIN  Dihydrolipoamide dehydrogen...   53   6e-07
gi|12644432|sp|Q39243|TRB1_ARATH  THIOREDOXIN REDUCTASE 1 (N...   52   8e-07
gi|114049|sp|P19480|AHPF_SALTY  ALKYL HYDROPEROXIDE REDUCTAS...   52   1e-06
gi|2507291|sp|P35340|AHPF_ECOLI  ALKYL HYDROPEROXIDE REDUCTA...   51   1e-06
gi|11135078|sp|O07212|STHA_MYCTU  Probable soluble pyridine ...   51   1e-06
gi|18271667|sp|P77212|YKGC_ECOLI  Probable pyridine nucleoti...   51   2e-06
gi|11135195|sp|P57112|STHA_PSEAE  Soluble pyridine nucleotid...   50   3e-06
gi|3023936|sp|P77650|HCAD_ECOLI  3-PHENYLPROPIONATE DIOXYGEN...   50   3e-06
gi|140037|sp|P23160|R34K_CLOPA  34.2 kDa protein in rubredox...   50   4e-06
gi|544165|sp|P35484|DLDH_ACHLA  DIHYDROLIPOAMIDE DEHYDROGENA...   50   4e-06
gi|731028|sp|P39050|TYTR_LEIDO  TRYPANOTHIONE REDUCTASE (TR)...   50   4e-06
gi|14916998|sp|P00390|GSHR_HUMAN  Glutathione reductase, mit...   50   4e-06
gi|7531098|sp|O50286|DLDH_VIBPA  DIHYDROLIPOAMIDE DEHYDROGEN...   49   5e-06
gi|115088|sp|P08087|BNZD_PSEPU  BENZENE 1,2-DIOXYGENASE SYST...   49   8e-06
gi|3915897|sp|P94284|TRXB_BORBU  THIOREDOXIN REDUCTASE (TRXR)     49   1e-05
```

Figure 14.5 (continued)

How is this output to be read?

Every hit in the list is aligned to the query sequence and is revealed below the hit list (not shown).

The hit list immediately tells one whether the query sequence has similarity to something already in the database and how many similar sequences exist. It also assesses the trustability of a good hit by evaluating the bit score and the *E* value. The bit score is a measure of the statistical significance of the alignment; the higher the bit score, the more similar are the two sequences, and matches below 50 are very unreliable. The *E* value (expectation value) is another statistical tool, which estimates the number of times such a match could be expected by chance alone. The lower the *E* value, the better the hit and the more confidence can be put into the evaluation of a similarity. Sequences which are identical have an *E* value close to zero. Reliable homologies all range below *E* values of 10^{-4} (0.0001).

Blasting nucleotide sequences is not always that easy, because there is more ambiguity to the nucleotide sequence, and good hits have to have a 70% homology over the whole sequence to be reliable, compared to 25% with proteins. If the coding region is known, it is better to translate the nucleotide sequence into protein and then start a blast search.

14.4.2 Aligning Several Protein Sequences using ClustalW

Another very useful strategy in bioinformatics is to create a multiple alignment between sequences of interest. The most common device used for this is ClustalW, found on the EBI server. It is useful for predicting protein structures and for predicting the function of proteins, and indispensable for generating phylogenetic trees (see Section 14.4.3). Here again, as for the BLAST searches, the better the alignment turns out to be, the better the predictions from it are going to be. Sequence alignment does not work very well for assembling sequence pieces or for ESTs; it is better used for whole protein sequences or domains. The smaller the sequence piece, the less useful is the information achieved. In sequence alignment the amino acids that show either structural or sequence similarity are put into the column. In the default position of the program, column structure is based upon sequence similarity: the more homologous the sequences, the more can be depicted from them regarding functional or evolutional similarity as well, because they then become equivalent.

More useful applications that can be derived from a multiple sequence alignment are:

- extrapolation of a yet unknown member into a certain family;
- phylogenetic analysis (see Section 14.4.3);
- pattern identification (very conserved regions);
- domain identification;
- secondary structure prediction.

```
Mycopl. pneumoniae P75389|NAOX_MYCP --------------MKKVIVIGVNHAGTSFIRTLLSKSKDFQVNAYDRNT      36
Mycopl. genitalis  Q49408|NAOX_MYCG --------------MKKVIVIGINHAGTSFIRTLLSKSKDFKVNAYDRNT      36
Enterococ.faecalis P37061|NAOX_ENTFA ---------------MKVVVVGCTHAGTSAVKSILANHPEAEVTVYERND     35
Lactococcus lactis NP_266547.1|      ---------------MKIVVIGTNHAGIATANTLIDRYPGHEIVMIDRNS     35
Borrelia burgdorferi AE001172        --------------MMKIIIIGGTSAGTSAAAKANRLNKKLDITIYEKTN     36
Methanoc.japanicus Q58065|NAOX_METJ MVNRKPNNPNKNGEEMRAIIIGSGAAGLTTASTIRKYNKDMEIVVITKEK      50

Mycopl. pneumoniae P75389|NAOX_MYCP NISFLGCGIALAVSGVVKNTEDLFYSTPEELKA-MGANVFMAHDVVGLDL      85
Mycopl. genitalis  Q49408|NAOX_MYCG NISFLGCGIALAVSGVVKNTDDLFYSNPEELKQ-MGANIFMSHDVTNIDL      85
Enterococ.faecalisP37061|NAOX_ENTFA NISFLSCGIALYVGGVVKNAADLFYSNPEELAS-LGATVKMEHNVEEINV      84
Lactococcus lactis |NP_266547.1|     NMSYLGCGTAIWVGRQIEKPDELFYAKAEDFEK-KGVKILTETEVSEIDF      84
Borrelia burgdorferi AE001172        IVSFGTCGLPYFVGGFFDNPNTMISRTQEEFEK-TGISVKTNHEVIKVDA      85
Methanoc.japanicus Q58065|NAOX_METJ EIAYSPCAIPYVIEGAIKSFDDIIMHTPEDYKRERNIDILTETTVIDVDS     100

Mycopl. pneumoniae P75389|NAOX_MYCP DKKQVIVKDLATGKETVDHYDQLVVASGAWPICMNVENEVTHTQLQFNHT     135
Mycopl. genitalis  Q49408|NAOX_MYCG IKKQVTVRDLTSNKEFTDQFDQLVIASGAWPICMNVENKVTHKPLEFNYT     135
Enterococ.faecalisP37061|NAOX_ENTFA DDKTVTAKNLQTGATETVSYDKLVMTTGSWPIIPPIP-------------     121
Lactococcus lactis  NP_266547.1|     TNKMIYAKS-KTGEKITESYDKLVLATGSRPIIPNLP-------------     120
Borrelia burgdorferi AE001172        KNNTIVIKNQKTGTIFNNTYDQLMIATGAKPIIPPIN-------------     122
Methanoc.japanicus Q58065|NAOX_METJ KNNKIKCVD-KDGNEFEMNYDYLVLATGAEPFIPPIE-------------     136

Mycopl. pneumoniae P75389|NAOX_MYCP DKYCGNIKNLISCKLYQHALTLIDSFRHDKSIKSVAIVGSGYIGLELAEA     185
Mycopl. genitalis Q49408|NAOX_MYCG  DKYCGNVKNLISCKLYQHALTLIDSFRKDKTIKSVAIVGSGYIGLELAEA     185
Enterococ.faecalisP37061|NAOX_ENTFA --GIDAENILLCKNYSQANVIIE--KAKDAKRVVVVGGGYIGIELVEA       165
Lactococcus lactis  NP_266547.1|     --GKDLKGIHFLKLFQEGQAIDEEFAKN-DVKRIAVIGAGYIGTEIAEA      166
Borrelia burgdorferi AE001172        --NINLENFHTLKNLEDGQKIKKLMDRE-EIKNIVIIGGGYIGIEMVEA      168
Methanoc.japanicus Q58065|NAOX_METJ --GKDLDGVFKVRTIEDGRAILKYIEEN-GCKKVAVVGAGAIGLEMAYG      182

Mycopl. pneumoniae P75389|NAOX_MYCP AWQCGKQVTVIDMLDKPAGNNFDEEFTNELEKAMKKAGINLMMGSAVKGF     235
Mycopl. genitalis  Q49408|NAOX_MYCG AWLCKKQVTVIDLLDKPAGNNFDHEFTDELEKVMQKDGLKLMMGCSVKGF     235
Enterococ.faecalisP37061|NAOX_ENTFA FVESGKQVTLVDGLDRILNKYLDKPFTDVLEKELVDRGVNLALGENVQQF     215
Lactococcus lactis  NP_266547.1|     AKRRGKEVLLFDAESTSLASYYDEEFAKGMDENLAQHGIELHFGELAQEF     216
Borrelia burgdorferi AE001172        AKNKRKNVRLIQLDKHILIDSFDEEIVTIMEEELTKKGVNLHTNEFVKSL     218
Methanoc.japanicus Q58065|NAOX_METJ LKCRGLDVLVVEMAPQVLPRFLDPDMAEIVQKYLEKEGIKVMLSKPLEKI     232

Mycopl. pneumoniae P75389|NAOX_MYCP IVDADKNVVKGVETDKGRVDADLVIQSIGFRPNTQFVPKDRQFEFNRNGS     285
Mycopl. genitalis Q49408|NAOX_MYCG  VVDSTNNVVKGVETDKGIVNADLVNQSIGFRPSTKFVPKDQNFEFIHNGS     285
Enterococ.faecalisP37061|NAOX_ENTFA VADEQGKVAK-VITPSQEFEADMVIMCVGFRPNTELLKD-KVDMLPNGA      262
Lactococcus lactis  NP_266547.1|     KANEKGHVSQ-IVTNKSTYDVDLVINCIGFTANSALAGE-HLETFKNGA      263
Borrelia burgdorferi AE001172        IGEKK--AEGVVTNKNTYQADAVILATGIKPDTEFLEN-QLKTTKNGA       263
Methanoc.japanicus Q58065|NAOX_METJ VGKEK--VEAVYVDGKLYDVDMVIMATGVRPNIELAKK-AGCKIGKFA      277

Mycopl. pneumoniae P75389|NAOX_MYCP IKVNEYLQALNHENVYVIGGAAAIYDAASEQYENIDLATNAVKSGLVAAM     335
Mycopl. genitalis Q49408|NAOX_MYCG  IKVNEFLQALNHKDVYVIGGCAAIYNAASEQYENIDLATNAVKSGLVAAM     335
Enterococ.faecalisP37061|NAOX_ENTFA IEVNEYMQTSN-PDIFAAGDSAVVHYNPSQTKNYIPLATNAVRQGMLVGR     311
Lactococcus lactis  NP_266547.1|     IKVDKHQQSSD-PDVSAVGDVATIYSNALQDFTYIALASNAVRSGIVAG-     311
Borrelia burgdorferi AE001172        IIVNEYG-ETSIKNIFSAGDCATIYNIVSKKNEYIPLATTANKLGRIVG-     311
Methanoc.japanicus Q58065|NAOX_METJ IEVNEKMQTSI-PNIYAVGDCVEVIDFITGEKTLSPFGTAAVRQGKVAG-     325

Mycopl. pneumoniae P75389|NAOX_MYCP HMIGSKAVKLESIVGTNALHVFGLNLAATGLTEKRAKMNGFDVGVSIVDD     385
Mycopl. genitalis Q49408|NAOX_MYCG  HIIGSNQVKLQSIVGTNALHIFGLNLAACGLTEQRAKKLGFDVGISVVDD     385
Enterococ.faecalisP37061|NAOX_ENTFA NLT-EQKLAYRGTQGTSGLYLFGWKIGSTGVTKESAKLNGLDVEATVFED     360
Lactococcus lactis NP_266547.1|      HNIGGKSIESVGVQGSNGISIFGYNMTSTGLSVKAAKKIGLEVSFSDFED     361
Borrelia burgdorferi AE001172        ENLAGNHTAFKGTLGSASIKILSLEAARTGLTEKDAKKLQIKYKTIFVKD     361
Methanoc.japanicus Q58065|NAOX_METJ KNIAGVEAKFYPVLNSAVSKIGDLEIGGTGLTAFSANLKRIPIVIGRAKA     375

Mycopl. pneumoniae P75389|NAOX_MYCP NDRPEFMG-TFDKVRFKLIYDKKTLRLLGAQLLSWNTNHSEIIFYIALAV     434
Mycopl. genitalis Q49408|NAOX_MYCG  NDRPEFMG-SYDKVRFKLVYDKKTLRILGAQLLSWNTNHSEIIFYIALAI     434
Enterococ.faecalisP37061|NAOX_ENTFA NYRPEFMP-TTEKVLMELVYEKGTQRIVGGQLMS-KYDITQSANTLSLAV     408
Lactococcus lactis NP_266547.1|      KQKAWFLHENNDSVKIRIVYETKNRRIIGAQLASKSEIIAGNINMFSLAI     411
Borrelia burgdorferi AE001172        KNHTNYYP-GQEDLYIKLIYEENTKIILGAQAIGKNGAVIR-IHALSIAI     409
Methanoc.japanicus Q58065|NAOX_METJ LTRARYYP-GGKEIEIKMIFNEDG-KVVGCQIVG-GERVAERIDAMSIAI     422

Mycopl. pneumoniae P75389|NAOX_MYCP QKKMLISELGLVDVYFLPHYNKPFNFVLAAVLQALGFSYYTPKNK         479
Mycopl. genitalis Q49408|NAOX_MYCG  QKQMLLTELGLVDVYFLPHYNKPFNFVLATVLQALGFSYYIPKK-         478
Enterococ.faecalisP37061|NAOX_ENTFA QNKMTVEDLAISDFFFQPHFDRPWNYLNLLAQAALENM-------       446
Lactococcus lactis NP_266547.1|      QEKKTIDELALLDLFFLPHFNSPYNYMTVAALNAK----------      446
Borrelia burgdorferi AE001172        YSKLTTKELGMMDFSYSPPFSRTWDILNIAGNAAK----------      444
Methanoc.japanicus Q58065|NAOX_METJ FKKVSAEELANMEFCYAPPVSMVHEPLSLAAEDALKKLSNK----      463
```

Figure 14.6 Six water-forming NADH oxidases were aligned using ClustalW.
The color code provides information about the similarity of amino acids, and the boxes give examples of information gathering from the multiple alignment: the first box shows the FAD binding domain, the second box shows a highly conserved cysteine residue crucial for this family, and the third box shows a consensus as part of the NAD binding domain.

When several sequences are being aligned, choice of the right sequences is crucial to a good alignment. The sequences should be more than 30%, but not more than 90%, identical (identical sequences do no harm, but neither do they help), and one should try to choose sequences of almost identical length.

Online servers providing multiple alignments include:
• www.ebi.ac.uk/clustalw/index.html
• bibiserv.techfak.uni-bielefield.de/dialign
• igs-server-cnrs-mrs.fr/Tcoffee

Here again, water-forming bacterial NADH oxidases are chosen to provide an example of multiple alignment using ClustalW (Figure 14.6).

14.4.3 Task: Whole Genome Analysis

Bioinformatics tools can be useful in whole genome analysis for some of the following tasks:

• investigating divergence of two related species;
• evidence of lateral gene transfer (LGT) (Chapter 15, Section 15.4.3);
• modification of a particular ORF between three related species;
• generation of virtual ancestors.

The information gleaned from BLAST and CROSS comparisons make it possible to draw some conclusions, which might be:

• some genes are present in a single copy only;
• other genes are present in up to four copies;
• whole cluster are duplicated;
• internal gene duplication occurs in one cluster.

From these conclusions, researchers can begin to build hypotheses about gene function. In some areas there is evidence of duplication, whereas in other areas there is greater or lesser divergence. The more that is learned about genomes, the more researchers are coming to appreciate that they are constantly expanding and contracting (duplicating and deleting material). The whole-system, genomic approach to biology gives researchers a different perspective on basic questions. It complements more traditional approaches.

14.4.4 Phylogenetic Tree

To study the evolutionary relationships between organisms alive today, various methods can be employed to estimate when those organisms may have diverged from a common ancestor. Having this information for a group of living organisms would allow one to construct a *phylogenetic* tree. This tree depicts in graphic form these divergences over time and accounts for the various current species in question. Such a tree for six current species might look like the chart in Figure 14.7.

In the past, much of this work was done by making observations of anatomy and physiology, and with comparisons in fossil records. More recently, techniques have been developed in molecular biology for performing such comparisons. One of the newest and most quantitative methods is to compare the nucleotide sequence in a particular segment of DNA common to two or more living organisms. Often referred to as a "molecular clock", this method utilizes a predictable rate at which mutations occur in DNA sequences. Those organisms that show the greatest number of nucleotide sequence differences are considered to have diverged from a common ancestor (following separate evolutionary paths) the greatest number of years ago (e.g., nos. 1 and 6 in Figure 14.7). If two organisms have few nucleotide differences between them, but a large and approximately equal number of differences from some third organism, they would be closely related and likely be found on "twigs" of a branch that are far removed from that third species (e.g., nos. 3, 4, and 1).

The purpose of phylogeny is to reconstruct the history of homology and similarity and to get an idea of when and how diversity between species occurred, and how it can be explained. Phylogenetics involves determination of the closest relatives of the organisms or the sequence of interest, discovery of the function of an uncharacterized gene, or tracing the origin of the gene of interest.

Two genes are orthologous when they come from two different organisms and have a common ancestor gene, separation being due only to speciation events. Genes are paralogous when they arose from gene duplication. They are more likely to have different functions than the orthologous ones.

The problem in phylogenetic tree construction is to find the tree which best describes the relationship between a set of objects. We are mostly interested in the case where the objects are species or taxa. Again, a good phylogenetic tree is more likely to be achieved using protein sequences, although many of the large trees are constructed using 16 S RNA, for example, and the better the data set, the more suitable it will be to obtain a reasonable tree using different methods (phylogenetic trees should be checked by two different approaches). Using ClustalW to start construction of a phylogenetic tree, a multiple alignment is created. For a perfect tree, sequence fragments, recombinant sequences, and large, complex families are to be avoided.

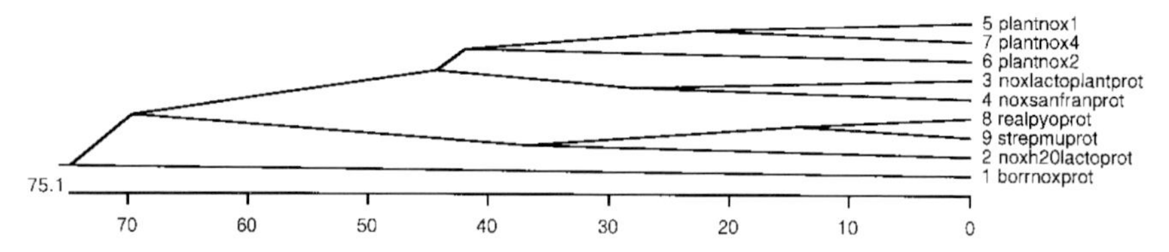

Figure 14.7 Phylogenetic tree of nine species, using the water-forming NADH oxidase protein as an example.

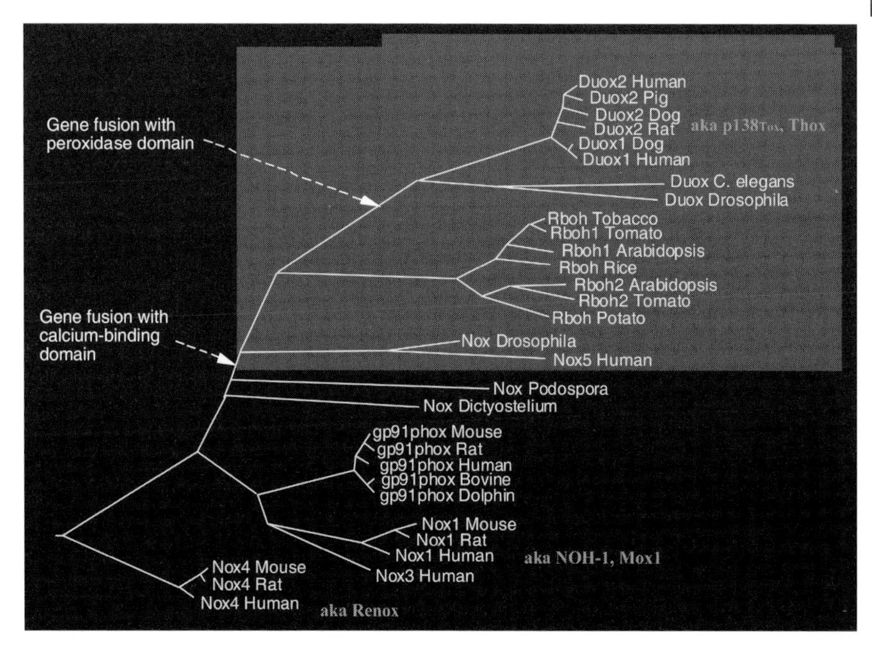

Figure 14.8 Phylogenetic tree of the gp91 phox subunit of NADPH oxidase (Nox), by courtesy of Dr. D. Lambeth, Pathology, Emory University, USA.

Phylogenetic tree construction is based on cladistics, the systematic classification of groups of organisms on the basis of shared characteristics thought to derive from a common ancestor. The basic assumptions of cladistics are:

1. The members of any group of organisms are related by descent from a common ancestor.
2. There is a bifurcation (binary) pattern of cladogenesis.
3. Changes in characteristics occur in lineages over time.

14.5 Bioinformatics for Structural Information on Enzymes

Even in the area of structural information, long regarded as one with access for experts only, today's databases can help in various ways:

- from PDB current holdings, statistics on solved 3D structures are available;
- with programs such as Swiss-3D-Image, viewing of existing 3D structures is possible, even from an advanced PC;
- modeling of novel structures on the basis of existing 3D structures can be performed with the program SwissModel (available on ExPASy) if the new structure features at least 50% sequence identity to the existing structure.

14.5.1
The Status of Predicting Protein Three-Dimensional Structure

Given the vast amounts of protein sequence information from the growing number of published genome sequences, there is growing interest in developing more efficient methods for determining a protein's structure from its amino acid sequence. The main goal of protein structure modeling is to generate structures that indicate something about function, and that can be used to help design experiments to test particular functions.

Although much progress has been made over the last 40 years, during which structural models of protein have been built, the problem is a long way from being solved. Improvements in computational methods, however, are opening up new possibilities in this field, particularly in the area of comparative modeling. There are three basic approaches to modeling protein structure, with the lines of distinction between the three approaches becoming increasingly blurred:

- *comparative modeling*: models for proteins where there is a clear sequence relationship between the sequence of interest and that of an already experimentally determined structure;
- *fold recognition modeling*: models for proteins where there is no easily detectable sequence relationship but the protein contains a fold that is already known, and either the protein is considered to be evolutionarily related to a known structure (homologous folds) or the protein has no detectable evolutionary relationship (analogous folds);
- *ab-initio modeling*: models that are developed for proteins without making any direct use of known structures.

While there is a great deal of information in published genome sequences, much more information can be gleaned when the sequences are annotated. A new discipline, structural genomics, has developed to address the challenge of providing the structural information for these annotations.

There are increasing numbers of protein sequences available as a result of the increasing numbers of genome sequences that are now accessible in public databanks. It is estimated that there were experimentally determined structures for 1% of the protein sequences known in 1999, meaning that inferences about structure had to rely on information from models for the remaining 99%.

The number of experimentally known structures is also increasing, and comparative modeling is likely to play an increasing role in protein structural modeling in the future.

14.6
Conclusion and Outlook

Some important potential trends until 2005 and beyond can already be discerned. Currently, projects are gearing up to increase throughput of 3D structures to 10–100 per day (structural genomics projects, synchroton use). Given that up to 90% of genes have no proven function there is an increasing need for assignment of function by machine-annotated DNA sequences. Lastly, one of the noblest goals of bioinformatics is the mission to improved understanding of the connection between the five main quests in biocatalysis:

sequence – structure – mechanism – function – stability

In summary, bioinformatics is important today and will be more important tomorrow for the following three reasons:

- Bioinformatics aids in closing the gap between the avalanche of sequences and the more steady advance of structural and functional assignments.
- Bioinformatics grows exponentially on the basis of similar growth of processor density and ability to sequence DNA.
- Bioinformatics is an increasingly important and necessary tool that is complementary to biochemistry and molecular biology techniques.

Suggested Further Reading

T. K. Attwood and D. J. Parry-Smith, *Introduction to Bioinformatics*, Prentice Hall, 1st edition, **2001**.

A. D. Baxevanis and B. F. F. Ouellette, *Bioinformatics: A Practical Guide to the Analysis of Genes and Proteins*, Wiley-Interscience; 2nd edition, **2001**.

Relevant URLs

Abbreviation: NCBI, National Center for Biotechnology Information.

NCBI homepage: http://www.ncbi.nlm.nih.gov/
NCBI education page on bioinformatics: http://www.ncbi.nlm.nih.gov/education/index.html
ExPASy Molecular Biology Server: http://www.expasy.ch/
Protein Data Bank (PDB): http://www.rcsb.org/pdb/
PDB current holdings: http://www.rcsb.org/pdb/holdings.html
Bioinformatics virtual newsletter: http://www.miljolare.no/virtue/newsletter/01_03/curr-dennis/index.php
Searching for amino acid modification: http://pir.georgetown.edu/pirwww/dbinfo/resid.html
Searching for protein motifs: http://scansite.mit.edu
Structural genomics: http://www.miljolare.no/virtue/newsletter/01_06/curr-lennart/index.php

Other useful websites for proteomics and genomics

http://ensembl.ebi.ac.uk
www.hgmp.mrc.ac.uk
http://wit.integratedgenomics.com/GOLD
www.sanger.ac.uk
www.tigr.org/tdb
http://genome.csc.ucsc.edu

References

H. M. Berman, J. Westbrook, Z. Feng, G. Gilliland, T. N. Bhat, H. Weissig, I. N. Shindyalov, and P. E. Bourne, The Protein Data Bank, *Nucleic Acids Res.* **2000**, *28*, 235-242.

A. S. Bommarius, Development of enzymatic process for the synthesis of chiral building blocks with emphasis on protein stability (in German), Habilitation thesis, RWTH Aachen, Aachen, Germany, **2000**.

N. Guex and M. C. Peitsch, Protein modelling via the Internet, *BioWorld* **1997**, *(5)*, 9–13.

J. Moult, Predicting protein three-dimensional structure, *Curr. Opin. Biotechnol.* **1999**, *10*, 583–588.

M. C. Peitsch, Protein modelling by E-mail, *Bio/Technology* **1995**, *13*, 658–660.

M. C. Peitsch, ProMod and SwissModel: internet-based tools for automated comparative protein modelling, *Biochem. Soc. Trans.* **1996**, *24*, 274–279.

G. A. Petsko, G. L. Kenyon, J. A. Gerlt, D. Ringe, und J. W. Kozarich, On the origin of enzymatic species, *Trends Biochem. Sci. (TIBS)* **1993**, *18*, 372–376.

T. Smidt-Schmidt, Banking on structures, *BioWorld* **2002**, *1(8)*.

J. Yin, Bio-informatics – a chemical engineering frontier?, *Chem. Eng. Progr.* **1999**, *95*, 65–74.

15
Systems Biology for Biocatalysis

Summary

The "reductionist" approach to science has been extremely successful for several centuries. Its goal was the postulation of a hypothesis, later of a model, and ultimately of a theory, from experimental data (induction) or the postulation of a model with subsequent experimental verification (deduction). Parallel to the availability of a hyper-exponentially growing amount of data, a complementary approach is taking shape, termed a "data-driven" or "systems" approach.

Data-driven or systems approaches have to rely on discovery science and hypothesis-driven research in individual fields of the life sciences and are complementary to them. Without knowledge gained with reductionist approaches, interpretation of data with the systems approach is not feasible.

The completion of several microbial genomes and the big project of the last decade, the sequencing of the human genome, have raised expectations of the speeding up of the discovery of new drug targets by means of high-throughput comparison of disease samples versus control samples. Several new areas of research have emerged:

- *genomics*, the study of the entirety of genes (*genome*) and their function in a cell;
- *proteomics*, the study of the entirety of the proteins in a cell as a whole (*proteome*);
- *transcriptomics*, the study of RNA transcripts as a whole occurring in a cell at a certain point in time; and
- *metabolomics*, the systematic investigation of the comprehensive profile of all the metabolites (*metabolome*) present in a biological sample across a range of organisms.

The practice of protein analysis of whole proteomes relies on (i) two-dimensional gel electrophoresis for separation; (ii) mass spectrometry for analysis; and (iii) protein arrays for achieving massively parallel analysis.

DNA array technologies provide rapid and cost-effective methods of sampling, testing, and identifying patterns of gene expression and genetic variations or mutations (genotyping).

Metabolic engineering allows directed improvement of cellular processes by modifying specific biochemical reactions or by introducing new genes through recombinant DNA technology. Metabolic engineering is the rational design of metabolic pathways in a cell to achieve a certain goal, such as synthesizing a new product, increasing the yield of desired products, or reducing undesirable by-products. Determination of cellular concentrations of metabolites as well as of metabolic fluxes and their mathematical modeling are at the heart of metabolic engineering.

15.1 Introduction to Systems Biology

15.1.1 Systems Approach versus Reductionism

For centuries, biologists as well as chemists, physicists, and other scientists approached scientific questions by successively dividing the system under investigation into smaller and smaller sub-systems until the problem was sufficiently reduced in complexity to be solved. This approach, termed *reductionism*, was extremely successful, and yielded monumental advances such as the structure of DNA, the mode of action of a gene, the mechanism of an enzyme, and many others. The goal of the reductionist approach was the postulation of a hypothesis and later of a model from the experimental data (induction) or the postulation of a model and subsequent efforts to seek experimental verification (deduction). With the advent of ever denser and thus faster chips to process bits of information and the concomitant hyper-exponentially growing amounts of available data (see Chapter 14), a novel, complementary approach is taking shape, termed a *data-driven* or *systems approach*. (Figure 15.1).

The conditions for the application of a data-based approach are twofold: (i) massive amounts of data are available, and (ii) analytical model building is either not pos-

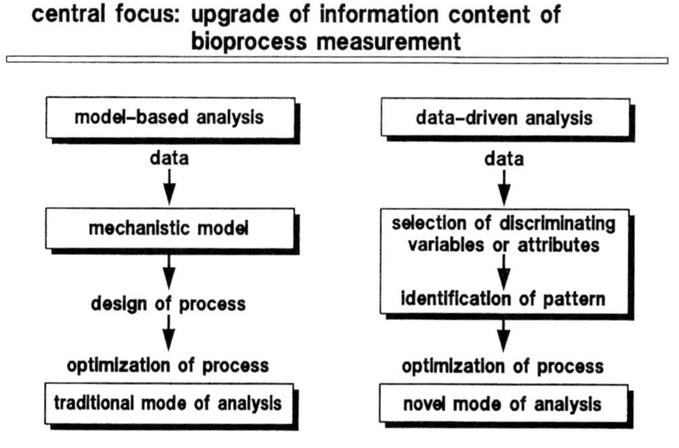

Figure 15.1 Data-based versus model-based approach.

sible or not practical. Several tools have been developed to upgrade the information content of bioprocess measurements, such as artificial intelligence, neural networks, or pattern recognition software. These tools have been applied to problems in metabolic fluxes and engineering (see below) (Kiss, 1992), fermentation process development (Kamimura, 1996; Stephanopoulos, 1997), and vaccine manufacture.

The data-driven approach does not render the model-driven approach obsolete or superfluous. In fact, data-driven approaches have to rely, and will do so for the foreseeable future, on model-driven advances in single fields of science (Aebersold, 2000). In addition, without knowledge gained with reductionist approaches interpretation of data with the systems approach is not possible. Novel tools such as pathway analysis and pattern recognition, and novel kinds of research questions such as the investigation of hypotheses based on the interaction of several system components, become possible with the systems approach.

15.1.2 Completion of Genomes: Man, Earthworm, and Others

The completion of probably *the* flagship project of the 1990s, the sequencing of the human genome, as well as of several microbial and eukaryotic genomes, gave an enormous impetus to the field of *genomics* and has raised expectations of acceleration of the discovery of new drug targets by means of high-throughput comparison of diseased samples versus control samples. As it is not genes but proteins that control the cause and action of many diseases, the requirement of thorough knowledge about drug targets (mainly proteins), including extensive validation, is now a widely accepted condition for successful drug development. Thus, *proteomics* is the upcoming field of high-throughput and large-scale protein target identification. Currently, the best-established applications of proteomics are in the clinical and biomedical fields: alterations in the proteome of body fluids or tissues are measured and compared using the samples from the control and disease state (Blackstock, 1999).

15.2 Genomics, Proteomics, and other -omics

15.2.1 Genomics

Genomics comprises the study of genes and their function. Genomics exploits the very fast-developing advances in DNA sequencing, which have been strongly supported from the private sector regarding the Human Genome Sequencing Project, initially started by Craig Venter and TIGR (Rockville/MD, USA) and being continued with the new company Craig Venter founded, Celera (also in Rockville).

Recent advances in genomics are bringing about a revolution in our understanding of the molecular mechanisms of disease, including the complex interplay of genetic and environmental factors. Genomics automatically leads to the question of the transcripts of those genes sequenced, which is named, as the summary of all transcribed genes, the "transcriptome". Not every transcriptome is identical to the proteome, a reason why scientists have to distinguish between the different steps a cell can go through, starting with the genome, which leads to selection of the transcriptome, which again leads to selection of the possible proteome.

Genomics has provided a vast amount of information linking gene activity with disease. However, some expressed genes lead to several different protein species, depending on splicing forms and cell-specific expression patterns. Genes cannot provide a complete and accurate profile of a protein's abundance or its final structure and state of activity.

After transcription from DNA to RNA in eukaryotic cells, the gene transcript can be spliced (eukaryotic genes consist of several exons, varying in length, divided by introns, which need to be spliced out) in different ways prior to translation into protein. Following translation, most proteins are chemically altered through post-translational modification, such as phosphorylation, proteolytic cleavage (both also occurring in prokaryotes), glycosylation, or lipid anchoring. Such modification plays a vital role in modulating the function of many proteins but is not directly coded by genes. As a consequence, the information from a single gene can encode as many as 50 different protein species. While awaiting the completion of the Human Genome Sequencing Project (it was 95.8% finished in January 2003), estimates of the number of genes in the human range between 35 000 and 100 000 (Pennington, 2001), which can lead up to 500 000 individual protein moieties due to modification and splicing (Pennington, 2001). As a consequence, the proteome is far more complex than the genome.

15.2.2
Proteomics

Proteomics is a scientific discipline which analyzes and documents proteins in biological samples as a whole. Among its aims are to establish the protein distribution pattern in relation to cell type or phenotype, and to characterize individual proteins of interest and their functional role and relationships with other proteins. Proteomics is regarded as a fairly open approach to specify known, altered, or new proteins of unknown function and it holds the potential to select for successful pharmaceuticals early in the drug discovery process, therefore saving significant resources for clinical studies (Koepke, 2001). It enables correlations to be drawn between the range of proteins produced by a cell or tissue and the initiation or progression of a disease state. Proteome research also permits the discovery of new protein markers for diagnostic purposes and of novel molecular targets for drug discovery.

The abundance of information provided by proteome research is entirely complementary to the genetic information being generated by genomic research. Proteomics will make a key contribution to the development of functional genomics

(Figure 15.2). The combination of proteomics and genomics will play a major role in biomedical research and will have a significant impact on the development of the diagnostic and therapeutic products of the future.

With the availability of DNA microarray analysis, permitting the expression of thousands of genes to be monitored simultaneously, one may ask why proteomics is so important. First, DNA or protein arrays can only reflect information covered on the chip. Secondly, the importance of the proteome cannot be overstated as it is the proteins within the cell that provide structure, produce energy, and allow communication, movement, and reproduction. Basically, proteins provide structural and functional frameworks for cellular life. Genetic information is static while the protein complement of a cell is dynamic. Proteomics got a huge push through introduction of two crucial changes: first, the improvements in protein mass analysis, and second, the growth of protein databases, especially for biomedical research the expressed sequence tag (EST) databases, though these are still incompletely covered (Blackstock, 1999).

Among several different descriptions of the nature and purpose of proteomics, one of the most useful is the separation into *expression proteomics* and *cell-map proteomics* (Blackstock. 1999):

- *expression proteomics* relies on 2D-PAGE or other orthogonal separation methods to study cellular pathways under different conditions, which is especially useful for disease-marker discovery or drug-target evaluation;
- *cell-map proteomics* determines the subcellular location of proteins and their protein–protein interactions or protein-complex identification.

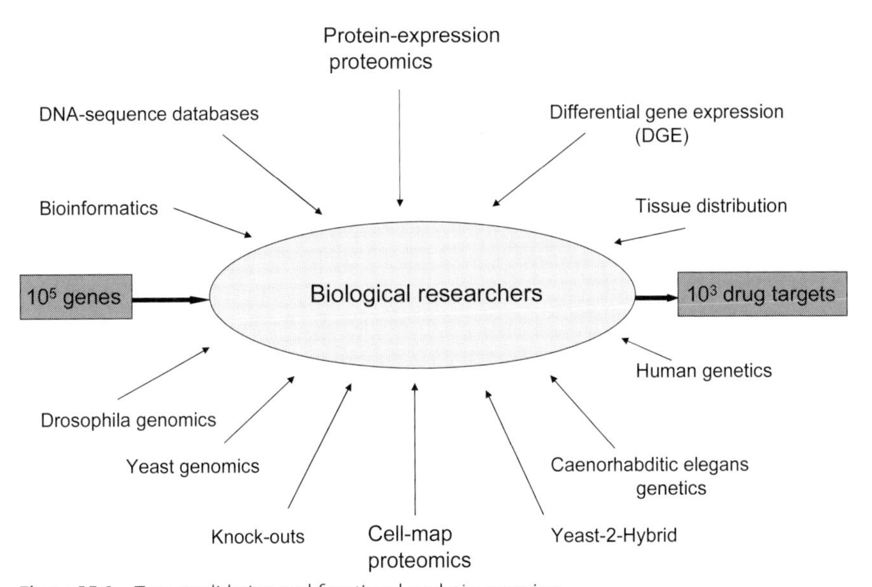

Figure 15.2 Target validation and functional analysis: overview of the methods currently available to biological researchers. Modified from Blackstock (1999).

15.3
Technologies for Systems Biology

The practice of protein analysis of whole proteomes relies on several techniques, in particular (i) 2D gel electrophoresis for separation; (ii) mass spectrometry for analysis; and (iii) protein arrays for achieving massively parallel analysis. Chapters 8 and 9 provide the basis of these techniques, each of which we discuss below.

15.3.1
Two-Dimensional Gel Electrophoresis (2D PAGE)

The most established and accepted method for separation of proteins from each other is two-dimensional gel electrophoresis (two-dimensional polyacrylamide gel electrophoresis or 2D PAGE). 2D PAGE can achieve the separation of several thousand different proteins in one gel. High-resolution 2D PAGE can resolve up to 10 000 protein spots per gel. Staining with dyes such as Coomassie Blue, silver, and SYPRO Ruby Red can be employed to visualize the proteins. 2D PAGE separates proteins according to their charge in the first dimension, followed by a second dimension where the proteins are separated on the basis of their molecular weights (classical discontinuous PAGE, after Laemmli, 1970) (remember 1D PAGE in Chapter 9, Section 9.3.3).

Preceding versions of this technology developed techniques for the first dimension, such as coupling IEF (isoelectrical focusing), that differed from the already established SDS-PAGE (Klose, 1975). Based on these developments, an alternative method was published for the first dimension, the NEPHGE technique, non-equilibrium pH gradient electrophoresis, (O'Farrell, 1977). Separation occurs on the basis of protein mobility [see Chapter 9, Eq. (16)] in the presence of a rapidly forming pH gradient using polyacrylamide gels with 8 M urea and 2% Nonidet P40, a non-ionic surfactant.

However, a challenge of IEF is its reproducibility because the ampholytes employed for creating the pH gradient are manufactured by a complex synthetic process which is difficult to control (Goerg, 2000); therefore, the ampholytes are not fixed within the gel, which leads to a phenomenon called the cathodic drift: during IEF, the build-up of electroendoosmotic flow of water leads the ampholytes to migrate towards the cathode, resulting in pH gradient instability (Westermeier, 1990). This problem was solved with the development of immobilized pH gradients, IPG, based on the use of immobilines (Amersham–Pharmacia Biotech): polyacrylamide IEF gels are poured using a series of eight immobilines (CH_2=CH-CO-NH-R) creating a pH gradient, the widest between pH 3 and 10. During polymerization the buffering groups of the immobilines are covalently attached via vinyl groups to the polyacrylamide matrix and are therefore immune to endoosmosis during IEF. IPG-IEF is the current method for the first dimension of 2DE (Goerg, 2000). For the second dimension, urea and glycerol are added to the solution though the main buffer system is still Tris buffer, pH 8.8, containing SDS. After equilibration, the

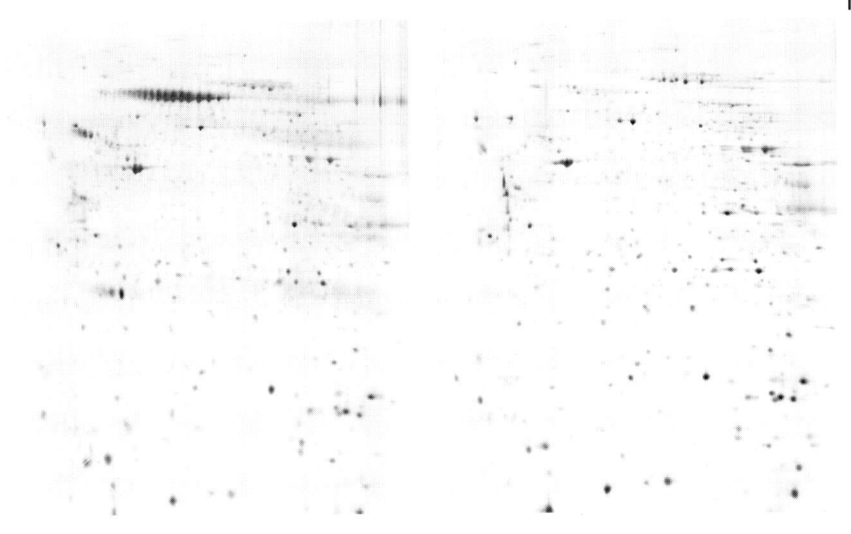

Figure 15.3 Comparison of two different sets of cell lysates, control (left) and disease state (right). Taken from: 2D PAGE training gels deposited on the ExPASy website.

samples are placed on the stacking gel of an SDS-PAGE apparatus to obtain the measurement.

High-resolution analytical 2D PAGE separates between 25 and 100 μg of total protein (equivalent to 10^5 cells or 1 mg of tissue) to ≤ 3 mg of protein (equivalent to 10^6 cells or 10 mg of tissue) for a preparative gel. In summary, the IPG technology offers high reproducibility due to standardized protocols and ready-to-use gels that are commercially available, whereas the NEPHGE technology offers higher resolution at the cost of being more difficult and labor-intensive (Koepke, 2001). Examples of 2D PAGE are given in Figure 15.3 and in the Inlet Figure at the end of this chapter.

Alternatives to two-dimensional electrophoresis are currently under development: in these approaches, liquid chromatography or capillary electrophoresis is used for easier automation and capability of high-throughput application.

15.3.1.1 **Separation by Chromatography or Capillary Electrophoresis**

Link et al. have used a two-dimensional chromatographic separation approach to characterize yeast ribosome complex proteins (Link, 1999). This technique employs a cation exchange column (SCX) for the first separation and, subsequently, two parallel reversed-phase HPLC columns (Figure 15.4), and thus works extremely rapidly and efficiently. While the first column loads, the second elutes using an acetonitrile gradient. The flow from the column is directed to parallel online MS detectors as well as to offline fraction collection with UV detectors.

The ribosomes, which after all are proteins, were proteolytically digested and applied to a cationic exchange chromatography column in the first dimension (an

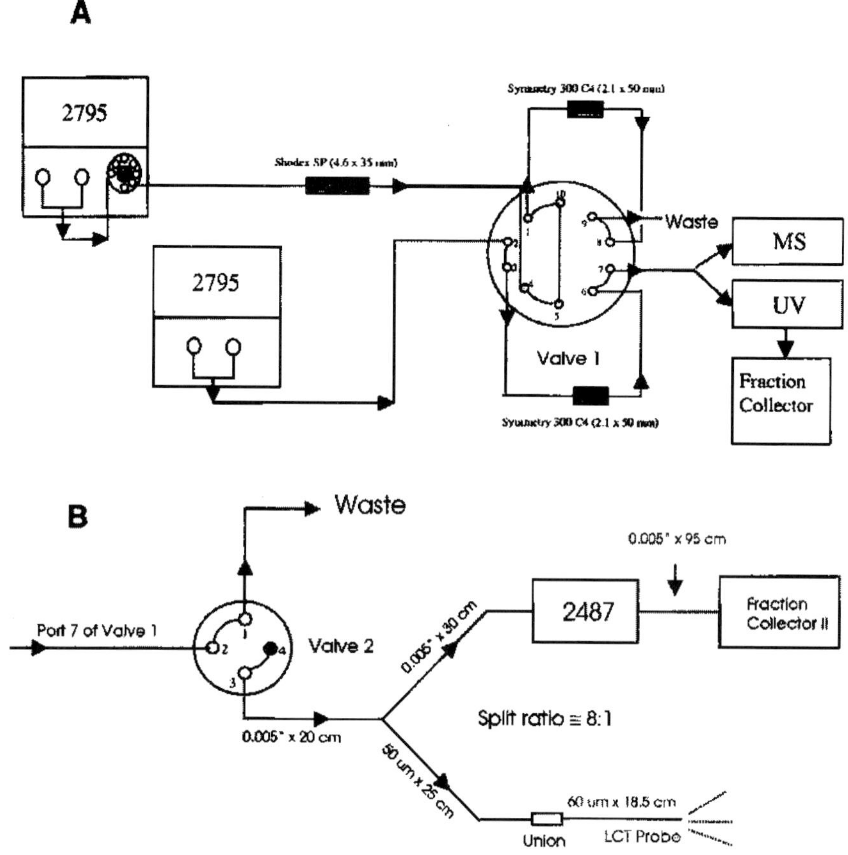

Figure 15.4 Schematic overview of the two-dimensional separation system (A) and a detailed view of the post-column configuration (B) according to Link (1999).

SCX column on HPLC), the eluates were directly transferred onto a reversed-phase HPLC column, and subsequently the eluates from the second column were validated by tandem MS. Of 78 existing proteins, 75 were correctly identified. The analysis of an enriched yeast ribosomal fraction identified over 80% of the subunits in the complex (Lui, 2002). According to the described method of validation and optimization, the potential of this method lies within practical application due to automation and up-scaling.

15.3.1.2 **Separation by Chemical Tagging**

For proteins of low abundance as well as membrane proteins, 2D PAGE suffers from a lack of resolving power. Another possibility, besides chromatography, to circumvent the limitations of 2D PAGE is to use chemically labeled tags, such as two different isotopes (ICAT, isotope-coded affinity tag) (Adam, 2002). A suitable tag consists of at least three elements: (i) a reactive group, such as iodacetamide for cysteine side chains, capable of labeling the protein at a specific site; (ii) an isotope-coded linker; and (iii) a tag, such as biotin, for affinity purification and isolation of the target. ICATs can be used in solution as well as on solid phases such as tagged beads. Cells are grown in or exposed to the desired environment, then lysed, and labeled with two ICAT probes, one with a light isotope, the other with a heavy one. Subsequently, the two cell lysates are combined and proteolyzed, for instance with trypsin. After proteolytic digestion, the ICAT-labeled peptides are affinity-purified according to their tags (for biotin, streptavidin beads would be used). After elution from the column, the samples can be analyzed by LC-MS (liquid chromatography mass spectrometry). For quantification, the ratio of signal intensities of differently labeled peptide pairs (heavy and light isotopes) is compared and the relative levels of the corresponding proteins in the samples are determined. This method is believed to be capable of determining quantities of low-abundance proteins such as phosphorylated kinases within cell samples, using phosphoramidate modification as the ICAT labeling (Zhou, 2001).

15.3.2 Mass Spectroscopy

Mass spectroscopy (MS) methods can be used to analyze complex mixtures of proteins and peptides in minutes and with mass accuracy several orders of magnitude better than that obtained from electrophoretic methods. For proteins with a known sequence, the mass measurement accuracies are generally sufficient to identify the products resistant to proteolysis precisely and to define compact domains within proteins effectively.

The history of MS began with Sir J. J. Thomson of the Cavendish Laboratory, University of Cambridge (Nobel prize for physics, 1906), whose studies on electrical discharges in gases led to the discovery of the electron in 1897. In the first decade of the 20th century, Thomson went on to construct the first mass spectrometer (then called a parabola spectrograph) for the determination of the mass-to-charge ratios of ions. Ions generated in discharge tubes were passed into electric and magnetic fields, which made them move through parabolic trajectories. The rays were then detected on a fluorescent screen or photographic plate. Thomson's protégé, Francis W. Aston of the University of Cambridge (Nobel prize for chemistry, 1922), designed a mass spectrometer in which ions were dispersed by mass and focused by velocity, improving MS resolving power by an order of magnitude. Around 1920, A. J. Dempster (University of Chicago) developed a magnetic deflection instrument with direction focusing as well as the first electron impact source, which ionized volatilized molecules with a beam of electrons from a hot wire filament. Direction focusing and electron impact ion sources are still very widely used in modern mass spectrometers today. (Adapted from: http://masspec.scripps.edu/information/history/tandem.html.) Figure 15.5 depicts a timeline of important steps in the progress of mass spectroscopy.

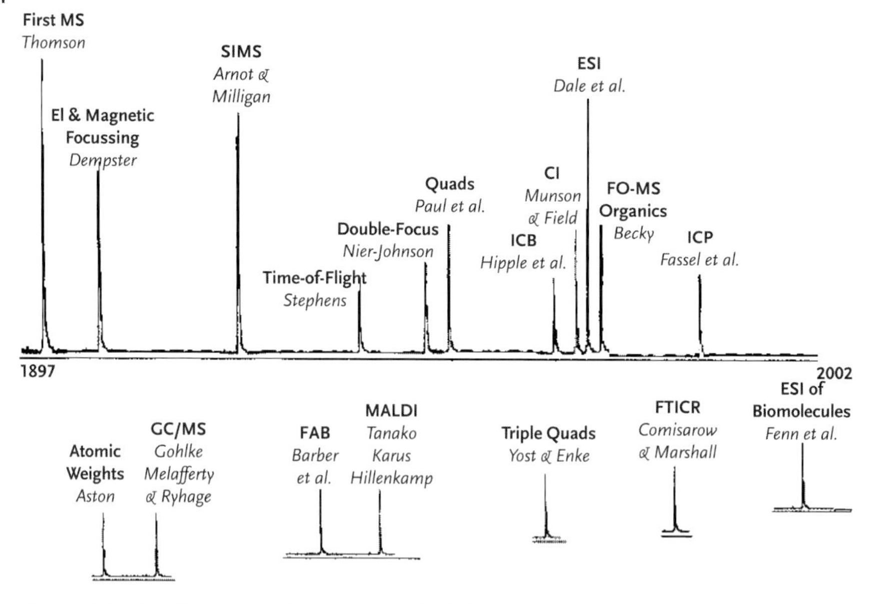

Figure 15.5 Timeline of crucial inventions in the field of mass spectrometry, taken from: http://masspec.scripps.edu/information/history/tandem.html.

Table 15.1 Useful internet sites for mass spectrometric protein identification.

Resource	*WWW Uniform Resource Locator (http://...)*	*Features*
EMBL, Heidelberg, Germany	www.mann.embl-heidelberg.de/Services/PeptideSearch/PeptideSearchIntro.html	peptide mass and fragment ion search programs
ExPASy	www.expasy.ch/tools/proteome	peptide mass and fragment ion search programs
Mascot	www.matrix-science.com/cgi/index.pl?page=/home.html	peptide mass and fragment ion search programs
Rockefeller University, NY	prwol.rockefeller.edu	peptide mass and fragment ion search programs
UCSF	Prospector.ucsf.edu	peptide mass and fragment ion search programs
University of Washington, DC	Thompson.mbt.washington.edu/sequest	instruction on how to get SEQUEST(fragment ion search program)

Taken from: Pennington (2001).

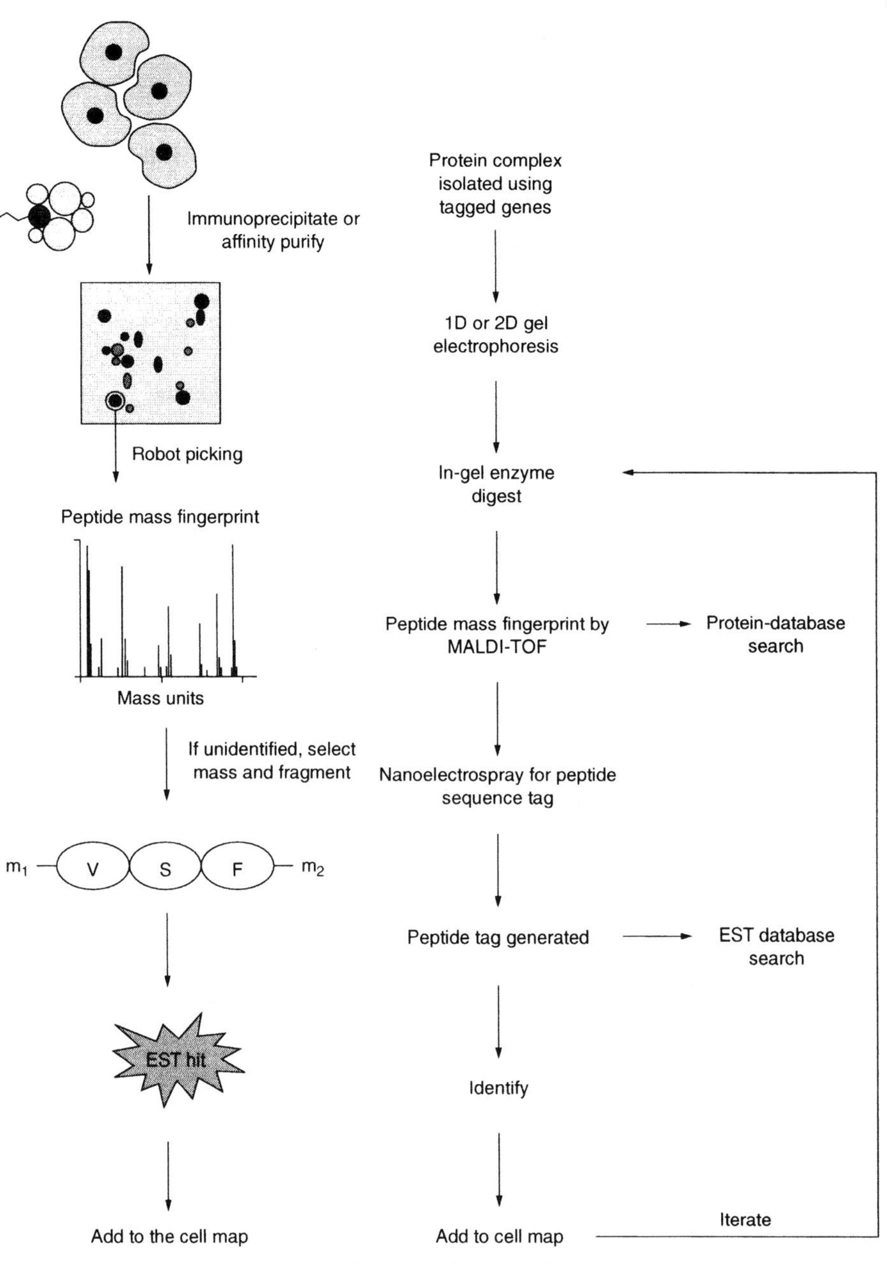

Figure 15.6 Schematic depiction of one example of protein identification by mass spectrometry. Genes of interest are tagged, then transfected into mammalian cells, and proteins associated with the cognate tagged protein are purified by affinity methods. Separation of the complex is carried out by 2D SDS-PAGE. Identification of the proteins is by MALDI and ESI (Blackwell, 1999).

Among the parameters determining the identification of a specific protein, mass determination is particularly attractive because of its accuracy, speed, and sensitivity. As MS measures mass-to-charge ratios (m/z) of ions under vacuum, conventional MS is limited to those molecules that can be ionized and transferred to vacuum. For biological material, soft ionization methods such as electrospray ionization MS, ESI-MS, and matrix-assisted laser desorption and ionization MS, MALDI-MS, are most relevant because of less fragmentation of the corresponding ions and therefore fewer degradation products that hinder identification. Mass spectrometers contain three key components: the ionization source, the mass analyzer, and the detector. Both single- and double-focusing MS instruments use high-power ionization which, irrespective of their extremely high resolution power compared to the newer developments, causes excessive degradation of large ions. The most important variants of the ionization source and analyzer will be discussed within the three types of commercially available instruments common nowadays, the MALDI-TOF, ESI-triple-quadrupole and ESI ion trap. For sources of more information, see Table 15.1.

An example of protein identification by MS is shown in Figure 15.6.

15.3.2.1 MALDI-TOF-MS (Matrix-Assisted Laser Desorption/Ionization Time-of-Flight MS)

MALDI refers to the ionization source and TOF to the type of analyzer used. Ions are produced by MALDI and their m/z ratios are measured in the analyzer. Accuracy always depends on the mass/charge (m/z) ratio, ranging from 0.01 to 0.05% for MALDI and 0.01 to 0.001% for ESI-MS. MALDI is also the instrument of choice when biological buffers are used as solvents. Samples are mixed with the matrices (small aromatic molecules) and placed on a metallic support. Under vacuum a high voltage is applied to the matrix to generate ions and a short laser pulse is directed towards the matrix. Laser energy is absorbed, resulting in rapid heating and thus in sublimation of the sample and expansion of molecules into the gas phase of the MS where ionization occurs. The ions are directed towards the analyzer and arrive with the same kinetic energy. Their velocity is inversely correlated with their m/z-ratio (Patterson, 2001). In the TOF analyzer, ions are separated by velocity differences as they move in a straight path towards a collector in order of increasing mass/charge ratio. TOF MS is fast, it is applicable to chromatographic detection, and it is now the method of choice for the determination of large biomolecules.

15.3.2.2 ESI-triple-quadrupole MS

In ESI, the analyte-containing solution is pressed through a microfine needle at high voltage, causing the solution to disperse into a fine spray of highly charged droplets. The spray is directed across a small orifice at a lower voltage than the initial application. The orifice is the interface between the ion source at atmospheric pressure and the MS under vacuum. The droplets are carried on a flow of inert gas which causes them to desolvate, resulting in an increase in the charge density of the ions. The increase causes a Coulombic explosion, transferring the ions into the gas phase (Patterson, 2001).

In a quadrupole device, not as accurate and precise as double-focusing instruments but fast, a quadrupolar electrical field comprising radio-frequency (RF) and direct-current components is used to separate ions. Quadrupole instruments as mass analyzers are used together with ESI as the ion source; the configuration employing a three-dimensional quadrupolar RF electric field (Wolfgang Paul, University of Bonn, 1989 Nobel prize for physics) is referred to as an ion trap analyzer (see below).

A triple-quadrupole MS consists of three sets of quadrupole analyzers. As ions drift through the space between quadrupoles, pairs of rods of opposite polarity, the voltage can be varied so that only certain ions of specific m/z-ratios pass through the filter, whereas others are diverted. The applied voltage can be varied over time (scanned) and the number of ions exiting the filter can be analyzed as a function of a selected m/z value. In a triple-quadrupole MS, the second quadrupole is run in RF (transmitting all the ions present in the mixture) mode only, without preselection of m/z ratios, to induce fragmentation of ions by collisions with inert gas (CID, collision-induced dissociation).

ESI-triple-quadrupole MS is one example of the often used term tandem MS (MS-MS), where by definition a precursor ion is mass-selected and typically fragmented by CID, followed by mass analysis of the resulting product ions. The technique requires two mass analyzers (in the case above, three are used) in series (or a single mass analyzer that can be used sequentially) to analyze the precursor and product ions. Tandem MS provides structural information by establishing relationships between precursor ions and their fragmentation products.

Usually, owing to insolubility or unacceptable buffers, large protein fragments are not beneficially analyzed by ESI-MS. However, the identity of stable domains of protein molecules can still be inferred from the accurate mass measurements of the smaller polypeptide fragments obtained from limited proteolysis. This provides a basis for resolving the compact domains for proteins in situations where matrix-assisted laser desorption ionization (MALDI) data are ambiguous and ESI-MS of the large protein may be precluded.

15.3.2.3 **ESI-MS Using an Ion Trap Analyzer**

ESI-MS is also a tandem MS technique, in which ions can be accumulated and stored prior to analysis. The ion trap is not only the mass analyzer but also the collision cell. RF-only mode quadrupoles are used to guide ions from the ESI source into the ion trap, which consists of two types of electrodes, a ring electrode and two end-cap electrodes. Ions enter and leave the trap through holes in the end-cap electrodes and are trapped in the electrical field created by the ring electrode. The magnitude of the applied voltage to create the field determines the frequency and motion of the trapped ions. A linear increase in the RF voltage in the ring electrode versus small voltages across the end-cap electrodes generates the mass spectrum, causing ions to become unstable in the trap and to be ejected axially through the end-cap electrodes towards the detector. Through this process, ions of specific m/z-ratios can be isolated in the trap whereas all other ions of higher or lower m/z-ratios are ejected. After isolation, the remaining ions are fragmented through CID and then scanned with the quadrupole analyzer as described above.

15.3.3
DNA Microarrays

Alterations either in a DNA sequence (mutation) or in gene expression patterns can have profound effects on biological function. Variations in gene expression are probably the main reason for altered processes in the cell and pathological diseases. DNA array technologies provide rapid and cost-effective methods of sampling, testing, and identifying patterns of gene expression and genetic variations or mutations (genotyping). DNA microarrays typically consist of thousands of immobilized DNA sequences present on a miniaturized surface the size of a business card or less. Microarrays are distinguished from macroarrays by their smaller DNA spot size, allowing the presence of thousands of DNA sequences instead of the hundreds on macroarrays.

How DNA are microarrays produced

A fully automated system uses either cDNA fragments of known genes or libraries of oligonucleotides, which are then spotted in duplicate onto hybridization filters (nylon membranes or PVDF) with a grid for guidance. When nylon membranes are used, these filters in microarrays can be used several times. There are two possible ways of applying the DNA to the filter (Cahill, 2001):

- needle or pin spotting, where the liquid is delivered through stainless steel pins;
- piezo-ink-jet method, which sprays miniature droplets onto the surface without contact.

Well-developed PCR techniques are of key importance for DNA arrays. Commercially available, fully automated cycling devices can handle up to 1536 samples in parallel (Cahill, 2001).

Two possible DNA sources can be applied as samples on the array:
- cDNA, for testing gene expression, obtained through reverse transcriptase PCR from the corresponding mRNA from cells or tissues of interest;
- genomic DNA, for the purpose of genotyping, compared to the random oligonucleotides on the array.

Using DNA–DNA hybridization, the single-stranded DNA will stick to a corresponding match on the chip; prior radioactive or, more commonly today, fluorescent labeling allows precise recognition of position and sequence. Quantification of the signal enables not only detection of the presence of a certain DNA sequence, but also determination of its relative abundance compared to controls.

15.3.4 Protein Microarrays

In protein microarrays, proteins are detected through the use of antibodies or specifically binding peptides that are immobilized on a small surface. Protein arrays may be engineered for a particular experiment (Figure 15.7). Thus, an array might contain all the combinatorial variants of a bioactive peptide or all the splicing variants of a single protein (Emili, 2000). There exist living and non-living arrays (see Table 15.2). Non-living arrays, using synthetic peptides or affinity-purified proteins, are the more common (Emili, 2000).

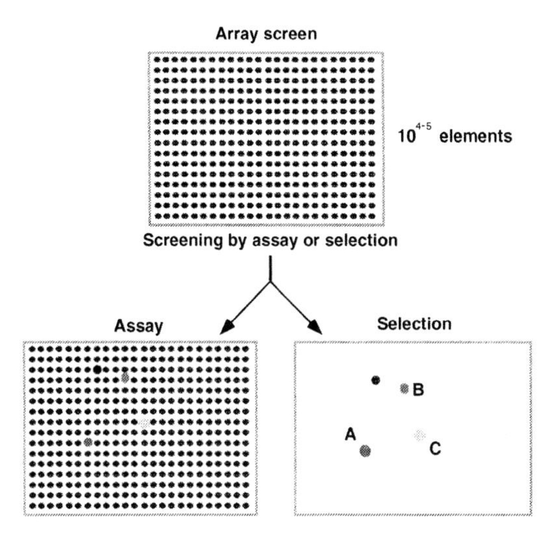

Figure 15.7 Example of a protein array setup. Elements can be screened either using a specific assay or through selection for certain characteristics. Arrays can also be screened in parallel, comparing two different sets of protein expressions of one specimen under two different conditions (Emili, 2000).

Table 15.2 Reported applications of peptide and protein arrays.

Array	*Applications*
Non-living arrays	
Synthetic peptides on plastic	protein interactions, epitope mapping
Synthetic peptides on paper	protein interactions, epitope mapping, and amino acid specificities of glycation reactions
Affinity-purified proteins	biochemical activities
Living arrays	
Yeast colonies expressing GAL4 fusion proteins	protein–protein and protein–DNA/RNA interactions
Yeast colonies with transposon disruptions	phenotype analysis

Antibodies against synthetic peptides as well as specifically binding peptides can be produced inexpensively and in large amounts through phage display (Koepke, 2001). However, unlike the situation with DNA microarrays, where only one molecular structure is involved so that all genes have the same properties, it is difficult to achieve the same binding conditions for all antibodies or peptides on one chip, so protein microarrays can be labor- and maintenance-intensive. Proteins are much too divergent in their properties to feature consistent protocols for antibody or peptide binding. Identification of protein changes in a cell will be limited to known proteins with pre-characterized properties, and used for establishing the binding protocol. Moreover, since several steps between translation and mature protein occur, it will be difficult to detect post-translational modifications or splicing forms, since these variations are not revealed by the synthetic peptides.

15.3.5
Applications of Genomics and Proteomics in Biocatalysis

At the time of writing this book, publications regarding applications of genomics and proteomics in biocatalysis are still exceedingly rare. However, one particular pertinent example is discussed below, and many more examples are certain to be published in the next few years, and the field of *metabolic engineering*, which has already achieved a number of successes and is discussed in the next section, is very likely to use more and more genomics and proteomics methods to achieve its future goals.

15.3.5.1 Lactic Acid Bacteria and Proteomics

Lactic acid bacteria are of extreme importance in the food industry (just consider all the milk fermentations), but also in the field of investigating human pathogens. Lactic acid bacteria are a bacterial group encompassing species of the genera *Carnobacterium, Enterococcus, Leuconostoc, Pediococcus, Lactobacillus, Lactococcus,* and *Streptococcus.* Some of those species are gaining importance in biocatalysis as certain enzymes can produce a set of interesting chiral building blocks.

Proteomics has been employed on lactic acid bacteria to obtain hints on what kinds of genes are induced or repressed under certain stress responses compared to the "normal" state of the cell (Champomier-Verges, 2002). Protein patterns of a given strain were compared after being exposed to different environmental conditions. The results were based on 2D PAGE and either Edman degradation of the spots or MS. Different stress conditions such as heat, acid, alkali, cold, oxidation, salt or starvation, among others, were tested. Some proteins, such as the heat-shock proteins GroEL and GroES, also termed chaperones, were found to be induced under almost every stress condition; others responded only to certain stress conditions, e.g., the *Csp* genes, coding for cold-shock proteins, under cold shock.

Analysis of a whole group of related species under industrially important conditions to identify potential target genes for metabolic engineering or isolation is promising, although proteomics can only be as good as the underlying genomics. Only a few genomes of lactic acid bacteria, such as the genome of *Lactococcus lactis,* have been sequenced so far, so only a little reliable information for protein identification is deposited in the databases. It is therefore important not only to improve proteomic tools but also enhance the speed of genomic sequencing and develop even more sophisticated bioinformatics.

15.4 Metabolic Engineering

According to one definition, metabolic engineering is the directed improvement of product formation or cellular properties through the modification of specific biochemical reactions or introduction of new characteristics through recombinant DNA technology (Stephanopoulos, 1999). Metabolic engineering allows directed improvement of cellular processes by modifying specific intracellular biochemical reactions or by introducing new genes through recombinant DNA technology. It is the rational design of metabolic pathways in a cell to achieve a certain goal, such as synthesizing a new product, increasing the yield of desired products, or reducing undesirable by-products. Metabolic engineering can be used for a broad range of applications, from the microbial production of fine chemicals such as amino acids to gene therapy in biomedical research. Although the main impetus for this field has come from recent advances in molecular biological techniques, metabolic engineering is an interdisciplinary field requiring coordinated inputs from biochemistry, molecular biology, cell physiology, and chemical engineering. The metabolic engineering goal of identifying genes that confer a particular phenotype is conceptually and methodically congruent with central issues in drug discovery and functional genomics (Bailey, 1999).

15.4.1 Concepts of Metabolic Engineering

In this context, only a brief outline of metabolic engineering can be given and the reader is referred to the "Suggested Further Reading" section at the end of this chapter.

Recently there have been discussions on whether single gene mutation or overexpression (heterologous or not) can be as efficient as more global, systems-oriented, approaches relying on mathematically modeled biochemical reaction networks (Bailey, 1999). One of the challenges in metabolic engineering for industrial applications is to coax the organism to produce large amounts of the desired product or to use the preferred substrate. A good example for a single gene cloning has been shown by Zaslavakaia et al. (Zaslavakaia, 2001). The photoautotroph microalga *Phaeodactylum tricornutum* was forced to use exogenous glucose as an energy source in the absence of light. The human glucose transporter gene (*glut1*) was heterologously expressed in this alga.

Despite several good examples of single gene alteration (either deletion, mutation, or homologous or heterologous overexpression) a shift from optimizing single enzyme steps to metabolic networks and systems biology is required, because kinetic control is distributed over the entire network, not only in a single enzymatic step (Sanford, 2002).

A good example is the effort to enhance ethanol production in yeast by overexpression of each enzyme in the pathway, but this has failed to increase the rate significantly (Bailey, 1999).

The increasing availability of knock-out animals, which more often than expected lack any phenotype that can be associated with the knock-out gene, gives rise to the realization that single events might not have any influence at all on many flux ratios (Bailey, 1999).

It is probably impossible to do any experiment in vivo that will change the level of only one protein or transcript; rather, such an experiment most often changes more than can be accessed by the phenotype.

Some of the most important concepts and tools of metabolic engineering include the following (Stephanopoulos, 1999):

- *Metabolic flux*: this is defined as the amount of the unit of interest, usually the mass of a metabolite in moles (often micromoles, rather) per unit time per unit area or volume, or often grams dry cell weight (dcw) [μmol (h g dcw)$^{-1}$] passing between components A and B of the metabolic system. A metabolic rate is equivalent dimensionally to a specific reaction rate. Metabolic flux is the basic unit of observation and modeling in metabolic engineering.

- *Metabolic pathway analysis*: A metabolic pathway is any sequence of feasible and observable biochemical reaction steps connecting a specified set of input and output metabolites. Unless a metabolite is observable *independently*, it is better lumped in with other metabolites and pathways.

- *Metabolic flux analysis*: Cellular metabolites and metabolic fluxes can be combined into a series of balance equations, not unlike a series of (bio)chemical reactions in a kinetic model. Metabolic flux analysis is the description of the components and their connections in a metabolic network.

- *Metabolic control analysis*: This produces a quantitative representation of the degree of flux control exercised by specific enzymes, metabolites, or effectors in the metabolic pathway.

Components are commonly represented as nodes in the metabolic network. These nodes can act as branch points if the number of input and output fluxes is not equivalent. Non-essential reactions around a node can be collected into reaction groups; the coefficients of their fluxes, in general termed the "metabolic flux coefficients" (in analogy to rate coefficients), can be rearranged as *group control coefficients*.

Methods of measuring cellular concentrations of metabolites include sensors (such as electrodes), tracers (such as radioactive isotope tracers) (Yanagimachi,

2001), off-gas analysis, and offline techniques such as spectroscopy and fluorimetry.

In all likelihood, the future in metabolic engineering will be driven by systems biology methods, involving functional genomics and proteomics, dissecting the gene–environment interactions that generate certain phenotype changes, and starting to improve rather than assessing the genetic alterations first and then searching for a phenotype. Kinetics and metabolic flux evaluation are essential in industrial applications in terms of optimizing yield and expenses, and new computational tools to achieve these goals have to be further developed (Bailey, 1999).

15.4.2
Examples of Metabolic Engineering

Example 1: *E. coli* pathway engineering: reduction of acetate production for enhanced recombinant protein production

Fermentation of recombinant *E. coli* regularly results in acetate accumulation owing to too little oxygen being available, coupled with high cell densities. The amount of acetate accumulated in the reactor increases to a level that can have a detrimental effect on both cell and protein yields, in contrast to the goal. In several attempts to remedy this situation, the glucose metabolic pathways have been manipulated to minimize acetate production (Yang, 1998). By supplying a glucose analog, which is taken up by but inhibits the phosphotransferase system (PTS), together with glucose, the inhibition resulted in increased recombinant protein production (Chou, 1994). The same results were achieved by mutating *ptsG* (the corresponding gene) which resulted in slower growth upon glucose feed and therefore in higher biomass and recombinant protein production due to lower acetate secretion.

Another approach targeted the heterologous expression of a *Bacillus subtilis* gene, acetolactate synthase (ALS) (see Chapter 7, Section 7.2.3.2), which converts pyruvate to acetoin (Aristidou, 1995). Pyruvate, as the central position of metabolic flux in glycolysis, is a good target for redirection of metabolic fluxes. Acetate levels in this recombinant strain were drastically reduced, and the by-product acetoin produced proved to be influencing yield and biomass production less negatively.

Example 2: Genome shuffling

What are the properties of an ideal cell that converts a substrate to a valuable product? Such a cell (i.e., biocatalyst) would grow rapidly to high density, then be sustained at high viability under no or minimal growth, utilize an inexpensive substrate, and secrete the product of interest at high rates, and – for maximal yield – make little or no other side product. In addition, an optimal cell would be able to grow and produce under conditions that are optimal for the overall process, including media preparation and downstream steps; such conditions may differ from those that are optimal for growth.

Two major advances describe the construction of mutants with an apparently significantly higher probability of improved phenotype. The method involves whole-genome shuffling – with protoplast fusion – of a small number of parental strains exhibiting subtle improvements in the phenotype of interest obtained by classical strain-improvement methods. The method was demonstrated on two cases of industrial importance: (i) the improvement of tylosin (a complex polyketide antibiotic) production by *Streptomyces fradiae* (Zhang, 2002), and (ii) the isolation of acid-tolerant strains of *Lactobacillus* suitable for the production of lactic acid in low-pH fermentations (Patnaik, 2002). In both cases, significant phenotypic improvements were obtained after only a few rounds of genome shuffling, at least comparable to those resulting from multi-year strain improvement efforts by mutation and screening/selection. Although random mutation and selection methods have succeeded in generating many industrial strains, today they must compete with recombinant technologies and direct genetic manipulation of strains through the introduction and control of specific genes.

However, the profile of an ideal cell depends on a multitude of genes that are rather poorly understood, mostly unknown, and broadly distributed throughout the genome. This raises serious obstacles in the direct application of genetic engineering for strain improvement and invites combinatorial approaches for the determination of the optimal genetic configuration in microbes for industrial applications. In combinatorial approaches, cells most useful with respect to the trait of interest are screened or selected from a genotypically diverse collection (or library) of candidates. Such libraries have been generated by random mutation of genes, successive rounds of mutation, or recombination of gene homologs from related strains combined with screening or selection of individual genes. This approach, termed directed evolution, has been described in detail in Chapter 11.

The success of combinatorial approaches depends on the initial selection of variants, the efficiency of creating genetic diversity, and the power of the selection method. Protoplast fusion seems to allow recombination events to occur throughout the genome, at much longer range relative to other methods. The method also allows simultaneous examination of several mutations by evaluating phenotypes, incorporating gene diversity, and improvement of several genes in one step. As such, this method signifies a reorientation of attention from a directed evolution of single enzymes to optimization of pathways. Still, however, the key feature of this technology is the formation of a library of candidate cells by shuffling the genomes of mutants with improved phenotypes. This apparently yields many improved cells; however, lack of a growth advantage associated with the desired phenotype dramatically reduces the probability of identifying the truly optimal cell. Another potential drawback could be the necessity to evaluate more than one phenotype: while acid tolerance in the lactobacilli mentioned above might be improved, other features might decline, such as cell growth or utilization of a wide range of nutrients.

Several important examples of metabolic engineering, ranging from applications in basic chemicals, such as the manufacture of propanediol from glucose, to the synthesis of chiral pharmaceutical intermediates, such as (2*R*)-indanediol, a building block of the HIV protease inhibitor Crixivan (Indinavir®, Merck; see Chapter 13, Section 13.3.3.30.), are presented in Chapter 20.

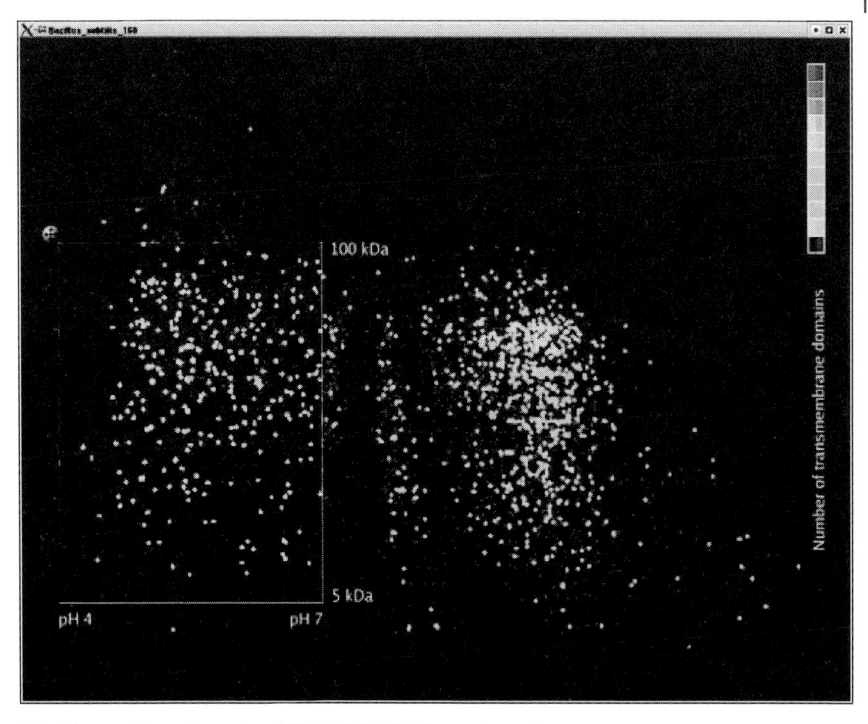

Inlet Figure Two-dimensional IPG-IEF PAGE electrophoresis, after modification for a specific feature, in this case transmembrane domains. From: http://www.mips.biochem.mpg.de/projects/proteomics.

Suggested Further Reading

S. R. Pennington and M. J. Dunn (eds.), *Proteomics: from Protein Sequence to Function*, BIOS Scientific Publishers, New York, **2001**.

G. Stephanopoulos, A. Aristidou, and J. Nielsen, *Metabolic Engineering: Principles and Methodologies*, Academic Press, New York, **1998**.

E. O. Voit, *Computational Analysis of Biochemical Systems: A Practical Guide for Biochemists and Molecular Biologists*, Cambridge University Press, Cambridge, **2000**.

References

G. C. Adam, E. J. Sorensen, and B. F. Cravatt, Chemical stratagies for functional proteomics, *Mol. Cell. Proteomics.* **2002**, *1*, 781–790.

R. Aebersold, L. E. Hood, and J. D. Watts, Equipping scientists for the new biology, *Nature Biotechnol.* **2000**, *18*, 359.

A. A. Aristidou, G. N. Bennett, and K. Y. San, Metabolic engineering of *Escherichia coli* to enhance recombinant protein production through acetate reduction, *Biotechnol. Progr.* **1995**, 375–478.

J. E. Bailey, Lessons from metabolic engineering functional genomics and drug discovery, *Nature Biotechnol.* **1999**, *17*, 616–618.

W. P. Blackstock and M. P. Weir, Proteomics: quantitative and physical mapping of cellular proteins, *TIPTECH* **1999**, *17*, 121–127.

B. C. Buckland, S. W. Drew, N. C. Conors, M. M. Chartrain, C. Lee, P. M. Salmon, K. Gbewonyo, W. Zhou, P. Gaillot, R. Singhvi, R. C. Olewinski, Jr., W.-J. Sun, J. Reddy, J. Zhang, B. A. Jackey, C. Taylor, K. E. Goklen, B. Junker, and R. L. Greasham, Microbial conversion of indene to indandiol: A key intermediate in the synthesis of CRIXIVAN, *Metabolic Eng.* **1999**, *1*, 63–74.

B. C. Buckland, D. K. Robinson, and M. Chartrain, Biocatalysis for pharmaceuticals – status and prospects for a key technology, *Metabolic Eng.* **2000**, *2*, 42–48.

D. J. Cahill, E. Nordhoff, J. O'Brien, J. Klose, H. Eickhoff and H. Lehrach, Bridging genomics and proteomics, in: *Proteomics, from Protein to Punction* (S. R. Pennington and M. J. Dunn), BIOS Scientific Publishers, pp. 1–22, **2001**.

M.-C. Champomier-Verges, E. Maguin, M.-Y. Mistou, P. Anglade, and J.-F. Chich, Lactic acid bacteria and proteomics: current knowledge and perspectives, *J. Chromatogr. B* **2002**, *771*, 329–342.

C.-H. Chou, G. N. Bennett, and K. Y. San, Effect of modified glucose uptake using genetic engineering techniques on high-level recombinant protein production in *Escherichia coli* dense cultures, *Biotechnol. Bioeng.* **1994**, *44*, 952–960.

A. Q. Emili and G. Cagney, Large-scale functional analysis using peptide or protein arrays, *Nature Biotechol.* **2000**, *18*, 393–397.

A. Goerg, C. Obermaier, G. Boguth, A. Harder, B. Scheibe, R. Wildgruber, and W. Weiss, The current state of two-dimensional electrophoresis with immobilized pH gradients, *Electrophoresis* **2000**, *21*, 1037–1053.

R. Kamimura, K. Konstantinov, and G. N. Stephanopoulos, Knowledge-based systems, artificial neural networks and pattern recognition: applications to biotechnological processes, *Curr. Opin. Biotechnol.* **1996**, *7*, 231–234.

R. D. Kiss and G. N. Stephanopoulos, Metabolic characterization of a L-lysine-producing strain by continuous culture, *Biotechnol. Bioeng.* **1992**, *39*, 565–574.

J. Klose, Protein mapping by combined isoelectric focusing and electrophoresis of mouse tissues. A novel approach to testing for induced point mutations in mammals, *Humangenetik* **1975**, *26*, 231–243.

A. Koepke, Proteomics – a new drug discovery tool, *Bioworld* **2001** (March), *3*, 14–15.

U. K. Laemmli, Cleavage of structural proteins during the assembly of the head of bacteriophage T_4, *Nature* **1970**, *227*, 680–685.

A. J. Link, J. Eng, D. M. Schieltz, E. Carmack, G. J. Mize, D. R. Morris, B. M. Garvik, and J. R. Yates, Direct analysis of protein complexes using mass spectrometry, *Nature Biotechnol.* **1999**, *17*, 676–682.

H. Liu , S. J. Berger, A. B. Chakrabort, R. S. Plumb, and S. A. Cohen, Multidimensional chromatography coupled to electrospray ionization time-of-flight mass spectrometry as an alternative to two-dimensional gels for the identification and analysis of complex mixtures of intact proteins, *J. Chromatogr. B* **2002**, *782*, 267–289.

P. H. O'Farell, High resolution two-dimensional electrophoresis of proteins, *J. Biol. Chem.* **1975**, *250*, 4007–4021.

R. Patnaik, S. Louie, V. Gavrilovic, K. Perry, W. P. Stemmer, C. M. Ryan, and S. Cardayre, Genome shuffling of *Lactobacillus* for improved acid tolerance, *Nat. Biotechnol.* **2002**, *20*, 707–712.

S. D. Patterson, R. Aebersold, and D. R. Goodlett, Mass spectrometry-based methods for protein identification and phosphorylation site analysis, in: *Proteomics, from Protein to Function* (eds: S. R. Pennington and M. J. Dunn), BIOS Scientific Publishers, pp. 97–124, **2001.**

S. R. Pennington and M. J. Dunn (eds.), *Proteomics: from protein sequence to function*, BIOS Scientific Publishers, New York, **2001**.

K. Sanford, P. Soucaille, G. Whited and G. Chotani, Genomics to fluxomics and physiomics – pathway engineering, *Curr. Opin. Microbiol.* **2002**, *5*, 318–322.

G. N. Stephanopoulos, G. Locher, M. J. Duff, R. Kamimura, and Ge. Stephanopoulos, Fermentation database mining by pattern recognition, *Biotechnol. Bioeng.* **1997**, *53*, 443–452.

G. N. Stephanopoulos, Metabolic fluxes and metabolic engineering, *Metabol Eng* **1999**, *1*, 1–11.

R. Westermeier, *Electrophorese Praktikum*, VCH, Weinheim, **1990**.

K. S. Yanagimachi, D. E. Stafford, A. F. Dexter, A. J. Sinskey, S. Drew, and G. N. Stephanopoulos, Application of radiolabeled tracers to biocatalytic flux analysis, *Eur. J. Biochem.* **2001**, *268*, 4950–4960.

Y.-T. Yang, G. N. Bennett, and K.-Y. San, Genetic and metabolic engineering, *J. Biotechnol.* **1998**, *1*, 134–141.

L. A. Zaslavakaia, J. C. Lippmeier, C. Shih, D. Ehrhardt, A. R. Grossman, and K. E. Apt, Trophic conversion of an obligate photoautotrophic organism through metabolic engineering, *Science* **2001**, *292*, 2073–2075.

Y. X. Zhang, K. Perry, V. A. Vinci, K. Powell, W. P. Stemmer, and S. B. del Cardayre, Genome shuffling leads to rapid phenotypic improvement in bacteria, *Nature* **2002**, *415*, 644–646.

H. Zhou, J. D. Watts, and R. Aebersold, A systematic approach to the analysis of protein phosphorylation, *Nat. Biotechnol.* **2001**, *19*, 375–378.

16
Evolution of Biocatalytic Function

Summary

This chapter attempts to shed some light on the question of the origin and the evolution of biocatalytic function. How does nature generate new enzyme functions? Is an existing protein mutated until the catalytic function is up to the right level (as in the generation and enhancement of activity in catalytic antibodies), or is an enzyme chosen with an already low level of activity which, by further mutation, is selected towards the target (in analogy to protein engineering)? Is it the binding or the chemical mechanism that is the decisive starting point for the development of novel enzymatic function?

Relatedness of proteins can be expressed by the terms *homologs* (enzymes derived from a common ancestor and thus related), *orthologs* (homologs in different species that catalyze the same reaction), *paralogs* (homologs in the same species that have diverged from each other), and *analogs* (enzymes catalyzing the same reaction but structurally unrelated). In nature, several possible mechanisms for the development of novel function have been observed: (i) dual-functionality proteins ("moonlighting proteins" and catalytic promiscuity); (ii) gene duplication; (iii) horizontal gene transfer; and (iv) circular permutation of structure.

There are three dominant principles of enzyme evolution:

1. *Retained substrate specificity (binding), changed chemistry*: A new enzyme evolves to supply substrate from an available precursor by evolution of enzyme using the substrate. The underlying hypothesis states that metabolic pathways evolve backwards: A $\rightarrow$ B $\rightarrow$ C, $E_{A \rightarrow B}$ is new, $E_{B \rightarrow C}$ is old enzyme;
2. *Retained chemistry, changed substrate specificity (binding)*: Nature selects protein from a pool of enzymes whose mechanism provide a partial reaction or stabilization strategy for intermediates or transition states. Evolution decreases the proficiency of the reaction catalyzed by the progenitor. The underlying hypothesis states that chemical mechanism dominance starts with a low level of promiscuous activity and that once evolved it is beneficial for nature to utilize it over and over again.
3. *Active-site architecture retained*: In this case, the active site is able to support alternative reactions with shared functional groups in a different mechanistic or metabolic context.The study of α/β-barrel proteins, also termed $(\beta\alpha)_8$-barrel

or TIM barrel proteins, is important because (i) the structure is found in 10% of all known structures, 68% of them enzymes; (ii) residues responsible for stability are separate from residues responsible for function; and (iii) the general scaffold catalyzes several classes of catalysis: oxidoreductases, hydrolases, lyases, and isomerases.

The mandelate and β-ketoadipate pathways serve as an example of gene duplication, as there is strong evidence pointing to the former evolving from the latter. Evolution of mandelate racemase from muconate lactonizing enzyme points to the relevance of the enzyme mechanism for catalytic reactivity.

The earliest currently discernible ancestor in the histidine biosynthesis pathway was a $(\beta\alpha)_4$-half-barrel protein which gene-duplicated into two initially identical half-barrels which fused to form an ancestral α/β-barrel protein. The second gene duplication step led to diversification into two enzymes with distinct catalytic activities.

16.1 Introduction

How does nature generate new enzyme functions? How is a protein selected which already binds a desired substrate? How is the protein mutated until the catalytic function is up to the right level (this procedure corresponds to the generation and enhancement of activity in catalytic antibodies), or is an enzyme chosen that by accident already features a low level of activity and that, by further mutation, is selected towards the target (in analogy to protein engineering)? Ultimately, is it binding or is it chemical mechanism that is the decisive starting point for the development of novel enzymatic function?

As already discussed in Chapter 11, there are more than 10 000 protein structures known but only about 30 3D structure types. This might be traced to a limited number of possible stable polypeptide structures but most probably reflects the evolutionary history of the diversity of proteins. There are structural motifs which repeat themselves in a multitude of enzymes which are otherwise neither structurally nor functionally related, such as TIM barrel proteins, four-helix bundle proteins, Rossmann folds, or α/β-folds of hydrolases (Figure 16.1).

Since comparatively few structural types are found among proteins it appears likely that the evolutionary history of an enzyme can be traced through its structure, even if mutations have clouded the primary structure during the course of evolutionary time.

As already demonstrated in Ch. 11, Section 11.1.3 (Figure 11.2), many different enzymes can evolve from a common protein structure. The enzymes in Figure 11.2 feature a highly conserved α/β-barrel structure, but neither the amino acid sequence nor the function of these proteins is constant over the timescale of evolution. Conversely, as mentioned in Section 11.1.3 as well, an enzyme catalyzing a certain activity can otherwise feature very different properties in different organisms, such as a melting point T_m for GAPDH of between 50 °C in the American lobster and 98 °C in an archaeum from a deep-sea vent. While the essential structural components of GAPDH are conserved, the detailed amino acid sequence is not.

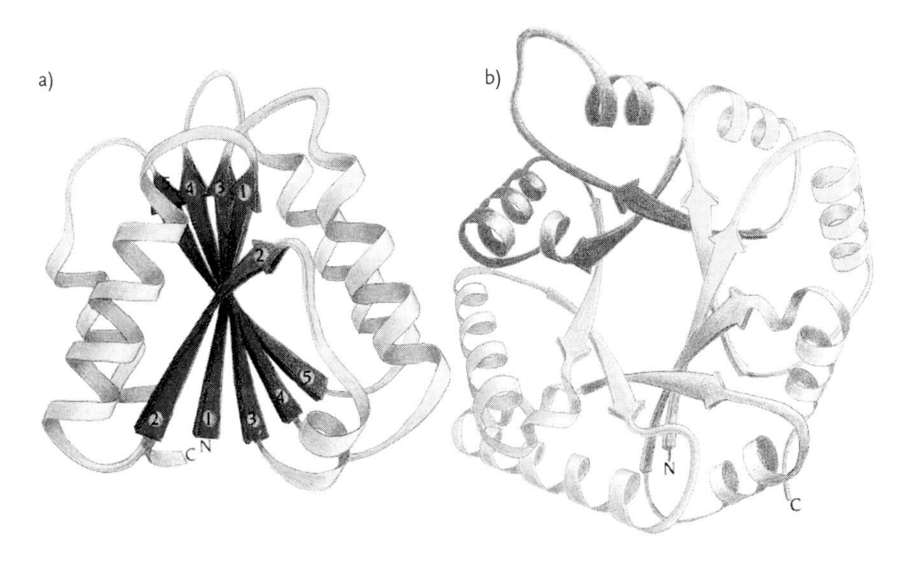

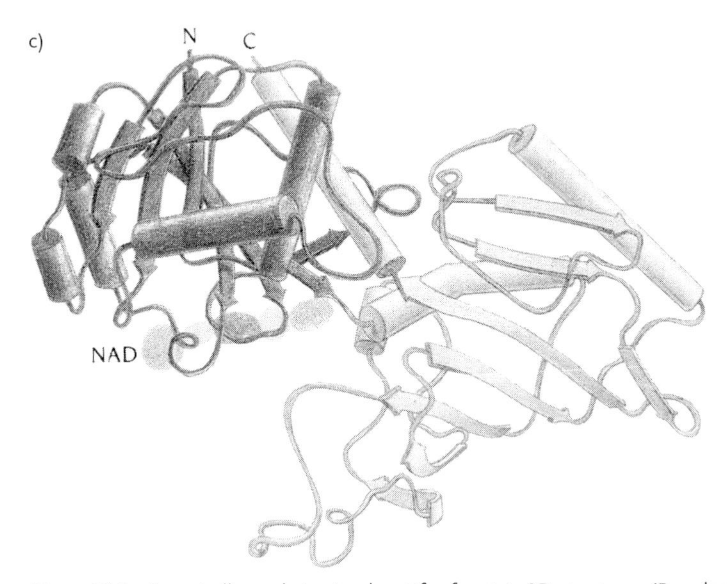

Figure 16.1 Repeatedly used structural motifs of protein 3D structures (Branden, 1999):
a) α/β–fold, for the redox protein flavodoxin;
b) closed α/β barrel structure, for triose phosphate isomerase;
c) Rossmann fold between substrate-binding domain (left) and dinucleotide-binding domain (right).

16.1.2
Congruence of Sequence, Function, Structure, and Mechanism

What determines protein function? Is it determined by the amino acid sequence of the protein? Or by three-dimensional structure? Or is the chemical mechanism most important? The probable answer is that protein function is not determined by any one factor alone but a combination of factors that, as yet, cannot be predicted well at all. Therefore, all the elucidated cases which are discussed in this chapter should not be treated as generalizable. The following examples illustrate the lack of general congruence between different factors impacting on protein function:

- *Lack of congruence of structure and mechanism*: Common structure does not imply a common mechanism! The β-barrel structures triosephosphate isomerase and xylose isomerase function by hydride transfer through enol, whereas aldose reductase performs hydride transfer through a metal ion.
- *Lack of congruence of sequence and structure with function*: Common sequence and structure, indeed identity of the protein itself, do not imply a unique function! Each pair, *o*-succinylbenzoate synthase (OSBS)–*N*-acetylamino acid racemase, and lens crystallin–lactate dehydrogenase, share sequence and structure but differ in function.
- *Lack of congruence of function and structure*: A common reaction catalyzed by two enzymes does not imply a common structure or sequence! Dehydrogenation is the reaction common to alcohol DH from horse liver (HLADH) and alcohol DH from yeast (YADH), but they have neither a common structure nor a common sequence.
- *Lack of congruence of structure and structure*: In several cases, no single structure of an enzyme exists. Several (at least two) structures can be verified during the catalytic cycle! An example is distinct lipases, with an open (hydrophilic) and with a closed lid (hydrophobic).

Especially interesting is the question of congruence with respect to sequence and structure: while it was accepted long ago that each protein sequence folds into a unique 3D structure, the implicit assumption was that similar sequences are congruent with similar structures. Indeed, for sequences with more than 30% identical residues, a similar structure is predicted (Sander, 1991). This high robustness of structures with respect to residue exchanges explains the scope available for variety in evolution. Surprisingly, most structurally similar proteins have less than 45% pairwise sequence identity, similar folds can often arise from protein pairs below 25–30% of pairwise sequence identity and, in one study, most pairs of structurally similar proteins even clustered around 8–9% sequence identity, a level expected for two random, unrelated sequences (Rost, 1997). In conclusion, most pairs of similar structures have sequence identity as low as randomly related sequences while on average only 3–4% of all residues are "anchor" residues crucial for maintaining the structure.

The position of protein function has undergone a radical change since the advent of conveniently accessible databases in the 1990s with a plethora of sequence (and increasingly, also structural) information. As Figure 16.2 illustrates, protein function before the advent of databases was the *starting point* for characterizing a protein in most cases. Information about function was available before sequence, mechanism, or biochemical details of the protein were known. With the available databases, assignment of protein function is now often the *endpoint* of the investigation, which often starts as the investigation of an annotated function (Chapter 14, Section 14.4) (Firestine, 1996). Certainly, both DNA and amino acid sequences are known before the assignment of (experimentally verified) function and, lastly, biochemical studies.

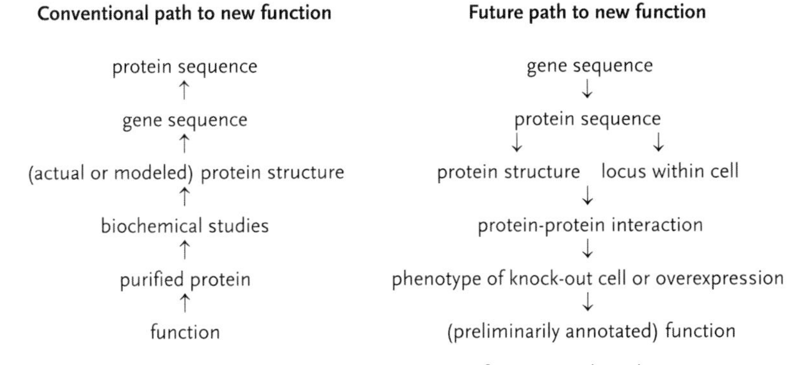

Figure 16.2 Congruence between sequence, structure, function, and mechanism.

16.2 Search Characteristics for Relatedness in Proteins

16.2.1 Classification of Relatedness of Proteins: the -log Family

The simplest measures for comparison of two proteins, the degree of sequence identity (identical amino acids in a certain position in the sequence) and similarity (similar amino acids in a certain position), do not specify the relationship between the two proteins. The following terms help to characterize such relationships; examples for each term are provided in Table 16.1:

- *Homologs*: enzymes which derive from a common ancestor and thus are related. A high degree of sequence homology allows identification from sequence databases, whereas highly divergent homologs with respect to sequence need additional proof from common 3D structures.
- *Orthologs*: homologs in different species that catalyze the same reaction.
- *Paralogs*: homologs in the same species that have diverged from each other.
- *Analogs*: enzymes catalyzing the same reaction but that are structurally unrelated.

Table 16.1 Examples of different relatedness in proteins.

Category	*Examples*	*References*
Homologs	subtilisin in *B. subtilis, S. carlsberg, B. liquifaciens* MLE and MR in *Pseudomonas putida* D-lactate DH and FDH in yeast	v.d. Osten, 1993 Hasson, 1998 Vinals, 1995
Orthologs	subtilisin in *B. subtilis, S. carlsberg, B. liquifaciens*	v.d. Osten, 1993
Paralogs	MLE and MR in *Pseudomonas putida*	Hasson, 1998
Analogs	HL-ADH and YADH	many

HL-ADH: horse liver alcohol dehydrogenase; Y-ADH: yeast alcohol dehydrogenase;
FDH: formate dehydrogenase; D-Lactate DH: D-lactate dehydrogenase;
MLE: muconate-lactonizing enzyme; MR: mandelate racemase.

The terms above imply the mode of relationship but not a timeline. Evolution can be thought of as divergent or convergent in time:

Divergent evolution refers to two or more species that originate from a common ancestor but that are becoming more and more dissimilar over time. Examples include:

- even-toed (with an even number of toes on each hoof) hoofed mammals. During the Eocene epoch, only one species of even-toed hoofed mammal existed. However, by the Pleistocene epoch, multiple species had diverged from this original species, such as the bison, pronghorn, chevrotain, peccary, wild boar, hippopotamus, giraffe, deer, and camel.

- many, possibly all, α/β-barrel or TIM barrel enzymes. This fold functions as a generic scaffold catalyzing 15 different enzymatic functions. The numbers in Figure 16.3 of the α/β-barrel fold indicate the different locations of the active site in four proteins that have this fold. These four proteins – xylose isomerase, aldose reductase, enolase, and adenosine deaminase – carry out very different enzymatic functions, in four of the main E.C. classes (1.–.–, 3.–.–, 4.–.–, and 5.–.–).

Convergent evolution occurs when species with different ancestors evolve to look similar. This happens as a result of adaptation to living in similar environments. Examples of convergent evolution include:

- the body shape of dolphins, sharks, and penguins; red tube-shaped flowers of plants pollinated by hummingbirds; the long bills of hummingbirds and sunbirds designed to reach the nectar at the base of flowers.
- the structures of two carbonic anhydrases with the same enzymatic function (E.C. 4.2.1.1; Chapter 9, Section 9.4) but with different folds.

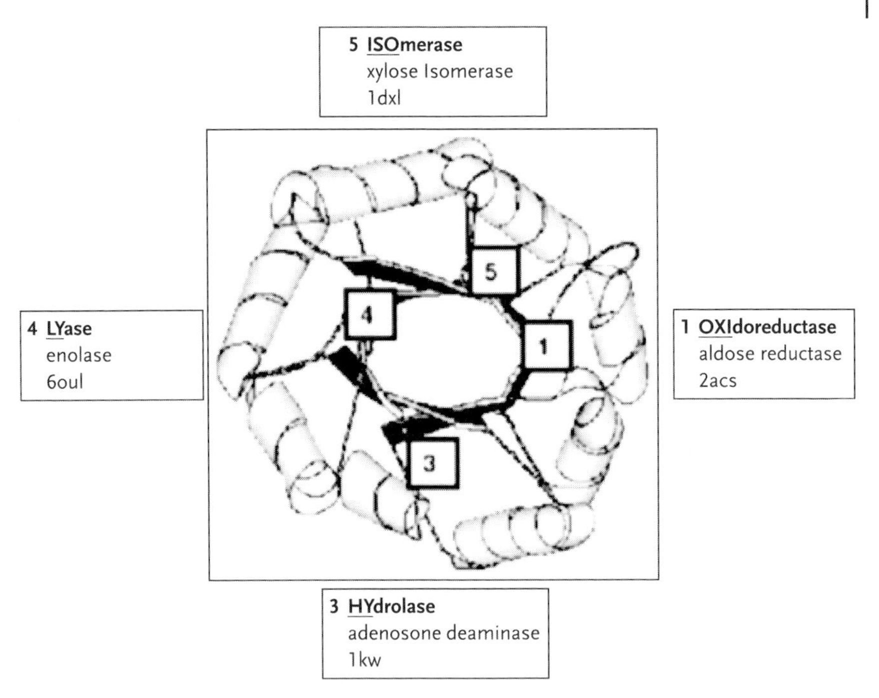

Figure 16.3 Specific example of divergent evolution: α/β-barrel enzymes (Hegyi, 1998).

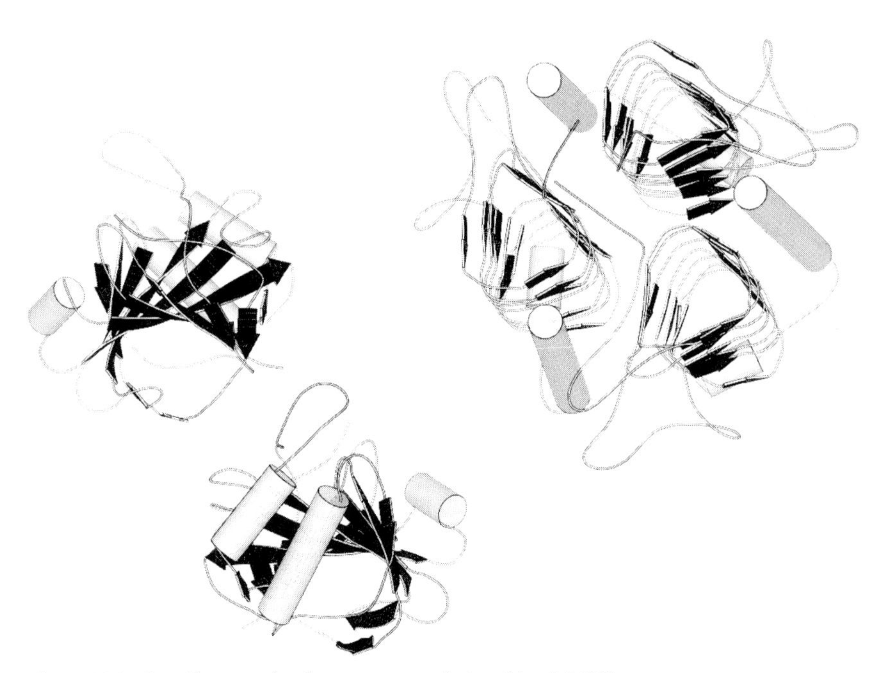

Figure 16.4 Specific example of convergent evolution (Hegyi, 1998).

16.2.2
Classification into Protein Families

The protein sequence and structure databases are now sufficiently representative for the strategies nature uses to evolve new catalytic functions to be identified. Another way to classify related enzymes concerns grouping into different levels of structural similarity:

- *Fold*: a common fold between proteins is indicated by the same major secondary structures in the same spatial arrangement and the same topological connection.
- *Families*: a group of homologous enzymes catalyzing the same reaction (mechanism and substrate specificity); there is often ≥ 30% sequence identity.
- *Superfamilies*: a group of homologous enzymes that catalyze either (i) the same reaction with differing substrate specificity, or (ii) different reactions with a shared mechanistic attribute (transition state, intermediate); there is often ≤ 20% sequence identity.
- *Suprafamilies*: a group of homologous enzymes that catalyze different overall reactions which do not share common mechanistic attributes and do not share a high level of sequence identity (≤ 20%) but are metabolically linked such as in successive reactions. Active-site residues may be conserved but perform different functions in family members.Tables 16.2 and 16.3 list known superfamilies and suprafamilies with some of their family members.

Two types of functionally distinct suprafamilies have been identified (Gerlt, 2001):
- suprafamilies whose members catalyze successive transformations in the tryptophan and histidine biosynthetic pathways, and
- suprafamilies whose members catalyze different reactions in different metabolic pathways.

Table 16.2 Known superfamilies and some of their members (Gerlt, 1999).

Superfamily	*Examples*	*References*
Enolases	mandelate racemase (MR), muconate-lactonizing enzyme (MLE), *N*-acetylamino acid racemase (NAAR)	Hasson, 1998; Palmer, 1999
Amidohydrolases	urease, amidase, carbamoylase, phosphotriesterase, chlorohydrolase	Holm, 1997
α/β-hydrolases	hydroxynitrile lyase (HNL), haloalkane dehalogenase	
crotonases	crotonase, enoyl-CoA hydratase, β-hydroxyisobutyryl-CoA hydrolase	Muller-Newen, 1995
vicinal-oxygen chelates (VOC)	glyoxalase I, dioxygenase	Babbitt, 1997
Fe-dependent oxidase/oxygenases	cephalosporin synthase, isopenicillin *N*-synthase (IPNS)	

Table 16.3 Known suprafamilies and some of their members (Gerlt, 2001).

Suprafamily	*Examples*	*References*
His biosynthesis	PFACRI and ImGPS	Jürgens, 2000
	OMPDC, R5PE, HUMPS	Gerlt, 2001

PFACRI: *N'*-[(5'-phosphoribosyl)formimino]-5-aminoimidazole-4-carboxamide ribonucleotide isomerase; ImGPS: imidazole 3-glycerol phosphate synthase; OMPDC: orotidine 5'-phosphate decarboxylase; R5PE: ribulose 5-phosphate epimerase; HUMPS: hex-3-ulose monophosphate synthase.

An understanding of the structural bases for the catalytic diversity observed in super- and suprafamilies may provide the basis for discovering the functions of proteins and enzymes in new genomes as well as providing guidance for in-vitro evolution/engineering of new enzymes.

16.2.3
Dominance of Different Mechanisms

Independently of the mode of evolution of new function, several different mechanisms can be dominant, in either substrate specificity, or chemical mechanisms, or active-site architecture.

1. *Substrate specificity dominance*: The first and best known dominant mechanism in evolution concerns substrate specificity. The hypothesis underlying this mechanism states that metabolic pathways evolve backwards [Eq. (16.1), where E_1 is the new enzyme and E_2 is the old enzyme].

$$A \xrightarrow{E_1} B \xrightarrow{E_2} C \qquad (16.1)$$

   New enzyme evolves to supply substrate from an available precursor by evolution of enzyme using the substrate. Old and new enzymes are members of the same *suprafamily*. This hypothesis was first espoused by Horowitz (1945).

2. *Chemical mechanism dominance*: This second mechanism assumes that it is not binding but chemical mechanism that is most difficult to create anew. Once a chemical mechanism has been evolved it is beneficial for nature to utilize it over and over again. Nature selects protein from a pool of enzymes whose mechanism provides partial reaction or stabilization strategy for intermediates or transition states. Evolution decreases the proficiency of the reaction catalyzed by the progenitor. The underlying hypothesis of chemical mechanism dominance is that it starts with a low level of promiscuous activity [Eq. (16.2)].

$$(k_{cat}/k_{uncat})_{progenitor}/(k_{cat}/k_{uncat})_{successor} \gg 1 \text{ (old)}; < 1 \text{ (new)} \qquad (16.2)$$

Old and new enzymes developed by this mechanism are members of the same *superfamily*. The main champions of this mechanism are Gerlt and Babbitt (Gerlt, 1999; Babbitt, 2000). Readers are also encouraged to consult the reference on catalytic promiscuity by O'Brien and Herschlag (O'Brien, 1999).

3. *Dominance of active-site architecture*: In some cases, the intuitive notion that the active-site architecture is rarely the dominant influence will play the most important role. In such a case, the active site is able to support an alternative reaction with shared functional groups in a different mechanistic or metabolic context. Old and new enzymes are going to be members of a single *suprafamily*. Binding of proteins to putative substrate molecules was already claimed more than 50 years ago to be the most important property of future enzymes (Horowitz, 1945); catalysis was described as a consequence of evolution from an existing binding situation and was thought to be the simpler step. However, after analysis of more than 100 000 confirmed or predicted enzyme sequences and characterization of more than 4000 3D structures of enzymes, only about 30 enzyme 3D structure types are known (see above). The work on mechanistic similarity of superfamily members demonstrates that acquisition of novel substrate specificity on the basis of existing catalytic mechanisms seems to be the more relevant way of generating new enzymatic activity. Catalysis, even in its sometimes initial rudimentary way, seems to be at least as important as a basis for new enzymes as binding to a specific substrate, which seems to be acquired more easily in foreseeable timeframes.

16.3 Evolution of New Function in Nature

In nature, several possible mechanisms for the development of novel function have been observed. They can be summarized as:

- dual functionality proteins ("moonlighting proteins" and catalytic promiscuity),
- gene duplication,
- horizontal gene transfer, and
- circular permutation of structure.

Evolution counteracts simplistic notions of a static world in which all effects have one cause and where linear thinking and operations are dominant. In Chapter 4, we have encountered the central dogma of biology, according to which all information flows from DNA to RNA to protein. As reverse transcriptases exist and transform RNA back into DNA, we know the central dogma is not strictly correct. In this chapter, we will encounter several situations which collectively challenge the notion of

one gene–one protein–one structure–one mechanism–one function

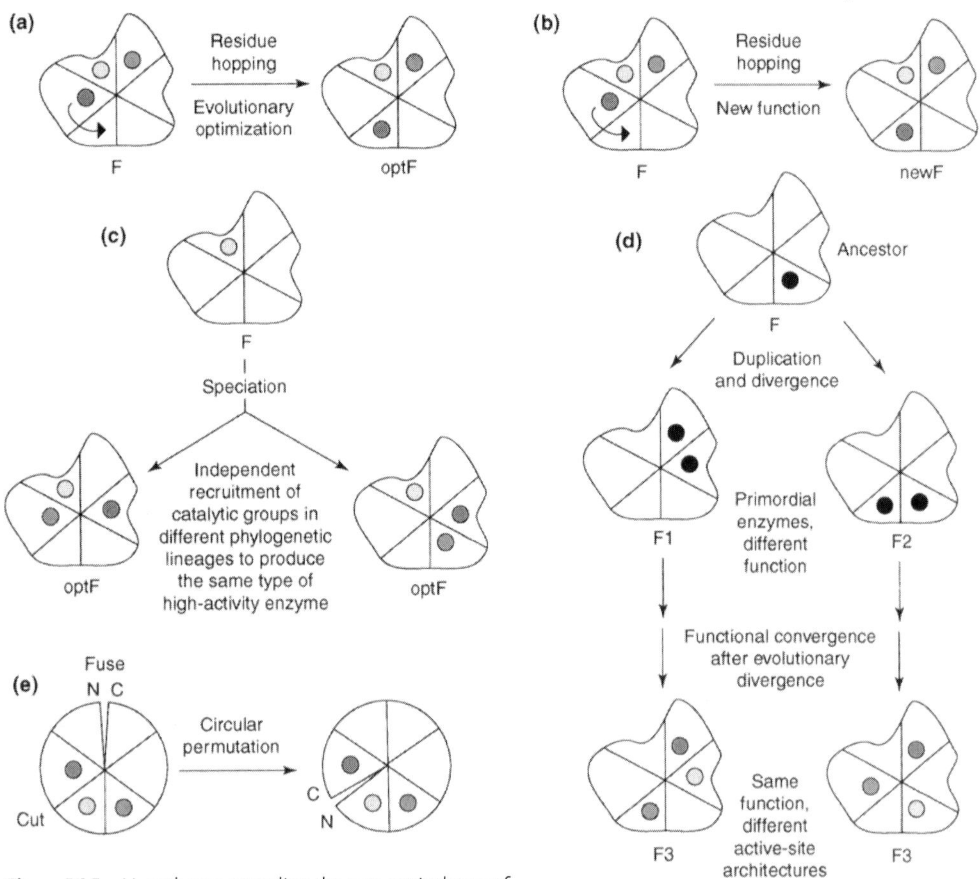

Figure 16.5 Hypotheses regarding the non-equivalence of functional residues that play the same catalytic role in enzyme homologs (Todd, 2002). Each shape represents a protein structure, and each "segment" in the protein structure represents a position in the fold at which a functional residue can be situated. Circles denote functional residues. "F" and "F1", etc., denote particular functions, and "optF" denotes the same function as "F", but optimized. "newF" denotes a new function., "N" and "C" denote the N- and C-termini, respectively.

Figure 16.5 summarizes possible options for evolution of function (Todd, 2002). Evolutionary optimization can occur by residue hopping of a functional residue to a different domain, strand, or subunit of the protein (a). Alternatively, residue hopping results in a new function (b). Recruitment of catalytic groups from different proteins for activity optimization might result in two novel and different enzymes (c, speciation). As discussed below, gene duplication can occur and different functions can develop on the proteins coded for by the genes (d, duplication and divergence). Lastly, in a circular protein the N- and C-termini can fuse and open up new

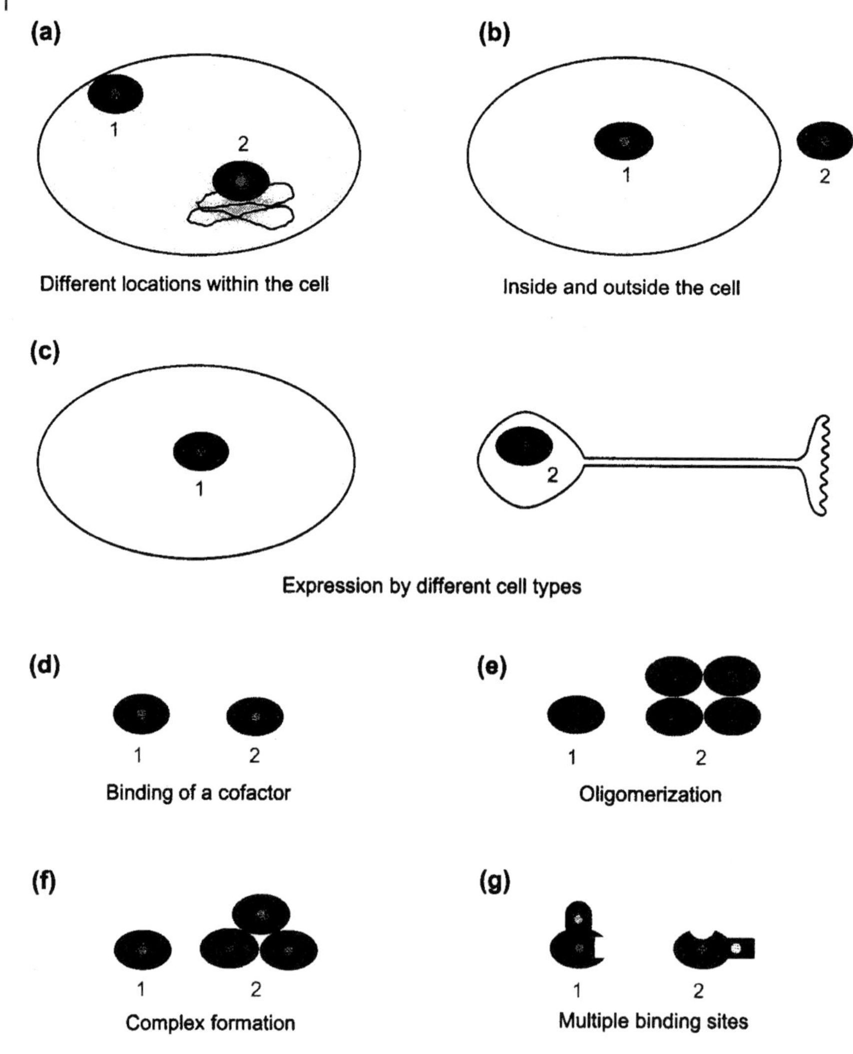

Figure 16.6 Examples of mechanisms of switching between one function and the other (Jeffery, 1999):
a) different functions in different locations of the cell;
b) proteins with enzymatic activity in cell cytoplasm but growth factor function when secreted;
c) proteins with different functions when expressed by different cell types;
d) different activity of protein caused by binding of substrate, product, or cofactor;
e) protein with different function in monomer than in multimer;
f) different function through interaction with different polypeptides; and
g) proteins with different binding sites for different substrates.

termini in a different position on the protein while conserving the catalytic residues (e, circular permutation).

One example, although not yet for optimization, is the shift of the catalytic lysine from one β strand to a neighboring strand: catalytic competency is maintained despite spatial reorganization of the active site with respect to substrate; this case could cover a) or b) (Wymer, 2001). Duplication and divergence (Section 16.4.2) and circular permutation (Section 16.3.4) are covered below.

16.3.1 Dual-Functionality Proteins

16.3.1.1 Moonlighting Proteins

Proteins with more than one function, also called moonlighting proteins, are much more frequent than expected. Such proteins challenge the chain of assumptions of "one gene–one protein–one function". The state of affairs has been summarized in a review article (Jeffery, 1999). Examples of moonlighting protein include:

- serum albumin: no activity vs. Kemp elimination;
- lens crystallins vs. lactate DH;
- thymidylate synthase vs. translation inhibitor;
- thrombin (protease) vs. cell surface receptor ligand;
- aconitase vs. iron-responsive element binding protein; and
- uracil-DNA-glycosidase vs. glyceraldehyde-3-phosphate DH.

How can a cell differentiate which of the two functions a protein molecule is supposed to exert? There are several modes of differentiation, all depicted in Figure 16.6.

16.3.1.2 Catalytic Promiscuity

In contrast to, and possibly as a subset of, moonlighting proteins, other proteins exist which feature not just two functions but even two different catalytic functions. Such proteins challenge the assumption of "one protein–one mechanism–one function".

In contrast to moonlighting proteins, which often are unknown to a researcher working in biocatalysis, proteins demonstrating catalytic promiscuity are fairly common and have also, in fact, been discussed in previous chapters of this book. Table 16.3 lists some examples of promiscuous enzymes, for a review, readers are referred to O'Brien (1999).

In cases of promiscuous activity, one species in the pathway often acts as a common intermediate for both mechanisms leading to different activities. An enamine between PLP and the substrate branches off into a decarboxylation or a transamination, a TPP-bound intermediate can react either to decarboxylation or to carboligation, and the triad Ser-His-Asp in hydroxynitrile lyase is responsible for both the main function, oxynitrilation, and the subordinate function, ester hydrolysis. It can be assumed that many more promiscuous functionalities will be discovered in the coming years.

Table 16.4 Examples of promiscuous enzymes.

Enzyme	*First function*	*Second function*	*Comment*	*Reference*
Subtilisin	peptidase	esterase	common to serine and cysteine proteases	many
Aspartate aminotransferase (AAT)	transamination	decarboxylation	in wt, transamination dominates	Yano, 1998; Oue, 1999
Phytase	phosphomonoesterase	sulfoxidation	sulfoxidation aided by vanadate addition	van de Velde, 1998
Hydroxynitrile lyase (HNL)	oxynitrilase	esterase	catalytic triad Ser-His-Asp	Wagner, 1996
Rubisco	carboxylase (addition of CO_2)	Oxygenase (addition of O_2)		Storroe, 1983
Decarboxylases (PDC, ADC)	decarboxylase	carboligation	thiamine pyrophosphate (TPP) catalysis	Bruhn, 1995
Chymotrypsin	amidation	phosphotriesterase		many
O-Succinylbenzoate synthase (OSBS)	synthase (OSBS)	racemization (NAAR)	identical protein: racemization fortuitous?	Palmer, 1999

PDC: pyruvate decarboxylase; ADC: acetoacetate decarboxylase;
NAAR: *N*-acetylamino acid racemase; OSBS: *O*-succinylbenzoate synthase.

16.3.2
Gene Duplication

Possibly the most important source of novel enzyme activity is divergence of genetic material from a common ancestral source by gene duplication, twofold incorporation of the same DNA material upon duplication of the cell. Gene duplication can be viewed from the perspectives of post-duplication and of the extent of the genomic region involved. The former encompasses (i) duplications of single structural genes followed by divergent evolution, (ii) genes which exist in several copies within each genome but which remain essentially identical to each other in DNA sequence and in function, and (iii) in eukaryotes only, relatively short sequences of DNA of various lengths that are each repeated many times (10^6) and which are not identical. The latter can be classified as (i) partial or internal gene duplications, (ii) complete gene duplications, (iii) partial chromosomal duplications (regional duplications), (iv) complete chromosomal duplications and, lastly, (v) polyploidy or whole-genome duplications.

The idea of gene duplications as a major means of creating new genes and ultimately new function is about 70 years old and connected with the name of John B. S. Haldane, who suggested in 1932 that a redundant duplicate of a gene may acquire divergent mutations and eventually emerge as a new gene and whom we have met already during the discussion of enzyme kinetics (Chapter 5).

16.3.3 Horizontal Gene Transfer (HGT)

Comparison of gene sequences is the primary way to study evolution of microbes. Most early studies assumed species evolved primarily by *vertical inheritance*, normal chromosomal inheritance, from parent or parents. We have been thinking of inheritance as vertical inheritance but we now know that *horizontal gene transfer*, gene transfer from species to species or genus to genus, has played a significant role in evolution. Drawing gene-based family trees for microbes offers a challenge owing to horizontal gene transfer. Bacteria and other single-celled creatures regularly pick up genetic material from organisms outside their species, even distant relatives. However, the recent sequencing of microbial genomes reveals that horizontal transfers have occurred far more often than most researchers had appreciated. In a study of the common gut bacterium *E. coli*, investigators calculate that nearly 20% of its DNA originated in other microbes.

In 1859, the great Charles Darwin in his seminal work *The Origin of Species* discussed the universal phylogenetic tree, a taxonomy based on genes and phenotype. Figure 16.7, incidentally the only Figure in *The Origin of Species*, illustrates Darwin's idea. More than a century later, Erwin Zuckerkandl and Linus Pauling at Caltech (Pasadena/CA, USA) in 1965 came up with the idea of molecular phylogenetics, a taxonomy based on molecular sequences rather than mostly on phenotype. A few years later, in the 1970s, Carl Woese (University of Illinois, Urbana-Champaign/IL, USA) developed what today is called the standard model, a taxonomy based on the sequence relationships of small-subunit ribosomal RNA (ssu RNA) (Figure 16.8). Since then, the standard model has been called into question; W. Ford Doolittle (Dalhousie University, Halifax, Nova Scotia, Canada) in 1999 advanced the notion of chimerism. Chimerism renders the role of the universal tree unclear (Figure 16.9).

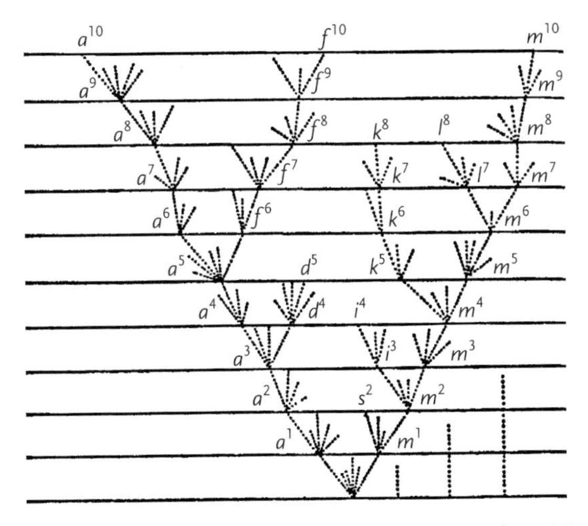

Figure 16.7 Universal phylogenetic tree (Darwin, *The Origin of Species*, 1859).

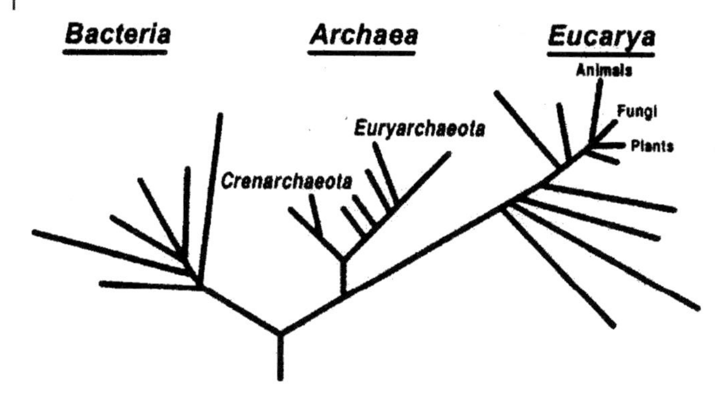

Figure 16.8 Standard model of taxonomy (Woese, 2002).

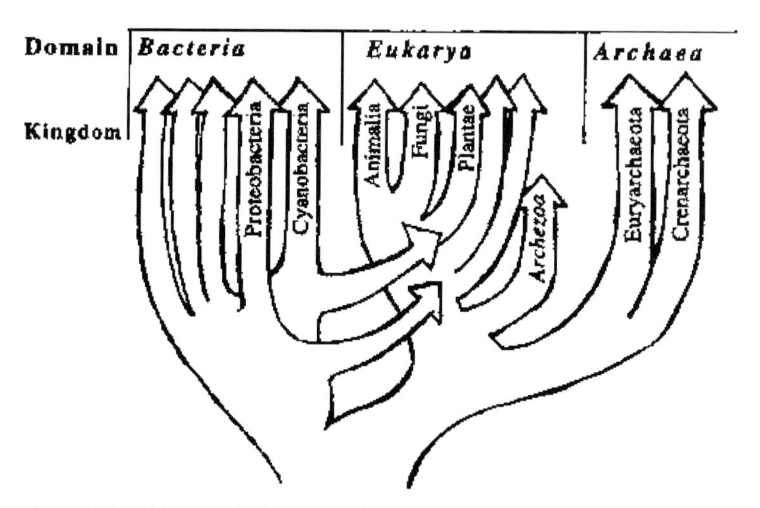

Figure 16.9 Chimerism in the "tree of life" (Doolittle, 1999).

At the beginning of evolution, cellular entities (progenotes) were very simple and information processing systems inaccurate. Enhanced levels of horizontal gene transfer were responsible for the evolutionary dynamic rather than vertical inheritance caused by the mutation rates (themselves high). Organismal lineages, and so organisms as we know them, did not exist at these early stages. With increasingly complex and precise biological structures and processes evolving, the mutation rate and the scope and level of lateral gene transfer decreased. One after the other, the various subsystems of the cell became refractory to horizontal gene transfer, with the translation apparatus probably crystallizing first and the evolutionary dynamic gradually becoming that characteristic of modern cells. The universal ancestor at the base of the universal phylogenetic tree, therefore, is not a discrete

Table 16.5 Levels and mechanisms of horizontal gene transfer.

Level	*Mechanism*
Recombination within species	passive absorption
Hybridization between species	mating/conjunction
Interspecies exchange	scavenging from environment
Plasmids and extrachromosomal elements	plasmid exchange
Organelle to nucleus	organelle to nucleus exchange
Viruses	viruses and viroids
Transposons	transposable elements fusion of cells

entity but a diverse community of cells that survives and evolves as a biological unit (Woese, 2002). It is the community as a whole, the ecosystem, which evolves. As a cell design becomes more complex and interconnected, a critical point is reached where a more integrated cellular organization emerges, a critical point, termed the "Darwinian threshold", is reached at which vertically generated novelty can and does assume greater importance.

How does horizontal gene transfer occur? Table 16.5 describes the levels and mechanisms of HGT.

Horizontal gene transfer (HGT) is a natural genetic transformation that is believed to be the essential mechanism for the attainment of genetic plasticity in many species. Kanamycin, streptomycin, and tetracycline resistance genes found in Gram-negative and Gram-positive bacteria are essentially identical in nucleotide sequence. Thus, the development of resistance against antibiotics in organisms can act as proof of the existence of horizontal gene transfer. Bacteria of the genus *Agrobacterium* have developed a special mechanism that allows transfer of genes from the bacterium to higher plant chromosomes. Horizontal gene transfer is suspected of contributing up to 50% of bacterial genetic material, which dramatically underscores the importance of this pathway for the development of evolution. As much as HGT affects the concept of a "tree of life" and even calls into question the notion of a prokaryote species, current thinking argues that each organism has a core of genes, responsible for critical functions, which is not transferred.

The surprisingly large extent of gene swapping has researchers asking whether comparisons of genes indicate how closely related different microbes are and if such analyses can truly point the way back to a universal ancestor. Sequence comparisons suggest the recent horizontal transfer of many genes among diverse species, including transfer across the boundaries of phylogenetic "domains". Thus determining the phylogenetic history of a species cannot be done conclusively by determining evolutionary trees for single genes. The wide transfer of genes that must have occurred naturally weakens the argument that human intervention through genetic engineering is placing genes where they have never been before.

16.3.4
Circular Permutation

In proteins with a symmetric structure, circular permutation can account for the shift of active-site residues over the course of evolution. A very good model of symmetric proteins are the $(\beta\alpha)_8$-barrel enzymes with their typical eightfold symmetry. Circular permutation is characterized by fusion of the N and C termini in a protein ancestor followed by cleavage of the backbone at an equivalent locus around the circular structure. Both fructose-bisphosphate aldolase class I and transaldolase belong to the aldolase superfamily of $(\alpha\beta)_8$-symmetric barrel proteins; both feature a catalytic lysine residue required to form the Schiff base intermediate with the substrate in the first step of the reaction (Chapter 9, Section 9.6.2). In most family members, the catalytic lysine residue is located on strand 6 of the barrel, but in transaldolase it is not only located on strand 4 but optimal sequence and structure alignment with aldolase class I necessitates rotation of the structure and thus circular permutation of the β-barrel strands (Jia, 1996).

16.4
α/β-Barrel Proteins as a Model for the Investigation of Evolution

16.4.1
Why Study α/β-Barrel Proteins?

The study of α/β-barrel proteins, also called TIM barrels after the first enzyme investigated in this class, triose phosphate isomerase, or $(\beta\alpha)_8$-barrel enzymes after the maximally eightfold symmetry axis, is important for the following three reasons:

1. the structure has been found in 10% of all known structures, and 68% of these known structures are enzymes;
2. the residues responsible for stability are separate from the residues responsible for function (Figure 16.10); and
3. the general scaffold catalyzes all classes of catalysis: oxidoreductases, hydrolases, lyases, and isomerases (Figure 16.3).

The main characteristics of α/β-barrel proteins are:

- a *high degree of symmetry*: they consist of a circular or elliptical eightfold repeat of (βα) units, with eight β–strands are surrounded by eight α–helices, each strand being H-bonded to two neighboring strands;
- *separation of purpose*: the β/α-loops (follow each β-strand) are important for function, the α/β-loops (follow each α-helix) are important for stability; the active site is found at the C-terminal end of the β-strands; and
- *evolutionary relationship*: evolution has been divergent for most α/β-barrel families, from a less efficient and specific common ancestor for enzymes involved in central metabolism (Farber, 1990).

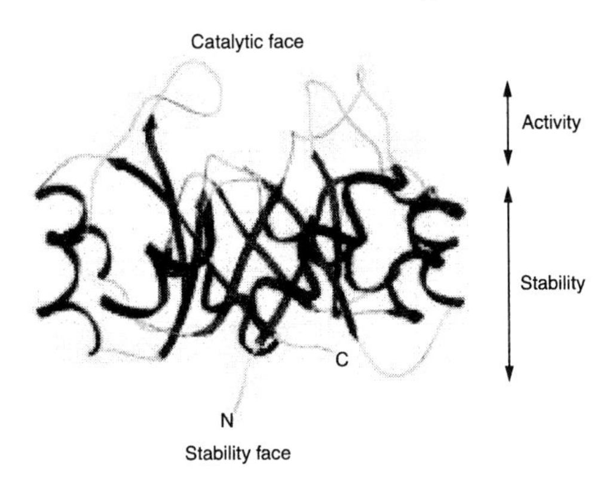

Figure 16.10 Separation of catalysis and stability domains in barrel proteins (Höcker, 2001).

α/β-Barrel enzymes can be classified in six families (listed in Table 16.7 for simplicity; Reardon, 1995) or even 21 superfamilies with 76 sequence families (Nagano, 2002), determined by (i) the domains in the β/α-loops, (ii) the shape of the barrel, and (iii) the location of extra sheets or helices such as enolases with their $\beta\beta\alpha\alpha(\beta\alpha)_6$ topology instead of the $(\beta\alpha)_8$-barrel fold.

Owing to the importance of the fold and rapid progress, the subject of α/β-barrel proteins has been reviewed extensively. Structural evidence for evolution of the β/α barrel scaffold by gene duplication and fusion is provided by Lang et al. (Lang, 2000). Aspects of stability, catalytic versatility, and evolution of the $(\beta\alpha)_8$-barrel fold have been covered by Höcker (Höcker, 2001). The prevalence of divergent evolution in $(\beta\alpha)_8$-barrel enzymes has been argued (Henn-Sax, 2001). Recent work has examined natural, designed, and directed evolution of function in several superfamilies of $(\beta\alpha)_8$-barrel-containing enzymes (Gerlt, 2003).

Lastly, α/β-barrel proteins are interesting because there exists at least one very good example with this class of fold for all modes of evolution of biocatalytic function: gene duplication, horizontal gene transfer, and circular permutation. The α/β-barrel template allows enzyme improvement through both protein engineering by site-directed mutagenesis (Chapter 10) and directed evolution by random mutagenesis (Chapter 11). Several studies have been performed with the goal of enzyme improvement, most notably by Jürgens (2000), who changed substrate specificity of HisA to that of PRAI (see Figure 16.13) although the two enzymes feature only 10% sequence identity (for an example outside the realm of α/β-barrels, see MacBeath, 1998).

16.4.2 Example of Gene Duplication: Mandelate and α-Ketoadipate Pathways

Mandelate racemase (MR) enables some strains of the common soil bacterium *Pseudomonas putida* to utilize mandelate from decomposing plant matter as a carbon source. MR is the first of five enzymes in the bacterial pathway that converts mandelate to benzoate. Then benzoate is broken down by another set of five enzymes of the ensuing β-ketoadipate pathway to compounds that can be used to generate ATP, the cell's major source of chemical energy.

Table 16.6 α/β Barrel proteins (Reardon, 1995).

Family	*Name*	*Cofactor*	*Chemical reaction*
A	flavocytochrome b_2	FMN, heme	oxidation
A	glycolate oxidase	FMN	oxidation
A	trimethylamine dehydrogenase	FMN, 4Fe–4S	oxidative demethylation
A	ribulose-1,5-biphosphate carboxylase/oxygenase	Mg^{2+}	formation of an enediol(ate) addition of either CO_2 or O_2/C–C bond cleavage
B	mandelate racemase	Mg^{2+}	proton abstraction to generate a carbanionic intermediate reprotonation of the intermediate
B	muconate cycloisomerase	Mn^{2+}	cycloisomerization
B	chloromuconate cycloisomerase	Mn^{2+}	cycloisomerization
B	*α*-amylase	Ca^{2+}	endohydrolysis of 1,4-*α*-D -glucoside bonds
B	*β*-amylase	none	hydrolysis of 1,4-*α*-D -glucoside bonds
B	oligo-1,6-glucosidase	none	hydrolysis of 1,6-*α*-D-glucoside bonds
B	cyclodextrin glycosyl-transferase	Ca^{2+}	formation (and hydrolysis) of 1,4-*α*-D-glucoside bonds
B	pyruvate kinase	Mg^{2+}, K^+	proton abstraction phosphorylation
B	xylose isomerase	Mg^{2+}	hydride transfer between adjacent carbons during aldose/ketose isomerization
C	*N*-(5′-phosphoribosyl) anthranilate isomerase	none	Amadori rearrangement
C	indole-3-glycerol phosphate synthase	none	decarboxylative closure of the indole ring
C	tryptophan synthase (*α*-unit)	none	aldol cleavage
C	triose phosphate isomerase	none	proton transfer between two adjacent carbons during aldose/ketose isomerization
C	narbonin		no known enzymatic activity
D	fructose bisphosphate aldolase	none	aldol cleavage
D	2-dehydro-3-deoxyphos-phogluconate aldolase	none	aldol cleavage
D	*N*-acetylneuraminate lyase	none	aldol cleavage
E	aldose reductase	NADPH	reduction

Table 16.5 (continued)

Family	*Name*	*Cofactor*	*Chemical reaction*
E	3α-hydroxysteroid dehydrogenase	NAD(P)H	reduction
F	adenosine deaminase	Zn^{2+}	hydrolysis
F	phosphotriesterase	Zn^{2+}	hydrolysis
*	(1→3)-β-glucanase	none	hydrolysis
*	(1→3,1→4)-β-glucanase	none	hydrolysis
PSR	endo-β-*N*-acetyl-glucoseaminidase		
PSR	urease	Ni^{2+}	
PSR	old yellow enzyme	FMN, NADP	

Table 16.7 α/β Barrel enzyme families (Reardon, 1995).*

Family	*Characteristics*	*Examples*
A	barrels nearly circular small helix between β-strand 8 and α-helix 8 domain covering N-terminus of the barrel	glycolate oxidase, trimethylamine dehydrogenase
B	major axis near β-strand 1 MR, MLE are missing final α-helix domains cover the C-terminal end of the barrel domain blocks N-terminus of the barrel	mandelate racemase, muconate cycloisomerase, xylose (glucose) isomerase
C	composed of a single domain major axis near β-strand 3 small helix between β-strand 8 and α-helix 8	tryptophan synthase, triose phosphate isomerase (TIM), PRAI, IGPS (see Figure 16.13)
D	have an α-helix preceding the α/β barrel this helix blocks the N-terminal end of the barrel	fructose-biphosphate aldolase, 2-keto-3-deoxy-6-phospho-gluconate
E	major axis of ellipse points towards β-strand large loops at the carboxyl-terminal end of the barrel after β-strands 4 and 7 bind NAD	aldose reductase, 3α-hydroxysteroid dehydrogenase
F	long thin barrel structure group of α-helices in the carboxyl-terminal loop between β-strand 1 and α-helix 1 require divalent transition metal ion	phosphotriesterase, adenosine deaminase

* Not all the characteristics refer to each member of the family.
There are other α/β-barrel enzymes that are in a family.

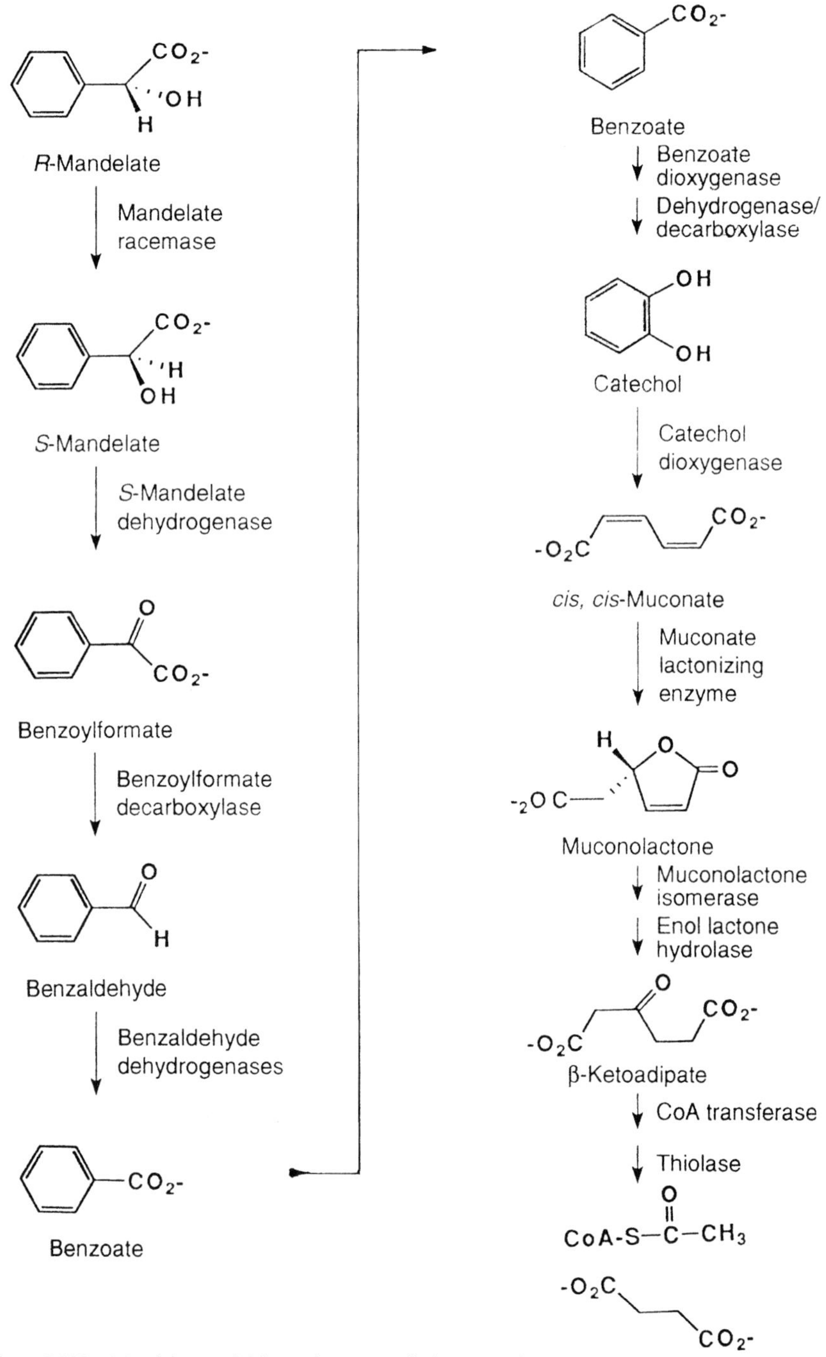

Figure 16.11 Mandelate and β-ketoadipate metabolism (Petsko, 1993).

During the investigation of the mandelate metabolism (Figure 16.11), a team of eminent bioscientists collaborating from several independent research groups, among them George Kenyon from the University of California at San Francisco (UCSF), John Gerlt from the University of Illinois at Urbana-Champaign/IL, John Kozarich from Merck (Rahway/NJ), and Greg Petsko from MIT (Cambridge/MA), later Brandeis University (Waltham/MA), all USA, noticed several surprises when characterizing the enzymes of the mandelate and β-ketoadipate pathways (Petsko, 1993):

- MR, an α/β-barrel protein, features a high degree of similarity with muconate-lactonizing enzyme (MLE) of the β-ketoadipate metabolism: 26% sequence identity, the same symmetry and organization of subunits (octamer), a divalent metal ion requirement (Mn^{2+} for MLE, Mg^{2+} for MR). As MLE is almost ubiquitous among *Pseudomonas*, compared to 5% frequency of MR, it can be concluded that MLE is the older enzyme and probably the precedessor of MR.
- As all enzymes of the mandelate path are located on one operon, the other four enzymes could also be identified, cloned, and sequenced. (*S*)-Mandelate dehydrogenase [(*S*)-ManDH] features similarities to glycolate oxidase and flavocytochrome b_2 (a lactate-DH): both are α/β-barrel proteins and require FMN as cofactor as does (*S*)-ManDH. It is reasonable to conclude that in this case nature has also sequestered a template that already catalyzes a chemistry similar to the desired reaction.
- Benzoylformate decarboxylase (BFD), a TPP-dependent enzyme, features a high degree of sequence similarity with a whole family of other TPP-dependent decarboxylases, among them pyruvate decarboxylase (PDC), one of the most important enzymes in metabolism.
- After comparison of DNA sequences of an open reading frame (orf) of the mandelate pathway with databases it was found that there exists sequence similarity of an otherwise unannotated piece of sequence with some amide hydrolases. It is suspected that this DNA sequence piece codes for a mandelate amidase which catalyzes hydrolysis of mandelamide into mandelate and ammonium ion. This hypothesis recently was borne out and the mandelamide hydrolase found (MacLeish, 2003).

The following conclusions can be drawn from these findings:

- The mandelate and β-ketoadipate pathways are a good example of *gene duplication*, with strong evidence of the former evolving from the latter: the congruence of MR and MLE, (*S*)-ManDH and benzoate dihydrodiol dehydrogenase, and possibly benzoyl formate decarboxylase and protocatechuate decarboxylase.
- Evolution of MR from MLE and possibly (*S*)-ManDH from benzoate dihydrodiol dehydrogenase points to the relevance of the *enzyme mechanism for catalytic reactivity*. If the chemistry is right, the remaining necessary enhancement of specificity, i.e., the enhancement of binding of substrate, is the simpler task, in contrast to the procedure for design of catalytic antibodies.

16.4.2.1 **Description of Function**

MR from *Pseudomonas putida* ATCC 12633 catalyzes the reversible interconversion of (*R*)-mandelate and (*S*)-mandelate, by removing a hydrogen ion from one side of mandelate and attaching it to the other. In the mandelate pathway, (*S*)-mandelate dehydrogenase is specific for the (*S*)-enantiomer, thus requiring the racemase to allow (*R*)-mandelate oxidation to benzoate. MR therefore allows utilization of either enantiomer of mandelate. There is persuasive evidence that the reaction catalyzed by MR proceeds via a two-base mechanism in which Lys166 abstracts the α proton from (*S*)-mandelate (Kallarakal, 1995) and His297 the α proton from (*R*)-mandelate (Landro, 1991). MR is not inhibited by pathway intermediates, indicating that it has no regulatory role. Reversible inhibition was, however, observed to occur when carboxylic acids having an aromatic substituent were incubated with the enzyme prior to addition of mandelic acid.

So what is the source of the power of MR catalysis? Many enzymes catalyze the abstraction of a proton in the α-position to a carbonyl group, such as in a carboxylic acid. The enzymatic rates of abstraction, however, are much higher than could be inferred from the ΔpK_a between substrate in solution and the base in the active center [Eq. (16.3)] ($k_BT/h = 6.2 \times 10^{-12}\ s^{-1}$).

$$k = (k_BT/h)\ \exp\{-[(\Delta G^{\neq}/RT) + 2.303 \cdot \Delta pK_a]\} \tag{16.3}$$

For carboxylic acids, $\Delta G^{\neq} > 30\ kJ\ mol^{-1}$, so $\Delta pK_a < 2–5$. The catalytic rate constants of the reaction, however, are often between 10^2 and $10^5\ s^{-1}$. Through analysis of known and well estimated pK_a data, *electrophilic catalysis* can be demonstrated with a cycle process (Figure 16.12) to suffice as a quantitative explanation for the observed rates.

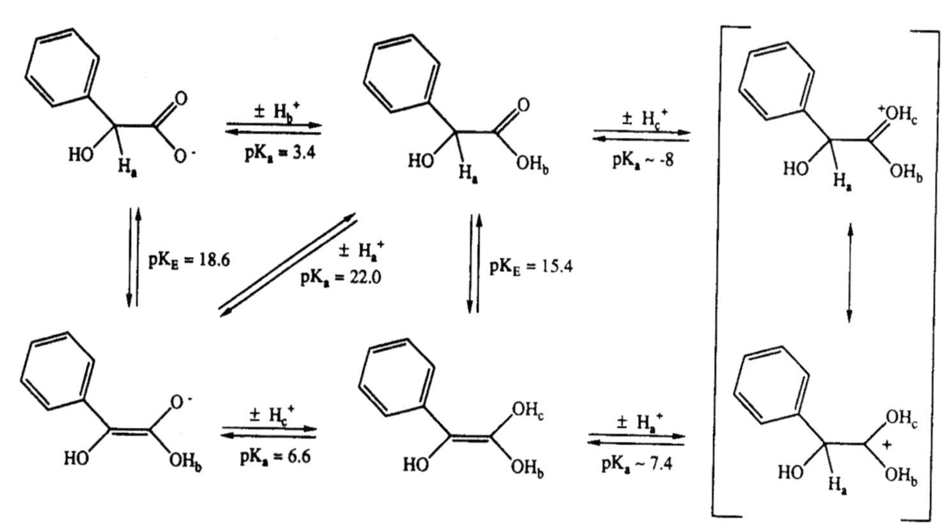

Figure 16.12 Cycle process: pK_a values of mandelic acid at different degrees of protonation (Gerlt, 1991).

If K_E signifies the concentrations of enol and keto tautomers in a carboxylic acid, then pK_E is the difference between the pK_a values of the α-protons of the keto tautomer and the hydroxyl group of the enol tautomer. pK_E, however, is also the difference in pK_a values between the α-protons and the proton of the carbonyl group of the carbonyl-protonated acid, that is deemed to be decisive (Gerlt, 1991) for the kinetics of abstraction of a proton, rather than the pK_a value of the substrate in solution. The pK_E of mandelic acid (15.4) links the pK_a of the α-proton of the keto tautomer (22.0) with the pK_a value of the enol tautomer (6.6). The pK_E value also links the pK_a value of the α-proton with that of the carbonyl-bound proton of the protonated mandelic acid. If the pK_a value of the carbonyl-bound proton of the protonated mandelic acid is assumed to be about –8.0 then the pK_a value of the α-proton is about 7.4. This value matches well with the pK_a -values of Lys and His residues which have been assigned recently in the active center of mandelate racemase, so electrophilic catalysis alone is able to explain the catalytic power of mandelate racemase.

16.4.3
Exchange of Function in the Aromatic Biosynthesis Pathways: Trp and His Pathways

In the biosynthesis pathways of histidine and tryptophan, the enzymes HisA (*N'*-[(5'-phosphoribosyl)formimino]-5-aminoimidazole-4-carboxamide ribonucleotide isomerase) and TrpF (PRAI, *p*hospho*r*ibosyl*a*nthranilate *i*somerase) (Figure 16.13), both of which are $(\beta\alpha)_8$-barrels, both catalyze Amadori rearrangements of a thermolabile aminoaldose into the corresponding aminoketose (Figure 16.13).

All four enzymes in Figure 16.13, His A, His F, TrpF, and TrpC, have a similar $(\beta\alpha)_8$-barrel structure. Site-directed mutagenesis of conserved residues at the active sites of both HisA and TrpF and steady-state enzyme kinetics of variants pointed to the following residues as catalytically important (Henn-Sax, 2002): aspartate-8, aspartate-127, and threonine-164 for the HisA reaction, cysteine-7 and aspartate-126 for the TrpF reaction. In conjunction with a three-dimensional structure of an enzyme–product analog conjugate, a similar reaction mechanism involving general acid–base catalysis and a Schiff base intermediate is postulated for both enzymes.

With the goal of creating an ancestor-like enzyme for *N'*-[(5'-phosphoribosyl)formimino]-5-aminoimidazole-4-carboxamide ribonucleotide isomerase (HisA; E.C. 5.3.1.16) and phosphoribosylanthranilate isomerase (TrpF; E.C. 5.3.1.24), HisA variants were generated by random mutagenesis and selection via an auxotroph (Jürgens, 2000). Several variants catalyzed the TrpF reaction and one variant even retained significant HisA activity, the quadruple mutant Asn24Asp, Arg97Gly, His175Tyr, and Met236Ile. A single amino acid exchange, Asp127Val, close to the active site was found to be able to establish TrpF activity on the HisA scaffold.

According to these results, HisA and TrpF may indeed have evolved from an ancestral enzyme of broader substrate specificity.

HisA and HisF both consist of α/β-barrels with a twofold repeat pattern. It is likely that these proteins evolved by twofold gene duplication and gene fusion from a common half-barrel ancestor (Lang, 2000). The strongest evidence for this idea, besides the twofold sequence repeat pattern, stems from the catalytic aspartate residue which is located at the end of β-strand 1 in the HisA-N-terminal half (HisA-N) and HisF-N, and on β-strand 5 in HisA-C and HisF-C. Furthermore, the individual purified half-barrel proteins HisF-N and HisF-C, catalytically inactive but featuring a well-defined secondary and tertiary structure, assemble into the catalytically fully active full barrel HisF-NC (Höcker, 2001).

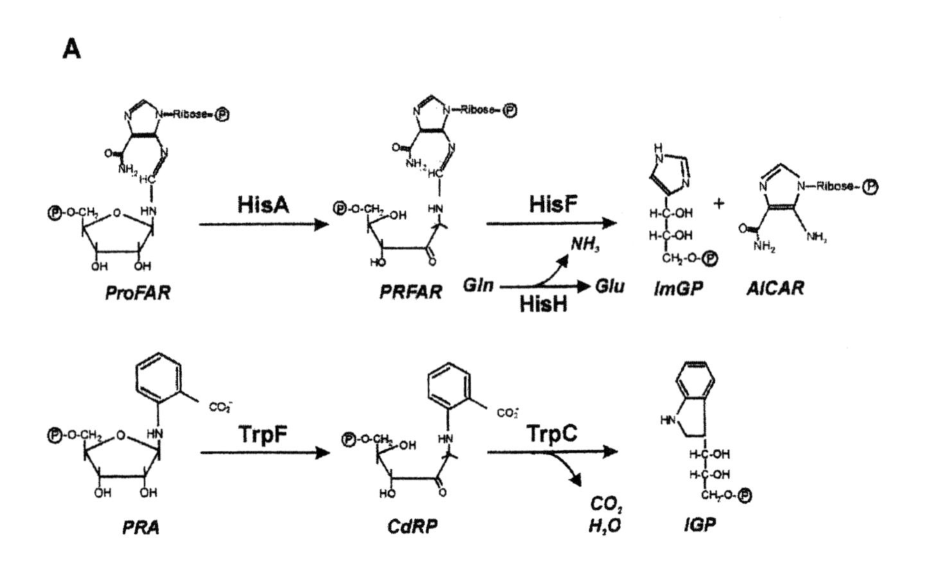

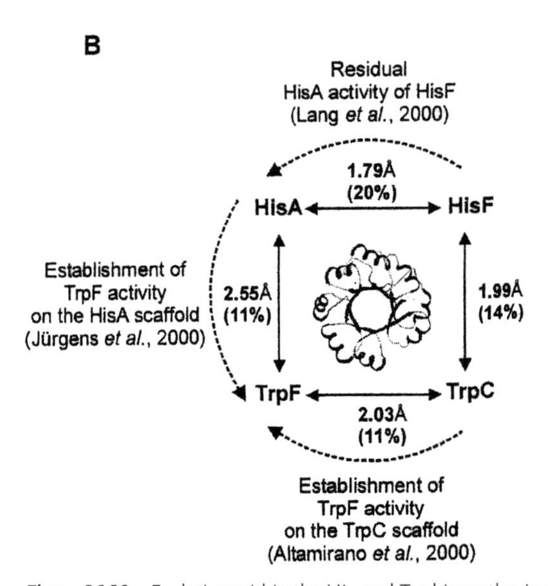

Figure 16.13 Evolution within the His and Trp biosynthesis pathways (Jürgens, 2000).

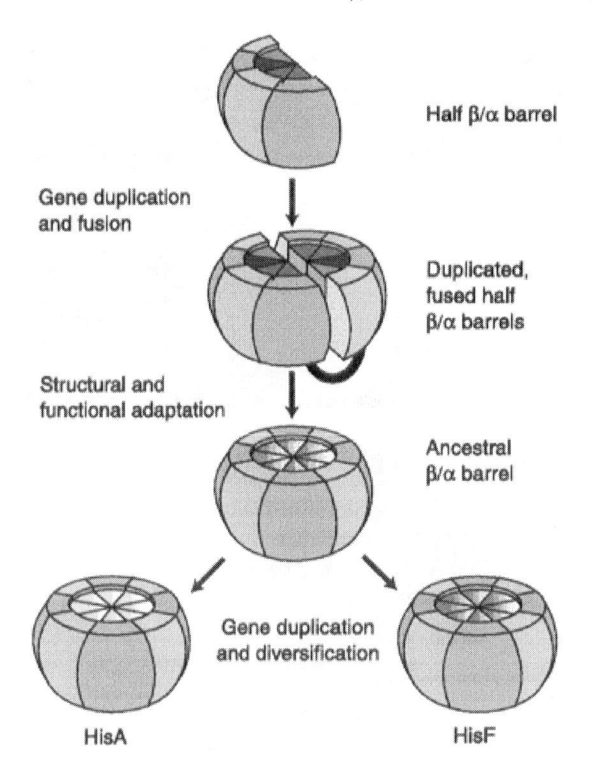

Figure 16.14 Evolution of α/β-barrel protein scaffold (Lang, 2000).

As a summary of all the current results, the picture in Figure 16.14 emerges.

The earliest currently discernible ancestor was a $(\beta\alpha)_4$-half barrel protein which, through gene duplication, generates two initially identical half-barrels which fuse and form an ancestral α/β-barrel protein. This full-barrel ancestor already featured or developed catalytic activity on intermediates of the histidine biosynthesis pathway. The second gene duplication step leads to the diversification of the ancestral α/β-barrel protein into two enzymes with distinct catalytic activities, in this case into HisA and HisF catalyzing successive reaction in the histidine biosynthesis pathway.

Therefore, the Horowitz hypothesis seems to have been demonstrated on the example of HisA and HisF as one possibility for the development of novel enzymatic activity.

Suggested Further Reading

C. C. Darwin, *On the Origin of Species by Means of Natural Selection or the Preservation of Favored Races in the Struggle for Life (The Origin of Species)*, John Murray, London, **1859.**

J. A. Gerlt and F. M. Raushel, Evolution of function in $(\beta/\alpha)_{(8)}$-barrel enzymes, *Curr. Opin. Chem. Biol.* **2003**, *7*, 252–264.

B. Höcker, C. Jürgens, M. Wilmanns, and R. Sterner, **2001**, Stability, catalytic versatility and evolution of the (β α)(8)-barrel fold, *Curr. Opin. Biotechnol. 12*, 376–381.

References

G. Alagona, C. Ghio, and P. A. Kollman, Do enzymes stabilize transition states by electrostatic interactions or pK_a balance: the case of triose phosphate isomerase (TIM)? *J. Am. Chem. Soc.* **1995**, *117*, 9855–9862.

F. H. Arnold, Directed evolution: creating biocatalysts for the future, *Chem. Eng. Sci.* **1996**, *51*, 5091–5102.

P. C. Babbitt, M. S. Hasson, J. E. Wedekind, D. R. Palmer, W. C. Barrett, G. H. Reed, I. Raymond, D. Ringe, G. L. Kenyon, and J. A. Gerlt, The enolase superfamily: a general strategy for enzyme-catalyzed abstraction of the α-protons of carboxylic acids, *Biochemistry* **1996**, *35*, 16489–16501.

P. C. Babbitt and J. A. Gerlt, Understanding enzyme superfamilies: chemistry as the fundamental determinant in the evolution of new catalytic activities, *J. Biol. Chem.* **1997**, *272*, 30591–30594.

P. C. Babbitt and J. A. Gerlt, Understanding enzyme superfamilies, *Biochemistry* **1999**, *36*, 000.

S. Borman, Enzyme's activity designed to order, *Chem. Eng. News* **2000**, Feb. 21, 35–36.

C. Branden and J. Tooze, *Introduction to Protein Science,* Garland Publishing, London, New York, **1991**.

H. Bruhn, M. Pohl, J. Groetzinger, and M.-R. Kula, The replacement of Trp392 by alanine influences the decarboxylase/carboligase activity and stability of pyruvate decarboxylase from *Zymomonas mobilis*, *Eur. J. Biochem.* **1995**, *234*, 650–655.

C. C. Darwin, *On the Origin of Species by Means of Natural Selection or the Preservation of Favored Races in the Struggle for Life (The Origin of Species)*, John Murray, London, **1859.**

W. F. Doolittle, Phylogenetic classification and the universal tree, *Science* **1999**, *284*, 2124–2128.

G. K. Farber and G. A. Petsko, The evolution of α/β-barrel enzymes, *Trends Biochem. Sci.* 228–234. **1990**,

S. M. Firestine, A. E. Nixon, and S. J. Benkovic, Threading your way to protein function, *Chem. Biol.* **1996**, *3*, 779–783.

J. A. Gerlt, J. W. Kozarich, G. L. Kenyon, and P. G. Gassman, Electrophilic catalysis can explain the unexpected acidity of carbon acids in enzyme-catalyzed reactions, *J. Am. Chem. Soc* **1991**, *113*, 9667–9669.

J. A. Gerlt and P. C. Babbitt, Mechanistically diverse enzyme superfamilies: the importance of chemistry in the evolution of catalysis, *Curr. Opin. Chem. Biol.* **1998**, *2*, 607–612.

J. A. Gerlt and P. C. Babbitt, Divergent evolution of enzymatic function: mechanistically diverse superfamilies and functionally diverse suprafamilies, *Annu. Rev. Biochem.* **2001**, *70*, 209–246.

J. A. Gerlt and F. M. Raushel, Evolution of function in $(\beta/\alpha)_{(8)}$-barrel enzymes, *Curr. Opin. Chem. Biol.* **2003**, *7*, 252–264.

M. S. Hasson, I. Schlichting, J. Moulai, K. Taylor, W. Barrett, G. L. Kenyon, P. C. Babbitt, J. A. Gerlt, G. A. Petsko, and D. Ringe, Evolution of an enzyme active site: the structure of a new crystal form of muconate lactonizing enzyme compared with mandelate racemase and enolase, *Proc. Natl. Acad. Sci. USA* **1998**, *95*, 10396–10401.

H. Hegyi and M. Gerstein, http://bioinfo.mbb.yale.edu/genome/foldfunc/Fig1-struct.html), **1998**.

M. Henn-Sax, B. Hocker, M. Wilmanns, and R. Sterner, Divergent evolution of $(\beta\alpha)_8$-barrel enzymes, *Biol. Chem.* **2001**, *382*, 1315–1320.

M. Henn-Sax, R. Thoma, S. Schmidt, M. Hennig, K. Kirschner, and R. Sterner, Two $(\beta\alpha)_{(8)}$-barrel enzymes of histidine and tryptophan biosynthesis have similar reaction mechanisms and common strategies for protecting their labile substrates, *Biochemistry* **2002**, *41*, 12032–120342.

B. Höcker, S. Beismann-Driemeyer, S. Hettwer, A. Lustig, and R. Sterner, Dissection of a $(\beta\alpha)_8$-barrel enzyme into two folded halves, *Nature Struct. Biol.* **2001**, *8*, 32–36.

B. Höcker, C. Jürgens, M. Wilmanns, and R. Sterner, Stability, catalytic versatility and evolution of the (beta alpha)(8)-barrel fold, *Current Opin. Biotechmol.* **2001**, *12*, 376–381.

L. Holm and C. Sanders, An evolutionary treasure; unification of a broad set of amidehydrolases related to urease, *Protein: Structure, Function, and Genetics* **1997**, *28*, 72–82.

N. H. Horowitz, On the evolution of biochemical syntheses, *Proc. Natl. Acad. Sci. USA* **1945**, *31*, 153–157.

C. J. Jeffery, Moonlighting proteins, *Trends Biochem. Sci. (TIBS)*, **1999**, *24*, 8–11.

J. Jia, W. Huang, U. Schoerken, H. Sahm, G. A. Sprenger, Y. Lindqvist, G. Schneider, Crystal structure of transaldolase B from *Escherichia coli* suggests a circular permutation of the α/β-barrel within the class I aldolase family, *Structure* **1996**, *4*, 715–724.

C. Jürgens, A. Strom, D. Wegener, S. Hettwer, M. Wilmanns, and R. Sterner, Directed evolution of a (beta alpha)(8)-barrel enzyme to catalyze related reactions in two different metabolic pathways, *Proc. Natl. Acad. Sci.* **2000**, *97*, 18, 9925–9930.

A. T. Kallarakal, B. Mitra, J. W. Kozarich, J. A. Gerlt, J. G. Clifton, G. A. Petsko, and G. L. Kenyon, Mechanism of the reaction catalyzed by mandelate racemase: structure and mechanistic properties of the K166R mutant, *Biochemistry* **1995**, *34*, 2788–2797.

J. A. Landro, A. T. Kallarakal, S. C. Ransom, J. A. Gerlt, J. W. Kozarich, D. J. Neidhart, and G. L. Kenyon, Mechanism of the reaction catalyzed by mandelate racemase. 3. Asymmetry in reactions catalyzed by the H297N mutant, *Biochemistry* **1991**, *30*, 9274–9281.

D. Lang, R. Thoma, M. Henn-Sax, R. Sterner, and M. Wilmanns, Structural evidence for evolution of the beta/alpha barrel scaffold by gene duplication and fusion, *Science* **2000**, *289*, 1546–1550.

G. MacBeath, P. Kast and D. Hilvert, Redesigning enzyme topology by directed evolution, *Science* **1998**, *279*, 1958–1961.

M. J. McLeish, M. M. Kneen, K. N. Gopalakrishna, C. W. Koo, P. C. Babbitt, J. A. Gerlt, G. L. Kenyon, Identification and characterization of a mandelamide hydrolase and an $NAD(P)^+$-dependent benzaldehyde dehydrogenase from *Pseudomonas putida* ATCC 12633, *J Bacteriol.* **2003**, *185*, 2451–2456.

G. Muller-Newen, U. Janssen, and W. Stoffel, Enoyl-CoA hydratase and isomerase form a superfamily with a common active-site glutamate residue, *Eur. J. Biochem.* **1995**, *228*, 68–73.

N. Nagano, C. A. Orengo, and J. M. Thornton, One fold with many functions: The evolutionary relationships between TIM barrel families based on their sequences, structures and functions, *J. Mol. Biol.* **2002**, *321*, 741–765.

P. J. O'Brien and D. Herschlag, Catalytic promiscuity and the evolution of new enzymatic activities, *Chem. Biol.* **1999**, *6*, R91–R105.

S. Oue, A. Okamoto, T. Yano, and H. Kagamiyama, Redesigning the substrate specificity of an enzyme by cumulative effects of the mutations of non-active site residues, *J. Biol. Chem.* **1999**, *274*, 2344–2349.

D. R. Palmer, J. B. Garret, V. Sharma, R. Meganathan, P. C. Babbitt, and J. A. Gerlt, Unexpected divergence of enzyme function and sequence: '*N*-Acetylamino Acid Racemase' is *o*-succinylbenzoate synthase, *Biochemistry* **1999**, *38*, 4252–4258.

G. A. Petsko, G. L. Kenyon, J. A. Gerlt, D. Ringe, and J. W. Kozarich, On the

origin of enzymatic species, *Trends Biochem. Sci. (TIBS)* **1993**, *18*, 372–376.
G. A. Petsko, Design by necessity, *Nature* **2000**, *403*, 606–607.
E. Quéméneur, M. Moutiez, J.-B. Charbonnier and A. Ménez, Engineering cyclophilin into a proline-specific endopeptidase, *Science* **1998**, *391*, 301–304.
D. Reardon, and G. K. Farber, Protein motifs. 4. The structure and evolution of alpha/beta barrel proteins, *FASEB J.* **1995**, *9*, 497–503.
B. Rost, Protein structures sustain evolutionary drift, *Fold. Des.* **1997**, *2*, S19–S24.
C. Sander and R. Schneider, Database of homology-derived protein structures and the structural meaning of sequence alignment, *Proteins: Struct. Funct. Genet.* **1991**, *9*, 56–68.
J. D. Stevenson, S. Lutz, and S. J. Benkovic, Retracing enzyme evolution in the $(\beta\alpha)_8$-barrel scaffold, *Angew. Chem. Int. Ed.* **2001**, *40*, 1854–1856.
I. Storroe, and B. R. McFadden, Ribulose diphosphate carboxylase/oxygenase in toluene-permeabilized *Rhodospirillum rubrum*, *Biochem. J.* **1983**, *212*, 45–54.
A. E. Todd, C. A. Orengo, and J. M. Thornton, Plasticity of enzyme active sites, *Trends Biochem. Sci.* **2002**, *27*, 419–426.
F. van de Velde, L. Konemann, F. van Rantwijk, and R. A. Sheldon, Enantioselective sulfoxidation mediated by vanadium-incorporated phytase: a hydrolase acting as a peroxidase, *Chem. Commun.* **1998**, *(17)*, 1891–1892.
C. von der Osten, S. Branner, S. Hastrup, L. Hedegaard, M. D. Rasmussen, H. Bisgard-Frantzen, S. Carlsen, and J. M. Mikkelsen, Protein engineering of subtilisins to improve stability in detergent formulations, *J. Biotechnol.* **1993**, *28*, 55–68.
C. Vinals, X. De Bolle, E. Depiereux, and E. Feytmans, Knowledge-based modeling of the D-lactate dehydrogenase three-dimensional structure, *Proteins* **1995**, *21*, 307–318.
U. G. Wagner, M. Hasslacher, H. Griengl, H. Schwab, and C. Kratky, Mechanism of cyanogenesis: the crystal structure of hydroxynitrile lyase from *Hevea brasiliensis*, *Structure* **1996**, *4*, 811–822.
R. K. Wierenga, The TIM-barrel fold: a versatile framework for efficient enzymes, *FEBS Lett.* **2001**, *492*, 193–198.
M. K. Winson and D. B. Kell, Going places: forces and natural molecular evolution, *TIBTECH* **1996**, *14*, 323–325.
C. Woese, The universal ancestor, *Proc. Natl. Acad. Sci. USA* **1998**, *95*, 6854–6859.
C. R. Woese, On the evolution of cells, *Proc. Natl. Acad. Sci. USA* **2002**, *99*, 8742–8747.
N. J. Wymer, L. V. Buchanan, D. Henderson, N. Mehta, C. H. Botting, L. Pocivavsek, C. A. Fierke, E. J. Toone, and J. H. Naismith, Directed evolution of a new catalytic site in 2-keto-3-deoxy-6-phosphogluconate aldolase from *Escherichia coli*, *Structure* **2001**, *9*, 1–10.
T. Yano, S. Oue, and H. Kagamiyama, Directed evolution of an aspartate aminotransferase with new substrate specificities, *Proc. Natl. Acad. Sci. USA*, **1998**, *95*, 5511–5515.

17
Stability of Proteins

Summary

Protein folding cannot occur in a random fashion but must follow distinct pathways: a protein of 100 amino acids with diffusional control of encounter between different parts of the chain would take $(2^{100}/2)/10^{10}\ s^{-1} = 10^{33}\ s = 5 \times 10^{24}$ years, or longer than the age of the universe, to fold. Levinthal's paradoxon reaches the same conclusion: a protein of 100 amino acids with side chains in three different states can feature 100!/3 states; diffusion of those states would take even longer than 10^{24} years. Proteins cannot be stable without hydrophobic interactions; convergence temperatures for unfolding enthalpies and entropies (ΔH_m and ΔS_m) are around 112 °C, the presumed T_{max} for proteins in aqueous solution.

Protein unfolding and deactivation is described most commonly by the Lumry–Eyring model: $N \Leftrightarrow U \rightarrow D$, where N denotes the native or active, U the unfolded, and D the irreversibly deactivated form. As $k_{d,obs} \approx k_d/(1 + K)$ $(K = [U]/[N])$, K has to be known when obtaining k_d from $k_{d,obs}$. Measuring $k_{d,obs}$ during a reaction can influence biocatalyst stability because (i) the enzyme molecule unfolds more slowly in the presence of a bound molecule, particularly if binding is tight (low K_M, K_P, or K_I), and (ii) at high concentrations, independently of active site interaction, many compounds can exert a chaotropic or kosmotropic effect.

An enzyme can deactivate irreversibly for two kinds of reasons: (i): *conformational processes*, such as aggregation (intermolecular), or incorrect structure formation (intramolecular), such as scrambled disulfide bond formation between wrong side chains, and (ii): *covalent processes*, such as reduction and thus destruction of disulfide bonds, deamidation of asparagine (Asn) or glutamine (Gln) side chains, or hydrolysis of (usually) labile asp–X bonds in the protein sequence.

The formation of aggregates during protein folding or unfolding can be viewed as kinetic competition between correct folding and aggregate formation reactions. Aggregation results in the formation of inclusion bodies, amyloid fibrils, and folding aggregates. Substantial data support the hypotheses that (i) partially folded intermediates are key precursors to aggregates; (ii) aggregation involves specific intermolecular interactions; and (iii) most aggregates involve β-sheets. Protein aggregation can be a major problem, not just in biocatalysis but also:

- when considering the shelf-life of protein pharmaceuticals,
- during fermentation and expression of proteins, as they can form inclusion bodies, and lastly,
- in misfolding diseases such as Alzheimers, Type II diabetes, Creutzfeldt–Jakob disease (CJD) and "mad cow" disease (BSE, bovine spongiform encephalopathy).

Overexpression of a recombinant protein often leads to the production of *inclusion bodies*, insoluble aggregates of misfolded protein. These inclusion bodies can easily be purified. However, the expressed protein can usually be solubilized only by using strongly denaturing conditions, and a major problem is then to achieve an efficient folding in vitro. Options to reduce the formation of inclusion bodies include (i) decreasing the rate of expression by limiting induction of gene expression or lowering the temperature, (ii) coexpressing the target protein with a hydrophilic fusion protein partner, or (iii) modifying the target protein's sequence to improve in-vivo and in-vitro folding.

17.1
Summary: Protein Folding, First-Order Decay, Arrhenius Law

17.1.1
The Protein Folding Problem

The most central problem of protein science is the protein folding problem, the challenge to predict the three-dimensional folding pattern of a protein and thus its 3D dimensional structure from its given primary sequence. Up to this day this has not been possible. One reason is the lack of speed of contemporary computers: even with the assumed continual increase in storage density and processing speed (doubling every 18 months: "Moore's law") it would be about another 25 years until prediction of protein folding seems possible with mean-field calculations using Newtonian mechanics.

Protein folding cannot occur as a random process. Assuming a chain length of 100 amino acids and diffusional control of encounter between different parts of the chain to probe folding occurring over a time constant of $10^{10}\ s^{-1}$, the process of sampling of all configurations would take

$$2^{100}/2 \div 10^{10}\ s^{-1} = 10^{33}\ s = 5 \times 10^{24}\ \text{years}$$

or considerably longer than the age of the universe. An alternative approach with a similar calculation is known as Levinthal's paradoxon (Levinthal, 1969): a protein of 100 amino acids with side chains in three different conformational states can attain a total of 100!/3 ($= 3.1 \times 10^{157}$) different configurational states; diffusion of those states would take even longer.

As a consequence of either calculation, there must be shortcuts to arrive at the final and thus active conformation of a protein (usually only one conformation). It is known that the formation of secondary structure elements such as α-helices and ß-sheets proceeds preferentially and much faster than other parts of the protein; the rapid formation of such elements pre-orders the amino acid chain and sets it up for faster formation of whole domains and then of the complete, native, catalytically active protein.

17.1.2
Why do Proteins Fold?

As early as the 1930s it was believed (by Linus Pauling, among others) that hydrogen bonds provide the main driving force for folding. However, as was pointed out in the 1950s, hydrogen bonding would not favor the folded state strongly because the unfolded conformation can form hydrogen bonds with water which compare in strength with the bonds in the folded protein. The next attempt to explain protein folding focused on hydrophobic interactions. Non-polar, oil-like molecules do not like to mix with water; the mixing process is accompanied by a large positive free energy and is disfavored entropically.

The three-dimensional structure of a protein does not depend on environmental factors within the cell but is determined solely by the amino acid sequence information (Epstein, 1963); this is called the Anfinsen dogma. Model systems of enzymes converted into linear polypeptide chains devoid of disulfide bonds and of secondary and tertiary structures support the theory that the folding process is essentially a thermodynamic one and that no genetic information other than that present in the amino acid sequence of the protein is required.

So what drives an amino acid chain to fold? The chain contains sufficient hydrophobic residues to favor minimization of area by folding. However, as there are fewer possible conformational states in the folded state than in the unfolded one, folding is still entropically disfavored, and a positive configurational entropy has to be considered.

The balance of favorable minimization of hydrophobic area and unfavorable reduction of conformational states upon folding will determine the stability of the protein. As these forces tend to be large and comparable in magnitude, the free enthalpy of formation of a protein is the sum of two large opposing forces and may thus be negative or positive. In any case, a folded and catalytically active protein is always just a few kilojoules away from instability.

The current picture of protein folding is nicely summarized in Figure 17.1. There are a multitude of pathways from the state of unfolded protein at the top of the figure to the native protein at the bottom. Without folding intermediates, the walls of the funnel to the bottom would be smooth. If the energy trough around a non-native protein is too deep, the misfolded protein cannot exit this local minimum.

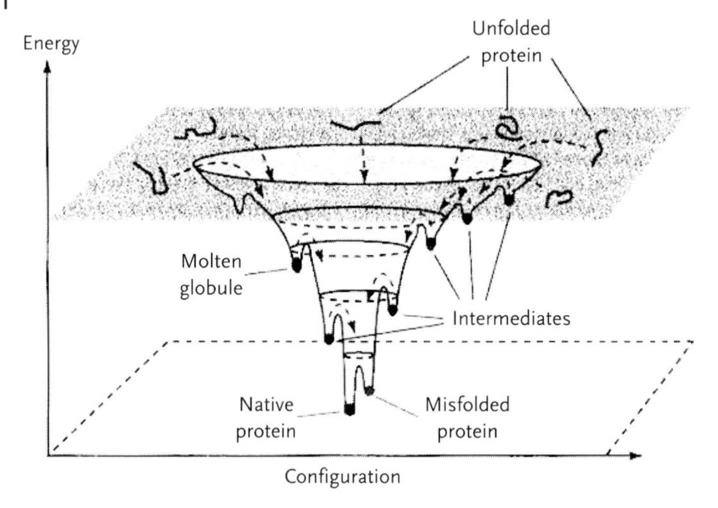

Figure 17.1 Schematic diagram of a folding energy landscape (Radford, 2000; from Schultz, 2000).

Maximum temperature for proteins?

Is there a maximum temperature at which enzymes are stable? At such a temperature, proteins would unfold in any solvent system. In a study, the temperature dependence of the thermodynamics of unfolding for cytochrome *c* [ΔH_m and ΔS_m as $f(T)$] has been determined as a function of low concentrations of methanol (Fu, 1992). As expected, the heat capacity change (ΔC_p) for unfolding decreased with increased methanol concentrations, indicating a higher solvent hydrophobicity which, at constant melting temperature T_m, results in higher experimental enthalpy (ΔH_m) and entropy (ΔS_m) changes with increased methanol concentrations. When the data sets for ΔH_m and ΔS_m at different methanol concentrations were plotted as a function of temperature, each set converged to a common value, around 100 °C for ΔH_m and around 112 °C for ΔS_m. These convergence temperatures approximately match those obtained for ΔH_m and ΔS_m for other proteins in aqueous solution, once ΔH_m and ΔS_m are normalized with respect to the number of residues in each protein. The convergence temperatures for ΔH_m and ΔS_m have been identified with the point at which hydrophobic contributions to ΔH_m and ΔS_m become zero. This viewpoint suggests that proteins cannot be stable without hydrophobic interactions and thus the convergence temperatures can be interpreted as the maximum temperature for proteins to exist in aqueous solution; this temperature then is around 112 °C.

17.2 Two-State Model: Thermodynamic Stability of Proteins (Unfolding)

17.2.1 Protein Unfolding and Deactivation

Often, it is convenient to ignore the process of folding or unfolding of a protein altogether and just consider the result at both ends of the folding or unfolding process: the completely folded protein possessing catalytic activity is termed the *native* protein, designated "N"; the state at the other end of the folding process, which is assumed reversible, is aptly termed the *unfolded* form, designated "U", and is assumed to be completely inactive. This picture [Eq. (17.1) with $K_u = [U]/[N]$ as the unfolding equilibrium constant] is known as the *two-state model.*

$$N \overset{K_u}{\Leftrightarrow} U \tag{17.1}$$

The unfolded state U is modeled as a single conformation even though this is by no means obvious and usually cannot be verified. The unfolding equilibrium between N and U can be shifted to the unfolded form by measures such as an increase in temperature or an increase in the concentration of "structure breakers", termed chaotropes, such as urea or guanidinium hydrochloride. When either temperature or the concentration of chaotropes is decreased (or the concentration of "structure formers", kosmotropes, is increased), the folding equilibrium reverts to the native state N. The two-state model is an approximation; very often, analytically verified folding intermediates render the two-state model incomplete.

The two-state model according to Eq. (17.1) will be enhanced during the following discussion of deactivation, aggregation and formation of inclusion bodies.

17.2.2 Thermodynamics of Proteins

The free enthalpy of formation of a protein can be written as a difference in the Gibbs free enthalpy between folded and unfolded states, G_u and G_f [Eq. (17.2)].

$$\Delta G = \Delta H - T \cdot \Delta S = G_u - G_f \tag{17.2}$$

The difference in heat capacity between unfolded and folded protein is assumed, based on good experimental data, to be positive and constant [Eq. (17.3)].

$$C_p(U) - C_p(N) = \Delta C_p > 0 \tag{17.3}$$

Equations (17.4) and (17.5) then give expressions for enthalpy and entropy, and yield Eq. (17.6) when inserted into Eq. (17.2).

$$\Delta H = \Delta H_0 + \Delta C_p\ (T - T_0) \tag{17.4}$$

$$\Delta S = \Delta S_0 + \Delta C_p \ln\ (T/T_0) \tag{17.5}$$

$$\Delta G = \Delta H_0 - T \cdot \Delta S_0 + \Delta C_p \cdot [(T - T_0) - T \cdot \ln\ (T/T_0)] \tag{17.6}$$

The stability curve of the protein can be inferred from the behavior of the free enthalpy ΔG over the temperature T. From the other important equation for the free enthalpy ΔG, Eq. (17.7), the limit of stability at $\Delta G = 0$ is reached when $K = 1$ or when the concentration of native and unfolded species are equal.

$$\Delta G = -\ RT \ln K = -\ RT \cdot \ln\ [F_d/(1 - F_d)] \tag{17.7}$$

From Eqs. (17.7) and (17.2) it can be seen that, at that point, $K = 1$, $F_n = F_{d,}$ and $T = T_m$; T_m is the melting temperature of the protein. Eq. (17.2) than gives Eqs. (17.8).

$$\Delta G_m = 0 = \Delta H_m - T_m \cdot \Delta S_m \tag{17.8a}$$

or

$$T_m = \Delta H_m / \Delta S_m \tag{17.8b}$$

If Eq. (17.8b) is inserted into Eq. (17.6) and T_0 is identified with T_m Eq. (17.9) results.

$$\Delta G(T) = \Delta H_m\ (1 - T/T_m) - \Delta C_p\ [(T_m - T) + T \cdot \ln\ (T/T_m)] \tag{17.9}$$

To find the temperature of optimal stability, Eq. (17.9) needs to be differentiated with respect to T and the derivative set equal to zero [Eq. (17.10), or with some algebra, Eq. (17.11)].

$$[\partial G/\partial T]_{T = Ts} = 0 = \Delta H_m / T_m + \Delta C_p \cdot \ln\ (T_s/T_m) \tag{17.10}$$

$$T_s = T_m \exp\ [-\Delta H_m/(T_m \cdot \Delta C_p)] \tag{17.11}$$

T_s is the temperature of maximum stability, at which the $\Delta G(T)$ versus T curve passes through its maximum, i.e., $\Delta G(T_s) > 0$. At T_m, the maximum temperature of stability, $\Delta G(T) = 0$. By necessity, $T_s < T_m$. There is another special point in the temperature stability ($\Delta G(T)$) curve, however: as the temperature stability curve resembles a inverted parabola, there must be a second intersection point with the line of minimum stability, $\Delta G(T) = 0$. This intersection point is the minimum of the range of temperature stability and is called *cold denaturation temperature*, T_c. The cold denaturation temperature is practically never taken into consideration in the discussion of temperature stability of biocatalysts. One reason lies in its frequent inaccessibility: often, T_c is below 0 °C and thus cannot be measured in water or any other mostly aqueous medium.

Experimental evidence for the two-state model

Hen egg-white lysozyme, lyophilized from aqueous solutions of different pH from pH 2.5 to 10.0 and then dissolved in water and in anhydrous glycerol, exhibits a cooperative conformational transition in both solvents occurring between 10 and 100°C (Burova, 2000). The thermal transition in glycerol is reversible and equilibrium follows the classical two-state mechanism. The transition enthalpies ΔH_m in glycerol are substantially lower than in water, while transition temperatures T_m are similar to values in water, but follow similar pH dependences. The transition heat capacity increment ΔC_p in glycerol does not depend on the pH and is 1.25 ± 0.31 kJ (mol K)$^{-1}$ compared to 6.72 ± 0.23 kJ (mol K)$^{-1}$ in water. Thermodynamic analysis of the calorimetric data reveals that the stability of the folded conformation of lysozyme in glycerol is similar to that in water at 20–80°C but exceeds it at lower and higher temperatures.

17.3 Three-State Model: Lumry–Eyring Equation

17.3.1 Enzyme Deactivation

The two-state model does not capture irreversible deactivation effects which lead to decrease of enzyme activity over time. To capture irreversible decrease in enzyme activity, we build on the two-state model and include an irreversible first-order deactivation term (U → D) from "U" to a new state, the (irreversibly) inactivated state, designated "D" [Eq. (17.12)].

$$\mathrm{N} \underset{k_{-1}}{\overset{k_1}{\Leftrightarrow}} \mathrm{U} \overset{k_d}{\rightarrow} \mathrm{D} \tag{17.12}$$

As before, "N" denotes the native, catalytically active state of the enzyme, "U" an inactive, reversibly unfolded state, and "D" the irreversibly inactivated state, which we assume is only accessible from the unfolded state U. A three-state model kinetic representation of protein unfolding and deactivation as in Eq. (17.12) is the most common representation and is referred to as the *Lumry–Eyring model* (Lumry, 1954). As mentioned above, the unfolded form U can be refolded into the native form N by lowering the temperature below the melting temperature T_m, or by the addition of kosmotropic agents, or by withdrawal, through dilution or dialysis, of chaotropes. Operationally, an enzyme can be deactivated irreversibly from the unfolded state U by further raising the temperature beyond the range of unfolding the protein.

In the following paragraphs, we will investigate the relationship between the observed deactivation rate constant $k_{d,obs}$ and the intrinsic deactivation rate constant k_d.

If $[E]_0 = [N] + [U]$, then the rate of deactivation is given by Eq. (17.13), but according to Eq. (17.12) the rate must also obey Eq. (17.14).

$$v = k_{d,obs} \cdot [E]_0 = k_{d,obs} \cdot ([N] + [U]) \tag{17.13}$$

$$v = k_d \cdot [U] \tag{17.14}$$

With the definition of the folding equilibrium constant $K = [U]/[N]$, Eq. (17.13) now reads according to Eq. (17.15) or (17.16).

$$v = k_d \cdot K \cdot N = k_{d,obs} \cdot [N] \cdot (1 + K) \tag{17.15}$$

or

$$k_{d,obs} = k_d \cdot (K/(1 + K)) \tag{17.16}$$

Three separate cases [Eqs. (17.17)] can be distinguished in the application of Eq. (17.16).

High T, or $T > T_m$: then $[U] \gg [N]$, thus $K \ll 1$, and $k_{d,obs} \approx k_d \cdot K$ (17.17a)

Moderate T, or $T < T_m$: then $[U] < [N]$, thus $K > 1$, and $k_{d,obs} \approx k_d \cdot K/(1 + K)$ (17.17b)

Low T, or $T \ll T_m$: then $[U] \ll [N]$, thus $K \gg 1$, and $k_{d,obs} \approx k_d$ (17.17c)

Equations (17a–c) indicate that extreme care is required when trying to determine k_d by calculating $k_{d,obs}$.

Evidence for the three-state Lumry–Eyring model
Thermal denaturation of thermolysin from *Bacillus thermoproteolyticus rokko*, important for the manufacture of aspartame (Chapter 7, Section 7.3.2), investigated by differential scanning calorimetry (DSC), was found to be irreversible and highly dependent on the scan rate (rate of temperature increase) (Sanchez-Ruiz, 1988). Both findings point towards a deactivation rate law according to Eq. (17.1), $N \rightarrow D$, with a rate constant $k_{d,obs}$, or even more appropriately, to a rate law according to the Lumry–Eyring model, $N \Leftrightarrow U \rightarrow D$ [Eq. (17.12)]. On the basis of this model, the value of the rate constant, as a function of temperature, and the activation energy have been calculated.

17.3.2 Empirical Deactivation Model

Empirically, the following equation often holds:

$$N \xrightarrow{k_{d,obs}} D$$

and thus one obtains Eq. (17.18), which yields Eq. (17.19) upon integration.

$$d[N]/dt = k_{d,obs} \cdot [N] \quad (17.18)$$

$$[N]_t = [N]_0 \cdot \exp(-k_{d,obs} \cdot t) \quad (17.19)$$

If we now write the equation for a batch reactor (Chapter 5, Section 5.2.2) with deactivating enzyme over time, we obtain Eq. (17.20), or after separation of variables Eq. (17.21), and after integration Eq. (17.22).

$$r = -d[S]/dt = k_{cat} \cdot [N]_t \cdot [S]/(K_M + [S]) = k_{cat} \cdot [N]_0 \cdot \exp(-k_{d,obs} \cdot t) \cdot [S]/(K_M + [S]) \quad (17.20)$$

$$\{(K_M + [S])/[S]\}\, d[S] = [N]_0 \cdot k_{cat} \cdot \exp(-k_{d,obs} \cdot t)\, dt \quad (17.21)$$

$$([S]_0 - [S]) - K_M \ln([S]/[S]_0) = [N]_0 \cdot (k_{cat}/k_{d,obs}) \{1 - \exp(-k_{d,obs} \cdot t)\} \quad (17.22)$$

With the definition of the degree of conversion x as: $x = 1 - ([S]/[S]_0)$ we obtain Eq. (17.23).

$$x \cdot [S]_0 - K_M \cdot \ln(1 - x) = [N]_0 \cdot (k_{cat}/k_{d,obs}) \{1 - \exp(-k_{d,obs} \cdot t)\} \quad (17.23)$$

Deactivation of the enzyme catalyst often goes unnoticed, especially if the biocatalyst is fairly stable and/or the time of observation is short, such as during an initial rate measurement. If we analyze the short-time behavior of Eq. (17.23) we can replace the exponential by a linear term deviating from unity. At short times, $\exp(-k_{d,obs} \cdot t) \approx 1 - k_{d,obs} \cdot t$, and Eq. (17.24) is recovered.

$$x \cdot [S]_0 - K_M \cdot \ln(1 - x) = [N]_0 \cdot k_{cat} \cdot t \quad (17.24)$$

which corresponds exactly to the equation for the batch reactor with non-deactivating enzyme already encountered in Chapter 5, Section 5.2.3 [Eq. (17.14b)]. As a consequence, at short times the equations for deactivating and non-deactivating enzyme behave identically, which is the reason why such short-time behavior is not suitable for analyzing deactivation.

At short times, as again typically encountered during an initial-rate experiment, the degree of conversion is low, especially in the case of a highly soluble substrate with high $[S]_0$, so we can neglect the logarithmic term in Eq. (17.23) as $\ln(1 - x) \approx \ln 1 \approx 0$, and Eq. (17.23) becomes Eq. (17.25) or (17.26).

$$x \cdot [S]_0 \approx [N]_0 \cdot (k_{cat}/k_{d,obs}) \{1 - \exp(-k_{d,obs} \cdot t)\} \quad (17.25)$$

or

$$x \cdot ([S]_0 \cdot k_{d,obs})/([N]_0 \cdot k_{cat}) \approx \{1 - \exp(-k_{d,obs} \cdot t)\} \quad (17.26)$$

At short times ($t \rightarrow 0$) or stable enzyme ($k_{d,obs} \rightarrow 0$), we again replace the term ($\exp(-k_{d,obs} \cdot t)$) by ($1 - k_{d,obs}\, t$) and obtain Eq. (17.27), which explains the often observed linear relationship between time and degree of conversion.

$$x \cdot [S]_0/([N]_0 \cdot k_{cat}) \approx t \tag{17.27}$$

Resting versus operational stability revisited

Regardless of whether it is the observed deactivation constant $k_{d,obs}$ or the intrinsic deactivation constant k_d, that is being analyzed, when running a biocatalytic process it is often of great importance to keep deactivation over time as small as possible. In this context, readers are reminded of the difference between resting stability and operational or process stability as already discussed in Chapter 5, Section 5.6. Any numerical values of $k_{d,obs}$ or k_d from the literature should be examined critically for the conditions in which they were obtained. As discussed in Section 5.6, deactivation constants in resting, quiescent solution, often under storage conditions, differ from those under reacting conditions, even if the internal mass transfer is not limiting ($\eta_I \to 1$) and the biocatalytic reaction is operated under saturation conditions ($[S] \gg K_M$). The reaction can influence the stability of the enzyme catalyst in several ways:

- In the presence of substrate, product, or an inhibitor, stabilization of the biocatalyst is often observed because the enzyme molecule unfolds more slowly or in harsher conditions in the presence of a bound molecule, particularly if binding is tight (low K_M, K_P, or K_I).

- At a high concentration of substrate, product, or inhibitor, these compounds can exert a chaotropic or kosmotropic effect, regardless of their ability to interact with the active site of the enzyme.

17.4 Four-State Model: Protein Aggregation

17.4.1 Folding, Deactivation, and Aggregation

In some situations, the three-state Lumry–Eyring model, which includes unfolding and (first-order) denaturation, is insufficient to describe experimental results. A four-state model describes the situation better. Causes necessitating a four-state model can be aggregation of the biocatalyst or folding intermediates between the native and the fully unfolded state.

Protein aggregation can be a major problem, not just in biocatalysis but also
- when considering the shelf-life of protein pharmaceuticals,
- during fermentation and expression of proteins, as they can form inclusion bodies, and lastly,
- in misfolding diseases such as Alzheimer's, Type II diabetes, Creutzfeldt–Jakob disease (CJD) and "mad cow" disease (BSE, bovine spongiform encephalopathy).

Aggregation results in the formation of inclusion bodies, amyloid fibrils, and folding aggregates (Figure 17.2) (Fink, 1998). Substantial data support the hypotheses that

- partially folded intermediates are key precursors to aggregates,
- aggregation involves specific intermolecular interactions, and
- most aggregates involve β-sheets.

If aggregation is important, the Lumry–Eyring model has to be extended by introducing a folding intermediate I, which can lead reversibly either to the native protein N, or the fully unfolded protein U, or irreversibly to a protein aggregate A_n, assumed catalytically inactive. The kinetic representation is thus expressed by Eq. (17.28).

$$\begin{array}{ccccccc} & k_1 & & & & k_d & \\ N & \Leftrightarrow & I & \Leftrightarrow & U & \rightarrow & D \\ & k_{-1} & \downarrow k_A & & & & \\ & & A_n & & & & \end{array} \tag{17.28}$$

Aggregation of partially folded intermediates during protein folding is widely regarded as leading to termination of further protein folding. In some cases, at a high protein or salt concentration, association as soluble aggregates supports further secondary structure formation (Uversky, 1998) but this seems to be an exception rather than the rule. In general, formation of folding intermediates has to be avoided.

Proteins are both colloids and polymers. Therefore, attempts have been made to understand the phenomenon of protein aggregation with the help of models from the polymer and colloid fields such as DLVO theory, describing the stability of colloidal particles, or phase behavior and attraction–repulsion models from polymers (De Young, 1993). For faster progress, more phase diagrams for equilibrium protein precipitation, in both the crystalline and the non-crystalline state, as well as more data on observations of defined protein oligomers or polymers, are required.

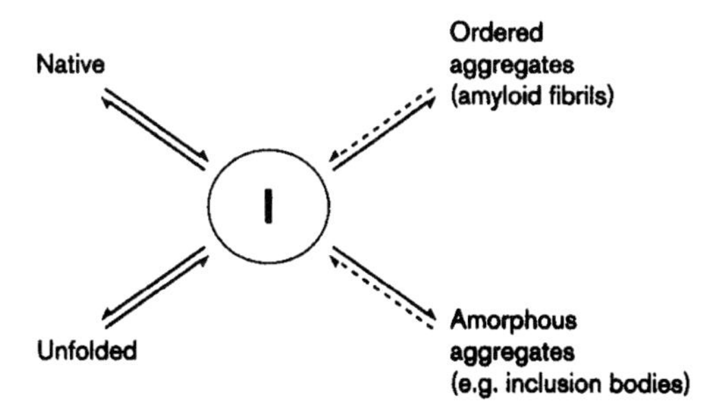

Figure 17.2 Basic model of protein aggregation with the partially folded intermediate in the center (Fink, 1998).

17.4.2
Model to Account for Competition between Folding and Inclusion Body Formation

The formation of aggregates during protein folding or unfolding can be viewed as kinetic competition between correct folding and aggregate formation reactions. We present a kinetic model originally published by Kiefhaber et al. (Kiefhaber, 1991) which operates on the basis of the kinetic scheme in Eq. (17.28) and takes into account three different processes with, as it turns out, different reaction order (Figure 17.3):

- *folding* from the intermediate to the correct native structure, a reversible *first-order* process;
- *formation of aggregates* ultimately resulting in inclusion bodies, a higher-order process modeled here as a *second-order* process; and
- formation of the folding intermediate from *protein synthesis*; as ribosomal protein is assumed completely saturated with tRNA amino acids, the substrates, protein synthesis is modeled as a *zeroth-order* process.

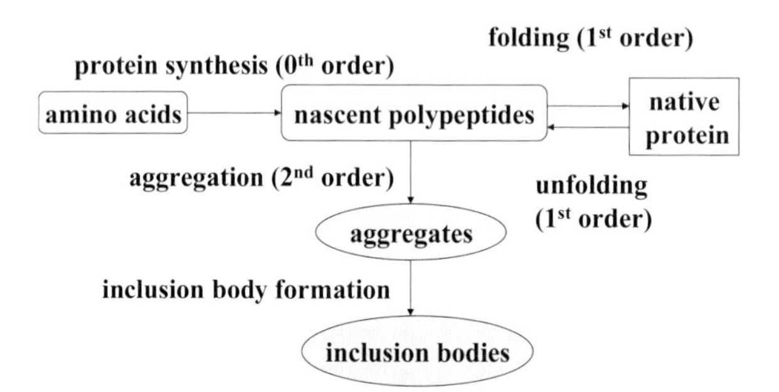

Figure 17.3 Protein synthesis, folding to native structure, and aggregate formation.

The aggregation rate is modeled as second-order because dimerization is often the initial step to the formation of aggregates of aggregation number *A*, with initial dimerization as the rate-limiting step.

17.4.2.1 Case 1: In Vitro – Protein Synthesis Unimportant

The molar concentration [I] over time can be expressed by Eq. (17.29), with k_1 as the rate of folding and k_2 as the rate of dimerization.

$$d[\mathrm{I}]/dt = -(k_1 \cdot [\mathrm{I}] + k_2 \cdot A \cdot [\mathrm{I}]^2) \tag{17.29}$$

The product $k_2 \cdot A$ is the apparent rate constant of aggregation. The solution to the differential Eq. (17.29) is Eq. (17.30), with $[\mathrm{I}]_0$ as the initial concentration of the folding intermediate I.

$$[\mathrm{I}]/[\mathrm{I}]_0 = \frac{\exp(-k_1 \cdot t)}{1 + \{[\mathrm{I}]_0 \cdot (k_2 \cdot A/k_1) \cdot (1 - \exp(-k_1 \cdot t)\}} \tag{17.30}$$

The concentration of N, the native species, and after all the concentration one wishes to optimize, is related to [I] according to Eq. (17.31), with $[N]_0 = 0$.

$$[N] = k_1 \cdot \int [I](\tau) \cdot d\tau \tag{17.31}$$

Substitution of Eq. (17.30) into Eq. (17.31) and integration yields Eq. (17.32).

$$[N]/[I]_0 = \{k_1/(k_2 \cdot A \cdot [I]_0)\} \ln \{1 + (k_2 \cdot A \cdot [I]_0/k_1)(1 - \exp(-k_1 \cdot t)\} \tag{17.32}$$

Equation (17.32) indicates that increasing the initial concentration of denatured protein $[I]_0$ results in a decrease in the relative amount of native protein $[N]/[I]_0$ at time t. If we define the yield of native protein y_N as the ratio $[N]^\infty/[I]_0$ we have to examine the long-time asymptotic solution of Eq. (17.32), when $t \to \infty$ and $[N] \to [N]^\infty$, to obtain the ratio y_N [Eq. (17.33)].

$$y_N = k_1/\{(k_2 \cdot A \cdot [I]_0)\} \ln \{1 + (k_2 \cdot A \cdot [I]_0/k_1)\} \tag{17.33}$$

From Eq. (17.33) we discern that y_N just depends monotonically on the ratio $k_1/(k_2 \cdot A \cdot [I]_0)$ (Figure 17.4).

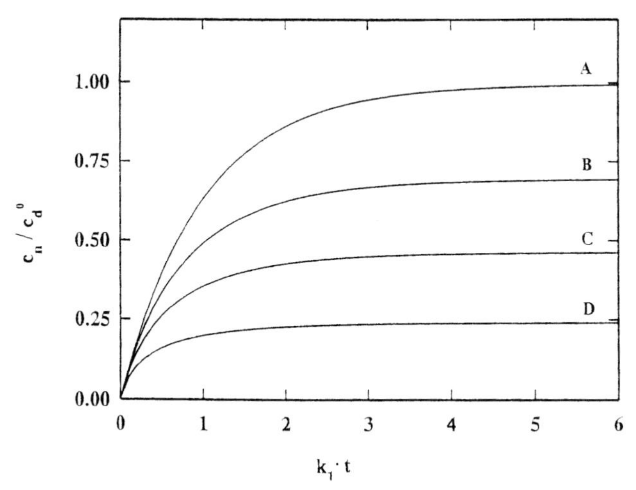

Figure 17.4 Normalized time course of the relative amount of renatured protein ($[N]/[I]_0 \equiv$ "c_n/c_d^0") during refolding. Values of $(k_2 \cdot A \cdot [I]_0)/k_1$: A): 0.01; B) 1; C) 3; D) 10 (Kiefhaber, 1991).

17.4.2.2 **Case 2: In Vivo – Protein Synthesis Included**

This case is particularly relevant in regard to the formation of inclusion bodies (see Section 17.6 below) as overexpression of a target protein is a mode of obtaining the protein. With the assumption that folding and aggregation behave as in case 1, and nascent protein is expressed with a zeroth-order rate constant k_0 with respect to the amount of polypeptide per volume and time, its mass n_p changes according to Eq. (17.34), where k_1, k_2, and A have the same meaning as above, and [P] replaces [I].

$$dn_p/dt = \{k_0 - (k_1 \cdot [P] + k_2 \cdot A \cdot [P]^2)\} \cdot V \qquad (17.34)$$

Substituting $n_p = [P] \cdot V$, Eq. (17.35) is obtained, where $\mu = (1/V)(dV/dt)$ is the specific cell growth rate, also acting as a dilution rate, which was encountered in Chapter 8, Section 8.2.6.

$$d[P]/dt = k_0 - \{(k_1 + \mu) \cdot [P] + k_2 \cdot A \cdot [P]^2\} \qquad (17.35)$$

As $\mu = \ln 2/\tau_2$ and the doubling time τ_2 is hardly less than 20 min, $\mu \leq 6 \times 10^{-4}$ s^{-1}. As long as $k_1 > 10^{-3}$ s^{-1}, the contribution of μ can be neglected.

Under steady-state conditions, $t \to \infty$ and $d[P]/dt = 0$, Eq. (17.35) reduces to just a quadratic equation in $[P]^\infty$. The steady-state yield y_0 of native protein is then found to be given by Eq. (17.36).

$$y_0 = k_1 \cdot [P]^\infty / k_0 = \frac{k_1 \cdot (k_1 + \mu)}{2 \cdot k_0 \cdot k_2 \cdot A} \left[\left(1 + \frac{4 \cdot k_0 \cdot k_2 \cdot A}{(k_1 + \mu)^2} \right)^{0.5} - 1 \right] \qquad (17.36)$$

If the contribution of μ can be neglected, this becomes Eq. (17.37), and y_0 only depends monotonically on the ratio $(k_0 \cdot k_2 \cdot A)/k_1^2$ in an inverse way.

$$y_0 = k_1 \cdot [P]^\infty / k_0 = k_1^2/(2 \cdot k_0 \cdot k_2 \cdot A) \, [\{1 + (4 \cdot k_0 \cdot k_2 \cdot A)/k_1^2\}^{0.5} - 1] \qquad (17.37)$$

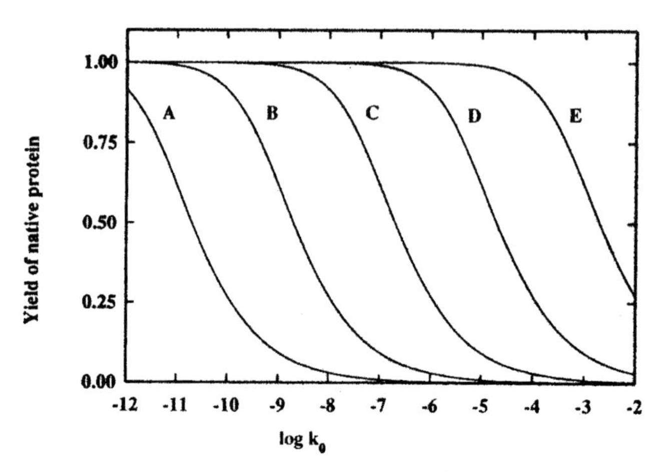

Figure 17.5 Effect of the rate of expression (k_0) on the yield of native protein in vivo (Kiefhaber, 1991). The lines are calculated from Eq. (17.37) with $k_2 \cdot A = 10^5$ M^{-1} and the following values for k_1 [s^{-1}]: A) 0.001; B) 0.01; C) 0.1; D) 1; E) 10 (Kiefhaber, 1991).

According to Eq. (17.37) the yield of native protein increases with the second power of the folding rate but decreases inversely with increasing rate of expression. In Figure 17.5, the yield of protein is plotted against log k_0.

At low rates of expression, even slowly folding proteins can reach quantitative yields. As an upper bound for k_0, Kiefhaber et al. (1991) calculated the rate of expression to be 4×10^{-6} M s^{-1} for 20% total cellular protein content, a molecular weight of 40 kDa and a doubling time τ_2 of 20 min. The actual rate of expression is much lower, about 2.5×10^{-8} M s^{-1} if 10^9 cells produce 0.1 mg of recombinant protein in 2 h, again for a protein of 40 kDa and a cell diameter of 3 μm. In the latter case, even a folding constant as low as $k_1 = 0.05$ s^{-1} would still give a yield of 77% native protein whereas in the former case $k_1 = 1$ s^{-1} would be required for that yield and $k_1 = 0.01$ s^{-1} would only result in a yield of 1.6%.

17.5 Causes of Instability of Proteins: $\Delta G < 0$, $\gamma(t)$, A

In the preceding sections, we have covered the kinetic behavior of enzyme deactivation. We now discuss the molecular basis for enzyme deactivation. Alexander Klibanov's group at MIT (Cambridge/MA, USA) analyzed the mechanisms of deactivation for ribonuclease A. They drew the conclusion that an enzyme can deactivate irreversibly for two kinds of reasons (Ahern, 1985, 1988):

- Deactivation by *conformational processes*, such as *aggregation*, an intermolecular process, or through *incorrect structure formation*, an intramolecular process. The latter can happen if the wrong cysteine side chains form disulfide bonds, which renders the protein structure inactive.

- Deactivation by *covalent processes*, such as
  - reduction and thus destruction of disulfide bonds to two cysteine bonds,
  - deamidation of asparagine or glutamine side chains to aspartate or glutamate, or
  - hydrolysis of (usually) labile Asp–X bonds in the protein sequence via intermittent and intramolecular formation of isoaspartate.

In the following sections, we will discuss the influence of several reaction parameters such as temperature, degree of stirring, and the presence of gas/liquid surfaces on the deactivation behavior of proteins. Mechanisms of enzyme inactivation and aggregation are still poorly understood so, in the best case, the rate of activity loss can be correlated with system parameters such as surface area, power input, or collision frequency. However, the ability to control the resistance of an enzyme to inactivation due to exposure to elevated temperatures is essential for the understanding of thermophilic behavior and for developing rational approaches to enzyme stabilization.

17.5.1 Thermal Inactivation

In a classic study on bovine pancreatic ribonuclease A at 90°C and pH conditions relevant for catalysis, irreversible deactivation behavior was found to be a function of pH (Zale, 1986): at pH 4, enzyme inactivation is caused mainly by hydrolysis of peptide bonds at aspartic acid residues as well as deamidation of asparagine and/or glutamine residues, whereas at pH 6–8, enzyme inactivation is caused mainly by thiol–disulfide interchange but also by β-elimination of cystine residues, and deamidation of asparagine and/or glutamine residues.

To improve its thermostability, point mutations were engineered by site-directed mutagenesis in the dimeric enzyme yeast triosephosphate isomerase (Ahern, 1987). The mutant proteins were expressed in a strain of *E. coli* lacking the bacterial isomerase and purified by ion-exchange and immunoadsorption chromatography (Casal, 1987). In an attempt to model the deleterious effects of deamidation, interfacial Asn78 was replaced by aspartic acid which, as expected, led to a strong decrease of the melting temperature T_m, an increase in the irreversible deactivation rate constant k_d, and a reduction of the stability against dilution-induced dissociation as well as lowered resistance to denaturants, tetrabutylammonium bromide, alkaline pH, and proteases.

As expected, the pI of the Asn78Asp mutant was equivalent to that of wild-type yeast TIM that had undergone a single, heat-induced deamidation. Cumulative replacement of asparagine residues at the subunit interface by residues approximating the geometry of asparagine but resistant to heat-induced deterioration, Asn14Thr and Asn78Ile, nearly doubled the half-life of the enzyme at pH 6 and 100°C but had little effect on the kinetic constants. In contrast, the single mutant Asn78Asp and the double mutant Asn14Thr/Asn78Ile registered appreciably lower k_{cat} for the substrate D-glyceraldehyde 3-phosphate.

To ascertain the upper limit of protein thermostability and to evaluate the effect of additional disulfide bridges on the enhancement of protein thermostability, additional cysteine residues were introduced into several unrelated proteins by site-directed mutagenesis and deactivation behavior tested at 100°C (Volkin, 1987). All the proteins investigated underwent heat-induced beta-elimination of cystine residues in the pH 4–8 range with first-order kinetics and similar deactivation constants k_d that just depended on pH: $0.8 \pm 0.3\ h^{-1}$ at pH 8.0 and $0.06 \pm 0.02\ h^{-1}$ at pH 6.0. These results indicate that beta-elimination is independent of both primary amino acid sequence and the presence of secondary structure elements. Elimination of disulfides produces free thiols that cause yet another deleterious reaction in proteins, heat-induced disulfide interchange, which can be much faster than beta-elimination.

Disulfide interchange was also found to dominate mostly in another case: the difference in half-lives $\tau_{1/2}$ at 90°C and pH 6.5 of *Bacillus* α-amylases extended two orders of magnitude from *Bacillus amyloliquefaciens* through *B. stearothermophilus* to *B. licheniformis* (Tomazic, 1988). For *B. licheniformis* α-amylase, deamidation of Asn/Gln residues was the main cause of inactivation. The cause of thermostability of the *B. amyloliquefaciens* enzyme was found in additional salt bridges at a few specific lysine residues; they reduce unfolding at high temperatures, thus lowering the rate of incorrect structure formation, and thus also increasing the overall rate of irreversible thermoinactivation.

17.5.2
Deactivation under the Influence of Stirring

The combined effects of inactivation and aggregation on lysozyme were investigated in a stirred tank reactor. The inactivation kinetics was found to follow a first-order law. For stirring speeds N in the range of 0–700 rpm or impeller tip speeds of 0 to 0.77 m s^{-1}, the rate constant of deactivation $k_{d,obs}$ was found to be proportional to the power imparted to the impeller (Colombie, 2000). This finding suggested that inactivation depends on collisions between enzyme molecules. Efficient collisions between native and inactive molecules are thought to induce the conversion of native molecules into inactive molecules, accompanied by lysozyme aggregation as evidenced by light scattering measurements. The structure of aggregates was investigated on samples treated by chemical denaturation and reduction (Colombie, 2001). The aggregates were found to resemble either supramolecular entities, mainly made up of inactivated enzymes linked by weak forces, or to be made up of dimers (18%) and trimers (5%) of lysozyme which are linked by disulfide bridges. Activity could not be recovered upon stopping of stirring or cooling of the solution, so deactivation must be considered irreversible.

The first-order dependence of the deactivation constant was found to be proportional not only to the power imparted by the impeller but also to the area between the liquid and the glass wall, air surface, or poly(tetrafluoroethylene) (PTFE) surface (Colombie, 2001). Hydrophobic PTFE and air interfaces increased lysozyme inactivation fourfold over glass. In addition, the number and thus the molecular surface of inactivated enzymes, which are more hydrophobic than native enzymes, enhanced lysozyme inactivation and aggregation.

The effect of agitation rate, interfacial interactions, and insulin concentration on the overall aggregation rate of insulin in aqueous solutions, an factor important to be minimized during production, storage, and delivery, was investigated (Sluzky, 1991). Aggregation rate and formation of intermediate species were monitored by UV spectroscopy, HPLC, and quasi-elastic light scattering (QELS). The reaction proceeded in two stages; insulin stability was enhanced at higher concentrations. Any kinetic model would have to be able to predict this fact. Such a model was built based on the following known facts, observations, and assumptions (Sluzky, 1991):

- Insulin monomers are in equilibrium with more stable dimers and hexamers with known association constants: $K_{dimer} = 1.1 \times 10^5$ M^{-1}, $K_{hexamer} = 2.89 \times 10^8$ M^{-2}.
- The monomer is assumed to be the denaturing species at the interface whereas the dimer has stabilized hydrophobic surface parts sufficiently well to adsorb but not to denature at the interface, and the hexamer even more so ($K_{adsorption} = 1.2 \times 10^5$ M^{-1}, $K_{denaturation} = 1.2 \times 10^5$ M^{-1}).

- The unfolded species either refolded back to the native conformation or initiated nucleation with other unfolded species. At a critical size U_n, the intermediates start reacting with native species to form large aggregates of 170 nm.

Equations (a)–(e) (θ = interfacial site) summarize the situation.

$$6\,\mathrm{N} \Leftrightarrow 3\,\mathrm{N}_2 \Leftrightarrow \mathrm{N}_6 \tag{a}$$

$$\mathrm{N}_2 + \theta \Leftrightarrow \mathrm{N}_2\theta\,, \qquad \mathrm{N}_6 + \theta \Leftrightarrow \mathrm{N}_6\theta \tag{b1, b2}$$

$$\mathrm{N} + \theta \rightarrow \mathrm{U} + \theta \text{ with } k_{\text{unfold}}, \qquad \mathrm{U} \rightarrow \mathrm{N} \text{ with } k_{\text{refold}} \tag{c1, c2}$$

$$\mathrm{U} + \mathrm{U} \Leftrightarrow \mathrm{U}_2, \qquad \mathrm{U}_2 + \mathrm{U} \Leftrightarrow \mathrm{U}_3, \qquad \mathrm{U}_{n-1} + \mathrm{U} \Leftrightarrow \mathrm{U}_3, \text{ with } k_{\text{U-association}} \tag{d}$$

$$\mathrm{U}_n + \mathrm{N} \rightarrow \text{aggr.}, \quad \mathrm{U}_n + \mathrm{N}_2 \rightarrow \text{aggr.}, \quad \mathrm{U}_n + \mathrm{N}_6 \rightarrow \text{aggr.}, \text{ with } k_{\text{aggregation}} \tag{e}$$

Owing to the complexity of this model, Eqs.(a)–(e) cannot be solved analytically; instead, the parameters for nuclei formation, $k_{\text{U-association}}$ and $k_{\text{U-dissociation}}$, and aggregation, $k_{\text{aggregation}}$, were fitted to the experimentally obtained curves of insulin concentration over time.

17.5.3
Deactivation under the Influence of Gas Bubbles

Investigating enzyme inactivation at gas/liquid interfaces, it was observed that bubbling nitrogen into an aqueous enzyme solution strongly enhances enzyme inactivation (Caussette, 1998a,b, 1999): at a flow rate of 150 mL min^{-1} the half-life $\tau_{1/2}$ of lysozyme was determined to be 12 min, in comparison with no loss of activity after 8 h without bubbling. Whereas the inactivation kinetics of lysozyme and lipase was found to behave linearly with specific interfacial area, thus demonstrating the presence of an interfacial mechanism, no dependence on specific interfacial area could be observed for pectin methylesterase. The inactivation process was first-order with respect to time for lysozyme and pectin methylesterase but biphasic for lipase: a rapid initial decline in activity was followed by a decrease in deactivation rate. Potentially such behavior is explained by competition between open-lid (active) and closed-lid (inactive) lipase species at the interface. The rate of deactivation induced by nitrogen bubbling strongly depends on temperature and pH value, in all likelihood by influencing the adsorption process of the enzyme at the interface. As a conclusion, there should be careful controls for gas/liquid interfaces in bioreactors, whether for enzyme reactions or fermentations. In a case where deactivation of an enzyme is desired, gas/liquid interfaces can be employed towards this goal.

17.5.4 Deactivation under the Influence of Aqueous/Organic Interfaces

A liquid/liquid bubble column was employed to investigate and quantify interfacial inactivation of enzymes by organic liquids (Ross, 2000). The amount of enzyme inactivated was found to be proportional to the area of organic solvent exposed, as is characteristic of the interfacial mechanism. Tests were made with a series of 12 solvents with log P close to 4.0 but with different functional groups. With α- and δ-chymotrypsin, inactivation was much less severe with amphiphilic molecules such as decanol than with less polar compounds such as heptane. This corresponds to a correlation with aqueous/organic interfacial tension, and presumably reflects a more polar interface as seen by the enzyme adsorbing from the aqueous phase. A 1 : 1 mixture of decanol and heptane behaved similarly to pure decanol, which would be expected to accumulate at the interface. With pig-liver esterase, the correlation was rather weak, however. Accumulated data for interfacial inactivation by alkanes were examined for the above enzymes, and also papain, trypsin, urease and RNase. The differing sensitivities did not show a clear correlation with any enzyme property, although there was some relationship with adiabatic compressibility, thermal denaturation temperature, and mean hydrophobicity. Inactivation rates can differ considerably even between similar enzymes, and, unlike adsorption, do not show a maximum at the isoelectric point (pI) (Halling, 1998).

17.5.5 Deactivation under the Influence of Salts and Solvents

The refolding of unfolded hen egg-white lysozyme, i.e., the reoxidation of reduced enzyme, was investigated in a variety of predominantly non-aqueous media consisting of protein-dissolving organic solvents and water (Rariy, 1999). LiCl and other common salts dramatically increased the refolding yield of lysozyme in non-aqueous systems, up to more than 100-fold, while reducing it in water. The mechanism of this surprising phenomenon appears to involve salt-induced suppression of non-specific lysozyme aggregation during refolding due to enhanced protein solubility.

17.6 Biotechnological Relevance of Protein Folding: Inclusion Bodies

It is often difficult to obtain soluble and active proteins from expression in prokaryotes. Often, overexpression leads to the production of inclusion bodies, which are insoluble aggregates of misfolded protein. These inclusion bodies can be purified easily. However, the solubilization of the expressed protein can usually only be obtained using strongly denaturing conditions, and a major problem is then to achieve an efficient folding in vitro.

Insoluble, inactive inclusion bodies are frequently formed upon recombinant protein production in transformed microorganisms. These inclusion bodies, which contain the recombinant protein in a highly enriched though inactive form, can be isolated by solid/liquid separation. After solubilization, native proteins can be generated from the inactive material by in-vitro folding techniques. New folding procedures have been developed for efficient in-vitro reconstitution of complex hydrophobic, multi-domain, oligomeric, or highly disulfide-bonded proteins (Rudolph, 1996; Lilie, 1998).

The challenge is mainly the lowering of the denaturant concentration to allow folding while at the same time preventing aggregation. The often published suggestion of eliminating the denaturing agent using dialysis is usually a mistake. Dialysis exposes the protein solution to a decreasing denaturant gradient over a few hours, so that proteins will remain exposed for an extended period of time to an intermediate denaturant concentration (2–4 M urea or guanidine) when they are not yet folded but no longer denatured and thus extremely prone to aggregation. Thus, the folding yield of tryptophanase was demonstrated to reach a minimum when the folding mixture was incubated at intermediate denaturant concentration (London, 1974). It is important to limit the aggregation using mild solubilizing agents during the refolding steps. Non-detergent sulfobetaines have been shown to be particularly efficient for the purpose of refolding nascent polypeptides (Vuillard, 1995).

There are several options for reducing formation of inclusion bodies (see also the discussion in Chapter 4, Section 4.5.3):

- decrease the rate of expression by (i) offering a C-source of less value to the organism, (ii) limiting induction of gene expression, e.g., by lowering the amount of inducer, or (iii) lowering the temperature of expression;
- enhance the solubility by coexpressing the target protein with a hydrophilic fusion protein partner such as maltose-binding protein or glutathione-S-transferase (Chapter 4, Section 4.5.3);
- modification of the target protein sequence to improve in-vitro folding.

17.7 Summary: Stabilization of Proteins

Proteins can be stabilized through a series of measures which can be classified as physical, biological, or chemical:

- *Physical methods*:
  - optimization of parameters such as temperature and ionic strength
  - immobilization or matrix entrapment
  - reduction of amount of water (by use of organic solvents or high ionic strength)

- *Biological methods*:
  - strain development
  - use of thermophilic organisms
  - application of rational or combinatorial methods of protein design

- *Chemical methods*:
  - addition of metal ions
  - addition of stabilizing agents (kosmotropes or compatible solutes) such as polyols or amino acid derivatives

17.7.1
Correlation between Stability and Structure

It is a long sought-after goal of biocatalysis research to find a general correlation between the stability of an enzyme and its structure, especially its primary structure (amino acid sequence).

Some good hints to help answer this question have been obtained by studying thermophilic proteins which are in turn obtained from thermophilic organisms. The optimum growth temperature for thermophilic organisms is between 40° and 65°C (moderate thermophiles) and 70° and 105°C (extreme thermophiles). Their respective enzymes have catalytically indistinguishable reactivity and catalytic sites from those isolated from mesophilic organisms.

The main reasons for thermostability of a protein seem to include the following:

- *Reduction of conformational entropy*: Crosslinking, such as that achieved by forming disulfide bonds, reduces conformational entropy according to Eq. (17.38).

$$\Delta S = 0.75 \cdot \sigma \cdot R \cdot \ln(n' + 3) \tag{17.38}$$

  Here σ is the number of connected chains (twice the number of crosslinks) and n' is the number of amino acids in each loop.

- *Hydrophobic interactions*: This is probably the biggest contribution. Proteins are not usually tightly packed but contain around 25% of their volume in cavities. If these cavities are filled and the contact area is thus increased, the protein becomes more stable against denaturation. Creation of a cavity with the size of a $-CH_2-$ group destabilizes the enzyme by 4.5 kJ mol^{-1}.

- *Paired charges*: Paired charges such as salt bridges are much rarer than hydrophobic interactions but do contribute about 12–14 kJ mol^{-1} to stabilization of a protein.

Several investigations have yielded a connection between enzyme stability to temperature, to organic solvents and to other denaturing agents (Suzuki, 1987).

One group of kosmotropic stabilizers of proteins are the osmolytes, organic osmoprotectant solutes, mostly polyhydric alcohols and amino acids or their derivatives. Osmolyte compatibility of organisms is thought to result from absence of osmolyte interactions with substrates and cofactors, and the non-perturbing or favorable effects on macromolecular–solvent interactions (Yancey, 1982).

A study of two of the most prominent and widespread osmolytes, betaine and beta-hydroxyectoine, by differential scanning calorimetry (DSC) on bovine ribonuclease A (RNase A) revealed an increase in the melting temperature T_m of RNase A of more than 12 K and of protein stability ΔG of 10.6 kJ mol^{-1} at room temperature at a 3 M concentration of beta-hydroxyectoine. The heat capacity difference ΔC_p between the folded and unfolded state was significantly increased. In contrast, betaine stabilized RNase A only at concentrations less than 3 M. When enzymes are applied in the presence of denaturants or at high temperature, beta-hydroxyectoine should be an efficient stabilizer.

Suggested Further Reading

L. R. De Young, A. L. Fink, and K. A. Dill, Aggregation of globular proteins, *Acc. Chem. Res.* **1993**, *26*, 614–620.

S. E. Radford, Protein folding: progress made and promises ahead, *Trends Biochem. Sci.* **2000**, *25*, 611–618.

References

T. J. Ahern and A. M. Klibanov, The mechanisms of irreversible enzyme inactivation at 100C, *Science* **1985**, *228*, 1280–1284.

T. J. Ahern, J. I. Casal, G. A. Petsko, and A. M. Klibanov, Control of oligomeric enzyme thermostability by protein engineering, *Proc. Natl. Acad. Sci. USA* **1987**, *84*, 675–679.

T. J. Ahern and A. M. Klibanov, Analysis of processes causing thermal inactivation of enzymes, *Methods Biochem. Anal.* **1988**, *33*, 91–127.

J. Broos, A. J. W. G. Visser, J. F. J. Engbersen, W. Verboom, A. van Hoek, and D. N. Reinhoudt, Flexibility of enzymes suspended in organic solvents probed by time-resolved fluorescence anisotropy. Evidence that enzyme activity and enantioselecticity are directly related to enzyme flexibility, *J. Am. Chem. Soc.* **1995**, *117*, 12657–12663.

T. V. Burova, N. V. Grinberg, V. Y. Grinberg, R. V. Rariy, and A. M. Klibanov, Calorimetric evidence for a native-like conformation of hen egg-white lysozyme dissolved in glycerol, *Biochim. Biophys. Acta* **2000**, *1478*, 309–317.

J. I. Casal, T. J. Ahern, R. C. Davenport, G. A. Petsko, and A. M. Klibanov, Subunit interface of triosephosphate isomerase: site-directed mutagenesis and characterization of the altered enzyme, *Biochemistry* **1987**, *26*, 1258–1264.

M. Caussette, A. Gaunand, H. Planche, P. Monsan, and B. Lindet, Inactivation of enzymes by inert gas bubbling: a kinetic study, *Ann. N. Y. Acad. Sci.* **1998a**, *864* (*Enz. Eng. XIV*), 228–233.

M. Caussette, A. Gaunand, H. Planche, and B. Lindet, Enzyme inactivation by inert gas bubbling, *Prog. Biotechnol.* **1998b**, *15* (*Stability and Stabilization of Biocatalysis*), 393–398.

M. Caussette, A. Gaunand, H. Planche, S. Colombie, P. Monsan, and B. Lindet, Lysozyme inactivation by inert gas bubbling: kinetics in a bubble column reactor, *Enz. Microb. Technol.* **1999**, *24*, 412–418.

H. S. Chan and K. A. Dill, The protein folding problem, *Physics Today* **1993**, 24–32.

S. Colombie, A. Gaunand, M. Rinaudo, and B. Lindet, Irreversible lysozyme inactivation and aggregation induced by stirring: kinetic study and aggregates characterization, *Biotechnol. Lett.* **2000**, *22*, 277–283.

S. Colombie, A. Gaunand, and B. Lindet, Lysozyme inactivation and aggregation in stirred-reactor, *J. Mol. Cat. B: Enzymatic* **2001a**, *11*, 559–565.

S. Colombie, A. Gaunand, and B. Lindet, Lysozyme inactivation under mechanical stirring: effect of physical and molecular interfaces, *Enz. Microb. Technol.* **2001**, *28*, 820–826.

L. R. De Young, A. L. Fink, and K. A. Dill, Aggregation of globular proteins, *Acc. Chem. Res.* **1993**, *26*, 614–620.

C. J. Epstein, R. F. Goldberger, and C. B. Anfinsen, The genetic control of tertiary protein structure. Model systems, *Cold Spring Harbor Symp. Quant. Biol.* **1963**, *28*, 439–449.

A. L. Fink, Protein aggregation: folding aggregates, inclusion bodies and amyloid, *Folding & Design*, **1998**, *3*, R9–R23.

L. Fu and E. Freire, On the origin of the enthalpy and entropy convergence temperatures in protein folding, *Proc. Natl. Acad. Sci. USA* **1992**, *89*, 9335–9338.

P. J. Halling, A. C. Ross, and G. Bell, Inactivation of enzymes at the aqueous–organic interface, *Prog. Biotechnol.* **1998**, *15* (*Stability and Stabilization of Biocatalysis*), 365–372.

D. T. Haynie and E. Freire, Estimation of the folding/unfolding energetics of marginally stable proteins using differential scanning calorimetry, *Anal. Biochem.* **1994**, *216*, 33–41.

T. Kiefhaber, R. Rudolph, H. H. Kohler, and J. Buchner, Protein aggregation in vitro and in vivo: a quantitative model of the kinetic competition between folding and aggregation, *Bio/Technology* **1991**, *9*, 825–829.

S. Knapp, R. Ladenstein, and E. A. Galinski, Extrinsic protein stabilization by the naturally occurring osmolytes beta-hydroxyectoine and betaine, *Extremophiles* **1999**, *3*, 191–198.

C. Levinthal, How to fold graciously, in: *Mossbauer Spectroscopy in Biological Systems: Proceedings of a Meeting held at Allerton House, Monticello, Illinois*, J. T. P. DeBrunner and E. Munck (eds.), University of Illinois Press, Chicago/IL, **1969**, pp. 22–24.

H. Lilie, E. Schwarz, and R. Rudolph, Advances in refolding of proteins produced in *E. coli*, *Curr. Opin. Biotechnol.* **1998**, *9*, 497–501.

J. London, C. Skrzynia, and M. E. Goldberg, Renaturation of *Escherichia coli* tryptophanase after exposure to 8 M urea. Evidence for the existence of nucleation centers, *Eur. J. Biochem.* **1974**, *47*, 409–415.

R. Lumry and H. Eyring, Conformation changes of proteins, *J. Phys. Chem.* **1954**, *58*, 110–120.

C. N. Pace, Measuring and increasing protein stability, *TIBTECH* **1990**, *8*, 93–98.

I. M. Plaza del Pino and J. M. Sanchez-Ruiz, An osmolyte effect on the heat capacity change for protein folding, *Biochemistry* **1995**, *34*, 8621–8630.

S. E. Radford, Protein folding: progress made and promises ahead, *Trends Biochem. Sci.* **2000**, *25*, 611–618.

R. V. Rariy and A. M. Klibanov, Protein refolding in predominantly organic media markedly enhanced by common salts, *Biotechnol. Bioeng.* **1999**, *62*, 704–710.

A. C. Ross, G. Bell, and P. J. Halling, Organic solvent functional group effect on enzyme inactivation by the interfacial mechanism, *J. Mol. Cat. B: Enzymatic* **2000**, *8*, 183–192.

R. Rudolph and H. Lilie, In vitro folding of inclusion body proteins, *FASEB J.* **1996**, *10*, 49–56.

A. Sadana, *Biocatalysis – Fundamentals of Enzyme Deactivation Kinetics*, Prentice-Hall, Englewood Cliffs/NJ, **1991**.

J. M. Sanchez-Ruiz, J. L. Lopez-Lacomba, M. Cortijo, and P. L. Mateo, Differential scanning calorimetry of the irreversible thermal denaturation of thermolysin, *Biochemistry* **1988**, *27*, 1648–1652.

C. P. Schultz, Illuminating folding intermediates, *Nature Struct. Biol.* **2000**, *7*, 7–10.

V. Sluzky, J. A. Tamada, A. M. Klibanov, and R. Langer, Kinetics of insulin aggregation in aqueous solutions upon agitation in the presence of hydrophobic surfaces, *Proc. Natl. Acad. Sci. USA* **1991**, *88*, 9377–9381.

Y. Suzuki, K. Oishi, H. Nakano, and T. Nagayama, A strong correlation between the increase in number of proline residues and the rise in thermostability of five *Bacillus* oligo-1,6-glucosidases, *Appl. Microbiol. Biotechnol.* **1987**, *26*, 546–551.

S. J. Tomazic and A. M. Klibanov, Why is one *Bacillus* alpha-amylase more resistant against irreversible thermoinactivation than another? *J. Biol. Chem.* **1988**, *263*, 3092–3096.

V. N. Uversky, D. J. Segel, S. Doniach, and A. L. Fink, Association-induced folding of globular proteins, *Proc. Natl. Acad. Sci. USA* **1998**, *95*, 5480–5483.

D. B. Volkin and A. M. Klibanov, Thermal destruction processes in proteins involving cystine residues, *J. Biol. Chem.* **1987**, *262*, 2945–2950.

L. Vuillard, C. Braun-Breton, and T. Rabilloud, Non-detergent sulphobetaines: a new class of mild solubilization agents for protein purification, *Biochem. J.* **1995**, *305*, 337–343.

P. H. Yancey, M. E. Clark, S. C. Hand, R. D. Bowlus, and G. N. Somero, Living with water stress: evolution of osmolyte systems, *Science* **1982**, *217*, 1214–1222.

S. E. Zale and A. M. Klibanov, **1986**, Why does ribonuclease irreversibly inactivate at high temperatures? *Biochemistry*, *25*, 5432–5444.

18
Artificial Enzymes

Summary

Catalytic antibodies, predicted by Jencks in 1969 and first discovered in 1986, can now be raised against a wide variety of haptens covering nearly every reaction. Catalytic antibodies are regarded as the best enzyme mimics, with very good selectivity, but almost always their catalytic efficiency is by far insufficient. Some natural RNA molecules act as catalysts with intrinsic enzyme-like activity which permits them to catalyze chemical reactions in the complete absence of protein cofactors. In addition, ribozymes identified through in-vitro selection have extended the repertoire of RNA catalysis. This versatility has lent credence to the idea that RNA molecules may have been central to the early stages of life on Earth.

For certain reactions, such as acid–base catalysis, reactivity of up to 10^5 over background was demonstrated with serum albumins, ordinary "off-the-shelf" proteins without any designated catalytic activity but with a lysine side chain group acting as a general base.

Criteria for calling a compound a "synthetic enzyme" are: (i) completion of at least one catalytic cycle; (ii) its presence after the catalytic cycle in unchanged form; and (iii) a saturation kinetics behavior such as is manifested by Michaelis–Menten kinetics. There is a tetrameric helical peptide that catalyzes the decarboxylation of oxaloacetate with Michaelis–Menten kinetics and accelerates the reaction 10^3–10^4-fold faster than *n*-butylamine as control, a record for a chemically derived artificial enzyme.

The various technologies for obtaining the heterogenization of enantioselective catalysts for asymmetric syntheses include the following approaches:

1. immobilization of chiral complexes on organic polymers and inorganic oxides and encapsulation in zeolites, mesoporous silicates, and poly(dimethylsiloxane) membranes;
2. attachment of chiral catalysts to soluble polymeric ligands, including dendritic polymers, to combine the advantages of catalysis in a homogeneous phase with facile separation by ultrafiltration in a membrane reactor; and
3. use of liquid–liquid biphasic systems with the catalyst residing in one phase, which is either aqueous or organic, e.g., fluorous, with the substrate and product in a second phase, often a hydrocarbon medium.

Among the different approaches to immobilization, main-chain chiral polymer catalysts are different from the traditional polymer catalysts prepared by anchoring monomeric chiral catalysts to an achiral polymer backbone. The three classes of synthetic main-chain chiral polymers include: (i) helical polymers, such as polypeptides; (ii) polymers with flexible chiral chains, such as polyesters and polyamides; and (iii) polymers of rigid, sterically regular chiral chains, represented by chiral conjugated polybinaphthyls. Covalent binding of homogeneous catalysts and/or their ligands to soluble or insoluble polymers with the purpose of catalyst recovery has been applied to catalysts for a wide variety of reactions.

Molecular-weight-enhanced homogeneous catalysts can be retained in a chemzyme-membrane reactor (CMR). Enantioselectivities often compare with the non-heterogenized case and volumetric productivities, with space–time yields reported up to more than 1 kg $(L \cdot d)^{-1}$, are often favorable for such a reactor configuration. Currently, the biggest technological challenge is to improve long-term stability of the catalysts. Such operational stability would allow the very promising data on enantioselectivities and volumetric productivity to be translated into a serious alternative to biological catalysts.

18.1 Catalytic Antibodies

18.1.1 Principle of Catalytic Antibodies: Connection between Chemistry and Immunology

Much groundwork was required until in 1980, the first successful reports with catalytic antibodies appeared simultaneously by the groups of Richard Lerner and Peter Schultz:

- In the first half of the 20th century antibodies were discovered and researched by Karl Landsteiner (Austria).
- Building on an idea of Haldane (Chapter 5) advanced in the 1920s, in 1944 Pauling and Pressmann formulated the hypothesis of complementarity of the catalyst with the transition state of a reaction. Until then it was assumed that the ground state of a substrate was bound most tightly to the enzyme molecule.
- In 1969, the enzymologist William Jencks (Brandeis University, Waltham/MA, USA) described the possibility of causing enzyme-like catalytic effects with antibodies.
- In 1975, George Köhler and Caesar Milstein (then in Cambridge, UK) discovered the principle of monoclonal antibodies (Nobel prize 1981).
- In 1986, Richard Lerner (Scripps Institute, La Jolla/CA, USA) and Peter Schultz (then at the University of California Berkeley, now also at the Scripps Institute), independently of each other, delivered proof of the concept that catalytic antibodies exist.

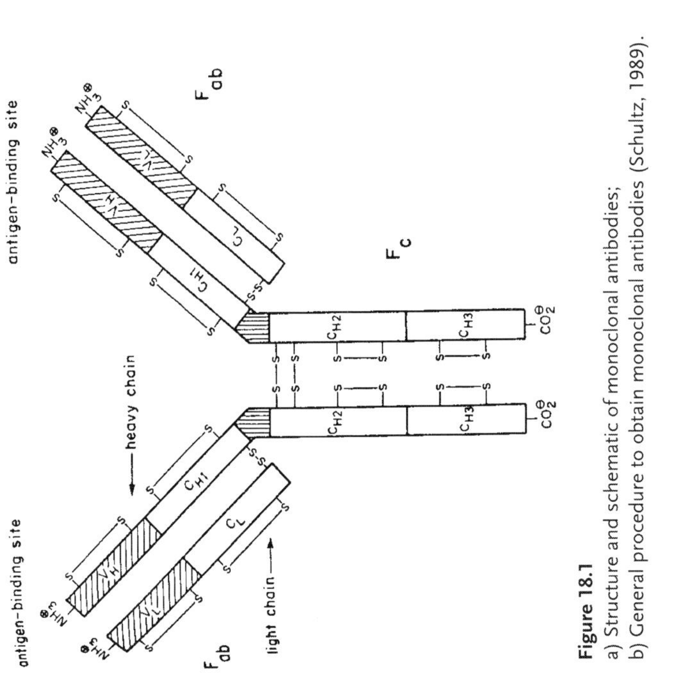

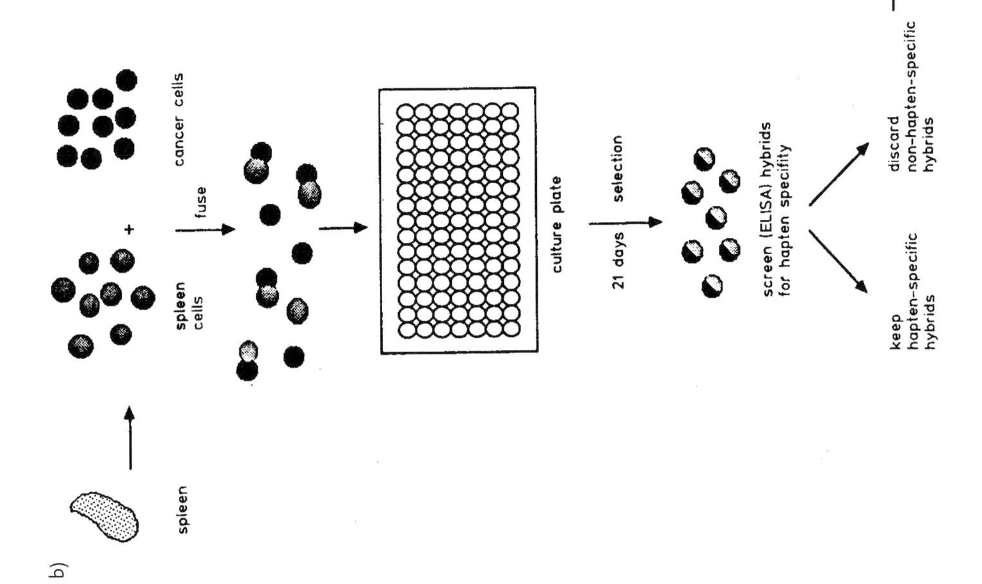

Figure 18.1
a) Structure and schematic of monoclonal antibodies;
b) General procedure to obtain monoclonal antibodies (Schultz, 1989).

Antibodies protect an organism through their capability to recognize molecules foreign to the body. Selective recognition rests on weak interactions such as hydrogen-bond bridges, van der Waals forces, and solvent effects, which are not sufficiently known and researched. Through X-ray structural analysis an antibody structure of about 150 kDa was found which in immunoglobulin G (IgG) consists of two identical pairs of chains, the light and heavy chains (V_L and V_H). The first 110 amino acids of the light and heavy chain form the variable region, which is responsible for the variability of antibodies (Figure 18.1).

Usually, antibodies feature association constants of 10^{-4} to 10^{4} M^{-1} and specificities which can surpass those of enzymes. The two greatest differences between enzymes and antibodies are:

- enzymes stabilize the transition state, whereas antibodies stabilize the ground state of a reaction;
- the specificity of enzymes in nature develops over a timeframe of millions of years (see, however, Chapter 11), but in contrast, development of the specificity of antibodies takes just over a couple of weeks.

The idea that antibodies also can stabilize the transition state of a reaction, that they feature a sterically or electronically complementary active center to the rate-determining transition state just like enzymes, has existed since 1969. This concept of catalytic antibodies could be investigated only after the advent of monoclonal antibodies. The capability to raise molecularly uniform antibodies instead of polyclonal sera

- facilitates kinetic, mechanistic, and structural characterization,
- spares large numbers of experiments as well as purifications and characterization steps, and
- allows reproducible production of large amounts of corresponding antibodies.

Three strategies are possible for raising molecular antibodies:
- the use of antibodies for the stabilization of negatively and positively charged transition states;
- the use of antibodies as "entropy traps"; and
- the raising of antibodies with catalytic or cofactor groups in the variable region.

18.1.2
Test Reaction Selection, Haptens, Mechanisms, Stabilization

Which reactions have been chosen as test reactions for the concept of catalytic antibodies? The following criteria were applied:

- the organic reactions serving as a base should be well understood;

- the steric and electronic nature of the transition state of this reaction should be

  - sufficiently different from the ground state of the substrate so that specific antibodies selectively stabilize the ground state, and
  - sufficiently different from the ground state of the product so as not to cause product inhibition in the production of the respective antibodies;

- inhibitors of a reaction should be able to serve as the starting point for a search of transition-state analogs because such low-molecular-weight antigens (haptens) based on the same mechanism stand a good chance of inducing antibody binding sites which are sterically and electronically complementary to the transition state.

As the first test reaction, the hydrolysis of carbonates and esters was selected according to the above criteria (Figure 18.2). Corresponding antibodies were indeed found. Typically, such antibodies catalyze reactions by a factor of 10^3–10^4 over back-

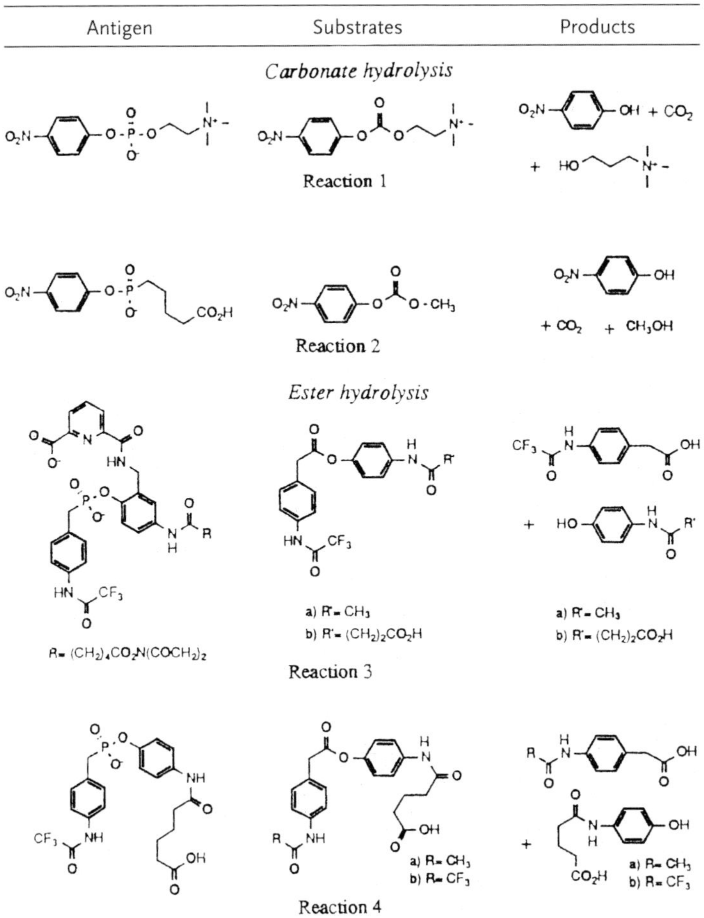

Figure 18.2 Hydrolysis of carbonates and esters by catalytic antibodies (Lerner, 1991).

ground, feature ordinary Michaelis–Menten kinetics and substrate specificity, and bind much more strongly to the transition state than to the ground state, consistently with Pauling's postulate. Carbonate and ester hydrolysis by biological catalysts is of such interest because stereoselective hydrolysis of non-activated esters with enantiomeric ratios of more than 100 still is an interesting problem in chemical synthesis.

18.1.2.1 Mechanism of Antibody-Catalyzed Reactions

Although the structure of haptens exerts an important influence on catalysis by antibodies, stabilization of the transition state by catalytic antibodies cannot always and cannot fully explain all observations. Sometimes, in analogy to the evolution of enzyme function, antibodies can take mechanistically unexpected detours. As an example, the corresponding antibody catalyzes the hydrolysis of arylamides according to Figure 18.3 by a factor of 2.5×10^5 over background.

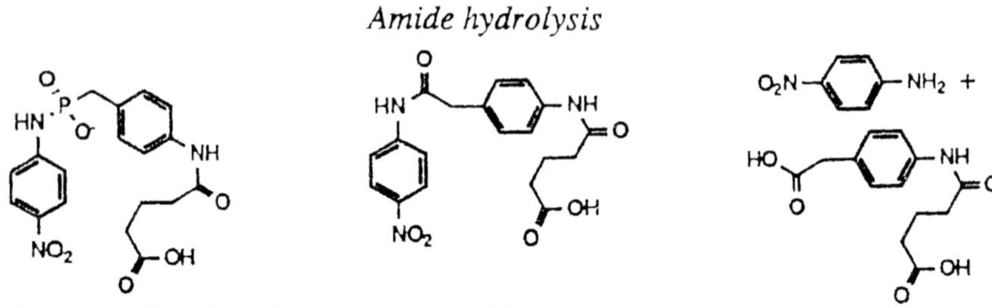

Figure 18.3 Hydrolysis of an amide (Lerner, 1991).

Kinetic investigation with $H_2{}^{18}O$ tracer studies and measurements of isotope effects with D_2O yielded a mechanism for reaction (1) consistent with the intermediate formation of an acyl–antibody complex (Figure 18.4).

$$\mathrm{Ab + S} \underset{k_{-1}}{\overset{k_1}{\rightleftharpoons}} \mathrm{AbS} \underset{k_{-2}}{\overset{k_2}{\rightleftharpoons}} \mathrm{AbIP_1} \xrightarrow{k_3\,[\mathrm{OH^-}]} \mathrm{AbP_1P_2} \xrightarrow[\mathrm{P_1}]{k_4} \mathrm{AbP_2} \xrightarrow[\mathrm{P_2}]{k_5} \mathrm{Ab}$$

Figure 18.4 Reaction (1) scheme with an intermediate acyl–antibody complex (Lerner, 1991).

This mechanism is similar to that of the serine protease α-chymotrypsin. In comparison to α-chymotrypsin, however, which features a general acid–base mechanism, the antibody requires a hydroxide ion (OH^-) to initiate the deacylation step. The enzyme liberates both products P_1 and P_2 early in the catalytic cycle and very rapidly whereas the antibody liberates both products at the end of the catalytic cycle only. Even given all the similarities between enzymes and antibodies, one thus cannot expect complete analogy down to mechanistic details. However, the specificity of enzymes and antibodies can be very similar, as Table 18.1 corroborates.

Table 18.1 Comparison of hydrolysis of esters and amides for enzymes and antibodies.

Catalyst	*Rate constant k_{cat} [s^{-1}]*	
	N-Ac-L-Phe-*p*-NO_2-Ph-ester	*N*-Ac-L-Tyr-*p*-NO_2-anilide
α-CT	77	0.051
Antibody	40	0.002

18.1.2.2 **Stabilization of Charged Transition States**

Phosphonates and phosphonamidates were selected as transition-state analogs for carbonate and ester hydrolysis: they are relatively stable molecules, they are known inhibitors of acyl transfer enzymes and they imitate the initial negative charge of the tetrahedral oxyanion of acyl transfer enzymes (such as serine proteases) in the dipole of the P=O bond.

18.1.2.3 **Effect of Antibodies as Entropy Traps**

Besides the stabilization of charged transition states, the effect as entropy traps is decisive for the effect as a catalyst. "Entropy trap" refers to the ability of a catalyst to bind substrates in a favorable orientation and thus to freeze out translational and rotational degrees of freedom of the substrate (Chapter 2, Section 2.2.3). The result corresponds to a drastic increase in the *effective molarity* of the substrate, to a concentration level which can never be reached in free solution. The transesterification reaction in Figure 18.5 (Schultz, 1993) would be difficult to conduct in water under ordinary circumstances as the hydrolysis side reaction is aided by a water concentration of 55.5 M. A comparison of k_{cat}/K_M [$(\text{M s})^{-1}$] with the uncatalyzed reaction in water [s^{-1}] (strictly speaking, [H_2O] at 55.5 M has to be considered as well, so a dimension of [$(\text{M s})^{-1}$] again results) yields an effective molarity of 10^6 M.

To prove that an antibody-catalyzed reaction indeed preferentially stabilizes the transition state, the ratio of dissociation constants for substrate K_S and transition-state K_T is frequently compared to the ratio of the antibody-catalyzed rate constant k_{Ab} and the rate constant of the uncatalyzed reaction k_{un} [Eq. (18.1), which is equivalent to Eq. (2.8)].

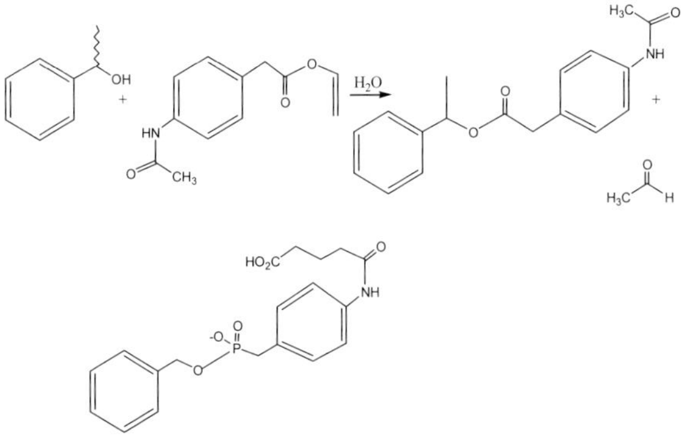

Figure 18.5 Transesterification reaction with entropy trap (Schultz, 1993).

$$K_{Ab}^{\ddagger}/K_{un}^{\ddagger} = K_S/K_T = k_{Ab}/k_{un} \tag{18.1}$$

For unimolecular lactonization reactions, hydrolyses of aryl esters, Diels–Alder reactions, and porphyrin metalization reactions the validity of Eq. (18.2) has been demonstrated.

18.1.3
Breadth of Reactions Catalyzed by Antibodies

18.1.3.1 Fastest Antibody-Catalyzed Reaction in Comparison with Enzymes

Ferrochelatase, the terminal enzyme of the heme biosynthesis pathway, catalyzes the metalization of mesoporphyrin with Zn^{II} with a k_{cat} value of 800 h^{-1}. An antibody was found whose reaction according to Figure 18.6 features a k_{cat} value of 80 h^{-1}, the relatively highest value for antibodies in comparison to enzymes (Cochran, 1990).

18.1.3.2 Antibody-Catalyzed Reactions without Corresponding Enzyme Equivalent

After an enzyme was finally found for the long-elusive Diels–Alder reaction (Oikawa, 1995), the Cope rearrangement is the last reaction for which no enzyme is known. Thus, efforts were made to find catalytic antibodies for this reaction as well as for the then-unsolved case of the Diels–Alder reaction. The transition state of a Diels–Alder reaction is a highly ordered six-membered ring with interacting orbitals; therefore, an unfavorable entropic interaction of –30 to –40 eu is observed. Through the design of a hapten, by whose antibody reaction SO_2 is liberated irreversibly after the Diels–Alder reaction and which features a dichloromethylene bridge in the hapten but not in the substrate or product, an effective entropy trap was constructed and product inhibition effects were avoided (Fig. 18.7., top reaction). The bottom reaction solves the problem in a more general way, by forcing both Diels–Alder reaction partners into a transition-state-like conformation.

18.1.3.3 Example of a Pericyclic Reaction: Claisen Rearrangement

A thoroughly investigated reaction on the biosynthetic pathway to aromatics is the [3+3]-sigmatropic Claisen rearrangement from chorismic acid to prephenic acid (Figure 18.8).

The reaction, which proceeds via a conformationally tight chair-type transition state, is clearly entropically dominated, with a $\Delta S^{\ddagger}$ of –13 eu. Whereas the known enzyme chorismate mutase from *E. coli* achieves a 3×10^6-fold accelerated catalysis, the antibody reaches a 10^4-fold enhancement. A decrease of $\Delta S^{\ddagger}$ to almost 0 eu points to the presence of an entropy trap.

18.1.3.4 Antibody Catalysts with Dual Activities

Recently, an antibody has been described which catalyzes not just a Diels–Alder reaction between an N-substituted maleimide and acetoxybutadiene ($k_{cat} = 0.055\ s^{-1}$, $K_M = 8.3$ mM, $k_{cat}/K_M = 6.6\ (M\ s)^{-1}$, but also the subsequent hydrolysis of the acetoxy group ($k_{cat} = 9.2 \times 10^{-4}\ s^{-1}$, $K_M = 1.1$ mM, $k_{cat}/K_M = 0.9\ (M\ s)^{-1}$), which is about 1.5% as fast as the Diels–Alder reaction itself (Figure 18.9).

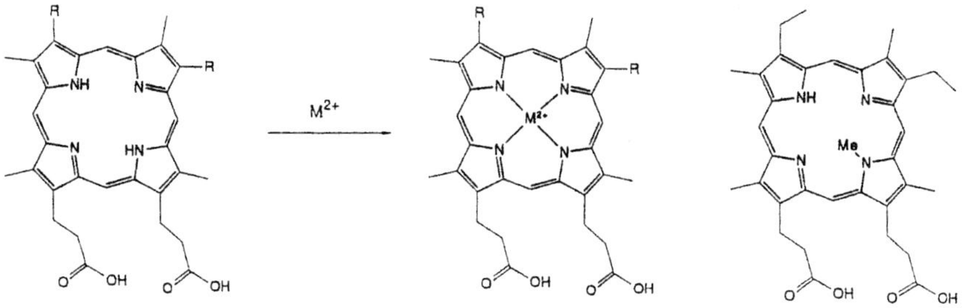

Figure 18.6 Ferrochelatase (Cochran, 1990).

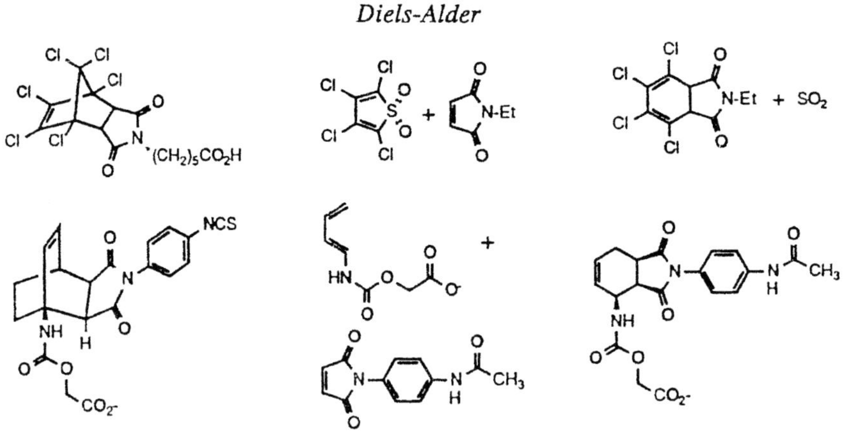

Diels-Alder

Figure 18.7 Diels–Alder reactions catalyzed by antibodies (Lerner, 1991).

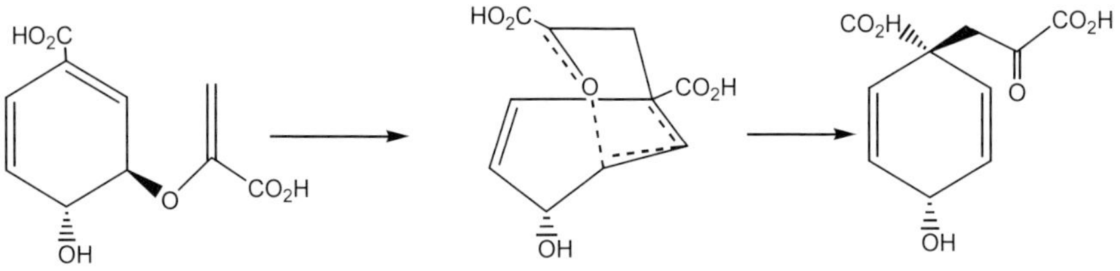

Figure 18.8 Claisen rearrangement from chorismate to prephenate catalyzed by antibodies.

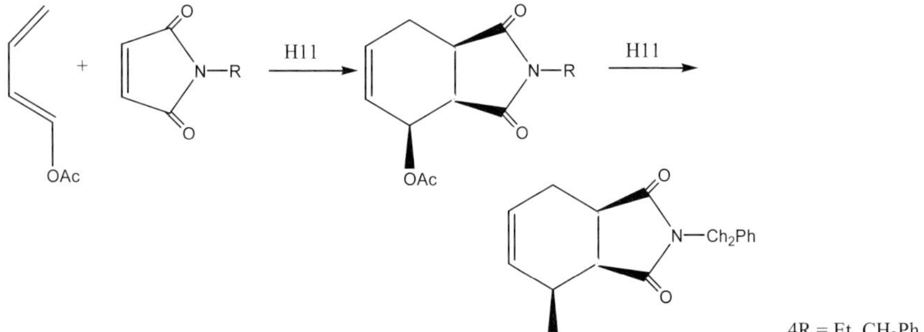

Figure 18.9 Dual-reactivity antibody.

4R = Et, CH_2Ph

18.1.3.5 **Scale-Up of an Antibody-Catalyzed Reaction**

Antibody-catalyzed reactions, too, can be conducted on a gram scale (Reymond, 1994); the biocatalyst was separated by dialysis in a cellulose membrane (12–14 kDa, antibody 150 kDa) and reused. The enantioselective hydrolysis of the enol ether (**4**) to the ketone (**5**) ran with an e.e. of 89–91% (Figure 18.10).

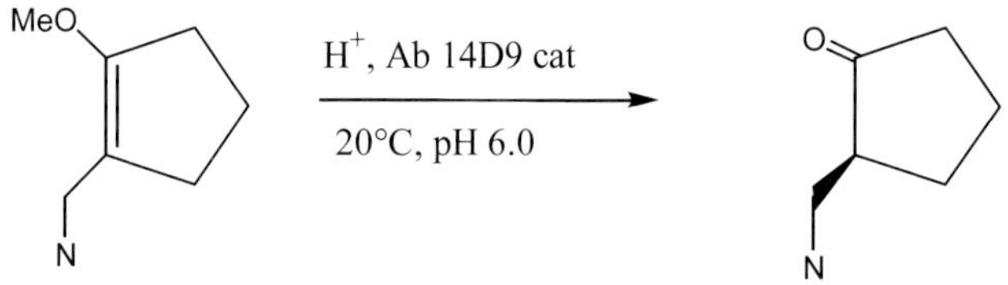

MeO ... N $\xrightarrow[20°C,\ pH\ 6.0]{H^+,\ Ab\ 14D9\ cat}$ O ... N

Figure 18.10 Scaled-up antibody reaction.

After five cycles, a total of 2.55 g could be isolated (yield: 62%). The turnover number of the antibody was 4700. The main losses seemed to have been caused by inaccurate transfer between cycles and and by instabilities of the purified antibody in comparison with the crude extract.

18.1.3.6 **Perspective for Catalytic Antibodies**

The idea that antibodies raised against transition-state analogs should show specific catalytic activity is beautiful and seductive. In the tenth year since the idea became an experimental reality, a preliminary assessment of their potential was made (Kirby, 1996). It was concluded that their high stereoselectivity makes abzymes excellent prospects for asymmetric synthesis, though their practical usefulness is currently limited by their catalytic efficiency.

Antibody molecules elicited with rationally designed transition-state analogs catalyze numerous reactions, including many that cannot be achieved by standard chemical methods (Hilvert, 2000). Although relatively primitive when compared with natural enzymes, these catalysts are valuable tools for probing the origins and evolution of biological catalysis. Mechanistic and structural analyses of representative antibody catalysts, generated with a variety of strategies for several different reaction types, suggest that their modest efficiency is a consequence of imperfect hapten design and indirect selection. Development of improved transition-state analogs, refinements in immunization and screening protocols, and elaboration of general strategies for augmenting the efficiency of first-generation catalytic antibodies are identified as evident, but difficult, challenges for this field. Rising to these challenges and more successfully integrating programmable design with the selective forces of biology will enhance our understanding of enzymatic catalysis. Further, it should yield useful protein catalysts for an enhanced range of practical applications in chemistry and biology.

18.2 Other Proteinaceous Catalysts: Ribozymes and Enzyme Mimics

18.2.1 Ribozymes: RNA World before Protein World?

In 1982, Cech and his co-workers reported the first catalytic RNA or ribozyme, the self-splicing intron of *Tetrahymena* pre-rRNA (Kruger, 1982). Although RNA is generally thought to be a passive genetic blueprint, some RNA molecules act as catalysts with intrinsic enzyme-like activity which permits them to catalyze chemical reactions in the complete absence of protein cofactors (Doubna, 2002). In addition to the well-known small ribozymes that cleave phosphodiester bonds, we now know that RNA catalysts probably affect a number of key cellular reactions. This versatility has lent credence to the idea that RNA molecules may have been central to the early stages of life on Earth. Additional examples of natural ribozymes were soon found, and research in the field focused on their enzymatic mechanism and secondary and tertiary structure. Ribozymes identified through in-vitro selection extended the repertoire of RNA catalysis. Because RNA is chemically and structurally dissimilar to protein, the finding of catalytic activity in RNA was initially surprising (Cech, 1993a). Quantitative measurements of reaction rates show that RNA can be as efficient a catalyst as protein. On the other hand, the potential versatility of RNA to catalyze diverse types of reactions has only begun to be explored. Understanding the efficiency and versatility of RNA as a catalyst helps the authors evaluate origin-of-life scenarios involving self-replicating RNA, and may explain why RNA catalysis remains important in contemporary cells. Two directions of current and future interest are the determination of atomic-resolution structures of large ribozymes by X-ray crystallography, and the structural and mechanistic analysis of complexes of ribozymes with protein facilitators of their activity (Cech, 2002).

18.2.2 Proteinaceous Enzyme Mimics

When attempting to mimic the efficiency of enzymes it is useful to remind oneself that enzymes bind to and stabilize the transition state for a particular reaction. There are several artificial mimics known that bind the ground state for such a reaction very efficiently. Even rate enhancements comparable to those of enzymes have been achieved as long as the substrate groups and the catalytic site were brought together in an intramolecular reaction (Kirby, 1996). Intermolecular catalysis of a similar performance, however, requiring selective binding and efficient catalysis including precise orientation of substrate and catalytically active group, remains elusive. Given the elusiveness of precise orientation of catalytic groups for intermolecular catalysis, the community reacted with excitement to the announcement of a catalytic antibody that efficiently catalyzes the eliminative ring-opening of a benzisoxazole (Kemp elimination) using carboxylate as the general base, raising the intriguing possibility that this high efficiency derives from precise positioning of catalytic and substrate groups (Thorn, 1995). A while later, however, a similar reactivity was demonstrated with serum albumins, ordinary "off-the-shelf" proteins without any designated catalytic activity but with a lysine side chain group acting as a general base (Hollfelder, 1996).

Kemp elimination was used as a probe of catalytic efficiency in antibodies, in non-specific catalysis by other proteins, and in catalysis by enzymes. Several simple reactions were found to be catalyzed by the serum albumins with Michaelis–Menten kinetics and could be shown to involve substrate binding and catalysis by local functional groups (Kirby, 2000). Known binding sites on the protein surface were found to be involved. In fact, formal general base catalysis seems to contribute only modestly to the efficiency of both the antibody and the non-specific albumin system, whereas antibody catalysis seems to be boosted by a non-specific medium effect.

Medium effects at the active sites of enzymes differ dramatically from those in bulk solvents, both in diversity (the presence of several "solvent" types) and in spatial arrangement, so that comparing reaction rates in different solvents is not particularly revealing. Kemp elimination of benzisoxazoles, the test reaction mentioned above, was employed to test and describe medium effects with synzyme catalysts, obtained by systematic alkylation of a polymeric scaffold bearing amine groups, rendering a remarkably efficient catalyst (Hollfelder, 2001). In contrast to the non-specific desolvation activation effects exhibited by solvents, medium effects with such polymeric synzymes is driven primarily by specific, localized enzyme-like effects not available in isotropic bulk solvents. Ligand-binding studies indicate that the synzyme active sites provide localized microenvironments affording a combination of hydrophobic and apolar regions on one hand and dipolar, protic, and positively charged regions on the other. A Brønsted (leaving group) analysis indicates that, in comparison to solvent catalysis, the efficiency of synzyme catalysis shows little sensitivity to leaving group pK_a. Even in the absence of efficient positioning of the catalytic amine base relative to the substrate, rate accelerations of up to 10^5 (!) can be achieved, for both activated and non-activated substrates. Such accidental identification of active sites on the surfaces of non-catalytic proteins and the promiscuous activities found in many enzymes suggest that the interfaces of protein surfaces and their hydrophobic cores provide a microenvironment that is intrinsically active and may serve as a basis for further evolutionary improvements to give proficient and selective enzymes.

18.3
Design of Novel Enzyme Activity: Enzyme Models (Synzymes)

18.3.1
Introduction

With chemical model reactions, valuable insights can be gained about enzymatic reactions and their mechanisms. In most cases, modeling of the active center is the main target, so that simple model compounds for the decisive groups in and around the active site are sought, with which mechanism, substrate breadth, or other reaction parameters can be tested.

In the early days of biocatalysis, the purpose of such investigations with model compounds was to gain insight about unknown active sites, as 3D structures were unavailable for most enzymes. These days, however, with 3D structures available through X-ray crystallography or NMR, and site-directed mutagenesis available as a tool to probe the influence of individual amino acid residues, studies with model compounds increasingly serve the purpose of mimicking the whole enzyme.

Not just the enzyme but also whole enzyme catalytic systems can be mimicked: as an example, if the influence of interaction of an enzyme with its environment is the target of the investigation, frequently either model membrane surfaces or aggregates formed with surfactant molecules such as micelles or vesicles are employed.

There are two approaches to the discovery of enzyme mimics, i.e., to identify molecules that are able to bind substrate(s) and then catalyze reactions (Rowan, 1997). The first approach, often inspired by enzymes themselves, utilizes chemical knowledge and experience to design the catalyst. The other approach is to create a library and select the best host for a transition-state analog of the required reaction. The second approach, i.e., the combinatorial search, is covered in Chapter 11; the first one is discussed below. This first approach, the finding of enzyme mimics, has been reviewed by Kirby (1996) and Reichwein (1994), among others.

Criteria for counting a compound as a synthetic enzyme are the following:
- completion of at least one catalytic cycle,
- its presence after the catalytic cycle in unchanged form, and
- a saturation kinetics behavior such as is manifested by Michaelis–Menten kinetics.

18.3.2
Enzyme Models on the Basis of the Binding Step: Diels–Alder Reaction

A good example of a synthetic model compound with significantly tight binding is the (substrate-free) porphyrin trimer depicted in Figure 18.11 (Kirby, 1996a; Bonar-Law, 1994). This host compound binds di- or tripyridyl compounds very strongly, and substituted monomeric pyridines well enough, so that the Diels–Alder reaction between pyridyl-substituted substrates is accelerated if they are bound inside the porphyrin trimer. The stereochemical course of the Diels–Alder reaction is

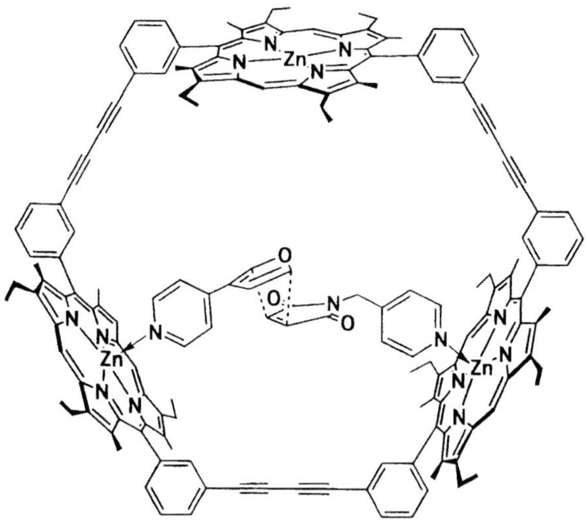

Figure 18.11 Catalysis by artificial porphyrins (Kirby 1996b).

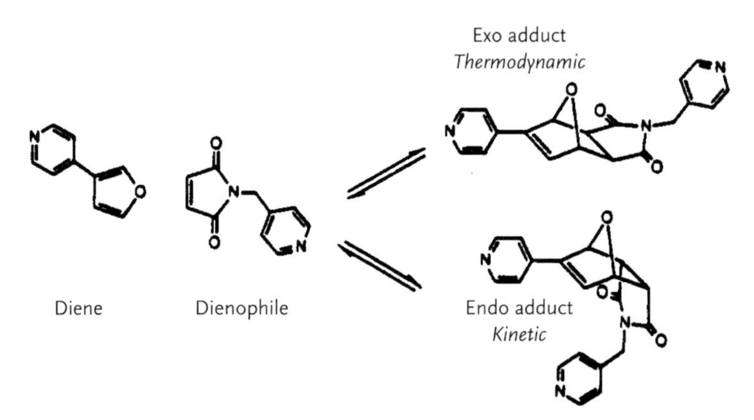

Figure 18.12 Diels–Alder adducts: exo versus endo.

markedly influenced by the binding inside the trimer (Figure 18.12): whereas the endo isomer is formed under kinetic control without the presence of the macrocycle, the exo adduct is formed about 1000-fold faster than the endo adduct with orientation of the substrates by the porphyrin trimer, corresponding to an effective molarity (Chapter 2, Section 2.2.3) of 20 M. Despite all these remarkable features the reaction is not catalytic, as product inhibition prevents multiple turnover.

Product inhibition occurs if the catalyst binds to the product with a similar tightness to the substrate. In a similar geometric approach of reactants, however, catalysis was successful (Mackay, 1994). Besides the above-mentioned Diels–Alder reaction, the substrate-free porphyrin trimer also catalyzes the acetylation of 4-hydroxymethylpyridine with *N*-acetylimidazole. Although both reactants are bound to the porphyrin trimer, their ground states are bound less tightly than the transition states and intermediate so that a complete catalytic cycle results. While the catalytic performance is not exactly earth-shattering, with an effective molarity of 2 M and a turnover number in toluene at 70 °C of 25/week, i.e., $4.1 \times 10^{-5}\ s^{-1}$, the results nevertheless prove catalysis. However, the competitive inhibitor to the acyl transfer, 1,3-bis-(4′-pyridyl)propane, binds with an association constant $K_{ass} = 2.3 \times 10^7\ M^{-1}$ (in comparison: for pyridine, $K_{ass} = 1.8 \times 10^3\ M^{-1}$ and for N-acetylimidazole, $K_{ass} = 3.6 \times 10^2\ M^{-1}$), which demonstrates the potential for selective binding of the transition state. Insofar as the porphyrin trimer catalyzes the transacetylation reaction, the compound already belongs to the catagory of catalytic systems discussed in the next section.

18.3.3 Enzyme Models with Binding and Catalytic Effects

The known properties of proteins and the known mechanism of amine-catalyzed decarboxylation of β-oxo acids (Laursen, 1966) were combined to generate a synthetic decarboxylase in the form of two tetradecameric peptides, termed Oxaldie 1 and 2, with leucine and lysine as the main amino acids (Johnsson, 1993).

Oxaldie 1: H_2N-Leu-Ala-**Lys**-Leu-Leu-**Lys**-Ala-Leu-Ala-**Lys**-Leu-Leu-**Lys**-**Lys**-$CONH_2$
Oxaldie 2: Ac-HN-Leu-Ala-**Lys**-Leu-Leu-**Lys**-Ala-Leu-Ala-**Lys**-Leu-Leu-**Lys**-**Lys**-$CONH_2$

The peptide sequence was chosen to allow formation of an α-helix in water with the hydrophobic leucine residues pointing towards one side of the helix and the hydrophilic lysine residues towards the other. As expected, Oxaldies 1 and 2 catalyzed the decarboxylation of oxaloacetate. Both carboxyl groups are believed to interact with lysine residues, whereas acetoacetate, with only one carboxyl group, is not a substrate. The reaction shows Michaelis–Menten kinetics and accelerates decarboxylation 10^3–10^4-fold faster than n-butylamine as control, a record for a chemically derived artificial enzyme. In addition, the degree of reactivity seems to correlate with helix formation of the peptide.

Table 18.2 provides some data on the performance of enzyme mimics.

Table 18.2 Reactivity and binding data for synthetic enzyme mimics (Kirby, 1996).

Host	*Reaction*	k_{cat} [s^{-1}]	*Eff. molarity* [M]	k_{cat}/k_{uncat} [–]
Porphyrin trimer	Diels–Alder	5.8×10^{-4}	20	–
Porphyrin trimer	acetyl transfer	6.0×10^{-5}	2	–
Porphyrin trimer	ester aminolysis	3.5×10^{-3}	0.035	160
Substd. barbiturates	intramolecular esterthiolysis	8.8×10^{-2}	6760	$> 10^4$ *
Oxaldie	decarboxylation	6.7×10^{-3}	–	10^3–10^4

* No catalysis, receptor in excess of substrate.

18.4
Heterogenized/Immobilized Chiral Chemical Catalysts

18.4.1
Overview of Different Approaches

The various technologies for the heterogenization of enantioselective catalysts for asymmetric syntheses include the following approaches (Fodor, 1999):

1. immobilization of chiral complexes on organic polymers and inorganic oxides, and encapsulation in zeolites, mesoporous silicates, and poly(dimethylsiloxane) membranes;
2. attachment of chiral catalysts to soluble polymeric ligands, including dendritic polymers, to combine the advantages of catalysis in a homogeneous phase with facile separation by ultrafiltration in a membrane reactor; and
3. liquid–liquid biphasic systems with the catalyst residing in one phase, which is either aqueous or organic, e.g., fluorous, and the substrate and product in a second phase, often a hydrocarbon medium.

Among the different approaches to immobilization, main chain *chiral* polymer catalysts are different from the traditional polymer catalysts prepared by anchoring monomeric chiral catalysts to an *achiral* polymer backbone (Pu, 1998). The three classes of synthetic main chain chiral polymers include:

- helical polymers such as those represented by polypeptides;
- polymers with flexible chiral chains, such as polyesters and polyamides; and
- polymers of rigid, sterically regular chiral chains represented by chiral conjugated polybinaphthyls.

Covalent binding of homogeneous catalysts and/or their ligands to soluble or insoluble polymers with the purpose of catalyst recovery has been applied to catalysts for a wide variety of reactions, including alkene hydrogenation, carbonyl and imine reduction, carbon–carbon bond formation, carbonyl alkylation, Diels–Alder reactions, enolate chemistry, Strecker chemistry, asymmetric dihydroxylation, epoxidation and epoxide ring opening, and acylation catalysts, among others (Bergbreiter, 2000a).

18.4.2
Immobilization with Polyamino Acids as Chiral Polymer Catalysts

One topic where biocatalysis has not scored big successes (yet?) is the epoxidation reaction of alkenes. Against this background, the finding that poly-L-alanine catalyzes the epoxidation of enones such as chalcone, $Ar_1C(O)CH{=}CHAr_2$, raised plenty of interest (Julia, 1980). By employing hydrogen peroxide and NaOH in a biphasic water–toluene mixture together with poly-L-alanine, 2*R*,3*S*-epoxychalcone was obtained in 85% yield with 93% e.e. after 24 h. Over time, the main focus of the Julia–Colonna epoxidation, asymmetric epoxidation with hydrogen peroxide in the presence of poly(amino acid)s, is still on α,β-unsaturated ketones. The state of the art in Julia–Colonna epoxidation can currently be summarized as follows (Porter, 1999):

- the three phases, solid catalyst, aqueous and organic phases, have to be present in optimum proportions: 1 g catalyst per 10 mL water and 17 mL toluene;
- the reaction displays a strong solvent effect; on α,β-unsaturated ketones, toluene and carbon tetrachloride give the highest e.e. values while *n*-hexane and cyclohexane give poor values and methanol gives a racemic product;
- poly-L-leucine, -alanine, and -isoleucine give the best result whereas poly-L-phenylalanine and -proline result in racemic product.

Unnatural amino acid backbones can also act as catalysts. Poly-β-leucines have been evaluated as catalysts for the Julia–Colonna asymmetric epoxidation of enones; the β-3-isomer was found to be an effective catalyst for the epoxidation of chalcone (70% e.e.) and some analogs (Coffey, 2001).

All these findings serve to illustrate that the Julia–Colonna epoxidation is not completely understood yet. More importantly in the context of this book, there is not much indication that poly-(amino acid)s as catalysts bear much resemblance to proteins.

18.4.3
Immobilization on Resins or other Insoluble Carriers

Highly porous silica gel served as a support for the TADDOL moiety derived from inexpensive and readily available L-tartaric acid, which provided access to titanium-based Lewis acid catalysts (Heckel, 2000). Such entities are employed successfully for enantioselective reactions. TADDOLs were covalently attached to the trimethylsilyl-hydrophobized silica gel, controlled-pore glass (CPG) at about 300 $m^2\ g^{-1}$, at a loading of 0.3–0.4 mmol g^{-1} (Heckel, 2002). In a carefully monitored multi-step immobilization procedure, the TADDOLs were titanated to yield dichloro-, diisopropyl-, or ditosyl-TADDOLates. These catalysts were employed in dialkylzinc addition to benzaldehydes and diphenyl nitrone addition to 3-crotonyloxazolidinone, a [3+2] cycloaddition.

Among the results, the following are worth mentioning:
- Conversions and enantioselectivities of CPG-immobilized Ti-TADDOLates match the results observed with unsupported catalysts under similar but homogeneous conditions, a finding also observed in other situations (for references, see the next section);
- in cases where rates and enantioselectivities decrease after several rounds of repeated batch use, an aqueous HCl/acetone wash and reloading with titanate leads to full restoration of performance;

- for the [3+2] cycloaddition catalyzed by dichloro-Ti-TADDOLate, seasoning of catalyst is observed: whereas the immobilized catalyst requirement initially is 0.5 equiv, it drops to 0.1 equiv after acidic washing during runs.

Enantiomerically pure (2*S*,3*S*)-2,3-epoxy-3-phenylpropanol has been anchored to Merrifield resins with different degrees of crosslinking and functionalization (Vidal-Ferran, 1998). The resulting epoxy-functionalized resins have been submitted to completely regioselective (C3 attack) and stereospecific ring-opening with secondary amines such as piperidine, *N*-methylpiperazine, and *cis*-2,6-dimethylpiperidine in the presence of lithium perchlorate to afford (2*R*,3*R*)-3-(dialkylamino)-2-hydroxy-3-phenylpropoxy resin ethers. Analogously, (2*R*,3*R*)-3-(*cis*-2,6-dimethylpiperidino)-3-phenyl-1,2-propanediol has been anchored to a 2-chlorotrityl chloride resin (Barlos resin) in dichloromethane in the presence of diisopropylethylamine. The resulting resin has been employed as a chiral ligand in the catalytic addition of diethylzinc to a family of 14 aromatic and aliphatic aldehydes at 0 °C to afford the corresponding (*S*)-1-substituted 1-propanols with a mean enantiomeric excess of 92%; best results were obtained with the *cis*-2,6-dimethylpiperidine-containing ligand.

A crosslinked polystyrene resin containing poly(amino acid) bound through amide linkages to *p*-methylene-substituted aromatic rings is a useful regenerable chiral catalyst for the enantioselective epoxidation of α,β-unsaturated ketones (Itsuno, 1990). A series of polymer-supported poly(amino acids) were synthesized by polymerization of amino acid *N*-carboxyanhydrides with aminomethylated polystyrene beads. The poly(amino acid) moiety catalyzes the epoxidation of benzalacetophenone derivatives leading to optically active epoxides having enantiomeric purity of up to 99%. In a similar approach, an oxazaborolidine catalyst, derived from (*S*)-α,α-diphenyl-2-pyrrolidinemethanol and poly-*p*-styrene boronic acid (1% crosslinked), was bound to a poly-*p*-styrene-supported backbone and employed in the enantioselective borane reduction of two model ketones (Franot, 1995). After reduction of acetophenone, the reaction mixture was quenched and the polymer-bound catalyst was recovered by simple filtration. Whereas subsequent use of this catalyst “as is” for a second batch resulted in only a marginal loss in selectivity, selectivity began to diminish as of the third batch.

18.4.4
Heterogenization with Dendrimers

To compare dendritic and non-dendritic carriers for enantioselective epoxidation and hetero-Diels–Alder reactions, salen derivatives from either (*R*,*R*)-diphenylethylenediamine or (*R*,*R*)-cyclohexanediamine carrying two to eight styryl groups for crosslinking copolymerization with styrene were prepared (Sellner, 2001). First, the salicylic aldehyde moieties were attached to the styryl groups, then subsequent condensation with the diamines provided the chiral salens. [Salens lacking peripheral vinyl groups were prepared for comparison of catalytic activity in homogeneous solution and served as controls.] Crosslinking radical suspension copolymerization of styrene and the styryl salens yielded beads ($d \sim 400$ μm) which were loaded with Mn or Cr [~0.2 mmol complex (g polymer)$^{-1}$]; more than 95% of the salen incorporated was found to be accessible for complexation.

The polymer-bound Mn and Cr complexes were used as catalysts for epoxidations of six phenyl-substituted olefins with *m*-CPBA/NMO and for dihydropyranone formation from the Danishefsky diene and aldehydes. There are several remarkable features of the novel immobilized salens:

- dendritic branches do not slow down the catalytic activity of the complexes in solution;
- reactions with salen catalysts gave essentially products of similar enantiopurity, whether incorporated in polystyrene, dendritically substituted, or in homogeneous solutions
- some Mn-loaded beads have been stored for a year, without loss of activitys and
- the biphenyl- and the acetylene-linked salen polymers especially give Mn complexes of excellent performance: after ten uses (without recharging with Mn!) there is no loss of enantioselectivity or degree of conversion under the standard conditions.

The efficient reversible functionalization of the periphery of urea adamantyl poly(propyleneimine) dendrimers with catalytic sites using non-covalent interactions is described (de Groot, 2001). Phosphine ligands equipped with urea acetic groups, a binding motif complementary to that of the dendrimer host, have been prepared and assembled on the dendrimer support. The resulting supramolecular complex has been used as a multi-dentate ligand system in the palladium-catalyzed allylic amination reaction in a batch process and in a continuous-flow membrane reactor. Activity and selectivity of the dendrimeric complex is similar to that of the monomer complex, which indicates that the catalytic centers act as independent sites. The supramolecular system is sufficiently large and the binding of the guests is strong, enabling a good separation of the catalyst components from the reaction mixture by nanofiltration techniques.

While a number of dendritic catalysts have been described, catalyst recyclization in most cases is an unsolved problem. Diaminopropyl-type dendrimers bearing Pd–phosphine complexes have been retained by ultra- or nanofiltration membranes, and the constructs have been used as catalysts for allylic substitution in a continuously operating chemzyme membrane reactor (CMR) (Brinkmann, 1999). Retention rates were found to be higher than 99.9%, resulting in a sixfold increase in the total turnover number (TTN) for the Pd catalyst.

18.4.5
Retention of Heterogenized Chiral Chemical Catalysts in a Membrane Reactor

In analogy to the enzyme membrane reactor, discussed in Chapters 5 (Section 5.4.3) and 19 (Section 19.2), a membrane reactor for the retention of heterogenized organometallic catalysts has been developed, the chemzyme membrane reactor (CMR) (Kragl, 1996; Woltinger, 2001; Woeltinger, 2001). The main advantage of a

CMR over other catalyst retention techniques is its potentially infinitely high "partition" coefficient between the retentate, the retained phase in the reactor, and the filtrate, the stream leaving the CMR; retention by phase change such as precipitation or extraction can never achieve such partition coefficients.

The variety of reactions that have been catalyzed in a CMR (Table 18.3) comprises dialkylzinc addition to aldehydes (Kragl, 1996), borane reduction of ketones (Giffels, 1998), transfer hydrogenation to acetophenone (Laue, 2001), Sharpless dihydroxylation (Woltinger, 2001b), allylic amination (de Groot, 2001), Julia–Colonna asymmetric epoxidation (Tsogoeva, 2002), and opening of *meso*-anhydrides (Woltinger, 2002). Relevant performance data are listed in Table 18.3.

Inspection of Table 18.3 reveals that in principle any reaction catalyzed by organometallic catalysts is amenable to running in a CMR. Further, enantioselectivities seem to be just as high in the non-heterogenized case. The space–time yields achieved, often in excess of 1 kg $(L\ d)^{-1}$, point to the potentially very favorable volumetric productivities of a CMR configuration.

Retention of homogeneous catalysts can be achieved by binding the low-molecular-weight catalyst to a dendrimer, to an already formed polymeric backbone, or to a polymerizable monomeric unit which is polymerized subsequently. Currently, the disadvantages concern the durability of the nanofiltration or ultrafiltration membranes, which, after all, in most cases have not been slated for use in

Table 18.3 Reactions and their performance data in a chemzyme membrane reactor (CMR).

Reaction	*Catalyst*	*Backbone*	*Substrate*	*TTN* [-]	*S.t.y.* [g $(L\ d)^{-1}$]	*E.e.* [%]	*Reference*
Alkylation	Ph_2-L-Pro-ol	meth-acrylates	PhCHO, R_2Zn			20 (*R*) –50 (*S*)	Kragl, 1996
Borane reduction	oxazaboro-lidine		acetophenone	560	1400	≤ 99	Giffels, 1998
Allylic substitution	Pd dendrimer	$(Ph_2PCH_2)_2N$	cinnamyl carbonate	95 (Pd)	262	n.a.	Brinkmann, 1996
Borane reduction	Ph_2-L-Pro-ol	(Me) siloxane	α-tetralone	1400	1420	96	Woeltinger, 2001a
Hydrogen transfer	Gao–Noyori		acetophenone		578	94	Laue, 2001
Allylic amination	$Pd(PPh_3)_4$	urea–ada-mantyl-PPI					De Groot, 2001
Epoxidation	oligo-L-leucine		chalcone				Tsogoeva, 2001
Sharpless di-hydroxylation	Os		ethyl cinnamate				Woltinger, 2001b
Hydrolysis	cinchona alkaloid		*meso*-anhydrides			decr. to 60	Woltinger, 2002

TTN: total turnover number (= product/catalyst [mol mol^{-1}]);
s.t.y. = space–time-yield (volumetric productivity); n.a. = not applicable

organic solvents, as well as the problem of moisture in the CMR, as often even traces of water affect the catalyst detrimentally, especially with respect to enantioselectivity (Woltinger, 2001a). Reduction with borane compounds belongs to the class of very moisture-sensitive reactions. In the case of transfer hydrogenation with a Gao–Noyori catalyst in a continuously operated reactor, a continuous and steady dosage of isopropoxide is required to compensate for deactivation caused by water residues in the feed stream. Membrane materials tested so far encompass polyaramide (Kragl, 1996), silicone, and ceramic membranes (Karau, 2002); the latter of these certainly are sufficiently stable against organic solvents but suffer from occasional holes and lack of large-scale availability.

During the investigation of Sharpless dihydroxylation in the CMR, certainly one of the more sophisticated organometallic catalytic reactions, leaching of the osmium component was recognized to be a major problem. The catalytically decisive moiety in the Sharpless dihydroxylation is an intermittent osmate ester moiety which apparently is not strongly associated with the remainder of the catalyst complex. Tightness of binding of the osmium moiety to the remainder of the catalytic unit has been correlated with enantioselectivity even in the monomeric homogeneous catalyst (Kolb, 1994), corroborating the findings with the CMR.

In the alkylation of benzaldehyde and in the borane reduction of ketones, interesting dependencies of enantioselectivity on the ratio of substrates were observed. Depending on which substrate was present in excess, the alkylation of benzaldehyde yielded α-ethylbenzenemethanol product of about 20% (*R*) configuration (excess of benzaldehyde) or around 50% (*S*) configuration (excess of diethylzinc) (Kragl, 1996). Similarly, a higher amount of catalyst, as expected, turned out to accelerate the catalytic reaction over the uncatalyzed and enantiounspecific chemical background reaction (Giffels, 1998). Different ratios of substrate to catalyst can be investigated most easily in a continuous reactor.

Continuous reactors are not always beneficial to achievement of good reactor performance (Woltinger, 2002): in the asymmetric opening of *meso*-anhydrides, due to product inhibition of the cinchona alkaloid catalyst the conversion and enantiomeric excess decreased rapidly during a continuous reaction. Even optimization of reaction parameters to decrease residence time to a very low value (1 h) did not improve the situation sufficiently. In contrast, performing the reaction in repetitive batch mode allowed a modest 60% e.e. to be sustained over 18 cycles.

Currently, the technological challenges mentioned above seem to affect the long-term stability of a CMR configuration. Such long-term operational stability would allow the very promising performance regarding enantioselectivity and volumetric productivity to be translated into a serious alternative to biological catalysts.

18.4.6
Recovery of Organometallic Catalysts by Phase Change: Liquid–Liquid Extraction

A hydrolase-type reaction is otherwise rare in homogeneous catalysis, but the opening of prochiral cyclic anhydrides mediated by cinchona alkaloid in the presence of methanol leads to optically active hemiesters (Bolm, 2000). Very structurally diverse anhydrides are converted into their corresponding methyl monoesters, and either enantiomer can be obtained with up to 99% e.e. by using quinine or quinidine as directing additive. After the reaction, the alkaloids can be recovered almost quantitatively and re-used without loss of enantioselectivity by first extracting the catalysts into an acidic aqueous phase, from which they are back-extracted after neutralization into chloroform, and finally obtained by evaporating the solvent.

Liquid–liquid biphasic systems that exhibit an increase in phase miscibility at elevated temperature in connection with soluble polymer-bound catalysts that have a strong phase preference at ambient temperature can be utilized for product isolation and catalyst recovery by a liquid/liquid separation. Such a thermomorphic catalyst recovery system has been employed for palladium-catalyzed carbon–carbon bond-forming reactions (Bergbreiter, 2000b). Air-stable tridentate SCS-$Pd^{II}Cl$ catalysts bound to poly(*N*-isopropylacrylamide) (PNIPAM) or polyethylene glycol (PEG) are efficient catalysts in allylic substitution reactions, such as Heck or Suzuki reactions, and sp–sp^2 cross-coupling reactions. A particular advantage of these SCS catalysts over conventional phosphine ligands with a Pd^0 catalyst is the avoidance of time-consuming solvent purification and degassing protocols, as no precautions against adventitious catalyst oxidation need be taken.

18.5
Tandem Enzyme Organometallic Catalysts

While enzymes and chiral chemical catalysts compete for best performance in a variety of situations, they have also been used jointly to afford a desired reaction result (Choi, 1999). By far the most frequent application of this concept, termed an "enzyme–metal combi reaction (EMCR)", is the dynamic kinetic resolution (DFR) of a racemic mixture with a lipase and an organometallic complex to afford in-situ racemization.

Racemic resolution of α-hydroxy esters was achieved with *Pseudomonas cepacia* lipase (PCL) and a ruthenium catalyst (for a list, see Figure 18.13) as well as 4-chlorophenyl acetate as an acyl donor in cyclohexane, with high yields and excellent enantiomeric excesses (Huerta, 2000) (Figure 18.14). Combining dynamic kinetic resolution with an aldol reaction yielded β-hydroxy ester derivatives in very high enantiomeric excesses (≤ 99% e.e.) in a one-pot synthesis (Huerta, 2001).

The same concept is applicable to allylic alcohols, ketones, or ketoximes. Enol acetates or ketones were successfully converted in multi-step reactions to chiral acetates in high yields and optical yields through catalysis by *Candida antarctica* lipase B (CALB, Novozyme® 435) and a ruthenium complex. 2,6-Dimethylheptan-4-ol served as a hydrogen donor and 4-chlorophenyl acetate as an acyl donor for the conversion of the ketones (Jung, 2000a).

A similar lipase/ruthenium combination of an enzyme with an organometallic compound was employed to resolve racemates of allylic alcohols to allylic acetates of high optical purity and more than 80% yield in the presence of an acyl donor (Lee, 2000).

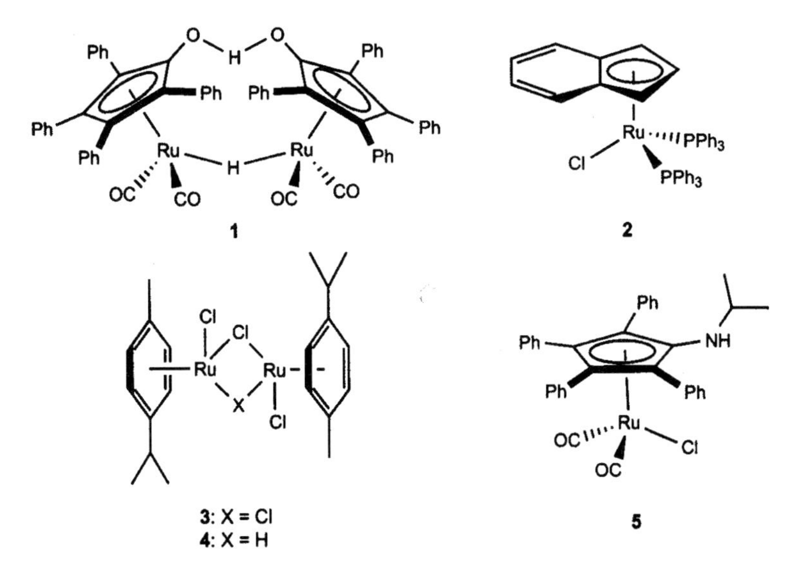

Figure 18.13 Asymmetric ruthenium-based catalysts for enzyme–metal combi reaction (Kim, 2002).

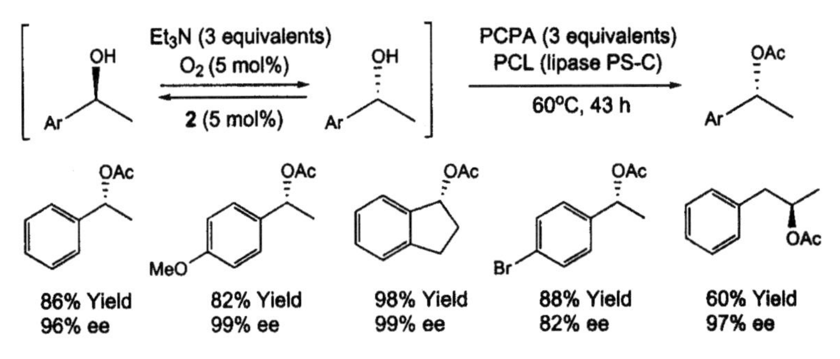

Figure 18.14 Racemic resolution of α-hydroxy esters with a tandem enzyme metallocomplex (Kim, 2002).

After the initial successes, a series of improvements and simplifications were introduced to the technique:

- In a one-pot reaction, a series of ketones were converted to chiral acetates with the help of an achiral ruthenium complex and CALB at 1 atm of hydrogen gas in ethyl acetate. Molecular hydrogen was equally effective in the transformation of enol acetates to chiral acetates in the same catalyst system without addition of additional acyl donors (Jung, 2000b).
- Lipase-catalyzed transesterifications proceed in ionic liquids with an up to 25-fold increased enantioselectivity in comparison with conventional organic solvents (Kim, 2001).

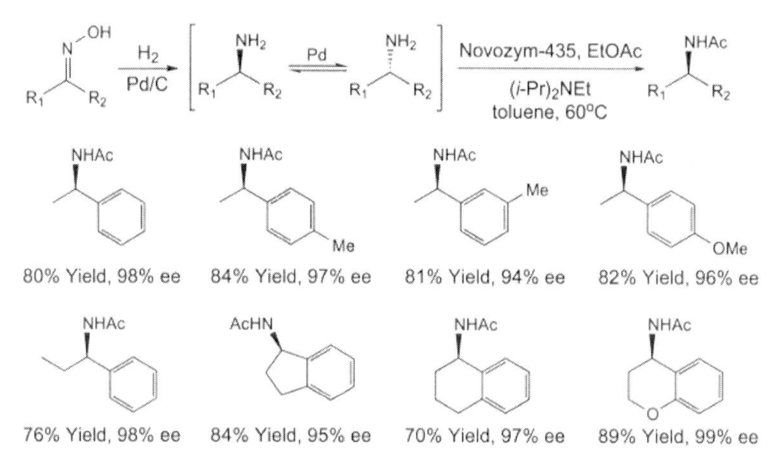

Figure 18.15 Resolution of ketoximes to amine acetates via EMCR (Kim, 2002).

The EMCR has been extended from obtaining enantiomerically pure alcohols to obtaining such amines. Prochiral ketoximes were transformed to optically active amine acetates in a coupled CALB/palladium catalysis in the presence of an acyl donor at 1 atm hydrogen (Figure 18.15) (Choi, 2001).

The standard in the field was set with the following racemic resolution of secondary alcohols: the transesterification of vinyl acetate to (*S*)-*O*-acetyl phenylethanol was catalyzed at room temperature by (isopropylamino)cyclopentadienylruthenium chloride and CALB with 97% yield and > 99% e.e. A reaction temperature of 25 °C, a reaction time of 30 h, and use of molecular sieve 4 Å as a water trap, sodium carbonate as base, and vinyl acetate as acyl donor, are all major improvements over the previous state-of-the-art (Choi, 2002).

Suggested Further Reading

D. Hilvert, Critical analysis of antibody catalysis, *Annu. Rev. Biochem.* **2000**, *69*, 751–793.

A. J. Kirby, Enzyme mechanisms, models, and mimics, *Angew. Chem., Int. Ed. Engl.* **1996**, *35*, 707–724; *Angew. Chem.* **1996**, *108*, 770–790.

References

D. E. Bergbreiter, Organic polymers as a catalyst recovery vehicle, in: *Chiral Catalyst Immobilization and Recycling*, D. E. De Vos, I. F. J. Vankelecom, and P. A. Jacobs (eds.), Wiley-VCH, Weinheim, **2000a**, pp. 43–80.

D. E. Bergbreiter, P. L. Osburn, A. Wilson, and E. M. Sink, Palladium-catalyzed C–C coupling under thermomorphic conditions, *J. Am. Chem. Soc.* **2000b**, *122*, 9058–9064.

C. Bolm, I. Schiffers, C. L. Dinter, and A. Gerlach, Practical and highly enantioselective ring opening of cyclic *meso*-anhydrides mediated by cinchona alkaloids, *J. Org. Chem.* **2000**, *65*, 6984–6991.

R. P. Bonar-Law, L. G. Mackay, C. J. Walter, V. Marvaud, and J. K. M. Sanders, Towards synthetic enzymes based on porphyrins and steroids, *Pure Appl. Chem.* **1994**, *66*, 803–810.

S. Borman, Chemical engineering focuses increasingly on the biological, *Chem. Eng. News* **1993**, January 11, 26–36.

N. Brinkmann, D. Giebel, G. Lohmer, M. T. Reetz, and U. Kragl, Allylic substitution with dendritic palladium catalysts in a continuously operating membrane reactor, *J. Catal.* **1999**, *183*, 163–168.

T. R. Cech, The efficiency and versatility of catalytic RNA: implications for an RNA world, *Gene* **1993a**, *135*, 33–36.

T. R. Cech, RNA. fishing for fresh catalysts, *Nature* **1993b**, *365*, 204–205.

T. R. Cech, Ribozymes, the first 20 years, *Biochem. Soc. Trans.* **2002**, *30*, 1162–1166.

Y. K. Choi, J. H. Suh, D. Lee, I. T. Lim, J. Y. Jung, and M. J. Kim, Dynamic kinetic resolution of acyclic allylic acetates using lipase and palladium, *J. Org. Chem.* **1999**, *64(22)*, 8423–8424.

Y. K. Choi, M. J. Kim, Y. Ahn, and M. J. Kim, Lipase/palladium-catalyzed asymmetric transformations of ketoximes to optically active amines, *Org. Lett.* **2001**, *3(25)*, 4099–4101.

J. H. Choi, Y. H. Kim, S. H. Nam, S. T. Shin, M. J. Kim, and J. Park, Aminocyclopentadienyl ruthenium chloride: catalytic racemization and dynamic kinetic resolution of alcohols at ambient temperatures, *Angew. Chem. Int. Ed.* **2002**, *41(13)*, 2373–2376.

A. G. Cochran and P. G. Schultz, Antibody-catalyzed porphyrin metallation, *Science* **1990**, *249*, 781–783.

P. E. Coffey, K. H. Drauz, S. M. Roberts, J. Skidmore, and J. A. Smith, beta-Peptides as catalysts: poly-beta-leucine as a catalyst for the Julia–Colonna asymmetric epoxidation of enones, *Chem. Commun.* **2001**, *(22)*, 2330–2331.

D. de Groot, B. F. de Waal, J. N. Reek, A. P. Schenning, P. C. Kamer, E. W. Meijer, and P. W. van Leeuwen, Noncovalently functionalized dendrimers as recyclable catalysts, *J. Am. Chem. Soc.* **2001**, *123*, 8453–8458.

J. A. Doudna and T. R. Cech, The chemical repertoire of natural ribozymes, *Nature* **2002**, *418*, 222–228.

M. Felder, G. Giffels, and C. Wandrey, A polymer-enlarged homogeneously soluble oxazaborolidine catalyst for the asymmetric reduction of ketones by borane, *Tetrahedron Asymm.* **1997**, *8*, 1975–1977.

K. Fodor, S. G. A. Kolmschot, and R. A. Sheldon, Heterogeneous enantioselective catalysis: state of the art, *Enantiomer* **1999**, *4*, 497–511.

C. Franot, G. B. Stone, P. Engeli, C. Spondlin, and E. Waldvogel, A polymer-bound oxazaborolidine catalyst: enantioselective borane reductions of ketones, *Tetrahedron: Asymm.* **1995**, *6*, 2755–2766.

G. Giffels, J. Beliczey, M. Felder, and U. Kragl, Polymer enlarged oxazaborolidines in a membrane reactor: enhancing effectivity by retention of the homogeneous catalyst, *Tetrahedron: Asymm.* **1998**, *9*, 691–696.

A. Heckel and D. Seebach, Immobilization of TADDOL with a high degree of loading on porous silica gel and first applications in enantioselective catalysis, *Angew. Chem., Int. Ed.* **2000**, *39*, 163–165.

A. Heckel and D. Seebach, Preparation and characterization of TADDOLs immobilized on hydrophobic controlled-pore-glass silica gel and their use in enantioselective heterogeneous catalysis, *Chem – Eur. J.* **2002**, *8*, 559–572.

D. Hilvert, K. W. Hill, K. D. Narel, and M.-T. M. Auditor, Antibody catalysis of the Diels–Alder reaction, *J. Am. Chem. Soc.* **1989**, *111*, 9261–9262.

D. Hilvert, Critical analysis of antibody catalysis, *Annu. Rev. Biochem.* **2000**, *69*, 751–793.

F. Hollfelder, A. J. Kirby, and D. S. Tawfik, Off-the-shelf proteins that rival tailor-made antibodies as catalysts, *Nature* **1996**, *383*, 60–62 (comment in: *Nature* **1996**, *383(6595)*, 23–24).

F. Hollfelder, A. J. Kirby, and D. S. Tawfik, On the magnitude and specificity of medium effects in enzyme-like catalysts for proton transfer, *J. Org. Chem.* **2001**, *66*, 5866–5874.

F. F. Huerta, Y. R. Laxmi, and J. E. Bäckvall, Dynamic kinetic resolution of alpha-hydroxy acid esters, *Org. Lett.* **2000**, *2(8)*, 1037–1040.

F. F. Huerta and J. E. Bäckvall, Enantioselective synthesis of beta-hydroxy acid derivatives via a one-pot aldol reaction – dynamic kinetic resolution, *Org. Lett.* **2001**, *3(8)*, 1209–1212.

S. Itsuno, M. Sakakura, and K. Ito, Polymer-supported poly(amino acids) as new asymmetric epoxidation catalyst of α,β-unsaturated ketones, *J. Org. Chem.* **1990**, *55*, 6047–6049.

K. Johnsson, R. K. Allemann, H. Widmer, and S. A. Benner, Synthesis, structure and activity of artificial, rationally designed catalytic polypeptides, *Nature* **1993**, *365*, 530–532.

Julia, **1980**.

H. M. Jung, J. H. Koh, M. J. Kim, and J. Park, Concerted catalytic reactions for conversion of ketones or enol acetates to chiral acetates, *Org. Lett.* **2000a**, *2(3)*, 409–411.

H. M. Jung, J. H. Koh, M. J. Kim, and J. Park, Practical ruthenium/lipase-catalyzed asymmetric transformations of ketones and enol acetates to chiral acetates, *Org. Lett.* **2000b**, *2(16)*, 2487–2490.

Karau, **2002**.

K. W. Kim, B. Song, M. Y. Choi, and M. J. Kim, Biocatalysis in ionic liquids: markedly enhanced enantioselectivity of lipase, *Org. Lett.* **2001**, *3(10)*, 1507–1509.

M. J. Kim, Y. Ahn, and J. Park, Dynamic kinetic resolutions and asymmetric transformations by enzymes couples with metal catalysis, *Curr. Opin. Biotechnol.* **2002**, *13*, 578–587.

A. J. Kirby, Simulation of enzymes, *Angew. Chem. Int. Ed. Engl.* **1994**, *33*, 551–553; *Angew. Chem.* **1994**, *106*, 573–576.

A. J. Kirby, Enzyme mechanisms, models, and mimics, *Angew. Chem., Int. Ed. Engl.* **1996a**, *35*, 707–724; *Angew. Chem.* **1996a**, *108*, 770–790.

A. J. Kirby, The potential of catalytic antibodies, *Acta Chem. Scand.* **1996b**, *50*, 203–10.

A. J. Kirby, F. Hollfelder, and D. S. Tawfik, Nonspecific catalysis by protein surfaces, *Appl. Biochem. Biotechnol.* **2000**, *83*, 173–180 (discussion in: *Appl. Biochem. Biotechnol.* **2000**, *83*, 180–181, 297–313).

H. C. Kolb, P. G. Andersson, and K. B. Sharpless, Toward an understanding of the high enantioselectivity in the osmium-catalyzed asymmetric dihydroxylation (AD). 1. Kinetics, *J. Am. Chem. Soc.* **1994**, *116*, 1278–1291.

U. Kragl and C. Dreisbach, Continuous asymmetric synthesis in a membrane reactor, *Angew. Chem., Int. Ed. Engl.* **1996**, *35*, 642–644.

K. Kruger, P. J. Grabowski, A. J. Zaug, J. Sands, D. E. Gottschling, and T. R. Cech, Self-splicing RNA: autoexcision and autocyclization of the ribosomal RNA intervening sequence of *Tetrahymena*, *Cell* **1982**, *31*, 147–157.

S. Laue, L. Greiner, J. Woltinger, and A. Liese, Continuous application of chemzymes in a membrane reactor: asymmetric transfer hydrogenation of acetophenone, *Adv. Synth. Catal.* **2001**, *343*, 711–720.

R. A. Laursen and F. H. Westheimer, The active site of acetoacetate decarboxylase, *J. Am. Chem. Soc.* **1966**, *88*, 3426–3430.

D. Lee, E. A. Huh, M. J. Kim, H. M. Jung, J. H. Koh, and J. Park, Dynamic kinetic resolution of allylic alcohols mediated by ruthenium- and lipase-based catalysts, *Org. Lett.* **2000**, *2(15)*, 2377–2379.

R. A. Lerner, S. J. Benkovic, and P. G. Schultz, At the crossroads of chemistry and immunology: catalytic antibodies, *Science* **1991**, *252*, 659–667.

L. G. Mackay, R. S. Wylie, and J. K. M. Sanders, Catalytic acyl transfer by a cyclic porphyrin trimer: efficient turnover without product inhibition, *J. Am. Chem. Soc.* **1994**, *116*, 3141–3142.

H. Oikawa, K. Katayama, Y. Suzuki, and A. Ichihara, Enzymic activity catalyzing *exo*-selective Diels–Alder reaction in solanapyrone biosynthesis, *J. Chem. Soc. Chem. Commun.* **1995**, 1321–1322.

S. J. Pollack, J. W. Jacobs, and P. G. Schultz, Selective chemical catalysis by an antibody, *Science* **1986**, *234*, 1570–1573.

M. J. Porter, S. M. Roberts, and J. Skidmore, Polyamino acids as catalysts in asymmetric synthesis, *Bioorgan. Med. Chem.* **1999**, *7*, 2145–2156.

L. Pu, Recent developments in asymmetric catalysis using synthetic polymers with main chain chirality, *Tetrahedron: Asymm.* **1998**, *9*, 1457–1477.

A. M. Reichwein, W. Verboom, and D. N. Reinhoudt, Enzyme models, *Rec. Trav. Chim. Pays-Bas* **1994**, *113*, 343–349.

J.-L. Reymond, J.-L. Reber, and R. A. Lerner, Antibody-catalyzed, enantioselective synthesis on the gram scale, *Angew. Chem.* **1994**, *106*, 485–486; *Angew. Chem., Int. Ed. Engl.* **1994**, *33*, 475–477.

S. J. Rowan and J. K. Sanders, Enzyme models: design and selection, *Curr. Opin. Chem. Biol.* **1997**, *1*, 483–490.

P. G. Schultz, Catalytic antibodies, *Angew. Chem., Int. Ed. Engl.* **1989**, *28*, 1283–1295.

P. G. Schultz and R. A. Lerner, Antibody catalysis of difficult chemical transformations, *Acc. Chem. Res.* **1995**, *26*, 391–395.

H. Sellner, J. K. Karjalainen, and D. Seebach, Preparation of dendritic and non-dendritic styryl-substituted Salens for cross-linking suspension copolymerization with styrene and multiple use of the corresponding Mn and Cr complexes in enantioselective epoxidations and hetero-Diels–Alder reactions, *Chem – Eur. J.* **2001**, *7*, 2873–2887.

S. Stinson, Studies probe catalytic antibody mechanisms, *Chem. Eng. News* **1998**, May 18, 44–45.

C. J. Suckling, M. C. Tedford, L. M. Bence, J. I. Irvine, and W. H. Stimson, An antibody with dual catalytic activity, *Bioorg. Med. Chem.* **1992**, *2*, 49–52.

S. N. Thorn, R. G. Daniels, M.-T. M. Auditor, and D. Hilvert, Large rate accelerations in antibody catalysis by strategic use of haptenic charge, *Nature* **1995**, *373*, 228–230.

A. Tramontano, K. D. Janda, and R. A. Lerner, Catalytic antibodies, *Science* **1986a**, *234*, 1566–1570.

A. Tramontano, K. D. Janda, and R. A. Lerner, Chemical reactivity at an antibody binding site elicited by mechanistic design of a synthetic antigen, *Proc. Natl. Acad. Sci. USA* **1986b**, *83*, 6736–40.

S. B. Tsogoeva, J. Woltinger, C. Jost, D. Reichert, A. Kuhnle, H.-P. Krimmer, and K. Drauz, Julia–Colonna asymmetric epoxidation in a continuously operated chemzyme membrane reactor, *Synlett* **2002**, *(5)*, 707–710.

A. Vidal-Ferran, N. Bampos, A. Moyano, M. A. Pericas, A. Riera, and J. K. Sanders, High catalytic activity of chiral amino alcohol ligands anchored to polystyrene resins, *J. Org. Chem.* **1998**, *63*, 6309–6318.

J. Woeltinger, A. S. Bommarius, K. Drauz, and C. Wandrey, The chemzyme membrane reactor in the fine chemicals industry, *Org. Proc. Res. Dev.* **2001**, 5, 241–248.

J. Woltinger, K. Drauz, and A. S. Bommarius, The membrane reactor in the fine chemicals industry, *Appl. Catal. A: General* **2001a**, *221*, 171–185.

J. Woltinger, H. Henniges, H.-P. Krimmer, A. S. Bommarius, and K. Drauz, Application of the continuous Sharpless dihydroxylation, *Tetrahedron: Asymm.* **2001b**, *12*, 2095–2098.

J. Woltinger, H.-P. Krimmer, and K. Drauz, The potential of membrane reactors in the asymmetric opening of *meso*-anhydrides, *Tetrahedr. Lett.* **2002**,*43*, 8531–8533.

19 Design of Biocatalytic Processes

Summary

The manufacture of HFCS (high-fructose corn syrup) is certainly the largest and one of the most important enzyme processes. The large scale necessitates low manufacturing costs and thus very efficient processing. For this reason, we will investigate in more depth the common process option for the GI-catalyzed reaction to HFCS, the plug-flow packed-bed reactor (GI = glucose isomerase).

First-order kinetics can be assumed both for the enzymatic reaction, as $[S] \ll K_M$, and for enzyme deactivation $[k = k_0 \cdot \exp(-k_d \cdot t)]$. The productivity p of an enzyme catalyst is the total amount of glucose syrup converted during the lifetime of that catalyst and is calculated by integrating the flow rate over all times of operation from 0 to t. As strong diffusion limitations are assumed, the Thiele modulus ϕ becomes a function of time t: $\phi = \phi_0 \cdot \exp(-\frac{1}{2}k_d \cdot t)$. The analytical solution for the optimum productivity is a function of geometry (A, H, L), pore structure and diffusion (D_{eff}, ε), enzyme deactivation (k_d), and the Thiele moduli ϕ_0 and ϕ. At an arbitrarily long time, productivity is highest at the lowest possible temperature. At 50 °C compared to 65 °C, it is higher by a factor of seven. At a possible production time of 50 days, maximum productivity is 30% higher at 60 °C than at 65 °C. In the daily production routine, both enzyme costs and operating costs play decisive roles. If realistic numbers are known for both contributions the costs of HFCS production can be optimized.

Among retention methods for biocatalysts (heterogenization on supports, recovery through phase change, and membrane filtration of a homogeneous catalyst), the enzyme–membrane reactor (EMR) is an established mode for running continuous biocatalytic processes. Productivity is influenced by retention of the enzyme catalyst as much as by its deactivation behavior. In the case of modeling activity decline by leakage through the membrane only (no deactivation), the observed first-order decay constant can be identified as $k_{d,obs} = (1 - R)/\tau$, whereas in the case of leakage and parallel deactivation, $k_{d,obs}$ depends in complex fashion on R, k_d and τ.

Enantiomerically pure alcohols belong to the most sought-after building blocks for pharmaceuticals, crop-protection agents, and food and nutraceuticals. There are a variety of process configurations available for their manufacture:

- whole-cell, enzymatically catalyzed,or homogeneously catalyzed, process routes;
- reduction by several possible reductants, such as borane, formate, or molecular hydrogen;
- different reactor configurations, such as membrane reactors, two-phase systems, organic solvents, or combinations thereof.

Isolated enzymes are advantageous over whole cells owing to the absence of side activities compromising selectivity and the greater ease of repeated re-use to achieve high total turnover numbers (TTNs). Principal disadvantages are the requirement for a purification protocol, the yield loss upon purification, and the requirement for external addition of cofactors.

Turnover numbers (TONs) and substrate/catalyst ratios ([S]/[C] ratios) seem the preferred quantities in homogeneous catalysis, in contrast to biocatalyst loading (units L^{-1}) and TTNs in biocatalysis. In the case of slow homogeneous chemical catalysts, the [S]/[C] ratio can approach unity (stoichiometric conditions). In the limit of no recycle, the values for TTN and TON are identical; upon re-use of catalyst, TTN increases correspondingly. Whereas recycling is very important in biocatalysis, it does not seem to be common practice in homogeneous chemical asymmetric catalysis.

19.1 Design of Enzyme Processes: High-Fructose Corn Syrup (HFCS)

19.1.1 Manufacture of HFCS from Glucose with Glucose Isomerase (GI): Process Details

The GI process from glucose to a mixture of glucose and fructose, the largest biocatalytic process with an annual worldwide production volume in excess of 10 million tons, has already been discussed in Chapter 7, Section 7.3.1. In this section, we cover a more in-depth model of the packed-bed reactor and provide model equations for its output and productivity.

As a reminder, Figure 19.1 shows the process configuration originally provided in Figure 7.24: parallel trains of packed-bed reactors are operated with differently deactivated enzyme batches in tube bundles. The age and temperature of each enzyme batch are adapted to regulate and stabilize production capacity: a high capacity is achieved at 63–65°C with fresh enzyme, whereas for a lower capacity 55–57°C and partially already-deactivated enzyme suffice. A typical system (Figure 19.1) runs with parallel rows of columns, each producing a 58% glucose/42% fructose mixture, termed HFCS 42.

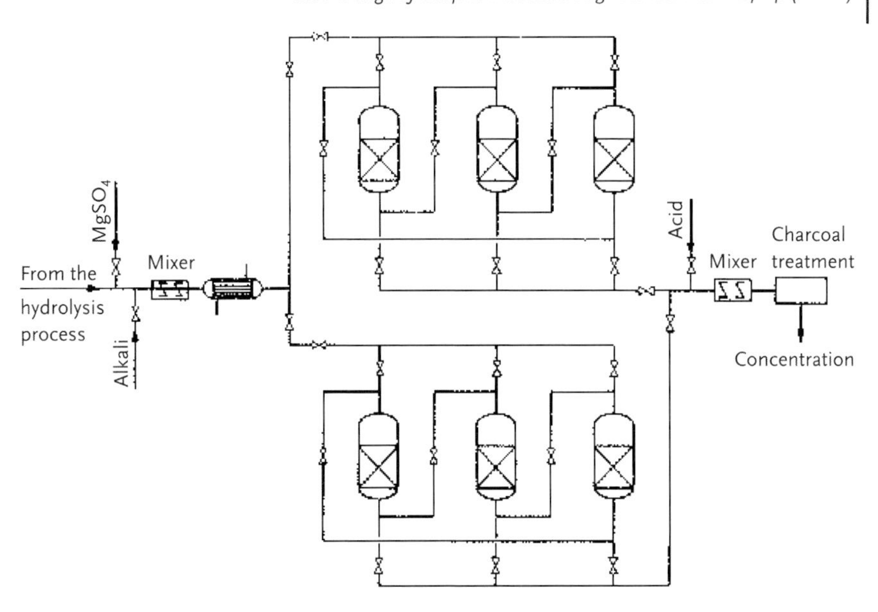

Figure 19.1 Process scheme for the glucose isomerase process (Crueger, 1989).

19.1.2
Mathematical Model for the Description of the Enzyme Kinetics of Glucose Isomerase (GI)

The large scale of the manufacture of HFCS necessitates low manufacturing costs and thus very efficient processing. For this reason, we will investigate in more depth the common process option for the GI-catalyzed reaction to HFCS, the plug-flow packed-bed reactor (Roels, 1979).

As the isomerization of glucose and fructose is reversible, the kinetics of the reaction can be represented by Eq. (19.1) (G = glucose; E = enzyme, F = fructose).

$$\mathrm{G} + \mathrm{E} \Leftrightarrow \mathrm{EX} \Leftrightarrow \mathrm{E} + \mathrm{F} \tag{19.1}$$

Accordingly, the rate expression, a standard Michaelis–Menten equation, has to be based not on the overall glucose concentration [G] but on the difference in concentration from the equilibrium glucose concentration [G]*, both in mol m^{-3}:

$$r = v_{max} \cdot [\mathrm{E}] \cdot ([\mathrm{G}] - [\mathrm{G}]^*)/\{K_{M,S} + ([\mathrm{G}] - [\mathrm{G}]^*)\} \tag{19.2}$$

The remaining symbols are:

r = reaction rate of glucose to fructose [mol m^{-3} s^{-1}]

v_{max} = maximum specific rate of fructose formation [mol (kg dry biomass)$^{-1}$$s^{-1}$]

[E] = concentration of the organism [kg dry biomass m^{-3}].

However, as $K_{M,S}$ is around 5000 mol m^{-3}, the difference ([G] – [G]*) commonly lies between only 3000 mol m^{-3} (column entrance) and 20 mol m^{-3} (column exit), first-order kinetics can be assumed and Eq. (19.2) can be simplified to Eq. (19.3), where k' is a pseudo first-order rate constant given by Eq. (19.4).

$$r = k' \cdot ([G] - [G]^*) \tag{19.3}$$

$$k' = \{k_2 (K^* + 1) \cdot [E] \cdot [E_{intr}]\}/\{K^* [(1 + K^* \cdot (K_{M,G}/K_{M,S})) \cdot [G]^* + K_{M,G}]\} \tag{19.4}$$

As k' is dependent on [G]* and thus also on $[G]_0$, k' cannot represent a true kinetic constant; $[E_{intr}]$ is the intrinsic enzyme concentration [mol (kg dry biomass)$^{-1}$], and K^* is the equilibrium constant.

From Chapter 5, Section 5.5.2, we recall the definition for the internal effectiveness factor η ((Eq. (19.5)), with the solution for first-order kinetics and spherical geometry in Eq. (19.6) (which is identical to Eq. (5.59) in Chapter 5), where $\phi = L \cdot (k'/D_{eff})^{1/2}$ represents the dimensionless Thiele modulus, and L is the characteristic length, i.e., $r/3$ for a sphere.

$$\eta = \frac{\text{reaction rate with consideration of internal diffusional effects}}{\text{intrinsic reaction rate without diffusional effects}} \tag{19.5}$$

$$\eta = 3/(\phi \cdot \tan \phi) - 1/\phi \tag{19.6}$$

It is assumed that deactivation of the enzyme occurs according to first-order kinetics (Eq. (19.7)).

$$k = k_0 \cdot \exp(-k_d \cdot t) \tag{19.7}$$

The effect of operating parameters on k' and k_d can be captured with an Arrhenius equation each (Eqs. (19.8) and (19.9)).

$$k' = k_0' \cdot \exp(-\Delta H'/RT) \tag{19.8}$$

$$k_d = k_{d,0} \cdot \exp(-\Delta H_d/RT) \tag{19.9}$$

Figure 19.2 depicts the mass balance for the fixed-bed reactor (Eq. (19.10)), with the parameters ε = porosity of fixed bed [–], A = area of fixed bed [m^2], and h = height of fixed bed [m].

$$F \cdot d[G] = -\eta (1 - \varepsilon) A \cdot k' ([G] - [G]^*) \, dh \tag{19.10}$$

The solution to Eq. (19.10) is provided by Eq. (19.11) with the two boundary conditions listed below ($[G]_0$ = entrance and $[G]_f$ = exit concentration of glucose, both in mol m^{-3}).

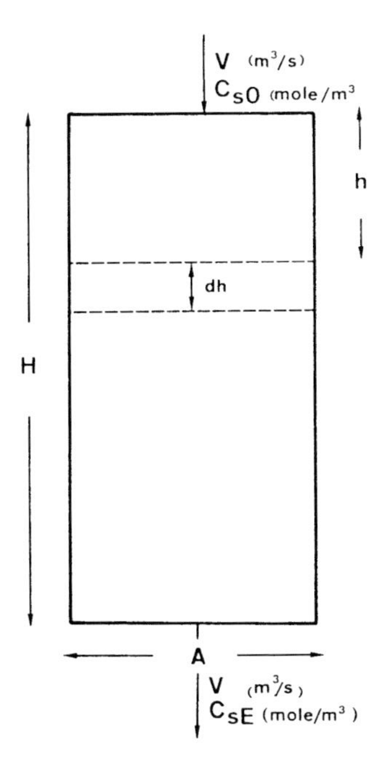

Figure 19.2 Model for fixed-bed isomerization (Roels, 1979).

$$F = \eta\,(1-\varepsilon)\,A \cdot k' \cdot H/\ln\{([G]_0-[G]^*)/([G]_f-[G]^*)\} \qquad (19.11)$$

Boundary condition 1: $[G]=[G]_0$ at $h=0$ and
Boundary condition 2: $[G]=[G]_f$ at $h=H$

Equation (19.11) contains a number of assumptions:

- external mass transfer can be neglected;
- the glucose syrup flows through the column in a plug-flow configuration;
- pseudo-stationarity with respect to k' as a function of t is assumed, i.e., k' is assumed to be constant for the duration of a residence time in the column. (As k_d features a timescale of about 100 days in comparison to a residence time of 1 h, this assumption is justified.)

19.1.3
Evaluation of the Model of the GI Reaction in the Fixed-Bed Reactor

At pH 7.5, with [G] = 3 M and additions of 3 mM $MgSO_4$ and 100 ppm SO_2 as activators, Eq. (19.11) was validated experimentally at Gist–Brocades with Maxazyme®, the GI from Gist–Brocades. Runs with constant substrate flow at varying degrees of conversion (constant-flow policy) were conducted as well as runs with constant conversion but varying substrate flow (constant-conversion policy). The temperature was varied between 55 °C and 75 °C, and 65 °C was picked as the reference temperature for most data points. The data obtained with multivariate curve fitting by non-linear regression are listed in Table 19.1

Table 19.1 Validation data for GI reaction model.

Parameter	***Value***	***Units***
K^*	$28.8 \cdot \exp(-1100/RT)$	(–)
ΔH_d	203	$kJ\ mol^{-1}$
$\Delta H'$	79	$kJ\ mol^{-1}$
D_{eff}	6.7×10^{-11}	$m^2\ s^{-1}$
$k_{0,T}'$	$4.83 \times 10^9 \exp(-9500/T)$	s^{-1}
$k_{0,65}'$	3.0×10^{-3}	s^{-1}
k_d	$1.51 \times 10^{30} \exp(-24\,400/T)$	d^{-1}
L	0.6	mm

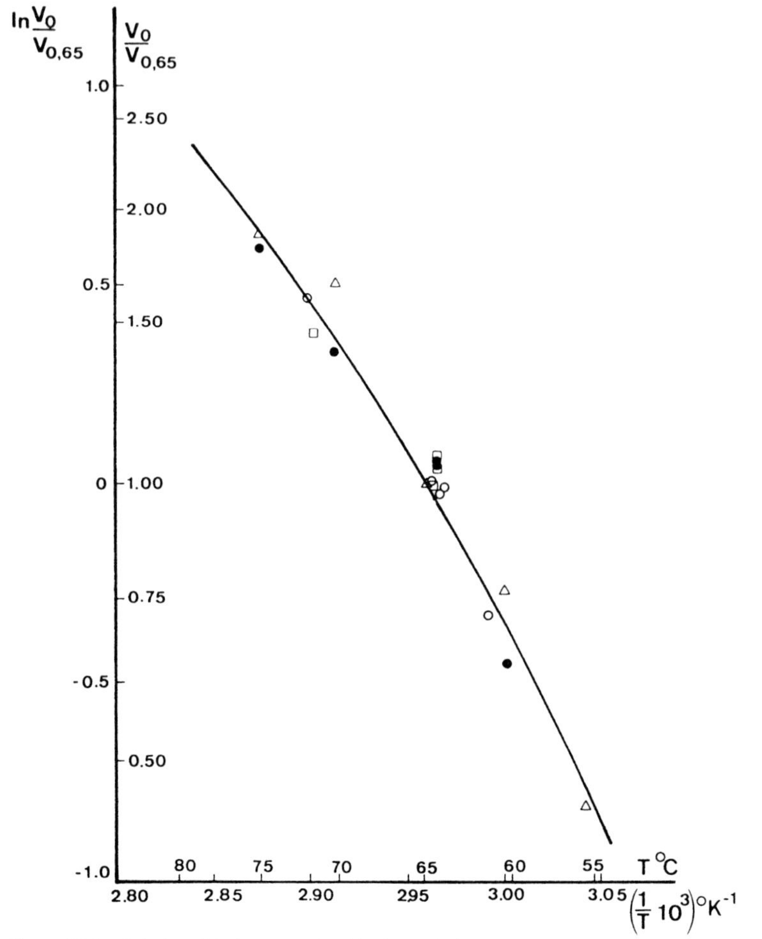

Figure 19.3 Relationship between fixed-bed initial flow rate and temperature (Roels, 1979).

Figure 19.3 demonstrates the Arrhenius plot, normalized to $T = 65$ °C. The curve is not a straight line, so the Arrhenius equation is not valid. In Figure 19.4, it can be discerned that the gradient of the Arrhenius plot in the present case is about half as great at high temperature as at low temperature and half that in the case of soluble homogeneous enzyme. As a decrease in apparent activation energy is known to occur in the case of strong diffusional limitations, it can be surmised that in the present case at high temperature the enzyme reaction is slowed down by the diffusion of substrate or product.

The present model allows the determination of the flow rate at constant conversion and operating time in a fixed-bed reactor. In Figure 19.5 it can be seen that the actually measured deactivation curve is located between the assumed exponential

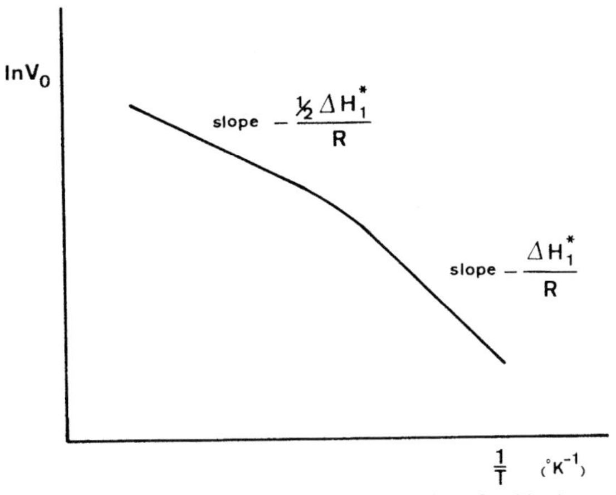

Figure 19.4 Characteristic Arrhenius plot of the fixed-bed initial flow rate (Roels, 1979).

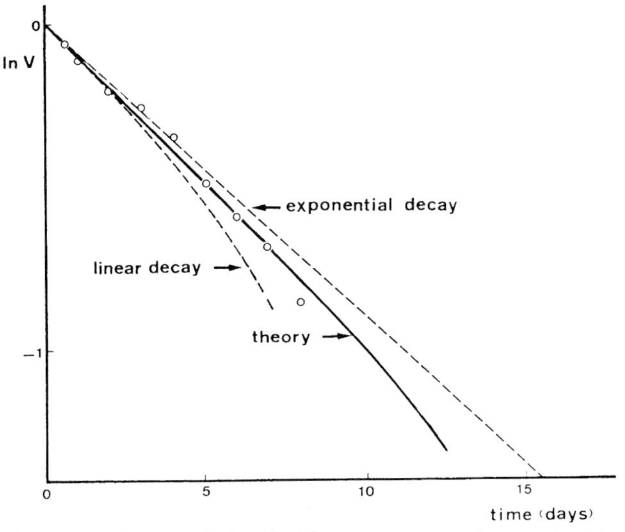

Figure 19.5 Decrease in fixed-bed flow rate at constant conversion (Roels, 1979).

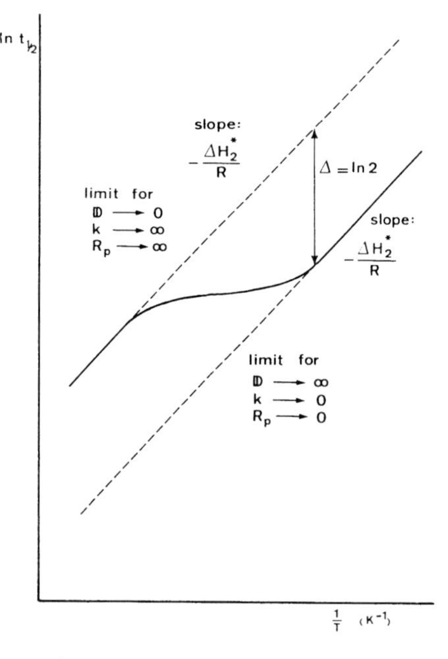

Figure 19.6 Characteristic Arrhenius plot of half-life at constant conversion (Roels, 1979).

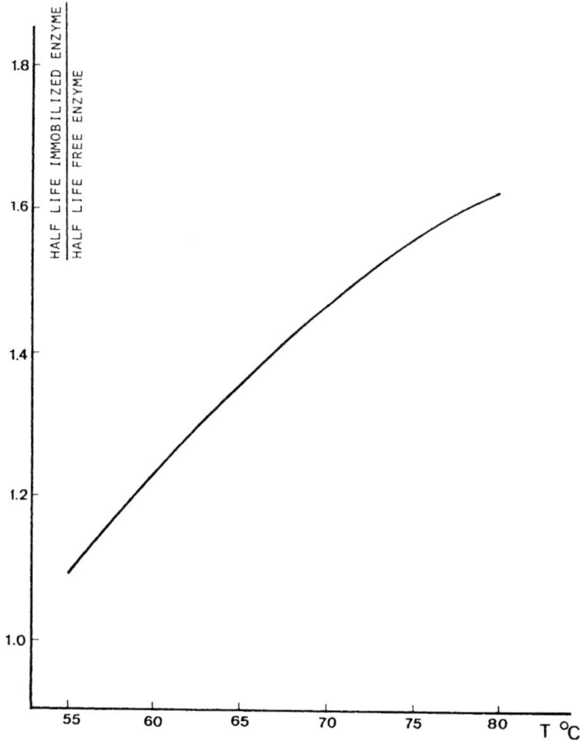

Figure 19.7 Theoretical relationship between activity half-life of free and immobilized enzyme (Roels, 1979).

deactivation over time and a linearly decreasing deactivation. The assumption of an exponential deactivation overestimates the actually determined half-life of the immobilized enzyme; in contrast, the assumption of a linear deactivation underestimates the half-life.

Initially, i.e., at low Thiele moduli, the Arrhenius plot in Figure 19.6 is linear with a slope equal to the deactivation enthalpy of the free, dissolved enzyme. At higher temperatures, the slope decreases, only to adopt the same value for the deactivation enthalpy as the dissolved enzyme at even higher temperatures and thus also at high Thiele moduli. However, the half-life is now twice as high as in the case of the free enzyme (see the discussion in Chapter 5, Section 5.6.3).

From the previous paragraph it can be discerned that the half-life of a biocatalyst is determined not only by the deactivation properties of the free enzyme but also by the mass-transfer properties of an immobilisate. Figure 19.7 reveals the factor by which the dependence on temperature increases operating stability in the case of immobilized enzyme.

19.1.4 Productivity of a Fixed-Bed Enzyme Reactor

The productivity p of an enzyme catalyst is the total amount of glucose syrup converted during the lifetime of that catalyst. It is calculated according to Eq. (19.12).

$$p(t) = \int_0^t F(t)\,\mathrm{d}t \tag{19.12}$$

$F(t)$ is calculated according to Eq. (19.11), the values for k' as a function of operating time according to Eq. (19.7) and the effectiveness factor according to Eq. (19.6). As strong diffusion limitations are assumed, the Thiele modulus ϕ as a function of time behaves as described by Eq. (19.13), where Eq. (19.14) holds for ϕ_0.

$$\phi = \phi_0 \cdot \exp\left(-\tfrac{1}{2}\, k_\mathrm{d} \cdot t\right) \tag{19.13}$$

$$\phi_0 = L \cdot (k_0'/D_\mathrm{eff})^{1/2} \tag{19.14}$$

Equations (19.13) and (19.7) yield Eq. (19.15).

$$k' = k_0'\,(\phi/\phi_0)^2 \tag{19.15}$$

From Eqs. (19.11), (19.12), and (19.14) Eq. (19.16) follows for Eq. (19.12).

$$p(t) = \int_0^t \{D_\mathrm{eff}\,(1-\varepsilon)\,A \cdot H\,(3\phi\tanh\phi - 3)/L^2 \cdot \ln[([\mathrm{G}]_0 - [\mathrm{G}]^*)/([\mathrm{G}]_\mathrm{f} - [\mathrm{G}]^*)]\}\,\mathrm{d}t \tag{19.16}$$

For the integration, Eq. (19.16) has to be simplified. With Eq. (19.17), Eq. (19.13) can be transformed into Eq. (19.18).

$$dt = (dt/d\phi) \cdot d\phi \tag{19.17}$$

$$dt/d\phi = -2/(k_d \cdot \phi) \tag{19.18}$$

With Eqs. (19.17) and (19.18), Eq. (19.16) can be transformed into Eq. (19.19), for which Eq. (19.20) is the solution.

$$p(\phi) = \int_{\phi_0}^{\phi} \{-6D_{eff}(1-\varepsilon)\, A \cdot H\,(1/\tanh\phi - 1)/[k_d \cdot L^2]\}\, d\phi \tag{19.19}$$

$$p\,(\phi) = \{-6\, D_{eff}(1-\varepsilon)\, A \cdot H/[k_d \cdot L^2]\} \cdot \{\ln[(\sinh\phi_0)/\phi_0] - \ln[(\sinh\phi)/\phi]\} \tag{19.20}$$

With the help of Eq. (19.20) productivities can be calculated as a function of operating temperature (see Figures 19.8 and 19.9). Figure 19.8 demonstrates that at an arbitrarily long time productivity is highest at the lowest possible temperature. At 50 °C it is higher by a factor of seven than at 65 °C. If one only has 100 days at one's disposal, the maximum productivity is achieved at 55 °C and it is twice as high as at 65 °C. At a possible production time of 50 days, maximum productivity is 30% higher at 60 °C than at 65 °C.

In the daily production routine, both enzyme costs and operating costs play decisive roles. In addition, the amount of production is likely to be adapted to demand the load for the product, here HFCS, so that at high demand the optimal production temperature is higher than at lower demand. From Figure 19.9 it becomes apparent how optimal operating times and temperatures can be estimated if the operating costs of a column as well as the costs for the immobilized enzyme

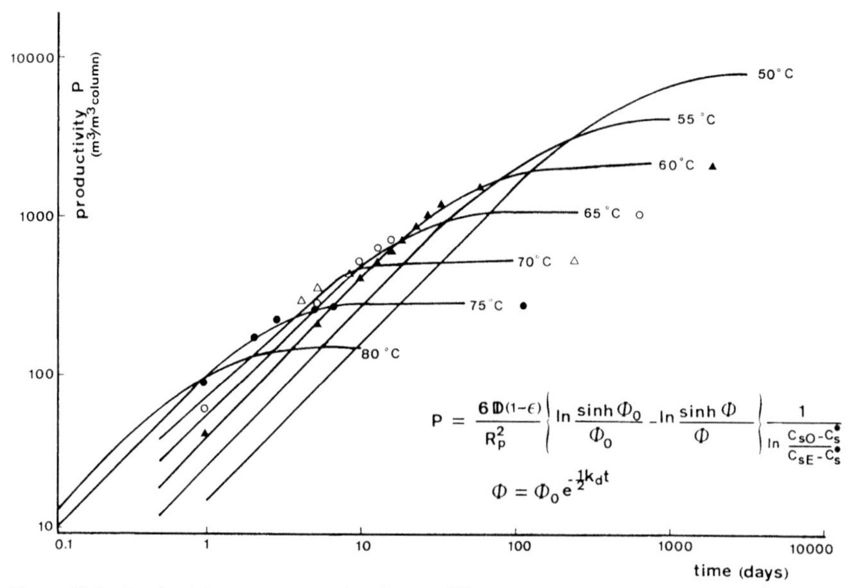

Figure 19.8 Productivity versus operating time at different operating temperatures (Roels, 1979).

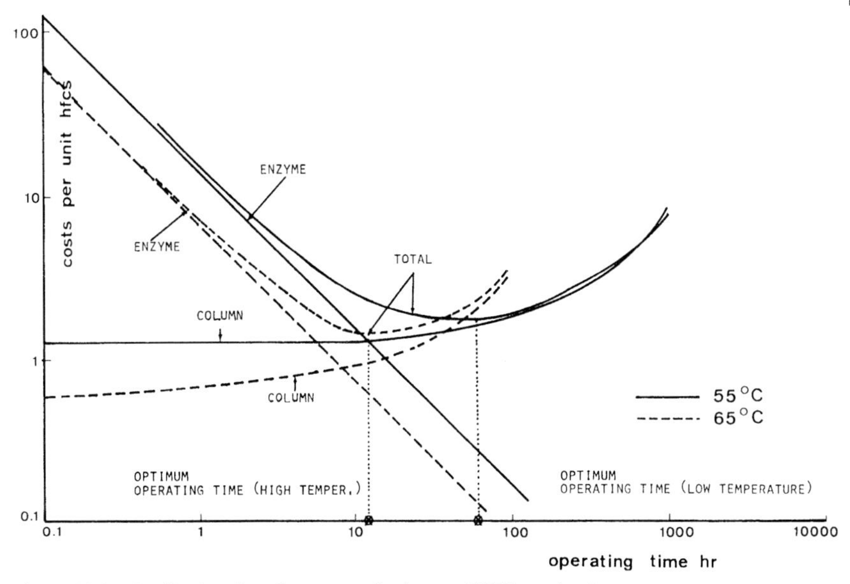

Figure 19.9 Qualitative plot of cost contributions to HFCS production at different operating temperatures (Roels, 1979).

are known. If realistic values are known for both, the costs of HFCS production can be optimized. The graph readily reveals that a variable temperature increasing with deactivation yields better results than a constant temperature.

19.2
Processing of Fine Chemicals or Pharmaceutical Intermediates in an Enzyme Membrane Reactor

19.2.1
Introduction

In the preceding section, we analyzed an immobilized enzyme process and calculated some important parameters such as productivity. In this section, we investigate another process configuration for retaining biocatalysts, the membrane reactor. The advantages and disadvantages of immobilization and membrane retention have already been discussed in Chapter 5. As in the case of immobilization, retention of catalyst by a membrane vastly improves biocatalyst productivity, a feature important on a processing scale but usually not on a laboratory scale.

Methods for (bio)catalyst retention include: (i) heterogenization on supports; (ii) recovery through phase change, such as precipitation or extraction of the catalyst; and (iii) membrane filtration of a homogeneous catalyst. All methods, in principle, enable repeated use of a chiral catalyst without much loss of activity or selectivity. In a recent review (Kragl, 2001), examples are given from laboratory and industrial processes, including hydrogenations, ketone reductions, epoxidations, dihydroxylations, diethylzinc additions, and Diels–Alder reactions catalyzed by chemo- or biocatalysts. Data from some industrial processes include the production of metalochlor and the production of (–)-menthol.

The enzyme membrane reactor (EMR) is an established mode for running continuous biocatalytic processes, ranging from laboratory units of 3 mL volume via pilot-scale units (0.5–500 L) to full-scale industrial units of several cubic meters volume and production capacities of hundreds of tons per year (Woltinger, 2001; Bommarius, 1996). The analogous chemzyme membrane reactor (CMR) concept, discussed in Chapter 18, Section 18.4.5, is based on the same principles as the EMR but is far less developed yet.

19.2.2
Determination of Process Parameters of a Membrane Reactor

In this section, we build on concepts covered in several earlier chapters; membrane reactors were discussed in Chapter 5 and membrane filtration was covered in Chapter 8, Section 8.5.1. A typical membrane reactor configuration is depicted in Figure 19.10, comprising a pump working into the membrane reactor, pre-filtering units, the membrane reactor itself, often a recycle reactor with a separate pump, and the membrane unit itself. Hollow-fiber membrane units are the most common, as they allow tangential filtration minimizing concentration polarization (Ch. 8, Section 8.5.1.), and a space-saving configuration.

We will calculate the reactor performance itself as well as the productivity over time; we will see that productivity is influenced by retention of the enzyme catalyst as much as by its deactivation behavior. In the schematic of an enzyme membrane reactor

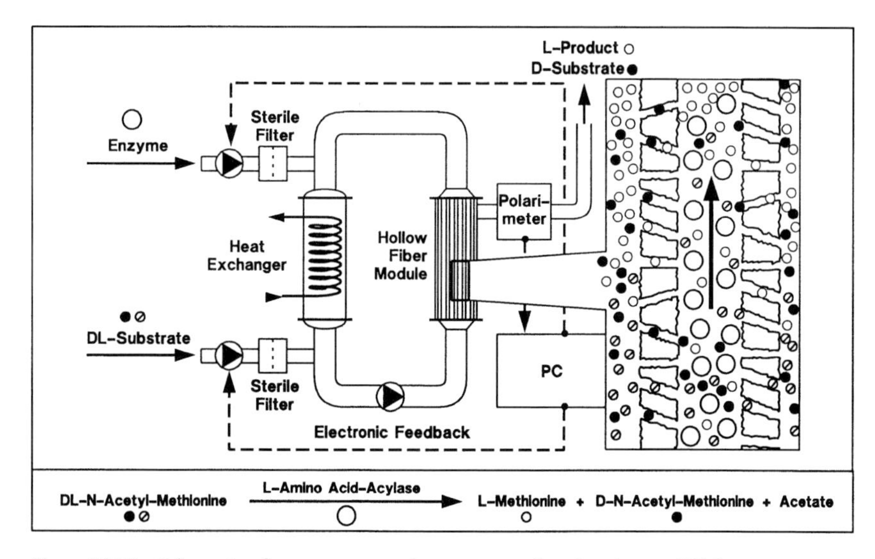

Figure 19.10 Schematic of an enzyme membrane reactor (Leuchtenberger, 1984).

(EMR) in Figure 19.10, the solvent (usually water) with flux J passes through the membrane of area A_m after a residence time τ in the reactor of volume V; the active enzyme concentration E in the reactor at any time can be diminished by leakage through the membrane and can then be found in the filtrate with a concentration E_f.

19.2.2.1 **Case 1: Leakage through Membrane, no Deactivation**

If only leakage through the membrane can diminish enzyme concentration, the rate of decrease is given by Eq. (19.21).

$$-V \cdot (dE/dt) = J \cdot A_m \cdot E_f \tag{19.21}$$

The enzyme concentration E_f in the filtrate and the residence time τ can be expressed as Eqs. (19.22) and (19.23) respectively.

$$E_f = (1 - R)\ E \tag{19.22}$$

$$\tau = V/Q = V/(J \cdot A_m) \tag{19.23}$$

Inserting Eqs, (19.21) and (19.22) into Eq. (19.23) yields Eq. (19.24), which upon integration from time t_0 to t and enzyme concentration E_0 to E_t yields Eq. (19.25).

$$-V \cdot (dE/dt) = (V/\tau) \cdot (1 - R) \cdot E \tag{19.24}$$

$$\ln (E_t/E_0) = -(1 - R)(t - t_0)/\tau \tag{19.25}$$

As it often happens that $t_0 = 0$, Eq. (19.25) is reduced to Eq. (19.26).

$$\ln (E_t/E_0) = -(1 - R)(t/\tau) \tag{19.26}$$

If the enzyme leaks strongly through the membrane, i.e., low values of R ($R \to 0$), Eq. (19.26) becomes Eq. (19.27); in other words, a negative exponential function with parameter (t/τ) results.

$$(E_t/E_0) = \exp -(t/\tau) \tag{19.27}$$

In the case of very good retention of enzyme, $R \to 1$; thus $x = (1 - R)(t/\tau) \to 0$, so that $\exp(-x) \approx 1 - x$ and hence Eq. (19.28) is obtained, so that straight-line behavior over time results.

$$(E_t/E_0) = 1 - (t/\tau) \tag{19.28}$$

When observing a decline in enzyme activity in the reactor, one assumes deactivation of the enzyme. If this deactivation phenomenon is modeled according to Eq. (19.29) [which is equivalent to Chapter 2, Eq. (19.19) and Chapter 5, Eq. (5.74)], then $k_{d,obs}$ can be identified with Eq. (19.26), as Eq. (19.30) states.

$$\ln (E_t/E_0) = -k_{d,obs} \cdot t \tag{19.29}$$

$$k_{d,obs} = (1 - R)/\tau \tag{19.30}$$

19.2.2.2 Case 2: Leakage through the Membrane and Deactivation of Enzyme

If we now assume that the enzyme deactivates according to the first-order rate law, $\ln (E_t/E_0) = -k_d \cdot t$ and permeates the membrane so that the retention is $(1 - R)$, Eq. (19.24) is modified to give Eq. (19.31), which upon integration from time 0 to t and enzyme concentration E_0 to E_t (Eq. (1932)) yields Eq. (19.33) or, upon rearrangement, Eq. (19.34).

$$-V \cdot (dE/dt) = (V/\tau) \cdot (1 - R) \cdot E_0 \cdot \exp(-k_d \cdot t) \tag{19.31}$$

$$\int_{E_0}^{E_t} dE = \{-E_0 \cdot (1 - R)/\tau\} \int \exp(-k_d \cdot t)\, dt \tag{19.32}$$

$$(E_t - E_0) = \{E_0 (1 - R)/(k_d \cdot \tau)\} \{1 - \exp(-k_d \cdot t)\} \tag{19.33}$$

$$E_t/E_0 = 1 - \{(1 - R)/(k_d \cdot \tau)\} \{1 - \exp(-k_d \cdot t)\} \tag{19.34}$$

If now Eq. (19.34) is compared with Eq. (19.29), k_{obs} no longer can be expressed in simple terms of R, k_d, and τ. Only if k_d and $k_{d,obs}$ are small, $k_{d,obs}$ can be identified with Eq. (19.35).

$$k_{d,obs} = (1 - R)\, k_d/(k_d \cdot \tau) = (1 - \mathrm{R})/\tau \tag{19.35}$$

and Eq. (19.30) is reavered.

19.2.2.3 Design Criterion for EMRs

Under certain conditions, scale-up of membrane reactors is straightforward. Provided that (i) the reactor contents are well mixed so that the reactor is operated as a CSTR, and that (ii) the membrane is configured for filtration in the tangential mode, the pertinent design criterion, besides constant residence time τ in the reactor, is constant fluidity F of the substrate/product solution through the membrane at all reactor scales. Fluidity is defined by Eq. (19.36) (V = ultrafiltered volume, ΔP = transmembrane pressure, t = filtration time, and A = membrane area).

$$F = V/(\Delta P \cdot t \cdot A) \quad [\mathrm{L\ (bar\ h\ m^2)^{-1}}] \tag{19.36}$$

Fluidity can be assumed to stay constant when scaling up an enzyme–substrate system, provided that solutions of identical composition are used for the laboratory-scale model and the full-size plant design. Application of the design criterion in Eq. (19.36) assumes operation in the linear regime of transmembrane pressure ΔP up to about 1–2 bar, as described in Chapter, Section 8.5.1, Eq. (8.78), so that the volume-to-surface ratio *V*/*A* should be kept constant. The validity of fluidity *F* as the appropriate design criterion has been verified during operation over six orders of magnitude, for reactor volumes ranging from 10 mL on laboratory scale via pilot-scale models between 300 cm^2 and 0.5 m^2 up to production-scale reactors with a volume of several cubic meters and more than 100 m^2 membrane area (Bommarius, 1996).

19.2.3
Large-Scale Applications of Membrane Reactors

19.2.3.1 Enantiomerically Pure L-Amino Acids for Infusion Solutions and as Building Blocks for New Drugs

The acylase-catalyzed resolution of *N*-acetyl-D,L-amino acids to obtain enantiomerically pure L-amino acids (see Chapter 7, Section 7.2.1) has been scaled up to the multi-hundred ton level. For the immobilized-enzyme reactor (Takeda, 1969) as well as the enzyme membrane reactor technology (Degussa, 1980) the acylase process was the first to be scaled up to industrial levels. Commercially acylase has broad substrate specificity and sufficient stability during both storage and operation. The process is fully developed and allowed major market penetration for its products, mainly pharmaceutical-grade L-methionine and L-valine.

Acylase (acylase I; aminoacylase; *N*-acetyl amino acid amidohydrolase; E.C. 3.5.1.14), is one of the best-known enzymes as far as substrate specificity (Chenault, 1989) or use in immobilized (Takahashi, 1989) or membrane reactors (Wandrey, 1977, 1979; Leuchtenberger, 1984; Bommarius, 1992a) is concerned; however, its exact mechanism or 3D structure is still not known (Gentzen, 1979; 1980). Acylase is available in large, process-scale quantities from two sources, porcine kidney and the mold *Aspergillus oryzae*.

As in all catalytic processes, catalyst stability is an essential feature. We have investigated the stability of acylase in conditions that are pertinent for large-scale processes. Instead of just determining thermal stability, which can be done by measuring storage stability of the enzyme in particular conditions of temperature and pH, we have also determined operational stability. The relevant parameter for operational stability studies of enzymes is the product of active enzyme concentration [E] and residence time *t*, $[E] \cdot t$. For a CSTR, the quantities [E] and *t* are linked by Eq. (19.37), where $[S_0]$ = initial substrate concentration, x = degree of conversion, and $r(x)$ = reaction rate (Wandrey, 1977; Bommarius, 1992b).

$$([E] \cdot t)/[S_0] = x/r(x) \tag{19.37}$$

On the pilot and full plant scale, the reactor is usually operated properly because it has been studied and designed well in the laboratory. However, the greatest amount of capital is invested not in the upstream reactor part but in the downstream part, in chromatography columns, crystallizers, and dryers. Even though the study of these unit operations should be commensurate with their importance, this is often not taken sufficiently into account.

19.2.3.2 Aqueous–Organic Membrane Reactors

In a biphasic hollow-fiber membrane reactor, the membrane can be used as a phase contactor, phase separator, and interfacial catalyst (McConville, 1990). Such a reactor was used by Sepracor (Marlborough/MA, USA) to contact sparingly soluble substrate, such as esters of racemic propanoic acids, with an immobilized enzyme, such as lipase from *Candida rugosa*, to produce 2-chloropropanoic acid, naproxen, or ibuprofen (Matson, 1989). Significant process improvements were realized by taking advantage of the compartmentalization properties of the membrane to create two aqueous environments in the reactor system. One environment was maintained in optimum conditions for the enzyme (low pH to minimize poisoning) and the other for the substrate (high pH to strip away the product). Increasing the aqueous solubility of the substrate ester has made the substrate more available to the immobilized enzyme, thus reducing diffusional limitations. This system has been scaled up to the ton level. As an example, ibuprofen sulfomethyl ester Na salt was reacted at 20–22°C in 0.2 M phosphate buffer, pH 7.0, and gave a conversion of 39% and an enantiomeric excess (e.e.) of > 99%.

19.2.3.3 Other Processes in Enzyme Membrane Reactors

A variety of processes have been developed in an enzyme membrane reactor system. Table 19.2 lists a selection of such developments:

- An enzyme membrane reactor allows continuous transketolase-catalyzed production of L-erythrulose from hydroxypyruvate and glycolaldehyde with high conversion, stable operational points, and good productivity (space–time yield) of 45 g $(L\ d)^{-1}$, thus best overcoming transketolase deactivation by substrates (Bongs, 1997).
- (*R*)-2-Hydroxy-4-phenylbutyric acid was produced continuously in an enzyme membrane reactor by enzymatic reductive amination of the α-keto acid with D-lactate dehydrogenase coupled with formate dehydrogenase (FDH) for regeneration of NADH. Reactor performance data matched a kinetic reactor model (Schmidt, 1992).
- *N*-Acetylneuraminic acid (NANA) is produced on a kilogram scale in an enzyme membrane reactor by using *N*-acetylglucosamine 2-epimerase to convert *N*-acetylglucosamine to *N*-acetylmannosamine, which is reacted with pyruvic acid in presence of the *N*-acetylneuraminic acid aldolase to give NANA. Feeding of *N*-acetylmannosamine (300 nM) with a twofold excess of pyruvate was found to be optimum with 86% conversion (Kragl, 1995).
- The highest space–time yield (120 g $L^{-1}d^{-1}$) was achieved in a continuously operated enzyme membrane reactor for the chloroperoxidase-catalyzed oxidation of indole to oxindole with H_2O_2 in aqueous *t*-BuOH, whereas a fed-batch reactor obtained the highest total turnover number (TTN: 860 000) (Seelbach, 1997).
- Optimization of a continuous enzymic reaction yielding (*R*)-phenylacetylcarbinol (PAC), an L-ephedrine intermediate (Chapter 7, Section 7.5.2), from acetaldehyde and benzaldehyde with PDC from *Zymomonas mobilis* demonstrated that

conducting the reaction with the mutant enzyme PDCW392M in a continuous reaction system with an equimolar substrate concentration was most efficient (Goetz, 2001).

- The enzymatic synthesis of 2-keto-3-deoxy-D-glycero-D-galacto-nonopyranulosonic acid (KDN) starting from D-mannose and pyruvic acid using Neu5Ac-aldolase has been scaled up to 100 g in repetitive batch ultrafiltration mode and, alternatively, to a 440 mL pilot-scale enzyme membrane reactor (EMR) for continuous production of KDN (Salagnad, 1997).

Table 19.2 Selection of processes developed in an enzyme membrane reactor system.

Product	***Enzyme***	***Reaction***	***Volume, yield***	***S.t.y. [g (L d)$^{-1}$]***	***Reference***
L-Erythrulose	transketolase	C–C bond formation	10 mL, [S] = 50 mM	45	Bongs, 1997
(*R*)-2-OH-4-phenylbutyric acid	D-LDH/FDH	reductive amination	220 mL, 3.1 kg	165	Schmidt, 1992
NANA	NANA aldolase/ Nac-epimerase	aldol reaction		45	Kragl, 1995
Oxindole	chloroperoxidase	oxidation	TTN ≤ 860 000	120	Seelbach, 1997
(*R*)-PAC	PDC, mutant W392M	carboligation	[S] = 50 mM	81	Goetz, 2001
KDN	Neu5Ac-aldolase	aldol reaction	440 mL, x = 0.75, o.y. = 75%	375	Salagnad, 1997
Naproxen ester	lipase	hydrolysis			Sakaki, 2001
Polyfructan (inulin)	fructosyl-transferase (FTF)	nucl. substn.: fructose transfer			Hicke, 1999
(*R*)-1-Aminotetralin	(*S*)-ω-transaminases (ω-TA)	transamination	39 mL, c.y. = 97%, [S] = 0.2 M	2.8	Shin, 2001
UDPGA	β-glucuronidase	glucuronidation	100–200 mg	0.07	Pfaar, 1999
dTDP-glucose	sucrose synthase	nucl. substn.: glucose transfer	10 mL	98	Elling, 1995

PDC: pyruvate decarboxylase; (*R*)-PAC: (*R*)-phenylacetylcarbinol;
KDN: 2-keto-3-deoxy-D-glycero-D-galacto-nonopyranulosonic acid;
UDPGA: β-D-uridine diphosphoglucuronic acid;
dTDP: deoxythymine diphosphate;
NANA aldolase: *N*-acetylneuraminic acid aldolase;
Nac-epimerase: *N*-acetylglucosamine 2-epimerase;
D-LDH: D-lactate dehydrogenase;
FDH: formate dehydrogenase.

- A lipase-immobilized membrane reactor was applied for the optical resolution of racemic naproxen. Lipase stability was enhanced by the EMR set-up to > 200 h in comparison with a half-life of 2 h in a stirred tank. Only pure lipase gave the best enantioselectivity (Sakaki, 2001).
- Polyfructan (inulin) of MW 30–50 million was synthesized from sucrose with fructosyltransferase (FTF, inulin sucrase from *Streptococcus mutans*) immobilized by photoinitiated graft copolymerization with poly(2-aminoethyl methacrylate) in the pores of a 3.0 μm microfiltration membrane (Hicke, 1999).
- Chiral amines, here (*R*)-1-aminotetralin, were obtained from racemic amine and pyruvate in a 39 mL hollow-fiber membrane reactor with (*S*)-ω-transaminases (ω-TA) (Shin, 2001). The substrates were recirculated until the e.e. value exceeded 95%. Simulations suggested residence times should be short to minimize product inhibition.
- 2*O*-Glucuronides were produced continuously in an enzyme membrane reactor from aglycones and β-D-uridine diphosphoglucuronic acid (UDPGA) in the presence of a guinea-pig liver preparation of β-glucuronidase for between 24 and 118 h (Pfaar, 1999).
- A higher space–time yield (98 versus 59 g L^{-1} d^{-1}) was reached in an EMR than in a batch reactor for the synthesis of dTDP-glucose and by-product D-fructose from sucrose and dTDP (Elling, 1995).

19.3
Production of Enantiomerically Pure Hydrophobic Alcohols: Comparison of Different Process Routes and Reactor Configurations

Enantiomerically pure alcohols belong to the most sought-after building blocks for pharmaceuticals, crop-protection agents, and food and nutraceuticals. There are a variety of process configurations available for their manufacture:

- whole-cell, enzymatically catalyzed,or homogeneously catalyzed process routes;
- reduction by several possible reductants, such as borane, formate, or molecular hydrogen;
- different reactor configurations, such as membrane reactors, two-phase systems, organic solvents, or combinations thereof.

19.3.1
Isolated Enzyme Approach

Isolated enzymes belong to the conceptually easiest catalyst configurations. Their principal disadvantages are the requirement to devise a purification protocol and the costs of running the purification scheme before use, including the yield loss incurred. Furthermore, regeneration enzymes have to be employed when there is a requirement for cofactor regeneration. The advantages are absence of the side activities plaguing whole-cell approaches and the much greater ease in repetitive-batch or continuous use with the corresponding advantages for total turnover number (TTN). Numerous reductases have been developed for the synthesis of enantiomerically pure alcohols; Table 19.3 provides an overview.

One of the most important goals is the enhancement of volumetric productivities. As high volumetric productivity correlates with high solubility of substrates (Chapter 2; Bommarius, 2001), enhancement of substrate solubility is an excellent measure for improving volumetric productivity.

One option for achieving this goal in the case of less-soluble substrates is the utilization of an aqueous–organic two-phase medium. The organic phase acts as a reservoir for substrate. Aromatic ketones could be reduced with moderate to good conversions and very high e.e., in most cases > 99%, to the (*S*)-alcohols (Groger, 2003). Both enzymes, the (*S*)-ADH from *Rhodococcus erythropolis* and the FDH from *Candida boidinii*, were found to be stable over many days in a 4 : 1 aqueous buffer/hexane (or heptane) system. Additions of even 10% to the buffer of several other organic solvents ranging from isopropanol and acetone to ethyl acetate, toluene, or chloroform proved to be very detrimental to the activity of FDH, the more susceptible enzyme, after one day or less. The overall concentration of aromatic ketone substrates can be strongly enhanced over purely aqueous systems in such systems: acetophenone can be reacted in overall concentrations up to at least 200 mM without a significant drop in conversion, in comparison to aqueous systems with a solubility limit of about 10 mM. Thus, volumetric productivities of 22 mol $(L\ d)^{-1}$ were found, a 250-fold enhancement over the purely aqueous case (aqueous case not optimized).

Another concept to overcome low substrate solubility is the use of an emulsion enzyme membrane reactor consisting of an organic phase acting as the substrate reservoir and an aqueous phase containing the enzyme; the two phases are separated by a hydrophilic ultrafiltration membrane (Liese, 1998). A gentle transmembrane pressure causes the phase boundary to be located within the membrane pores. As product is transported away by the organic phase, the aqueous phase is recharged by substrate partitioning to the aqueous phase. The concept was demonstrated by reducing 2-octanone with the help of carbonyl reductase from *Candida parapsilosis* to (*S*)-2-octanol with > 99.5% e.e. and NAD^+/NADH cofactor regeneration with FDH from *Candida boidinii*. Whereas the space–time yield in the monophasic aqueous reactor over four months was a modest 21.2 g $(L\ d)^{-1}$, the total turnover number could be enhanced by a factor of nine in the emulsion membrane reactor.

In previous versions of this concept, continuous extraction was employed to alleviate product inhibition during the oxidation of 1-phenyl-1,2-ethanediol with glycerol dehydrogenase (GDH); the cofactor NAD^+ required was regenerated by coupling with a lactate dehydrogenase (LDH)-catalyzed reduction of pyruvate to lactate (Liese, 1996). The product, hydroxyacetophenone, causes strong product inhibition. In a differential recycle reactor, the small amount of product formed per cycle is extracted with the help of a hydrophobic hollow-fiber membrane module. At a given degree of conversion, the dehydrogenase is considerably more active. In addition, the reaction could be run to completion (50% conversion), and the remaining (*S*)-1-phenyl-1,2-ethanediol was > 99% e.e.

Table 19.3 Processes with isolated enzyme for synthesis of enantiomerically pure alcohols.

Catalyst	*Substrate specificity*	*Solvents*	*Selectivity* [% e.e.]	*[S]/[C]* [–]	*[P],[S]** [mol L^{-1}]	*Vol. productivity* [g $(L\ d)^{-1}$]	*Process stability* TTN [–]	*References*
ADH/FDH *R. erythropolis*	acetophenone, PhO-acetone	biphasic H_2O/org.	(*S*), > 99 NAD(H)	20 U $mmol^{-1}$	0.2	22.2	–	Groger, 2003
ADH/FDH *R. erythropolis*	broad	H_2O, pH 7	(*S*), > 99 NAD(H)	20 U $mmol^{-1}$	0.005	0.09	–	Hummel, 2003
CPCR/FDH	2-octanone, sulcatone	2-octanone/H_2O, mono/biphasic	(*S*), > 99.5 NAD(H)	7 U $mmol^{-1}$	0.007/neat[b]	21.2/11[b]	13.6/124[b]	Liese, 1998
CPCR/FDH	acetophenone	biphasic isooctane/H_2O	(*S*), > 99.5 NAD(H)	6×10^{6}[b]	0.040[b]	88[b]	2.4×10^{8}[b]	Rissom, 1999
ADH/FDH *L. brevis* wild type G37D	broad	H_2O, pH 7	(*R*), > 99 NADP(H) NAD(H)					Riebel, 2002; Hummel, 1999
ADH/FDH *L. lactis*	broad	H_2O, pH 7	(*R*), > 99 NADP(H)					Riebel, unpublished
CR/GDH *E. coli*	COBE	biphasic H_2O/BA	(*S*), > 99 NADP(H)		0.2[b]	37.5[b]	650[a]	Yasohara, 2001

R. erythropolis: Rhodococcus erythropolis; CPCR: *Candida parapsilosis* carbonyl reductase;
L. brevis: Lactobacillus brevis; L. lactis: Lactococcus lactis; Candida magnoliae;
CR: carbonyl reductase; GDH: glucose dehydrogenase; COBE: ethyl 4-chloro-3-oxobutanoate;
BA: butyl acetate.

[a] mol product (mol cofactor)$^{-1}$; [b] two-phase system (total volume).

Similarly, enantiomerically pure hydrophobic (*S*)-1-phenyl-2-propanol, (*S*)-4-phenyl-2-butanol, and (*S*)-6-methylhept-5-en-2-ol (sulcatol) were obtained with high purities in a dual-loop enzyme membrane reactor unit with separate membrane extraction unit (Kruse, 1996). Whereas substrate concentrations were low at 9–12 mM, space–time yields higher than 100 g $(L\ d)^{-1}$ as well as concentrated product solutions were obtained.

19.3.2
Whole-Cell Approach

As the use of dehydrogenases requires regeneration of the cofactor, employment of whole cells instead of isolated enzymes appears attractive: growing or resting cells can recycle the cofactor internally so that no external addition is necessary. Table 19.4 lists some results of the use of whole cells in the reduction of keto compounds.

The different whole-cell approaches are at very different stages of development. The simplest and most direct is the use of reductase in the tried environment of the native organism; the work with *Rhodococcus ruber* (Stampfer, 2002; 2003a,b) and with *Candida magnoliae* (Yasohara, 1999) fall into that category. The reductase gene of *Saccharomyces cerevisiae* (Stewart, 2001) was cloned and subsequently overexpressed back in its own host. These recombinant yeasts, termed "designer yeasts", are capable of reducing a wide range of keto functions without external addition of cofactor. Whereas some reductases have been found and preliminarily optimized in the native whole-cell environment, ultimately they have been cloned and overexpressed in recombinant systems; the work on the reduction of COBE to (*S*)-CHBE with *Candida magnoliae* serves as an example.

In a whole-cell approach using both native activity in a cell (Yasohara, 1999) and cloned genes expressed in recombinant *E. coli* cells (Kataoka, 1999), completely enantiomerically pure ethyl (*S*)-4-chloro-3-hydroxybutanoate [(*S*)-CHBE], a potential precursor of (*S*)-carnitine, was obtained from its *ß*-keto ester precursor ethyl 4-chloro-3-oxobutanoate (COBE). Cloning the carbonyl reductase gene from *Candida magnoliae* and the glucose dehydrogenase gene of *Bacillus megaterium* on a common plasmid into *E. coli* HB 101 cells, 1.25 M (*S*)-CHBE product could be accumulated in the totally aqueous system by continuous feeding of COBE substrate, which is unstable in aqueous medium (Kizaki, 2001). In contrast, in the two-phase medium butyl acetate : water (1 : 1), a concentration of 2.58 M could be achieved with a molar yield on substrate of 85%. No external GDH or cofactor needs to be added to this system.

A further simplification beyond co-cloning of two genes is manifested by conducting either an asymmetric reduction of a ketone in the presence of up to 20% isopropanol as hydrogen donor or the complementary enantioselective oxidation of a racemic secondary alcohol in the presence of a similar amount of acetone as hydrogen acceptor, catalyzed by a secondary alcohol dehydrogenase in whole lyophilized cells of *Rhodococcus ruber* DSM 44541 (Stampfer, 2002). External addition of $NADH/NAD^{+}$ can be omitted. Owing to tolerance of high levels of substrates,

Table 19.4 Whole-cell approaches for the reduction of keto compounds.

Catalyst	*Substrate specificity*	*Solvents*	*Selectivity*	*[S]/[C]*	*[P],[S]**	*Vol. productivity*	*Process stability*	*References*
			[% e.e.]	*[–]*	*[mol L^{-1}]*	*[g $(L\ d)^{-1}$]*	*TTN [–]*	
L. kefir ADH	(2*R*,5*R*) hexanediol	water	(*S*), > 99 no extl cof	30 g/g DCW	fed-batch	0.54	–	Haberland, 2002
CR *S. cerevisiae*	WM and 4-Me-HP	water	(*S*), no ext. cofactor				–	Stewart, 2001
CR *C. magnoliae*	COBE	water, 2-phase	100% (*S*), NADP(H)		2.58*, 1.25		21,600[&]	Kataoka, 1999
ADH *T. brockii*	sec. alcs to ketones	water	(*S*), NADP(H)					
ADH *C. beijerinckii*	sec. alcs to ketones	water	(*S*), NADP(H)					
DH *T. et-hanolicus*		water						Holt, 2000
ADH *R. ruber*	broad	H_2O/ace H_2O/iPr	(*S*), > 99, no extl cof.	2.5 mmol g^{-1} dcw	0.15	0.17	–	Stampfer, 2003a

Organisms: *Lactobacillus kefir* DSM 20587, *Saccharomyces cerevisiae, Candida magnoliae, Bacillus megaterium, Thermoanaerobium brockii, Clostridium beijerinckii, Thermoanaerobacter ethanolicus, Rhodococcus ruber* DSM 44541. Solvents: ace = acetone; iPr = i-PrOH.
Substrates: WM: Wieland–Miescher ketone; 4-Me-HP: 4-methyl Hajos–Parrish ketone; COBE: ethyl 4-chloro-3-oxobutanoate.

co-substrates, products, and solvents, the process features high productivities. The system displays thermostability of up to 60 °C and pH stability of up to pH 11 (Stampfer, 2003). Regarding stability, the operational half-life of the redox system is 29 h in 20% v/v acetone and 37 h in 30% v/v 2-propanol (Stampfer, 2003b).

While enantioselectivity during reduction of ethyl 3-oxobutanoate by baker's yeast (*Saccharomyces cerevisiae*) to ethyl (*S*)-3-hydroxybutanoate was found to exceed 99%, yields did not exceed 50–70% (Chin-Joe, 2000). Elimination of two of three causes, evaporation of substrate and product esters and absorption or adsorption of the two esters by the yeast cells, increased the yield to 85%. Alleviation of hydrolysis of the two esters by yeast enzymes could increase the yield even more. Low supply rates of glucose as an electron donor provided the most efficient strategy for electron donor provision and yielded a high enantiomeric excess of ethyl (*S*)-3-hydroxybutanoate, low by-product formation and biomass increase, with a low oxygen requirement(Chin-Joe, 2001).

19.3.3
Organometallic Catalyst Approach

There are probably few areas of catalysis where there is a more intense and interesting competition between biocatalytic and homogeneously catalyzed approaches than in the topic of this section, the reduction of keto functionalities. Hydrogenation is a core technology in chemical synthesis so it is not surprising that corresponding catalytic technologies have been developed. It is therefore very appropriate, even in a treatise on biocatalysis, to appreciate the abilities of organometallic catalysts in this field and to compare the performance of organometallic and biological catalysts. The work to mention first and foremost is that of Ryoji Noyori (Chemistry Nobel prize 2000), who developed a range of options for the reduction of keto functions (Noyori, 2001).

Table 19.5 illustrates pertinent organometallic catalysts and their performance. When scanning the literature on organometallic catalysts, one searches in vain for numbers regarding catalyst process stability, as embodied by TTNs. Upon closer inspection, none of the catalysts is recovered and re-used, so the TTN was never determined. Instead, catalyst productivity is a key dimension of merit in homogeneous catalysis, so often the turnover number, TON [mol product $(\text{mol catalyst})^{-1}$], or its equivalent, the frequently used [S]/[C] ratio, is referenced in the literature and thus also in Table 19.5.

The vast majority of catalyst systems in Table 19.5 utilize ruthenium as a metal; examples include *trans*-RuH(η^1-BH_4)(binap)(1,2-diamine) (Ohkuma, 2002), *trans*-$RuCl_2$[(*R*)-xylbinap][(*R*)-daipen] (Ohkuma, 2000b), and *trans*-$RuCl_2[P(C_6H_4\text{-}4\text{-}CH_3)_3]_2$(ethylenediamine) (Ohkuma, 2000a).

High rates and selectivities are attainable only by the coordination of structurally well-designed catalysts and suitable reaction conditions. The base system employs a [$RuCl_2(\text{phosphane})_2$(1,2-diamine)] complex as precatalyst, isopropanol serving both as solvent and hydrogen donor and in the presence of base, typically *t*-BuOK. Use of chiral diphosphanes, particularly BINAP compounds, and/or chiral diamines

Table 19.5 Organometallic catalyst approaches to the reduction of keto compounds.

Catalyst	***Substrate specificity***	***Solvents***	***Selectivity*** [% e.e.]	***[S]/[C], TON*** [–]	***[P],[S]**** [mol L^{-1}]	***Vol. productivity*** [mol $(L\ d)^{-1}$]	***References***
$[RuCl_2(p\text{-cymene})]_2$,	EAA, 2-Ac-pyridine, MeO-acetone	i-PrOH	56–89 (*X*), 100% c.y.	100	0.1*	0.0625–0.2	Everaere, 1999
(*S,S*)-DPBPD	acetophenone	i-PrOH	91 (*S*)	100	0.01*	0.005	Gao, 1999
(1*S*,2*R*)-ADPE, $[RuCl_2(p\text{-cymene})]_2$	ketoisophorone	i-PrOH	97 (*X*), 92% c.y.	192	0.096*	0.002	Hennig, 2000
(*S,S*)-DPBPD	acetophenone	i-PrOH	97 (*X*), 95% c.y.	200	0.1	0.014	Gao, 1996
Oxazaborolidine	acetophenone	THF	94 (*R*), 93% c.y.	560	0.25	2.69	Rissom, 1999
[(*R*)-Xylbinap], [(*R*)-daipen]	γ-amino ketones	i-PrOH	92 (*S*), 99% c.y.	2000	1.0	0.167	Ohkuma, 2000c
t-$RuCl_2$ $[P(C_6H_4\text{-}4\text{-}Me)_3]_2$ (ethylenediamine)	*o*-benzophenones, *o*-benzoylferrocene	i-PrOH/*t*-BuOK	*x*-*y*%e.e., 8 atm, 23–35 °C	2000–20 000			Ohkuma, 2000a
(*R*)-Phanephos, (*S,S*)-DPEN	substituted acetophenones	i-PrOH	98–99 (*R*), 100% c.y.	≤ 20 000	2–2.5	48–60	Burk, 2000
[(*R*)-Xylbinap], [(*R*)-daipen]	heteroaromatic ketones	i-PrOH/*t*-BuOH	92–99 (*S*)	1000–40 000	4.5	4.5	Ohkuma, 2000b
(*S,S*)-DPEN, [(*R*)-tolbinap] [(*S*)-daipen], [(*S*)-xylbinap]	acetophenone substituted acetophenone	i-PrOH	–80 (*R*), 97% c.y.; –99 (*R*), 97–100 c.y.	100 000–2.4×10^6	3.33	0.07	Ohkuma, 1998

(*S,S*)-DPBPD: (*S,S*)-*N,N'*-bis[*o*-(diphenylphosphino)benzyl]propane-1,2-diamine-$RuCl_2$; (*S,S*)-DPEN: 1,2-diphenylethylenediamine; daipen: 1,1-di(4-anisyl)-2-isopropyl-1,2-ethylenediamine; (1*S*,2*R*)-ADPE: (1*S*,2*R*)-amino-1,2-diphenylethanol-N(*R*)-phanephos: (*R*)-4,12-bis(diphenylphosphino)[2.2]paracyclophane; (*R*)-xylbinap: (*R*)-2,2'-bis(di-3,5-xylylphosphino)-1,1'-binaphthyl; (*R*)-tolbinap: (*R*)-2,2'-bis(di-4-tolylphosphino)-1,1'-binaphthyl.

results in rapid and productive asymmetric hydrogenation of a range of aromatic and heteroaromatic ketones and gives high enantioselectivity. Substrate specificity currently encompasses cyclic and acyclic ketones as well as certain amino and alkoxy ketones. The hydrogenation tolerates many substituents, including F, Cl, Br, I, CF_3, OCH_3, $OCH_2C_6H_5$, $COOCH(CH_3)_2$, NO_2, NH_2, and NRCOR as well as various electron-rich and -deficient heterocycles. As this catalyst system preferentially reduces C=O functions over coexisting C=C linkages, cyclic and acyclic α,β-unsaturated ketones can be converted into chiral allyl alcohols of high enantiomeric purity.

Reviewing the status of this technology (Noyori, 2001), the high rate and carbonyl selectivity seem to be based on non-classical metal–ligand bifunctional catalysis involving an 18-electron aminorunium hydride complex and a 16-electron amidoruthenium species (Noyori, 2001). The simple ketones substrates lack any functionality capable of interacting with the metal center. Stereoselectivity can be controlled by the electronic and steric properties (bulkiness and chirality) of the ligands as well as the reaction conditions.

trans-$RuCl_2[P(C_6H_4\text{-}4\text{-}CH_3)_3]_2$(ethylenediamine), a chiral BINAP/chiral diamine Ru complex, effects asymmetric hydrogenation of various ortho-substituted benzophenones and benzoylferrocene to chiral diarylmethanols smoothly at 8 atm and 23–35 °C with consistently high e.e. values (Ohkuma, 2000a).

Use of a BINAP/chiral diamine Ru complex *trans*-$RuCl_2$[(*R*)-xylbinap][(*R*)-daipen] or the *S,S* complex efficiently catalyzes asymmetric hydrogenation of heteroaromatic ketones, with little influence of the ring properties on the enantioselectivity. Duloxetine, an inhibitor of serotonin and norepinephrine uptake carriers, has been synthesized in this way (Ohkuma, 2000b).

PhanePhos–ruthenium–diamine complexes catalyze the asymmetric hydrogenation of a wide range of aromatic, heteroaromatic, and α,β-unsaturated ketones with high activity and excellent enantioselectivity (Burk, 2000).

19.3.4
Comparison of Different Catalytic Reduction Strategies

- *Enantioselectivity*: The enantiomeric purity of the product is often higher in the case of enzymatic catalysis than in homogeneous chemical catalysis: whereas enzymes in many reported cases reach in excess of 99% selectivity, chemical catalysts rarely exceed 95%. However, the design principle of homogeneous chemical catalysts in most cases allows for both enantiomers of the chiral moiety to serve as part of the catalyst to generate both enantiomers of the product. This feature is a distinct advantage of organometallic catalysts.

- *Catalyst stability and productivity*: Interestingly, a comparison of catalyst stability and productivity for the different technologies is not straightforward, as practitioners of homogeneous catalysis seem to prefer to reference turnover numbers (TONs) and substrate/catalyst ratios ([S]/[C] ratios) whereas researchers in biocatalysis quote biocatalyst loading (Units L^{-1}) and total turnover numbers

(TTNs). In the case of slow homogeneous chemical catalysts, the [S]/[C] ratio can even be unity (i.e., stoichiometric conditions). The numerical values for the total turnover number and the turnover number are driven apart by the average number of recycles a catalyst experiences. In the limit of no recycle, the values are identical (TTN ≡ TON). Whereas recycling is very important in biocatalysis, it is apparently not high on the agenda in chemical asymmetric catalysis (Blaser, 2001).

- *Solvent systems*: One of the potential drawbacks of biocatalysts is their limitation to water as a solvent. Reductases, dependent on a regeneration enzyme, whether in isolated form or intracellularly, tend not to be active in organic solvents. As many substrates of interest in synthesis do not dissolve well in water, other solvent systems are required. Homogeneous chemical asymmetric catalysts were thought to answer that challenge. As Table 19.5 reveals, however, transfer hydrogenation catalysts tend to function only in isopropanol or similar solvents, so the problem of finding a suitable broad range of solvent systems remains.
  In a direct comparison of the reduction of acetophenone to highly enantio-enriched (*R*)-phenylethanol (94% e.e.) by heterogenized (*S*)-diphenyloxazaborolidine (Corey–Itsuno catalyst) or to enantiomerically pure (*S*)-phenylethanol (> 99% e.e.) by *Candida parapsilosis* carbonyl reductase (CPCR), the superior solubility of acetophenone in THF (0.25 M) versus water (0.04 M) leads to a vastly superior space–time yield of 290 g $(L\ d)^{-1}$ in THF with the Corey–Itsuno catalyst in comparison with 27 g $(L\ d)^{-1}$ in water with CPCR (Rissom, 1999). Conversely, the turnover frequencies (tofs) of 0.3 min^{-1} (Corey–Itsuno catalyst) versus 2.3×10^4 min^{-1} (CPCR) portend the difference in total turnover number (TTNs) of 2.4×10^8 versus 560.

In summary, in the case of well-developed technologies, both homogeneous chemical catalysis and biocatalysis tend to provide solutions to challenges in synthesis. Depending on the requirements for the reaction, either selectivity (in the case of pharma compounds) or both performance and cost (in the case of fine chemical applications) are paramount criteria for decisions about a process route. Biocatalysts often more easily hit the selectivity target and high total turnover numbers. Homogeneous catalysts often provide high space–time yields and still can be designed more rapidly. While the trend has favored biocatalysts over the last decade it is performance that will decide the use of a catalyst in most cases in the future.

Suggested Further Reading

H. U. Blaser, F. Spindler, and M. Studer, Enantioselective catalysis in fine chemicals production, *Appl. Catal. A: General* **2001**, *221*, 119–143.

R. Noyori and T. Ohkuma, Asymmetric catalysis by architectural and functional molecular engineering: practical chemo- and stereoselective hydrogenation of ketones, *Angew. Chem. Int. Ed. Engl.* **2001**, *40*, 40–73.

R. A. Sheldon, *Chirotechnology: Industrial Synthesis of Optically Active Compounds*, Marcel Dekker, Amsterdam, **1993**.

References

H. U. Blaser, F. Spindler, and M. Studer, Enantioselective catalysis in fine chemicals production, *Appl. Catal. A: General* **2001**, *221*, 119–143.

A. S. Bommarius, K. Drauz, U. Groeger, and C. Wandrey, Membrane bioreactors for the production of enantiomerically pure α-amino acids, in: *Chirality in Industry*, A. N. Collins, G. N. Sheldrake and J. Crosby (eds.), Wiley & Sons Ltd., London, **1992a**, Chapter 20, pp. 371–397.

A. S. Bommarius, K. Drauz, H. Klenk and C. Wandrey, Operational stability of enzymes – acylase-catalyzed resolution of *N*-acetyl amino acids to enantiomerically pure L-amino acids, *Ann. N. Y. Acad. Sci. (Enzyme Eng. XI)* **1992b**, *672*, 126–136.

A. S. Bommarius, M. Schwarm, and K. Drauz, **1996**, The membrane reactor for the acylase process – from laboratory results to commercial scale, *Chimica Oggi 14(10)*, 61–64.

A. S. Bommarius, M. Schwarm, and K. Drauz, **2001**, Factors for process design towards enantiomerically pure L- and D-amino acids, *Chimia 55*, 50–59.

J. Bongs, D. Hahn, U. Schoerken, G. A. Sprenger, U. Kragl, and C. Wandrey, Continuous production of erythrulose using transketolase in a membrane reactor, *Biotechnol. Lett.* **1997**, *19*, 213–215.

M. J. Burk, W. Hems, D. Herzberg, C. Malan, and A. Zanotti-Gerosa, **2000**, A catalyst for efficient and highly enantioselective hydrogenation of aromatic, heteroaromatic, and α,β-unsaturated ketones, *Org. Lett. 2*, 4173–4176.

K. Chenault, J. Dahmer, and G. M. Whitesides, Kinetic resolution of unnatural and rarely occurring amino acids: enantioselective hydrolysis of *N*-acyl amino acids catalyzed by acylase I, *J. Am. Chem. Soc.* **1989**, *111*, 6354–6364.

I. Chin-Joe, P. M. Nelisse, A. J. Straathof, J. A. Jongejan, J. T. Pronk, and J. J. Heijnen, Hydrolytic activity in baker's yeast limits the yield of asymmetric 3-oxo ester reduction, *Biotechnol. Bioeng.* **2000**, *69*, 370–3706.

I. Chin-Joe, A. J. Straathof, J. T. Pronk, J. A. Jongejan, and J. J. Heijnen, Influence of the ethanol and glucose supply rate on the rate and enantioselectivity of 3-oxo ester reduction by baker's yeast, *Biotechnol. Bioeng.* **2001**, *75*, 29–38.

W. Crueger and A. Crueger, *Biotechnology: A Textbook of Industrial Microbiology*, Science Tech Publishers, Madison/WI, **1989**.

L. Elling, M. Grothus, A. Zervosen, and M.-R. Kula, Application of sucrose synthase from rice grains for the synthesis of carbohydrates, *Ann. N. Y. Acad. Sci.* USA **1995**, *750*, 329–331.

K. Evetraere, J.-F. Carpentier, A. Morteux, and M. Bulliard, *N*-Benzoyl-norephedrine derivatives as new, efficient ligands for ruthenium-catalyzed asymmetric transfer hydrogenation of functionalized ketones, *Tetrahedron: Asymm.* **1999**, *10*, 4083–4086.

J. Gao, T. Ikariya, and R. Noyori, A ruthenium(ii) complex with a C_2-symmetric diphosphine/diamine tetradentate ligand for asymmetric transfer hydrogenation of aromatic ketones, *Organometallics* **1996**, *15*, 1087–1089.

J. Gao, P. Xu, X. Yi, C. Yang, H. Zhang, S. Cheng, H. Wan, K. Tsai, and T. Ikariya, Asymmetric transfer hydrogenation of prochiral ketones catalyzed by chiral ruthenium complexes with aminophosphine ligands, *J. Mol. Cat. A: Chemical* **1999**, *147*, 105–111.

I. Gentzen, H.-G. Löffler, and F. Schneider, Purification and partial characterization of amino-acylase from *Aspergillus oryzae*, in: *Metalloproteins*, U. Eser (ed.), Thieme, Stuttgart, **1979**, pp. 270–274.

I. Gentzen, H.-G. Löffler, and F. Schneider, Aminoacylase from *Aspergillus oryzae*. Comparison with the pig kidney enzyme, *Z. Naturforsch.* **1980**, *35c*, 544–550.

G. Goetz, P. Iwan, B. Hauer, M. Breuer, and M. Pohl, Continuous production of (*R*)-phenylacetylcarbinol in an enzyme membrane reactor using a potent mutant of pyruvate decarboxylase from *Zymomonas mobilis*, *Biotechnol. Bioeng.* **2001**, *74*, 317–325.

H. Groger, W. Hummel, S. Buchholz, K. Drauz, T. V. Nguyen, C. Rollmann, H. Husken, and K. Abokitse, Practical asymmetric enzymatic reduction through discovery of a dehydrogenase-compatible biphasic reaction media, *Org. Lett.* **2003**, *5*, 173–176.

J. Haberland, A. Kriegesmann, E. Wolfram, W. Hummel, and A. Liese, **2002**, Diastereoselective synthesis of optically active (2*R*,5*R*)-hexanediol, *Appl Microbiol Biotechnol. 58*, 595–599.

M. Hennig, K. Püntener, and M. Scalone, Synthesis of (*R*)- and (*S*)-4-hydroxy-isophorone by ruthenium-catalyzed asymmetric transfer of hydrogenation of ketoisophorone, *Tetrahedron: Asymm.* **2000**, *11*, 1849–1858.

H.-G. Hicke, M. Ulbricht, M. Becker, S. Radosta, and A. G. Heyer, Novel enzyme membrane reactor for poly-saccharide synthesis, *J. Membrane Sci.* **1999**, *161*, 239–245.

P. J. Holt, R. E. Williams, K. N. Jordan, C. R. Lowe, and N. C. Bruce, Cloning, sequencing and expression in *Escherichia coli* of the primary alcohol dehydrogenase gene from *Thermoanaerobacter ethanolicus* JW200, *FEMS Microbiol. Lett.* **2000**, *190*, 57–62.

W. Hummel and B. Riebel, Dehydrogenase mutants with improved NAD-dependence, their manufacture with recombinant microorganisms, and their use for chiral hydroxy compound preparation, PCT Int. Appl. WO 9947684, **1999.**

W. Hummel, K. Abokitse, K. Drauz, C. Rollmann, and H. Groger, Towards a large-scale asymmetric reduction process with isolated enzymes: Expression of an (*S*)-alcohol dehydrogenase in *E. coli* and studies on the synthetic potential of this biocatalyst, *Adv. Synth. Catal.* **2003**, *345*, 153–159.

M Kataoka, K. Yamamoto, H. Kawabata, M. Wada, K. Kita, H. Yanase, and S. Shimizu, Stereoselective reduction of ethyl 4-chloro-3-oxobutanoate by *Escherichia coli* transformant cells coexpressing the aldehyde reductase and glucose dehydrogenase genes, *Appl. Microbiol Biotechnol.* **1999**, *51*, 486–490.

N. Kizaki, Y. Yasohara, J. Hasegawa, M. Wada, M. Kataoka, and S. Shimizu, Synthesis of optically pure ethyl (S)-4-chloro-3-hydroxybutanoate by *Escherichia coli* transformant cells coexpressing the carbonyl reductase and glucose dehydrogenase genes, *Appl. Microbiol. Biotechnol.* **2001**, *55*, 590–595.

U. Kragl, M. Kittelmann, O. Ghisalba, and C. Wandrey, *N*-Acetylneuraminic acid: from a rare chemical from natural sources to a multikilogram enzymic synthesis for industrial application, *Ann. N. Y. Acad. Sci. USA* **1995**, *750*, 300–305.

U. Kragl and T. Dwars, The development of new methods for the recycling of chiral catalysts, *Trends Biotechnol.* **2002**, *19*, 442–449 (erratum in: *Trends Biotechnol.* **2002**, *20*, 45).

W. Kruse, W. Hummel, and U. Kragl, Alcohol-dehydrogenase-catalyzed production of chiral hydrophobic alcohols. A new approach leading to a nearly waste-free process, *Rec. Trav. Chim. Pays-Bas* **1996**, *115*, 239–243.

W. Leuchtenberger, M. Karrenbauer, and U. Ploecker, Scale-up of an enzyme membrane reactor process for the manufacture of L-enantiomeric compounds, *Ann. N. Y. Acad. Sci. USA* **1984**, 434, 78–86.

A. Liese, M. Karutz, J. Kamphuis, C. Wandrey, and U. Kragl, Enzymic resolution of 1-phenyl-1,2-ethanediol by enantioselective oxidation: overcoming product inhibition by continuous extraction, *Biotechnol. Bioeng.* **1996**, *51*, 544–550.

A Liese, T. Zelinski, M.-R. Kula, H. Kierkels, M. Karutz, U. Kragl, and C. Wandrey, **1998**, A novel reactor concept for the enzymatic reduction of poorly soluble ketones, *J. Mol. Catal. B: Enzymatic, 4*, 91–99.

S. L. Matson, S. A. Wald, C. M. Zepp, and D. R. Dodds, Process for the preparation of chiral propanoic acids via enantioselective enzymic hydrolysis of their water-soluble esters, PCT Int. Appl. WO 8909765, **1989.**

F. X. McConville, J. L. Lopez, and S. A. Wald, Enzymic resolution of ibuprofen in a multiphase membrane reactor, in: *Biocatalysis*, D. A. Abramowicz (ed.), Van Nostrand Reinhold, New York, **1990**, pp. 167–177.

R. Noyori and T. Ohkuma, Asymmetric catalysis by architectural and functional molecular engineering: practical chemo- and stereoselective hydrogenation of ketones, *Angew. Chem. Int. Ed. Engl.* **2001**, *40*, 40–73.

T. Ohkuma, M. Koizumi, H. Doucet, T. Pham, M. Kozawa, K. Murata, E. Katayama, T. Yokozawa, T. Ikariya, and R. Noyori, Asymmetric hydrogenation of alkenyl, cyclopropyl, and aryl ketones. $RuCl_2$(xylbinap)(1,2-diamine) as a precatalyst exhibiting a wide scope, *J. Am. Chem. Soc.* **1998**, *120*, 13529–13530.

T. Ohkuma, M. Koizumi, H. Ikehira, T. Yokozawa, and R. Noyori, Selective hydrogenation of benzophenones to benzhydrols. Asymmetric synthesis of unsymmetrical diarylmethanols, *Org. Lett.* **2000a**, *2*, 659–662.

T. Ohkuma, M. Koizumi, M. Yoshida, and R. Noyori, General asymmetric hydrogenation of hetero-aromatic ketones, *Org. Lett.* **2000b**, *2*, 1749–1751.

T. Ohkuma, D. Ishii, H. Takeno, and R. Noyori, Asymmetric hydrogenation of amino ketones using chiral $RuCl_2$(diphosphine)(1,2-diamine) complexes, *J. Am. Chem. Soc.* **2000c**, *122*, 6510–6511.

T. Ohkuma, M. Koizumi, K. Muniz, G. Hilt, C. Kabuto, and R. Noyori, *trans*-RuH(eta1-BH_4)(binap)(1,2-diamine): a catalyst for asymmetric hydrogenation of simple ketones under base-free conditions, *J. Am. Chem. Soc.* **2002**, *124*, 6508–6509.

U. Pfaar, D. Gygax, W. Gertsch, T. Winkler, and O. Ghisalba, Enzymatic synthesis of β-D-glucuronides in an enzyme membrane reactor, *Chimia* **1999**, *53*, 590–593.

B. Riebel, W. Hummel, and A. Bommarius, Dehydrogenase mutants capable of using NAD as coenzyme and their preparation and use for chiral hydroxy compound preparation, Eur. Pat. Appl. EP 1176203, **2002**.

S. Rissom, J. Beliczey, G. Giffels, U. Kragl, and C. Wandrey, Asymmetric reduction of acetophenone in membrane reactors: comparison of oxazaborolidine and alcohol dehydrogenase, *Tetrahedron Asymm.* **1999**, *10*, 923–928.

J. A. Roels and R. van Tilberg, Temperature dependence of the stability and the activity of immobilized glucose isomerase, *ACS Symp. Series* **1979**, *106*, 147–172.

K. Sakaki, L. Giorno, and E. Drioli, Lipase-catalyzed optical resolution of racemic naproxen in biphasic enzyme membrane reactors, *J. Membrane Sci.* **2001**, *184*, 27–38.

C. Salagnad, A. Godde, B. Ernst, and U. Kragl, Enzymatic large-scale production of 2-keto-3-deoxy-D-glycero-D-galacto-nopyranulosonic acid in enzyme membrane reactors, *Biotechnol. Prog.* **1997**, *13*, 810–813.

E. Schmidt, O. Ghisalba, D. Gygax, and G. Sedelmeier, Optimization of a process for the production of (*R*)-2-hydroxy-4-phenylbutyric acid – an intermediate for inhibitors of angiotensin converting enzyme, *J. Biotechnol.* **1992**, *24*, 315–327.

K. Seelbach, M. P. J. van Deurzen, F. van Rantwijk, R. A. Sheldon, and U. Kragl, Improvement of the total turnover number and space–time yield for chloroperoxidase catalyzed oxidation, *Biotechnol. Bioeng.* **1997**, *55*, 283–288.

J.-S. Shin, B.-G. Kim, A. Liese, and C. Wandrey, Kinetic resolution of chiral amines with ω-transaminase using an enzyme membrane reactor, *Biotechnol. Bioeng.* **2001**, *73*, 179–187.

W. Stampfer, B. Kosjek, C. Moitzi, W. Kroutil, and K. Faber, Biocatalytic asymmetric hydrogen transfer, *Angew. Chem. Int. Ed.* **2002**, *41*, 1014–1017.

W. Stampfer, B. Kosjek, W. Kroutil, and K. Faber, On the organic solvent and thermostability of the biocatalytic redox system of *Rhodococcus ruber* DSM 44541, *Biotechnol. Bioeng.* **2003a**, *81*, 865–869.

W. Stampfer, B. Kosjek, K. Faber, and W. Kroutil, Biocatalytic asymmetric hydrogen transfer employing *Rhodococcus ruber* DSM 44541, *J. Org. Chem.* **2003b**, *68*, 402–406.

J. D. Stewart, Dehydrogenases and transaminases in asymmetric synthesis, *Curr. Opin. Chem. Biol.* **2001**, *5*, 120–129.

T. Takahashi, O. Izumi, and K. Hatano, Takeda Chemical Industries, Ltd., Eur. Patent EP 0 304 021, **1989**

C. Wandrey, Habilitation thesis, TU Hannover, Hannover, Germany, **1977**.

C. Wandrey and E. Flaschel, Process development and economic aspects in enzyme engineering. Acylase L-methionine system, *Adv. Biochem. Eng.* **1979**, *12*, 147–218.

J. Woltinger, K. Drauz, and A. S. Bommarius, The membrane reactor in the fine chemicals industry, *Appl. Catal. A: General* **2001**, *221*, 171–185.

Y. Yasohara, N. Kizaki, J. Hasegawa, S. Takahashi, M. Wada, M. Kataoka, and S. Shimizu, Synthesis of optically active ethyl 4-chloro-3-hydroxybutanoate by microbial reduction, *Appl. Microbiol. Biotechnol.* **1999**, *51*, 847–51.

Y. Yasohara, N. Kizaki, J. Hasegawa, M. Wada, M. Kataoka, and S. Shimizu, Molecular cloning and overexpression of the gene encoding an NADPH-dependent carbonyl reductase from *Candida magnoliae*, involved in stereoselective reduction of ethyl 4-chloro-3-oxobutanoate, *Biosci. Biotechnol. Biochem.* **2000**, *64*, 1430–1436.

Y. Yasohara, N. Kizaki, J. Hasegawa, M. Wada, M. Kataoka, and S. Shimizu, Stereoselective reduction of alkyl 3-oxobutanoate by carbonyl reductase from *Candida magnoliae*, *Tetrahedron Asymm.* **2001**, *12*, 1713–1718.

20
Comparison of Biological and Chemical Catalysts for Novel Processes

Summary

A direct comparison of catalysis of olefin epoxidation with a homogeneous chemical catalyst (Mn salen), an enzyme (CPO), and an antibody resulted in sufficiently high enantioselectivity for all three catalysts, a higher *turnover number* for the enzyme, and a slightly higher *substrate/catalyst ratio* for the homogenous catalyst. *Criteria* for comparison should be quantitative and include catalyst lifetime as well as volumetric productivities, but have been found to depend on the different needs of *laboratory synthetic chemists,* who need a broadly specific catalyst quickly, versus those of *process chemists,* who often control catalyst availability and can allow narrow specificity (provided their substrate is accepted) but need high productivity.

Key trends for novel, environmentally more benign, processes are: (i) replacement of stoichiometric processes by catalytic ones, (ii) process steps with 100% conversion and 100% selectivity; (iii) the highest possible substrate concentration; (iv) environmentally acceptable solvents with few solvent changes; (v) avoidance of buffers and salt production, but instead use of solid acids and bases and pH-stat techniques. Examples of novel processes included in the discussion are:

- The novel BHC ibuprofen process with its three catalytic steps and 80% atom utilization (99% with recycle) replaces a process with six stoichiometric steps at < 40% atom utilization.
- The Rh-BINAP-catalyzed process from myrcene to L-(–)-menthol includes a chiral isomerization and ratios of desirable (–)-isopulegol to other isopulegol isomers of 99.7 : 0.3 as well as space–time yields of 10 kmol enamine (mol Rh)$^{-1}$ d^{-1}, and total turnover numbers of 400 000.
- The Genencor–Eastman process to ascorbic acid (vitamin C) ferments glucose in one step to 2-ketogulonic acid, with subsequent chemical conversion to ascorbic acid. It replaces the old Reichstein–Grüssner process with its 55% overall yield, 3 d cycle time, five chemical steps, 17–20 different downstream processing steps, and at least seven different solvent systems.

Successful production of chemicals through pathway engineering requires integration of fermentation technologies and metabolic engineering with the goals of higher yield on C-source, turnover per cell, and volumetric productivity, and thus lower production costs:

- DuPont's process to the fiber intermediate 1,3-propanediol, after expressing all genes in *E. coli*, converts glucose directly to 1,3-propanediol. Process improvements, such as a yield increase by preventing equilibration of TIM-catalyzed intermediates or fourfold reduction of the vitamin B_{12} requirement of glycerol dehydratase, render the process competitive. Coupling a fuel infrastructure with the manufacture of chemicals would further reduce costs and might open a new era of chemical processing.
- The key goals of the reaction network from indene to *cis*-(1*S*,2*R*)-indanediol, an intermediate via *cis*-aminoindanol to indinavir (Crixivan®), Merck's HIV protease inhibitor, are to enhance toluene dioxygenase over monooxygenase activity, and to avoid degradation of *cis*-(1*S*,2*R*)-indanediol. Still to achieve are a low enough by-product spectrum and commercially attractive yields.

20.1 Criteria for the Judgment of (Bio-)catalytic Processes

20.1.1 Discussion: Jacobsen's Five Criteria

In the example of the asymmetric epoxidation of olefins, enzymes, synthetic catalysts, and catalytic antibodies have been compared side-by-side with respect to performance in chemical synthesis (Jacobsen, 1994). Epoxidation of olefins is a reaction of considerable industrial interest where, historically, enzymes have not performed extremely well. One reason is the dependence of the enantiomeric purity of the diol and epoxide products on the regiospecificity of the attack on the racemic epoxide by a water molecule (Figure 20.1).

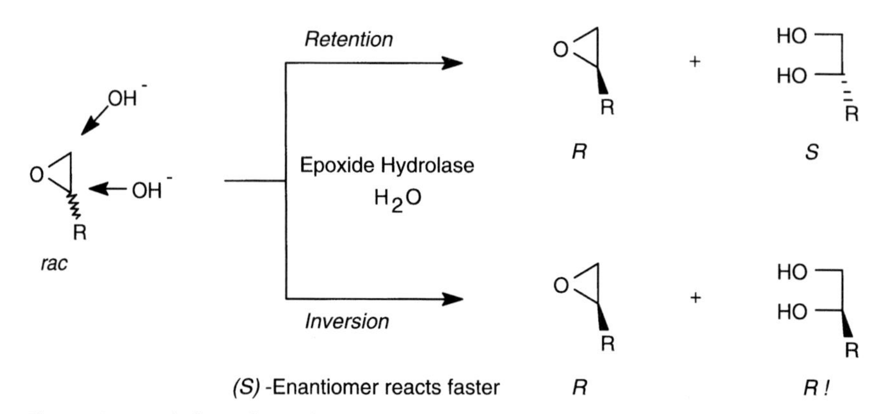

Figure 20.1 Hydrolysis of epoxides proceeding with retention or inversion (Drauz, 2002).

After presenting the criteria for performance evaluation of the three different techniques, we explain their meaning and significance. In the next section, we will comment on the appropriateness of the criteria and the reference points. This discussion relates to coverage of enzyme performance criteria in Chapter 2. Jacobsen and Finney suggest the following five criteria for comparison of performance ("Jacobsen's five criteria"):

1. enantioselectivity of the product,
2. amount of product obtained per amount of catalyst consumed,
3. availability and costs of the catalyst,
4. substrate specificity (breadth of substrates), and
5. comparison of the method with alternative strategies.

ad 1) *Enantioselectivity*: Jacobsen and Finney regard an e.e. value of 90% for the product as sufficient for the presence of an enantioselective reaction. According to their statement, enantiomeric purity can be lifted without difficulty to 95% by crystallization, albeit with the commensurate loss of yield. The three epoxidations compared were:

- reaction of hept-2-ene and H_2O_2 to (*2R*)-heptene oxide with chloroperoxidase (CPO; Allain, 1993);
- epoxidation of a cyclic ether with hypochlorite (NaOCl) with the help of a salen-Mn catalyst (Lee, 1991);
- oxidation of a benzyl-substituted cyclopentene in acetonitrile/H_2O_2 with the antibody 20B11 to the (*1R*)-epoxide (Koch, 1994).

The enantioselectivity of all three is very high. In the case of the antibody, however, the chemical epoxidation as a background reaction had to be taken into consideration for the experimental measured overall enantiomeric purity of 71% e.e. for the product. As the background reaction was about 30% of the total rate, the selectivity of the enzymatic epoxidation reaches a very high value with 98% e.e.

ad 2) *Amount of product obtained per unit of catalyst consumed*: The authors concede the difficulty in the simple use of this criterion; after all, terminology such as product yield, degree of conversion, product inhibition, total turnover number, molecular mass of the catalyst, and volumetric productivity of the process has to be standardized. One problem is the difference in definitions of the turnover number for enzymatic and chemical catalysts (Eqs. (20.1) and (20.2)).

Biocatalysis: turnover number $\equiv 1/k_{cat}$, obtained from $v = k_{cat} \cdot [E] \cdot [S]/(K_M + [S])$ (20.1)

Chemistry:
$$\text{turnover number} \equiv \frac{\text{total moles of converted substrate}}{\text{total moles of catalyst used}} \quad (20.2)$$

Table 20.1 Performance data on three catalysts for the epoxidation of olefins.

Criterion	*Enzyme (CPO)*	*Organomet. catalyst (Mn-salen)*	*Antibody (20B11)*
Turnover number (TON), acc. to Eq. (20.2)	2000	33	3.5
[S]/[C] ratio [–]	5	8	0.01

CPO: chloroperoxidase; [C]: concentration of catalyst [mol L^{-1}].

The values in Table 20.1 were taken as a basis for evaluation of the three catalytic technologies.

In comparing the amount of product produced per amount of catalyst used, Jacobsen and Finney found a draw between enzyme and organometallic complex because the authors interpreted the criterion as the optimum of the [S]/[C] ratio (5 for CPO, 8 for Mn-salen).

ad 3) *Availability and costs of the catalyst*: The authors point out (correctly) that data for this criterion cannot be obtained from the literature. For comparison of costs, the price of chloroperoxidase from *Caldariomyces fumago* from Sigma (then: $ 10/mg crude product, $ 100/mg pure enzyme) was compared with the price of a commercially available salen–manganese complex of below US $ 1/g in bulk. The availability of the antibody was judged to be difficult; as it is not available commercially, the costs cannot be estimated.

ad 4) *Substrate specificity (breadth of substrate specificity)*: In connection with enantioselectivity on key substrates, substrate specificity is usually the most well-known parameter of a catalyst. In many cases, however, investigations of substrate specificity have been limited to a few good substrates so that applicability to and performance evaluation of a novel, not yet investigated, substrate is often not possible. The authors are of the opinion that synthetic chemical catalysts feature broader substrate specificity than enzymes, which, however, convert good substrates fast. Antibodies allegedly feature a potentially broader specificity than enzymes when optimized for desired substrates.

ad 5) *Comparison of the method with alternative strategies*: On this point the authors do not give any facts, but limit themselves to general statements.

20.1.2
Comment on Jabobsen's Five Criteria

Jacobsen and Finney are to be credited with a major contribution to the comparison of different catalytic technologies by explicitly listing their evaluation criteria and by comparing quantitative data on a common basis for catalysts of different types and sources. Good criteria for catalyst evaluation should allow a clear and rapid decision in cases where all the data are available. Excellent performance with respect to all the criteria applied should be sufficient for an optimized process for the synthesis of a specific compound of interest, and should not require additional measures for evaluation. When viewed from this angle, three comments are particularly pertinent:

- Jacobsen's five criteria were picked for a *laboratory situation* in which a synthetic target compound, sometimes not even optimized for its purpose, was needed quickly and probably just once or at best a few times, and on a laboratory scale. The criteria were not developed for the situation of *large-scale, continual production* of a narrowly defined but well-known product. One-time laboratory synthesis protocols have to be concerned about instant availability and instant cost of catalysts (usually not under the influence of the user) but re-usability of catalyst and process performance criteria such as volumetric productivity are not and do not have to be considered. For large-scale continual production the situation is often reversed.

- Whereas the criteria (i) enantioselectivity, (ii) amount of product obtained per amount of catalyst used, and (iv) substrate specificity are of a *quantitative nature*, the point (iii) availability (though not cost) of a catalyst is only a semi-quantitative criterion, and (v) comparison of a method with alternative strategies is even redundant, as the process of comparison of options and their resulting evaluation and ranking are merely different parts of the decision-making process, and are not undertaken in parallel with performance measurements along dimensions of merit.

- If large-scale or continual production of a compound is the goal, as so often in industry, criteria for the *evaluation of the process* have to be added, such as lifetime catalyst stability and volumetric productivity.

We now turn away from evaluating the criteria themselves to turn towards commenting on the way in which they are applied and the appropriateness of their application. Each of the criteria deserves comments along those lines:

1) *Enantioselectivity* is indeed the best and most important quantity for the assessment of a process to enantiomerically pure compounds. Enantiomeric excess, the e.e. value, is sufficient when evaluating the optical quality of a chiral product. For a process, however, supplemental dimensions of merit need to be included, such as regioselectivity and product yield, or at least the final degree of conversion for a process. The enantiomeric value E, an alternative criterion to enantiomeric excess for the evaluation of a process (Eq. (5.90)), has the advantage of encompassing both the e.e. value and the degree of conversion (Eqs. (5.98)–(5.101)); however, the E value is fraught with its own challenges, such as in the case of mutual inhibition of the two enantiomers. If an e.e. value is determined on an isolated product, purification steps work-up can alter the value (this might not interest the user of such a product).

Lastly, a comment is warranted on the cut-off of 90% e.e. employed by Jacobsen and Finney. By today's standards, such a level of enantioselectivity is much too low; considering that the FDA, for new processes to enantiomerically pure pharmaceutical intermediates, demands a separate toxicological investigation for every by-product above the level of 1%, even the undesired enantiomer, an e.e. value of 98% should represent the lower bound. A value of 99% would be preferable and good biocatalytic processes feature an enantiomeric excess of > 99.5%. While such performance levels should serve as standard, many homogeneous chiral catalysts show enantioselectivities of 90–95% e.e., so a cut-off of 90% serves as a bias towards such catalysts.

2) The interpretation of the criterion of *mass of product obtained per mass of catalyst used* as the determination of the optimum [S]/[C] ratio is biased in two different ways:

- The [S]/[C] ratio is calculated on the basis of the molar mass of the catalyst. Typically, homogeneous chemical catalysts feature a molecular mass of about 1000 in comparison with that of an enzyme of 30–50 kDa. At equal turnover number (catalytic events per active site per unit time), such a view yields 30–50 times higher [S]/[C] ratios for homogeneous catalysts. As, however, the cost of a unit of activity or the ease of synthesis should be the guiding dimension and not the molar mass, such homogeneous catalysts are unduly favored over enzymes and antibodies.

- While an [S]/[C] ratio serves as a good dimension of merit for the productivity of one batch, it does not yield any information on the stability of the catalyst over its lifetime. As already discussed in Chapter 19, Section 19.3.4, the homogeneous catalysis community typically does not feel the need to recycle catalysts (Blaser, 2001), so the turnover number (TON) equals the total turnover number (TTN). The true utility and productivity advantage of biocatalysts is captured upon reuse of the catalyst, achieving catalyst lifetime productivities far in excess of catalysts used only once.

3) *The availability and cost of the catalyst* cannot be addressed within the framework of a *technological* assessment of a process. However, for a *logistical* assessment of a process not only are the availability (and delivery times) and price of the catalyst very important, but so are questions of proper formulation and storage as well as safety. As mentioned above, a laboratory chemist has no control over the price of a catalyst offered whereas a company desiring to produce a certain product can plan ahead to supply the catalyst at the right time and price.

4) The relevance of the *substrate specificity of a catalyst* as an evaluation criterion will be answered differently by a preparative (bio-)chemist and process (bio-)chemist. A preparative (bio-)chemist depends on a known and broad substrate specificity for a catalyst to be useful. The process researcher in the vast majority of cases

already knows the (almost exclusively affirmative) answer as to the appropriateness of a catalyst for a desired reaction. If the catalyst has the right specificity for a process application it might just as well be narrow; broad reactivity is not only irrelevant but often detrimental. (In analytical applications, narrow substrate specificity is often actually essential to exclude unwanted cross-reactivities.) On average, however, narrow substrate specificity is an obstacle to a synthetic chemist.

One remark regarding the state of knowledge on the substrate specificity of even well-researched enzymes: in our experience, many investigations of substrate specificity are conducted with a conservative view of its breadth. Risky substrates covering a large parameter space or those outside the narrowly defined range of interest are not tested. Too many researchers often just include standard substrates and are afraid of negative results although it is obvious how the reporting of even negative results can help other researchers.

5) As discussed above, *comparison with alternative strategies* should be part of the overall evaluation process but should not be regarded as a separate criterion.

20.2 Position of Biocatalysis in Comparison to Chemical Catalysts for Novel Processes

20.2.1 Conditions and Framework for Processes of the Future

Owing to economic pressures resulting from increased worldwide competition for market share and from increased costs from enhanced safety and environmental conditions (Chapter 1, Section 1.2.3), urgent requirements for revamping current industrial processes exist. In many cases, however, revamping will be insufficient to meet either cost pressures or environmental standards and therefore completely new processes will need to be developed.

Some of the most important trends of (bio-)chemical processing are (Sheldon, 1994):

- replacement of stoichiometric processes by catalytic ones,
- choice of process steps with 100% conversion and 100% selectivity if possible,
- utilization of the highest possible substrate concentration,
- utilization of environmentally acceptable solvents with as few solvent changes as possible,
- avoidance of the production of by-products during pH adjustment; avoidance of buffers, and use of solid acids and bases (ion exchange resins, zeolites) and of pH-stat techniques.

Adherence to these criteria results in a much better starting position when switching processes with the goal of *environmentally benign manufacturing* (Chapter 1, Section 1.2.3).

The procedure towards an environmentally benign process starts with the selection of raw materials: in addition to conventional raw materials from the petrochemical industry based on low- to medium-boiling aliphatic and aromatic hydrocarbons, replenishable raw materials from nature are increasingly available nowadays. In the simplest cases, these can be carbon sources, such as glucose and sucrose for fermentation, but also more complex molecules, frequently obtained from the "chiral pool" or through inexpensive fermentation from carbon sources, such as glutamic acid or citric acid. Table 20.2 lists a selection of raw materials from the chiral pool, with their estimated costs per kilogram.

Legal restrictions also increasingly influence the choice of solvent for (bio-)chemical processes. Halogenated solvents, which have very favorable solvating and processing properties, have been banned in the European Union since 1997. On the other hand, the use of methyl *tert*-butyl ether (MTBE) in the chemical industry is growing strongly. Good solvating properties, low boiling point and low propensity for explosion (owing to the inability to form peroxides because of the *tert*-butyl group) complement a low price and almost infinite availability. The last property is owed to large capacities but decreasing demand as an antiknock additive to unleaded gasoline after the ban in California, as of 2003, for causing pollution of groundwater reservoirs.

Novel environmentally benign processes can be based on three separate technological platforms: (i) processes utilizing heterogeneous catalysts; (ii) processes with homogeneous catalysts; and (iii) processes using biological catalysis. As discussed in Chapter 1, the latter encompass a continuum ranging from fermentation to reactions with purified enzyme. In many cases, novel processes based on different catalytic technologies compete against each other. Either economic criteria such as capacity or profit margin, or external criteria such as maximally permitted emissions, must decide on the question of which technology to adopt. The following examples illustrate the competitive situation for novel processes replacing older and established ones.

Table 20.2 Raw materials and their costs from biological sources and the chiral pool.

Product	*Production quantity [tons per year (tpy)]*	*Product price [\$ kg^{-1} or • kg^{-1}]*
Ethanol	15 000 000	0.8
Glucose–fructose (HFCS)	5 000 000	0.9
Monosodium glutamate (MSG)	1 000 000	1.1
Citric acid	400 000	1.2
L-Lysine, feed grade	115 000	2.0
Gluconic acid	50 000	30
6-Aminopenicillanic acid (6-APA)	6 000	
L-Aspartic acid		
L-Phenylalanine		
L-Tartaric aicd		
D-Shikimic acid		
Kainic acid		
L-Malic acid	500	25

20.2.2
Ibuprofen (Painkiller)

Ibuprofen, a well-known, non-steroidal, anti-inflammatory drug (NSAID) used as a common painkiller and marketed under brand names such as Advil™ and Motrin™, provides an example of the switching of processes from one chemical route to another.

There have been many commercial and laboratory publications on the synthesis of ibuprofen. Two of the most popular ways to obtain ibuprofen are the Boots process and the Hoechst process. The Boots process is an older commercial process developed by the Boots Pure Drug Company, the discoverers of ibuprofen in the 1960s, and the Hoechst process is a newer process developed by the Hoechst Company. Most of these routes to Ibuprofen begin with isobutylbenzene and use Friedel–Crafts acylation. The Boots process requires six steps, while the Hoechst process, with the assistance of catalysts, is completed in only three steps (Figure 20.2).

In 1992 BHC (Boots Hoechst–Celanese) Company commercialized a new synthetic process to manufacture ibuprofen in BHC's 3500 metric-ton-per-year facility in Bishop/TX, USA, which was cited as an industry model of environmental excellence in chemical processing technology. For its innovation, BHC was the recipient of the 1997 Alternative Synthetic Pathways Award of the Presidential Green Chemistry Challenge.

The BHC ibuprofen process is an innovative, efficient technology that revolutionized bulk pharmaceutical manufacturing. The new technology with its three catalytic steps achieves approximately 80% atom utilization, which is increased to almost 99% if recovery of the by-product acetic acid is included. It replaces a pro-

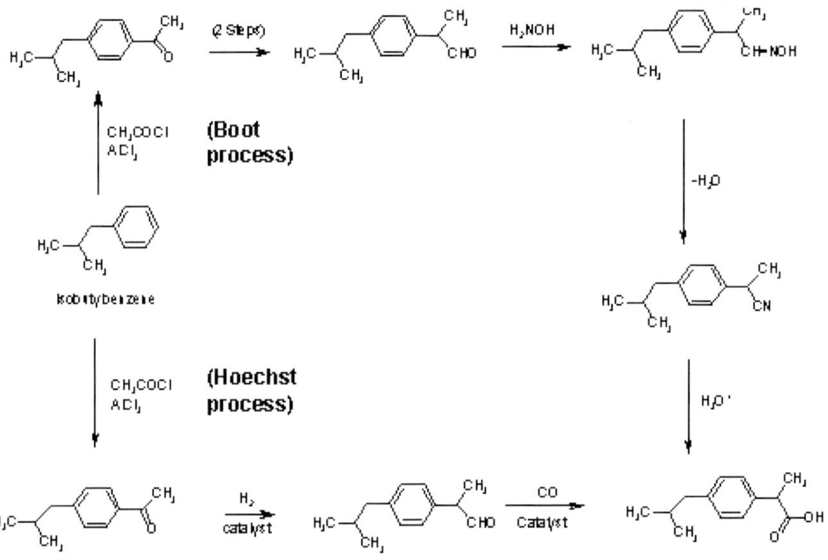

Figure 20.2 Ibuprofen processes (http://www.chm.bris.ac.uk/motm/ibuprofen/synthesisc.htm).

cess with six stoichiometric steps and less than 40% atom utilization. The use of anhydrous hydrogen fluoride as both a catalyst of Friedel–Crafts acylation and a solvent offers important advantages in both reaction selectivity and waste reduction, as HF is recycled with more than 99.9% efficiency. No other solvent is needed in the process, simplifying product recovery, minimizing fugitive emissions, and eliminating large volumes of aqueous salt waste solutions at the source. The nearly complete atom utilization of the starting materials either by conversion to product or by recycling renders this streamlined process a waste-minimizing, environmentally friendly technology.

Cheminor Drugs (part of Dr. Reddy's Group, Mumbai, India) have developed a process based on a chiral synthesis to (*S*)-ibuprofen. (*S*)-Ibuprofen is 160 times more active than the (*R*) form, so with an enantiomerically pure drug the dosage could be cut in half. As the human body has been reported to racemize (*R*)-ibuprofen partially into the (*S*) form and as (*R*)-ibuprofen is not harmful, nearly all the ibuprofen taken becomes active. The process discovered by Cheminor is therefore unlikely to have commercial significance.

20.2.3
Indigo (Blue Dye)

The industrial synthesis of indigo, a blue dye obtained from plants and used since antiquity, in 1891 was one of the early successes of industrial chemistry. Its chemical structure had been discovered only eight years earlier by Adolf von Baeyer. Indigo is produced by coupling two molecules of sodium phenylglycinate, obtained industrially from aniline which in turn is isolated from coal tar, in a melt of KOH–NaOH at 900 °C with $NaNH_2$, a reagent newly introduced for this process (Figure 20.3). The intermittently produced water-soluble dye indoxyl reacts to indigo upon exposure to air. The classic process has remained largely unchanged from the early 1900s until now. While economical, its ecological impact has spurred the search for other routes.

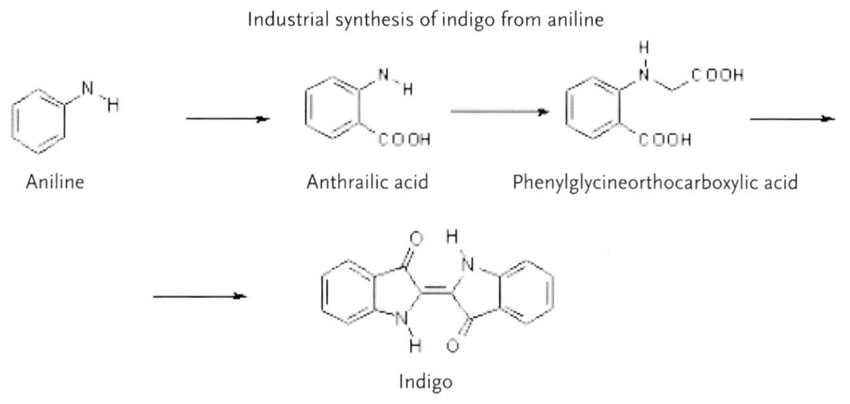

Figure 20.3 Classic indigo process.

An estimated world market of 16 000 tpy for indigo is totally dominated by its use to dye denim, the cloth in jeans. Just when many plants had been scheduled to be shut down in the 1960s, the denim boom brought a renaissance of demand and interest awoke again in developing a more contemporary process.

Genencor (Palo Alto/CA, USA) developed an alternative, biocatalytic, process. *E. coli* bacteria, using sugars as the main raw material, converted the carbon source into tryptophan, a suitable precursor for indigo because it already contained the ring-structure at the core of the indigo molecule. Metabolic engineering modified the tryptophan pathway to cause high-level indole production by adding the *Pseudomonas putida* genes encoding naphthalene dioxygenase (NDO). In comparison to a tryptophan-overproducing strain, the first indigo-producing strain made less than half of the expected amount of indigo. Severe inactivation of the first enzyme of aromatic biosynthesis, 3-deoxy-D-*arabino*-heptulosonate 7-phosphate (DAHP) synthase (the *aroG*^{fbr} gene product), was observed in cells collected from indigo fermentations. Subsequent in-vitro experiments revealed that DAHP synthase was inactivated by exposure to the spontaneous chemical conversion of indoxyl to indigo. Indigo production was thereafter improved by increasing the gene dosage of *aroG*^{fbr} or by increasing substrate availability to DAHP synthase in vivo by either amplifying the *tktA* (transketolase) gene or inactivating both isozymes of pyruvate kinase. By combining all three strategies for enhancing DAHP formation in the cell, a 60% increase in indigo production was achieved. As the last step, the indigo precursor indoxyl was spontaneously oxidatively dimerized into indigo when exposed to air (Figure 20.4). Metabolic engineering was then further applied to eliminate a by-product, isatin, of the spontaneous conversion of indoxyl to indigo, thereby solving a serious problem with the use of bio-indigo in the final denim dyeing application.

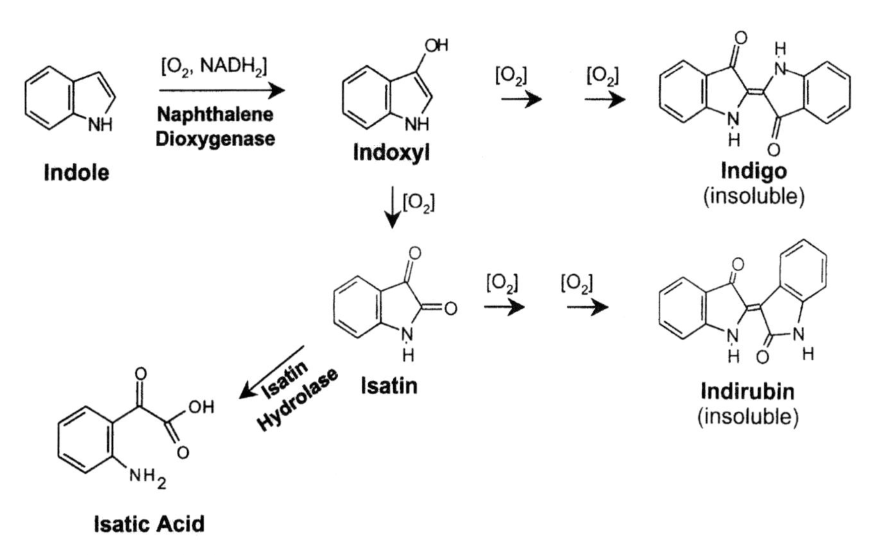

Figure 20.4 Biocatalytic route to indigo.

At first, however, a trace by-product, indirubin, rendered jeans an unfashionable shade of red. Genetic engineering removed the gene for the red pigment. The final color is "indistinguishable" from the globally popular deep blue of the chemically made dye. To be competitive with the established process, however, the efficiency of the process still needs to be improved. The process, however, is much more environmentally benign and produces much less waste.

To keep everything in perspective, it is important to mention that, in addition to the biological route, there is a novel chemical route under development using homogeneous catalysis.

20.2.4
Menthol (Peppermint Flavoring Agent)

L-(–)-menthol is responsible for peppermint flavor. The classic source of menthol is peppermint oil but the supply is insufficient to satisfy today's demand of about 4000 tpy worldwide. Besides the food industry (tea additive, sweets) the main customer is the tobacco industry: peppermint-flavored cigarettes are very popular in several countries, most notably in southeast Asia (*kretek* in Indonesia).

Three processes have been developed for L-(–)-menthol, which has three asymmetric centers: (i) separation of diastereomeric salt pairs; (ii) homogeneous catalysis with Rh-BINAP; and (iii) lipase resolution of menthol benzoates.

20.2.4.1 Separation of Diastereomeric Salt Pairs

The racemate is straightforwardly produced from *m*-cresol via Friedel–Crafts acylation to thymol and catalytic hydrogenation. Hydrogenation of thymol yields four pairs of racemic diastereomers: menthol, isomenthol, neomenthol, and neoisomenthol. In the classic synthetic process, separation of the eight isomers is afforded by formation of diastereomeric salt pairs. However, researchers at Haarmann & Reimer (H & R, Holzminden, Germany) found unexpectedly that the racemic menthyl benzoates can be separated by crystallization induced by seeding the mixture with one of the optically active menthyl benzoate enantiomers. After notable engineering development, this discovery resulted in a process of more than 90% overall yield and made H&R the market leader in synthetic L-(–)-menthol (Fleischer, 1972) (Figure 20.5).

20.2.4.2 Homogeneous Catalysis with Rh-BINAP

In the early 1980s, Takasago (Tokyo, Japan) developed an elegant, homogeneously catalyzed process to L-(–)-menthol from myrcene (Figure 20.6a). Myrcene is converted to diethylgeranylamine by the lithium-catalyzed addition of diethylamine, which in turn is catalytically isomerized with Rh-BINAP (Figure 20.6b) to the chiral (3*R*)-citronellal enamine with 96–99% e.e. Hydrolysis gives (3*R*)-(+)-citronellal of higher chiral purity than citronellal from citronella oil. The remainder of the synthesis employs classical chemistry: zinc bromide catalyzes the cyclization of (3*R*)-(+)-citronellal to (–)-isopulegol with a ratio of (–)-isopulegol to the other isopulegol isomers of 94 : 6. However, in a recent patent Takasago workers have shown that

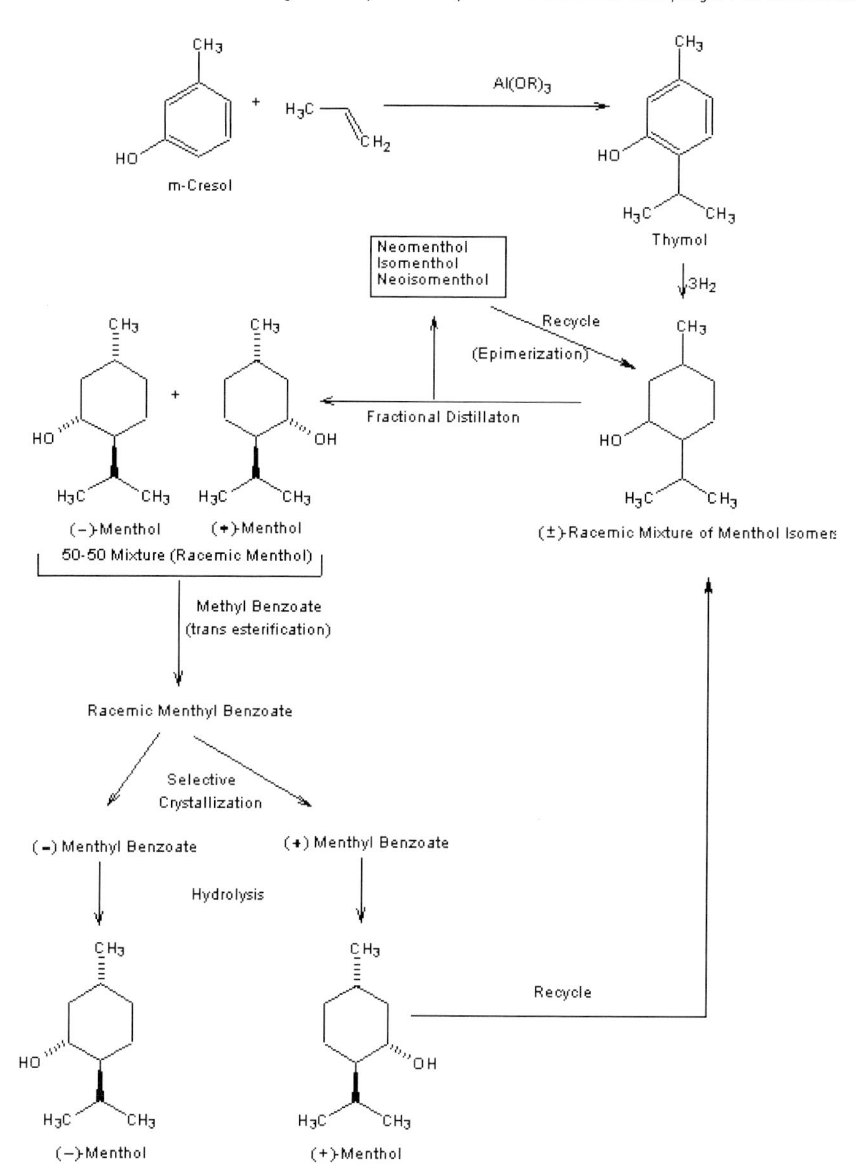

Figure 20.5 Classic synthetic route to L-(–)-menthol.

tris(2,6-diarylphenoxy)aluminum catalysts can give a ratio of (–)-isopulegol to the other isopulegol isomers of 99.7 : 0.3 (Iwata, 2002).

This is the second major commercial route to L-(–)-menthol and the largest application of homogeneous *asymmetric* catalysis. Yearly, about 1500 t (–)-menthol and other terpenes are produced by this route. The space–time yield could be improved to about 8000 mol enamine (mol Rh)$^{-1}$ in 18 h. The total turnover number (TTN) was improved to an impressive 400 000.

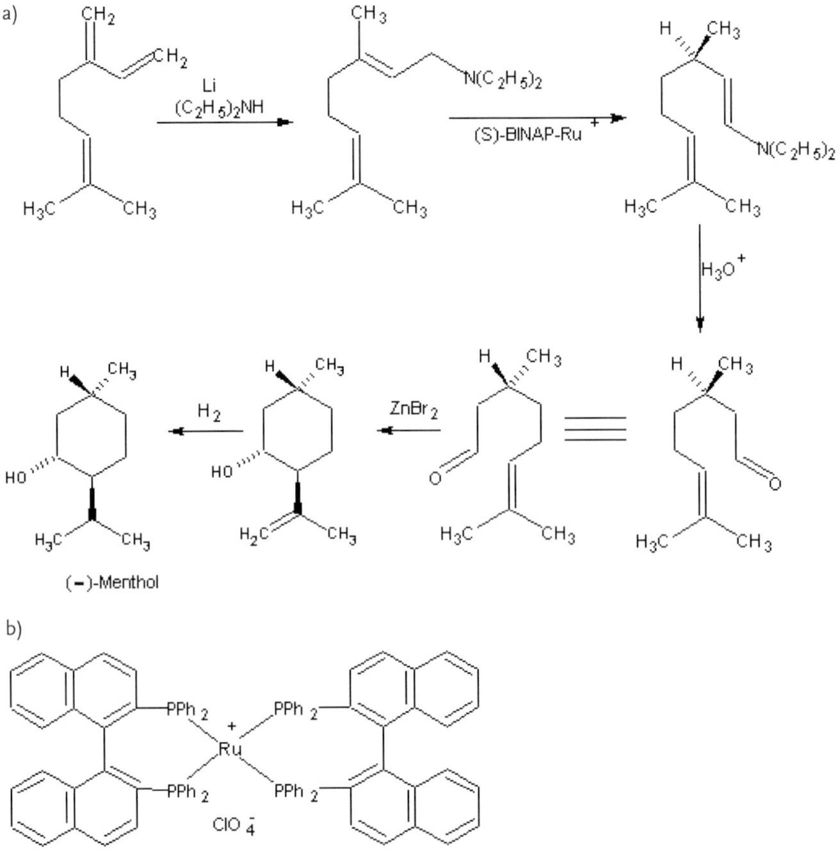

a) Li, $(C_2H_5)_2NH$; (S)-BINAP-Ru$^+$; H_3O^+; $ZnBr_2$; H_2; (−)-Menthol

b) ClO_4^-

Figure 20.6
a) Homogeneously catalyzed process to L-(–)-menthol from myrcene;
b) Takasago catalyst for the process to L-(–)-menthol.

20.2.4.3 **Lipase-Catalyzed Resolution of Racemic Menthol Esters**

For decades, biocatalytic processes to (–)-menthol have been sought. Takasago researchers were the first to achieve hydrolysis of racemic esters of menthol using an enzyme carboxylic acid hydrolase (Moroe, 1968). Currently, two much more recently developed enzymatic processes vie for realization on a commercial scale:

AECI (Gallo Manor, South Africa) employs stereoselective lipase-catalyzed acylation of the mixture of four pairs of racemic diastereomers mentioned above to obtain L-menthyl acetate in ≥ 96% e.e. Menthyl acetate is separated from the unreacted isomers by distillation and hydrolyzed to L-(–)-menthol (Figure 20.7). From a 2002 patent issued to AECI (Chaplin, 2002), the enzyme half-life is considerably longer than 2000 h. The remaining mixture of isomers is isomerized or racemized to yield the original composition which can then again be subjected to lipase catalysis. The overall yield on thymol over several cycles is near quantitative. According to the patent, the continuous process has been demonstrated on a 1 kg h^{-1} scale.

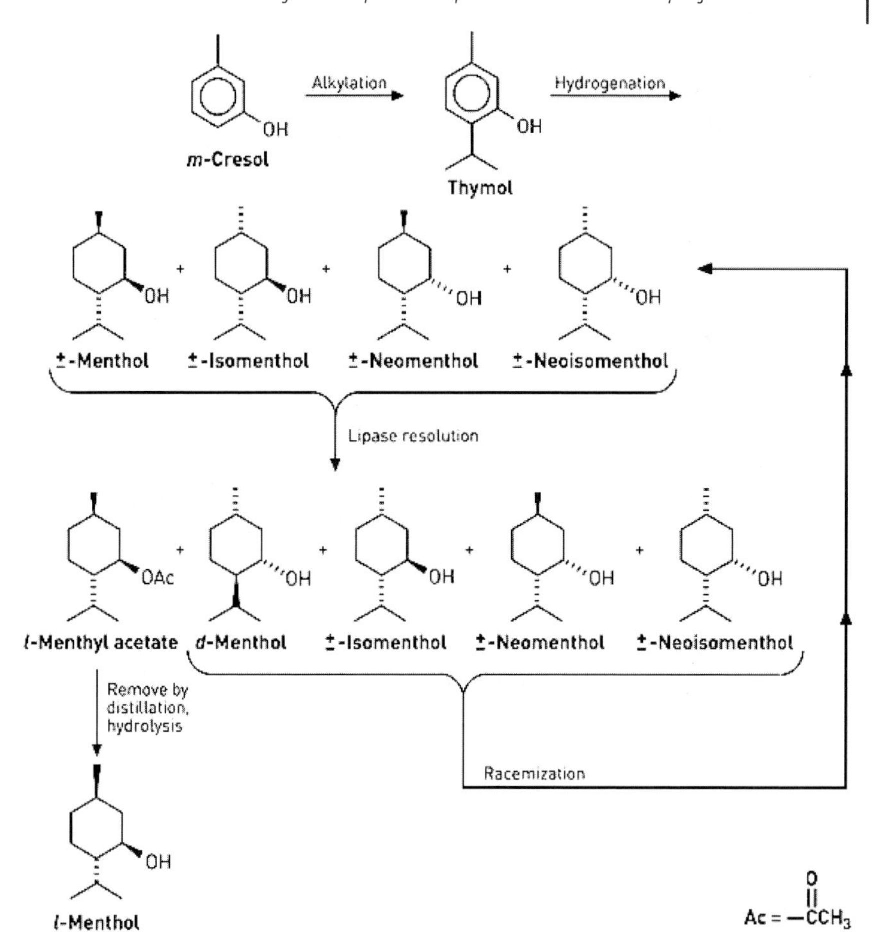

Figure 20.7 Lipase-catalyzed process to L-(–)-menthol.

Haarmann & Reimer (Holzminden, Germany) in collaboration with Rolf Schmid's group at the University of Stuttgart developed a process based on the resolution of racemic menthol esters, such as acetate or benzoate esters. The team employed several lipases such as *Candida cyclindracea* which resulted in enantioselectivities up to 100% e.e. (Bornscheuer, 2002).

20.2.5 Ascorbic Acid (Vitamin C)

Ascorbic acid as a water-soluble vitamin (vitamin C) is an essential component in the human diet. As one of many anti-oxidants (vitamin E and β-carotene are examples of fat-soluble anti-oxidants), ascorbic acid is required for the growth and repair of tissues in all parts of the body. It is necessary to form collagen, an important protein used to make skin, scar tissue, tendons, ligaments, and blood vessels.

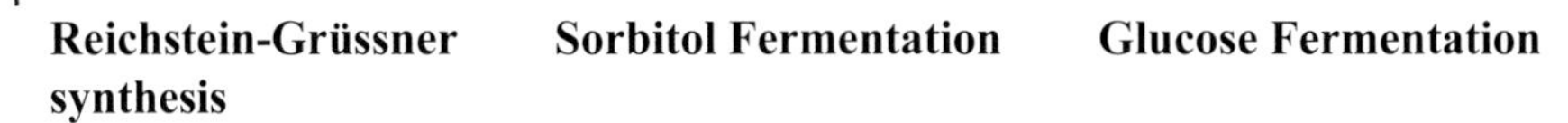

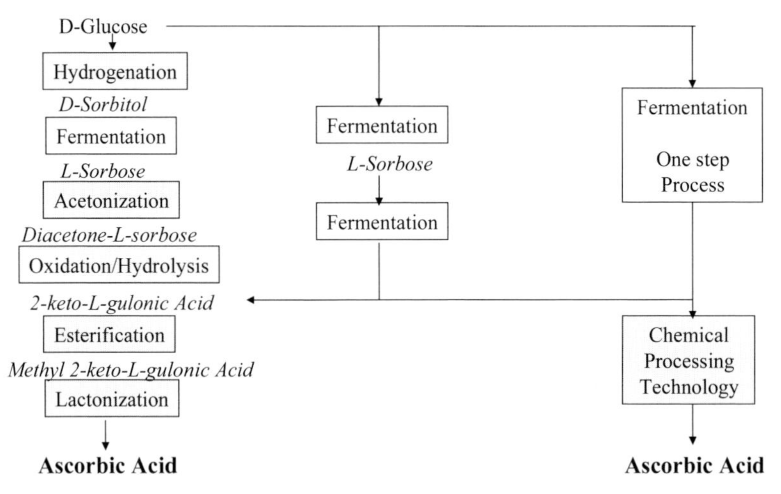

Figure 20.8 Comparison of ascorbic acid processes.

Vitamin C is essential for the healing of wounds, and for the repair and maintenance of cartilage, bones, and teeth.

There are a total of three ascorbic acid processes either in place or in late stages of development with firm expectations of commercialization: (i) the traditional chemical Reichstein–Grüssner synthesis; (ii) the two-step fermentation process to 2-ketogulonic acid with subsequent chemical esterification/lactonization to ascorbic acid; and (iii) the one-step fermentation to 2-ketogulonic acid with the same last chemical step. Figure 20.8 and Table 20.3 provide an overview of the three processes.

20.2.5.1 The Traditional Reichstein–Grüssner Synthesis

The Reichstein–Grüssner process, developed in 1934, takes in all five chemical steps (hydrogenation, fermentative oxidation, acetonization, oxidation, and hydrolysis/rearrangement) (Table 20.3). Over the course of the whole synthesis there are 17–20 different downstream processing steps, six solid-handling steps, and at least seven different solvent systems to handle. The overall yield is about 55%, and overall cycle time is around three days. Such values clearly suggest possible improvement in the process towards ascorbic acid.

20.2.5.2 Two-Step Fermentation Process to 2-Ketogulonic Acid with Chemical Step to Ascorbic Acid

In a process design towards a biological process, the enzymes for the first two steps, E1 – glucose dehydrogenase and E2 – gluconic acid dehydrogenase, were cocloned into *E. coli* and thus yielded L-sorbose directly. The next fermentative step was performed to obtain 2-keto-L-gulonic acid, which was then treated as in the conventional process.

Table 20.3 Features of the steps in the Reichstein–Grüssner synthesis.

Step	*Yield [%]*	*Cycle time [h]; T, p; catalyst, solvent*	*Work-up steps*	*[S] [g L^{-1}]*	*Biggest challenges*
Hydro-genation	95	2; 140 °C, 80–125 bar; Ra–Ni, H_2O/MeOH	hot filtration, ion exchange, filtration		
Sorbitol oxidation	90	24; 30 °C, 2 atm; pH 5–6 → 2, H_2O	centrifugation, deionization, crystallization	200	sterility and 2 atm O_2 required; Ni tank material toxic
Acetonization	85	24; 30 °C, 135 °C/3 Torr; acetone/H_2SO_4, ether	2 × distillation, filtration, vac. distillation	50	
Oxidation	90	6; 50 °C; Pd/C; pH H_2O, acid, acetone	precipitation, filtration, drying		
Hydrolysis/rearrange-ment	85	2; 100 °C; pH 2: HCl/MeOH, ClHC, EtOH	distillation, recrystallization, evaporation		N_2/CO_2 atm. required

ClHC: chlorinated hydrocarbon.

20.2.5.3 One-Step Fermentation to 2-Ketogulonic Acid with Chemical Step to Ascorbic Acid

In a collaboration between Genencor (Palo Alto/CA, USA) and Eastman Chemical (Kingsport/TN, USA) co-sponsored by the US federal government through the Advanced Technology Program, the most direct process so far was developed (Figure 20.9): all four enzymes to the intermediate 2-keto-L-gulonic acid, glucose dehydrogenase (E1, to gluconic acid), gluconic acid dehydrogenase (E2, to sorbose),

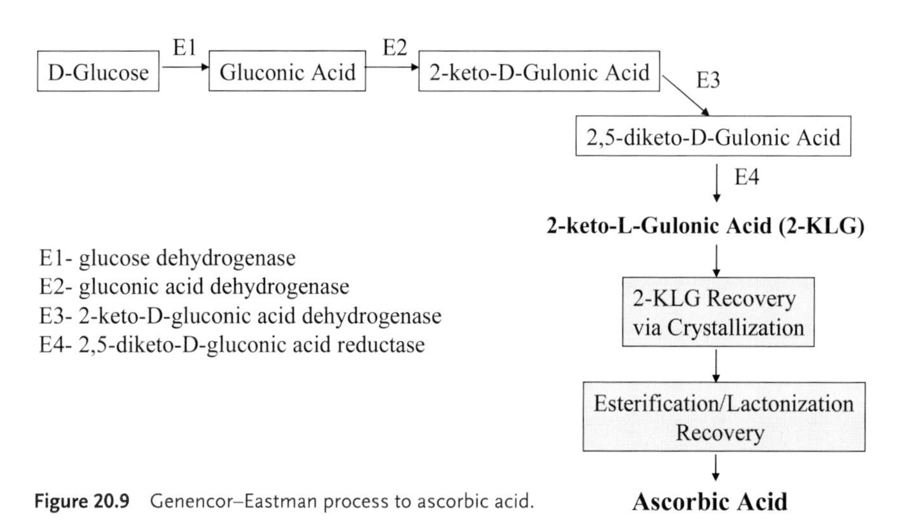

Figure 20.9 Genencor–Eastman process to ascorbic acid.

2-keto-D-gluconic acid dehydrogenase (E3, to 2,5-diketo-D-gluconic acid) and 2,5-diketo-D-gluconic acid reductase (E4, to 2-keto-L-gulonic acid) were co-cloned into *E. coli* to produce the intermediate directly. The last step is parallel to the conventional chemical synthesis.

20.3 Pathway Engineering through Metabolic Engineering

Essential for the successful production of chemicals through pathway engineering will be the integration of fermentation technologies and metabolic engineering (Chotani, 2000). Key to the application of metabolic engineering is the ability to measure multiple metabolic fluxes quantitatively, especially through intracellularly applied microanalytical methods. Pathway engineering is also aided by advances in genomics and proteomics (Chapter 15). Successful application will achieve improvements in several aspects of a process such as higher product turnover per cell, more efficient channeling of carbon, higher productivity and thus smaller factories, and lastly, minimized product development cycle times to bring products to market. Progress in industrially applicable metabolic pathway engineering driven by the coalescence of flux analysis ("fluxomics"), genomics, proteomics, and computational modeling has been reviewed recently (Sanford, 2002).

20.3.1 Pathway Engineering for Basic Chemicals: 1,3-Propanediol

The superior properties of poly(propylene terephthalate) (PPT) polymer and fibers over the chemically analogous poly(ethylene terephthalate) (PET, used for soda bottles) and poly(butylene terephthalate) (PBT) have been well known for several decades: PPT fibers are much more elastic and less brittle than PET and offer better recovery from stretching than PBT; they are also easier to dye than either PET or PBT. Compared to the intermediate for PET, ethylene glycol, which is available inexpensively from ethylene oxide, and to that for PBT, butanediol, likewise available inexpensively from butene or butadiene, the intermediate for PPT, 1,3 propanediol (1,3-PPD or PDO), was not – and on a large scale is still not – available. Three processes, two chemical ones and one biotechnological, compete to change this situation (Figure 20.10).

1,3-Propanediol is produced either from the reductive hydration of acrolein (Degussa–DuPont process), or through reductive carbonylation of ethylene oxide (Shell process), or through fermentation of glucose via glycerol (DuPont–Genencor process).

The DuPont–Genencor process follows the glycolysis pathway to the equilibrium ($K = 1.1$) between glyceraldehyde-3-phosphate (GAP) and dihydroxyacetone phosphate (DHAP) catalyzed by triose phosphate isomerase (TIM, Chapter 10, Section 10.3.2) (Figure 20.11). DHAP then is reduced to glycerol-3-phosphate with $NADP^+$-dependent glycerol dehydrogenase, followed by hydrolysis to glycerol with

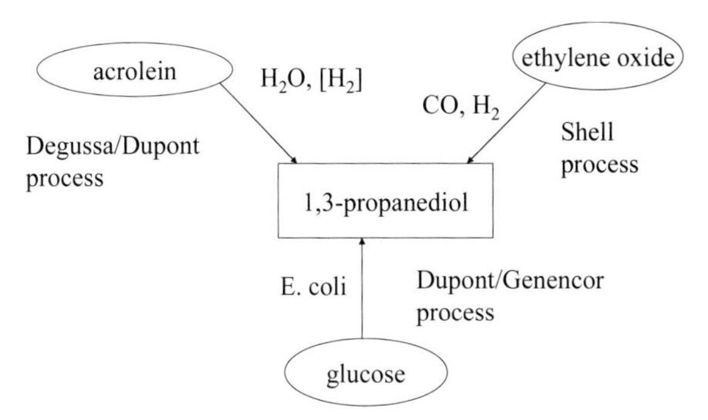

Figure 20.10 Competing processes to 1,3-propanediol.

glycerol-3-phosphate phosphatase. The last two steps are a dehydration to 3-hydroxypropionaldehyde and another $NADP^+$-dependent reduction to the end-product, 1,3-propanediol.

The DuPont process originally consisted of two separate fermentations, the first in yeast leading from glucose to glycerol, and the second in *E. coli* from glycerol to the end-product 1,3-propanediol. A major advance, especially in terms of lower investment costs, was the cloning and expression of all genes of the pathway into *E. coli* to lead straight from glucose to the end-product.

Triose phosphate isomerase, TIM, as a nearly-diffusion-limited enzyme (Ch. 2, Section 2.2.3), catalyzes the equilibration of GAP and DHAP very efficiently. However, equilibrium concentrations of GAP were metabolized to pyruvate and further to ethanol or acetate, so the stoichiometric yield on glucose to 1,3-PPD of 42.5% was lower than the target of 50%. Thus, TIM was cloned out to prevent equilibration between the desired DHAP and the undesired side-product GAP, which successfully increased the yield beyond the minimum target.

The penultimate enzyme in the pathway, glycerol dehydratase (E. C. 4.2.1.30), catalyzing the dehydration of glycerol to 3-hydroxypropionaldehyde, is a vitamin B_{12}-dependent, cyanocobalamin-containing enzyme, which employs a radical

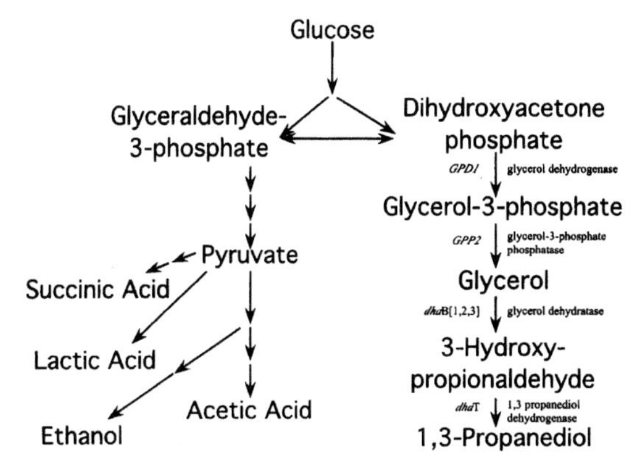

Figure 20.11 Schematic of the reaction network from glucose to 1,3-propanediol (Chotani, 2000).

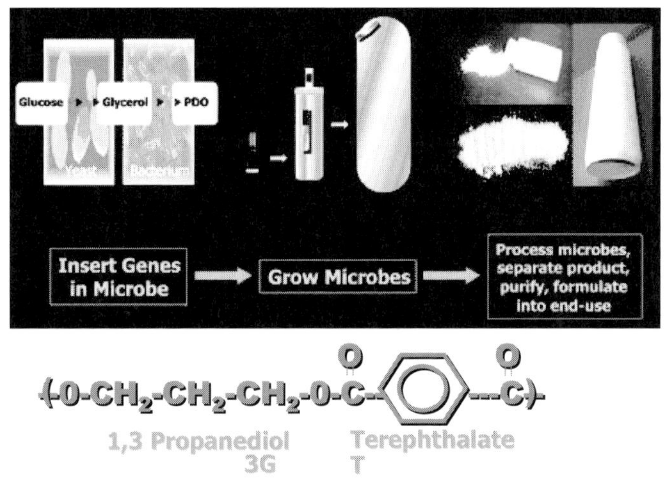

Figure 20.12 Overview of the DuPont process to poly(propylene terephthalate) (3-GT, Sorona®) (Ray Miller, DuPont).

mechanism based on hemolytic cleavage of the cobalt–carbon bond (Kajiura, 2001; Toraya, 2002). Combining a variety of protein engineering techniques, the cofactor requirement per unit mass of product could be decreased fourfold so that the cost of vitamin B_{12} no longer seems to be limiting.

On 5–10 000 ton pilot-plant level, chemical and now also biological 1,3-PPD have been incorporated into PPT fibers. The fiber product, Sorona®, is synthesized in a continuous process via transesterification from dimethyl terephthalate ("T") and 1,3-propanediol ("3-G") with the help of a catalyst to the polymer ("3-GT"), finished under vacuum, and pelletized. Figure 20.12 summarizes the process.

Should the biological route to 1,3-propanediol be competitive vis-à-vis the two chemical processes (and many signs points towards the fact that it will), fermentation of a biological carbon source with an optimized recombinant strain would supply a "non-biological" product, i.e., a product with no life science use. While chemical raw materials such as acetone, butanol, or ethanol have been obtained through fermentation, 1,3-propanediol would furnish the first case of a chemical company utilizing recombinant DNA as well as fermentation technology to make a product, or at least a chemical intermediate. For this work, Dupont won the Presidential Green Challenge Award in 2003. Coupling a fuel infrastructure with the manufacture of chemicals would further reduce costs and might usher in a new era of industrial processing.

20.3.2
Pathway Engineering for Pharmaceutical Intermediates: cis-Aminoindanol

In Chapter 13, Section 13.3.3, we discussed indinavir (Crixivan®), Merck's HIV protease inhibitor, which is currently manufactured via a chemical synthesis route. An alternative biocatalytic route, at least to intermediates, briefly discussed in Section 13.3.3, is sought to increase efficiency of the current process and to decrease costs. At Merck Research Laboratories in Rahway/NJ, USA, Barry Buckland's group applied metabolic engineering to construct new pathways for the biosynthesis of *cis*-aminoindanol, a useful intermediate towards indinavir (Buckland, 1999).

The group employed several microorganisms, such as two different *Pseudomonas putida* strains and a *Rhodococcus* sp. strain to achieve the original goal of producing *cis*-aminoindanol. Biotransformation in the original *Pseudomonas* strain did result in the desired *cis*-(1*S*,2*R*)-indanediol with over 99% e.e. The high e.e. value was achieved by further metabolism exclusively of the undesired enantiomer, *cis*-(1*R*,2*S*)-indanediol, within the cell, thereby resolving the racemic solution and resulting in the high e.e. value, albeit at the expense of overall yield.

The indene bioconversion in *Rhodococcus* sp. leads to the desired product of *cis*-(1*S*,2*R*)-indanediol as well, but also to several by-products which disrupt the overall yield and enantiomeric excess of the desired product (see Figure 20.13).

A toluene dioxygenase (TOD) operon was isolated from a *Pseudomonas* strain which can grow on toluene through expression of the TOD operon, and cloned into *E. coli*. The TOD complex can convert both toluene and indene to the corresponding *cis*-dihydrodiols but has to be induced by toluene, which is toxic to the microorganism above a certain concentration. One measure to reduce toxicity is addition of soybean oil to the biotransformation to form a second, hydrophobic solvent phase to partition the toxic toluene from the aqueous biocatalyst solution. Expression of the operon in *E. coli* and process optimization lead to a titer of 1 g L^{-1} product with over 99% e.e. after 24 h of fermentation. However, the titer was not high enough for commercially attractive levels.

More than 1 kg of purified *cis*-(1*S*,2*R*)-indanediol was produced and incorporated into indinavir sulfate, the active ingredient of Crixivan®. The undesired necessity to induce the dioxygenase with toluene was by-passed through mutation rounds to screen for a constitutive producer of the TDO enzyme system solely

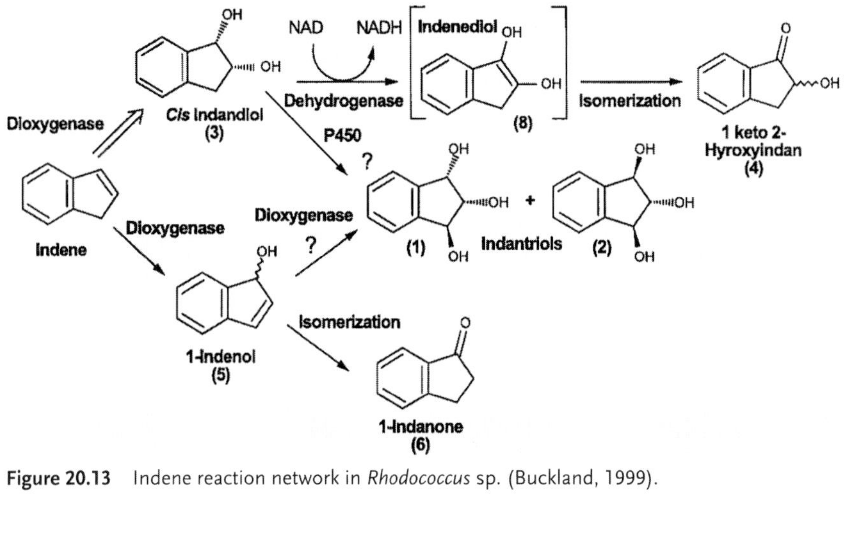

Figure 20.13 Indene reaction network in *Rhodococcus* sp. (Buckland, 1999).

growing on indene. Mutants were screened by the production of blue-colored indigo, since indole can be converted by this enzyme complex first to *cis*-indole-2,3-dihydrodiol and then, after water elimination, to indoxyl, which air-oxidizes to indigo. The successful constitutive mutant strain converted indene to *cis*-(1*S*,2*R*)-indanediol with > 90% e.e. owing to two *cis*-glycol dehydrogenases, which prefer *cis*-(1*R*,2*S*)-indanediol and thus resolve the racemate kinetically. Further experiments demonstrated that each product emerging from the bioconversion inhibited formation of all reaction products, so that end-product inhibition limits the productivity of this approach.

Perspective for a process to cis-aminoindanol

While bioconversion of indene to *cis*-(1*S*,2*R*)-indanediol features the correct enantiomeric specificity, the conversion is part of a degradation pathway resulting in by-products and degradation of the product. Metabolic engineering is being applied to reduce side reactions and avoid the further degradation of *cis*-indanediol, pushing synthesis even further to the next step, the desired amine derivate, *cis*-aminoindanol (Figure 20.14). This approach is still in development, indicating the difficulties and obstacles researchers have to overcome to tune a biocatalyst into producing a desired product with (i) a good enantiomeric excess; (ii) straightforward metabolic pathways to simplify fermentation; (iii) as few by-products as possible to facilitate product purification; and (iv) commercially attractive yields.

The indene bioreaction network that catalyzed the bioconversion to (2*R*)-indanediol was also investigated in *Rhodococcus* sp. (Stafford, 2001). Metabolic engineering methodology was applied step-by-step:

1. Steady-state physiology of the *Rhodococcus* sp. 124 strain was investigated by a chemostat with a novel indene air delivery system.
2. Prolonged cultivation of this organism in a continuous flow system actually resulted in the evolution of a mutant strain, designated KY1, with improved

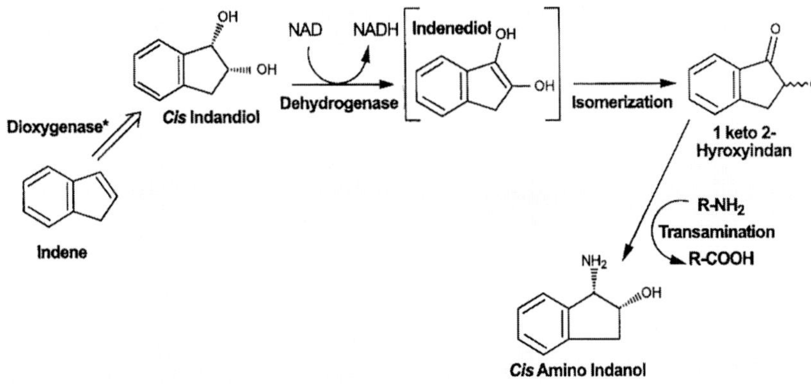

Figure 20.14 Potential future metabolically engineered pathway in *Rhodococcus* to achieve the desired product *cis*-aminoindanol (Buckland, 1999).

bioconversion properties, in particular a twofold increase in yield of (2*R*)-indanediol relative to the strain 124. (Before the advent of protein engineering (Chapter 10) and directed evolution (Chapter 11), a chemostat was employed to observe mutation rates over time.)
3. Induction studies with both strains revealed the lack of a toluene-inducible dioxygenase (TOD) activity present in 124 but lacking in KY1, which is responsible for the formation of several undesired by-products.
4. In KY1, more than 95% of indene substrate is oxidized by a monooxygenase to indane oxide, which in turn is hydrolyzed to undesired *trans*-(1*R*,2*R*)-indanediol and *cis*-(1*S*,2*R*)-indanediol, as was verified by steady-state metabolite balancing and labeling with ^{14}C-tracers.

As a result of the flux analysis, the most promising target for pathway modification to augment the yield of desired *trans*-(1*R*,2*R*)-indanediol is enhanced selective hydrolysis of indane oxide to *trans*-(1*R*,2*R*)-indanediol, either by modification of the culture conditions or by overexpression of an epoxide hydrolase.

Metabolic engineering efforts and tools are nicely summarized and reviewed by Stafford et al. (Stafford, 2002) for the example of 1,2-indanediol synthesis from indene in *Rhodococcus*.

Suggested Further Reading

R. A. Sheldon, *Chirotechnology*, Marcel Dekker, New York, **1993**.

References

E. J. Allain, L. P. Hager, L. Deng, and E. N. Jacobsen, Highly enantioselective epoxidation of disubstituted alkenes with hydrogen peroxide catalyzed by chloroperoxidase, *J. Am. Chem. Soc.* **1993**, *115*, 4415–4416.

A. Berry, T. C. Dodge, M. Pepsin, and W. Weyler, Application of metabolic engineering to improve both the production and use of biotech indigo, *J. Ind. Microbiol. Biotechnol.* **2002**, *28*, 127–133.

H. U. Blaser, F. Spindler, and M. Studer, Enantioselective catalysis in fine chemicals production, *Appl. Catal. A: General* **2001**, *221*, 119–143.

U. T. Bornscheuer, I.-L. Gatfield, J.-M. Hilmer, S. Vorlova, and R. Schmid, Method for preparing D- or L-menthol, Eur. Patent. EP 1223223, **2002** (July 17).

B. C. Buckland, S. W. Drew, N. C. Conors, M. M. Chartrain, C. Lee, P. L. Salmon, K. Gbewonyo, W. Zhou, P. Gaillot, R. Singhri, R. C. Oleminski jr., W.-J. Sun, J. Reddy, J. Zhang, B. A. Sachly, C. Taylor, K. E. Goklen, B. Junker, and R. L. Greasham, Microbial conversion of indene to indanediol: A key intermediate in the synthesis of CRIXIVAN, *Metabolic Eng.* **1999**, *1*, 63–74.

J. A. Chaplin, M. D. E. Dickson, S. F. Marais, R. K. Mitra, S. Reddy, D. Brady, M. Portwig, N. S. Gardiner, B. A. Mboniswa, and C. J. Parkinson, Process for preparing (–)-menthol and similar compounds, World Patent WO 02/36795 A2, **2002**(May 10).

G. Chotani, T. Dodge, A. Hsu, M. Kumar, R. LaDuca, D. Trimbur, W. Weyler, and

K. Sanford, The commercial production of chemicals using pathway engineering, *Biochim. Biophys. Acta* **2000**, *1543*, 434–455.

K. Drauz and H. Waldmann (eds.), *Enzyme Catalysis in Organic Synthesis*, Wiley-VCH, Weinheim, 2nd edition, **2002**, Chapter 11.2, pp. 579–605.

N. S. Finney, Enantioselective epoxide hydrolysis: catalysis involving microbes, mammals and metals, *Chem. Biol.* **1998**, *5*, R73–R79.

J. Fleischer, K. Bauer, and R. Hopp, Resolution of DL-menthyl benzoates, Ger. Offen. DE 2109456, **1972**.

T. Iwata, Y. Okeda, and Y. Hori, Process for producing isopulegol by citronellal selective cyclization over tris(2,6-diarylphenoxy)aluminum catalysts, Eur. Pat. Appl. EP 1225163, **2002**.

E. N. Jacobsen and N. S. Finney, Synthetic and biological catalysts in chemical synthesis: how to assess practical utility, *Chem. Biol.* **1994**, *1*, 85–90.

H. Kajiura, K. Mori, T. Tobimatsu, and T. Toraya, Characterization and mechanism of action of a reactivating factor for adenosylcobalamin-dependent glycerol dehydratase, *J. Biol. Chem.* **2001**, *276*, 36514–36519.

J. R. Knowles, Enzyme catalysis: not different, just better, *Nature* **1991**, *350*, 121–124.

A. Koch, J. L. Reymond, and R. A. Lerner, Antibody-catalyzed activation of unfunctionalized olefins for highly enantioselective asymmetric epoxidation, *J. Am. Chem. Soc.* **1994**, *116*, 803–804.

G. Langrand, J. Baratti, G. Buono, and C. Triantaphylides, Lipase catalyzed reactions and strategy for alcohol resolution, *Tetrahedron Lett.* **1986**, *27*, 29–32.

G. Langrand, M. Secchi, G. Buono, J. Baratti, and C. Triantaphylides, Lipase-catalyzed ester formation in organic solvents. An easy preparative resolution of α-substituted cyclohexanols, *Tetrahedron Lett.* **1985**, *26*, 1857–1860.

N. H. Lee, A. R. Muci, and E. N. Jacobsen, Enantiomerically pure epoxychromans via asymmetric catalysis, *Tetrahedron Lett.* **1991**, *32*, 5055–5058.

T. Moroe, S. Hattori, A. Komatsu, and Y. Yuzo, Method for the biochemical isolation of l-menthol, US Patent 3607651 (to Takasago Perfumery Co.), **1971**.

C. G. Rabiller, K. Koenigsberger, K. Faber, and H. Griengl, Enzymatic recognition of diastereomeric esters, *Tetrahedron* **1990**, *46*, 4231–4240.

K. Sanford, P. Soucaille, G. Whited, and G. Chotani, Genomics to fluxomics and physiomics – pathway engineering, *Curr. Opin. Microbiol.* **2002**, *5*, 318–322.

R. A. Sheldon, Consider the environmental quotient, *Chemtech* **1994**, *(3)*, 38–47.

L. H. Shu and Y. Shi, An efficient ketone-catalyzed epoxidation using hydrogen peroxide as oxidant, *J. Org. Chem.* **2000**, *65*, 8807–8810.

D. E. Stafford, K. S. Yanagimachi, and G. N. Stephanopoulos, Metabolic engineering of indene bioconversion in *Rhodococcus* sp., *Adv. Biochem. Eng./Biotechnol.* **2001**, *73*, 85–101.

D. E. Stafford, K. S. Yanagimachi, P. A. Lessard, S. K. Rijhwani, A. J. Sinskey, and G. N. Stephanopoulos, Optimizing bioconversion pathways through systems analysis and metabolic engineering, *Proc. Natl. Acad. Sci. USA* **2002**, *99*, 1801–1806.

H. Q. Tian, X. G. She, and Y. Shi, Electronic probing of ketone catalysts for asymmetric epoxidation. Search for more robust catalysts, *Org. Lett.* **2001**, *3*, 715–718.

T. Toraya, Enzymatic radical catalysis: coenzyme B_{12}-dependent diol dehydratase, *Chem. Record* **2002**, *2*, 352–366.

A. R. Vaino, Sodium percarbonate as an oxygen source for MTO catalyzed epoxidations, *J. Org. Chem.* **2000**, *65*, 4210–4212.

Index

Biocatalysis. A. S. Bommarius, B. R. Riebel

ISBN 3-527-30344-8

I

Q

R